SCHAUM'S OUTLINE OF

THEORY AND PROBLEMS

OF

LINEAR
ALGEBRA

Second Edition

•

SEYMOUR LIPSCHUTZ, Ph.D.

Professor of Mathematics
Temple University

•

SCHAUM'S OUTLINE SERIES

McGRAW-HILL, INC.

New York   St. Louis   San Francisco   Auckland   Bogotá   Caracas   Hamburg
Lisbon   London   Madrid   Mexico   Milan   Montreal   New Delhi   Paris
San Juan   São Paulo   Singapore   Sydney   Tokyo   Toronto

SEYMOUR LIPSCHUTZ, who is presently on the mathematics faculty of Temple University, formerly taught at the Polytechnic Institute of Brooklyn and was a visiting professor in the Computer Science Department of Brooklyn College. He received his Ph.D. in 1960 at the Courant Institute of Mathematical Sciences of New York University. Some of his other books in the Schaum's Outline Series include *Programming with Fortran* (with Arthur Poe), *Discrete Mathematics*, and *Probability*.

Schaum's Outline of Theory and Problems of
LINEAR ALGEBRA

1 2 3 4 5 6 7 8 9 10 11 12 13 14 15 16 17 18 19 20 SHP SHP 9 2 1

ISBN 0-07-038007-4

Sponsoring Editor, David Beckwith
Production Supervisor, Annette Mayeski
Editing Supervisors, Meg Tobin, Maureen Walker
Cover Design by Amy E. Becker.

Library of Congress Cataloging-in-Publication Data

Lipschutz, Seymour.
   Schaum's outline of theory and problems of linear algebra /
Seymour Lipschutz.—2nd ed.
     p. cm.—(Schaum's outline series)
   Includes index.
   ISBN 0-07-038007-4
   1. Algebras, Linear—Outlines, syllabi, etc. I. Title.
QA188.L57 1991
512′.5—dc20

90-42357
CIP

Preface

Linear algebra has in recent years become an essential part of the mathematical background required of mathematicians, engineers, physicists and other scientists. This requirement reflects the importance and wide applications of the subject matter.

This book is designed for use as a textbook for a formal course in linear algebra or as a supplement to all current standard texts. It aims to present an introduction to linear algebra which will be found helpful to all readers regardless of their fields of specialization. More material has been included than can be covered in most first courses. This has been done to make the book more flexible, to provide a useful book of reference, and to stimulate further interest in the subject.

Each chapter begins with clear statements of pertinent definitions, principles and theorems together with illustrative and other descriptive material. This is followed by graded sets of solved and supplementary problems. The solved problems serve to illustrate and amplify the theory, bring into sharp focus those fine points without which the student continually feels himself on unsafe ground, and provide the repetition of basic principles so vital to effective learning. Numerous proofs of theorems are included among the solved problems. The supplementary problems serve as a complete review of the material of each chapter.

The first chapter treats systems of linear equations. This provides the motivation and basic computational tools for the subsequent material. After vectors and matrices are introduced, there are chapters on vector spaces and subspaces and on inner products. This is followed by chapters covering determinants, eigenvalues and eigenvectors, and diagonalizing matrices (under similarity) and quadratic forms (under congruence). The later chapters cover abstract linear maps and their canonical forms, specifically the triangular, Jordan and rational canonical forms. The last chapter treats abstract linear maps on inner product spaces.

The main changes in the second edition have been for pedagogical reasons (form) rather than in content. Here, the notion of a matrix mapping is introduced early in the text, and inner products are introduced right after the chapter on vector spaces and subspaces. Also, algorithms for row reduction, matrix inversion computing determinants, and diagonalizing matrices and quadratic forms are presented using algorithmic notation. Furthermore, such topics as elementary matrices, LU factorization, Fourier coefficients, and various norms in $\mathbf{R}^n$ are introduced directly in the text, rather than in the problem sections. Lastly, by treating the more advanced abstract topics in the latter part of the text, we make the text easier to be used for an elementary course or for a two semester course in linear algebra.

I wish to thank the staff at McGraw Hill-Schaum Series, especially, John Aliano, David Beckwith and Margaret Tobin, for invaluable suggestions and for their very helpful cooperation. Lastly, I want to express my gratitude to Wilhelm Magnus, my teacher, advisor and friend, who introduced me to the beauty of mathematics.

Temple University   SEYMOUR LIPSCHUTZ
January, 1991

Contents

Chapter 1   SYSTEMS OF LINEAR EQUATIONS   1

1.1 Introduction. 1.2 Linear Equations, Solutions. 1.3 Linear Equations in Two Unknowns. 1.4 Systems of Linear Equations, Equivalent Systems, Elementary Operations. 1.5 Systems in Triangular and Echelon Form. 1.6 Reduction Algorithm. 1.7 Matrices. 1.8 Row Equivalence and Elementary Row Operations. 1.9 Systems of Linear Equations and Matrices. 1.10 Homogeneous Systems of Linear Equations.

Chapter 2   VECTORS IN R^n AND C^n, SPATIAL VECTORS   39

2.1 Introduction. 2.2 Vectors in R^n. 2.3 Vector Addition and Scalar Multiplication. 2.4 Vectors and Linear Equations. 2.5 Dot (Scalar) Product. 2.6 Norm of a Vector. 2.7 Located Vectors, Hyperplanes, and Lines in R^n. 2.8 Spatial Vectors, **ijk** Notation in R^3. 2.9 Complex Numbers. 2.10 Vectors in C^n.

Chapter 3   MATRICES   74

3.1 Introduction. 3.2 Matrices. 3.3 Matrix Addition and Scalar Multiplication. 3.4 Matrix Multiplication. 3.5 Transpose of a Matrix. 3.6 Matrices and Systems of Linear Equations. 3.7 Block Matrices.

Chapter 4   SQUARE MATRICES, ELEMENTARY MATRICES   89

4.1 Introduction. 4.2 Square Matrices. 4.3 Diagonal and Trace, Identity Matrix. 4.4 Powers of Matrices, Polynomials, and Matrices. 4.5 Invertible (NonSingular) Matrices. 4.6 Special Types of Square Matrices. 4.7 Complex Matrices. 4.8 Square Block Matrices. 4.9 Elementary Matrices and Applications. 4.10 Elementary Column Operations, Matrix Equivalence. 4.11 Congruent Symmetric Matrices, Law of Inertia. 4.12 Quadratic Forms. 4.13 Similarity. 4.14 LU Factorization.

Chapter 5   VECTOR SPACES   141

5.1 Introduction. 5.2 Vector Spaces. 5.3 Examples of Vector Spaces. 5.4 Subspaces. 5.5 Linear Combinations, Linear Spans. 5.6 Linear Dependence and Independence. 5.7 Basis and Dimension. 5.8 Linear Equations and Vector Spaces. 5.9 Sums and Direct Sums. 5.10 Coordinates. 5.11 Change of Basis.

Chapter 6   INNER PRODUCT SPACES, ORTHOGONALITY   202

6.1 Introduction. 6.2 Inner Product Spaces. 6.3 Cauchy–Schwarz Inequality, Applications. 6.4 Orthogonality. 6.5 Orthogonal Sets and Bases, Projections. 6.6 Gram–Schmidt Orthogonalization Process. 6.7 Inner Products and Matrices. 6.8 Complex Inner Product Spaces. 6.9 Normed Vector Spaces.

Chapter 7   DETERMINANTS   246
7.1 Introduction. 7.2 Determinants of Orders One and Two. 7.3 Determinants of Order Three. 7.4 Permutations. 7.5 Determinants of Arbitrary Order. 7.6 Properties of Determinants. 7.7 Minors and Cofactors. 7.8 Classical Adjoint. 7.9 Applications to Linear Equations, Cramer's Rule. 7.10 Submatrices, General Minors, Principal Minors. 7.11 Block Matrices and Determinants. 7.12 Determinants and Volume. 7.13 Multilinearity and Determinants.

Chapter 8   EIGENVALUES AND EIGENVECTORS, DIAGONALIZATION   280
8.1 Introduction. 8.2 Polynomials and Matrices. 8.3 Characteristic Polynomial, Cayley–Hamilton Theorem. 8.4 Eigenvalues and Eigenvectors. 8.5 Computing Eigenvalues and Eigenvectors, Diagonalizing Matrices. 8.6 Diagonalizing Real Symmetric Matrices. 8.7 Minimum Polynomial.

Chapter 9   LINEAR MAPPINGS   312
9.1 Introduction. 9.2 Mappings. 9.3 Linear Mappings. 9.4 Kernel and Image of a Linear Mapping. 9.5 Singular and Nonsingular Linear Mappings, Isomorphisms. 9.6 Operations with Linear Mappings. 9.7 Algebra $A(V)$ of Linear Operators. 9.8 Invertible Operators.

Chapter 10   MATRICES AND LINEAR MAPPINGS   344
10.1 Introduction. 10.2 Matrix Representation of a Linear Operator. 10.3 Change of Basis and Linear Operators. 10.4 Diagonalization of Linear Operators. 10.5 Matrices and General Linear Mappings.

Chapter 11   CANONICAL FORMS   369
11.1 Introduction. 11.2 Triangular Form. 11.3 Invariance. 11.4 Invariant Direct-Sum Decompositions. 11.5 Primary Decomposition. 11.6 Nilpotent Operators. 11.7 Jordan Canonical Form. 11.8 Cyclic Subspaces. 11.9 Rational Canonical Form. 11.10 Quotient Spaces.

Chapter 12   LINEAR FUNCTIONALS AND THE DUAL SPACE   397
12.1 Introduction. 12.2 Linear Functionals and the Dual Space. 12.3 Dual Basis. 12.4 Second Dual Space. 12.5 Annihilators. 12.6 Transpose of a Linear Mapping.

Chapter 13   BILINEAR, QUADRATIC, AND HERMITIAN FORMS   409
13.1 Introduction. 13.2 Bilinear Forms. 13.3 Bilinear Forms and Matrices. 13.4 Alternating Bilinear Forms. 13.5 Symmetric Bilinear Forms, Quadratic Forms. 13.6 Real Symmetric Bilinear Forms, Law of Inertia. 13.7 Hermitian Forms.

Chapter *14*   **LINEAR OPERATORS ON INNER PRODUCT SPACES**   425

14.1 Introduction.   14.2 Adjoint Operators.   14.3 Analogy Between $A(V)$ and **C,** Special Operators.   14.4 Self-Adjoint Operators.   14.5 Orthogonal and Unitary Operators.   14.6 Orthogonal and Unitary Matrices.   14.7 Change of Orthonormal Basis.   14.8 Positive Operators.   14.9 Diagonalization and Canonical Forms in Euclidean Spaces.   14.10 Diagonalization and Canonical Forms in Unitary Spaces.   14.11 Spectral Theorem.

Appendix   **POLYNOMIALS OVER A FIELD**   446

A.1 Introduction.   A.2 Divisibility; Greatest Common Divisor.   A.3 Factorization.

INDEX   449

Chapter 1

Systems of Linear Equations

1.1 INTRODUCTION

The theory of linear equations plays an important and motivating role in the subject of linear algebra. In fact, many problems in linear algebra are equivalent to studying a system of linear equations, e.g., finding the kernel of a linear mapping and characterizing the subspace spanned by a set of vectors. Thus the techniques introduced in this chapter will be applicable to the more abstract treatment given later. On the other hand, some of the results of the abstract treatment will give us new insights into the structure of "concrete" systems of linear equations.

This chapter investigates systems of linear equations and describes in detail the Gaussian elimination algorithm which is used to find their solution. Although matrices will be studied in detail in Chapter 3, matrices, together with certain operations on them, are also introduced here, since they are closely related to systems of linear equations and their solution.

All our equations will involve specific numbers called *constants* or *scalars*. For simplicity, we assume in this chapter that all our scalars belong to the real field $\mathbf{R}$. The solutions of our equations will also involve n-tuples $u = (k_1, k_2, \ldots, k_n)$ of real numbers called *vectors*. The set of all such n-tuples is denoted by $\mathbf{R}^n$.

We note that the results in this chapter also hold for equations over the complex field $\mathbf{C}$ or over any arbitrary field K.

1.2 LINEAR EQUATIONS, SOLUTIONS

By a *linear equation* in unknowns $x_1, x_2, \ldots, x_n$, we mean an equation that can be put in the *standard form*:

$$a_1 x_1 + a_2 x_2 + \cdots + a_n x_n = b \qquad (1.1)$$

where $a_1, a_2, \ldots, a_n, b$ are constants. The constant a_k is called the *coefficient* of x_k and b is called the *constant* of the equation.

A *solution* of the above linear equation is a set of values for all the unknowns, say $x_1 = k_1$, $x_2 = k_2, \ldots, x_n = k_n$, or simply an n-tuple $u = (k_1, k_2, \ldots, k_n)$ of constants, with the property that the following statement (obtained by substituting each k_i for x_i in the equation) is true:

$$a_1 k_1 + a_2 k_2 + \cdots + a_n k_n = b$$

This set of values is then said to *satisfy* the equation.

The set of all such solutions is called the *solution set* or *general solution* or, simply, the *solution* of the equation.

> **Remark:** The above notions implicitly assume there is an ordering of the unknowns. In order to avoid subscripts, we will usually use variables x, y, z, as ordered, to denote three unknowns, x, y, z, t, as ordered, to denote four unknowns, and x, y, z, s, t, as ordered, to denote five unknowns.

1

Example 1.1

(a) The equation $2x - 5y + 3xz = 4$ is not linear since the product xz of two unknowns is of second degree.

(b) The equation $x + 2y - 4z + t = 3$ is linear in the four unknowns x, y, z, t.
     The 4-tuple $u = (3, 2, 1, 0)$ is a solution of the equation since

$$3 + 2(2) - 4(1) + 0 = 3 \qquad \text{or} \qquad 3 = 3$$

is a true statement. However, the 4-tuple $v = (1, 2, 4, 5)$ is not a solution of the equation since

$$1 + 2(2) - 4(4) + 5 = 3 \qquad \text{or} \qquad -6 = 3$$

is not a true statement.

Linear Equations in One Unknown

The following basic result is proved in Problem 1.5.

Theorem 1.1: Consider the linear equation $ax = b$.

    (i)   If $a \neq 0$, then $x = b/a$ is a unique solution of $ax = b$.

    (ii) If $a = 0$, but $b \neq 0$, then $ax = b$ has no solution.

    (iii) If $a = 0$ and $b = 0$, then every scalar k is a solution of $ax = b$.

Example 1.2

(a) Solve $4x - 1 = x + 6$.
     Transpose to obtain the equation in standard form: $4x - x = 6 + 1$ or $3x = 7$. Multiply by 1/3 to obtain the unique solution $x = \frac{7}{3}$ [Theorem 1.1(i)].

(b) Solve $2x - 5 - x = x + 3$.
     Rewrite the equation in standard form: $x - 5 = x + 3$, or $x - x = 3 + 8$, or $0x = 8$. The equation has no solution [Theorem 1.1(ii)].

(c) Solve $4 + x - 3 = 2x + 1 - x$.
     Rewrite the equation in standard form: $x + 1 = x + 1$, or $x - x = 1 - 1$, or $0x = 0$. Every scalar k is a solution [Theorem 1.1(iii)].

Degenerate Linear Equations

A linear equation is said to be *degenerate* if it has the form

$$0x_1 + 0x_2 + \cdots + 0x_n = b$$

that is, if every coefficient is equal to zero. The solution of such an equation is as follows:

Theorem 1.2: Consider the degenerate linear equation $0x_1 + 0x_2 + \cdots + 0x_n = b$.

    (i) If the constant $b \neq 0$, then the equation has no solution.

    (ii) If the constant $b = 0$, then every vector $u = (k_1, k_2, \ldots, k_n)$ is a solution.

Proof. (i) Let $u = (k_1, k_2, \ldots, k_n)$ be any vector. Suppose $b \neq 0$. Substituting u in the equation we obtain:

$$0k_1 + 0k_2 + \cdots + 0k_n = b \qquad \text{or} \qquad 0 + 0 + \cdots + 0 = b \qquad \text{or} \qquad 0 = b$$

This is not a true statement since $b \neq 0$. Hence no vector u is a solution.

(ii) Suppose $b = 0$. Substituting u in the equation we obtain:

$$0k_1 + 0k_2 + \cdots + 0k_n = 0 \qquad \text{or} \qquad 0 + 0 + \cdots + 0 = 0 \qquad \text{or} \qquad 0 = 0$$

which is a true statement. Thus every vector u in $\mathbf{R}^n$ is a solution, as claimed.

Example 1.3. Describe the solution of $4y - x - 3y + 3 = 2 + x - 2x + y + 1$.

Rewrite in standard form by collecting terms and transposing:

$$y - x + 3 = y - x + 3 \qquad \text{or} \qquad y - x - y + x = 3 - 3 \qquad \text{or} \qquad 0x + 0y = 0$$

The equation is degenerate with a zero constant; thus every vector $u = (a, b)$ in $\mathbf{R}^2$ is a solution.

Nondegenerate Linear Equations, Leading Unknown

This subsection covers the solution of a single nondegenerate linear equation in one or more unknowns, say

$$a_1 x_1 + a_2 x_2 + \cdots + a_n x_n = b$$

By the *leading unknown* in such an equation, we mean the first unknown with a nonzero coefficient. Its position p in the equation is therefore the smallest integral value of j for which $a_j \neq 0$. In other words, x_p is the leading unknown if $a_j = 0$ for $j < p$, but $a_p \neq 0$.

Example 1.4. Consider the linear equation $5y - 2z = 3$. Here y is the leading unknown. If the unknowns are x, y, and z, then $p = 2$ is its position; but if y and z are the only unknowns, then $p = 1$.

The following theorem, proved in Problem 1.9, applies.

Theorem 1.3: Consider a nondegenerate linear equation $a_1 x_1 + a_2 x_2 + \cdots + a_n x_n = b$ with leading unknown x_p.

(i) Any set of values for the unknowns x_j with $j \neq p$ will yield a unique solution of the equation. (The unknowns x_j are called *free variables* since one can assign any values to them.)

(ii) Every solution of the equation is obtained in (i).

(The set of all solutions is called the *general solution* of the equation.)

Example 1.5

(a) Find three particular solutions to the equation $2x - 4y + z = 8$.

Here x is the leading unknown. Accordingly, assign any values to the free variables y and z, and then solve for x to obtain a solution. For example:

(1) Set $y = 1$ and $z = 1$. Substitution in the equation yields

$$2x - 4(1) + 1 = 8 \qquad \text{or} \qquad 2x - 4 + 1 = 8 \qquad \text{or} \qquad 2x = 11 \qquad \text{or} \qquad x = \tfrac{11}{2}$$

Thus $u_1 = (\tfrac{11}{2}, 1, 1)$ is a solution.

(2) Set $y = 1$, $z = 0$. Substitution yields $x = 6$. Hence $u_2 = (6, 1, 0)$ is a solution.

(3) Set $y = 0$, $z = 1$. Substitution yields $x = \frac{7}{2}$. Thus $u_3 = (\frac{7}{2}, 0, 1)$ is a solution.

(b) The general solution of the above equation $2x - 4y + z = 8$ is obtained as follows.

First, assign arbitrary values (called *parameters*) to the free variables, say, $y = a$ and $z = b$. Then substitute in the equation to obtain

$$2x - 4a + b = 8 \quad \text{or} \quad 2x = 8 + 4a - b \quad \text{or} \quad x = 4 + 2a - \tfrac{1}{2}b$$

Thus

$$x = 4 + 2a - \tfrac{1}{2}b, \; y = a, \; z = b \quad \text{or} \quad u = (4 + 2a - \tfrac{1}{2}b, a, b)$$

is the general solution.

1.3 LINEAR EQUATIONS IN TWO UNKNOWNS

This section considers the special case of linear equations in two unknowns, x and y, that is, equations that can be put in the standard form

$$ax + by = c$$

where a, b, c are real numbers. (We also assume that the equation is nondegenerate, i.e., that a and b are not both zero.) Each solution of the equation is a pair of real numbers, $u = (k_1, k_2)$, which can be found by assigning an arbitrary value to x and solving for y, or vice versa.

Every solution $u = (k_1, k_2)$ of the above equation determines a point in the cartesian plane $\mathbf{R}^2$. Since a and b are not both zero, all such solutions correspond precisely to the points on a straight line (whence the name "linear equation"). This line is called the *graph* of the equation.

Example 1.6. Consider the linear equation $2x + y = 4$. We find three solutions of the equation as follows. First choose any value for either unknown, say $x = -2$. Substitute $x = -2$ into the equation to obtain

$$2(-2) + y = 4 \quad \text{or} \quad -4 + y = 4 \quad \text{or} \quad y = 8$$

Thus $x = -2$, $y = -8$ or the point $(-2, 8)$ in $\mathbf{R}^2$ is a solution. Now find the y-intercept, that is, substitute $x = 0$ in the equation to get $y = 4$; hence $(0, 4)$ on the y axis is a solution. Next find the x-intercept, that is, substitute $y = 0$ in the equation to get $x = 2$; hence $(2, 0)$ on the x axis is a solution.

To plot the graph of the equation, first plot the three solutions, $(-2, 8)$, $(0, 4)$, and $(2, 0)$, in the plane $\mathbf{R}^2$ as pictured in Fig. 1-1. Then draw the line L determined by two of the solutions and note that the third solution also lies on L. (Indeed, L is the set of all solutions of the equation.) The line L is the graph of the equation.

System of Two Equations in Two Unknowns

This subsection considers a system of two (nondegenerate) linear equations in the two unknowns x and y:

$$\begin{aligned} a_1 x + b_1 y = c_1 \\ a_2 x + b_2 y = c_2 \end{aligned} \tag{1.2}$$

(Thus a_1 and b_1 are not both zero, and a_2 and b_2 are not both zero.) This simple system is treated separately since it has a geometrical interpretation, and its properties motivate the general case.

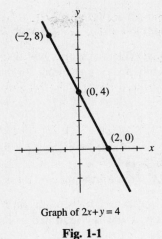

Graph of $2x + y = 4$

Fig. 1-1

A pair $u = (k_1, k_2)$ of real numbers which satisfies both equations is called a simultaneous solution of the given equations or a solution of the system of equations. There are three cases, which can be described geometrically.

(1) The system has exactly one solution. Here the graphs of the linear equations intersect in one point, as in Fig. 1-2(a).

(2) The system has no solutions. Here the graphs of the linear equations are parallel, as in Fig. 1-2(b).

(3) The system has an infinite number of solutions. Here the graphs of the linear equations coincide, as in Fig. 1-2(c).

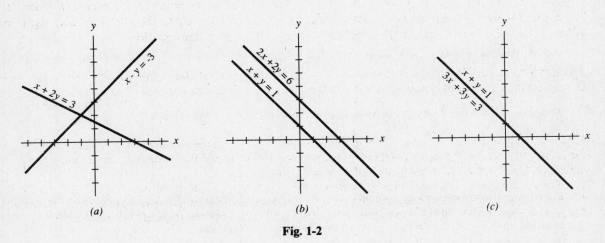

(a)   (b)   (c)

Fig. 1-2

The special cases (2) and (3) can only occur when the coefficients of x and y in the two linear equations are proportional; that is,

$$\frac{a_1}{a_2} = \frac{b_1}{b_2} \qquad \text{or} \qquad \begin{vmatrix} a_1 & b_1 \\ a_2 & b_2 \end{vmatrix} = a_1 b_2 - a_2 b_1 = 0$$

Specifically, case (2) or (3) occurs if

$$\frac{a_1}{a_2} = \frac{b_1}{b_2} \neq \frac{c_1}{c_2} \qquad \text{or} \qquad \frac{a_1}{a_2} = \frac{b_1}{b_2} = \frac{c_1}{c_2}$$

respectively. Unless otherwise stated or implied, we assume we are dealing with the general case (1).

Remark: The expression $\begin{vmatrix} a_1 & b_1 \\ a_2 & b_2 \end{vmatrix}$, which has the value $a_1 b_2 - a_2 b_1$, is called a determinant of order two. Determinants will be studied in Chapter 7. Thus the system has a unique solution when the determinant of the coefficients is not zero.

Elimination Algorithm

The solution to system (*1.2*) can be obtained by the process known as elimination, whereby we reduce the system to a single equation in only one unknown. Assuming the system has a unique solution, this elimination algorithm consists of the following two steps:

Step 1. Add a multiple of one equation to the other equation (or to a nonzero multiple of the other equation) so that one of the unknowns is eliminated in the new equation.

Step 2. Solve the new equation for the given unknown, and substitute its value in one of the original equations to obtain the value of the other unknown.

Example 1.7

(*a*) Consider the system

$$L_1: \quad 2x + 5y = \quad 8$$
$$L_2: \quad 3x - 2y = -7$$

We eliminate x from the equations by forming the new equation $L = 3L_1 - 2L_2$; that is, by multiplying L_1 by 3 and multiplying L_2 by -2 and adding the resultant equations:

$$
\begin{array}{rl}
3L_1: & 6x + 15y = 24 \\
-2L_2: & -6x + 4y = 14 \\
\hline
\text{Addition:} & 19y = 38
\end{array}
$$

Solving the new equation for y yields $y = 2$. Substituting $y = 2$ into one of the original equations, say L_1, yields

$$2x + 5(2) = 8 \quad \text{or} \quad 2x + 10 = 8 \quad \text{or} \quad 2x = -2 \quad \text{or} \quad x = -1$$

Thus $x = -1$ and $y = 2$, or the pair $(-1, 2)$, is the unique solution to the system.

(*b*) Consider the system

$$L_1: \quad x - 3y = 4$$
$$L_2: \quad -2x + 6y = 5$$

Eliminate x from the equations by multiplying L_1 by 2 and adding it to L_2; that is, by forming the equation $L = 2L_1 + L_2$. This yields the new equation $0x + 0y = 13$. This is a degenerate equation which has a nonzero constant; therefore, the system has no solution. (Geometrically speaking, the lines are parallel.)

(*c*) Consider the system

$$L_1: \quad x - 3y = \quad 4$$
$$L_2: \quad -2x + 6y = -8$$

Eliminate x by multiplying L_1 by 2 and adding it to L_2. This yields the new equation $0x + 0y = 0$ which is a degenerate equation where the constant term is also zero. Hence the system has an infinite number of solutions, which correspond to the solutions of either equation. (Geometrically speaking, the lines coincide.) To find the general solution, let $y = a$ and substitute in L_1 to obtain $x - 3a = 4$ or $x = 3a + 4$. Accordingly, the general solution to the system is

$$(3a + 4, a)$$

where a is any real number.

1.4 SYSTEMS OF LINEAR EQUATIONS, EQUIVALENT SYSTEMS, ELEMENTARY OPERATIONS

This section considers a system of m linear equations, say $L_1, L_2, \ldots, L_m$, in n unknowns $x_1, x_2, \ldots, x_n$ which can be put in the *standard form*

$$
\begin{aligned}
a_{11}x_1 + a_{12}x_2 + \cdots + a_{1n}x_n &= b_1 \\
a_{21}x_1 + a_{22}x_2 + \cdots + a_{2n}x_n &= b_2 \\
&\cdots\cdots\cdots\cdots\cdots\cdots\cdots\cdots\cdots\cdots \\
a_{m1}x_1 + a_{m2}x_2 + \cdots + a_{mn}x_n &= b_m
\end{aligned}
\tag{1.3}
$$

where the a_{ij}, b_i are constants.

A *solution* (or a *particular solution*) of the above system is a set of values for the unknowns, say $x_1 = k_1, x_2 = k_2, \ldots, x_n = k_n$, or an n-tuple $u = (k_1, k_2, \ldots, k_n)$ of constants, which is a solution of each of the equations of the system. The set of all such solutions is called the *solution set* or the *general solution* of the system.

Example 1.8.   Consider the system

$$
\begin{aligned}
x_1 + 2x_2 - 5x_3 + 4x_4 &= 3 \\
2x_1 + 3x_2 + x_3 - 2x_4 &= 1
\end{aligned}
$$

Determine whether $x_1 = -8$, $x_2 = 4$, $x_3 = 1$, $x_4 = 2$ is a solution of the system.
  Substitute in each equation to obtain

(1) $-8 + 2(4) - 5(1) + 4(2) = 3$ or $-8 + 8 - 5 + 8 = 3$ or $3 = 3$

(2) $2(-8) + 3(4) + 1 - 2(2) = 1$ or $-16 + 12 + 1 - 4 = 1$ or $-7 = 3$

No, it is not a solution since it is not a solution of the second equation.

Equivalent Systems, Elementary Operations

Systems of linear equations in the same unknowns are said to be *equivalent* if the systems have the same solution set. One way of producing a system which is equivalent to a given system, with linear equations $L_1, L_2, \ldots, L_m$, is by applying a sequence of the following operations called *elementary operations*:

[E_1]   Interchange the ith equation and the jth equation: $L_i \leftrightarrow L_j$.

[E_2]   Multiply the ith equation by a nonzero scalar k: $kL_i \rightarrow L_i$, $k \neq 0$.

[E_3]   Replace the ith equation by k times the jth equation plus the ith equation: $(kL_j + L_i) \rightarrow L_i$.

In actual practice, we apply [E_2] and then [E_3] in one step, that is, the operation

[E]   Replace the ith equation by k' times the jth equation plus k (nonzero) times the ith equation:

$$
(k'L_j + kL_i) \rightarrow L_i, \; k \neq 0.
$$

The above is formally stated in the following theorem proved in Problem 1.46.

Theorem 1.4:   Suppose a system (#) of linear equations is obtained from a system (∗) of linear equations by a finite sequence of elementary operations. Then (#) and (∗) have the same solution set.

Our method for solving the system (*1.3*) of linear equations consists of two steps:

Step 1.   Use the above elementary operations to reduce the system to an equivalent simpler system (in triangular or echelon form).

Step 2.   Use back-substitution to find the solution of the simpler system.

The two steps are illustrated in Example 1.9. However, for pedagogical reasons, we first discuss Step 2 in detail in Section 1.5 and then we discuss Step 1 in detail in Section 1.6.

Example 1.9.   The solution of the system

$$\begin{aligned} x + 2y - 4z &= -4 \\ 5x + 11y - 21z &= -22 \\ 3x - 2y + 3z &= 11 \end{aligned}$$

is obtained as follows:

Step 1.   First we eliminate x from the second equation by the elementary operation $(-5L_1 + L_2) \to L_2$, that is, by multiplying L_1 by -5 and adding it to L_2; and then we eliminate x from the third equation by applying the elementary operation $(-3L_1 + L_3) \to L_3$, i.e., by multiplying L_1 by -3 and adding it to L_3:

$$\begin{array}{ll}
-5 \times L_1: & -5x - 10y + 20z = 20 \\
L_2: & \underline{5x + 11y - 21z = -22} \\
\text{new } L_2: & y - z = -2
\end{array}
\qquad
\begin{array}{ll}
-3 \times L_1: & -3x - 6y + 12z = 12 \\
L_3: & \underline{3x - 2y + 3z = 11} \\
\text{new } L_3: & -8y + 15z = 23
\end{array}$$

Thus the original system is equivalent to the system

$$\begin{aligned} x + 2y - 4z &= -4 \\ y - z &= -2 \\ -8y + 15z &= 23 \end{aligned}$$

Next we eliminate y from the third equation by applying $(8L_2 + L_3) \to L_3$ that is, by multiplying L_2 by 8 and adding it to L_3:

$$\begin{array}{ll}
8 \times L_2: & 8y - 8z = -16 \\
L_3: & \underline{-8y + 15z = 23} \\
\text{new } L_3: & 7z = 7
\end{array}$$

Thus we obtain the following equivalent triangular system:

$$\begin{aligned} x + 2y - 4z &= -4 \\ y - z &= -2 \\ 7z &= 7 \end{aligned}$$

Step 2.   Now we solve the simpler triangular system by back-substitution. The third equation gives $z = 1$. Substitute $z = 1$ into the second equation to obtain

$$y - 1 = -2 \quad \text{or} \quad y = -1$$

Now substitute $z = 1$ and $y = -1$ into the first equation to obtain

$$x + 2(-1) - 4(1) = -4 \quad \text{or} \quad x - 2 - 4 = -4 \quad \text{or} \quad x - 6 = -4 \quad \text{or} \quad x = 2$$

Thus $x = 2$, $y = -1$, $z = 1$, or, in other words, the ordered triple $(2, -1, 1)$, is the unique solution to the given system.

The above two-step algorithm for solving a system of linear equations is called *Gaussian elimination*. The following theorem will be used in Step 1 of the algorithm.

Theorem 1.5:   Suppose a system of linear equations contains the degenerate equation

$$L: \quad 0x_1 + 0x_2 + \cdots + 0x_n = b$$

  (a)   If $b = 0$, then L may be deleted from the system without changing the solution set.

  (b)   If $b \neq 0$, then the system has no solution.

Proof. The proof follows directly from Theorem 1.2, that is, Part (*a*) follows from the fact that every vector in $\mathbf{R}^n$ is a solution to L, and Part (*b*) follows from the fact that L has no solution and hence the system has no solution.

1.5 SYSTEMS IN TRIANGULAR AND ECHELON FORM

This section considers two simple types of systems of linear equations: systems in triangular form and the more general systems in echelon form.

Triangular Form

A system of linear equations is in *triangular form* if the number of equations is equal to the number of unknowns and if x_k is the leading unknown of the kth equation. Thus a triangular system of linear equations has the following form:

$$
\begin{aligned}
a_{11}x_1 + a_{12}x_2 + \cdots + \quad a_{1,n-1}x_{n-1} + \quad & a_{1n}x_n = b_1 \\
a_{22}x_2 + \cdots + \quad a_{2,n-1}x_{n-1} + \quad & a_{2n}x_n = b_2 \\
\cdots\cdots\cdots\cdots\cdots\cdots\cdots\cdots\cdots\cdots\cdots\cdots\cdots & \\
a_{n-1,n-1}x_{n-1} + a_{n-1,n}x_n & = b_{n-1} \\
a_{nn}x_n & = b_n
\end{aligned} \tag{1.4}
$$

where $a_{11} \neq 0, a_{22} \neq 0, \ldots, a_{nn} \neq 0$.

The above triangular system of linear equations has a unique solution which may be obtained by the following process known as back-substitution. First, we solve the last equation for the last unknown, x_n:

$$
x_n = \frac{b_n}{a_{nn}}
$$

Second, we substitute this value for x_n in the next-to-last equation and solve it for the next-to-last unknown, x_{n-1}:

$$
x_{n-1} = \frac{b_{n-1} - a_{n-1,n}(b_n/a_{nn})}{a_{n-1,n-1}}
$$

Third, we substitute these values for x_n and x_{n-1} in the third-from-last equation and solve it for the third-from-last unknown, x_{n-2}:

$$
x_{n-2} = \frac{b_{n-2} - (a_{n-2,n-1}/a_{n-1,n-1})[b_{n-1} - a_{n-1,n}(b_n/a_{nn})] - (a_{n-2,n}/a_{nn})b_n}{a_{n-2,n-2}}
$$

In general, we determine x_k by substituting the previously obtained values of $x_n, x_{n-1}, \ldots, x_{k+1}$ in the kth equation:

$$
x_k = \frac{b_k - \sum\limits_{m=k+1}^{n} a_{km}x_m}{a_{kk}}
$$

The process ceases when we have determined the first unknown, x_1. The solution is unique since, at each step of the algorithm, the value of x_k is, by Theorem 1.1(i), uniquely determined.

Example 1.10. Consider the system

$$
\begin{aligned}
2x + 4y - \quad z &= 11 \\
5y + \quad z &= 2 \\
3z &= -9
\end{aligned}
$$

Since the system is in triangular form it may be solved by back-substitution.

(i)   The last equation yields $z = -3$.

(ii) Substitute in the second equation to obtain $5y - 3 = 2$ or $5y = 5$ or $y = 1$.

(iii) Substitute $z = -3$ and $y = 1$ in the first equation to obtain

$$2x + 4(1) - (-3) = 11 \qquad \text{or} \qquad 2x + 4 + 3 = 11 \qquad \text{or} \qquad 2x = 4 \qquad \text{or} \qquad x = 2$$

Thus the vector $u = (2, 1, -3)$ is the unique solution of the system.

Echelon Form, Free Variables

A system of linear equations is in *echelon form* if no equation is degenerate and if the leading unknown in each equation is to the right of the leading unknown of the preceding equation. The paradigm is:

$$a_{11}x_1 + a_{12}x_2 + a_{13}x_3 + a_{14}x_4 + \cdots + a_{1n}x_n = b_1$$
$$a_{2j_2}x_{j_2} + a_{2,\,j_2+1}x_{j_2+1} + \cdots + a_{2n}x_n = b_2$$
$$\cdots\cdots\cdots\cdots\cdots\cdots\cdots\cdots\cdots\cdots\cdots\cdots\cdots\cdots\cdots\cdots \qquad (1.5)$$
$$a_{rj_r}x_{j_r} + a_{r,\,j_r+1}x_{j_r+1} + \cdots + a_{rn}x_n = b_r$$

where $1 < j_2 < \cdots < j_r$, and where $a_{11} \neq 0, a_{2j_2} \neq 0, \ldots, a_{rj_r} \neq 0$. Note that $r \leq n$.

An unknown x_k in the above echelon system (1.5) is called a *free variable* if x_k is not the leading unknown in any equation, that is, if $x_k \neq x_1, x_k \neq x_{j_2}, \ldots, x_k \neq x_{j_r}$.

The following theorem, proved in Problem 1.13, describes the solution set of an echelon system.

Theorem 1.6: Consider the system (1.5) of linear equations in echelon form. There are two cases.

(i) $r = n$. That is, there are as many equations as unknowns. Then the system has a unique solution.

(ii) $r < n$. That is, there are fewer equations than unknowns. Then we can arbitrarily assign values to the $n - r$ free variables and obtain a solution of the system.

Suppose the echelon system (1.5) does contain more unknowns than equations. Then the system has an infinite number of solutions since each of the $n - r$ free variables may be assigned any real number. The general solution of the system is obtained as follows. Arbitrary values, called parameters, say $t_1, t_2, \ldots, t_{n-r}$, are assigned to the free variables, and then back-substitution is used to obtain values of the nonfree variables in terms of the parameters. Alternatively, one may use back-substitution to solve for the nonfree variables $x_1, x_{j_2}, \ldots, x_{j_r}$ directly in terms of the free variables.

Example 1.11. Consider the system

$$x + 4y - 3z + 2t = 5$$
$$z - 4t = 2$$

The system is in echelon form. The leading unknowns are x and z; hence the free variables are the other unknowns y and t.

To find the general solution of the system, we assign arbitrary values to the free variables y and t, say $y = a$ and $t = b$, and then use back-substitution to solve for the nonfree variables x and z. Substituting in the last equation yields $z - 4b = 2$ or $z = 2 + 4b$. Substitute in the first equation to get

$$x + 4a - 3(2 + 4b) + 2b = 5 \qquad \text{or} \qquad x + 4a - 6 - 12b + 2b = 5 \qquad \text{or} \qquad x = 11 - 4a + 10b$$

Thus

$$x = 11 - 4a + 10b, \; y = a, \; z = 2 + 4b, \; t = b \qquad \text{or} \qquad (11 - 4a + 10b, \, a, \, 2 + 4b, \, b)$$

is the general solution in parametric form. Alternatively, we can use back-substitution to solve for the nonfree variables x and z directly in terms of the free variables y and t. The last equation gives $z = 2 + 4t$. Substitute in the first equation to obtain

$$x + 4y - 3(2 + 4t) + 2t = 5 \qquad \text{or} \qquad x + 4y - 6 - 12t + 2t = 5 \qquad \text{or} \qquad x = 11 - 4y + 10t$$

Accordingly,

$$x = 11 - 4y + 10t$$
$$z = \ \ 2 + 4t$$

is another form for the general solution of the system.

1.6 REDUCTION ALGORITHM

The following algorithm (sometimes called row reduction) reduces the system (1.3) of m linear equations in n unknowns to echelon (possibly triangular) form, or determines that the system has no solution.

Reduction algorithm

Step 1. Interchange equations so that the first unknown, x_1, appears with a nonzero coefficient in the first equation; i.e., arrange that $a_{11} \neq 0$.

Step 2. Use a_{11} as a pivot to eliminate x_1 from all the equations except the first equation. That is, for each $i > 1$, apply the elementary operation (Section 1.4)

$$[E_3]: \ -(a_{i1}/a_{11})L_1 + L_i \rightarrow L_i \qquad \text{or} \qquad [E]: \ -a_{i1}L_1 + a_{11}L_i \rightarrow L_i$$

Step 3. Examine each new equation L:

(a) If L has the form $0x_1 + 0x_2 + \cdots + 0x_n = 0$ or if L is a multiple of another equation, then delete L from the system.

(b) If L has the form $0x_1 + 0x_2 + \cdots + 0x_n = b$ with $b \neq 0$, then exit from the algorithm. The system has no solution.

Step 4. Repeat Steps 1, 2, and 3 with the subsystem formed by all the equations, excluding the first equation.

Step 5. Continue the above process until the system is in echelon form or a degenerate equation is obtained in Step 3(b).

The justification of Step 3 is Theorem 1.5 and the fact that if $L = kL'$ for some other equation L' in the system, the operation $-kL' + L \rightarrow L$ replaces L by $0x_1 + 0x_2 + \cdots + 0x_n = 0$, which again may be deleted by Theorem 1.5.

Example 1.12

(a) The system

$$2x + \ \ y - 2z = 10$$
$$3x + 2y + 2z = \ \ 1$$
$$5x + 4y + 3z = \ \ 4$$

is solved by first reducing it to echelon form. To eliminate x from the second and third equations, apply the operations $-3L_1 + 2L_2 \rightarrow L_2$ and $-5L_1 + 2L_3 \rightarrow L_3$:

$-3L_1$:	$-6x - 3y + \ 6z = -30$	$-5L_1$:	$-10x - 5y + 10z = -50$
$2L_2$:	$6x + 4y + \ 4z = \ \ \ 2$	$2L_3$:	$10x + 8y + \ 6z = \ \ \ \ 8$
$-3L_1 + 2L_2$:	$y + 10z = -28$	$-5L_1 + 2L_3$:	$3y + 16z = -42$

This yields the following system, from which y is eliminated from the third equation by the operation $-3L_2 + L_3 \to L_3$:

$$\left.\begin{array}{rcl} 2x + y - 2z &=& 10 \\ y + 10z &=& -28 \\ 3y + 16z &=& -42 \end{array}\right\} \to \left\{\begin{array}{rcl} 2x + y - 2z &=& 10 \\ y + 10z &=& -28 \\ -14z &=& 42 \end{array}\right.$$

The system is now in triangular form. Therefore, we can use back-substitution to obtain the unique solution $u = (1, 2, -3)$.

(b) The system

$$\begin{array}{rcl} x + 2y - 3z &=& 1 \\ 2x + 5y - 8z &=& 4 \\ 3x + 8y - 13z &=& 7 \end{array}$$

is solved by first reducing it to echelon form. To eliminate x from the second and third equations, apply $-2L_1 + L_2 \to L_2$ and $-3L_1 + L_3 \to L_3$ to obtain

$$\begin{array}{rcl} x + 2y - 3z &=& 1 \\ y - 2z &=& 2 \\ 2y - 4z &=& 4 \end{array} \qquad \text{or} \qquad \begin{array}{rcl} x + 2y - 3z &=& 1 \\ y - 2z &=& 2 \end{array}$$

(The third equation is deleted since it is a multiple of the second equation.) The system is now in echelon form, with free variable z.

To obtain the general solution, let $z = a$ and solve by back-substitution. Substitute $z = a$ into the second equation to obtain $y = 2 + 2a$. Then substitute $z = a$ and $y = 2 + 2a$ into the first equation to obtain $x + 2(2 + 2a) - 3a = 1$ or $x = -3 - a$. Thus the general solution is

$$x = -3 - a, \; y = 2 + 2a, \; z = a \qquad \text{or} \qquad (-3 - a, 2 + 2a, a)$$

where a is the parameter.

(c) The system

$$\begin{array}{rcl} x + 2y - 3z &=& -1 \\ 3x - y + 2z &=& 7 \\ 5x + 3y - 4z &=& 2 \end{array}$$

is solved by first reducing it to echelon form. To eliminate x from the second and third equations, apply the operations $-3L_1 + L_2 \to L_2$ and $-5L_1 + L_3 \to L_3$ to obtain the equivalent system

$$\begin{array}{rcl} x + 2y - 3z &=& -1 \\ -7y + 11z &=& 10 \\ -7y + 11z &=& 7 \end{array}$$

The operation $-L_2 + L_3 \to L_3$ yields the degenerate equation

$$0x + 0y + 0z = -3$$

Thus the system has no solution.

The following basic result was indicated previously.

Theorem 1.7: Any system of linear equations has either: (i) a unique solution, (ii) no solution, or (iii) an infinite number of solutions.

Proof. Applying the above algorithm to the system, we can either reduce it to echelon form or determine that it has no solution. If the echelon form has free variables, then the system has an infinite number of solutions.

Remark: A system is said to be *consistent* if it has one or more solutions [Case (i) or (iii) in Theorem 1.7], and is said to be *inconsistent* if it has no solutions [Case (ii) in Theorem 1.7]. Figure 1-3 illustrates this situation.

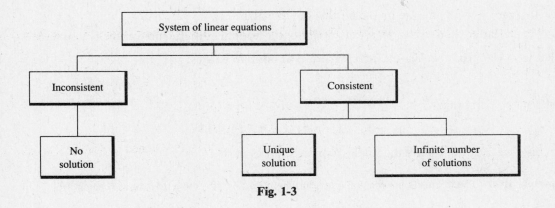

Fig. 1-3

1.7 MATRICES

Let A be a rectangular array of numbers as follows:

$$A = \begin{pmatrix} a_{11} & a_{12} & \cdots & a_{1n} \\ a_{21} & a_{22} & \cdots & a_{2n} \\ \cdots\cdots\cdots\cdots\cdots\cdots \\ a_{m1} & a_{m2} & \cdots & a_{mn} \end{pmatrix}$$

The array A is called a *matrix*. Such a matrix may be denoted by writing $A = (a_{ij})$, $i = 1, \ldots, m$, $j = 1, \ldots, n$, or simply $A = (a_{ij})$. The m horizontal n-tuples

$$(a_{11}, a_{12}, \ldots, a_{1n}), (a_{21}, a_{22}, \ldots, a_{2n}), \ldots, (a_{m1}, a_{m2}, \ldots, a_{mn})$$

are the *rows* of the matrix, and the n vertical m-tuples

$$\begin{pmatrix} a_{11} \\ a_{21} \\ \cdots \\ a_{m1} \end{pmatrix}, \begin{pmatrix} a_{12} \\ a_{22} \\ \cdots \\ a_{m2} \end{pmatrix}, \ldots, \begin{pmatrix} a_{1n} \\ a_{2n} \\ \cdots \\ a_{mn} \end{pmatrix}$$

are its *columns*. Note that the element a_{ij}, called the *ij-entry* or *ij-component*, appears in the ith row and the jth column. A matrix with m rows and n columns is called an m by n matrix, or $m \times n$ matrix; the pair of numbers (m, n) is called its *size*.

Example 1.13. Let $A = \begin{pmatrix} 1 & -3 & 4 \\ 0 & 5 & -2 \end{pmatrix}$. Then A is a 2×3 matrix. Its rows are $(1, -3, 4)$ and $(0, 5, -2)$; its columns are $\begin{pmatrix} 1 \\ 0 \end{pmatrix}, \begin{pmatrix} -3 \\ 5 \end{pmatrix}$, and $\begin{pmatrix} 4 \\ -2 \end{pmatrix}$.

The first nonzero entry in a row R of a matrix A is called the leading nonzero entry of R. If R has no leading nonzero entry, i.e., if every entry in R is 0, then R is called a zero row. If all the rows of A are zero rows, i.e., if every entry of A is 0, then A is called the zero matrix, denoted by 0.

Echelon Matrices

A matrix A is called an *echelon matrix*, or is said to be in *echelon form* if the following two conditions hold:

(i)   All zero rows, if any, are on the bottom of the matrix.

(ii) Each leading nonzero entry is to the right of the leading nonzero entry in the preceding row.

That is, $A = (a_{ij})$ is an echelon matrix if there exist nonzero entries

$$a_{1j_1}, a_{2j_2}, \ldots, a_{rj_r}, \qquad \text{where} \qquad j_1 < j_2 < \cdots < j_r$$

with the property that

$$a_{ij} = 0 \qquad \text{for } i \leq r, j < j_i, \qquad \text{and for} \qquad i > r$$

In this case, $a_{1j_1}, \ldots, a_{rj_r}$ are the leading nonzero entries of A.

Example 1.14.   The following are echelon matrices whose leading nonzero entries have been circled:

$$\begin{pmatrix} ② & 3 & 2 & 0 & 4 & 5 & -6 \\ 0 & 0 & ① & 1 & -3 & 2 & 0 \\ 0 & 0 & 0 & 0 & 0 & ⑥ & 2 \\ 0 & 0 & 0 & 0 & 0 & 0 & 0 \end{pmatrix} \begin{pmatrix} ① & 2 & 3 \\ 0 & 0 & ① \\ 0 & 0 & 0 \end{pmatrix} \begin{pmatrix} 0 & ① & 3 & 0 & 0 & 4 \\ 0 & 0 & 0 & ① & 0 & -3 \\ 0 & 0 & 0 & 0 & ① & 2 \end{pmatrix}$$

An echelon matrix A is said to be in *row canonical form* if it has the following two additional properties:

(iii) Each leading nonzero entry is 1.

(iv)   Each leading nonzero entry is the only nonzero entry in its column.

The third matrix above is an example of a matrix in row canonical form. The second matrix is not in row canonical form since the leading nonzero entry in the second row is not the only nonzero etnry in its column, there is a 3 above it. The first matrix is not in row canonical form since some leading nonzero entries are not 1.

The zero matrix 0, for any number of rows or columns, is also an example of a matrix in row canonical form.

1.8   ROW EQUIVALENCE AND ELEMENTARY ROW OPERATIONS

A matrix A is said to be *row equivalent* to a matrix B, written $A \sim B$, if B can be obtained from A by a finite sequence of the following operations called elementary row operations:

[E_1]   Interchange the ith row and the jth row: $R_i \leftrightarrow R_j$.

[E_2]   Multiply the ith row by a nonzero scalar k: $kR_i \rightarrow R_i$, $k \neq 0$.

[E_3]   Replace the ith row by k times the jth row plus the ith row: $kR_j + R_i \rightarrow R_i$.

In actual practice, we apply [E_2] and then [E_3] in one step, i.e., the operation

[E]   Replace the ith row by k' times the jth row plus k (nonzero) times the ith row:

$$k'R_j + kR_i \rightarrow R_i, \ k \neq 0.$$

The reader no doubt recognizes the similarity of the above operations and those used in solving systems of linear equations.

The following algorithm row reduces a matrix A into echelon form. (The term "row reduce" or simply "reduce" shall mean to transform a matrix by row operations.)

Algorithm A

Here $A = (a_{ij})$ is an arbitrary matrix.

Step 1. Find the first column with a nonzero entry. Suppose it is the j_1 column.

Step 2. Interchange the rows so that a nonzero entry appears in the first row of the j_1 column, that is, so that $a_{1j_1} \neq 0$.

Step 3. Use a_{1j_1} as a pivot to obtain 0s below a_{1j_1}; that is, for each $i > 1$, apply the row operation $-a_{ij_1}R_1 + a_{1j_1}R_i \to R_i$ or $(-a_{ij_1}/a_{1j_1})R_1 + R_i \to R_i$.

Step 4. Repeat Steps 1, 2, and 3 with the submatrix formed by all the rows, excluding the first row.

Step 5. Continue the above process until the matrix is in echelon form.

Example 1.15.   The matrix $A = \begin{pmatrix} 1 & 2 & -3 & 0 \\ 2 & 4 & -2 & 2 \\ 3 & 6 & -4 & 3 \end{pmatrix}$ is reduced to echelon form by Algorithm A as follows:

Use $a_{11} = 1$ as a pivot to obtain 0s below a_{11}, that is, apply the row operations $-2R_1 + R_2 \to R_2$ and $-3R_1 + R_3 \to R_3$ to obtain the matrix

$$\begin{pmatrix} 1 & 2 & -3 & 0 \\ 0 & 0 & 4 & 2 \\ 0 & 0 & 5 & 3 \end{pmatrix}$$

Now use $a_{23} = 4$ as a pivot to obtain a 0 below a_{23}, that is, apply the row operation $-5R_2 + 4R_3 \to R_3$ to obtain the matrix

$$\begin{pmatrix} 1 & 2 & -3 & 0 \\ 0 & 0 & 4 & 2 \\ 0 & 0 & 0 & 2 \end{pmatrix}$$

The matrix is now in echelon form.

The following algorithm row reduces an echelon matrix into its row canonical form.

Algorithm B

Here $A = (a_{ij})$ is in echelon form, say with leading nonzero entries

$$a_{1j_1}, a_{2j_2}, \dots, a_{rj_r}$$

Step 1. Multiply the last nonzero row R_r by $1/a_{rj_r}$ so that the leading nonzero entry is 1.

Step 2. Use $a_{rj_r} = 1$ as a pivot to obtain 0s above the pivot; that is, for $i = r-1, r-2, \dots, 1$, apply the operation

$$-a_{ir_i}R_r + R_i \to R_i$$

Step 3. Repeat Steps 1 and 2 for rows $R_{r-1}, R_{r-2}, \dots, R_2$.

Step 4. Multiply R_1 by $1/a_{1j_1}$.

Example 1.16. Using Algorithm B, the echelon matrix

$$A = \begin{pmatrix} 2 & 3 & 4 & 5 & 6 \\ 0 & 0 & 3 & 2 & 5 \\ 0 & 0 & 0 & 0 & 4 \end{pmatrix}$$

is reduced to row canonical form as follows:

Multiply R_3 by $\frac{1}{4}$ so that the leading nonzero entry equals 1; and then use $a_{35} = 1$ as a pivot to obtain 0s above it by applying the operations $-5R_3 + R_2 \to R_2$ and $-6R_3 + R_1 \to R_1$:

$$A \sim \begin{pmatrix} 2 & 3 & 4 & 5 & 6 \\ 0 & 0 & 3 & 2 & 5 \\ 0 & 0 & 0 & 0 & 1 \end{pmatrix} \sim \begin{pmatrix} 2 & 3 & 4 & 5 & 0 \\ 0 & 0 & 3 & 2 & 0 \\ 0 & 0 & 0 & 0 & 1 \end{pmatrix}$$

Multiply R_2 by $\frac{1}{3}$ so that the leading nonzero entry equals 1; and then use $a_{23} = 1$ as a pivot to obtain 0 above with the operation $-4R_2 + R_1 \to R_1$:

$$A \sim \begin{pmatrix} 2 & 3 & 4 & 5 & 0 \\ 0 & 0 & 1 & \frac{2}{3} & 0 \\ 0 & 0 & 0 & 0 & 1 \end{pmatrix} \sim \begin{pmatrix} 2 & 3 & 0 & \frac{7}{3} & 0 \\ 0 & 0 & 1 & \frac{2}{3} & 0 \\ 0 & 0 & 0 & 0 & 1 \end{pmatrix}$$

Finally, multiply R_1 by $\frac{1}{2}$ to obtain

$$\begin{pmatrix} 1 & \frac{3}{2} & 0 & \frac{7}{6} & 0 \\ 0 & 0 & 1 & \frac{2}{3} & 0 \\ 0 & 0 & 0 & 0 & 1 \end{pmatrix}$$

This matrix is the row canonical form of A.

Algorithms A and B show that any matrix is row equivalent to at least one matrix in row canonical form. In Chapter 5 we prove that such a matrix is unique, that is,

Theorem 1.8:   Any matrix A is row equivalent to a unique matrix in row canonical form (called the *row canonical form* of A).

> **Remark:**   If a matrix A is in echelon form, then its leading nonzero entries will be called *pivot entries*. The term comes from the above algorithm which row reduces a matrix to echelon form.

1.9   SYSTEMS OF LINEAR EQUATIONS AND MATRICES

The augmented matrix M of the system (*1.3*) of m linear equations in n unknowns is as follows:

$$M = \begin{pmatrix} a_{11} & a_{12} & \dots & a_{1n} & b_1 \\ a_{21} & a_{22} & \dots & a_{2n} & b_2 \\ \multicolumn{5}{c}{\dotfill} \\ a_{m1} & a_{m2} & \dots & a_{mn} & b_m \end{pmatrix}$$

Observe that each row of M corresponds to an equation of the system, and each column of M correspond to the coefficients of an unknown, except the last row which correspond to the constants of the system.

The coefficient matrix A of the system (*1.3*) is

$$A = \begin{pmatrix} a_{11} & a_{12} & \cdots & a_{1n} \\ a_{21} & a_{22} & \cdots & a_{2n} \\ \cdots\cdots\cdots\cdots\cdots \\ a_{m1} & a_{m2} & \cdots & a_{mn} \end{pmatrix}$$

Note that the coefficient matrix A may be obtained from the augmented matrix M by omitting the last column of M.

One way to solve a system of linear equations is by working with its augmented matrix M, specifically, by reducing its augmented matrix to echelon form (which tells whether the system is consistent) and then reducing it to its row canonical form (which essentially gives the solution). The justification of this process comes from the following facts:

(1) Any elementary row operation on the augmented matrix M of the system is equivalent to applying the corresponding operation on the system itself.

(2) The system has a solution if and only if the echelon form of the augmented matrix M does not have a row of the form $(0, 0, \ldots, 0, b)$ with $b = 0$.

(3) In the row canonical form of the augmented matrix M (excluding zero rows) the coefficient of each nonfree variable is a leading nonzero entry which is equal to one and is the only nonzero entry in its respective column; hence the free variable form of the solution is obtained by simply transferring the nonfree variable terms to the other side.

This process is illustrated in the following example.

Example 1.17

(*a*) The system

$$\begin{aligned} x + y - 2z + 4t &= 5 \\ 2x + 2y - 3z + t &= 3 \\ 3x + 3y - 4z - 2t &= 1 \end{aligned}$$

is solved by reducing its augmented matrix M to echelon form and then to row canonical form as follows:

$$M = \begin{pmatrix} 1 & 1 & -2 & 4 & 5 \\ 2 & 2 & -3 & 1 & 3 \\ 3 & 3 & -4 & -2 & 1 \end{pmatrix} \sim \begin{pmatrix} 1 & 1 & -2 & 4 & 5 \\ 0 & 0 & 1 & -7 & -7 \\ 0 & 0 & 2 & -14 & -14 \end{pmatrix} \sim \begin{pmatrix} 1 & 1 & 0 & -10 & -9 \\ 0 & 0 & 1 & -7 & -7 \end{pmatrix}$$

[The third row (in the second matrix) is deleted since it is a multiple of the second row and will result in a zero row.] Thus the free variable form of the general solution of the system is as follows:

$$\begin{aligned} x + y \quad\quad - 10t &= -9 \\ z - 7t &= -7 \end{aligned} \quad \text{or} \quad \begin{aligned} x &= -9 - y + 10t \\ z &= -7 + 7t \end{aligned}$$

Here the free variables are y and t, and the nonfree variables are x and z.

(*b*) The system

$$\begin{aligned} x_1 + x_2 - 2x_3 + 3x_4 &= 4 \\ 2x_1 + 3x_2 + 3x_3 - x_4 &= 3 \\ 5x_1 + 7x_2 + 4x_3 + x_4 &= 5 \end{aligned}$$

is solved as follows. First we reduce its augmented matrix to echelon form:

$$M = \begin{pmatrix} 1 & 1 & -2 & 3 & 4 \\ 2 & 3 & 3 & -1 & 3 \\ 5 & 7 & 4 & 1 & 5 \end{pmatrix} \sim \begin{pmatrix} 1 & 1 & -2 & 3 & 4 \\ 0 & 1 & 7 & -7 & -5 \\ 0 & 2 & 14 & -14 & 15 \end{pmatrix} \sim \begin{pmatrix} 1 & 1 & -2 & 3 & 4 \\ 0 & 1 & 7 & -7 & -5 \\ 0 & 0 & 0 & 0 & -5 \end{pmatrix}$$

There is no need to continue to find the row canonical form of the matrix since the echelon matrix already tells us that the system has no solution. Specifically, the third row of the echelon matrix corresponds to the degenerate equation

$$0x_1 + 0x_2 + 0x_3 + 0x_4 = -5$$

which has no solution.

(c)   The system

$$\begin{aligned} x + 2y + z &= 3 \\ 2x + 5y - z &= -4 \\ 3x - 2y - z &= 5 \end{aligned}$$

is solved by reducing its augmented matrix M to echelon form and then to row canonical form as follows:

$$M = \begin{pmatrix} 1 & 2 & 1 & 3 \\ 2 & 5 & -1 & -4 \\ 3 & -2 & -1 & 5 \end{pmatrix} \sim \begin{pmatrix} 1 & 2 & 1 & 3 \\ 0 & 1 & -3 & -10 \\ 0 & -8 & -4 & -4 \end{pmatrix} \sim \begin{pmatrix} 1 & 2 & 1 & 3 \\ 0 & 1 & -3 & -10 \\ 0 & 0 & -28 & -84 \end{pmatrix}$$

$$\sim \begin{pmatrix} 1 & 2 & 1 & 3 \\ 0 & 1 & -3 & -10 \\ 0 & 0 & 1 & 3 \end{pmatrix} \sim \begin{pmatrix} 1 & 2 & 0 & 0 \\ 0 & 1 & 0 & -1 \\ 0 & 0 & 1 & 3 \end{pmatrix} \sim \begin{pmatrix} 1 & 0 & 0 & 2 \\ 0 & 1 & 0 & -1 \\ 0 & 0 & 1 & 3 \end{pmatrix}$$

Thus the system has the unique solution $x = 2$, $y = -1$, $z = 3$ or $u = (2, -1, 3)$. (Note that the echelon form of M already indicated that the solution was unique since it corresponded to a triangular system.)

1.10   HOMOGENEOUS SYSTEMS OF LINEAR EQUATIONS

The system (*1.3*) of linear equations is said to be *homogeneous* if all the constants are equal to zero, that is, if the system has the form

$$\begin{aligned} a_{11}x_1 + a_{12}x_2 + \cdots + a_{1n}x_n &= 0 \\ a_{21}x_1 + a_{22}x_2 + \cdots + a_{2n}x_n &= 0 \\ \cdots\cdots\cdots\cdots\cdots\cdots\cdots\cdots\cdots\cdots\cdots\cdots \\ a_{m1}x_1 + a_{m2}x_2 + \cdots + a_{mn}x_n &= 0 \end{aligned} \qquad (1.6)$$

In fact, the system (*1.6*) is called the homogeneous system associated with the system (*1.3*).

The homogeneous system (*1.6*) always has a solution, namely the zero n-tuple $0 = (0, 0, \ldots, 0)$ called the zero or trivial solution. (Any other solution, if it exists, is called a nonzero or nontrivial solution.) Thus it can always be reduced to an equivalent homogeneous system in echelon form:

$$\begin{aligned} a_{11}x_1 + a_{12}x_2 + a_{13}x_3 + \cdots + a_{1n}x_n &= 0 \\ a_{2j_2}x_{j_2} + a_{2, j_2+1}x_{j_2+1} + \cdots + a_{2n}x_n &= 0 \\ \cdots\cdots\cdots\cdots\cdots\cdots\cdots\cdots\cdots\cdots\cdots\cdots\cdots \\ a_{rj_r}x_{j_r} + a_{r, j_r+1}x_{j_r+1} + \cdots + a_{rn}x_n &= 0 \end{aligned} \qquad (1.7)$$

There are two possibilities:

(i) $r = n$. Then the system has only the zero solution.

(ii) $r < n$. Then the system has a nonzero solution.

Accordingly, if we begin with fewer equations than unknowns then, in echelon form, $r < n$ and hence the system has a nonzero solution. This proves the following important theorem.

Theorem 1.9: A homogeneous system of linear equations with more unknowns than equations has a nonzero solution.

Example 1.18

(a) The homogeneous system

$$x + 2y - 3z + w = 0$$
$$x - 3y + z - 2w = 0$$
$$2x + y - 3z + 5w = 0$$

has a nonzero solution since there are four unknowns but only three equations.

(b) We reduce the following system to echelon form:

$$x + y - z = 0 \qquad x + y - z = 0 \qquad x + y - z = 0$$
$$2x - 3y + z = 0 \qquad -5y + 3z = 0 \qquad -5y + 3z = 0$$
$$x - 4y + 2z = 0 \qquad -5y + 3z = 0$$

The system has a nonzero solution, since we obtained only two equations in the three unknowns in echelon form. For example, let $z = 5$; then $y = 3$ and $x = 2$. In other words, the 3-tuple $(2, 3, 5)$ is a particular nonzero solution.

(c) We reduce the following system to echelon form:

$$x + y - z = 0 \qquad x + y - z = 0 \qquad x + y - z = 0$$
$$2x + 4y - z = 0 \qquad 2y + z = 0 \qquad 2y + z = 0$$
$$3x + 2y + 2z = 0 \qquad -y + 5z = 0 \qquad 11z = 0$$

Since in echelon form there are three equations in three unknowns, the given system has only the zero solution $(0, 0, 0)$.

Basis for the General Solution of a Homogeneous System

Let W denote the general solution of a homogeneous system. Nonzero solution vectors $u_1, u_2, \ldots, u_s$ are said to form a *basis* of W if every solution vector w in W can be expressed uniquely as a linear combination of $u_1, u_2, \ldots, u_s$. The number s of such basis vectors is called the *dimension* of W, written $\dim W = s$. (If $W = \{0\}$, we define $\dim W = 0$.)

The following theorem, proved in Chapter 5, tells us how to find such a basis.

Theorem 1.10: Let W be the general solution of a homogeneous system, and suppose an echelon form of the system has s free variables. Let $u_1, u_2, \ldots, u_s$ be the solutions obtained by setting one of the free variables equal to one (or any nonzero constant) and the remaining free variables equal to zero. Then $\dim W = s$ and $u_1, u_2, \ldots, u_s$ form a basis of W.

Remark: The above term *linear combination* refers to multiplying vectors by scalars and adding, where such operations are defined by

$$(a_1, a_2, \ldots, a_n) + (b_1, b_2, \ldots, b_n) = (a_1 + b_1, a_2 + b_2, \ldots, a_n + b_n)$$
$$k(a_1, a_2, \ldots, a_n) = (ka_1, ka_2, \ldots, ka_n)$$

These operations are studied in detail in Chapter 2.

Example 1.19 Suppose we want to find the dimension and a basis for the general solution W of the homogeneous system

$$x + 2y - 3z + 2s - 4t = 0$$
$$2x + 4y - 5z + s - 6t = 0$$
$$5x + 10y - 13z + 4s - 16t = 0$$

First we reduce the system to echelon. Applying the operations $-2L_1 + L_2 \to L_2$ and $-5L_2 + L_3 \to L_3$, and then $-2L_2 + L_3 \to L_3$, yields:

$$x + 2y - 3z + 2s - 4t = 0$$
$$z - 3s + 2t = 0 \qquad \text{and} \qquad x + 2y - 3z + 2s - 4t = 0$$
$$2z - 6s + 4t = 0 \qquad\qquad\qquad\qquad\quad z - 3s + 2t = 0$$

In echelon form, the system has three free variables, y, s and t; hence dim $(W) = 3$. Three solution vectors which form a basis for W are obtained as follows:

(1)   Set $y = 1, s = 0, t = 0$. Back-substitution yields the solution $u_1 = (-2, 1, 0, 0, 0)$.
(2)   Set $y = 0, s = 1, t = 0$. Back-substitution yields the solution $u_2 = (7, 0, 3, 1, 0)$.
(3)   Set $y = 0, s = 0, t = 1$. Back-substitution yields the solution $u_3 = (-2, 0, -2, 0, 1)$.

The set $[u_1, u_2, u_3]$ is a basis for W.

Now any solution of the system can be written in the form

$$au_1 + bu_2 + cu_3 = a(-2, 1, 0, 0, 0) + b(7, 0, 3, 1, 0) + c(-2, 0, -2, 0, 1)$$
$$= (-2a + 7b - 2c, a, 3b - 2c, b, c)$$

where a, b, c are arbitrary constants. Observe that this is nothing other than the parametric form of the general solution under the choice of parameters $y = a, s = b, t = c$.

Nonhomogeneous and Associated Homogeneous Systems

The relationship between the nonhomogeneous system (*1.3*) and its associated homogeneous system (*1.6*) is contained in the following theorem whose proof is postponed until Chapter 3 (Theorem 3.5).

Theorem 1.11:   Let v_0 be a particular solution and let U be the general solution of a nonhomogeneous system of linear equations. Then

$$U = v_0 + W = \{v_0 + w : w \in W\}$$

where W is the general solution of the associated homogeneous system.
That is, $U = v_0 + W$ may be obtained by adding v_0 to each element of W.

The above theorem has a geometrical interpretation in the space $\mathbf{R}^3$. Specifically, if W is a line through the origin, then, as pictured in Fig. 1-4, $U = v_0 + W$ is the line parallel to W which can be obtained by adding v_0 to each element in W. Similarly, whenever W is a plane through the origin, then $U = v_0 + W$ is a plane parallel to W.

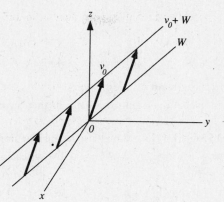

Fig. 1-4

Solved Problems

LINEAR EQUATIONS, SOLUTIONS

1.1. Determine whether each equation is linear:

    (a) $5x + 7y - 8yz = 16$ *(b)* $x + \pi y + ez = \log 5$ *(c)* $3x + ky - 8z = 16$

    (a)   No, since the product yz of two unknowns is of second degree.

    (b)   Yes, since π, e, and $\log 5$ are constants.

    (c)   As it stands, there are four unknowns: x, y, z, k. Because of the term ky it is not a linear equation. However, assuming k is a constant, the equation is linear in the unknowns x, y, z.

1.2. Consider the linear equation $x + 2y - 3z = 4$. Determine whether $u = (8, 1, 2)$ is a solution.

    Since x, y, z is the ordering of the unknowns, $u = (8, 1, 2)$ is short for $x = 8$, $y = 1$, $z = 2$. Substitute in the equation to obtain

$$8 + 2(1) - 3(2) = 4 \quad \text{or} \quad 8 + 2 - 6 = 4 \quad \text{or} \quad 4 = 4$$

Yes, it is a solution

1.3. Determine whether *(a)* $u = (3, 2, 1, 0)$ and *(b)* $v = (1, 2, 4, 5)$ are solutions of the equation $x_1 + 2x_2 - 4x_3 + x_4 = 3$.

    (a)   Substitute to obtain $3 + 2(2) - 4(1) + 0 = 3$, or $3 = 3$; yes, it is a solution.

    (b)   Substitute to obtain $1 + 2(2) - 4(4) + 5 = 3$, or $-6 = 3$; not a solution.

1.4. Is $u = (6, 4, -2)$ a solution of the equation $3x_2 + x_3 - x_1 = 4$?

    By convention, the components of u are ordered according to the subscripts on the unknowns. That is, $u = (6, 4, -2)$ is short for $x_1 = 6$, $x_2 = 4$, $x_3 = -2$. Substitute in the equation to obtain $3(4) - 2 - 6 = 4$, or $4 = 4$. Yes, it is a solution.

1.5. Prove Theorem 1.1.

    Suppose $a \neq 0$. Then the scalar b/a exists. Substituting b/a in $ax = b$ yields $a(b/a) = b$, or $b = b$; hence b/a is a solution. On the other hand, suppose x_0 is a solution to $ax = b$, so that $ax_0 = b$. Multiplying both sides by $1/a$ yields $x_0 = b/a$. Hence b/a is the unique solution of $ax = b$. Thus (i) is proved.

    On the other hand, suppose $a = 0$. Then, for any scalar k, we have $ak = 0k = 0$. If $b \neq 0$, then $ak \neq b$. Accordingly, k is not a solution of $ax = b$ and so (ii) is proved. If $b = 0$, then $ak = b$. That is, any scalar k is a solution of $ax = b$ and so (iii) is proved.

1.6.   Solve each equation:

$$(a) \;\; ex = \log 5 \qquad (c) \;\; 3x - 4 - x = 2x + 3$$
$$(b) \;\; cx = 0 \qquad\quad (d) \;\; 7 + 2x - 4 = 3x + 3 - x$$

(a)   Since $e \neq 0$, multiply by $1/e$ to obtain $x = (\log 5)/e$.

(b)   If $c \neq 0$, then $0/c = 0$ is the unique solution. If $c = 0$, then every scalar k is a solution [Theorem 1.1(iii)].

(c)   Rewrite in standard form, $2x - 4 = 2x + 3$ or $0x = 7$. The equation has no solution [Theorem 1.1(ii)].

(d)   Rewrite in standard form, $3 + 2x = 2x + 3$ or $0x = 0$. Every scalar k is a solution [Theorem 1.1(iii)].

1.7.   Describe the solutions of the equation $2x + y + x - 5 = 2y + 3x - y + 4$.

Rewrite in standard form by collecting terms and transposing:

$$3x + y - 5 = y + 3x + 4 \qquad \text{or} \qquad 0x + 0y = 9$$

The equation is degenerate with a nonzero constant; thus the equation has no solution.

1.8.   Describe the solutions of the equation $2y + 3x - y + 4 = x + 3 + y + 1 + 2x$.

Rewrite in standard form by collecting terms and transposing:

$$y + 3x + 4 = 3x + 4 + y \qquad \text{or} \qquad 0x + 0y = 0$$

The equation is degenerate with a zero constant; thus every vector $u = (a, b)$ in $\mathbf{R}^2$ is a solution.

1.9.   Prove Theorem 1.3.

First we prove (i). Set $x_j = k_j$ for $j \neq p$. Because $a_j = 0$ for $j < p$, substitution in the equation yields

$$a_p x_p + a_{p+1} k_{p+1} + \cdots + a_n k_n = b \qquad \text{or} \qquad a_p x_p = b - a_{p+1} k_{p+1} - \cdots - a_n k_n$$

with $a_p \neq 0$. By Theorem 1.1(i), x_p is uniquely determined as

$$x_p = \frac{1}{a_p}(b - a_{p+1} k_{p+1} - \cdots - a_n k_n)$$

Thus (i) is proved.

Now we prove (ii). Suppose $u = (k_1, k_2, \ldots, k_n)$ is a solution. Then

$$a_p k_p + a_{p+1} k_{p+1} + \cdots + a_n k_n = b \qquad \text{or} \qquad k_p = \frac{1}{a_p}(b - a_{p+1} k_{p+1} - \cdots - a_n k_n)$$

This, however, is precisely the solution

$$u = \left(k_1, \ldots, k_{p-1}, \frac{b - a_{p+1} k_{p+1} - \cdots - a_n k_n}{a_p}, k_{p+1}, \ldots, k_n\right)$$

obtained in (i). Thus (ii) is proved.

1.10.   Consider the linear equation $x - 2y + 3z = 4$. Find (a) three particular solutions and (b) the general solution.

(a)   Here x is the leading unknown. Accordingly, assign any values to the free variables y and z, and then solve for x to obtain a solution. For example:

(1)   Set $y = 1$ and $z = 1$. Substitution in the equation yields

$$x - 2(1) + 3(1) = 4 \qquad \text{or} \qquad x - 2 + 3 = 4 \qquad \text{or} \qquad x = 3$$

Thus $u_1 = (3, 1, 1)$ is a solution.

(2)   Set $y = 1, z = 0$. Substitution yields $x = 6$; hence $u_2 = (6, 1, 0)$ is a solution.

(3)   Set $y = 0, z = 1$. Substitution yields $x = 1$; hence $u_3 = (1, 0, 1)$ is a solution.

(b)   To find the general solution, assign arbitrary values to the free variables, say $y = a$ and $z = b$. (We call a and b parameters of the solution.) Then substitute in the equation to obtain

$$x - 2a + 3b = 4 \qquad \text{or} \qquad x = 4 + 2a - 3b$$

Thus $u = (4 + 2a - 3b, a, b)$ is the general solution.

SYSTEMS IN TRIANGULAR AND ECHELON FORM

1.11.   Solve the system

$$\begin{aligned} 2x - 3y + 5z - 2t &= 9 \\ 5y - z + 3t &= 1 \\ 7z - t &= 3 \\ 2t &= 8 \end{aligned}$$

The system is in triangular form; hence we solve by back-substitution.

(i)   The last equation gives $t = 4$.

(ii)   Substituting in the third equation gives $7z - 4 = 3$, or $7z = 7$, or $z = 1$.

(iii) Substituting $z = 1$ and $t = 4$ in the second equation gives

$$5y - 1 + 3(4) = 1 \quad \text{or} \quad 5y - 1 + 12 = 1 \quad \text{or} \quad 5y = -10 \quad \text{or} \quad y = -2$$

(iv)   Substituting $y = -2, z = 1, t = 4$ in the first equation gives

$$2x - 3(-2) + 5(1) - 2(4) = 9 \quad \text{or} \quad 2x + 6 + 5 - 8 = 9 \quad \text{or} \quad 2x = 6 \quad \text{or} \quad x = 3$$

Thus $x = 3, y = -2, z = 1, t = 4$ is the unique solution of the system.

1.12.   Determine the free variables in each system:

$$\begin{array}{lll} 3x + 2y - 5z - 6s + 2t = 4 & 5x - 3y + 7z = 1 & x + 2y - 3z = 2 \\ z + 8s - 3t = 6 & 4y + 5z = 6 & 2x - 3y + z = 1 \\ s - 5t = 5 & 4z = 9 & 5x - 4y - z = 4 \\ \qquad\qquad (a) & \qquad (b) & \qquad (c) \end{array}$$

(a)   In the echelon form, any unknown that is not a leading unknown is termed a free variable. Here, y and t are the free variables.

(b)   The leading unknowns are x, y, z. Hence there are no free variables (as in any triangular system).

(c)   The notion of free variable applies only to a system in echelon form.

1.13.   Prove Theorem 1.6.

There are two cases:

(i) $r = n$.   That is, there are as many equations as unknowns. Then the system has a unique solution.

(ii) $r < n$.   That is, there are fewer equations than unknowns. Then we can arbitrarily assign values to the $n - r$ free variables and obtain a solution of the system.

The proof is by induction on the number r of equations in the system. If $r = 1$, then we have a single, nondegenerate, linear equation, to which Theorem 1.3 applies when $n > r = 1$ and Theorem 1.1 applies when $n = r = 1$. Thus the theorem holds for $r = 1$.

Now assume that $r > 1$ and that the theorem is true for a system of $r - 1$ equations. We view the $r - 1$ equations

$$a_{2j_2} x_{j_2} + a_{2, j_2+1} x_{j_2+1} + \cdots + a_{2n} x_n = b_2$$
$$\cdots\cdots\cdots\cdots\cdots\cdots\cdots\cdots\cdots\cdots\cdots\cdots$$
$$a_{rj_r} x_{j_r} + a_{r, j_r+1} x_{j_r+1} + \cdots + a_{rn} x_n = b_r$$

as a system in the unknowns $x_{j_2}, \ldots, x_n$. Note that the system is in echelon form. By the induction hypothesis, we can arbitrarily assign values to the $(n - j_2 + 1) - (r - 1)$ free variables in the reduced system to obtain a solution (say, $x_{j_2} = k_{j_2}, \ldots, x_n = k_n$). As in case $r = 1$, these values together with arbitrary values for the additional $j_2 - 2$ free variables (say $x_2 = k_2, \ldots, x_{j_2-1} = k_{j_2-1}$) yield a solution of the first equation with

$$x_1 = \frac{1}{a_{11}}(b_1 - a_{12} k_2 - \cdots - a_{1n} k_n)$$

[Note that there are $(n - j_2 + 1) - (r - 1) + (j_2 - 2) = n - r$ free variables.] Furthermore, these values for $x_1, \ldots, x_n$ also satisfy the other equations since, in these equations, the coefficients of $x_1, \ldots, x_{j_2-1}$ are zero.

Now if $r = n$, then $j_2 = 2$. Thus by induction we obtain a unique solution of the subsystem and then a unique solution of the entire system. Accordingly, the theorem is proven.

1.14. Find the general solution of the echelon system

$$\begin{aligned} x - 2y - 3z + 5s - 2t &= 4 \\ 2z - 6s + 3t &= 2 \\ 5t &= 10 \end{aligned}$$

Since the equations begin with the unknowns x, z, and t, respectively, the other unknowns y and s are the free variables. To find the general solution, assign parameters to the free variables, say $y = a$ and $s = b$, and use back-substitution to solve for the nonfree variables x, z, and t.

(i)   The last equation yields $t = 2$.

(ii)   Substitute $t = 2$, $s = b$ in the second equation to obtain

$$2z - 6b + 3(2) = 2 \quad \text{or} \quad 2z - 6b + 6 = 2 \quad \text{or} \quad 2z = 6b - 4 \quad \text{or} \quad z = 3b - 2$$

(iii)   Substitute $t = 2$, $s = b$, $z = 3b - 2$, $y = a$ in the first equation to obtain

$$x - 2a - 3(3b - 2) + 5b - 2(2) = 4 \quad \text{or} \quad x - 2a - 9b + 6 + 5b - 4 = 4$$

$$\text{or} \quad x = 2a + 4b + 2$$

Thus

$$x = 2a + 4b + 2 \qquad y = a \qquad z = 3b - 2 \qquad s = b \qquad t = 2$$

or, equivalently,

$$u = (2a + 4b + 2, a, 3b - 2, b, 2)$$

is the parametric form of the general solution.

Alternately, solving for x, z, and t in terms of the free variables y and s yields the following free variable form of the general solution:

$$x = 2y + 4s + 2 \qquad z = 3s - 2 \qquad t = 2$$

SYSTEMS OF LINEAR EQUATIONS, GAUSSIAN ELIMINATION

1.15. Solve the system

$$\begin{aligned} x - 2y + z &= 7 \\ 2x - y + 4z &= 17 \\ 3x - 2y + 2z &= 14 \end{aligned}$$

Reduce to echelon form. Apply $-2L_1 + L_2 \to L_2$ and $-3L_1 + L_3 \to L_3$ to eliminate x from the second and third equations, and then apply $-4L_2 + 3L_3 \to L_3$ to eliminate y from the third equation. These operations yield

$$x - 2y + z = 7 \qquad\qquad x - 2y + z = 7$$
$$3y + 2z = 3 \quad \text{and} \qquad 3y + 2z = 3$$
$$4y - z = -7 \qquad\qquad\qquad -11z = -33$$

The system is in triangular form, and hence, after back-substitution, has the unique solution $u = (2, -1, 3)$.

1.16. Solve the system

$$2x - 5y + 3z - 4s + 2t = 4$$
$$3x - 7y + 2z - 5s + 4t = 9$$
$$5x - 10y - 5z - 4s + 7t = 22$$

Reduce the system to echelon form. Apply the operations $-3L_1 + 2L_2 \to L_2$ and $-5L_1 + 2L_3 \to L_3$, and then $-5L_2 + L_3 \to L_3$ to obtain

$$2x - 5y + 3z - 4s + 2t = 4 \qquad\qquad 2x - 5y + 3z - 4s + 2t = 4$$
$$y - 5z + 2s + 2t = 6 \quad \text{and} \qquad y - 5z + 2s + 2t = 6$$
$$5y - 25z + 12s + 4t = 24 \qquad\qquad\qquad 2s - 6t = -6$$

The system is now in echelon form. Solving for the leading unknowns, x, y, and s, in terms of the free variables, z and t, we obtain the free-variable form of the general solution:

$$x = 26 + 11z - 15t \qquad y = 12 + 5z - 8t \qquad s = -3 + 3t$$

From this follows at once the parametric form of the general solution (where $z = a$, $t = b$):

$$x = 26 + 11a - 15b \qquad y = 12 + 5a - 8b \qquad z = a \qquad s = -3 + 3b \qquad t = b$$

1.17. Solve the system

$$x + 2y - 3z + 4t = 2$$
$$2x + 5y - 2z + t = 1$$
$$5x + 12y - 7z + 6t = 7$$

Reduce the system to echelon form. Eliminate x from the second and third equations by the operations $-2L_1 + L_2 \to L_2$ and $-5L_1 + L_3 \to L_3$; this yields the system

$$x + 2y - 3z + 4t = 2$$
$$y + 4z - 7t = -3$$
$$2y + 8z - 14t = -3$$

The operation $-2L_2 + L_3 \to L_3$ yields the degenerate equation $0 = 3$. Thus the system has no solution (even though the system has more unknowns than equations).

1.18. Determine the values of k so that the following system in unknowns x, y, z has: (i) a unique solution, (ii) no solution, (iii) an infinite number of solutions.

$$x + y - z = 1$$
$$2x + 3y + kz = 3$$
$$x + ky + 3y = 2$$

Reduce the system to echelon form. Eliminate x from the second and third equations by the operations $-2L_1 + L_2 \to L_2$ and $-L_1 + L_3 \to L_3$ to obtain

$$x + y - z = 1$$
$$y + (k + 2)z = 1$$
$$(k - 1)y + 4z = 1$$

To eliminate y from the third equation, apply the operation $-(k-1)L_2 + L_3 \rightarrow L_3$ to obtain

$$
\begin{aligned}
x + y - \qquad\quad z &= 1 \\
y + \qquad (k+2)z &= 1 \\
(3+k)(2-k)z &= 2-k
\end{aligned}
$$

The system has a unique solution if the coefficient of z in the third equation is not zero; that is, if $k \neq 2$ and $k \neq -3$. In case $k = 2$, the third equation reduces to $0 = 0$ and the system has an infinite number of solutions (one for each value of z). In case $k = -3$, the third equation reduces to $0 = 5$ and the system has no solution. Summarizing: (i) $k \neq 2$ and $k \neq 3$, (ii) $k = -3$, (iii) $k = 2$.

1.19. What condition must be placed on a, b, and c so that the following system in unknowns x, y, and z has a solution?

$$
\begin{aligned}
x + 2y - 3z &= a \\
2x + 6y - 11z &= b \\
x - 2y + 7z &= c
\end{aligned}
$$

    Reduce to echelon form. Eliminating x from the second and third equation by the operations $-2L_1 + L_2 \rightarrow L_2$ and $-L_1 + L_3 \rightarrow L_3$, we obtain the equivalent system

$$
\begin{aligned}
x + 2y - 3z &= a \\
2y - 5z &= b - 2a \\
-4y + 10z &= c - a
\end{aligned}
$$

Eliminating y from the third equation by the operation $2L_2 + L_3 \rightarrow L_3$, we finally obtain the equivalent system

$$
\begin{aligned}
x + 2y - 3z &= a \\
2y - 5z &= b - 2a \\
0 &= c + 2b - 5a
\end{aligned}
$$

The system will have no solution if $c + 2b - 5a \neq 0$. Thus the system will have at least one solution if $c + 2b - 5a = 0$, or $5a = 2b + c$. Note, in this case, that the system will have infinitely many solutions. In other words, the system cannot have a unique solution.

MATRICES, ECHELON MATRICES, ROW REDUCTION

1.20. Interchange the rows in each of the following matrices to obtain an echelon matrix:

$$
\begin{pmatrix}
0 & 1 & -3 & 4 & 6 \\
4 & 0 & 2 & 5 & -3 \\
0 & 0 & 7 & -2 & 8
\end{pmatrix}
\qquad
\begin{pmatrix}
0 & 0 & 0 & 0 & 0 \\
1 & 2 & 3 & 4 & 5 \\
0 & 0 & 5 & -4 & 7
\end{pmatrix}
\qquad
\begin{pmatrix}
0 & 2 & 2 & 2 & 2 \\
0 & 3 & 1 & 0 & 0 \\
0 & 0 & 0 & 0 & 0
\end{pmatrix}
$$

$$
(a) \qquad\qquad\qquad\qquad\qquad (b) \qquad\qquad\qquad\qquad\qquad (c)
$$

(a) Interchange the first and second rows, i.e., apply the elementary row operation $R_1 \leftrightarrow R_2$.

(b) Bring the zero row to the bottom of the matrix, i.e., apply $R_1 \leftrightarrow R_2$ and then $R_2 \leftrightarrow R_3$.

(c) No amount of row interchanges can produce an echelon matrix.

1.21. Row reduce the following matrix to echelon form:

$$
A = \begin{pmatrix}
1 & 2 & -3 & 0 \\
2 & 4 & -2 & 2 \\
3 & 6 & -4 & 3
\end{pmatrix}
$$

Use $a_{11} = 1$ as a pivot to obtain 0s below a_{11}; that is, apply the row operations $-2R_1 + R_2 \rightarrow R_2$ and $-3R_1 + R_3 \rightarrow R_3$ to obtain the matrix

$$\begin{pmatrix} 1 & 2 & -3 & 0 \\ 0 & 0 & 4 & 2 \\ 0 & 0 & 5 & 3 \end{pmatrix}$$

Now use $a_{23} = 4$ as a pivot to obtain a 0 below a_{23}; that is, apply the row operation $-5R_2 + 4R_3 \rightarrow R_3$ to obtain the matrix

$$\begin{pmatrix} 1 & 2 & -3 & 0 \\ 0 & 0 & 4 & 2 \\ 0 & 0 & 0 & 2 \end{pmatrix}$$

which is in echelon form.

1.22. Row reduce the following matrix to echelon form:

$$B = \begin{pmatrix} -4 & 1 & -6 \\ 1 & 2 & -5 \\ 6 & 3 & -4 \end{pmatrix}$$

Hand calculations are usually simpler if the pivot element equals 1. Therefore, first interchange R_1 and R_2; then apply $4R_1 + R_2 \rightarrow R_2$ and $-6R_1 + R_3 \rightarrow R_3$; and then apply $R_2 + R_3 \rightarrow R_3$:

$$B \sim \begin{pmatrix} 1 & 2 & -5 \\ -4 & 1 & -6 \\ 6 & 3 & -4 \end{pmatrix} \sim \begin{pmatrix} 1 & 2 & -5 \\ 0 & 9 & -26 \\ 0 & -9 & 26 \end{pmatrix} \sim \begin{pmatrix} 1 & 2 & -5 \\ 0 & 9 & -26 \\ 0 & 0 & 0 \end{pmatrix}$$

The matrix is now in echelon form.

1.23. Describe the *pivoting* row reduction algorithm. Also, describe the advantages, if any, of using this pivoting algorithm.

The row reduction algorithm becomes a pivoting algorithm if the entry in column j of greatest absolute value is chosen as the pivot a_{1j_i} and if one uses the row operation

$$(-a_{ij_i}/a_{1j_i})R_1 + R_i \rightarrow R_i$$

The main advantage of the pivoting algorithm is that the above row operation involves division by the (current) pivot a_{1j_i} and, on the computer, roundoff errors may be substantially reduced when one divides by a number as large in absolute value as possible.

1.24. Use the pivoting algorithm to reduce the following matrix A to echelon form:

$$A = \begin{pmatrix} 2 & -2 & 2 & 1 \\ -3 & 6 & 0 & -1 \\ 1 & -7 & 10 & 2 \end{pmatrix}$$

First interchange R_1 and R_2 so that -3 can be used as the pivot, and then apply $(\frac{2}{3})R_1 + R_2 \rightarrow R_2$ and $(\frac{1}{3})R_1 + R_3 \rightarrow R_3$:

$$A \sim \begin{pmatrix} -3 & 6 & 0 & -1 \\ 2 & -2 & 2 & 1 \\ 1 & -7 & 10 & 2 \end{pmatrix} \sim \begin{pmatrix} -3 & 6 & 0 & -1 \\ 0 & 2 & 2 & \frac{1}{3} \\ 0 & -5 & 10 & \frac{5}{3} \end{pmatrix}$$

Now interchange R_2 and R_3 so that -5 may be used as the pivot, and apply $(\frac{2}{5})R_2 + R_3 \to R_3$:

$$A \sim \begin{pmatrix} -3 & 6 & 0 & -1 \\ 0 & -5 & 10 & \frac{5}{3} \\ 0 & 2 & 2 & \frac{1}{3} \end{pmatrix} \sim \begin{pmatrix} -3 & 6 & 0 & -1 \\ 0 & -5 & 10 & \frac{5}{3} \\ 0 & 0 & 6 & 1 \end{pmatrix}$$

The matrix has been brought to echelon form.

ROW CANONICAL FORM

1.25. Which of the following echelon matrices are in row canonical form?

$$\begin{pmatrix} 1 & 2 & -3 & 0 & 1 \\ 0 & 0 & 5 & 2 & -4 \\ 0 & 0 & 0 & 7 & 3 \end{pmatrix} \quad \begin{pmatrix} 0 & 1 & 7 & -5 & 0 \\ 0 & 0 & 0 & 0 & 1 \\ 0 & 0 & 0 & 0 & 0 \end{pmatrix} \quad \begin{pmatrix} 1 & 0 & 5 & 0 & 2 \\ 0 & 1 & 2 & 0 & 4 \\ 0 & 0 & 0 & 1 & 7 \end{pmatrix}$$

    The first matrix is not in row canonical form since, for example, two leading nonzero entries are 5 and 7, not 1. Also, there are nonzero entries above the leading nonzero entries 5 and 7. The second and third matrices are in row canonical form.

1.26. Reduce the following matrix to row canonical form:

$$B = \begin{pmatrix} 2 & 2 & -1 & 6 & 4 \\ 4 & 4 & 1 & 10 & 13 \\ 6 & 6 & 0 & 20 & 19 \end{pmatrix}$$

    First, reduce B to an echelon form by applying $-2R_1 + R_2 \to R_2$ and $-3R_1 + R_3 \to R_3$, and then $-R_2 + R_3 \to R_3$:

$$B \sim \begin{pmatrix} 2 & 2 & -1 & 6 & 4 \\ 0 & 0 & 3 & -2 & 5 \\ 0 & 0 & 3 & 2 & 7 \end{pmatrix} \sim \begin{pmatrix} 2 & 2 & -1 & 6 & 4 \\ 0 & 0 & 3 & -2 & 5 \\ 0 & 0 & 0 & 4 & 2 \end{pmatrix}$$

Next reduce the echelon matrix to row canonical form. Specifically, first multiply R_3 by $\frac{1}{4}$, so the pivot $b_{34} = 1$, and then apply $2R_3 + R_2 \to R_2$ and $-6R_3 + R_1 \to R_1$:

$$B \sim \begin{pmatrix} 2 & 2 & -1 & 6 & 4 \\ 0 & 0 & 3 & -2 & 5 \\ 0 & 0 & 0 & 1 & \frac{1}{2} \end{pmatrix} \sim \begin{pmatrix} 2 & 2 & -1 & 0 & 1 \\ 0 & 0 & 3 & 0 & 6 \\ 0 & 0 & 0 & 1 & \frac{1}{2} \end{pmatrix}$$

Now multiply R_2 by $\frac{1}{3}$, making the pivot $b_{23} = 1$, and apply $R_2 + R_1 \to R_1$:

$$B \sim \begin{pmatrix} 2 & 2 & -1 & 0 & 1 \\ 0 & 0 & 1 & 0 & 2 \\ 0 & 0 & 0 & 1 & \frac{1}{2} \end{pmatrix} \sim \begin{pmatrix} 2 & 2 & 0 & 0 & 3 \\ 0 & 0 & 1 & 0 & 2 \\ 0 & 0 & 0 & 1 & \frac{1}{2} \end{pmatrix}$$

Finally, multiply R_1 by $\frac{1}{2}$ to obtain the row canonical form

$$B \sim \begin{pmatrix} 1 & 1 & 0 & 0 & \frac{3}{2} \\ 0 & 0 & 1 & 0 & 2 \\ 0 & 0 & 0 & 1 & \frac{1}{2} \end{pmatrix}$$

1.27. Reduce the following matrix to row canonical form:

$$A = \begin{pmatrix} 1 & -2 & 3 & 1 & 2 \\ 1 & 1 & 4 & -1 & 3 \\ 2 & 5 & 9 & -2 & 8 \end{pmatrix}$$

First reduce A to echelon form by applying $-R_1 + R_2 \to R_2$ and $-2R_1 + R_3 \to R_3$, and then applying $-3R_2 + R_3 \to R_3$:

$$A \sim \begin{pmatrix} 1 & -2 & 3 & 1 & 2 \\ 0 & 3 & 1 & -2 & 1 \\ 0 & 9 & 3 & -4 & 4 \end{pmatrix} \sim \begin{pmatrix} 1 & -2 & 3 & 1 & 2 \\ 0 & 3 & 1 & -2 & 1 \\ 0 & 0 & 0 & 2 & 1 \end{pmatrix}$$

Now use back-substitution. Multiply R_3 by $\frac{1}{2}$ to obtain the pivot $a_{34} = 1$, and then apply $2R_3 + R_2 \to R_2$ and $-R_3 + R_1 \to R_1$:

$$A \sim \begin{pmatrix} 1 & -2 & 3 & 1 & 2 \\ 0 & 3 & 1 & -2 & 1 \\ 0 & 0 & 0 & 1 & \frac{1}{2} \end{pmatrix} \sim \begin{pmatrix} 1 & -2 & 3 & 0 & \frac{3}{2} \\ 0 & 3 & 1 & 0 & 2 \\ 0 & 0 & 0 & 1 & \frac{1}{2} \end{pmatrix}$$

Now multiply R_2 by $\frac{1}{3}$ to obtain the pivot $a_{22} = 1$, and then apply $2R_2 + R_1 \to R_1$:

$$A \sim \begin{pmatrix} 1 & -2 & 3 & 0 & \frac{3}{2} \\ 0 & 1 & \frac{1}{3} & 0 & \frac{2}{3} \\ 0 & 0 & 0 & 1 & \frac{1}{2} \end{pmatrix} \sim \begin{pmatrix} 1 & 0 & \frac{11}{3} & 0 & \frac{17}{6} \\ 0 & 1 & \frac{1}{3} & 0 & \frac{2}{3} \\ 0 & 0 & 0 & 1 & \frac{1}{2} \end{pmatrix}$$

Since $a_{11} = 1$, the last matrix is the desired row canonical form.

1.28. Describe the Gauss–Jordan elimination algorithm which reduces an arbitrary matrix A to its row canonical form.

The Gauss–Jordan algorithm is similar to the Gaussian elimination algorithm except that here the algorithm first normalizes a row to obtain a unit pivot and then uses the pivot to place 0s both below and above the pivot before obtaining the next pivot.

1.29. Use Gauss–Jordan elimination to obtain the row canonical form of the matrix of Problem 1.27.

Use the leading nonzero entry $a_{11} = 1$ as pivot to put 0s below it, applying $-R_1 + R_2 \to R_2$ and $-2R_1 + R_3 \to R_3$; this yields

$$A \sim \begin{pmatrix} 1 & -2 & 3 & 1 & 2 \\ 0 & 3 & 1 & -2 & 1 \\ 0 & 9 & 3 & -4 & 4 \end{pmatrix}$$

Multiply R_2 by $\frac{1}{3}$ to get the pivot $a_{22} = 1$ and produce 0s below and above a_{22} by applying $-9R_2 + R_3 \to R_3$ and $2R_2 + R_1 \to R_1$:

$$A \sim \begin{pmatrix} 1 & -2 & 3 & 1 & 2 \\ 0 & 1 & \frac{1}{3} & -\frac{2}{3} & \frac{1}{3} \\ 0 & 9 & 3 & -4 & 4 \end{pmatrix} \sim \begin{pmatrix} 1 & 0 & \frac{11}{3} & -\frac{1}{3} & \frac{8}{3} \\ 0 & 1 & \frac{1}{3} & -\frac{2}{3} & \frac{1}{3} \\ 0 & 0 & 0 & 2 & 1 \end{pmatrix}$$

Last, multiply R_3 by $\frac{1}{2}$ to get the pivot $a_{34} = 1$ and produce 0s above a_{34} by applying $(\frac{2}{3})R_3 + R_2 \to R_2$ and $(\frac{1}{3})R_3 + R_1 \to R_1$:

$$A \sim \begin{pmatrix} 1 & 0 & \frac{11}{3} & -\frac{1}{3} & \frac{8}{3} \\ 0 & 1 & \frac{1}{3} & -\frac{2}{3} & \frac{1}{3} \\ 0 & 0 & 0 & 1 & \frac{1}{2} \end{pmatrix} \sim \begin{pmatrix} 1 & 0 & \frac{11}{3} & 0 & \frac{17}{6} \\ 0 & 1 & \frac{1}{3} & 0 & \frac{2}{3} \\ 0 & 0 & 0 & 1 & \frac{1}{2} \end{pmatrix}$$

1.30. One speaks of "an" echelon form of a matrix A, "the" row canonical form of A. Why?

An arbitrary matrix A may be row equivalent to many echelon matrices. On the other hand, regardless of the algorithm that is used, a matrix A is row equivalent to a unique matrix in row canonical form. (The term "canonical" usually connotes uniqueness.) For example, the row canonical forms in Problems 1.27 and 1.29 are equal.

1.31. Given an $n \times n$ echelon matrix in triangular form,

$$A = \begin{pmatrix} a_{11} & a_{12} & a_{13} & \cdots & a_{1,\,n-1} & a_{1n} \\ 0 & a_{22} & a_{23} & \cdots & a_{2,\,n-1} & a_{2n} \\ 0 & 0 & a_{33} & \cdots & a_{3,\,n-1} & a_{3n} \\ \cdots\cdots\cdots\cdots\cdots\cdots\cdots\cdots\cdots\cdots\cdots \\ 0 & 0 & 0 & \cdots & 0 & a_{nn} \end{pmatrix}$$

with all $a_{ii} \neq 0$. Find the row canonical form of A.

Multiplying R_n by $1/a_{nn}$ and using the new $a_{nn} = 1$ as pivot, we obtain the matrix

$$\begin{pmatrix} a_{11} & a_{12} & a_{13} & \cdots & a_{1,\,n-1} & 0 \\ 0 & a_{22} & a_{23} & \cdots & a_{2,\,n-1} & 0 \\ 0 & 0 & a_{33} & \cdots & a_{3,\,n-1} & 0 \\ \cdots\cdots\cdots\cdots\cdots\cdots\cdots\cdots\cdots\cdots\cdots \\ 0 & 0 & 0 & \cdots & 0 & 1 \end{pmatrix}$$

Observe that the last column of A has been converted into a unit vector. Each succeeding back-substitution yields a new unit column vector, and the end result is

$$A \sim \begin{pmatrix} 1 & 0 & \cdots & 0 \\ 0 & 1 & \cdots & 0 \\ \cdots\cdots\cdots\cdots \\ 0 & 0 & \cdots & 1 \end{pmatrix}$$

i.e., A has the $n \times n$ *identity matrix* I as its row canonical form.

1.32. Reduce the following triangular matrix with nonzero diagonal elements to row canonical form:

$$C = \begin{pmatrix} 5 & -9 & 6 \\ 0 & 2 & 3 \\ 0 & 0 & 7 \end{pmatrix}$$

By Problem 1.31, C is row equivalent to the identity matrix. Alternately, by back-substitution,

$$C \sim \begin{pmatrix} 5 & -9 & 6 \\ 0 & 2 & 3 \\ 0 & 0 & 1 \end{pmatrix} \sim \begin{pmatrix} 5 & -9 & 0 \\ 0 & 2 & 0 \\ 0 & 0 & 1 \end{pmatrix} \sim \begin{pmatrix} 5 & -9 & 0 \\ 0 & 1 & 0 \\ 0 & 0 & 1 \end{pmatrix} \sim \begin{pmatrix} 5 & 0 & 0 \\ 0 & 1 & 0 \\ 0 & 0 & 1 \end{pmatrix} \sim \begin{pmatrix} 1 & 0 & 0 \\ 0 & 1 & 0 \\ 0 & 0 & 1 \end{pmatrix}$$

SYSTEMS OF LINEAR EQUATIONS IN MATRIX FORM

1.33. Find the augmented matrix M and the coefficient matrix A of the following system:

$$x + 2y - 3z = 4$$
$$3y - 4z + 7x = 5$$
$$6z + 8x - 9y = 1$$

First align the unknowns in the system to obtain

$$x + 2y - 3z = 4$$
$$7x + 3y - 4z = 5$$
$$8x - 9y + 6z = 1$$

Then

$$M = \begin{pmatrix} 1 & 2 & -3 & 4 \\ 7 & 3 & -4 & 5 \\ 8 & -9 & 6 & 1 \end{pmatrix} \quad \text{and} \quad A = \begin{pmatrix} 1 & 2 & -3 \\ 7 & 3 & -4 \\ 8 & -9 & 6 \end{pmatrix}$$

1.34. Solve, using the augmented matrix,

$$x - 2y + 4z = 2$$
$$2x - 3y + 5z = 3$$
$$3x - 4y + 6z = 7$$

Reduce the augmented matrix to echelon form:

$$\begin{pmatrix} 1 & -2 & 4 & 2 \\ 2 & -3 & 5 & 3 \\ 3 & -4 & 6 & 7 \end{pmatrix} \sim \begin{pmatrix} 1 & -2 & 4 & 2 \\ 0 & 1 & -3 & -1 \\ 0 & 2 & -6 & 1 \end{pmatrix} \sim \begin{pmatrix} 1 & -2 & 4 & 2 \\ 0 & 1 & -3 & -1 \\ 0 & 0 & 0 & 3 \end{pmatrix}$$

The third row of the echelon matrix corresponds to the degenerate equation $0 = 3$; hence the system has no solution.

1.35. Solve, using the augmented matrix,

$$x + 2y - 3z - 2s + 4t = 1$$
$$2x + 5y - 8z - s + 6t = 4$$
$$x + 4y - 7z + 5s + 2t = 8$$

Reduce the augmented matrix to echelon form and then to row canonical form:

$$\begin{pmatrix} 1 & 2 & -3 & -2 & 4 & 1 \\ 2 & 5 & -8 & -1 & 6 & 4 \\ 1 & 4 & -7 & 5 & 2 & 8 \end{pmatrix} \sim \begin{pmatrix} 1 & 2 & -3 & -2 & 4 & 1 \\ 0 & 1 & -2 & 3 & -2 & 2 \\ 0 & 2 & -4 & 7 & -2 & 7 \end{pmatrix} \sim \begin{pmatrix} 1 & 2 & -3 & -2 & 4 & 1 \\ 0 & 1 & -2 & 3 & -2 & 2 \\ 0 & 0 & 0 & 1 & 2 & 3 \end{pmatrix}$$

$$\sim \begin{pmatrix} 1 & 2 & -3 & 0 & 8 & 7 \\ 0 & 1 & -2 & 0 & -8 & -7 \\ 0 & 0 & 0 & 1 & 2 & 3 \end{pmatrix} \sim \begin{pmatrix} 1 & 0 & 1 & 0 & 24 & 21 \\ 0 & 1 & -2 & 0 & -8 & -7 \\ 0 & 0 & 0 & 1 & 2 & 3 \end{pmatrix}$$

Thus the free-variable form of the solution is

$$x + \quad z + 24t = 21 \qquad\qquad x = 21 - z - 24t$$
$$y - 2z - 8t = -7 \quad \text{or} \quad y = -7 + 2z + 8t$$
$$s + 2t = 3 \qquad\qquad s = 3 - 2t$$

where z and t are the free variables.

1.36. Solve, using the augmented matrix,

$$x + 2y - z = 3$$
$$x + 3y + z = 5$$
$$3x + 8y + 4z = 17$$

Reduce the augmented matrix to echelon form and then to row canonical form:

$$\begin{pmatrix} 1 & 2 & -1 & 3 \\ 1 & 3 & 1 & 5 \\ 3 & 8 & 4 & 17 \end{pmatrix} \sim \begin{pmatrix} 1 & 2 & -1 & 3 \\ 0 & 1 & 2 & 2 \\ 0 & 2 & 7 & 8 \end{pmatrix} \sim \begin{pmatrix} 1 & 2 & -1 & 3 \\ 0 & 1 & 2 & 2 \\ 0 & 0 & 3 & 4 \end{pmatrix}$$

$$\sim \begin{pmatrix} 1 & 2 & -1 & 3 \\ 0 & 1 & 2 & 2 \\ 0 & 0 & 1 & \frac{4}{3} \end{pmatrix} \sim \begin{pmatrix} 1 & 2 & 0 & \frac{13}{3} \\ 0 & 1 & 0 & -\frac{2}{3} \\ 0 & 0 & 1 & \frac{4}{3} \end{pmatrix} \sim \begin{pmatrix} 1 & 0 & 0 & \frac{17}{3} \\ 0 & 1 & 0 & -\frac{2}{3} \\ 0 & 0 & 1 & \frac{4}{3} \end{pmatrix}$$

The system has the unique solution $x = \frac{17}{3}, y = -\frac{2}{3}, z = \frac{4}{3}$ or $u = (\frac{17}{3}, -\frac{2}{3}, \frac{4}{3})$.

HOMOGENEOUS SYSTEMS

1.37. Determine whether each system has a nonzero solution.

$$\begin{array}{lll}
\begin{aligned}
x - 2y + 3z - 2w &= 0 \\
3x - 7y - 2z + 4w &= 0 \\
4x + 3y + 5z + 2w &= 0
\end{aligned}
&
\begin{aligned}
x + 2y - 3z &= 0 \\
2x + 5y + 2z &= 0 \\
3x - y - 4z &= 0
\end{aligned}
&
\begin{aligned}
x + 2y - z &= 0 \\
2x + 5y + 2z &= 0 \\
x + 4y + 7z &= 0 \\
x + 3y + 3z &= 0
\end{aligned}
\\
\qquad (a) & \qquad (b) & \qquad (c)
\end{array}$$

(a)　The system must have a nonzero solution since there are more unknowns than equations.

(b)　Reduce to echelon form:

$$\begin{array}{lcl}
\begin{aligned}
x + 2y - 3z &= 0 \\
2x + 5y + 2z &= 0 \\
3x - y - 4z &= 0
\end{aligned}
\quad \text{to} \quad
&
\begin{aligned}
x + 2y - 3z &= 0 \\
y + 8z &= 0 \\
-7y + 5z &= 0
\end{aligned}
\quad \text{to} \quad
&
\begin{aligned}
x + 2y - 3z &= 0 \\
y + 8z &= 0 \\
61z &= 0
\end{aligned}
\end{array}$$

In echelon form there are exactly three equations in the three unknowns; hence the system has a unique solution, the zero solution.

(c)　Reduce to echelon form:

$$\begin{array}{lcl}
\begin{aligned}
x + 2y - z &= 0 \\
2x + 5y + 2z &= 0 \\
x + 4y + 7z &= 0 \\
x + 3y + 3z &= 0
\end{aligned}
\quad \text{to} \quad
&
\begin{aligned}
x + 2y - z &= 0 \\
y + 4z &= 0 \\
2y + 8z &= 0 \\
y + 4z &= 0
\end{aligned}
\quad \text{to} \quad
&
\begin{aligned}
x + 2y - z &= 0 \\
y + 4z &= 0
\end{aligned}
\end{array}$$

In echelon form there are only two equations in the three unknowns; hence the system has a nonzero solution.

1.38. Find the dimension and a basis for the general solution W of the homogeneous system

$$\begin{aligned}
x + 3y - 2z + 5s - 3t &= 0 \\
2x + 7y - 3z + 7s - 5t &= 0 \\
3x + 11y - 4z + 10s - 9t &= 0
\end{aligned}$$

Show how the basis gives the parametric form of the general solution of the system.

Reduce the system to echelon form. Apply the operations $-2L_1 + L_2 \to L_2$ and $-3L_1 + L_3 \to L_3$, and then $-2L_2 + L_3 \to L_3$ to obtain

$$\begin{array}{lcl}
\begin{aligned}
x + 3y - 2z + 5s - 3t &= 0 \\
y + z - 3s + t &= 0 \\
2y + 2z - 5s &= 0
\end{aligned}
\quad \text{and} \quad
&
\begin{aligned}
x + 3y - 2z + 5s - 3t &= 0 \\
y + z - 3s + t &= 0 \\
s - 2t &= 0
\end{aligned}
\end{array}$$

In echelon form, the system has two free variables, z and t; hence dim $W = 2$. A basis $[u_1, u_2]$ for W may be obtained as follows:

(1) Set $z = 1$, $t = 0$. Back-substitution yields $s = 0$, then $y = -1$, and then $x = 5$. Therefore, $u_1 = (5, -1, 1, 0, 0)$.

(2) Set $z = 0$, $t = 1$. Back-substitution yields $s = 2$, then $y = 5$, and then $x = -22$. Therefore, $u_2 = (-22, 5, 0, 2, 1)$.

Multiplying the basis vectors by the parameters a and b, respectively, yields

$$au_1 + bu_2 = a(5, -1, 1, 0, 0) + b(-22, 5, 0, 2, 1) = (5a - 22b, -a + 5b, a, 2b, b)$$

This is the parametric form of the general solution.

1.39. Find the dimension and a basis for the general solution W of the homogeneous system

$$x + 2y - 3z = 0$$
$$2x + 5y + 2z = 0$$
$$3x - y - 4z = 0$$

Reduce the system to echelon form. From Problem 1.37(b) we have

$$x + 2y - 3z = 0$$
$$y + 8z = 0$$
$$61z = 0$$

There are no free variables (the system is in triangular form). Hence dim $W = 0$ and W has no basis. Specifically, W consists only of the zero solution, $W = \{0\}$.

1.40. Find the dimension and a basis for the general solution W of the homogeneous system

$$2x + 4y - 5z + 3t = 0$$
$$3x + 6y - 7z + 4t = 0$$
$$5x + 10y - 11z + 6t = 0$$

Reduce the system to echelon form. Apply $-3L_1 + 2L_2 \to L_2$ and $-5L_1 + 2L_3 \to L_3$, and then $-3L_2 + L_3 \to L_3$ to obtain

$$\begin{array}{l} 2x + 4y - 5z + 3t = 0 \\ z - t = 0 \\ 3z - 3t = 0 \end{array} \qquad \text{and} \qquad \begin{array}{l} 2x + 4y - 5z + 3t = 0 \\ z - t = 0 \end{array}$$

In echelon form, the system has two free variables, y and t; hence dim $W = 2$. A basis $[u_1, u_2]$ for W may be obtained as follows:

(1) Set $y = 1, t = 0$. Back-substitution yields the solution $u_1 = (-2, 1, 0, 0)$.

(2) Set $y = 0, t = 1$. Back-substitution yields the solution $u_2 = (1, 0, 1, 1)$.

1.41. Consider the system

$$x - 3y - 2z + 4t = 5$$
$$3x - 8y - 3z + 8t = 18$$
$$2x - 3y + 5z - 4t = 19$$

(a) Find the parametric form of the general solution of the system.

(b) Show that the result of (a) may be rewritten in the form given by Theorem 1.11.

(a) Reduce the system to echelon form. Apply $-3L_1 + L_2 \to L_2$ and $-2L_1 + L_3 \to L_3$, and then $-3L_2 + L_3 \to L_3$ to obtain

$$
\begin{array}{rl}
x - 3y - 2z + 4t = 5 & \\
y + 3z - 4t = 3 & \quad\text{and}\quad \\
3y + 9z - 12t = 9 &
\end{array}
\qquad
\begin{array}{l}
x - 3y - 2z + 4t = 5 \\
y + 3z - 4t = 3
\end{array}
$$

In echelon form, the free variables are z and t. Set $z = a$ and $t = b$, where a and b are parameters. Back-substitution yields $y = 3 - 3a + 4b$, and then $x = 14 - 7a + 8b$. Thus the parametric form of the solution is

$$x = 14 - 7a + 8b \qquad y = 3 - 3a + 4b \qquad z = a \qquad t = b \tag{$*$}$$

(b) Let $v_0 = (14, 3, 0, 0)$ be the vector of constant terms in $(*)$, let $u_1 = (-7, -3, 1, 0)$ be the vector of coefficients of a in $(*)$, and let $u_2 = (8, 4, 0, 1)$ be the vector of coefficients of b in $(*)$. Then the general solution $(*)$ may be rewritten in vector form as

$$(x, y, z, t) = v_0 + au_1 + bu_2 \tag{$**$}$$

We next show that $(**)$ is the general solution per Theorem 1.11. First note that v_0 is the solution of the inhomogeneous system obtained by setting $a = 0$ and $b = 0$. Consider the associated homogeneous system, in echelon form:

$$
\begin{array}{l}
x - 3y - 2z + 4t = 0 \\
y + 3z - 4t = 0
\end{array}
$$

The free variables are z and t. Set $z = 1$ and $t = 0$ to obtain the solution $u_1 = (-7, -3, 1, 0)$. Set $z = 0$ and $t = 1$ to obtain the solution $u_2 = (8, 4, 0, 1)$. By Theorem 1.10, $\{u_1, u_2\}$ is a basis for the solution space of the associated homogeneous system. Thus $(**)$ has the desired form.

MISCELLANEOUS PROBLEMS

1.42. Show that each of the elementary operations $[E_1]$, $[E_2]$, $[E_3]$ has an inverse operation of the same type.

[E_1]   Interchange the ith equation and the jth equation: $L_i \leftrightarrow L_j$.

[E_2]   Multiply the ith equation by a nonzero scalar k: $kL_i \to L_i$, $k \neq 0$.

[E_3]   Replace the ith equation by k times the jth equation plus the ith equation: $kL_j + L_i \to L_i$.

(a) Interchanging the same two equations twice, we obtain the original system; that is, $L_i \leftrightarrow L_j$ is its own inverse.

(b) Multiplying the ith equation by k and then by k^{-1}, or by k^{-1} and then k, we obtain the original system. In other words, the operations $kL_i \to L_i$ and $k^{-1}L_i \to L_i$ are inverses.

(c) Applying the operation $kL_j + L_i \to L_i$ and then the operation $-kL_j + L_i \to L_i$, or vice versa, we obtain the original system. In other words, the operations $kL_j + L_i \to L_i$ and $-kL_j + L_i \to L_i$ are inverses.

1.43. Show that the effect of applying the following operation $[E]$ can be obtained by applying $[E_2]$ and then $[E_3]$.

[E]   Replace the ith equation by k' times the jth equation plus k (nonzero) times the ith equation: $k'L_j + kL_i \to L_i$, $k \neq 0$.

Applying $kL_i \to L_i$ and then applying $k'L_j + L_i \to L_i$ has the same result as applying the operation $k'L_j + kL_i \to L_i$.

1.44. Suppose that each equation L_i in the system (1.3) is multiplied by a constant c_i, and that the resulting equations are added to yield

$$(c_1 a_{11} + \cdots + c_m a_{m1})x_1 + \cdots + (c_1 a_{1n} + \cdots + c_m a_{mn})x_n = c_1 b_1 + \cdots + c_m b_m \tag{1}$$

Such an equation is termed a linear combination of the equations L_i. Show that any solution of the system (1.3) is also a solution of the linear combination (1).

Suppose $u = (k_1, k_2, \ldots, k_n)$ is a solution of (1.3):

$$a_{i1}k_1 + a_{i2}k_2 + \cdots + a_{in}k_n = b_i \qquad (i = 1, \ldots, m) \tag{2}$$

To show that u is a solution of (1), we must verify the equation

$$(c_1 a_{11} + \cdots + c_m a_{m1})k_1 + \cdots + (c_1 a_{1n} + \cdots + c_m a_{mn})k_n = c_1 b_1 + \cdots + c_m b_m$$

But this can be rearranged into

$$c_1(a_{11}k_1 + \cdots + a_{1n}k_n) + \cdots + c_m(a_{m1} + \cdots + a_{mn}k_n) = c_1 b_1 + \cdots + c_m b_m$$

or, by (2)

$$c_1 b_1 + \cdots + c_m b_m = c_1 b_1 + \cdots + c_m b_m$$

which is clearly a true statement.

1.45. Suppose that a system ($\#$) of linear equations is obtained from a system ($*$) of linear equations by applying a single elementary operation—$[E_1]$, $[E_2]$, or $[E_3]$. Show that ($\#$) and ($*$) have all solutions in common (the two systems are equivalent).

Each equation in ($\#$) is a linear combination of the equations in ($*$). Therefore, by Problem 1.44, any solution of ($*$) will be a solution of all the equations in ($\#$). In other words, the solution set of ($*$) is contained in the solution set of ($\#$). On the other hand, since the operations $[E_1]$, $[E_2]$, and $[E_3]$ have inverse elementary operations, the system ($*$) can be obtained from ($\#$) by a single elementary operation. Accordingly, the solution set of ($\#$) is contained in the solution set of ($*$). Thus ($\#$) and ($*$) have the same solutions.

1.46. Prove Theorem 1.4.

By Problem 1.45, each step does not change the solution set. Hence the original system ($*$) and the final system ($\#$) (and any system in between) have the same solution set.

1.47. Prove that the following three statements about a system of linear equations are equivalent: (i) The system is consistent (has a solution). (ii) No linear combination of the equations is the equation

$$0x_1 + 0x_2 + \cdots + 0x_n = b \neq 0 \tag{*}$$

(iii)   The system is reducible to echelon form.

Suppose the system is reducible to echelon form. The echelon form has a solution, and hence the original system has a solution. Thus (iii) implies (i).

Suppose the system has a solution. By Problem 1.44, any linear combination of the equations also has a solution. But ($*$) has no solution; hence ($*$) is not a linear combination of the equations. Thus (i) implies (ii).

Finally, suppose the system is not reducible to echelon form. Then, in the Gaussian algorithm, it must yield an equation of the form ($*$). Hence ($*$) is a linear combination of the equations. Thus not-(iii) implies not-(ii), or, equivalently, (ii) implies (iii).

Supplementary Problems

SOLUTION OF LINEAR EQUATIONS

1.48. Solve:

$$\text{(a)}\quad \begin{array}{l} 2x + 3y = 1 \\ 5x + 7y = 3 \end{array} \qquad \text{(b)}\quad \begin{array}{l} 2x + 4y = 10 \\ 3x + 6y = 15 \end{array} \qquad \text{(c)}\quad \begin{array}{l} 4x - 2y = 5 \\ -6x + 3y = 1 \end{array}$$

1.49. Solve:

$$\text{(a)}\quad \begin{array}{l} 2x - y - 3z = 5 \\ 3x - 2y + 2z = 5 \\ 5x - 3y - z = 16 \end{array} \qquad \text{(b)}\quad \begin{array}{l} 2x + 3y - 2z = 5 \\ x - 2y + 3z = 2 \\ 4x - y + 4z = 1 \end{array} \qquad \text{(c)}\quad \begin{array}{l} x + 2y + 3z = 3 \\ 2x + 3y + 8z = 4 \\ 3x + 2y + 17z = 1 \end{array}$$

1.50. Solve:

$$\text{(a)}\quad \begin{array}{l} 2x + 3y = 3 \\ x - 2y = 5 \\ 3x + 2y = 7 \end{array} \qquad \text{(b)}\quad \begin{array}{l} x + 2y - 3z + 2t = 2 \\ 2x + 5y - 8z + 6t = 5 \\ 3x + 4y - 5z + 2t = 4 \end{array} \qquad \text{(c)}\quad \begin{array}{l} x + 2y - z + 3t = 3 \\ 2x + 4y + 4z + 3t = 9 \\ 3x + 6y - z + 8t = 10 \end{array}$$

1.51. Solve:

$$\text{(a)}\quad \begin{array}{l} x + 2y + 2z = 2 \\ 3x - 2y - z = 5 \\ 2x - 5y + 3z = -4 \\ x + 4y + 6z = 0 \end{array} \qquad \text{(b)}\quad \begin{array}{l} x + 5y + 4z - 13t = 3 \\ 3x - y + 2z + 5t = 2 \\ 2x + 2y + 3z - 4t = 1 \end{array}$$

HOMOGENEOUS SYSTEMS

1.52. Determine whether each system has a nonzero solution:

$$\text{(a)}\quad \begin{array}{l} x + 3y - 2z = 0 \\ x - 8y + 8z = 0 \\ 3x - 2y + 4z = 0 \end{array} \qquad \text{(b)}\quad \begin{array}{l} x + 3y - 2z = 0 \\ 2x - 3y + z = 0 \\ 3x - 2y + 2z = 0 \end{array} \qquad \text{(c)}\quad \begin{array}{l} x + 2y - 5z + 4t = 0 \\ 2x - 3y + 2z + 3t = 0 \\ 4x - 7y + z - 6t = 0 \end{array}$$

1.53. Find the dimension and a basis of the general solution W of each homogeneous system.

$$\text{(a)}\quad \begin{array}{l} x + 3y + 2z - s - t = 0 \\ 2x + 6y + 5z + s - t = 0 \\ 5x + 15y + 12z + s - 3t = 0 \end{array} \qquad \text{(b)}\quad \begin{array}{l} 2x - 4y + 3z - s + 2t = 0 \\ 3x - 6y + 5z - 2s + 4t = 0 \\ 5x - 10y + 7z - 3s + t = 0 \end{array}$$

ECHELON MATRICES AND ELEMENTARY ROW OPERATIONS

1.54. Reduce A to echelon form and then to its row canonical form, where

$$\text{(a)}\quad A = \begin{pmatrix} 1 & 2 & -1 & 2 & 1 \\ 2 & 4 & 1 & -2 & 3 \\ 3 & 6 & 2 & -6 & 5 \end{pmatrix} \qquad \text{(b)}\quad A = \begin{pmatrix} 2 & 3 & -2 & 5 & 1 \\ 3 & -1 & 2 & 0 & 4 \\ 4 & -5 & 6 & -5 & 7 \end{pmatrix}$$

1.55. Reduce A to echelon form and then to its row canonical form, where

$$(a) \quad A = \begin{pmatrix} 1 & 3 & -1 & 2 \\ 0 & 11 & -5 & 3 \\ 2 & -5 & 3 & 1 \\ 4 & 1 & 1 & 5 \end{pmatrix} \qquad (b) \quad A = \begin{pmatrix} 0 & 1 & 3 & -2 \\ 0 & 4 & -1 & 3 \\ 0 & 0 & 1 & 1 \\ 0 & 5 & -3 & 4 \end{pmatrix}$$

1.56. Describe all the possible 2×2 matrices which are in row reduced echelon form.

1.57. Suppose A is a square row reduced echelon matrix. Show that if $A \neq I$, the identity matrix, then A has a zero row.

1.58. Show that each of the following elementary row operations has an inverse operation of the same type.

$[E_1]$ Interchange the ith row and the jth row: $R_i \leftrightarrow R_j$.

$[E_2]$ Multiply the ith row by a nonzero scalar k: $kR_i \rightarrow R_i$, $k \neq 0$.

$[E_3]$ Replace the ith row by k times the jth row plus the ith row: $kR_j + R_i \rightarrow R_i$.

1.59. Show that row equivalence is an equivalence relation:

(i) A is row equivalent to A;

(ii) A row equivalent to B implies B row equivalent to A;

(iii) A row equivalent to B and B row equivalent to C implies A row equivalent to C.

MISCELLANEOUS PROBLEMS

1.60. Consider two general linear equations in two unknowns x and y over the real field **R**:

$$ax + by = e$$
$$cx + dy = f$$

Show that:

(i)   If $\dfrac{a}{c} \neq \dfrac{b}{d}$, i.e., if $ad - bc \neq 0$, then the system has the unique solution $x = \dfrac{de - bf}{ad - bc}$, $y = \dfrac{af - ce}{ad - bc}$;

(ii)   If $\dfrac{a}{c} = \dfrac{b}{d} \neq \dfrac{e}{f}$, then the system has no solution;

(iii)   If $\dfrac{a}{c} = \dfrac{b}{d} = \dfrac{e}{f}$, then the system has more than one solution.

1.61. Consider the system

$$ax + by = 1$$
$$cx + dy = 0$$

Show that if $ad - bc \neq 0$, then the system has the unique solution $x = d/(ad - bc)$, $y = -c/(ad - bc)$. Also show that if $ad - bc = 0$, $c \neq 0$ or $d \neq 0$, then the system has no solution.

1.62. Show that an equation of the form $0x_1 + 0x_2 + \cdots + 0x_n = 0$ may be added or deleted from a system without affecting the solution set.

1.63. Consider a system of linear equations with the same number of equations as unknowns:

$$
\begin{aligned}
a_{11}x_1 + a_{12}x_2 + \cdots + a_{1n}x_n &= b_1 \\
a_{21}x_1 + a_{22}x_2 + \cdots + a_{2n}x_n &= b_2 \\
&\cdots\cdots\cdots\cdots\cdots\cdots\cdots\cdots\cdots\cdots \\
a_{n1}x_1 + a_{n2}x_2 + \cdots + a_{nn}x_n &= b_n
\end{aligned}
\tag{1}
$$

(i) Suppose the associated homogeneous system has only the zero solution. Show that (*1*) has a unique solution for every choice of constants b_i.

(ii) Suppose the associated homogeneous system has a nonzero solution. Show that there are constants b_i for which (*1*) does not have a solution. Also show that if (*1*) has a solution, then it has more than one.

1.64. Suppose in a homogeneous system of linear equations the coefficients of one of the unknowns are all zero. Show that the system has a nonzero solution.

Answers to Supplementary Problems

1.48. (a) $x = 2, y = -1$;   (b) $x = 5 - 2a, y = a$;   (c) no solution

1.49. (a) $(1, -3, -2)$;   (b) no solution;   (c) $(-1 - 7a, 2 + 2a, a)$ or $\begin{cases} x = -1 - 7z \\ y = 2 + 2z \end{cases}$

1.50. (a) $x = 3, y = -1$;   (b) $(-a + 2b, 1 + 2a - 2b, a, b)$ or $\begin{cases} x = -z + 2t \\ y = 1 + 2z - 2t \end{cases}$

   (c) $(\frac{7}{2} - 5b/2 - 2a, a, \frac{1}{2} + b/2, b)$ or $\begin{cases} x = \frac{7}{2} - 5t/2 - 2y \\ z = \frac{1}{2} + t/2 \end{cases}$

1.51. (a) $(2, 1, -1)$;   (b) no solution

1.52. (a) yes;   (b) no;   (c) yes, by Theorem 1.8

1.53. (a) Dim $W = 3$; $u_1 = (-3, 1, 0, 0, 0)$, $u_2 = (7, 0, -3, 1, 0)$, $u_3 = (3, 0, -1, 0, 1)$;
   (b) Dim $W = 2$; $u_1 = (-2, 1, 0, 0, 0)$, $u_2 = (5, 0, -5, -3, 1)$

1.54. (a)
$$
\begin{pmatrix} 1 & 2 & -1 & 2 & 1 \\ 0 & 0 & 3 & -6 & 1 \\ 0 & 0 & 0 & -6 & 1 \end{pmatrix}
\quad\text{and}\quad
\begin{pmatrix} 1 & 2 & 0 & 0 & \frac{4}{3} \\ 0 & 0 & 1 & 0 & 0 \\ 0 & 0 & 0 & 1 & -\frac{1}{6} \end{pmatrix};
$$

   (b)
$$
\begin{pmatrix} 2 & 3 & -2 & 5 & 1 \\ 0 & -11 & 10 & -15 & 5 \\ 0 & 0 & 0 & 0 & 0 \end{pmatrix}
\quad\text{and}\quad
\begin{pmatrix} 1 & 0 & \frac{4}{11} & \frac{5}{11} & \frac{13}{11} \\ 0 & 1 & -\frac{10}{11} & \frac{15}{11} & -\frac{5}{11} \\ 0 & 0 & 0 & 0 & 0 \end{pmatrix}
$$

1.55. (a)
$$
\begin{pmatrix} 1 & 3 & -1 & 2 \\ 0 & 11 & -5 & 3 \\ 0 & 0 & 0 & 0 \\ 0 & 0 & 0 & 0 \end{pmatrix}
\quad\text{and}\quad
\begin{pmatrix} 1 & 0 & \frac{4}{11} & \frac{13}{11} \\ 0 & 1 & -\frac{5}{11} & \frac{3}{11} \\ 0 & 0 & 0 & 0 \\ 0 & 0 & 0 & 0 \end{pmatrix};
$$

   (b)
$$
\begin{pmatrix} 0 & 1 & 3 & -2 \\ 0 & 0 & -13 & 11 \\ 0 & 0 & 0 & 35 \\ 0 & 0 & 0 & 0 \end{pmatrix}
\quad\text{and}\quad
\begin{pmatrix} 0 & 1 & 0 & 0 \\ 0 & 0 & 1 & 0 \\ 0 & 0 & 0 & 1 \\ 0 & 0 & 0 & 0 \end{pmatrix}
$$

Chapter 2

Vectors in $\mathbf{R}^n$ and $\mathbf{C}^n$, Spatial Vectors

2.1 INTRODUCTION

In various physical applications there appear certain quantities, such as temperature and speed, which possess only "magnitude." These can be represented by real numbers and are called *scalars*. On the other hand, there are also quantities, such as force and velocity, which possess both "magnitude" and "direction." These quantities can be represented by arrows (having appropriate lengths and directions and emanating from some given reference point O) and are called *vectors*.

We begin by considering the following operations on vectors.

(i) *Addition:* The resultant $\mathbf{u} + \mathbf{v}$ of two vectors $\mathbf{u}$ and $\mathbf{v}$ is obtained by the so-called parallelogram law, i.e., $\mathbf{u} + \mathbf{v}$ is the diagonal of the parallelogram formed by $\mathbf{u}$ and $\mathbf{v}$ as shown in Fig. 2-1(a).

(ii) *Scalar multiplication:* The product $k\mathbf{u}$ of a real number k by a vector $\mathbf{u}$ is obtained by multiplying the magnitude of $\mathbf{u}$ by k and retaining the same direction if $k \geq 0$ or the opposite direction if $k < 0$, as shown in Fig. 2-1(b).

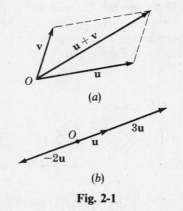

(a)

(b)

Fig. 2-1

Now we assume the reader is familiar with the representation of the points in the plane by ordered pairs of real numbers. If the origin of the axes is chosen at the reference point O above, then every vector is uniquely determined by the coordinates of its endpoint. The relationship between the above operations and endpoints follows.

(i) *Addition:* If (a, b) and (c, d) are the endpoints of the vectors $\mathbf{u}$ and $\mathbf{v}$, then $(a + c, b + d)$ will be the endpoint of $\mathbf{u} + \mathbf{v}$, as shown in Fig. 2-2(a).

(ii) *Scalar multiplication:* If (a, b) is the endpoint of the vector $\mathbf{u}$, then (ka, kb) will be the endpoint of the vector $k\mathbf{u}$, as shown in Fig. 2-2(b).

Mathematically, we identify the vector $\mathbf{u}$ with its endpoint (a, b) and write $\mathbf{u} = (a, b)$. In addition we call the ordered pair (a, b) of real numbers a point or vector depending upon its interpretation. We generalize this notion and call an n-tuple $(a_1, a_2, \ldots, a_n)$ of real numbers a vector. However, special notation may be used for spatial vectors in $\mathbf{R}^3$ (Section 2.8).

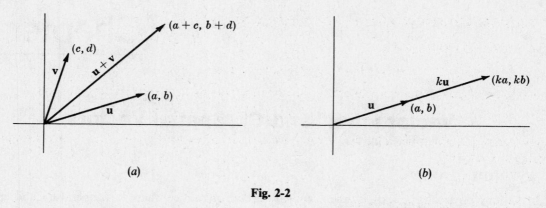

Fig. 2-2

We assume the reader is familiar with the elementary properties of the real number field which we denote by **R**.

2.2 VECTORS IN $\mathbf{R}^n$

The set of all *n*-tuples of real numbers, denoted by $\mathbf{R}^n$, is called *n-space*. A particular *n*-tuple in $\mathbf{R}^n$, say

$$u = (u_1, u_2, \ldots, u_n)$$

is called a *point* or *vector*; the real numbers u_i are called the *components* (or *coordinates*) of the vector u. Moreover, when discussing the space $\mathbf{R}^n$ we use the term *scalar* for the elements of **R**.

Two vectors u and v are *equal*, written $u = v$, if they have the same number of components, i.e., belong to the same space, and if corresponding components are equal. The vectors (1, 2, 3) and (2, 3, 1) are not equal, since corresponding elements are not equal.

Example 2.1

(a) Consider the following vectors

$$(0, 1) \qquad (1, -3) \qquad (1, 2, \sqrt{3}, 4) \qquad (-5, \tfrac{1}{2}, 0, \pi)$$

The first two vectors have two components and so are points in $\mathbf{R}^2$; the last two vectors have four components and so are points in $\mathbf{R}^4$.

(b) Suppose $(x - y, x + y, z - 1) = (4, 2, 3)$. Then, by definition of equality of vectors,

$$x - y = 4$$
$$x + y = 2$$
$$z - 1 = 3$$

Solving the above system of equations gives $x = 3$, $y = -1$, and $z = 4$.

Sometimes vectors in *n*-space are written vertically as columns rather than horizontally as rows, as above. Such vectors are called column vectors. For example,

$$\begin{pmatrix} 0 \\ 1 \end{pmatrix} \qquad \begin{pmatrix} 1 \\ -3 \end{pmatrix} \qquad \begin{pmatrix} 1 \\ 7 \\ -8 \end{pmatrix} \qquad \begin{pmatrix} 1.2 \\ -35 \\ 28 \end{pmatrix}$$

are column vectors with 2, 2, 3, and 3 components, respectively.

2.3 VECTOR ADDITION AND SCALAR MULTIPLICATION

Let u and v be vectors in $\mathbf{R}^n$:

$$u = (u_1, u_2, \ldots, u_n) \quad \text{and} \quad v = (v_1, v_2, \ldots, v_n)$$

The *sum* of u and v, written $u + v$, is the vector obtained by adding corresponding components:

$$u + v = (u_1 + v_1, u_2 + v_2, \ldots, u_n + v_n)$$

The *product* of a real number k by the vector u, written ku, is the vector obtained by multiplying each component of u by k:

$$ku = (ku_1, ku_2, \ldots, ku_n)$$

Observe that $u + v$ and ku are also vectors in $\mathbf{R}^n$. We also define

$$-u = -1u \quad \text{and} \quad u - v = u + (-v)$$

The sum of vectors with different numbers of components is not defined.

Basic properties of the vectors in $\mathbf{R}^n$ under the operations of vector addition and scalar multiplication are described in the following theorem (proved in Problem 2.4).

Theorem 2.1: For any vectors $u, v, w \in \mathbf{R}^n$ and any scalars $k, k' \in \mathbf{R}$,

(i) $(u + v) + w = u + (v + w)$ (v) $k(u + v) = ku + kv$

(ii) $u + 0 = u$ (vi) $(k + k')u = ku + k'u$

(iii) $u + (-u) = 0$ (vii) $(kk')u = k(k'u)$

(iv) $u + v = v + u$ (viii) $1u = u$

Suppose u and v are vectors in $\mathbf{R}^n$ for which $u = kv$ for some nonzero scalar $k \in \mathbf{R}$. Then u is called a *multiple* of v; and u is said to be in the *same direction* as v if $k > 0$, and in the *opposite direction* if $k < 0$.

2.4 VECTORS AND LINEAR EQUATIONS

Two important concepts involving vectors, linear combinations and linear dependence, are closely related to systems of linear equations as follows.

Linear Combinations

Consider a nonhomogeneous system of m equations in n unknowns:

$$\begin{aligned}
a_{11}x_1 + a_{12}x_2 + \cdots + a_{1n}x_n &= b_1 \\
a_{21}x_1 + a_{22}x_2 + \cdots + a_{2n}x_n &= b_2 \\
&\cdots \\
a_{m1}x_1 + a_{m2}x_2 + \cdots + a_{mn}x_n &= b_m
\end{aligned}$$

This system is equivalent to the following vector equation

$$x_1\begin{pmatrix}a_{11}\\a_{21}\\\vdots\\a_{m1}\end{pmatrix} + x_2\begin{pmatrix}a_{12}\\a_{22}\\\vdots\\a_{m2}\end{pmatrix} + \cdots + x_n\begin{pmatrix}a_{1n}\\a_{2n}\\\vdots\\a_{mn}\end{pmatrix} = \begin{pmatrix}b_1\\b_2\\\vdots\\b_m\end{pmatrix}$$

that is, the vector equation

$$x_1 u_1 + x_2 u_2 + \cdots + x_n u_n = v$$

where $u_1, u_2, \ldots, u_n, v$ are the above column vectors, respectively.

Now if the above system has a solution, then v is said to be a linear combination of the vectors u_i. We state this important concept formally.

Definition: A vector v is a *linear combination* of vectors $u_1, u_2, \ldots, u_n$ if there exist scalars $k_1, k_2, \ldots, k_n$ such that

$$v = k_1 u_1 + k_2 u_2 + \cdots + k_n u_n$$

that is, if the vector equation

$$v = x_1 u_1 + x_2 u_2 + \cdots + x_n u_n$$

has a solution where the x_i are unknown scalars.

The above definition applies to both column vectors and row vectors, although our illustration was in terms of column vectors.

Example 2.2. Suppose

$$v = \begin{pmatrix} 2 \\ 3 \\ -4 \end{pmatrix}, \qquad u_1 = \begin{pmatrix} 1 \\ 1 \\ 1 \end{pmatrix}, \qquad u_2 = \begin{pmatrix} 1 \\ 1 \\ 0 \end{pmatrix} \quad \text{and} \quad u_3 = \begin{pmatrix} 1 \\ 0 \\ 0 \end{pmatrix}$$

Then v is a linear combination of u_1, u_2, u_3 since the vector equation (or system)

$$\begin{pmatrix} 2 \\ 3 \\ -4 \end{pmatrix} = x \begin{pmatrix} 1 \\ 1 \\ 1 \end{pmatrix} + y \begin{pmatrix} 1 \\ 1 \\ 0 \end{pmatrix} + z \begin{pmatrix} 1 \\ 0 \\ 0 \end{pmatrix} \qquad \text{or} \qquad \begin{cases} 2 = x + y + z \\ 3 = x + y \\ -4 = x \end{cases}$$

has a solution $x = -4$, $y = 7$, $z = -1$. In other words,

$$v = -4u_1 + 7u_2 - u_3$$

Linear Dependence

Consider a homogeneous system of m equations in n unknowns:

$$a_{11} x_1 + a_{12} x_2 + \cdots + a_{1n} x_n = 0$$
$$a_{21} x_1 + a_{22} x_2 + \cdots + a_{2n} x_n = 0$$
$$\cdots\cdots\cdots\cdots\cdots\cdots\cdots\cdots\cdots\cdots\cdots\cdots$$
$$a_{m1} x_1 + a_{m2} x_2 + \cdots + a_{mn} x_n = 0$$

This system is equivalent to the following vector equation:

$$x_1 \begin{pmatrix} a_{11} \\ a_{21} \\ \vdots \\ a_{m1} \end{pmatrix} + x_2 \begin{pmatrix} a_{12} \\ a_{22} \\ \vdots \\ a_{m2} \end{pmatrix} + \cdots + x_n \begin{pmatrix} a_{1n} \\ a_{2n} \\ \vdots \\ a_{mn} \end{pmatrix} = \begin{pmatrix} 0 \\ 0 \\ \vdots \\ 0 \end{pmatrix}$$

that is, the vector equation

$$x_1 u_1 + x_2 u_2 + \cdots + x_n u_n = 0$$

where $u_1, u_2, \ldots, u_n$ are the above column vectors, respectively.

Now, if the above homogeneous system has a nonzero solution, the vectors $u_1, u_2, \ldots, u_n$ are said to be linearly dependent; on the other hand, if the equation has only the zero solution, then the vectors are said to be linearly independent. We state this important concept formally.

> **Definition:** Vectors $u_1, u_2, \ldots, u_n$ in $\mathbf{R}^n$ are linearly dependent if there exist scalars $k_1, k_2, \ldots, k_n$, not all zero, such that
>
> $$k_1 u_1 + k_2 u_2 + \cdots + k_n u_n = 0$$
>
> that is, if the vector equation
>
> $$x_1 u_1 + x_2 u_2 + \cdots + x_n u_n = 0$$
>
> has a nonzero solution where the x_i are unknown scalars. Otherwise, the vectors are said to be linearly independent.

The above definition applies to both column vectors and row vectors, although our illustration was in terms of column vectors.

Example 2.3

(a) The only solution to

$$x\begin{pmatrix}1\\1\\1\end{pmatrix} + y\begin{pmatrix}1\\1\\0\end{pmatrix} + z\begin{pmatrix}1\\0\\0\end{pmatrix} = \begin{pmatrix}0\\0\\0\end{pmatrix} \quad \text{or} \quad \begin{cases} x + y + z = 0 \\ x + y \quad\;\; = 0 \\ x \qquad\quad\; = 0 \end{cases}$$

is the zero solution $x = 0$, $y = 0$, $z = 0$. Hence the three vectors are linearly independent.

(b) The vector equation (or system of linear equations)

$$x\begin{pmatrix}1\\1\\1\end{pmatrix} + y\begin{pmatrix}2\\-1\\3\end{pmatrix} + z\begin{pmatrix}1\\-5\\3\end{pmatrix} = \begin{pmatrix}0\\0\\0\end{pmatrix} \quad \text{or} \quad \begin{cases} x + 2y + \;\;z = 0 \\ x - \;\;y - 5z = 0 \\ x + 3y + 3z = 0 \end{cases}$$

has a nonzero solution $(3, -2, 1)$, i.e., $x = 3$, $y = -2$, $z = 1$. Thus the three vectors are linearly dependent.

2.5 DOT (SCALAR) PRODUCT

Let u and v be vectors in $\mathbf{R}^n$:

$$u = (u_1, u_2, \ldots, u_n) \quad \text{and} \quad v = (v_1, v_2, \ldots, v_n)$$

The *dot*, *scalar*, or *inner product* of u and v, denoted by $u \cdot v$, is the scalar obtained by multiplying corresponding components and adding the resulting products:

$$u \cdot v = u_1 v_1 + u_2 v_2 + \cdots + u_n v_n$$

The vectors u and v are said to be orthogonal (or perpendicular) if their dot product is zero, that is, if $u \cdot v = 0$.

Example 2.4. Let $u = (1, -2, 3, -4)$, $v = (6, 7, 1, -2)$, and $w = (5, -4, 5, 7)$. Then

$$u \cdot v = 1 \cdot 6 + (-2) \cdot 7 + 3 \cdot 1 + (-4) \cdot (-2) = 6 - 14 + 3 + 8 = 3$$

$$u \cdot w = 1 \cdot 5 + (-2) \cdot (-4) + 3 \cdot 5 + (-4) \cdot 7 = 5 + 8 + 15 - 28 = 0$$

Thus u and w are orthogonal.

Basic properties of the dot product in $\mathbf{R}^n$ (proved in Problem 2.17) follow.

Theorem 2.2:   For any vectors $u, v, w \in \mathbf{R}^n$ and any scalar $k \in \mathbf{R}$,

   (i) $(u + v) \cdot w = u \cdot w + v \cdot w$ (iii) $u \cdot v = v \cdot u$

   (ii) $(ku) \cdot v = k(u \cdot v)$ (iv) $u \cdot u \geq 0$, and $u \cdot u = 0$ iff $u = 0$

> **Remark:**   The space $\mathbf{R}^n$ with the above operations of vector addition, scalar multiplication, and dot product is usually called Euclidean n-space.

2.6   NORM OF A VECTOR

Let $u = (u_1, u_2, \ldots, u_n)$ be a vector in $\mathbf{R}^n$. The *norm* (or *length*) of the vector u, written $\|u\|$, is defined to be the nonnegative square root of $u \cdot u$:

$$\|u\| = \sqrt{u \cdot u} = \sqrt{u_1^2 + u_2^2 + \cdots + u_n^2}$$

Since $u \cdot u \geq 0$, the square root exists. Also, if $u \neq 0$, then $\|u\| > 0$; and $\|0\| = 0$.

The above definition of the norm of a vector conforms to that of the length of a vector (arrow) in (Euclidean) geometry. Specifically, suppose $\mathbf{u}$ is a vector (arrow) in the plane $\mathbf{R}^2$ with endpoint $P(a, b)$ as shown in Fig. 2-3. Then $|a|$ and $|b|$ are the lengths of the sides of the right triangle formed by $\mathbf{u}$ and the horizontal and vertical directions. By the Pythagorean Theorem, the length $|\mathbf{u}|$ of $\mathbf{u}$ is

$$|\mathbf{u}| = \sqrt{a^2 + b^2}$$

This value is the same as the norm of u defined above.

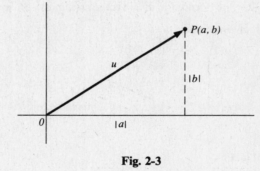

Fig. 2-3

Example 2.5.   Suppose $u = (3, -12, -4)$. To find $\|u\|$, we first find $\|u\|^2 = u \cdot u$ by squaring the components of u and adding:

$$\|u\|^2 = 3^2 + (-12)^2 + (-4)^2 = 9 + 144 + 16 = 169$$

Then $\|u\| = \sqrt{169} = 13$.

A vector u is a unit vector if $\|u\| = 1$ or, equivalently, if $u \cdot u = 1$. Now if v is any nonzero vector, then

$$\hat{v} \equiv \frac{1}{\|v\|} v = \frac{v}{\|v\|}$$

is a unit vector in the same direction as v. (The process of finding $\hat{v}$ is called *normalizing* v.) For example,

$$\hat{v} = \frac{v}{\|v\|} = \left(\frac{2}{\sqrt{102}}, \frac{-3}{\sqrt{102}}, \frac{8}{\sqrt{102}}, \frac{-5}{\sqrt{102}} \right)$$

is the unit vector in the direction of the vector $v = (2, -3, 8, -5)$.

We now state a fundamental relationship (proved in Problem 2.22) known as the Cauchy–Schwarz inequality.

Theorem 2.3 (Cauchy–Schwarz):   For any vectors u, v in $\mathbf{R}^n$, $|u \cdot v| \leq \|u\| \, \|v\|$.

Using the above inequality, we prove (Problem 2.33) the following result known as the triangle inequality or as Minkowski's inequality.

Theorem 2.4 (Minkowski):   For any vectors u, v in $\mathbf{R}^n$, $\|u + v\| \leq \|u\| + \|v\|$.

Distance, Angles, Projections

Let $u = (u_1, u_2, \ldots, u_n)$ and $v = (v_1, v_2, \ldots, v_n)$ be vectors in $\mathbf{R}^n$. The *distance* between u and v, denoted by $d(u, v)$, is defined as

$$d(u, v) \equiv \|u - v\| = \sqrt{(u_1 - v_1)^2 + (u_2 - v_2)^2 + \cdots + (u_n - v_n)^2}$$

We show that this definition corresponds to the usual notion of Euclidean distance in the plane $\mathbf{R}^2$. Consider $u = (a, b)$ and $v = (c, d)$ in $\mathbf{R}^2$. As shown in Fig. 2-4, the distance between the points $P(a, b)$ and $Q(c, d)$ is

$$d = \sqrt{(a - c)^2 + (b - d)^2}$$

On the other hand, by the above definition,

$$d(u, v) = \|u - v\| = \|(a - c, b - d)\| = \sqrt{(a - c)^2 + (b - d)^2}$$

Both give the same value.

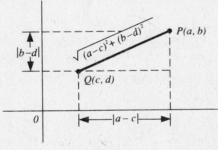

Fig. 2-4

Using the Cauchy–Schwarz inequality, we can define the angle θ between any two nonzero vectors u, v in $\mathbf{R}^n$ by

$$\cos \theta = \frac{u \cdot v}{\|u\| \, \|v\|}$$

Note that if $u \cdot v = 0$, then $\theta = 90°$ (or $\theta = \pi/2$). This then agrees with our previous definition of orthogonality.

Example 2.6.   Suppose $u = (1, -2, 3)$ and $v = (3, -5, -7)$. Then

$$d(u, v) = \sqrt{(1 - 3)^2 + (-2 + 5)^2 + (3 + 7)^2} = \sqrt{4 + 9 + 100} = \sqrt{113}$$

To find $\cos \theta$, where θ is the angle between u and v, we first find

$$u \cdot v = 3 + 10 - 21 = -8 \qquad \|u\|^2 = 1 + 4 + 9 = 14 \qquad \|v\|^2 = 9 + 25 + 49 = 83$$

Then

$$\cos \theta = \frac{u \cdot v}{\|u\|\|v\|} = -\frac{-8}{\sqrt{14}\sqrt{83}}$$

Let u and $v \neq 0$ be vectors in $\mathbf{R}^n$. The *(vector) projection of u onto v* is the vector

$$\mathrm{proj}\,(u, v) = \frac{u \cdot v}{\|v\|^2}\, v$$

We show how this definition conforms to the notion of vector projection in physics. Consider the vectors **u** and **v** in Fig. 2-5. The (perpendicular) projection of **u** onto **v** is the vector **u***, of magnitude

$$|\mathbf{u}^*| = |\mathbf{u}| \cos \theta = |\mathbf{u}| \frac{\mathbf{u} \cdot \mathbf{v}}{|\mathbf{u}||\mathbf{v}|} = \frac{\mathbf{u} \cdot \mathbf{v}}{|\mathbf{v}|}$$

To obtain **u***, we multiply its magnitude by the unit vector in the direction of **v**:

$$\mathbf{u}^* = |\mathbf{u}^*|\,\frac{\mathbf{v}}{|\mathbf{v}|} = \frac{\mathbf{u} \cdot \mathbf{v}}{|\mathbf{v}|^2}\,\mathbf{v}$$

This agrees with the above definition of proj (u, v).

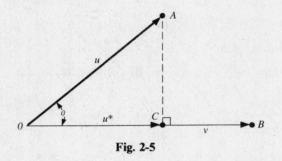

Fig. 2-5

Example 2.7.   Suppose $u = (1, -2, 3)$ and $v = (2, 5, 4)$. To find proj (u, v), we first find

$$u \cdot v = 2 - 10 + 12 = 4 \qquad \text{and} \qquad \|v\|^2 = 4 + 25 + 16 = 45$$

Then

$$\mathrm{proj}\,(u, v) = \frac{u \cdot v}{\|v\|^2}\, v = \frac{4}{45}\,(2, 5, 4) = \left(\frac{8}{45}, \frac{20}{45}, \frac{16}{45}\right) = \left(\frac{8}{45}, \frac{4}{9}, \frac{16}{45}\right)$$

2.7 LOCATED VECTORS, HYPERPLANES, AND LINES IN $\mathbf{R}^n$

This section distinguishes between an n-tuple $P(a_1, a_2, \ldots, a_n) \equiv P(a_i)$ viewed as a point in $\mathbf{R}^n$ and an n-tuple $v = [c_1, c_2, \ldots, c_n]$ viewed as a vector (arrow) from the origin O to the point $C(c_1, c_2, \ldots, c_n)$. Any pair of points $P = (a_i)$ and $Q = (b_i)$ in $\mathbf{R}^n$ defines the *located vector* or *directed line segment* from P to Q, written $\overrightarrow{PQ}$. We identify $\overrightarrow{PQ}$ with the vector

$$v = Q - P = [b_1 - a_1, b_2 - a_2, \ldots, b_n - a_n]$$

since $\overrightarrow{PQ}$ and v have the same magnitude and direction as shown in Fig. 2-6.

A hyperplane H in $\mathbf{R}^n$ is the set of points $(x_1, x_2, \ldots, x_n)$ that satisfy a nondegenerate linear equation

$$a_1 x_1 + a_2 x_2 + \cdots + a_n x_n = b$$

In particular, a hyperplane H in $\mathbf{R}^2$ is a line; and a hyperplane H in $\mathbf{R}^3$ is a plane. The vector $u = [a_1, a_2, \ldots, a_n] \neq 0$ is called a *normal* to H. This terminology is justified by the fact (Problem 2.33)

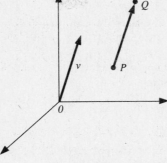

Fig. 2-6

that any directed line segment $\overrightarrow{PQ}$, where P, Q belong to H, is orthogonal to the normal vector u. This fact in $\mathbf{R}^3$ is shown in Fig. 2-7.

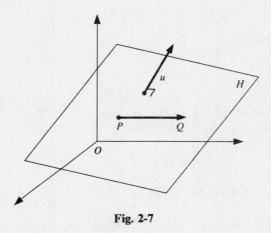

Fig. 2-7

The *line L* in $\mathbf{R}^n$ passing through the point $P(a_1, a_2, \ldots, a_n)$ and in the direction of the nonzero vector $u = [u_1, u_2, \ldots, u_n]$ consists of the points $X(x_1, x_2, \ldots, x_n)$ which satisfy

$$X = P + tu \qquad \text{or} \qquad \begin{cases} x_1 = a_1 + u_1 t \\ x_2 = a_2 + u_2 t \\ \cdots\cdots\cdots\cdots\cdots \\ x_n = a_n + u_n t \end{cases}$$

where the *parameter t* takes on all real numbers. (See Fig. 2-8.)

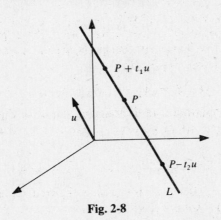

Fig. 2-8

Example 2.8

(a) Consider the hyperplane H in $\mathbf{R}^4$ which passes through the point $P(1, 3, -4, 2)$ and is normal to the vector $u = [4, -2, 5, 6]$. Such an equation has the form

$$4x - 2y + 5z + 6t = k$$

Substituting P into this equation, we obtain

$$4(1) - 2(3) + 5(-4) + 6(2) = k \quad\text{or}\quad 4 - 6 - 20 + 12 = k \quad\text{or}\quad k = -10$$

Thus $4x - 2y + 5z + 6t = -10$ is the equation of H.

(b) Consider the line L in $\mathbf{R}^4$ passing through the point $P(1, 2, 3, -4)$ in the direction of $u = [5, 6, -7, 8]$. A parametric representation of L is as follows:

$$\begin{aligned} x_1 &= 1 + 5t \\ x_2 &= 2 + 6t \\ x_3 &= 3 - 7t \\ x_4 &= -4 + 8t \end{aligned} \qquad\text{or}\qquad (1 + 5t, 2 + 6t, 3 - 7t, -4 + 8t)$$

Note $t = 0$ yields the point P on L.

Curves in $\mathbf{R}^n$

Let D be an interval (finite or infinite) in the real line $\mathbf{R}$. A continuous function $F: D \to \mathbf{R}^n$ is a curve in $\mathbf{R}^n$. Thus to each $t \in D$ there is assigned the following point (vector) in $\mathbf{R}^n$:

$$F(t) = [F_1(t), F_2(t), \ldots, F_n(t)]$$

Moreover, the derivative (if it exists) of $F(t)$ yields the vector

$$V(t) = dF(t)/dt = [dF_1(t)/dt, dF_2(t)/dt, \ldots, dF_n(t)/dt]$$

which is tangent to the curve and normalizing $V(t)$ yields

$$T(t) = \frac{V(t)}{\|V(t)\|}$$

which is the unit tangent vector to the curve.

Example 2.9.   Consider the following curve C in $\mathbf{R}^3$:

$$F(t) = [\sin t, \cos t, t]$$

Taking the derivative of $F(t)$ [or each component of $F(t)$] yields

$$V(t) = [\cos t, -\sin t, 1]$$

which is a vector tangent to the curve. We normalize $V(t)$. First we obtain

$$\|V(t)\|^2 = \cos^2 t + \sin^2 t + 1 = 1 + 1 = 2$$

Then

$$T(t) = \frac{V(t)}{\|V(t)\|} = \left[\frac{\cos t}{\sqrt{2}}, \frac{-\sin t}{\sqrt{2}}, \frac{1}{\sqrt{2}}\right]$$

which is the unit tangent vector to the curve.

2.8 SPATIAL VECTORS, ijk NOTATION IN $\mathbf{R}^3$

Vectors in $\mathbf{R}^3$, called spatial vectors, appear in many applications, especially in physics. In fact, a special notation is frequently used for such vectors as follows:

$\mathbf{i} = (1, 0, 0)$ denotes the unit vector in the x direction,

$\mathbf{j} = (0, 1, 0)$ denotes the unit vector in the y direction,

$\mathbf{k} = (0, 0, 1)$ denotes the unit vector in the z direction.

Then any vector $u = (a, b, c)$ in $\mathbf{R}^3$ can be expressed uniquely in the form

$$u = (a, b, c) = a\mathbf{i} + b\mathbf{j} + c\mathbf{k}$$

Since $\mathbf{i}, \mathbf{j}, \mathbf{k}$ are unit vectors and are mutually orthogonal, we have

$$\mathbf{i} \cdot \mathbf{i} = 1, \qquad \mathbf{j} \cdot \mathbf{j} = 1, \qquad \mathbf{k} \cdot \mathbf{k} = 1 \quad \text{and} \quad \mathbf{i} \cdot \mathbf{j} = 0, \qquad \mathbf{i} \cdot \mathbf{k} = 0, \qquad \mathbf{j} \cdot \mathbf{k} = 0$$

The various vector operations discussed previously may be expressed in the above notation as follows. Suppose $u = a_1\mathbf{i} + a_2\mathbf{j} + a_3\mathbf{k}$ and $v = b_1\mathbf{i} + b_2\mathbf{j} + b_3\mathbf{k}$. Then

$$u + v = (a_1 + b_1)\mathbf{i} + (a_2 + b_2)\mathbf{j} + (a_3 + b_3)\mathbf{k} \qquad u \cdot v = a_1 b_1 + a_2 b_2 + a_3 b_3$$

$$cu = ca_1\mathbf{i} + ca_2\mathbf{j} + ca_3\mathbf{k} \qquad \|u\| = \sqrt{u \cdot u} = \sqrt{a_1^2 + a_2^2 + a_3^2}$$

where c is a scalar.

Example 2.10.   Suppose $u = 3\mathbf{i} + 5\mathbf{j} - 2\mathbf{k}$ and $v = 4\mathbf{i} - 3\mathbf{j} + 7\mathbf{k}$.

(a)   To find $u + v$, add corresponding components yielding

$$u + v = 7\mathbf{i} + 2\mathbf{j} + 5\mathbf{k}$$

(b)   To find $3u - 2v$, first multiply the vectors by the scalars, and then add:

$$3u - 2v = (9\mathbf{i} + 15\mathbf{j} - 6\mathbf{k}) + (-8\mathbf{i} + 6\mathbf{j} - 14\mathbf{k}) = 4\mathbf{i} + 21\mathbf{j} - 20\mathbf{k}$$

(c)   To find $u \cdot v$, multiply corresponding components and then add:

$$u \cdot v = 12 - 15 - 14 = -17$$

(d)   To find $\|u\|$, square each component and then add to get $\|u\|^2$. That is,

$$\|u\|^2 = 9 + 25 + 4 = 38 \quad \text{and hence} \quad \|u\| = \sqrt{38}$$

Cross Product

There is a special operation for vectors u, v in $\mathbf{R}^3$, called the *cross product* and denoted by $u \times v$. Specifically, suppose

$$u = a_1\mathbf{i} + a_2\mathbf{j} + a_3\mathbf{k} \qquad \text{and} \qquad v = b_1\mathbf{i} + b_2\mathbf{j} + b_3\mathbf{k}$$

Then

$$u \times v = (a_2 b_3 - a_3 b_2)\mathbf{i} + (a_3 b_1 - a_1 b_3)\mathbf{j} + (a_1 b_2 - a_2 b_1)\mathbf{k}$$

Note $u \times v$ is a vector; hence $u \times v$ is also called *vector product* or *outer product* of u and v.

Using determinant notation (Chapter 7), where $\begin{vmatrix} a & b \\ c & d \end{vmatrix} = ad - bc$, the cross product may also be expressed as follows:

$$u \times v = \begin{vmatrix} a_2 & a_3 \\ b_2 & b_3 \end{vmatrix}\mathbf{i} - \begin{vmatrix} a_1 & a_3 \\ b_1 & b_3 \end{vmatrix}\mathbf{j} + \begin{vmatrix} a_1 & a_2 \\ b_1 & b_2 \end{vmatrix}\mathbf{k}$$

or, equivalently,

$$u \times v = \begin{vmatrix} \mathbf{i} & \mathbf{j} & \mathbf{k} \\ a_1 & a_2 & a_3 \\ b_1 & b_2 & b_3 \end{vmatrix}$$

Two important properties of the cross product follow (see Problem 2.56).

Theorem 2.5:   Let u, v, w be vectors in $\mathbf{R}^3$.

    (i)   The vector $w = u \times v$ is orthogonal to both u and v.

    (ii) The absolute value of the "triple product" $u \cdot v \times w$ represents the volume of the parallelepiped formed by the vectors $u, v,$ and w (as shown in Fig. 2-9).

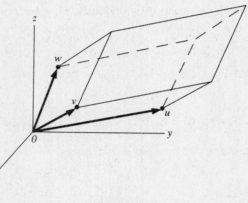

Fig. 2-9

Example 2.11

(a)   Suppose $u = 4\mathbf{i} + 3\mathbf{j} + 6\mathbf{k}$ and $v = 2\mathbf{i} + 5\mathbf{j} - 3\mathbf{k}$. Then

$$u \times v = \begin{vmatrix} 3 & 6 \\ 5 & -3 \end{vmatrix} \mathbf{i} - \begin{vmatrix} 4 & 6 \\ 2 & -3 \end{vmatrix} \mathbf{j} + \begin{vmatrix} 4 & 3 \\ 2 & 5 \end{vmatrix} \mathbf{k} = -39\mathbf{i} + 24\mathbf{j} + 14\mathbf{k}$$

(b) $$(2, -1, 5) \times (3, 7, 6) = \left(\begin{vmatrix} -1 & 5 \\ 7 & 6 \end{vmatrix}, \ - \begin{vmatrix} 2 & 5 \\ 3 & 6 \end{vmatrix}, \ \begin{vmatrix} 2 & -1 \\ 3 & 7 \end{vmatrix} \right) = (-41, 3, 17)$$

(Here we find the cross product without using the **ijk** notation.)

(c)   The cross products of the vectors $\mathbf{i}, \mathbf{j}, \mathbf{k}$ are as follows:

$$\mathbf{i} \times \mathbf{j} = \mathbf{k}, \qquad \mathbf{j} \times \mathbf{k} = \mathbf{i}, \qquad \mathbf{k} \times \mathbf{i} = \mathbf{j}, \qquad \text{and} \qquad \mathbf{j} \times \mathbf{i} = -\mathbf{k}, \qquad \mathbf{k} \times \mathbf{j} = -\mathbf{i}, \qquad \mathbf{i} \times \mathbf{k} = -\mathbf{j}$$

In other words, if we view the triple $(\mathbf{i}, \mathbf{j}, \mathbf{k})$ as a cyclic permutation, i.e., as arranged around a circle in the counterclockwise direction as in Fig. 2-10, then the product of two of them in the given direction is the third one but the product of two of them in the opposite direction is the negative of the third one.

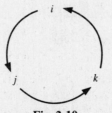

Fig. 2-10

2.9 COMPLEX NUMBERS

The set of complex numbers is denoted by $\mathbf{C}$. Formally, a complex number is an ordered pair (a, b) of real numbers; equality, addition, and multiplication of complex numbers are defined as follows:

$$(a, b) = (c, d) \quad \text{iff} \quad a = c \quad \text{and} \quad b = d$$
$$(a, b) + (c, d) = (a + c, b + d)$$
$$(a, b)(c, d) = (ac - bd, ad + bc)$$

We identify the real number a with the complex number $(a, 0)$:

$$a \leftrightarrow (a, 0)$$

This is possible since the operations of addition and multiplication of real numbers are preserved under the correspondence

$$(a, 0) + (b, 0) = (a + b, 0) \quad \text{and} \quad (a, 0)(b, 0) = (ab, 0)$$

Thus we view $\mathbf{R}$ as a subset of $\mathbf{C}$ and replace $(a, 0)$ by a whenever convenient and possible.

The complex number $(0, 1)$, denoted by i, has the important property that

$$i^2 = ii = (0, 1)(0, 1) = (-1, 0) = -1 \quad \text{or} \quad i = \sqrt{-1}$$

Furthermore, using the fact

$$(a, b) = (a, 0) + (0, b) \quad \text{and} \quad (0, b) = (b, 0)(0, 1)$$

we have

$$(a, b) = (a, 0) + (b, 0)(0, 1) = a + bi$$

The notation $z = a + bi$, where $a \equiv \text{Re } z$ and $b \equiv \text{Im } z$ are called, respectively, the *real* and *imaginary* *parts* of the complex number z, is more convenient than (a, b). For example, the sum and product of two complex numbers $z = a + bi$ and $w = c + di$ can be obtained by simply using the commutative and distributive laws and $i^2 = -1$:

$$z + w = (a + bi) + (c + di) = a + c + bi + di = (a + c) + (b + d)i$$

$$zw = (a + bi)(c + di) = ac + bci + adi + bdi^2 = (ac - bd) + (bc + ad)i$$

Warning: The above use of the letter i for $\sqrt{-1}$ has no relationship whatsoever to the vector notation $\mathbf{i} = (1, 0, 0)$ introduced in Section 2.8.

The conjugate of the complex number $z = (a, b) = a + bi$ is denoted and defined by

$$\bar{z} = \overline{a + bi} = a - bi$$

Then $z\bar{z} = (a + bi)(a - bi) = a^2 - b^2 i^2 = a^2 + b^2$. If, in addition, $z \neq 0$, then the inverse z^{-1} of z and division of w by z are given, respectively, by

$$z^{-1} = \frac{\bar{z}}{z\bar{z}} = \frac{a}{a^2 + b^2} + \frac{-b}{a^2 + b^2} i \quad \text{and} \quad \frac{w}{z} = wz^{-1}$$

where $w \in \mathbf{C}$. We also define

$$-z = -1z \quad \text{and} \quad w - z = w + (-z)$$

Just as the real numbers can be represented by the points on a line, the complex numbers can be represented by the points in the plane. Specifically, we let the point (a, b) in the plane represent the complex number $z = a + bi$, i.e., whose *real part* is a and whose imaginary part is b. (See Fig. 2-11.) The *absolute value* of z, written $|z|$, is defined as the distance from z to the origin:

$$|z| = \sqrt{a^2 + b^2}$$

Note that $|z|$ is equal to the norm of the vector (a, b). Also, $|z| = \sqrt{z\bar{z}}$.

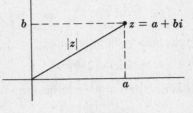

Fig. 2-11

Example 2.12. Suppose $z = 2 + 3i$ and $w = 5 - 2i$. Then

$$z + w = (2 + 3i) + (5 - 2i) = 2 + 5 + 3i - 2i = 7 + i$$

$$zw = (2 + 3i)(5 - 2i) = 10 + 15i - 4i - 6i^2 = 16 + 11i$$

$$\bar{z} = \overline{2 + 3i} = 2 - 3i \qquad \text{and} \qquad \bar{w} = \overline{5 - 2i} = 5 + 2i$$

$$\frac{w}{z} = \frac{5 - 2i}{2 + 3i} = \frac{(5 - 2i)(2 - 3i)}{(2 + 3i)(2 - 3i)} = \frac{4 - 19i}{13} = \frac{4}{13} - \frac{19}{13}i$$

$$|z| = \sqrt{4 + 9} = \sqrt{13} \qquad \text{and} \qquad |w| = \sqrt{25 + 4} = \sqrt{29}$$

> **Remark:** In Appendix *B* we define the algebraic structure called a *field*. We emphasize that the set $\mathbf{C}$ of complex numbers with the above operations of addition and multiplication is a field.

2.10 VECTORS IN $\mathbf{C}^n$

The set of all *n*-tuples of complex numbers, denoted by $\mathbf{C}^n$, is called *complex n-space*. Just as in the real case, the elements of $\mathbf{C}^n$ are called *points* or *vectors*, the elements of $\mathbf{C}$ are called *scalars*, and *vector addition* in $\mathbf{C}^n$ and *scalar multiplication* on $\mathbf{C}^n$ are given by

$$(z_1, z_2, \ldots, z_n) + (w_1, w_2, \ldots, w_n) = (z_1 + w_1, z_2 + w_2, \ldots, z_n + w_n)$$

$$z(z_1, z_2, \ldots, z_n) = (zz_1, zz_2, \ldots, zz_n)$$

where $z_i, w_i, z \in \mathbf{C}$.

Example 2.13

(a) $(2 + 3i, 4 - i, 3) + (3 - 2i, 5i, 4 - 6i) = (5 + i, 4 + 4i, 7 - 6i)$

(b) $2i(2 + 3i, 4 - i, 3) = (-6 + 4i, 2 + 8i, 6i)$

Now let u and v be arbitrary vectors in $\mathbf{C}^n$:

$$u = (z_1, z_2, \ldots, z_n) \qquad v = (w_1, w_2, \ldots, w_n) \qquad z_i, w_i \in \mathbf{C}$$

The *dot*, or *inner*, product of u and v is defined as follows:

$$u \cdot v = z_1 \bar{w}_1 + z_2 \bar{w}_2 + \cdots + z_n \bar{w}_n$$

Note that this definition reduces to the previous one in the real case, since $w_i = \bar{w}_i$ when w_i is real. The norm of u is defined by

$$\|u\| = \sqrt{u \cdot u} = \sqrt{z_1\bar{z}_1 + z_2\bar{z}_2 + \cdots + z_n\bar{z}_n} = \sqrt{|z_1|^2 + |z_2|^2 + \cdots + |z_n|^2}$$

Observe that $u \cdot u$ and so $\|u\|$ are real and positive when $u \neq 0$ and 0 when $u = 0$.

Example 2.14. Let $u = (2 + 3i, 4 - i, 2i)$ and $v = (3 - 2i, 5, 4 - 6i)$. Then

$$\begin{aligned}
u \cdot v &= (2 + 3i)\overline{(3 - 2i)} + (4 - i)\overline{(5)} + (2i)\overline{(4 - 6i)} \\
&= (2 + 3i)(3 + 2i) + (4 - i)(5) + (2i)(4 + 6i) \\
&= 13i + 20 - 5i - 12 + 8i = 8 + 16i \\
u \cdot u &= (2 + 3i)\overline{(2 + 3i)} + (4 - i)\overline{(4 - i)} + (2i)\overline{(2i)} \\
&= (2 + 3i)(2 - 3i) + (4 - i)(4 + i) + (2i)(-2i) \\
&= 13 + 17 + 4 = 34 \\
\|u\| &= \sqrt{u \cdot u} = \sqrt{34}
\end{aligned}$$

The space **C**^{n} with the above operations of vector addition, scalar multiplication, and dot product, is called *complex Euclidean n-space*.

> **Remark:** If $u \cdot v$ were defined by $u \cdot v = z_1 w_1 + \cdots + z_n w_n$, then it is possible for $u \cdot u = 0$ even though $u \neq 0$, e.g., if $u = (1, i, 0)$. In fact, $u \cdot u$ may not even be real.

Solved Problems

VECTORS IN **R**^{n}

2.1. Let $u = (2, -7, 1)$, $v = (-3, 0, 4)$, $w = (0, 5, -8)$. Find (a) $3u - 4v$, (b) $2u + 3v - 5w$.

First perform the scalar multiplication and then the vector addition.

(a) $3u - 4v = 3(2, -7, 1) - 4(-3, 0, 4) = (6, -21, 3) + (12, 0, -16) = (18, -21, -13)$

(b) $2u + 3v - 5w = 2(2, -7, 1) + 3(-3, 0, 4) - 5(0, 5, -8)$
$$= (4, -14, 2) + (-9, 0, 12) + (0, -25, 40)$$
$$= (4 - 9 + 0, -14 + 0 - 25, 2 + 12 + 40) = (-5, -39, 54)$$

2.2. Compute:

$$(a) \quad 2\begin{pmatrix} 1 \\ -1 \\ 3 \end{pmatrix} - 3\begin{pmatrix} 2 \\ 3 \\ -4 \end{pmatrix}, \qquad (b) \quad -2\begin{pmatrix} 5 \\ 3 \\ -4 \end{pmatrix} + 4\begin{pmatrix} -1 \\ 5 \\ 2 \end{pmatrix} - 3\begin{pmatrix} 3 \\ -1 \\ -1 \end{pmatrix}$$

First perform the scalar multiplication and then the vector addition.

$$(a) \quad 2\begin{pmatrix} 1 \\ -1 \\ 3 \end{pmatrix} - 3\begin{pmatrix} 2 \\ 3 \\ -4 \end{pmatrix} = \begin{pmatrix} 2 \\ -2 \\ 6 \end{pmatrix} + \begin{pmatrix} -6 \\ -9 \\ 12 \end{pmatrix} = \begin{pmatrix} -4 \\ -11 \\ 18 \end{pmatrix}$$

$$(b) \quad -2\begin{pmatrix} 5 \\ 3 \\ -4 \end{pmatrix} + 4\begin{pmatrix} -1 \\ 5 \\ 2 \end{pmatrix} - 3\begin{pmatrix} 3 \\ -1 \\ -1 \end{pmatrix} = \begin{pmatrix} -10 \\ -6 \\ 8 \end{pmatrix} + \begin{pmatrix} -4 \\ 20 \\ 8 \end{pmatrix} + \begin{pmatrix} -9 \\ 3 \\ 3 \end{pmatrix} = \begin{pmatrix} -23 \\ 17 \\ 19 \end{pmatrix}$$

2.3.   Find x and y if $(a)\ (x, 3) = (2, x + y)$; $(b)\ (4, y) = x(2, 3)$.

     (a)   Since the two vectors are equal, the corresponding components are equal to each other:

$$x = 2 \qquad 3 = x + y$$

            Substitute $x = 2$ into the second equation to obtain $y = 1$. Thus $x = 2$ and $y = 1$.

     (b)   Multiply by the scalar x to obtain $(4, y) = x(2, 3) = (2x, 3x)$. Set the corresponding components equal to each other:

$$4 = 2x \qquad y = 3x$$

            Solve the linear equations for x and y: $x = 2$ and $y = 6$.

2.4.   Prove Theorem 2.1.

       Let u_i, v_i, and w_i be the ith components of u, v, and w, respectively.

     (i)   By definition, $u_i + v_i$ is the ith component of $u + v$ and so $(u_i + v_i) + w_i$ is the ith component of $(u + v) + w$. On the other hand, $v_i + w_i$ is the ith component of $v + w$ and so $u_i + (v_i + w_i)$ is the ith component of $u + (v + w)$. But u_i, v_i, and w_i are real numbers for which the associative law holds, that is,

$$(u_i + v_i) + w_i = u_i + (v_i + w_i) \qquad \text{for} \qquad i = 1, \dots, n$$

           Accordingly, $(u + v) + w = u + (v + w)$ since their corresponding components are equal.

    (ii)   Here, $0 = (0, 0, \dots, 0)$; hence

$$u + 0 = (u_1, u_2, \dots, u_n) + (0, 0, \dots, 0)$$
$$= (u_1 + 0, u_2 + 0, \dots, u_n + 0) = (u_1, u_2, \dots, u_n) = u$$

   (iii)   Since $-u = -1(u_1, u_2, \dots, u_n) = (-u_1, -u_2, \dots, -u_n)$,

$$u + (-u) = (u_1, u_2, \dots, u_n) + (-u_1, -u_2, \dots, -u_n)$$
$$= (u_1 - u_1, u_2 - u_2, \dots, u_n - u_n) = (0, 0, \dots, 0) = 0$$

   (iv)   By definition, $u_i + v_i$ is the ith component of $u + v$, and $v_i + u_i$ is the ith component of $v + u$. But u_i and v_i are real numbers for which the commutative law holds, that is,

$$u_i + v_i = v_i + u_i \qquad i = 1, \dots, n$$

           Hence $u + v = v + u$ since their corresponding components are equal.

     (v)   Since $u_i + v_i$ is the ith component of $u + v$, $k(u_i + v_i)$ is the ith component of $k(u + v)$. Since ku_i and kv_i are the ith components of ku and kv, respectively, $ku_i + kv_i$ is the ith component of $ku + kv$. But k, u_i, and v_i are real numbers; hence

$$k(u_i + v_i) = ku_i + kv_i \qquad i = 1, \dots, n$$

           Thus $k(u + v) = ku + kv$, as corresponding components are equal.

   (vi)   Observe that the first plus sign refers to the addition of the two scalars k and k' whereas the second plus sign refers to the vector addition of the two vectors ku and $k'u$.

           By definition, $(k + k')u_i$ is the ith component of the vector $(k + k')u$. Since ku_i and $k'u_i$ are the ith components of ku and $k'u$, respectively, $ku_i + k'u_i$ is the ith component of $ku + k'u$. But k, k', and u_i are real numbers; hence

$$(k + k')u_i = ku_i + k'u_i \qquad i = 1, \dots, n$$

           Thus $(k + k')u = ku + k'u$, as corresponding components are equal.

   (vii)   Since $k'u_i$ is the ith component of $k'u$, $k(k'u_i)$ is the ith component of $k(k'u)$. But $(kk')u_i$ is the ith component of $(kk')u$ and, since k, k' and u_i are real numbers,

$$(kk')u_i = k(k'u_i) \qquad i = 1, \dots, n$$

           Hence $(kk')u = k(k'u)$, as corresponding components are equal.

  (viii) $1 \cdot u = 1(u_1, u_2, \dots, u_n) = (1u_1, 1u_2, \dots, 1u_n) = (u_1, u_2, \dots, u_n) = u$.

VECTORS AND LINEAR EQUATIONS

2.5. Convert the following vector equation to an equivalent system of linear equations and solve:

$$\begin{pmatrix} 1 \\ -6 \\ 5 \end{pmatrix} = x \begin{pmatrix} 1 \\ 2 \\ 3 \end{pmatrix} + y \begin{pmatrix} 2 \\ 5 \\ 8 \end{pmatrix} + z \begin{pmatrix} 3 \\ 2 \\ 3 \end{pmatrix}$$

Multiply the vectors on the right by the unknown scalars and then add:

$$\begin{pmatrix} 1 \\ -6 \\ 5 \end{pmatrix} = \begin{pmatrix} x \\ 2x \\ 3x \end{pmatrix} + \begin{pmatrix} 2y \\ 5y \\ 8y \end{pmatrix} + \begin{pmatrix} 3z \\ 2z \\ 3z \end{pmatrix} = \begin{pmatrix} x + 2y + 3z \\ 2x + 5y + 2z \\ 3x + 8y + 3z \end{pmatrix}$$

Set corresponding components of the vectors equal to each other, and reduce the system to echelon form:

$$\begin{array}{ccc}
x + 2y + 3z = 1 & x + 2y + 3z = 1 & x + 2y + 3z = 1 \\
2x + 5y + 2z = -6 & y - 4z = -8 & y - 4z = -8 \\
3x + 8y + 3z = 5 & 2y - 6z = 2 & 2z = 18
\end{array}$$

The system is triangular, and back-substitution yields the unique solution $x = -82$, $y = 28$, $z = 9$.

2.6. Write the vector $v = (1, -2, 5)$ as a linear combination of the vectors $u_1 = (1, 1, 1)$, $u_2 = (1, 2, 3)$, and $u_3 = (2, -1, 1)$.

We want to express v in the form $v = xu_1 + yu_2 + zu_3$ with x, y, and z as yet unknown. Thus we have

$$\begin{pmatrix} 1 \\ -2 \\ 5 \end{pmatrix} = x \begin{pmatrix} 1 \\ 1 \\ 1 \end{pmatrix} + y \begin{pmatrix} 1 \\ 2 \\ 3 \end{pmatrix} + z \begin{pmatrix} 2 \\ -1 \\ 1 \end{pmatrix} = \begin{pmatrix} x + y + 2z \\ x + 2y - z \\ x + 3y + z \end{pmatrix}$$

(It is more convenient to write the vectors as columns than as rows when forming linear combinations.) Setting corresponding components equal to each other we obtain:

$$\begin{array}{ccc}
x + y + 2z = 1 & x + y + 2z = 1 & x + y + 2z = 1 \\
x + 2y - z = -2 \quad \text{or} & y - 3z = -3 \quad \text{or} & y - 3z = -3 \\
x + 3y + z = 5 & 2y - z = 4 & 5z = 10
\end{array}$$

The unique solution of the triangular form is $x = -6$, $y = 3$, $z = 2$; thus $v = -6u_1 + 3u_2 + 2u_3$.

2.7. Write the vector $v = (2, 3, -5)$ as a linear combination of $u_1 = (1, 2, -3)$, $u_2 = (2, -1, -4)$, and $u_3 = (1, 7, -5)$.

Find the equivalent system of equations and then solve. First:

$$\begin{pmatrix} 2 \\ 3 \\ -5 \end{pmatrix} = x \begin{pmatrix} 1 \\ 2 \\ -3 \end{pmatrix} + y \begin{pmatrix} 2 \\ -1 \\ -4 \end{pmatrix} + z \begin{pmatrix} 1 \\ 7 \\ -5 \end{pmatrix} = \begin{pmatrix} x + 2y + z \\ 2x - y + 7z \\ -3x - 4y - 5z \end{pmatrix}$$

Setting corresponding components equal to each other we obtain

$$\begin{array}{ccc}
x + 2y + z = 2 & x + 2y + z = 2 & x + 2y + z = 2 \\
2x - y + 7z = 3 \quad \text{or} & -5y + 5z = -1 \quad \text{or} & -5y + 5z = -1 \\
-3x - 4y - 5z = -5 & 2y - 2z = 1 & 0 = 3
\end{array}$$

The third equation, $0 = 3$, indicates that the system has no solution. Thus v cannot be written as a linear combination of the vectors u_1, u_2 and u_3.

2.8. Determine whether the vectors $u_1 = (1, 1, 1)$, $u_2 = (2, -1, 3)$, and $u_3 = (1, -5, 3)$ are linearly dependent or linearly independent.

Recall that u_1, u_2, u_3 are linearly dependent or linearly independent according as the vector equation $xu_1 + yu_2 + zu_3 = 0$ has a nonzero solution or only the zero solution. Thus first set a linear combination of the vectors equal to the zero vector:

$$\begin{pmatrix} 0 \\ 0 \\ 0 \end{pmatrix} = x\begin{pmatrix} 1 \\ 1 \\ 1 \end{pmatrix} + y\begin{pmatrix} 2 \\ -1 \\ 3 \end{pmatrix} + z\begin{pmatrix} 1 \\ -5 \\ 3 \end{pmatrix} = \begin{pmatrix} x + 2y + z \\ x - y - 5z \\ x + 3y + 3z \end{pmatrix}$$

Set corresponding components equal to each other, and reduce the system to echelon form:

$$\begin{array}{lll}
x + 2y + z = 0 & x + 2y + z = 0 & x + 2y + z = 0 \\
x - y - 5z = 0 \quad \text{or} & -3y - 6z = 0 \quad \text{or} & y + 2z = 0 \\
x + 3y + 3z = 0 & y + 2z = 0 &
\end{array}$$

The system in echelon form has a free variable; hence the system has a nonzero solution. Thus the original vectors are linearly dependent. (We do not need to solve the system to determine linear dependence or independence; we only need to know if a nonzero solution exists.)

2.9. Determine whether or not the vectors $(1, -2, -3)$, $(2, 3, -1)$, and $(3, 2, 1)$ are linearly dependent.

Set a linear combination (with coefficients x, y, z) of the vectors equal to the zero vector:

$$\begin{pmatrix} 0 \\ 0 \\ 0 \end{pmatrix} = x\begin{pmatrix} 1 \\ -2 \\ -3 \end{pmatrix} + y\begin{pmatrix} 2 \\ 3 \\ -1 \end{pmatrix} + z\begin{pmatrix} 3 \\ 2 \\ 1 \end{pmatrix} = \begin{pmatrix} x + 2y + 3z \\ -2x + 3y + 2z \\ -3x - y + z \end{pmatrix}$$

Set corresponding components equal to each other, and reduce the system to echelon form:

$$\begin{array}{llll}
x + 2y + 3z = 0 & x + 2y + 3z = 0 & x + 2y + 3z = 0 & x + 2y + 3z = 0 \\
-2x + 3y + 2z = 0 \quad \text{or} & 7y + 8z = 0 \quad \text{or} & y + 2z = 0 \quad \text{or} & y + 2z = 0 \\
-3x - y + z = 0 & 5y + 10z = 0 & 7y + 8z = 0 & -6z = 0
\end{array}$$

The homogeneous system is in triangular form, with no free variables; hence it has only the zero solution. Thus the original vectors are linearly independent.

2.10. Prove: Any $n + 1$ or more vectors in $\mathbf{R}^n$ are linearly dependent.

Suppose $u_1, u_2, \ldots, u_q$ are vectors in $\mathbf{R}^n$ and $q > n$. The vector equation

$$x_1 u_1 + x_2 u_2 + \cdots + x_q u_q = 0$$

is equivalent to a homogeneous system of n equations in $q > n$ unknowns. By Theorem 1.9, this system has a nonzero solution. Therefore $u_1, u_2, \ldots, u_q$ are linearly dependent.

2.11. Show that any set of q vectors that includes the zero vector is linearly dependent.

Denoting the vectors as $0, u_2, u_3, \ldots, u_q$, we have $1(0) + 0u_2 + 0u_3 + \cdots + 0u_q = 0$.

DOT (INNER) PRODUCT, ORTHOGONALITY

2.12. Compute $u \cdot v$ where $u = (1, -2, 3, -4)$ and $v = (6, 7, 1, -2)$.

Multiply corresponding components and add: $u \cdot v = (1)(6) + (-2)(7) + (3)(1) + (-4)(-2) = 3$.

2.13. Suppose $u = (3, 2, 1)$, $v = (5, -3, 4)$, $w = (1, 6, -7)$. Find: (a) $(u + v) \cdot w$, (b) $u \cdot w + v \cdot w$.

(a)   First calculate $u + v$ by adding corresponding components:

$$u + v = (3 + 5, 2 - 3, 1 + 4) = (8, -1, 5)$$

Then compute

$$(u + v) \cdot w = (8)(1) + (-1)(6) + (5)(-7) = 8 - 6 - 35 = -33$$

(b)   First find $u \cdot w = 3 + 12 - 7 = 8$ and $v \cdot w = 5 - 18 - 28 = -41$. Then $u \cdot w + v \cdot w$ $8 - 41 = -33$. [Note: As expected from Theorem 2.2(i), both values are equal.]

2.14.   Let $u = (1, 2, 3, -4)$, $v = (5, -6, 7, 8)$, and $k = 3$. Find: (a) $k(u \cdot v)$, (b) $(ku) \cdot v$, (c) $u \cdot (kv)$.

(a)   First find $u \cdot v = 5 - 12 + 21 - 32 = -18$. Then $k(u \cdot v) = 3(-18) = -54$.

(b)   First find $ku = (3(1), 3(2), 3(3), 3(-4)) = (3, 6, 9, -12)$. Then

$$(ku) \cdot v = (3)(5) + (6)(-6) + (9)(7) + (-12)(8) = 15 - 36 + 63 - 96 = -54$$

(c)   First find $kv = (15, -18, 21, 24)$. Then

$$u \cdot (kv) = (1)(15) + (2)(-18) + (3)(21) + (-4)(24) = 15 - 36 + 63 - 96 = -54$$

2.15.   Let $u = (5, 4, 1)$, $v = (3, -4, 1)$, and $w = (1, -2, 3)$. Which pair of these vectors, if any, are perpendicular?

Find the dot product of each pair of vectors:

$$u \cdot v = 15 - 16 + 1 = 0 \qquad v \cdot w = 3 + 8 + 3 = 14 \qquad u \cdot w = 5 - 8 + 3 = 0$$

Hence vectors u and v and vectors u and w are orthogonal, but vectors v and w are not.

2.16.   Determine k so that the vectors u and v are orthogonal where $u = (1, k, -3)$ and $v = (2, -5, 4)$.

Compute $u \cdot v$, set it equal to 0, and solve for k. $u \cdot v = (1)(2) + (k)(-5) + (-3)(4) = 2 - 5k - 12 = 0$, or $-5k - 10 = 0$. Solving, $k = -2$.

2.17.   Prove Theorem 2.2.

Let $u = (u_1, u_2, \ldots, u_n)$, $v = (v_1, v_2, \ldots, v_n)$, $w = (w_1, w_2, \ldots, w_n)$.

(i)   Since $u + v = (u_1 + v_1, u_2 + v_2, \ldots, u_n + v_n)$,

$$\begin{aligned}
(u + v) \cdot w &= (u_1 + v_1)w_1 + (u_2 + v_2)w_2 + \cdots + (u_n + v_n)w_n \\
&= u_1 w_1 + v_1 w_1 + u_2 w_2 + v_2 w_2 + \cdots + u_n w_n + v_n w_n \\
&= (u_1 w_1 + u_2 w_2 + \cdots + u_n w_n) + (v_1 w_1 + v_2 w_2 + \cdots + v_n w_n) \\
&= u \cdot w + v \cdot w
\end{aligned}$$

(ii)   Since $ku = (ku_1, ku_2, \ldots, ku_n)$,

$$(ku) \cdot v = ku_1 v_1 + ku_2 v_2 + \cdots + ku_n v_n = k(u_1 v_1 + u_2 v_2 + \cdots + u_n v_n) = k(u \cdot v)$$

(iii) $u \cdot v = u_1 v_1 + u_2 v_2 + \cdots + u_n v_n = v_1 u_1 + v_2 u_2 + \cdots + v_n u_n = v \cdot v$

(iv)   Since u_i^2 is nonnegative for each i, and since the sum of nonnegative real numbers is nonnegative,

$$u \cdot u = u_1^2 + u_2^2 + \cdots + u_n^2 \geq 0$$

Furthermore, $u \cdot u = 0$ iff $u_i = 0$ for each i, that is, iff $u = 0$.

NORM (LENGTH) IN $\mathbf{R}^n$

2.18.   Find $\|w\|$ if $w = (-3, 1, -2, 4, -5)$.

$$\|w\|^2 = (-3)^2 + 1^2 + (-2)^2 + 4^2 + (-5)^2 = 9 + 1 + 4 + 16 + 25 = 55; \text{ hence } \|w\| = \sqrt{55}.$$

2.19. Determine k such that $\|u\| = \sqrt{39}$ where $u = (1, k, -2, 5)$.

$\|u\|^2 = 1^2 + k^2 + (-2)^2 + 5^2 = k^2 + 30$. Now solve $k^2 + 30 = 39$ and obtain $k = 3, -3$.

2.20. Normalize $w = (4, -2, -3, 8)$.

First find $\|w\|^2 = w \cdot w = 4^2 + (-2)^2 + (-3)^2 + 8^2 = 16 + 4 + 9 + 64 = 93$. Divide each component of w by $\|w\| = \sqrt{93}$ to obtain

$$\hat{w} = \frac{w}{\|w\|} = \left(\frac{4}{\sqrt{93}}, \frac{-2}{\sqrt{93}}, \frac{-3}{\sqrt{93}}, \frac{8}{\sqrt{93}} \right)$$

2.21. Normalize $v = (\frac{1}{2}, \frac{2}{3}, -\frac{1}{4})$.

Note that v and any positive multiple of v will have the same normalized form. Hence, first multiply v by 12 to "clear" fractions: $12v = (6, 8, -3)$. Then

$$\|12v\|^2 = 36 + 64 + 9 = 109 \qquad \text{and} \qquad \hat{v} = \widehat{12v} = \frac{12v}{\|12v\|} = \left(\frac{6}{\sqrt{109}}, \frac{8}{\sqrt{109}}, \frac{-3}{\sqrt{109}} \right)$$

2.22. Prove Theorem 2.3 (Cauchy–Schwarz).

We shall prove the following stronger statement: $|u \cdot v| \leq \sum_{i=1}^{n} |u_i v_i| \leq \|u\| \|v\|$. First, if $u = 0$ or $v = 0$, then the inequality reduces to $0 \leq 0 \leq 0$ and is therefore true. Hence we need only consider the case in which $u \neq 0$ and $v \neq 0$, i.e., where $\|u\| \neq 0$ and $\|v\| \neq 0$. Furthermore, because

$$|u \cdot v| = |\textstyle\sum u_i v_i| \leq \textstyle\sum |u_i v_i|$$

we need only prove the second inequality.

Now, for any real numbers $x, y \in \mathbf{R}$, $0 \leq (x - y)^2 = x^2 - 2xy + y^2$ or, equivalently,

$$2xy \leq x^2 + y^2 \tag{1}$$

Set $x = |u_i|/\|u\|$ and $y = |v_i|/\|v\|$ in (1) to obtain, for any i,

$$2 \frac{|u_i|}{\|u\|} \frac{|v_i|}{\|v\|} \leq \frac{|u_i|^2}{\|u\|^2} + \frac{|v_i|^2}{\|v\|^2} \tag{2}$$

But, by definition of the norm of a vector, $\|u\| = \sum u_i^2 = \sum |u_i|^2$ and $\|v\| = \sum v_i^2 = \sum |v_i|^2$. Thus summing (2) with respect to i and using $|u_i v_i| = |u_i| |v_i|$, we have

$$2 \frac{\sum |u_i v_i|}{\|u\| \|v\|} \leq \frac{\sum |u_i|^2}{\|u\|^2} + \frac{\sum |v_i|^2}{\|v\|^2} = \frac{\|u\|^2}{\|u\|^2} + \frac{\|v\|^2}{\|v\|^2} = 2$$

that is,

$$\frac{\sum |u_i v_i|}{\|u\| \|v\|} \leq 1$$

Multiplying both sides by $\|u\| \|v\|$, we obtain the required inequality.

2.23. Prove Theorem 2.4 (Minkowski).

By the Cauchy–Schwarz inequality (Problem 2.22) and the other properties of the inner product,

$$\|u + v\|^2 = (u + v) \cdot (u + v) = u \cdot u + 2(u \cdot v) + v \cdot v$$
$$\leq \|u\|^2 + 2\|u\| \|v\| + \|v\|^2 = (\|u\| + \|v\|)^2$$

Taking the square roots of both sides yields the desired inequality.

2.24. Prove that the norm in $\mathbf{R}^n$ satisfies the following laws:

(a) $[N_1]$ For any vector u, $\|u\| \geq 0$; and $\|u\| = 0$ iff $u = 0$.

(b) $[N_2]$ For any vector u and any scalar k, $\|ku\| = |k| \|u\|$.

(c) $[N_3]$ For any vectors u and v, $\|u + v\| \leq \|u\| + \|v\|$.

(a) By Theorem 2.2, $u \cdot u \geq 0$, and $u \cdot u = 0$ iff $u = 0$. Since $\|u\| = \sqrt{u \cdot u}$, $[N_1]$ follows.

(b) Suppose $u = (u_1, u_2, \ldots, u_n)$ and so $ku = (ku_1, ku_2, \ldots, ku_n)$. Then

$$\|ku\|^2 = (ku_1)^2 + (ku_2)^2 + \cdots + (ku_n)^2 = k^2(u_1^2 + u_2^2 + \cdots + u_n^2) = k^2\|u\|^2$$

Taking square roots gives $[N_2]$.

(c) $[N_3]$ was proved in Problem 2.23.

2.25. Let $u = (1, 2, -2)$, $v = (3, -12, 4)$, and $k = -3$. (a) Find $\|u\|$, $\|v\|$, and $\|ku\|$. (b) Verify that $\|ku\| = |k| \|u\|$ and $\|u + v\| \leq \|u\| + \|v\|$.

(a) $\|u\| = \sqrt{1 + 4 + 4} = \sqrt{9} = 3$, $\|v\| = \sqrt{9 + 144 + 16} = \sqrt{169} = 13$, $ku = (-3, -6, 6)$, and $\|ku\| = \sqrt{9 + 36 + 36} = \sqrt{81} = 9$.

(b) Since $|k| = |-3| = 3$, we have $|k| \|u\| = 3 \cdot 3 = 9 = \|ku\|$. Also $u + v = (4, -10, 2)$. Thus

$$\|u + v\| = \sqrt{16 + 100 + 4} = \sqrt{120} \leq 16 = 3 + 13 = \|u\| + \|v\|$$

DISTANCE, ANGLES, PROJECTIONS

2.26. Find the distance $d(u, v)$ between the vectors u and v where

(a) $u = (1, 7)$, $v = (6, -5)$,

(b) $u = (3, -5, 4)$, $v = (6, 2, -1)$,

(c) $u = (5, 3, -2, -4, -1)$, $v = (2, -1, 0, -7, 2)$.

In each case use the formula $d(u, v) = \|u - v\| = \sqrt{(u_1 - v_1)^2 + \cdots + (u_n - v_n)^2}$.

(a) $d(u, v) = \sqrt{(1 - 6)^2 + (7 + 5)^2} = \sqrt{25 + 144} = \sqrt{169} = 13$

(b) $d(u, v) = \sqrt{(3 - 6)^2 + (-5 - 2)^2 + (4 + 1)^2} = \sqrt{9 + 49 + 25} = \sqrt{83}$

(c) $d(u, v) = \sqrt{(5 - 2)^2 + (3 + 1)^2 + (-2 + 0)^2 + (-4 + 7)^2 + (-1 - 2)^2} = \sqrt{47}$

2.27. Find k such that $d(u, v) = 6$ where $u = (2, k, 1, -4)$ and $v = (3, -1, 6, -3)$.

First find

$$[d(u, v)]^2 = (2 - 3)^2 + (k + 1)^2 + (1 - 6)^2 + (-4 + 3)^2 = k^2 + 2k + 28$$

Now solve $k^2 + 2k + 28 = 6^2$ to obtain $k = 2, -4$.

2.28. From Problem 2.24, prove that the distance function $d(u, v)$ satisfies the following:

$[M_1]$ $d(u, v) \geq 0$; and $d(u, v) = 0$ iff $u = v$.

$[M_2]$ $d(u, v) = d(v, u)$.

$[M_3]$ $d(u, v) \leq d(u, w) + d(w, v)$ (triangle inequality).

$[M_1]$ follows directly from $[N_1]$. By $[N_2]$,

$$d(u, v) = \|u - v\| = \|(-1)(v - u)\| = |-1| \|v - u\| = \|v - u\| = d(v, u)$$

which is $[M_2]$. By $[N_3]$,

$$d(u, v) = \|u - v\| = \|(u - w) + (w - v)\| \le \|u - w\| + \|w - v\| = d(u, w) + d(w, v)$$

which is $[M_3]$.

2.29. Find $\cos \theta$, where θ is the angle between $u = (1, 2, -5)$ and $v = (2, 4, 3)$.

First find

$$u \cdot v = 2 + 8 - 15 = -5 \qquad \|u\|^2 = 1 + 4 + 25 = 30 \qquad \|v\|^2 = 4 + 16 + 9 = 29$$

Then

$$\cos \theta = \frac{u \cdot v}{\|u\| \|v\|} = -\frac{5}{\sqrt{30}\sqrt{29}}$$

2.30. Find proj (u, v) where $u = (1, -3, 4)$ and $v = (3, 4, 7)$.

First find $u \cdot v = 3 - 12 + 28 = 19$ and $\|v\|^2 = 9 + 16 + 49 = 74$. Then

$$\text{proj}\,(u, v) = \frac{u \cdot v}{\|v\|^2} v = \frac{19}{74}(3, 4, 7) = \left(\frac{57}{74}, \frac{76}{74}, \frac{133}{74}\right) = \left(\frac{57}{74}, \frac{38}{37}, \frac{133}{74}\right)$$

POINTS, LINES, AND HYPERPLANES

This section distinguishes between an n-tuple $P(a_1, a_2, \ldots, a_n) \equiv P(a_i)$ viewed as a point in $\mathbf{R}^n$ and an n-tuple $v = [c_1, c_2, \ldots, c_n]$ viewed as a vector (arrow) from the origin O to the point $C(c_1, c_2, \ldots, c_n)$.

2.31. Find the vector v identified with the directed line segment $\overrightarrow{PQ}$ for the points (a) $P(2, 5)$ and $Q(-3, 4)$, (b) $P(1, -2, 4)$ and $Q(6, 0, -3)$.

(a) $v = Q - P = [-3 - 2, 4 - 5] = [-5, -1]$

(b) $v = Q - P = [6 - 1, 0 + 2, -3 - 4] = [5, 2, -7]$

2.32. Consider points $P(3, k, -2)$ and $Q(5, 3, 4)$ in $\mathbf{R}^3$. Find k so that $\overrightarrow{PQ}$ is orthogonal to the vector $u = [4, -3, 2]$.

First find $v = Q - P = [5 - 3, 3 - k, 4 + 2] = [2, 3 - k, 6]$. Next compute

$$u \cdot v = 4 : 2 - 3(3 - k) + 2 \cdot 6 = 8 - 9 + 3k + 12 = 3k + 11$$

Lastly, set $u \cdot v = 0$ or $3k + 11 = 0$, from which $k = -11/3$.

2.33. Consider the hyperplane H in $\mathbf{R}^n$ which is the solution set of the linear equation

$$a_1 x_1 + a_2 x_2 + \cdots + a_n x_n = b \qquad\qquad (1)$$

where $u = [a_1, a_2, \ldots, a_n] \ne 0$. Show that the directed line segment $\overrightarrow{PQ}$ of any pair of points P, $Q \in H$ is orthogonal to the coefficient vector u; the vector u is said to be *normal* to the hyperplane H.

Let $w_1 = \overrightarrow{OP}$ and $w_2 = \overrightarrow{OQ}$; hence $v = w_2 - w_1 = \overrightarrow{PQ}$. By (1), $u \cdot w_1 = b$ and $u \cdot w_2 = b$. But then

$$u \cdot v = u \cdot (w_2 - w_1) = u \cdot w_2 - u \cdot w_1 = b - b = 0$$

Thus $v = \overrightarrow{PQ}$ is orthogonal to the normal vector u.

2.34. Find an equation of the hyperplane H in $\mathbf{R}^4$ which passes through $P(3, -2, 1, -4)$ and is normal to $u = (2, 5, -6, -2)$.

    An equation of H is of the form $2x + 5y - 6z - 2w = k$ since it is normal to u. Substitute P into this equation to obtain $k = -2$. Thus an equation of H is $2x + 5y - 6z - 2w = -2$.

2.35. Find an equation of the plane H in $\mathbf{R}^3$ which contains $P(1, -5, 2)$ and is parallel to the plane H' determined by $3x - 7y + 4z = 5$.

    H and H' are parallel if and only if their normals are parallel or antiparallel. Hence an equation of H is of the form $3x - 7y + 4z = k$. Substitute $P(1, -5, 2)$ into this equation to obtain $k = 46$. Thus the required equation is $3x - 7y + 4z = 46$.

2.36. Find a parametric representation of the line in $\mathbf{R}^4$ passing through $P(4, -2, 3, 1)$ in the direction $u = [2, 5, -7, 11]$.

    The line L in $\mathbf{R}^n$ passing through the point $P(a_i)$ and in the direction of the nonzero vector $u = [u_i]$ consists of the points $X = (x_i)$ which satisfy the equation

$$X = P + tu \qquad \text{or} \qquad x_i = a_i + u_i t \qquad (\text{for } i = 1, 2, \ldots, n) \qquad (1)$$

where the *parameter* t takes on all real numbers. Thus we obtain

$$\begin{cases} x = 4 + 2t \\ y = -2 + 5t \\ z = 3 - 7t \\ w = 1 + 11t \end{cases} \qquad \text{or} \qquad (4 + 2t,\ -2 + 5t,\ 3 - 7t,\ 1 + 11t)$$

2.37. Find a parametric equation of the line in $\mathbf{R}^3$ passing through the points $P(5, 4, -3)$ and $Q(1, -3, 2)$.

    First compute $u = \overrightarrow{PQ} = [1 - 5, -3 - 4, 2 - (-3)] = [-4, -7, 5]$. Then use Problem 2.36 to obtain

$$x = 5 - 4t \qquad y = 4 - 7t \qquad z = -3 + 5t$$

2.38. Give a *nonparametric* representation for the line of Problem 2.37.

    Solve each coordinate equation for t and equate the results

$$\frac{x - 5}{-4} = \frac{y - 4}{-7} = \frac{z + 3}{5}$$

or the pair of linear equations $7x - 4y = 19$ and $5x + 4z = 13$.

2.39. Find a parametric equation of the line in $\mathbf{R}^3$ perpendicular to the plane $2x - 3y + 7z = 4$ and intersecting the plane at the point $P(6, 5, 1)$.

    Since the line is perpendicular to the plane, the line must be in the direction of the normal vector $u = [2, -3, 7]$ of the plane. Hence

$$x = 6 + 2t \qquad y = 5 - 3t \qquad z = 1 + 7t$$

2.40. Consider the following curve C in $\mathbf{R}^4$ where $0 \le t \le 4$:

$$F(t) = (t^2, 3t - 2, t^3, t^2 + 5)$$

Find the unit tangent vector $\mathbf{T}$ when $t = 2$.

Take the derivative of (each component of) $F(t)$ to obtain a vector V which is tangent to the curve:

$$V(t) = \frac{dF(t)}{dt} = (2t, 3, 3t^2, 2t)$$

Now find V when $t = 2$. This yields $V = (4, 3, 12, 4)$. Normalize V to get the unit tangent vector T to the curve when $t = 2$. We have

$$\|V\|^2 = 16 + 9 + 144 + 16 = 185 \quad \text{or} \quad \|V\| = \sqrt{185}$$

Thus

$$\mathbf{T} = \left[\frac{4}{\sqrt{185}}, \frac{3}{\sqrt{185}}, \frac{12}{\sqrt{185}}, \frac{4}{\sqrt{185}} \right]$$

2.41. Let $\mathbf{T}(t)$ be the unit tangent vector to a curve C in $\mathbf{R}^n$. Show that $d\mathbf{T}(t)/dt$ is orthogonal to $\mathbf{T}(t)$.

We have $\mathbf{T}(t) \cdot \mathbf{T}(t) = 1$. Using $d(1)/dt = 0$, we have

$$d[\mathbf{T}(t) \cdot \mathbf{T}(t)]/dt = \mathbf{T}(t) \cdot d\mathbf{T}(t)/dt + d\mathbf{T}(t)/dt \cdot \mathbf{T}(t) = 2\mathbf{T}(t) \cdot d\mathbf{T}(t)/dt = 0$$

Thus $d\mathbf{T}(t)/dt$ is orthogonal to $\mathbf{T}(t)$.

SPATIAL VECTORS (VECTORS IN $\mathbf{R}^3$), PLANES, LINES, CURVES, AND SURFACES IN $\mathbf{R}^3$

The following formulas will be used in Problems 2.42–2.53.

The equation of a plane through the point $P_0(x_0, y_0, z_0)$ with normal direction $\mathbf{N} = a\mathbf{i} + b\mathbf{j} + c\mathbf{k}$ is

$$a(x - x_0) + b(y - y_0) + c(x - x_0) = 0 \qquad (2.1)$$

The parametric equation of a line L through a point $P_0(x_0, y_0, z_0)$ in the direction of the vector $v = a\mathbf{i} + b\mathbf{j} + c\mathbf{k}$ is

$$x = at + x_0 \qquad y = bt + y_0 \qquad z = ct + z_0$$

or, equivalently,

$$L(t) = (at + x_0)\mathbf{i} + (bt + y_0)\mathbf{j} + (ct + z_0)\mathbf{k} \qquad (2.2)$$

The equation of a normal vector $\mathbf{N}$ to a surface $F(x, y, z) = 0$ is

$$\mathbf{N} = F_x \mathbf{i} + F_y \mathbf{j} + F_z \mathbf{k} \qquad (2.3)$$

2.42. Find the equation of the plane with normal direction $\mathbf{N} = 5\mathbf{i} - 6\mathbf{j} + 7\mathbf{k}$ and containing the point $P(3, 4, -2)$.

Substitute P and $\mathbf{N}$ in equation (2.1) to get

$$5(x - 3) - 6(y - 4) + 7(z + 2) = 0 \quad \text{or} \quad 5x - 6y + 7z = -23$$

2.43. Find a normal vector $\mathbf{N}$ to the plane $4x + 7y - 12z = 3$.

The coefficients of x, y, z give a normal direction; hence $\mathbf{N} = 4\mathbf{i} + 7\mathbf{j} - 12\mathbf{k}$. (Any multiple of $\mathbf{N}$ also is normal to the plane.)

2.44. Find the plane H parallel to $4x + 7y - 12z = 3$ and containing the point $P(2, 3, -1)$.

H and the given plane have the same normal direction; that is, $\mathbf{N} = 4\mathbf{i} + 7\mathbf{j} - 12\mathbf{k}$ is normal to H. Substitute P and $\mathbf{N}$ in equation (a) to get

$$4(x - 2) + 7(y - 3) - 12(z + 1) = 0 \qquad \text{or} \qquad 4x + 7y - 12z = 41$$

2.45. Let H and K be, respectively, the planes $x + 2y - 4z = 5$ and $2x - y + 3z = 7$. Find $\cos \theta$ where ϕ is the angle between planes H and K.

The angle θ between H and K is the same as the angle between the normal $\mathbf{N}$ of H and the normal $\mathbf{N}'$ of K. We have

$$\mathbf{N} = \mathbf{i} + 2\mathbf{j} - 4\mathbf{k} \qquad \text{and} \qquad \mathbf{N}' = 2\mathbf{i} - \mathbf{j} + 3\mathbf{k}$$

Then

$$\mathbf{N} \cdot \mathbf{N}' = 2 - 2 - 12 = -12 \qquad \|\mathbf{N}\|^2 = 1 + 4 + 16 = 21 \qquad \|\mathbf{N}'\|^2 = 4 + 1 + 9 = 14$$

Thus

$$\cos \theta = \frac{\mathbf{N} \cdot \mathbf{N}'}{\|\mathbf{N}\| \|\mathbf{N}'\|} = -\frac{12}{\sqrt{21} \sqrt{14}} = -\frac{12}{7\sqrt{6}}$$

2.46. Derive equation (2.1).

Let $P(x, y, z)$ be an arbitrary point in the plane. The vector v from P_0 to P is

$$v = P - P_0 = (x - x_0)\mathbf{i} + (y - y_0)\mathbf{j} + (z - z_0)\mathbf{k}$$

Since v is orthogonal to $\mathbf{N} = a\mathbf{i} + b\mathbf{j} + c\mathbf{k}$ (Fig. 2-12), we get the required formula

$$a(x - x_0) + b(y - y_0) + c(z - z_0) = 0$$

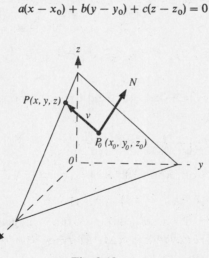

Fig. 2-12

2.47. Derive equation (2.2).

Let $P(x, y, z)$ be an arbitrary point on the line L. The vector w from P_0 to P is

$$w = P - P_0 = (x - x_0)\mathbf{i} + (y - y_0)\mathbf{j} + (z - z_0)\mathbf{k} \qquad (1)$$

Since w and v have the same direction (Fig. 2-13),

$$\mathbf{w} = t\mathbf{v} = t(a\mathbf{i} + b\mathbf{j} + c\mathbf{k}) = at\mathbf{i} + bt\mathbf{j} + ct\mathbf{k} \qquad (2)$$

Equations (1) and (2) give us our result.

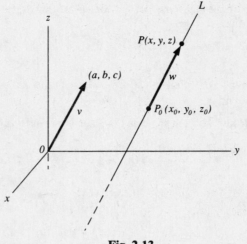

Fig. 2-13

2.48. Find the (parametric) equation of the line L through:

(a)   Point $P(3, 4, -2)$ and in the direction of $v = 5\mathbf{i} - \mathbf{j} + 3\mathbf{k}$,

(b)   Points $P(1, 3, 2)$ and $Q(2, 5, -6)$.

(a)   Substitute in equation (2.2) to get

$$L(t) = (5t + 3)\mathbf{i} + (-t + 4)\mathbf{j} + (3t - 2)\mathbf{k}$$

(b)   First find the vector v from P to Q: $v = Q - P = \mathbf{i} + 2\mathbf{j} - 8\mathbf{k}$. Then use (2.2) with v and one of the given points, say P, to get

$$L(t) = (t + 1)\mathbf{i} + (2t + 3)\mathbf{j} + (-8t + 2)\mathbf{k}$$

2.49. Let H be the plane $3x + 5y + 7z = 15$. Find the equation of the line L perpendicular to H and containing the point $P(1, -2, 4)$.

Since L is perpendicular to the plane H, L must be in the same direction as the normal $\mathbf{N} = 3\mathbf{i} + 5\mathbf{j} + 7\mathbf{k}$ to H. Thus use (2.2) with $\mathbf{N}$ and P to get

$$L(t) = (3t + 1)\mathbf{i} + (5t - 2)\mathbf{j} + (7t + 4)\mathbf{k}$$

2.50. Consider a moving body B whose position at time t is given by $R(t) = t^3\mathbf{i} + 2t^2\mathbf{j} + 3t\mathbf{k}$. [Then $V(t) = dR(t)/dt$ denotes the velocity of B and $A(t) = dV(t)/dt$ denotes the acceleration of B.]

(a)   Find the position of B when $t = 1$.   (c)   Find the speed s of B when $t = 1$.

(b)   Find the velocity v of B when $t = 1$.   (d)   Find the acceleration a of B when $t = 1$.

(a)   Substitute $t = 1$ into $R(t)$ to get $R(1) = \mathbf{i} + 2\mathbf{j} + 3\mathbf{k}$.

(b)   Take the derivative of $R(t)$ to get

$$V(t) = \frac{dR(t)}{dt} = 3t^2\mathbf{i} + 4t\mathbf{j} + 3\mathbf{k}$$

Substitute $t = 1$ in $V(t)$ to get $v = V(1) = 3\mathbf{i} + 4\mathbf{j} + 3\mathbf{k}$.

(c)   The speed s is the magnitude of the velocity v. Thus

$$s^2 = \|v\|^2 = 9 + 16 + 9 = 34 \qquad \text{and hence} \qquad s = \sqrt{34}$$

(d)   Take the second derivative of $R(t)$ or, in other words, the derivative of $V(t)$ to get

$$A(t) = \frac{dV(t)}{dt} = 6t\mathbf{i} + 4\mathbf{j}$$

Substitute $t = 1$ in $A(t)$ to get $a = A(1) = 6\mathbf{i} + 4\mathbf{j}$.

2.51.   Consider the surface $xy^2 + 2yz = 16$ in R^3. Find (a) the normal vector $\mathbf{N}(x, y, z)$ to the surface and (b) the tangent plane H to the surface at point $P(1, 2, 3)$.

(a)   Find the partial derivatives F_x, F_y, F_z where $F(x, y, z) = xy^2 + 2yz - 16$. We have

$$F_x = y^2 \qquad F_y = 2xy + 2z \qquad F_z = 2y$$

Thus, from equation (2.3), $\mathbf{N}(x, y, z) = y^2\mathbf{i} + (2xy + 2z)\mathbf{j} + 2y\mathbf{k}$.

(b)   The normal to the surface at point P is

$$\mathbf{N}(P) = \mathbf{N}(1, 2, 3) = 4\mathbf{i} + 10\mathbf{j} + 4\mathbf{k}$$

Thus $\mathbf{N} = 2\mathbf{i} + 5\mathbf{j} + 2\mathbf{k}$ is also a normal vector at P. Substitute P and $\mathbf{N}$ into equation (2.1) to get

$$2(x - 1) + 5(y - 2) + 2(z - 3) = 0 \qquad \text{or} \qquad 2x + 5y + 2z = 18$$

2.52.   Consider the ellipsoid $x^2 + 2y^2 + 3z^2 = 15$. Find the tangent plane H at point $P(2, 2, 1)$.

First find the normal vector [from equation (2.3)]

$$\mathbf{N}(x, y, z) = F_x\mathbf{i} + F_y\mathbf{j} + F_z\mathbf{k} = 2x\mathbf{i} + 4y\mathbf{j} + 6z\mathbf{k}$$

Evaluate the normal vector $\mathbf{N}(x, y, z)$ at P to get

$$\mathbf{N}(P) = \mathbf{N}(2, 2, 1) = 4\mathbf{i} + 8\mathbf{j} + 6\mathbf{k}$$

Thus $\mathbf{N} = 2\mathbf{i} + 4\mathbf{j} + 3\mathbf{k}$ is normal to the ellipsoid at P. Substitute P and $\mathbf{N}$ into (2.1) to obtain H:

$$2(x - 2) + 4(y - 2) + 3(z - 1) = 0 \qquad \text{or} \qquad 2x + 4y + 3z = 15$$

2.53.   Consider the function $f(x, y) = x^2 + y^2$ whose solution set $z = x^2 + y^2$ represents a surface S in $\mathbf{R}^3$. Find (a) the normal vector $\mathbf{N}$ to the surface S when $x = 2, y = 3$; and (b) the tangent plane H to the surface S when $x = 2, y = 3$.

(a)   Use the fact that, when $F(x, y, z) = f(x, y) - z$, we have $F_x = f_x, F_y = f_y$, and $F_z = -1$. Then

$$\mathbf{N} = (f_x, f_y, -1) = 2x\mathbf{i} + 2y\mathbf{j} - \mathbf{k} = 4\mathbf{i} + 6\mathbf{j} - \mathbf{k}$$

(b)   If $x = 2, y = 3$, then $z = 4 + 9 = 13$; hence $P(2, 3, 13)$ is the point on the surface S. Substitute P and $\mathbf{N} = 4\mathbf{i} + 6\mathbf{j} - \mathbf{k}$ into equation (2.1) to obtain H.

$$4(x - 2) + 6(y - 3) - (z - 13) = 0 \qquad \text{or} \qquad 4x + 6y - z = 13$$

CROSS PRODUCT

The cross product is defined only for vectors in $\mathbf{R}^3$.

2.54.   Find $u \times v$ where (a) $u = (1, 2, 3)$ and $v = (4, 5, 6)$, (b) $u = (7, 3, 1)$ and $v = (1, 1, 1)$, (c) $u = (-4, 12, 2)$ and $v = (6, -18, 3)$.

The cross product of $u = (a_1, a_2, a_3)$ and $v = (b_1, b_2, b_3)$ may be obtained as follows. Put the vector $v = (b_1, b_2, b_3)$ *under* the vector $u = (a_1, a_2, a_3)$ to form the array

$$\begin{pmatrix} a_1 & a_2 & a_3 \\ b_1 & b_2 & b_3 \end{pmatrix}$$

Then

$$u \times v = \left(\begin{vmatrix} \boxed{a_1} & a_2 & a_3 \\ \boxed{b_1} & b_2 & b_3 \end{vmatrix}, \; -\begin{vmatrix} a_1 & \boxed{a_2} & a_3 \\ b_1 & \boxed{b_2} & b_3 \end{vmatrix}, \; \begin{vmatrix} a_1 & a_2 & \boxed{a_3} \\ b_1 & b_2 & \boxed{b_3} \end{vmatrix} \right)$$

That is, cover the first column of the array and take the determinant to obtain the first component of $u \times v$; cover the second column and take the determinant backward to obtain the second component; and cover the third column and take the determinant to obtain the third component.

(a) $u \times v = \left(\begin{vmatrix} \boxed{1} & 2 & 3 \\ \boxed{4} & 5 & 6 \end{vmatrix}, \; -\begin{vmatrix} 1 & \boxed{2} & 3 \\ 4 & \boxed{5} & 6 \end{vmatrix}, \; \begin{vmatrix} 1 & 2 & \boxed{3} \\ 4 & 5 & \boxed{6} \end{vmatrix} \right) = (12 - 15, \; 12 - 6, \; 5 - 8) = (-3, 6, -3)$

(b) $u \times v = \left(\begin{vmatrix} \boxed{7} & 3 & 1 \\ \boxed{1} & 1 & 1 \end{vmatrix}, \; -\begin{vmatrix} 7 & \boxed{3} & 1 \\ 1 & \boxed{1} & 1 \end{vmatrix}, \; \begin{vmatrix} 7 & 3 & \boxed{1} \\ 1 & 1 & \boxed{1} \end{vmatrix} \right) = (3 - 1, \; 1 - 7, \; 7 - 3) = (2, -6, 4)$

(c) $u \times v = \left(\begin{vmatrix} \boxed{-4} & 12 & 2 \\ \boxed{6} & -18 & 3 \end{vmatrix}, \; -\begin{vmatrix} -4 & \boxed{12} & 2 \\ 6 & \boxed{-18} & 3 \end{vmatrix}, \; \begin{vmatrix} -4 & 12 & \boxed{2} \\ 6 & -18 & \boxed{-3} \end{vmatrix} \right)$

$= (36 + 36, \; 12 + 12, \; 72 - 72) = (72, 24, 0)$

2.55. Consider the vectors $u = 2\mathbf{i} - 3\mathbf{j} + 4\mathbf{k}$, $v = 3\mathbf{i} + \mathbf{j} - 2\mathbf{k}$, $w = \mathbf{i} + 5\mathbf{j} + 3\mathbf{k}$. Find: (a) $u \times v$, (b) $u \times w$, (c) $v \times w$.

Use

$$v_1 \times v_2 = \begin{vmatrix} \mathbf{i} & \mathbf{j} & \mathbf{k} \\ a_1 & a_2 & a_3 \\ b_1 & b_2 & b_3 \end{vmatrix} = \begin{vmatrix} a_2 & a_3 \\ b_2 & b_3 \end{vmatrix} \mathbf{i} - \begin{vmatrix} a_1 & a_3 \\ b_1 & b_3 \end{vmatrix} \mathbf{j} + \begin{vmatrix} a_1 & a_2 \\ b_1 & b_2 \end{vmatrix} \mathbf{k}$$

where $v_1 = a_1 \mathbf{i} + a_2 \mathbf{j} + a_3 \mathbf{k}$ and $v_2 = b_1 \mathbf{i} + b_2 \mathbf{j} + b_3 \mathbf{k}$.

(a) $u \times v = \begin{vmatrix} \mathbf{i} & \mathbf{j} & \mathbf{k} \\ 2 & -3 & 4 \\ 3 & 1 & -2 \end{vmatrix} = (6 - 4)\mathbf{i} + (12 + 4)\mathbf{j} + (2 + 9)\mathbf{k} = 2\mathbf{i} + 16\mathbf{j} + 11\mathbf{k}$

(*Remark*: Observe that the **j** component is obtained by taking the determinant "backwards." See Problem 2.54.)

(b) $u \times w = \begin{vmatrix} \mathbf{i} & \mathbf{j} & \mathbf{k} \\ 2 & -3 & 4 \\ 1 & 5 & 3 \end{vmatrix} = (-9 - 20)\mathbf{i} + (4 - 6)\mathbf{j} + (10 + 3)\mathbf{k} = -29\mathbf{i} - 2\mathbf{j} + 13\mathbf{k}$

(c) $v \times w = \begin{vmatrix} \mathbf{i} & \mathbf{j} & \mathbf{k} \\ 3 & 1 & -2 \\ 1 & 5 & 3 \end{vmatrix} = (3 + 10)\mathbf{i} + (-2 - 9)\mathbf{j} + (15 - 1)\mathbf{k} = 13\mathbf{i} - 11\mathbf{j} + 14\mathbf{k}$

2.56. Prove Theorem 2.5(i): The vector $u \times v$ is orthogonal to both u and v.

Suppose $u = (a_1, a_2, a_3)$ and $v = (b_1, b_2, b_3)$. Then

$$u \cdot (u \times v) = a_1(a_2 b_3 - a_3 b_2) + a_2(a_3 b_1 - a_1 b_3) + a_3(a_1 b_2 - a_2 b_1)$$
$$= a_1 a_2 b_3 - a_1 a_3 b_2 + a_2 a_3 b_1 - a_1 a_2 b_3 + a_1 a_3 b_2 - a_2 a_3 b_1 = 0$$

Thus $u \times v$ is orthogonal to u. Similarly, $u \times v$ is orthogonal to v.

2.57. Find a unit vector u orthogonal to $v = (1, 3, 4)$ and $w = (2, -6, 5)$.

First find $v \times w$ which is orthogonal to v and w. The array $\begin{pmatrix} 1 & 3 & 4 \\ 2 & -6 & -5 \end{pmatrix}$ gives

$$v \times w = (-15 + 24, \; 8 + 5, \; -6 - 6) = (9, 13, -12)$$

Now normalize $u \times w$ to get $u = (9/\sqrt{394}, \; 13/\sqrt{394}, \; -12/\sqrt{394})$.

2.58. Prove *Lagrange's identity*, $\|u \times v\|^2 = (u \cdot u)(v \cdot v) - (u \cdot v)^2$.

If $u = (a_1, a_2, a_3)$ and $v = (b_1, b_2, b_3)$, then

$$\|u \times v\|^2 = (a_2 b_3 - a_3 b_2)^2 + (a_3 b_1 - a_1 b_3)^2 + (a_1 b_2 - a_2 b_1)^2 \qquad (1)$$

$$(u \cdot u)(v \cdot v) - (u \cdot v)^2 = (a_1^2 + a_2^2 + a_3^2)(b_1^2 + b_2^2 + b_3^2) - (a_1 b_1 + a_2 b_2 + a_3 b_3)^2 \qquad (2)$$

Expansion of the right-hand sides of (1) and (2) establishes the identity.

COMPLEX NUMBERS

2.59. Suppose $z = 5 + 3i$ and $w = 2 - 4i$. Find: (a) $z + w$, (b) $z - w$, (c) zw.

Use the ordinary rules of algebra together with $i^2 = -1$ to obtain a result in the standard form $a + bi$.

(a) $z + w = (5 + 3i) + (2 - 4i) = 7 - i$

(b) $z - w = (5 + 3i) - (2 - 4i) = 5 + 3i - 2 + 4i = 3 + 7i$

(c) $zw = (5 + 3i)(2 - 4i) = 10 - 14i - 12i^2 = 10 - 14i + 12 = 22 - 14i$

2.60. Simplify: (a) $(5 + 3i)(2 - 7i)$, (b) $(4 - 3i)^2$, (c) $(1 + 2i)^3$.

(a) $(5 + 3i)(2 - 7i) = 10 + 6i - 35i - 21i^2 = 31 - 29i$

(b) $(4 - 3i)^2 = 16 - 24i + 9i^2 = 7 - 24i$

(c) $(1 + 2i)^3 = 1 + 6i + 12i^2 + 8i^3 = 1 + 6i - 12 - 8i = -11 - 2i$

2.61. Simplify: (a) i^0, i^3, i^4, (b) i^5, i^6, i^7, i^8, (c) $i^{39}, i^{174}, i^{252}, i^{317}$.

(a) $i^0 = 1, i^3 = i^2(i) = (-1)(i) = -i, i^4 = (i^2)(i^2) = (-1)(-1) = 1.$

(b) $i^5 = (i^4)(i) = (1)(i) = i, i^6 = (i^4)(i^2) = (1)(i^2) = i^2 = -1, i^7 = i^3 = -i, i^8 = i^4 = 1.$

(c) Using $i^4 = 1$ and $i^n = i^{4q+r} = (i^4)^q i^r = 1^q i^r = i^r$, divide the exponent n by 4 to obtain the remainder r:

$$i^{39} = i^{4(9)+3} = (i^4)^9 i^3 = 1^9 i^3 = i^3 = -i \qquad i^{174} = i^2 = -1 \qquad i^{252} = i^0 = 1 \qquad i^{317} = i^1 = i$$

2.62. Find the complex conjugate of each of the following:

(a) $6 + 4i, 7 - 5i, 4 + i, -3 - i$;   (b) $6, -3, 4i, -9i$.

(a) $\overline{6 + 4i} = 6 - 4i, \quad \overline{7 - 5i} = 7 + 5i, \quad \overline{4 + i} = 4 - i, \quad \overline{-3 - i} = -3 + i.$

(b) $\overline{6} = 6, \quad \overline{-3} = -3, \quad \overline{4i} = -4i, \quad \overline{-9i} = 9i.$

(Note that the conjugate of a real number is the original number, but the conjugate of a pure imaginary number is the negative of the original number.)

2.63. Find $z\bar{z}$ and $|z|$ when $z = 3 + 4i$.

For $z = a + bi$, use $z\bar{z} = a^2 + b^2$ and $z = \sqrt{z\bar{z}} = \sqrt{a^2 + b^2}$.

$$z\bar{z} = 9 + 16 = 25 \qquad |z| = \sqrt{25} = 5$$

2.64. Simplify $\dfrac{2 - 7i}{5 + 3i}$.

To simplify a fraction z/w of complex numbers, multiply both numerator and denominator by $\bar{w}$, the conjugate of the denominator.

$$\frac{2 - 7i}{5 + 3i} = \frac{(2 - 7i)(5 - 3i)}{(5 + 3i)(5 - 3i)} = \frac{-11 - 41i}{34} = -\frac{11}{34} - \frac{41}{34}i$$

2.65. Prove: For any complex numbers $z, w \in \mathbf{C}$, (i) $\overline{z + w} = \bar{z} + \bar{w}$, (ii) $\overline{zw} = \bar{z}\bar{w}$, (iii) $\bar{\bar{z}} = z$.

Suppose $z = a + bi$ and $w = c + di$ where $a, b, c, d \in \mathbf{R}$.

(i) $\overline{z + w} = \overline{(a + bi) + (c + di)} = \overline{(a + c) + (b + d)i}$
$= (a + c) - (b + d)i = a + c - bi - di$
$= (a - bi) + (c - di) = \bar{z} + \bar{w}$

(ii) $\overline{zw} = \overline{(a + bi)(c + di)} = \overline{(ac - bd) + (ad + bc)i}$
$= (ac - bd) - (ad + bc)i = (a - bi)(c - di) = \bar{z}\bar{w}$

(iii) $\bar{\bar{z}} = \overline{\overline{a + bi}} = \overline{a - bi} = a - (-b)i = a + bi = z$

2.66. Prove: For any complex numbers $z, w \in \mathbf{C}$, $|zw| = |z||w|$.

Suppose $z = a + bi$ and $w = c + di$ where $a, b, c, d \in \mathbf{R}$. Then

$$|z|^2 = a^2 + b^2 \qquad |w|^2 = c^2 + d^2 \qquad \text{and} \qquad zw = (ac - bd) + (ad + bc)i$$

Thus

$$|zw|^2 = (ac - bd)^2 + (ad + bc)^2$$

$$= a^2c^2 - 2abcd + b^2d^2 + a^2d^2 + 2abcd + b^2c^2$$
$$= a^2(c^2 + d^2) + b^2(c^2 + d^2) = (a^2 + b^2)(c^2 + d^2) = |z|^2|w|^2$$

The square root of both sides gives us the desired result.

2.67. Prove: For any complex numbers $z, w \in \mathbf{C}$, $|z + w| \le |z| + |w|$.

Suppose $z = a + bi$ and $w = c + di$ where $a, b, c, d \in \mathbf{R}$. Consider the vectors $u = (a, b)$ and $v = (c, d)$ in $\mathbf{R}^2$. Note that

$$|z| = \sqrt{a^2 + b^2} = \|u\| \qquad |w| = \sqrt{c^2 + d^2} = \|v\|$$

and

$$|z + w| = |(a + c) + (b + d)i| = \sqrt{(a + c)^2 + (b + d)^2} = \|(a + c, b + d)\| = \|u + v\|$$

By Minkowski's inequality (Problem 2.23), $\|u + v\| \le \|u\| + \|v\|$ and so

$$|z + w| = \|u + v\| \le \|u\| + \|v\| = |z| + |w|$$

2.68. Find the dot products $u \cdot v$ and $v \cdot u$ where (a) $u = (1 - 2i, 3 + i)$, $v = (4 + 2i, 5 - 6i)$, and (b) $u = (3 - 2i, 4i, 1 + 6i)$, $v = (5 + i, 2 - 3i, 7 + 2i)$.

Recall that the conjugates of the second vector appear in the dot product

$$(z_1, \ldots, z_n) \cdot (w_1, \ldots, w_n) = z_1\bar{w}_1 + \cdots + z_n\bar{w}_n$$

(a) $u \cdot v = (1 - 2i)\overline{(4 + 2i)} + (3 + i)\overline{(5 - 6i)}$
$= (1 - 2i)(4 - 2i) + (3 + i)(5 + 6i) = -10i + 9 + 23i = 9 + 13i$
$v \cdot u = (4 + 2i)\overline{(1 - 2i)} + (5 - 6i)\overline{(3 + i)}$
$= (4 + 2i)(1 + 2i) + (5 - 6i)(3 - i) = 10i + 9 - 23i = 9 - 13i$

(b) $u \cdot v = (3 - 2i)\overline{(5 + i)} + (4i)\overline{(2 - 3i)} + (1 + 6i)\overline{(7 + 2i)}$
$= (3 - 2i)(5 - i) + (4i)(2 + 3i) + (1 + 6i)(7 - 2i) = 20 + 35i$
$v \cdot u = (5 + i)\overline{(3 - 2i)} + (2 - 3i)\overline{(4i)} + (7 + 2i)\overline{(1 + 6i)}$
$= (5 + i)(3 + 2i) + (2 - 3i)(-4i) + (7 + 2i)(1 - 6i) = 20 - 35i$

In both cases, $v \cdot u = \overline{u \cdot v}$. This holds true in general, as seen in Problem 2.70.

2.69. Let $u = (7 - 2i, 2 + 5i)$ and $v = (1 + i, -3 - 6i)$. Find:

(a) $u + v$; (b) $2iu$; (c) $(3 - i)v$; (d) $u \cdot v$; (e) $\|u\|$ and $\|v\|$.

(a) $u + v = (7 - 2i + 1 + i, 2 + 5i - 3 - 6i) = (8 - i, -1 - i)$

(b) $2iu = (14i - 4i^2, 4i + 10i^2) = (4 + 14i, -10 + 4i)$

(c) $(3 - i)v = (3 + 3i - i - i^2, -9 - 18i + 3i + 6i^2) = (4 + 2i, -15 - 15i)$

(d) $u \cdot v = (7 - 2i)\overline{(1 + i)} + (2 + 5i)\overline{(-3 - 6i)}$

$\qquad = (7 - 2i)(1 - i) + (2 + 5i)(-3 + 6i) = 5 - 9i - 36 - 3i = -31 - 12i$

(e) $\|u\| = \sqrt{7^2 + (-2)^2 + 2^2 + 5^2} = \sqrt{82}, \|v\| = \sqrt{1^2 + 1^2 + (-3)^2 + (-6)^2} = \sqrt{47}$

2.70. Prove: For any vectors $u, v \in \mathbf{C}^n$ and any scalar $z \in \mathbf{C}$, (i) $u \cdot v = \overline{v \cdot u}$, (ii) $(zu) \cdot v = z(u \cdot v)$, (iii) $u \cdot (zv) = \bar{z}(u \cdot v)$.

Suppose $u = (z_1, z_2, \ldots, z_n)$ and $v = (w_1, w_2, \ldots, w_n)$.

(i) Using the properties of the conjugate,

$$\overline{v \cdot u} = \overline{w_1 \bar{z}_1 + w_2 \bar{z}_2 + \cdots + w_n \bar{z}_n} = \overline{w_1 \bar{z}_1} + \overline{w_2 \bar{z}_2} + \cdots + \overline{w_n \bar{z}_n}$$
$$= \bar{w}_1 z_1 + \bar{w}_2 z_2 + \cdots + \bar{w}_n z_n = z_1 \bar{w}_1 + z_2 \bar{w}_2 + \cdots + z_n \bar{w}_n = u \cdot v$$

(ii) Since $zu = (zz_1, zz_2, \ldots, zz_n)$,

$$(zu) \cdot v = zz_1 \bar{w}_1 + zz_2 \bar{w}_2 + \cdots + zz_n \bar{w}_n = z(z_1 \bar{w}_1 + z_2 \bar{w}_2 + \cdots + z_n \bar{w}_n) = z(u \cdot v)$$

(Compare with Theorem 2.2 on vectors in $\mathbf{R}^n$.)

(iii) **Method 1.**   Since $zv = (zw_1, zw_2, \ldots, zw_n)$,

$$u \cdot (zv) = z_1 \overline{zw_1} + z_2 \overline{zw_2} + \cdots + z_n \overline{zw_n} = z_1 \bar{z} \bar{w}_1 + z_2 \bar{z} \bar{w}_2 + \cdots + z_n \bar{z} \bar{w}_n$$
$$= \bar{z}(z_1 \bar{w}_1 + z_2 \bar{w}_2 + \cdots + z_n \bar{w}_n) = \bar{z}(u \cdot v)$$

Method 2.   Using (i) and (ii),

$$u \cdot (zv) = \overline{(zv) \cdot u} = z\overline{(v \cdot u)} = \bar{z}\overline{(v \cdot u)} = \bar{z}(u \cdot v)$$

Supplementary Problems

VECTORS IN $\mathbf{R}^n$

2.71. Let $u = (2, -1, 0, -3)$, $v = (1, -1, -1, 3)$, $w = (1, 3, -2, 2)$. Find: (a) $2u - 3v$; (b) $5u - 3v - 4w$; (c) $-u + 2v - 2w$; (d) $u \cdot v, u \cdot w$ and $v \cdot w$; (e) $\|u\|, \|v\|,$ and $\|w\|$.

2.72. Determine x and y if: (a) $x(3, 2) = 2(y, -1)$; (b) $x(2, y) = y(1, -2)$.

2.73. Find $d(u, v)$ and proj (u, v) when (a) $u = (1, -3), v = (4, 1)$ and (b) $u = (2, -1, 0, 1), v = (1, -1, 1, 2)$.

LINEAR COMBINATIONS, LINEAR INDEPENDENCE

2.74. Let

$$u_1 = \begin{pmatrix} 1 \\ 0 \\ 1 \end{pmatrix} \qquad u_2 = \begin{pmatrix} 1 \\ 1 \\ 1 \end{pmatrix} \qquad u_3 = \begin{pmatrix} 1 \\ 2 \\ 3 \end{pmatrix}$$

Express v as a linear combination of u_1, u_2, u_3 where

$$(a) \quad v = \begin{pmatrix} 1 \\ -2 \\ 4 \end{pmatrix}, \qquad (b) \quad v = \begin{pmatrix} 1 \\ 3 \\ 5 \end{pmatrix}, \qquad (c) \quad v = \begin{pmatrix} a \\ b \\ c \end{pmatrix}$$

2.75. Determine whether the following vectors u, v, w are linearly independent and, if not, express one of them as a linear combination of the others.

(a) $u = (1, 0, 1), v = (1, 2, 3), w = (3, 2, 5)$;

(b) $u = (1, 0, 1), v = (1, 1, 1), w = (0, 1, 1)$;

(c) $u = (1, 2), v = (1, -1), w = (2, 5)$;

(d) $u = (1, 0, 0, 1), v = (0, 1, 2, 1), w = (1, 2, 3, 4)$;

(e) $u = (1, 0, 0, 1), v = (0, 1, 2, 1), w = (1, 2, 4, 3)$.

2.76. Show that any $n + 1$ or more vectors in $\mathbf{R}^n$ are linearly dependent.

LOCATED VECTORS, HYPERPLANES, LINES, CURVES

2.77. Find the (located) vector v from (a) $P(2, 3, -7)$ to $Q(1, -6, -5)$; (b) $P(1, -8, -4, 6)$ to $Q(3, -5, 2, -4)$.

2.78. Find an equation of the hyperplane in $\mathbf{R}^3$ which:

(a) Passes through $(2, -7, 1)$ and is normal to $(3, 1, -11)$;

(b) Contains $(1, -2, 2), (0, 1, 3)$, and $(0, 2, -1)$;

(c) Contains $(1, -5, 2)$ and is parallel to $3x - 7y + 4z = 5$.

2.79. Find a parametric representation of the line which:

(a) Passes through $(7, -1, 8)$ in the direction of $(1, 3, -5)$;

(b) Passes through $(1, 9, -4, 5)$ and $(2, -3, 0, 4)$;

(c) Passes through $(4, -1, 9)$ and is perpendicular to the plane $3x - 2y + z = 18$.

SPATIAL VECTORS (VECTORS IN $\mathbf{R}^3$), PLANES, LINES, CURVES, AND SURFACES IN $\mathbf{R}^3$

2.80. Find the equation of the plane H:

(a) With normal $\mathbf{N} = 3\mathbf{i} - 4\mathbf{j} + 5\mathbf{k}$ and containing the point $P(1, 2, -3)$;

(b) Parallel to $4x + 3y - 2z = 11$ and containing the point $P(1, 2, -3)$.

2.81. Find a unit vector u which is normal to the plane:

(a) $3x - 4y - 12z = 11$;   (b) $2x - y - 2z = 7$.

2.82. Find $\cos \theta$ where θ is the angle between the planes:

(a) $3x - 2y - 4z = 5$ and $x + 2y - 6z = 4$;

(b) $2x + 5y - 4z = 1$ and $4x + 3y + 2z = 1$.

2.83. Find the (parametric) equation of the line L:

(a) Through the point $P(2, 5, -3)$ and in the direction of $v = 4\mathbf{i} - 5\mathbf{j} + 7\mathbf{k}$;

(b) Through the points $P(1, 2, -4)$ and $Q(3, -7, 2)$;

(c) Perpendicular to the plane $2x - 3y + 7z = 4$ and containing $P(1, -5, 7)$.

2.84. Consider the following curve where $0 \le t \le 5$:

$$F(t) = t^3\mathbf{i} - t^2\mathbf{j} + (2t - 3)\mathbf{k}$$

(a) Find $F(t)$ when $t = 2$.

(b) Find the endpoints of the curve.

(c) Find the unit tangent vector $\mathbf{T}$ to the curve when $t = 2$.

2.85. Consider the curve $F(t) = (\cos t)\mathbf{i} + (\sin t)\mathbf{j} + t\mathbf{k}$.

(a) Find the unit tangent vector $\mathbf{T}(t)$ to the curve.

(b) Find the unit *normal* vector $\mathbf{N}(t)$ to the curve by normalizing $U(t) = d\mathbf{T}(t)/dt$.

(c) Find the unit binormal vector $\mathbf{B}(t)$ to the curve using $\mathbf{B} = \mathbf{T} \times \mathbf{N}$.

2.86. Consider a moving body B whose position at time t is given by $\mathbf{R}(t) = t^2\mathbf{i} + t^3\mathbf{j} + 2t\mathbf{k}$. [Then $V(t) = dR(t)/dt$ denotes the velocity of B and $A(t) = dV(t)/dt$ denotes the acceleration of B.]

(a) Find the position of B when $t = 1$.

(b) Find the velocity v of B when $t = 1$.

(c) Find the speed s of B when $t = 1$.

(d) Find the acceleration a of B when $t = 1$.

2.87. Find the normal vector $\mathbf{N}$ and the tangent plane H to the surface at the given point:

(a) Surface: $x^2y + 3yz = 20$ and point $P(1, 3, 2)$;

(b) Surface: $x^2 + 3y^2 - 5z^2 = 16$ and point $P(3, -2, 1)$.

2.88. Given $z = f(x, y) = x^2 + 2xy$. Find the normal vector $\mathbf{N}$ and the tangent plane H when $x = 3$, $y = 1$.

CROSS PRODUCT

The cross product is defined only for vectors in $\mathbf{R}^3$.

2.89. Given $u = 3\mathbf{i} - 4\mathbf{j} + 2\mathbf{k}$, $v = 2\mathbf{i} + 5\mathbf{j} - 3\mathbf{k}$, $w = 4\mathbf{i} + 7\mathbf{j} + 2\mathbf{k}$. Find: (a) $u \times v$, (b) $u \times w$, (c) $v \times w$, (d) $v \times u$.

2.90. Find a unit vector w orthogonal to (a) $u = (1, 2, 3)$ and $v = (1, -1, 2)$; (b) $u = 3\mathbf{i} - \mathbf{j} + 2\mathbf{k}$ and $v = 4\mathbf{i} - 2\mathbf{j} - \mathbf{k}$.

2.91. Prove the following properties of the cross product:

(a) $u \times v = -(v \times u)$

(b) $u \times u = 0$ for any vector u

(c) $(ku) \times v = k(u \times v) = u \times (kv)$

(d) $u \times (v + w) = (u \times v) + (u \times w)$

(e) $(v + w) \times u = (v \times u) + (w \times u)$

(f) $(u \times v) \times w = (u \cdot w)v - (v \cdot w)u$

COMPLEX NUMBERS

2.92. Simplify: (a) $(4 - 7i)(9 + 2i)$; (b) $(3 - 5i)^2$; (c) $\dfrac{1}{4 - 7i}$; (d) $\dfrac{9 + 2i}{3 - 5i}$; (e) $(1 - i)^3$.

2.93. Simplify: (a) $\dfrac{1}{2i}$; (b) $\dfrac{2 + 3i}{7 - 3i}$; (c) i^{15}, i^{25}, i^{34}; (d) $\left(\dfrac{1}{3 - i}\right)^2$.

2.94. Let $z = 2 - 5i$ and $w = 7 + 3i$. Find: (a) $z + w$; (b) zw; (c) z/w; (d) $\bar{z}$, $\bar{w}$; (e) $|z|$, $|w|$.

2.95. Let $z = 2 + i$ and $w = 6 - 5i$. Find: (a) z/w; (b) $\bar{z}$, $\bar{w}$; (c) $|z|$, $|w|$.

2.96. Show that (a) $\mathrm{Re}\, z = \frac{1}{2}(z + \bar{z})$; (b) $\mathrm{Im}\, z = (z - \bar{z})/2i$.

2.97. Show that $zw = 0$ implies $z = 0$ or $w = 0$.

VECTORS IN $\mathbf{C}^n$

2.98. Prove: For any vectors $u, v, w \in \mathbf{C}^n$:

   (i) $(u + v) \cdot w = u \cdot w + v \cdot w$;   (ii) $w \cdot (u + v) = w \cdot u + w \cdot v$.

2.99. Prove that the norm in $\mathbf{C}^n$ satisfies the following laws:

   $[N_1]$ For any vector u, $\|u\| \geq 0$; and $\|u\| = 0$ iff $u = 0$.

   $[N_2]$ For any vector u and any complex number z, $\|zu\| = |z|\,\|u\|$.

   $[N_3]$ For any vectors u and v, $\|u + v\| \leq \|u\| + \|v\|$.

Answers to Supplementary Problems

2.71. (a) $2u - 3v = (1, 1, 3, -15)$ (d) $u \cdot v = -6$, $u \cdot w = -7$, $v \cdot w = 6$

   (b) $5u - 3v - 4w = (3, -14, 11, -32)$ (e) $\|u\| = \sqrt{14}$, $\|v\| = 2\sqrt{3}$, $\|w\| = 3\sqrt{2}$

   (c) $-u + 2v - 2w = (-2, -7, 2, 5)$

2.72. (a) $x = -1$, $y = -\frac{3}{2}$;   (b) $x = 0$, $y = 0$ or $x = -2$, $y = -4$

2.73. (a) $d = 5$, proj $(u, v) = (\frac{4}{17}, \frac{1}{17})$;   (b) $d = \sqrt{11}$, proj $(u, v) = (\frac{5}{7}, -\frac{5}{7}, \frac{5}{7}, \frac{10}{7})$

2.74. (a) $v = (\frac{9}{2})u_1 - 5u_2 + 3u_3$

   (b) $v = u_2 + 2u_3$

   (c) $v = [(a - 2b + c)/2]u_1 + (a + b - c)u_2 + [(c - a)/2]u_3$

2.75. (a) dependent;   (b) independent;   (c) dependent;   (d) independent;   (e) dependent.

2.76. The corresponding homogeneous system has more unknowns than equations and hence has a nonzero solution.

2.77. (a) $v = (-1, -9, 2)$;   (b) $v = (2, 3, 6, -10)$

2.78. (a) $3x + y - 11z = -12$;   (b) $13x + 4y + z = 7$;   (c) $3x - 7y + 4z = 46$

2.79. (a) $x = 7 + t$, $y = -1 + 3t$, $z = 8 - 5t$

   (b) $x_1 = 1 + t$, $x_2 = 9 - 12t$, $x_3 = -4 + 4t$, $x_4 = 5 - t$

   (c) $x = 4 + 3t$, $y = -1 - 2t$, $z = 9 + t$

2.80. (a) $3x - 4y + 5z = -20$;   (b) $4x + 3y - 2z = 16$

2.81. (a) $u = (3\mathbf{i} - 4\mathbf{j} - 12\mathbf{k})/13$;   (b) $u = (2\mathbf{i} - \mathbf{j} - 2\mathbf{k})/3$

2.82. (a) $23/(\sqrt{29}\sqrt{41})$;   (b) $15/(\sqrt{45}\sqrt{29})$

2.83. (a) $x = 2 + 4t, y = 5 - 5t, z = -3 + 7t$
(b) $x = 1 + 2t, y = 2 - 9t, z = -4 + 6t$
(c) $x = 1 + 2t, y = -5 - 3t, z = 7 + 7t$

2.84. (a) $8\mathbf{i} - 4\mathbf{j} + \mathbf{k}$;   (b) $-3\mathbf{k}$ and $125\mathbf{i} - 25\mathbf{j} + 7\mathbf{k}$;   (c) $\mathbf{T} = (6\mathbf{i} - 2\mathbf{j} + \mathbf{k})/\sqrt{41}$

2.85. (a) $\mathbf{T}(t) = (-\sin t)\mathbf{i} + (\cos t)\mathbf{j} + \mathbf{k})/\sqrt{2}$
(b) $\mathbf{N}(t) = (-\cos t)\mathbf{i} - (\sin t)\mathbf{j}$
(c) $\mathbf{B}(t) = (\sin t)\mathbf{i} - (\cos t)\mathbf{j} + \mathbf{k})/\sqrt{2}$

2.86. (a) $\mathbf{i} + \mathbf{j} + 2\mathbf{k}$;   (b) $2\mathbf{i} + 3\mathbf{j} + 2\mathbf{k}$;   (c) $\sqrt{17}$;   (d) $2\mathbf{i} + 6\mathbf{j}$

2.87. (a) $\mathbf{N} = 6\mathbf{i} + 7\mathbf{j} + 9\mathbf{k}, 6x + 7y + 9z = 45$
(b) $\mathbf{N} = 6\mathbf{i} - 12\mathbf{j} - 10\mathbf{k}, 3x - 6y - 5z = 16$

2.88. $\mathbf{N} = 8\mathbf{i} + 6\mathbf{j} - \mathbf{k}, 8x + 6y - z = 15$

2.89. (a) $2\mathbf{i} + 13\mathbf{j} + 23\mathbf{k}$ (c) $31\mathbf{i} - 16\mathbf{j} - 6\mathbf{k}$
(b) $-22\mathbf{i} + 2\mathbf{j} + 37\mathbf{k}$ (d) $-2\mathbf{i} - 13\mathbf{j} - 23\mathbf{k}$

2.90. (a) $(7, 1, -3)/\sqrt{59}$;   (b) $(5\mathbf{i} + 11\mathbf{j} - 2\mathbf{k})/\sqrt{150}$

2.92. (a) $50 - 55i$;   (b) $-16 - 30i$;   (c) $(4 + 7i)/65$;   (d) $(1 + 3i)/2$;   (e) $-2 - 2i$

2.93. (a) $-\frac{1}{2}i$;   (b) $(5 + 27i)/58$;   (c) $-i, i, -1$;   (d) $(4 + 3i)/50$

2.94. (a) $z + w = 9 - 2i$ (d) $\bar{z} = 2 + 5i, \bar{w} = 7 - 3i$
(b) $zw = 29 - 29i$ (e) $|z| = \sqrt{29}, |w| = \sqrt{58}$
(c) $z/w = (-1 - 41i)/58$

2.95. (a) $z/w = (7 + 16i)/61$;   (b) $\bar{z} = 2 - i, \bar{w} = 6 + 5i$;   (c) $|z| = \sqrt{5}, |w| = \sqrt{61}$

2.97. If $zw = 0$, then $|zw| = |z||w| = |0| = 0$. Hence $|z| = 0$ or $|w| = 0$; and so $z = 0$ or $w = 0$.

Chapter 3

Matrices

3.1 INTRODUCTION

Matrices were first introduced in Chapter 1 and their elements were related to the coefficients of systems of linear equations. Here we will reintroduce these matrices and we will study certain algebraic operations defined on them. The material here is mainly computational. However, as with linear equations, the abstract treatment presented later on will give us new insight into the structure of matrices.

The entries in our matrices shall come from some arbitrary, but fixed, field K. (See Appendix B.) The elements of K are called scalars. Nothing essential is lost if the reader assumes that K is the real field $\mathbf{R}$ or the complex field $\mathbf{C}$.

Last, we remark that the elements of $\mathbf{R}^n$ or $\mathbf{C}^n$ are conveniently represented by "row vectors" or "column vectors," which are special cases of matrices.

3.2 MATRICES

A *matrix over a field K* (or simply a *matrix* if K is implicit) is a rectangular array of scalars a_{ij} of the form

$$\begin{pmatrix} a_{11} & a_{12} & \dots & a_{1n} \\ a_{21} & a_{22} & \dots & a_{2n} \\ \hdotsfor{4} \\ a_{m1} & a_{m2} & \dots & a_{mn} \end{pmatrix}$$

The above matrix is also denoted by (a_{ij}), $i = 1, \dots, m$, $j = 1, \dots, n$, or simply by (a_{ij}). The m horizontal n-tuples

$$(a_{11}, a_{12}, \dots, a_{1n}), (a_{21}, a_{22}, \dots, a_{2n}), \dots, (a_{m1}, a_{m2}, \dots, a_{mn})$$

are the *rows* of the matrix, and the n vertical m-tuples

$$\begin{pmatrix} a_{11} \\ a_{21} \\ \dots \\ a_{m1} \end{pmatrix}, \begin{pmatrix} a_{12} \\ a_{22} \\ \dots \\ a_{m2} \end{pmatrix}, \dots, \begin{pmatrix} a_{1n} \\ a_{2n} \\ \dots \\ a_{mn} \end{pmatrix}$$

are its *columns*. Note that the element a_{ij}, called the *ij-entry* or *ij-component*, appears in the ith row and the jth column. A matrix with m rows and n columns is called an m by n matrix, or $m \times n$ matrix; the pair of numbers (m, n) is called its *size* or *shape*.

Matrices will usually be denoted by capital letters $A, B, \dots$, and the elements of the field K by lower-case letters $a, b, \dots$. Two matrices A and B are *equal*, written $A = B$, if they have the same shape and if corresponding elements are equal. Thus the equality of two $m \times n$ matrices is equivalent to a system of mn equalities, one for each pair of elements.

Example 3.1

(a) The following is a 2×3 matrix: $\begin{pmatrix} 1 & -3 & 4 \\ 0 & 5 & -2 \end{pmatrix}$.

Its rows are $(1, -3, 4)$ and $(0, 5, -2)$; its columns are $\begin{pmatrix} 1 \\ 0 \end{pmatrix}$, $\begin{pmatrix} -3 \\ 5 \end{pmatrix}$, and $\begin{pmatrix} 4 \\ -2 \end{pmatrix}$.

(b) The statement $\begin{pmatrix} x+y & 2z+w \\ x-y & z-w \end{pmatrix} = \begin{pmatrix} 3 & 5 \\ 1 & 4 \end{pmatrix}$ is equivalent to the following system of equations:

$$\begin{cases} x + y = 3 \\ x - y = 1 \\ 2z + w = 5 \\ z - w = 4 \end{cases}$$

The solution of the system is $x = 2, y = 1, z = 3, w = -1$.

> **Remark:** A matrix with one row is also referred to as a *row vector*, and with one column as a *column vector*. In particular, an element in the field K can be viewed as a 1×1 matrix.

3.3 MATRIX ADDITION AND SCALAR MULTIPLICATION

Let A and B be two matrices with the same size, i.e., the same number of rows and of columns, say, $m \times n$ matrices:

$$A = \begin{pmatrix} a_{11} & a_{12} & \dots & a_{1n} \\ a_{21} & a_{22} & \dots & a_{2n} \\ \dots\dots\dots\dots\dots\dots\dots \\ a_{m1} & a_{m2} & \dots & a_{mn} \end{pmatrix} \quad \text{and} \quad B = \begin{pmatrix} b_{11} & b_{12} & \dots & b_{1n} \\ b_{21} & b_{22} & \dots & b_{2n} \\ \dots\dots\dots\dots\dots\dots\dots \\ b_{m1} & b_{m2} & \dots & b_{mn} \end{pmatrix}$$

The *sum* of A and B, written $A + B$, is the matrix obtained by adding corresponding entries:

$$A + B = \begin{pmatrix} a_{11}+b_{11} & a_{12}+b_{12} & \dots & a_{1n}+b_{1n} \\ a_{21}+b_{21} & a_{22}+b_{22} & \dots & a_{2n}+b_{2n} \\ \dots\dots\dots\dots\dots\dots\dots\dots\dots\dots\dots \\ a_{m1}+b_{m1} & a_{m2}+b_{m2} & \dots & a_{mn}+b_{mn} \end{pmatrix}$$

The *product* of a scalar k by the matrix A, written $k \cdot A$ or simply kA, is the matrix obtained by multiplying each entry of A by k:

$$kA = \begin{pmatrix} ka_{11} & ka_{12} & \dots & ka_{1n} \\ ka_{21} & ka_{22} & \dots & ka_{2n} \\ \dots\dots\dots\dots\dots\dots\dots \\ ka_{m1} & ka_{m2} & \dots & ka_{mn} \end{pmatrix}$$

Observe that $A + B$ and kA are also $m \times n$ matrices. We also define

$$-A = -1 \cdot A \quad \text{and} \quad A - B = A + (-B)$$

The sum of matrices with different sizes is not defined.

Example 3.2. Let $A = \begin{pmatrix} 1 & -2 & 3 \\ 4 & 5 & -6 \end{pmatrix}$ and $B = \begin{pmatrix} 3 & 0 & 2 \\ -7 & 1 & 8 \end{pmatrix}$. Then

$$A + B = \begin{pmatrix} 1+3 & -2+0 & 3+2 \\ 4-7 & 5+1 & -6+8 \end{pmatrix} = \begin{pmatrix} 4 & -2 & 5 \\ -3 & 6 & 2 \end{pmatrix}$$

$$3A = \begin{pmatrix} 3\cdot1 & 3\cdot(-2) & 3\cdot3 \\ 3\cdot4 & 3\cdot5 & 3\cdot(-6) \end{pmatrix} = \begin{pmatrix} 3 & -6 & 9 \\ 12 & 15 & -18 \end{pmatrix}$$

$$2A - 3B = \begin{pmatrix} 2 & -4 & 6 \\ 8 & 10 & -12 \end{pmatrix} + \begin{pmatrix} -9 & 0 & -6 \\ 21 & -3 & -24 \end{pmatrix} = \begin{pmatrix} -7 & -4 & 0 \\ 29 & 7 & -36 \end{pmatrix}$$

The $m \times n$ matrix whose entries are all zero is called the zero matrix and will be denoted by $\mathbf{0}_{m,n}$ or simply $\mathbf{0}$. For example, the 2×3 zero matrix is

$$\begin{pmatrix} 0 & 0 & 0 \\ 0 & 0 & 0 \end{pmatrix}$$

The zero matrix is similar to the scalar 0, and the same symbol will be used for both. For any matrix A,

$$A + \mathbf{0} = \mathbf{0} + A = A$$

Basic properties of matrices under the operations of matrix addition and scalar multiplication follow.

Theorem 3.1:   Let V be the set of all $m \times n$ matrices over a field K. Then for any matrices $A, B, C \in V$ and any scalars $k_1, k_2 \in K$,

    (i) $(A + B) + C = A + (B + C)$ (v) $k_1(A + B) = k_1 A + k_1 B$

    (ii) $A + \mathbf{0} = A$ (vi) $(k_1 + k_2)A = k_1 A + k_2 A$

    (iii) $A + (-A) = \mathbf{0}$ (vii) $(k_1 k_2)A = k_1(k_2 A)$

    (iv) $A + B = B + A$ (viii) $1 \cdot A = A$ and $0A = \mathbf{0}$

        Using (vi) and (viii) above, we also have that $A + A = 2A$, $A + A + A = 3A$, ...

Remark: Suppose vectors in $\mathbf{R}^n$ are represented by row vectors (or by column vectors); say,

$$u = (a_1, a_2, \ldots, a_n) \quad \text{and} \quad v = (b_1, b_2, \ldots, b_n)$$

Then viewed as matrices, the sum $u + v$ and the scalar product ku are as follows:

$$u + v = (a_1 + b_1, a_2 + b_2, \ldots, a_n + b_n) \quad \text{and} \quad ku = (ka_1, ka_2, \ldots, ka_n)$$

But this corresponds precisely to the sum and scalar product as defined in Chapter 2. In other words, the above operations on matrices may be viewed as a generalization of the corresponding operations defined in Chapter 2.

3.4 MATRIX MULTIPLICATION

The product of matrices A and B, written AB, is somewhat complicated. For this reason, we first begin with a special case.

The product $A \cdot B$ of a row matrix $A = (a_i)$ and a column matrix $B = (b_i)$ with the same number of elements is defined as follows:

$$(a_1, a_2, \ldots, a_n)\begin{pmatrix} b_1 \\ b_2 \\ \cdots \\ b_n \end{pmatrix} = a_1 b_1 + a_2 b_2 + \cdots + a_n b_n = \sum_{k=1}^{n} a_k b_k$$

Note that $A \cdot B$ is a scalar (or a 1×1 matrix). The product $A \cdot B$ is not defined when A and B have different numbers of elements.

Example 3.3

$$(8, -4, 5)\begin{pmatrix} 3 \\ 2 \\ -1 \end{pmatrix} = 8 \cdot 3 + (-4) \cdot 2 + 5 \cdot (-1) = 24 - 8 - 5 = 11$$

Using the above definition, we now define matrix multiplication in general.

Definition: Suppose $A = (a_{ij})$ and $B = (b_{ij})$ are matrices such that the number of columns of A is equal to the number of rows of B; say, A is an $m \times p$ matrix and B is a $p \times n$ matrix. Then the *product AB* is the $m \times n$ matrix whose *ij*-entry is obtained by multiplying the *i*th row A_i of A by the *j*th column B^j of B:

$$AB = \begin{pmatrix} A_1 \cdot B^1 & A_1 \cdot B^2 & \ldots & A_1 \cdot B^n \\ A_2 \cdot B^1 & A_2 \cdot B^2 & \ldots & A_2 \cdot B^n \\ \cdots\cdots\cdots\cdots\cdots\cdots\cdots\cdots\cdots\cdots \\ A_m \cdot B^1 & A_m \cdot B^2 & \ldots & A_m \cdot B^n \end{pmatrix}$$

That is,

$$\begin{pmatrix} a_{11} & \ldots & a_{1p} \\ \cdot & \ldots & \cdot \\ a_{i1} & \ldots & a_{ip} \\ \cdot & \ldots & \cdot \\ a_{m1} & \ldots & a_{mp} \end{pmatrix} \begin{pmatrix} b_{11} & \ldots & b_{1j} & \ldots & b_{1n} \\ \cdot & \ldots & \cdot & \ldots & \cdot \\ \cdot & \ldots & \cdot & \ldots & \cdot \\ \cdot & \ldots & \cdot & \ldots & \cdot \\ b_{p1} & \ldots & b_{pj} & \ldots & b_{pn} \end{pmatrix} = \begin{pmatrix} c_{11} & \ldots & c_{1n} \\ \cdot & \ldots & \cdot \\ \cdot & c_{ij} & \cdot \\ \cdot & \ldots & \cdot \\ c_{m1} & \ldots & c_{mn} \end{pmatrix}$$

where $c_{ij} = a_{i1}b_{1j} + a_{i2}b_{2j} + \cdots + a_{ip}b_{pj} = \sum\limits_{k=1}^{p} a_{ik}b_{kj}$.

We emphasize that the product AB is not defined if A is an $m \times p$ matrix and B is a $q \times n$ matrix, where $p \neq q$.

Example 3.4

(a) $\begin{pmatrix} r & s \\ t & u \end{pmatrix}\begin{pmatrix} a_1 & a_2 & a_3 \\ b_1 & b_2 & b_3 \end{pmatrix} = \begin{pmatrix} ra_1 + sb_1 & ra_2 + sb_2 & ra_3 + sb_3 \\ ta_1 + ub_1 & ta_2 + ub_2 & ta_3 + ub_3 \end{pmatrix}$

(b) $\begin{pmatrix} 1 & 2 \\ 3 & 4 \end{pmatrix}\begin{pmatrix} 1 & 1 \\ 0 & 2 \end{pmatrix} = \begin{pmatrix} 1 \cdot 1 + 2 \cdot 0 & 1 \cdot 1 + 2 \cdot 2 \\ 3 \cdot 1 + 4 \cdot 0 & 3 \cdot 1 + 4 \cdot 2 \end{pmatrix} = \begin{pmatrix} 1 & 5 \\ 3 & 11 \end{pmatrix}$

$\begin{pmatrix} 1 & 1 \\ 0 & 2 \end{pmatrix}\begin{pmatrix} 1 & 2 \\ 3 & 4 \end{pmatrix} = \begin{pmatrix} 1 \cdot 1 + 1 \cdot 3 & 1 \cdot 2 + 1 \cdot 4 \\ 0 \cdot 1 + 2 \cdot 3 & 0 \cdot 2 + 2 \cdot 4 \end{pmatrix} = \begin{pmatrix} 4 & 6 \\ 6 & 8 \end{pmatrix}$

The above example shows that matrix multiplication is not commutative, i.e., the products AB and BA of matrices need not be equal.

Matrix multiplication does, however, satisfy the following properties:

Theorem 3.2:   (i) $(AB)C = A(BC)$ (associative law)

(ii) $A(B + C) = AB + AC$ (left distributive law)

(iii) $(B + C)A = BA + CA$ (right distributive law)

(iv) $k(AB) = (kA)B = A(kB)$, where k is a scalar

We assume that the sums and products in the above theorem are defined.

We remark that $0A = 0$ and $B0 = 0$ where 0 is the zero matrix.

3.5 TRANSPOSE OF A MATRIX

The *transpose* of a matrix A, denoted A^T, is the matrix obtained by writing the rows of A, in order, as columns:

$$\begin{pmatrix} a_{11} & a_{12} & \ldots & a_{1n} \\ a_{21} & a_{22} & \ldots & a_{2n} \\ \cdots\cdots\cdots\cdots\cdots\cdots \\ a_{m1} & a_{m2} & \ldots & a_{mn} \end{pmatrix}^T = \begin{pmatrix} a_{11} & a_{21} & \ldots & a_{m1} \\ a_{12} & a_{22} & \ldots & a_{m2} \\ \cdots\cdots\cdots\cdots\cdots\cdots \\ a_{1n} & a_{2n} & \ldots & a_{mn} \end{pmatrix}$$

In other words, if $A = (a_{ij})$ is an $m \times n$ matrix, then $A^T = (a_{ij}^T)$ is the $n \times m$ matrix where $a_{ij}^T = a_{ji}$, for all i and j.

Note that the transpose of a row vector is a column vector and vice versa.

The transpose operation on matrices satisfies the following properties:

Theorem 3.3: (i) $(A + B)^T = A^T + B^T$ (iii) $(kA)^T = kA^T$ (k a scalar)

(ii) $(A^T)^T = A$ (iv) $(AB)^T = B^T A^T$

Observe in (iv) that the transpose of a product is the product of transposes, but in the reverse order.

3.6 MATRICES AND SYSTEMS OF LINEAR EQUATIONS

Consider again a system of m linear equations in n unknowns:

$$\begin{aligned}
a_{11}x_1 + a_{12}x_2 + \cdots + a_{1n}x_n &= b_1 \\
a_{21}x_1 + a_{22}x_2 + \cdots + a_{2n}x_n &= b_2 \\
&\cdots \cdots \cdots \cdots \cdots \cdots \cdots \cdots \\
a_{m1}x_1 + a_{m2}x_2 + \cdots + a_{mn}x_n &= b_m
\end{aligned} \tag{3.1}$$

The above system is equivalent to the following matrix equation:

$$\begin{pmatrix} a_{11} & a_{12} & \dots & a_{1n} \\ a_{21} & a_{22} & \dots & a_{2n} \\ \vdots & & & \\ a_{m1} & a_{m2} & \dots & a_{mn} \end{pmatrix} \begin{pmatrix} x_1 \\ x_2 \\ x_3 \\ \cdots \\ x_n \end{pmatrix} = \begin{pmatrix} b_1 \\ b_2 \\ \cdots \\ b_m \end{pmatrix} \qquad \text{or simply} \qquad AX = B$$

where $A = (a_{ij})$ is the matrix of coefficients, called the *coefficient matrix*, $X = (x_j)$ is the column of unknowns, and $B = (b_i)$ is the column of constants. The statement that they are equivalent means that every solution of the system (3.1) is a solution of the matrix equation, and vice versa.

The *augmented matrix* of the system (3.1) is the following matrix:

$$\begin{pmatrix} a_{11} & a_{12} & \dots & a_{1n} & b_1 \\ a_{21} & a_{22} & \dots & a_{2n} & b_2 \\ \vdots & & & & \\ a_{m1} & a_{m2} & \dots & a_{mn} & b_m \end{pmatrix}$$

That is, the augmented matrix of the system $AX = B$ is the matrix which consists of the matrix A of coefficients followed by the column B of constants. Observe that the system (3.1) is completely determined by its augmented matrix.

Example 3.5.   The following are, respectively, a system of linear equations and its equivalent matrix equation:

$$\begin{aligned} 2x + 3y - 4z &= 7 \\ x - 2y - 5z &= 3 \end{aligned} \qquad \text{and} \qquad \begin{pmatrix} 2 & 3 & -4 \\ 1 & -2 & -5 \end{pmatrix} \begin{pmatrix} x \\ y \\ z \end{pmatrix} = \begin{pmatrix} 7 \\ 3 \end{pmatrix}$$

(Note that the size of the column of unknowns is not equal to the size of the column of constants.)

The augmented matrix of the system is

$$\begin{pmatrix} 2 & 3 & -4 & 7 \\ 1 & -2 & -5 & 3 \end{pmatrix}$$

In studying linear equations it is usually simpler to use the language and theory of matrices, as indicated by the following theorems.

Theorem 3.4: Suppose $u_1, u_2, \ldots, u_n$ are solutions of a homogeneous system of linear equations $AX = 0$. Then every linear combination of the u_i of the form $k_1 u_1 + k_2 u_2 + \cdots + k_n u_n$ where the k_i are scalars, is also a solution of $AX = 0$. Thus, in particular, every multiple ku of any solution u of $AX = 0$ is also a solution of $AX = 0$.

Proof. We are given that $Au_1 = 0, Au_2 = 0, \ldots, Au_n = 0$. Hence

$$A(ku_1 + ku_2 + \cdots + ku_n) = k_1 Au_1 + k_2 Au_2 + \cdots + k_n Au_n$$
$$= k_1 0 + k_2 0 + \cdots + k_n 0 = 0$$

Accordingly, $k_1 u_1 + \cdots + k_n u_n$ is a solution of the homogeneous system $AX = 0$.

Theorem 3.5: The general solution of a nonhomogeneous system $AX = B$ may be obtained by adding the solution space W of the homogeneous system $AX = 0$ to a particular solution v_0 of the nonhomogeneous system $AX = B$. (That is, $v_0 + W$ is the general solution of $AX = B$.)

Proof. Let w be any solution of $AX = 0$. Then

$$A(v_0 + w) = A(v_0) + A(w) = B + 0 = B$$

That is, the sum $v_0 + w$ is a solution of $AX = B$.

On the other hand, suppose v is any solution of $AX = B$ (which may be distinct from v_0). Then

$$A(v - v_0) = Av - Av_0 = B - B = 0$$

That is, the difference $v - v_0$ is a solution of the homogeneous system $AX = 0$. But

$$v = v_0 + (v - v_0)$$

Thus any solution of $AX = B$ can be obtained by adding a solution of $AX = 0$ to the particular solution v_0 of $AX = B$.

Theorem 3.6: Suppose the field K is infinite (e.g., if K is the real field **R** or the complex field C). Then the system $AX = B$ has no solution, a unique solution, or an infinite number of solutions.

Proof. It suffices to show that if $AX = B$ has more than one solution, then it has infinitely many. Suppose u and v are distinct solutions of $AX = B$; that is, $Au = B$ and $Av = B$. Then, for any $k \in K$,

$$A[u + k(u - v)] = Au + k(Au - Av) = B + k(B - B) = B$$

In other words, for each $k \in K$, $u + k(u - v)$ is a solution of $AX = B$. Since all such solutions are distinct (Problem 3.21), $AX = B$ has an infinite number of solutions as claimed.

3.7 BLOCK MATRICES

Using a system of horizontal and vertical (dashed) lines, we can partition a matrix A into smaller matrices called *blocks* (or *cells*) of A. The matrix A is then called a *block matrix*. Clearly, a given matrix may be divided into blocks in different ways; for example,

$$\begin{pmatrix} 1 & -2 & 0 & 1 & 3 \\ 2 & 3 & 5 & 7 & -2 \\ 3 & 1 & 4 & 5 & 9 \end{pmatrix} = \left(\begin{array}{cc:ccc} 1 & -2 & 0 & 1 & 3 \\ 2 & 3 & 5 & 7 & -2 \\ \hdashline 3 & 1 & 4 & 5 & 9 \end{array}\right) = \left(\begin{array}{ccc:cc} 1 & -2 & 0 & 1 & 3 \\ \hdashline 2 & 3 & 5 & 7 & -2 \\ \hdashline 3 & 1 & 4 & 5 & 9 \end{array}\right)$$

The convenience of the partition into blocks is that the result of operations on block matrices can be obtained by carrying out the computation with the blocks, just as if they were the actual elements of the matrices. This is illustrated below.

Suppose A is partitioned into blocks; say

$$A = \begin{pmatrix} A_{11} & A_{12} & \ldots & A_{1n} \\ A_{21} & A_{22} & \ldots & A_{2n} \\ \ldots\ldots\ldots\ldots\ldots\ldots\ldots\ldots \\ A_{m1} & A_{m2} & \ldots & A_{mn} \end{pmatrix}$$

Multiplying each block by a scalar k, multiplies each element of A by k; thus

$$kA = \begin{pmatrix} kA_{11} & kA_{12} & \ldots & kA_{1n} \\ kA_{21} & kA_{22} & \ldots & kA_{2n} \\ \ldots\ldots\ldots\ldots\ldots\ldots\ldots \\ kA_{m1} & kA_{m2} & \ldots & kA_{mn} \end{pmatrix}$$

Now suppose a matrix B is partitioned into the same number of blocks as A; say

$$B = \begin{pmatrix} B_{11} & B_{12} & \ldots & B_{1n} \\ B_{21} & B_{22} & \ldots & B_{2n} \\ \ldots\ldots\ldots\ldots\ldots\ldots\ldots \\ B_{m1} & B_{m2} & \ldots & B_{mn} \end{pmatrix}$$

Furthermore, suppose the corresponding blocks of A and B have the same size. Adding these corresponding blocks adds the corresponding elements of A and B. Accordingly,

$$A + B = \begin{pmatrix} A_{11} + B_{11} & A_{12} + B_{12} & \ldots & A_{1n} + B_{1n} \\ A_{21} + B_{21} & A_{22} + B_{22} & \ldots & A_{2n} + B_{2n} \\ \ldots\ldots\ldots\ldots\ldots\ldots\ldots\ldots\ldots\ldots\ldots\ldots\ldots\ldots \\ A_{m1} + B_{m1} & A_{m2} + B_{m2} & \ldots & A_{mn} + B_{mn} \end{pmatrix}$$

The case of matrix multiplication is less obvious but still true. That is, suppose matrices U and V are partitioned into blocks as follows

$$U = \begin{pmatrix} U_{11} & U_{12} & \ldots & U_{1p} \\ U_{21} & U_{22} & \ldots & U_{2p} \\ \ldots\ldots\ldots\ldots\ldots\ldots\ldots \\ U_{m1} & U_{m2} & \ldots & U_{mp} \end{pmatrix} \quad \text{and} \quad V = \begin{pmatrix} V_{11} & V_{12} & \ldots & V_{1n} \\ V_{21} & V_{22} & \ldots & V_{2n} \\ \ldots\ldots\ldots\ldots\ldots\ldots\ldots \\ V_{p1} & V_{22} & \ldots & V_{pn} \end{pmatrix}$$

such that the number of columns of each block U_{ik} is equal to the number of rows of each block V_{kj}. Then

$$UV = \begin{pmatrix} W_{11} & W_{12} & \ldots & W_{1n} \\ W_{21} & W_{22} & \ldots & W_{2n} \\ \ldots\ldots\ldots\ldots\ldots\ldots\ldots \\ W_{m1} & W_{m2} & \ldots & W_{mn} \end{pmatrix}$$

where

$$W_{ij} = U_{i1}V_{1j} + U_{i2}V_{2j} + \cdots + U_{ip}V_{pj}$$

The proof of the above formula for UV is straightforward, but detailed and lengthy. It is left as a supplementary problem.

Solved Problems

MATRIX ADDITION AND SCALAR MULTIPLICATION

3.1. Compute:

(a) $A + B$ for $A = \begin{pmatrix} 1 & 2 & 3 \\ 4 & 5 & 6 \end{pmatrix}$ and $B = \begin{pmatrix} 1 & -1 & 2 \\ 0 & 3 & -5 \end{pmatrix}$.

(b) $3A$ and $-5A$, where $A = \begin{pmatrix} 1 & -2 & 3 \\ 4 & 5 & -6 \end{pmatrix}$.

(a) Add corresponding elements:

$$A + B = \begin{pmatrix} 1+1 & 2+(-1) & 3+2 \\ 4+0 & 5+3 & 6+(-5) \end{pmatrix} = \begin{pmatrix} 2 & 1 & 5 \\ 4 & 8 & 1 \end{pmatrix}$$

(b) Multiply each entry by the given scalar:

$$3A = \begin{pmatrix} 3\cdot 1 & 3\cdot(-2) & 3\cdot 3 \\ 3\cdot 4 & 3\cdot 5 & 3\cdot(-6) \end{pmatrix} = \begin{pmatrix} 3 & -6 & 9 \\ 12 & 15 & -18 \end{pmatrix}$$

$$-5A = \begin{pmatrix} -5\cdot 1 & -5\cdot(-2) & -5\cdot 3 \\ -5\cdot 4 & -5\cdot 5 & -5\cdot(-6) \end{pmatrix} = \begin{pmatrix} -5 & 10 & -15 \\ -20 & -25 & 30 \end{pmatrix}$$

3.2. Find $2A - 3B$, where $A = \begin{pmatrix} 1 & -2 & 3 \\ 4 & 5 & -6 \end{pmatrix}$ and $B = \begin{pmatrix} 3 & 0 & 2 \\ -7 & 1 & 8 \end{pmatrix}$.

First perform the scalar multiplications, and then a matrix addition:

$$2A - 3B = \begin{pmatrix} 2 & -4 & 6 \\ 8 & 10 & -12 \end{pmatrix} + \begin{pmatrix} -9 & 0 & -6 \\ 21 & -3 & -24 \end{pmatrix} = \begin{pmatrix} -7 & -4 & 0 \\ 29 & 7 & -36 \end{pmatrix}$$

(Note that we multiply B by -3 and then add, rather than multiplying B by 3 and subtracting. This usually avoids errors.)

3.3. Find x, y, z, and w if $3\begin{pmatrix} x & y \\ z & w \end{pmatrix} = \begin{pmatrix} x & 6 \\ -1 & 2w \end{pmatrix} + \begin{pmatrix} 4 & x+y \\ z+w & 3 \end{pmatrix}$.

First write each side as a single matrix:

$$\begin{pmatrix} 3x & 3y \\ 3z & 3w \end{pmatrix} = \begin{pmatrix} x+4 & x+y+6 \\ z+w-1 & 2w+3 \end{pmatrix}$$

Set corresponding entries equal to each other to obtain the system of four equations,

$$\begin{array}{ll} 3x = x+4 & 2x = 4 \\ 3y = x+y+6 & 2y = 6+x \\ 3z = z+w-1 \qquad \text{or} \qquad & 2z = w-1 \\ 3w = 2w+3 & w = 3 \end{array}$$

The solution is: $x = 2$, $y = 4$, $z = 1$, $w = 3$.

3.4. Prove Theorem 3.1(v): Let A and B be $m \times n$ matrices and k a scalar. Then $k(A + B) = kA + kB$.

Suppose $A = (a_{ij})$ and $B = (b_{ij})$. Then $a_{ij} + b_{ij}$ is the ij-entry of $A + B$, and so $k(a_{ij} + b_{ij})$ is the ij-entry of $k(A + B)$. On the other hand, ka_{ij} and kb_{ij} are the ij-entries of kA and kB, respectively, and so $ka_{ij} + kb_{ij}$

is the ij-entry of $kA + kB$. But k, a_{ij} and b_{ij} are scalars in a field; hence

$$k(a_{ij} + b_{ij}) = ka_{ij} + kb_{ij} \qquad \text{for every } i, j$$

Thus $k(A + B) = kA + kB$, as corresponding entries are equal.

MATRIX MULTIPLICATION

3.5. Calculate: (a) $(3, 8, -2, 4)\begin{pmatrix} 5 \\ -1 \\ 6 \end{pmatrix}$ (b) $(1, 8, 3, 4)(6, 1, -3, 5)$

(a) The product is not defined when the row matrix and column matrix have different numbers of elements.

(b) The product of a row matrix and a row matrix is not defined.

3.6. Let $A = \begin{pmatrix} 1 & 3 \\ 2 & -1 \end{pmatrix}$ and $B = \begin{pmatrix} 2 & 0 & -4 \\ 3 & -2 & 6 \end{pmatrix}$. Find (a) AB, (b) BA.

(a) Since A is 2×2 and B is 2×3, the product AB is defined and is a 2×3 matrix. To obtain the entries in the first row of AB, multiply the first row $(1, 3)$ of A by the columns $\begin{pmatrix} 2 \\ 3 \end{pmatrix}$, $\begin{pmatrix} 0 \\ -2 \end{pmatrix}$, and $\begin{pmatrix} -4 \\ 6 \end{pmatrix}$ of B, respectively:

$$\begin{pmatrix} 1 & 3 \\ 2 & -1 \end{pmatrix}\begin{pmatrix} 2 & 0 & -4 \\ 3 & -2 & 6 \end{pmatrix} = \begin{pmatrix} 1 \cdot 2 + 3 \cdot 3 & 1 \cdot 0 + 3 \cdot (-2) & 1 \cdot (-4) + 3 \cdot 6 \end{pmatrix}$$

$$= \begin{pmatrix} 2 + 9 & 0 - 6 & -4 + 18 \end{pmatrix} = \begin{pmatrix} 11 & -6 & 14 \end{pmatrix}$$

To obtain the entries in the second row of AB, multiply the second row $(2, -1)$ of A by the columns of B, respectively:

$$\begin{pmatrix} 1 & 3 \\ 2 & -1 \end{pmatrix}\begin{pmatrix} 2 & 0 & -4 \\ 3 & -2 & 6 \end{pmatrix} = \begin{pmatrix} 11 & -6 & 14 \\ 4 - 3 & 0 + 2 & -8 - 6 \end{pmatrix}$$

Thus $AB = \begin{pmatrix} 11 & -6 & 14 \\ 1 & 2 & -14 \end{pmatrix}$

(b) Note that B is 2×3 and A is 2×2. Since the inner numbers 3 and 2 are not equal, the product BA is not defined.

3.7. Given $A = \begin{pmatrix} 2 & -1 \\ 1 & 0 \\ -3 & 4 \end{pmatrix}$ and $B = \begin{pmatrix} 1 & -2 & -5 \\ 3 & 4 & 0 \end{pmatrix}$, find (a) AB, (b) BA.

(a) Since A is 3×2 and B is 2×3, the product AB is defined and is a 3×3 matrix. To obtain the first row of AB, multiply the first row of A by each column of B, respectively:

$$\begin{pmatrix} 2 & -1 \\ 1 & 0 \\ -3 & 4 \end{pmatrix}\begin{pmatrix} 1 & -2 & -5 \\ 3 & 4 & 0 \end{pmatrix} = \begin{pmatrix} 2 - 3 & -4 - 4 & -10 + 0 \end{pmatrix} = \begin{pmatrix} -1 & -8 & -10 \end{pmatrix}$$

To obtain the second row of AB, multiply the second row of A by each column of B, respectively:

$$\begin{pmatrix} 2 & -1 \\ 1 & 0 \\ -3 & 4 \end{pmatrix}\begin{pmatrix} 1 & -2 & -5 \\ 3 & 4 & 0 \end{pmatrix} = \begin{pmatrix} -1 & -8 & -10 \\ 1 + 0 & -2 + 0 & -5 + 0 \end{pmatrix} = \begin{pmatrix} -1 & -8 & -10 \\ 1 & -2 & -5 \end{pmatrix}$$

To obtain the third row of AB, multiply the third row of A by each column of B, respectively:

$$\begin{pmatrix} 2 & -1 \\ 1 & 0 \\ -3 & 4 \end{pmatrix}\begin{pmatrix} 1 & -2 & -5 \\ 3 & 4 & 0 \end{pmatrix} = \begin{pmatrix} -1 & -8 & -10 \\ 1 & -2 & -5 \\ -3+12 & 6+16 & 15+0 \end{pmatrix} = \begin{pmatrix} -1 & -8 & -10 \\ 1 & -2 & -5 \\ 9 & 22 & 15 \end{pmatrix}$$

Thus

$$AB = \begin{pmatrix} -1 & -8 & -10 \\ 1 & -2 & -5 \\ 9 & 22 & 15 \end{pmatrix}$$

(b) Since B is 2×3 and A is 3×2, the product BA is defined and is a 2×2 matrix. To obtain the first row of BA, multiply the first row of B by each column of A, respectively:

$$\begin{pmatrix} 1 & -2 & -5 \\ 3 & 4 & 0 \end{pmatrix}\begin{pmatrix} 2 & -1 \\ 1 & 0 \\ -3 & 4 \end{pmatrix} = \begin{pmatrix} 2-2+15 & -1+0-20 \end{pmatrix} = \begin{pmatrix} 15 & -21 \end{pmatrix}$$

To obtain the second row of BA, multiply the second row of B by each column of A, respectively:

$$\begin{pmatrix} 1 & -2 & -5 \\ 3 & 4 & 0 \end{pmatrix}\begin{pmatrix} 2 & -1 \\ 1 & 0 \\ -3 & 4 \end{pmatrix} = \begin{pmatrix} 15 & -21 \\ 6+4+0 & -3+0+0 \end{pmatrix} = \begin{pmatrix} 15 & -21 \\ 10 & -3 \end{pmatrix}$$

Thus

$$BA = \begin{pmatrix} 15 & -21 \\ 10 & -3 \end{pmatrix}$$

Remark: Observe that in this problem both AB and BA are defined, but they are not equal; in fact they do not even have the same shape.

3.8. Find AB, where

$$A = \begin{pmatrix} 2 & 3 & -1 \\ 4 & -2 & 5 \end{pmatrix} \qquad B = \begin{pmatrix} 2 & -1 & 0 & 6 \\ 1 & 3 & -5 & 1 \\ 4 & 1 & -2 & 2 \end{pmatrix}$$

Since A is 2×3 and B is 3×4, the product is defined as a 2×4 matrix. Multiply the rows of A by the columns of B to obtain:

$$AB = \begin{pmatrix} 4+3-4 & -2+9-1 & 0-15+2 & 12+3-2 \\ 8-2+20 & -4-6+5 & 0+10-10 & 24-2+10 \end{pmatrix} = \begin{pmatrix} 3 & 6 & -13 & 13 \\ 26 & -5 & 0 & 32 \end{pmatrix}$$

3.9. Refer to Problem 3.8. Suppose that only the third column of the product AB were of interest. How could it be computed independently?

By the rule for matrix multiplication, the jth column of a product is equal to the first factor times the jth column vector of the second. Thus

$$\begin{pmatrix} 2 & 3 & -1 \\ 4 & -2 & 5 \end{pmatrix}\begin{pmatrix} 0 \\ -5 \\ -2 \end{pmatrix} = \begin{pmatrix} 0-15+2 \\ 0+10-10 \end{pmatrix} = \begin{pmatrix} -13 \\ 0 \end{pmatrix}$$

Similarly, the ith row of a product is equal to the ith row vector of the first factor times the second factor.

3.10. Let A be an $m \times n$ matrix, with $m > 1$ and $n > 1$. Assuming u and v are vectors, discuss the conditions under which (a) Au, (b) vA is defined.

(a) The product Au is defined only when u is a column vector with n components, i.e., an $n \times 1$ matrix. In such case, Au is a column vector with m components.

(b) The product vA is defined only when v is a row vector with m components, i.e., a $1 \times m$ matrix. In such case, vA is a row vector with n components.

3.11. Compute: (a) $\begin{pmatrix} 2 \\ 3 \\ -1 \end{pmatrix} (6 \quad -4 \quad 5)$ and (b) $(6 \quad -4 \quad 5) \begin{pmatrix} 2 \\ 3 \\ -1 \end{pmatrix}$.

(a) The first factor is 3×1 and the second factor is 1×3, so the product is defined as a 3×3 matrix:

$$\begin{pmatrix} 2 \\ 3 \\ -1 \end{pmatrix} (6 \quad -4 \quad 5) = \begin{pmatrix} (2)(6) & (2)(-4) & (2)(5) \\ (3)(6) & (3)(-4) & (3)(5) \\ (-1)(6) & (-1)(-4) & (-1)(5) \end{pmatrix} = \begin{pmatrix} 12 & -8 & 10 \\ 18 & -12 & 15 \\ -6 & 4 & -5 \end{pmatrix}$$

(b) The first factor is 1×3 and the second factor is 3×1, so the product is defined as a 1×1 matrix, which we frequently write as a scalar.

$$(6 \quad -4 \quad 5) \begin{pmatrix} 2 \\ 3 \\ -1 \end{pmatrix} = (12 - 12 - 5) = (-5) = -5$$

3.12. Prove Theorem 3.2(i): $(AB)C = A(BC)$.

Let $A = (a_{ij})$, $B = (b_{jk})$, and $C = (c_{kl})$. Furthermore, let $AB = S = (s_{ik})$ and $BC = T = (t_{jl})$. Then

$$s_{ik} = a_{i1}b_{1k} + a_{i2}b_{2k} + \cdots + a_{im}b_{mk} = \sum_{j=1}^{m} a_{ij}b_{jk}$$

$$t_{jl} = b_{j1}c_{1l} + b_{j2}c_{2l} + \cdots + b_{jn}c_{nl} = \sum_{k=1}^{n} b_{jk}c_{kl}$$

Now multiplying S by C, i.e., (AB) by C, the element in the ith row and lth column of the matrix $(AB)C$ is

$$s_{i1}c_{1l} + s_{i2}c_{2l} + \cdots + s_{in}c_{nl} = \sum_{k=1}^{n} s_{ik}c_{kl} = \sum_{k=1}^{n} \sum_{j=1}^{m} (a_{ij}b_{jk})c_{kl}$$

On the other hand, multiplying A by T, i.e., A by BC, the element in the ith row and lth column of the matrix $A(BC)$ is

$$a_{i1}t_{1l} + a_{i2}t_{2l} + \cdots + a_{im}t_{ml} = \sum_{j=1}^{m} a_{ij}t_{jl} = \sum_{j=1}^{m} \sum_{k=1}^{n} a_{ij}(b_{jk}c_{kl})$$

Since the above sums are equal, the theorem is proven.

3.13. Prove Theorem 3.2(ii): $A(B + C) = AB + AC$.

Let $A = (a_{ij})$, $B = (b_{jk})$, and $C = (c_{jk})$. Furthermore, let $D = B + C = (d_{jk})$, $E = AB = (e_{ik})$, and $F = AC = (f_{ik})$. Then

$$d_{jk} = b_{jk} + c_{jk}$$

$$e_{ik} = a_{i1}b_{1k} + a_{i2}b_{2k} + \cdots + a_{im}b_{mk} = \sum_{j=1}^{m} a_{ij}b_{jk}$$

$$f_{ik} = a_{i1}c_{1k} + a_{i2}c_{2k} + \cdots + a_{im}c_{mk} = \sum_{j=1}^{m} a_{ij}c_{jk}$$

Hence the element in the ith row and kth column of the matrix $AB + AC$ is

$$e_{ik} + f_{ik} = \sum_{j=1}^{m} a_{ij}b_{jk} + \sum_{j=1}^{m} a_{ij}c_{jk} = \sum_{j=1}^{m} a_{ij}(b_{jk} + c_{jk})$$

On the other hand, the element in the ith row and kth column of the matrix $AD = A(B + C)$ is

$$a_{i1}d_{1k} + a_{i2}d_{2k} + \cdots + a_{im}d_{mk} = \sum_{j=1}^{m} a_{ij}d_{jk} = \sum_{j=1}^{m} a_{ij}(b_{jk} + c_{jk})$$

Thus $A(B + C) = AB + AC$ since the corresponding elements are equal.

TRANSPOSE

3.14. Given $A = \begin{pmatrix} 1 & 3 & 5 \\ 6 & -7 & -8 \end{pmatrix}$, find A^T and $(A^T)^T$.

Rewrite the rows of A as columns to obtain A^T, and then rewrite the rows of A^T as columns to obtain $(A^T)^T$:

$$A^T = \begin{pmatrix} 1 & 6 \\ 3 & -7 \\ 5 & -8 \end{pmatrix} \qquad (A^T)^T = \begin{pmatrix} 1 & 3 & 5 \\ 6 & -7 & -8 \end{pmatrix}$$

[As expected from Theorem 3.3(ii), $(A^T)^T = A$.]

3.15. Show that the matrices AA^T and A^TA are defined for any matrix A.

If A is an $m \times n$ matrix, then A^T is an $n \times m$ matrix. Hence AA^T is defined as an $m \times m$ matrix, and A^TA is defined as an $n \times n$ matrix.

3.16. Find AA^T and A^TA, where $A = \begin{pmatrix} 1 & 2 & 0 \\ 3 & -1 & 4 \end{pmatrix}$.

Obtain A^T by rewriting the rows of A as columns:

$$A^T = \begin{pmatrix} 1 & 3 \\ 2 & -1 \\ 0 & 4 \end{pmatrix} \quad \text{whence} \quad AA^T = \begin{pmatrix} 1 & 2 & 0 \\ 3 & -1 & 4 \end{pmatrix}\begin{pmatrix} 1 & 3 \\ 2 & -1 \\ 0 & 4 \end{pmatrix} = \begin{pmatrix} 5 & 1 \\ 1 & 26 \end{pmatrix}$$

$$A^TA = \begin{pmatrix} 1 & 3 \\ 2 & -1 \\ 0 & 4 \end{pmatrix}\begin{pmatrix} 1 & 2 & 0 \\ 3 & -1 & 4 \end{pmatrix} = \begin{pmatrix} 1+9 & 2-3 & 0+12 \\ 2-3 & 4+1 & 0-4 \\ 0+12 & 0-4 & 0+16 \end{pmatrix} = \begin{pmatrix} 10 & -1 & 12 \\ -1 & 5 & -4 \\ 12 & -4 & 16 \end{pmatrix}$$

3.17. Prove Theorem 3.3(iv): $(AB)^T = B^T A^T$.

If $A = (a_{ij})$ and $B = (b_{kj})$, the ij-entry of AB is

$$a_{i1}b_{1j} + a_{i2}b_{2j} + \cdots + a_{im}b_{mj} \tag{1}$$

Thus (1) is the ji-entry (reverse order) of $(AB)^T$.

On the other hand, column j of B becomes row j of B^T, and row i of A becomes column i of A^T. Consequently, the ji-entry of $B^T A^T$ is

$$(b_{1j}, b_{2j}, \ldots, b_{mj})\begin{pmatrix} a_{i1} \\ a_{i2} \\ \cdots \\ a_{im} \end{pmatrix} = b_{1j}a_{i1} + b_{2j}a_{i2} + \cdots + b_{mj}a_{im}$$

Thus, $(AB)^T = B^T A^T$, since corresponding entries are equal.

BLOCK MATRICES

3.18. Compute AB using block multiplication, where

$$A = \begin{pmatrix} 1 & 2 & \vdots & 1 \\ 3 & 4 & \vdots & 0 \\ \hdashline 0 & 0 & \vdots & 2 \end{pmatrix} \quad \text{and} \quad B = \begin{pmatrix} 1 & 2 & 3 & \vdots & 1 \\ 4 & 5 & 6 & \vdots & 1 \\ \hdashline 0 & 0 & 0 & \vdots & 1 \end{pmatrix}$$

Here $A = \begin{pmatrix} E & F \\ 0_{1 \times 2} & G \end{pmatrix}$ and $B = \begin{pmatrix} R & S \\ 0_{1 \times 3} & T \end{pmatrix}$, where $E, F, G, R, S,$ and T are the given blocks. Hence

$$AB = \begin{pmatrix} ER & ES + FT \\ 0_{1 \times 3} & GT \end{pmatrix} = \begin{pmatrix} \begin{pmatrix} 9 & 12 & 15 \\ 19 & 26 & 33 \end{pmatrix} & \begin{pmatrix} 3 \\ 7 \end{pmatrix} + \begin{pmatrix} 1 \\ 0 \end{pmatrix} \\ (0 \quad 0 \quad 0) & (2) \end{pmatrix} = \begin{pmatrix} 9 & 12 & 15 & 4 \\ 19 & 26 & 33 & 7 \\ 0 & 0 & 0 & 2 \end{pmatrix}$$

3.19. Compute CD by block multiplication, where

$$C = \begin{pmatrix} 1 & 2 & \vdots & 0 & 0 & 0 \\ 3 & 4 & \vdots & 0 & 0 & 0 \\ \hdashline 0 & 0 & \vdots & 5 & 1 & 2 \\ 0 & 0 & \vdots & 3 & 4 & 1 \end{pmatrix} \quad \text{and} \quad D = \begin{pmatrix} 3 & -2 & \vdots & 0 & 0 \\ 2 & 4 & \vdots & 0 & 0 \\ \hdashline 0 & 0 & \vdots & 1 & 2 \\ 0 & 0 & \vdots & 2 & -1 \\ 0 & 0 & \vdots & -4 & 1 \end{pmatrix}$$

$$CD = \begin{pmatrix} \begin{pmatrix} 1 & 2 \\ 3 & 4 \end{pmatrix}\begin{pmatrix} 3 & -2 \\ 2 & 4 \end{pmatrix} & 0_{2 \times 2} \\ 0_{2 \times 2} & \begin{pmatrix} 5 & 1 & 2 \\ 3 & 4 & 1 \end{pmatrix}\begin{pmatrix} 1 & 2 \\ 2 & -3 \\ -4 & 1 \end{pmatrix} \end{pmatrix}$$

$$= \begin{pmatrix} \begin{pmatrix} 3+4 & -2+8 \\ 9+8 & -6+16 \end{pmatrix} & 0_{2 \times 2} \\ 0_{2 \times 2} & \begin{pmatrix} 5+2-8 & 10-3+2 \\ 3+8-4 & 6-12+1 \end{pmatrix} \end{pmatrix} = \begin{pmatrix} 7 & 6 & 0 & 0 \\ 17 & 10 & 0 & 0 \\ 0 & 0 & -1 & 9 \\ 0 & 0 & 7 & -5 \end{pmatrix}$$

MISCELLANEOUS PROBLEMS

3.20. Show: (a) If A has a zero row, then AB has a zero row. (b) If B has a zero column, then AB has a zero column.

(a) Let R_i be the zero row of A, and $B^1, \ldots, B^n$ the columns of B. Then the ith row of AB is

$$(R_i \cdot B^1, R_i \cdot B^2, \ldots, R_i \cdot B^n) = (0, 0, \ldots, 0)$$

(b) Let C_j be the zero column of B, and $A_1, \ldots, A_m$ the rows of A. Then the jth column of AB is

$$\begin{pmatrix} A_1 \cdot C_j \\ A_2 \cdot C_j \\ \ldots \\ A_m \cdot C_j \end{pmatrix} = \begin{pmatrix} 0 \\ 0 \\ \ldots \\ 0 \end{pmatrix}$$

3.21. Let u and v be distinct vectors. Show that, for each scalar $k \in K$, the vectors $u + k(u - v)$ are distinct.

It suffices to show that if

$$u + k_1(u - v) = u + k_2(u - v) \tag{1}$$

then $k_1 = k_2$. Suppose (*1*) holds. Then

$$k_1(u - v) = k_2(u - v) \qquad \text{or} \qquad (k_1 - k_2)(u - v) = 0$$

Since u and v are distinct, $u - v \neq 0$. Hence $k_1 - k_2 = 0$ and $k_1 = k_2$.

Supplementary Problems

MATRIX OPERATIONS

Problems 3.22–3.25 refer to the following matrices:

$$A = \begin{pmatrix} 1 & 2 \\ 3 & -4 \end{pmatrix} \qquad B = \begin{pmatrix} 5 & 0 \\ -6 & 7 \end{pmatrix} \qquad C = \begin{pmatrix} 1 & -3 & 4 \\ 2 & 6 & -5 \end{pmatrix}$$

3.22. Find $5A - 2B$ and $2A + 3B$.

3.23. Find: (*a*) AB and $(AB)C$, (*b*) BC and $A(BC)$. [Note $(AB)C = A(BC)$.]

3.24. Find A^T, B^T, and $A^T B^T$. [Note $A^T B^T \neq (AB)^T$.]

3.25. Find $AA = A^2$ and AC.

3.26. Suppose $e_1 = (1, 0, 0)$, $e_2 = (0, 0, 1)$, $e_3 = (0, 0, 1)$, and $A = \begin{pmatrix} a_1 & a_2 & a_3 & a_4 \\ b_1 & b_2 & b_3 & b_4 \\ c_1 & c_2 & c_3 & c_4 \end{pmatrix}$. Find $e_1 A$, $e_2 A$, and $e_3 A$.

3.27. Let $e_i = (0, \ldots, 0, 1, 0, \ldots, 0)$ where 1 is the ith component. Show the following:

    (*a*) $e_i A = A_i$, the ith row of a matrix A.

    (*b*) $B e_j^T = B^j$, the jth column of B.

    (*c*)   If $e_i A = e_i B$ for each i, then $A = B$.

    (*d*)   If $A e_j^T = B e_j^T$ for each j, then $A = B$.

3.28. Let $A = \begin{pmatrix} 1 & 2 \\ 3 & 6 \end{pmatrix}$. Find a 2×3 matrix B with distinct entries such that $AB = 0$.

3.29. Prove Theorem 3.2(iii): $(B + C)A = BA + CA$; (iv) $k(AB) = (kA)B = A(kB)$, where k is a scalar. [Parts (i) and (ii) were proven in Problems 3.12 and 3.13, respectively.]

3.30. Prove Theorem 3.3: (i) $(A + B)^T = A^T + B^T$; (ii) $(A^T)^T = A$; (iii) $(kA)^T = kA^T$, for k scalar. [Part (iv) was proven in Problem 3.17.]

3.31. Suppose $A = (A_{ik})$ and $B = (B_{kj})$ are block matrices for which AB is defined and the number of columns of each block A_{ik} is equal to the number of rows of each block B_{kj}. Show that

$$AB = (C_{ij}), \text{ where } C_{ij} = \sum_k A_{ik} B_{kj}.$$

Answers to Supplementary Problems

3.22. $\begin{pmatrix} -5 & 10 \\ 27 & -36 \end{pmatrix}, \begin{pmatrix} 17 & 4 \\ -12 & 13 \end{pmatrix}$

3.23. $(a) \begin{pmatrix} -7 & 14 \\ 39 & -28 \end{pmatrix}, \begin{pmatrix} 21 & 105 & -98 \\ -17 & -285 & 296 \end{pmatrix};$ $(b) \begin{pmatrix} 5 & -15 & 20 \\ 8 & 60 & -59 \end{pmatrix}, \begin{pmatrix} 21 & 105 & -98 \\ -17 & -285 & 296 \end{pmatrix}$

3.24. $\begin{pmatrix} 1 & 3 \\ 2 & -4 \end{pmatrix}, \begin{pmatrix} 5 & -6 \\ 0 & 7 \end{pmatrix}, \begin{pmatrix} 5 & 15 \\ 10 & -40 \end{pmatrix}$

3.25. $\begin{pmatrix} 7 & -6 \\ -9 & 22 \end{pmatrix}, \begin{pmatrix} 5 & 9 & -6 \\ -5 & -33 & 32 \end{pmatrix}$

3.26. $(a_1, a_2, a_3, a_4), (b_1, b_2, b_3, b_4), (c_1, c_2, c_3, c_4)$, the rows of A.

3.28. $\begin{pmatrix} 2 & 4 & 6 \\ -1 & -2 & -3 \end{pmatrix}$

Chapter 4

Square Matrices, Elementary Matrices

4.1 INTRODUCTION

Matrices with the same number of rows as columns are called *square matrices*. These matrices play a major role in linear algebra and will be used throughout the text. This chapter introduces us to these matrices and certain of their elementary properties.

This chapter also introduces us to the elementary matrices which are closely related to the elementary row operations in Chapter 1. We use these matrices to justify two algorithms—one which finds the inverse of a matrix, and the other which diagonalizes a quadratic form.

The scalars in this chapter are real numbers unless otherwise stated or implied. However, we will discuss the special case of complex matrices and some of their properties.

4.2 SQUARE MATRICES

A *square matrix* is a matrix with the same number of rows as columns. An $n \times n$ square matrix is said to be of *order n* and is called an *n-square matrix*.

Recall that not every two matrices can be added or multiplied. However, if we only consider square matrices of some given order n, then this inconvenience disappears. Specifically, the operations of addition, multiplication, scalar multiplication, and transpose can be performed on any $n \times n$ matrices and the result is again an $n \times n$ matrix.

Example 4.1. Let $A = \begin{pmatrix} 1 & 2 & 3 \\ -4 & -4 & -4 \\ 5 & 6 & 7 \end{pmatrix}$ and $B = \begin{pmatrix} 2 & -5 & 1 \\ 0 & 3 & -2 \\ 1 & 2 & -4 \end{pmatrix}$. Then A and B are square matrices of order 3.

Also,

$$A + B = \begin{pmatrix} 3 & -3 & 4 \\ -4 & -1 & -6 \\ 6 & 8 & 3 \end{pmatrix} \qquad 2A = \begin{pmatrix} 2 & 4 & 6 \\ -8 & -8 & -8 \\ 10 & 12 & 14 \end{pmatrix} \qquad A^T = \begin{pmatrix} 1 & -4 & 5 \\ 2 & -4 & 6 \\ 3 & -4 & 7 \end{pmatrix}$$

and

$$AB = \begin{pmatrix} 2+0+3 & -5+6+6 & 1-4-12 \\ -8+0-4 & 20-12-8 & -4+8+16 \\ 10+0+7 & -25+18+14 & 5-12-28 \end{pmatrix} = \begin{pmatrix} 5 & 7 & -15 \\ -12 & 0 & 20 \\ 17 & 7 & -35 \end{pmatrix}$$

are matrices of order 3.

> **Remark:** A nonempty collection **A** of matrices is called an *algebra* (of matrices) if **A** is closed under the operations of matrix addition, scalar multiplication of a matrix, and matrix multiplication. Thus the collection $\mathbf{M}_n$ of all *n*-square matrices forms an algebra of matrices.

Square Matrices as Functions

Let A be any *n*-square matrix. Then A may be viewed as a function $A : \mathbf{R}^n \to \mathbf{R}^n$ in two different ways:

(1) $A(u) = Au$ where u is a column vector;

(2) $A(u) = uA$ where u is a row vector.

This text adopts the first meaning of $A(u)$, that is, that the function defined by the matrix A will be $A(u) = Au$. Accordingly, unless otherwise stated or implied, vectors u in $\mathbf{R}^n$ are assumed to be column vectors (not row vectors). For typographical convenience, such column vectors u will often be presented as transposed row vectors.

Example 4.2. Let $A = \begin{pmatrix} 1 & -2 & 3 \\ 4 & 5 & -6 \\ 2 & 0 & -1 \end{pmatrix}$. If $u = (1, -3, 7)^T$, then

$$A(u) = Au = \begin{pmatrix} 1 & -2 & 3 \\ 4 & 5 & -6 \\ 2 & 0 & -1 \end{pmatrix} \begin{pmatrix} 1 \\ -3 \\ 7 \end{pmatrix} = \begin{pmatrix} 1 + 6 + 21 \\ 4 - 15 - 42 \\ 2 + 0 - 7 \end{pmatrix} = \begin{pmatrix} 28 \\ -53 \\ -5 \end{pmatrix}$$

If $w = (2, -1, 4)^T$, then

$$A(w) = Aw = \begin{pmatrix} 1 & -2 & 3 \\ 4 & 5 & -6 \\ 2 & 0 & -1 \end{pmatrix} \begin{pmatrix} 2 \\ -1 \\ 4 \end{pmatrix} = \begin{pmatrix} 2 + 2 + 12 \\ 8 - 5 - 24 \\ 4 + 0 - 4 \end{pmatrix} = \begin{pmatrix} 16 \\ -21 \\ 0 \end{pmatrix}$$

Warning! Some texts define the above function A by

$$A(u) = uA$$

where the vectors u in $\mathbf{R}^n$ are assumed to be row vectors. Clearly, if one only works with vectors in $\mathbf{R}^n$ and not matrices, then it does not matter which way the vectors are defined.

Commuting Matrices

Matrices A and B are said to *commute* if $AB = BA$, a condition that applies only for square matrices of the same order. For example, suppose

$$A = \begin{pmatrix} 1 & 2 \\ 3 & 4 \end{pmatrix} \quad \text{and} \quad B = \begin{pmatrix} 5 & 4 \\ 6 & 11 \end{pmatrix}$$

Then

$$AB = \begin{pmatrix} 5 + 12 & 4 + 22 \\ 15 + 24 & 12 + 44 \end{pmatrix} = \begin{pmatrix} 17 & 26 \\ 39 & 56 \end{pmatrix} \quad \text{and} \quad BA = \begin{pmatrix} 5 + 12 & 10 + 16 \\ 6 + 33 & 12 + 44 \end{pmatrix} = \begin{pmatrix} 17 & 26 \\ 39 & 56 \end{pmatrix}$$

Since $AB = BA$, the matrices commute.

4.3 DIAGONAL AND TRACE, IDENTITY MATRIX

Let $A = (a_{ij})$ be an n-square matrix. The *diagonal* (or *main diagonal*) of A consists of the elements $a_{11}, a_{22}, \ldots, a_{nn}$. The *trace* of A, written tr A, is the sum of the diagonal elements, that is,

$$\text{tr } A = a_{11} + a_{22} + \cdots + a_{nn} \equiv \sum_{i=1}^{n} a_{ii}$$

The n-square matrix with 1's on the diagonal and 0's elsewhere, denoted by I_n or simply I, is called the *identity* (or *unit*) matrix. The matrix I is similar to the scalar 1 in that, for any matrix A (of the same order),

$$AI = IA = A$$

More generally, if B is an $m \times n$ matrix, then $BI_n = B$ and $I_m B = B$ (Problem 4.9).

For any scalar $k \in K$, the matrix kI which contains k's on the diagonal and 0's elsewhere is called the *scalar matrix* corresponding to the scalar k.

Example 4.3.

(a) The Kronecker delta δ_{ij} is defined by

$$\delta_{ij} = \begin{cases} 0 & \text{if } i \neq j \\ 1 & \text{if } i = j \end{cases}$$

Thus the identity matrix may be defined by $I = (\delta_{ij})$.

(b) The scalar matrices of orders 2, 3, and 4 corresponding to the scalar $k = 5$ are, respectively,

$$\begin{pmatrix} 5 & 0 \\ 0 & 5 \end{pmatrix} \qquad \begin{pmatrix} 5 & 0 & 0 \\ 0 & 5 & 0 \\ 0 & 0 & 5 \end{pmatrix} \qquad \begin{pmatrix} 5 & & & \\ & 5 & & \\ & & 5 & \\ & & & 5 \end{pmatrix}$$

(It is common practice to omit blocks or patterns of 0's as in the third matrix.)

The following theorem is proved in Problem 4.10.

Theorem 4.1: Suppose $A = (a_{ij})$ and $B = (b_{ij})$ are n-square matrices and k is a scalar. Then

(i) $\operatorname{tr}(A + B) = \operatorname{tr} A + \operatorname{tr} B$,   (ii) $\operatorname{tr} kA = k \cdot \operatorname{tr} A$,   (iii) $\operatorname{tr} AB = \operatorname{tr} BA$

4.4 POWERS OF MATRICES, POLYNOMIALS, AND MATRICES

Let A be an n-square matrix over a field K. Powers of A are defined as follows:

$$A^2 = AA \qquad A^3 = A^2 A, \ldots, A^{n+1} = A^n A, \ldots \qquad \text{and} \qquad A^0 = I$$

Polynomials in the matrix A are also defined. Specifically, for any polynomial

$$f(x) = a_0 + a_1 x + a_2 x^2 + \cdots + a_n x^n$$

where the a_i are scalars, $f(A)$ is defined to be the matrix

$$f(A) = a_0 I + a_1 A + a_2 A^2 + \cdots + a_n A^n$$

[Note that $f(A)$ is obtained from $f(x)$ by substituting the matrix A for the variable x and substituting the scalar matrix $a_0 I$ for the scalar a_0.] In the case that $f(A)$ is the zero matrix, the matrix A is called a *zero* or *root* of the polynomial $f(x)$.

Example 4.4. Let $A = \begin{pmatrix} 1 & 2 \\ 3 & -4 \end{pmatrix}$. Then

$$A^2 = \begin{pmatrix} 1 & 2 \\ 3 & -4 \end{pmatrix}\begin{pmatrix} 1 & 2 \\ 3 & -4 \end{pmatrix} = \begin{pmatrix} 7 & -6 \\ -9 & 22 \end{pmatrix} \quad \text{and} \quad A^3 = A^2 A = \begin{pmatrix} 7 & -6 \\ -9 & 22 \end{pmatrix}\begin{pmatrix} 1 & 2 \\ 3 & -4 \end{pmatrix} = \begin{pmatrix} -11 & 38 \\ 57 & -106 \end{pmatrix}$$

If $f(x) = 2x^2 - 3x + 5$, then

$$f(A) = 2\begin{pmatrix} 7 & -6 \\ -9 & 22 \end{pmatrix} - 3\begin{pmatrix} 1 & 2 \\ 3 & -4 \end{pmatrix} + 5\begin{pmatrix} 1 & 0 \\ 0 & 1 \end{pmatrix} = \begin{pmatrix} 16 & -18 \\ -27 & 61 \end{pmatrix}$$

If $g(x) = x^2 + 3x - 10$, then

$$g(A) = \begin{pmatrix} 7 & -6 \\ -9 & 22 \end{pmatrix} + 3\begin{pmatrix} 1 & 2 \\ 3 & -4 \end{pmatrix} - 10\begin{pmatrix} 1 & 0 \\ 0 & 1 \end{pmatrix} = \begin{pmatrix} 0 & 0 \\ 0 & 0 \end{pmatrix}$$

Thus A is a zero of the polynomial $g(x)$.

The above map from the ring $K[x]$ of polynomials over K into the algebra $\mathbf{M}_n$ of n-square matrices defined by

$$f(x) \to f(A)$$

is called the *evaluation map at A*.

The following theorem (proved in Problem 4.11) applies.

Theorem 4.2:   Let $f(x)$ and $g(x)$ be polynomials and let A be an n-square matrix (all over K). Then

(i) $(f + g)(A) = f(A) + g(A)$,

(ii) $(fg)(A) = f(A)g(A)$,

(iii) $f(A)g(A) = g(A)f(A)$.

In other words, (i) and (ii) state that if we first add (multiply) the polynomials $f(x)$ and $g(x)$ and then evaluate the sum (product) at the matrix A, we get the same result as if we first evaluated $f(x)$ and $g(x)$ at A and then added (multiplied) the matrices $f(A)$ and $g(A)$. Part (iii) states that any two polynomials in A commute.

4.5   INVERTIBLE (NONSINGULAR) MATRICES

A square matrix A is said to be *invertible* (or *nonsingular*) if there exists a matrix B with the property that

$$AB = BA = I$$

where I is the identity matrix. Such a matrix B is unique; for

$$AB_1 = B_1 A = I \text{ and } AB_2 = B_2 A = I \quad \text{implies} \quad B_1 = B_1 I = B_1(AB_2) = (B_1 A)B_2 = IB_2 = B_2$$

We call such a matrix B the *inverse* of A and denote it by A^{-1}. Observe that the above relation is symmetric; that is, if B is the inverse of A, then A is the inverse of B.

Example 4.5

(a)   Suppose $A = \begin{pmatrix} 2 & 5 \\ 1 & 3 \end{pmatrix}$ and $B = \begin{pmatrix} 3 & -5 \\ -1 & 2 \end{pmatrix}$. Then

$$AB = \begin{pmatrix} 2 & 5 \\ 1 & 3 \end{pmatrix}\begin{pmatrix} 3 & -5 \\ -1 & 2 \end{pmatrix} = \begin{pmatrix} 6-5 & -10+10 \\ 3-3 & -5+6 \end{pmatrix} = \begin{pmatrix} 1 & 0 \\ 0 & 1 \end{pmatrix} = I$$

$$BA = \begin{pmatrix} 3 & -5 \\ -1 & 2 \end{pmatrix}\begin{pmatrix} 2 & 5 \\ 1 & 3 \end{pmatrix} = \begin{pmatrix} 6-5 & 15-15 \\ -2+2 & -5+6 \end{pmatrix} = \begin{pmatrix} 1 & 0 \\ 0 & 1 \end{pmatrix} = I$$

Thus A and B are invertible and are inverses of each other.

(b)   Suppose $A = \begin{pmatrix} 1 & 0 & 2 \\ 2 & -1 & 3 \\ 4 & 1 & 8 \end{pmatrix}$ and $B = \begin{pmatrix} -11 & 2 & 2 \\ -4 & 0 & 1 \\ 6 & -1 & -1 \end{pmatrix}$. Then

$$AB = \begin{pmatrix} -11+0+12 & 2+0-2 & 2+0-2 \\ -22+4+18 & 4+0-3 & 4-1-3 \\ -44-4+48 & 8+0-8 & 8+1-8 \end{pmatrix} = \begin{pmatrix} 1 & 0 & 0 \\ 0 & 1 & 0 \\ 0 & 0 & 1 \end{pmatrix} = I$$

By Problem 4.21, $AB = I$ if and only if $BA = I$; hence we do not need to test if $BA = I$. Thus A and B are inverses of each other.

Consider now a general 2×2 matrix

$$A = \begin{pmatrix} a & b \\ c & d \end{pmatrix}$$

We are able to determine when A is invertible and, in such a case, to give a formula for its inverse. First of all, we seek scalars x, y, z, t such that

$$\begin{pmatrix} a & b \\ c & d \end{pmatrix}\begin{pmatrix} x & y \\ z & t \end{pmatrix} = \begin{pmatrix} 1 & 0 \\ 0 & 1 \end{pmatrix} \qquad \text{or} \qquad \begin{pmatrix} ax + bz & ay + bt \\ cx + dz & cy + dt \end{pmatrix} = \begin{pmatrix} 1 & 0 \\ 0 & 1 \end{pmatrix}$$

which reduces to solving the following two systems

$$\begin{cases} ax + bz = 1 \\ cx + dz = 0 \end{cases} \qquad \begin{cases} ay + bt = 0 \\ cy + dt = 1 \end{cases}$$

where the original matrix A is the coefficient matrix of each system. Set $|A| = ad - bc$ (the determinant of A). By Problems 1.60 and 1.61, the two systems are solvable, and A is invertible, if and only if $|A| \neq 0$. In that case, the first system has the unique solution $x = d/|A|$, $z = -c/|A|$, and the second system has the unique solution $y = -b/|A|$, $t = a/|A|$. Accordingly,

$$A^{-1} = \begin{pmatrix} d/|A| & -b/|A| \\ -c/|A| & a/|A| \end{pmatrix} = \frac{1}{|A|}\begin{pmatrix} d & -b \\ -c & a \end{pmatrix}$$

In words: When $|A| \neq 0$, the inverse of a 2×2 matrix A is obtained by (i) interchanging the elements on the main diagonal, (ii) taking the negatives of the other elements, and (iii) multiplying the matrix by $1/|A|$.

Remark 1: The above property that A is invertible if and only if its determinant $|A| \neq 0$ is true for square matrices of any order. (See Chapter 7.)

Remark 2: Suppose A and B are invertible, then AB is invertible and $(AB)^{-1} = B^{-1}A^{-1}$. More generally, if $A_1, A_2, \ldots, A_k$ are invertible, then their product is invertible and

$$(A_1 A_2 \cdots A_k)^{-1} = A_k^{-1} \cdots A_2^{-1} A_1^{-1}$$

the product of the inverses in the reverse order.

4.6 SPECIAL TYPES OF SQUARE MATRICES

This section describes a number of special kinds of square matrices which play an important role in linear algebra.

Diagonal Matrices

A square matrix $D = (d_{ij})$ is *diagonal* if its nondiagonal entries are all zero. Such a matrix is frequently notated as $D = \text{diag}(d_{11}, d_{22}, \ldots, d_{nn})$, where some or all of the d_{ii} may be zero. For example,

$$\begin{pmatrix} 3 & 0 & 0 \\ 0 & -7 & 0 \\ 0 & 0 & 2 \end{pmatrix} \qquad \begin{pmatrix} 4 & 0 \\ 0 & -5 \end{pmatrix} \qquad \begin{pmatrix} 6 & & \\ & 0 & \\ & & -9 \\ & & & 1 \end{pmatrix}$$

are diagonal matrices which may be represented, respectively, by

$$\text{diag}(3, -7, 2) \qquad \text{diag}(4, -5) \qquad \text{and} \qquad \text{diag}(6, 0, -9, 1)$$

(Observe that patterns of 0s in the third matrix have been omitted.)

Clearly, the sum, scalar product, and product of diagonal matrices are again diagonal. Thus all the $n \times n$ diagonal matrices form an algebra of matrices. In fact, the diagonal matrices form a commutative algebra since any two $n \times n$ diagonal matrices commute.

Triangular Matrices

A square matrix $A = (a_{ij})$ is an *upper triangular matrix* or simply a *triangular matrix* if all entries below the main diagonal are equal to zero; that is, if $s_{ij} = 0$ for $i > j$. Generic upper triangular matrices of orders 2, 3, and 4 are, respectively,

$$
\begin{pmatrix} a_{11} & a_{12} \\ 0 & a_{22} \end{pmatrix}
\qquad
\begin{pmatrix} b_{11} & b_{12} & b_{13} \\ & b_{22} & b_{23} \\ & & b_{33} \end{pmatrix}
\qquad
\begin{pmatrix} c_{11} & c_{12} & c_{13} & c_{14} \\ & c_{22} & c_{23} & c_{24} \\ & & c_{33} & c_{34} \\ & & & c_{44} \end{pmatrix}
$$

(As in diagonal matrices, it is common practice to omit patterns of 0s.)

The upper triangular matrices also form an algebra of matrices. In fact,

Theorem 4.3:   Suppose $A = (a_{ij})$ and $B = (b_{ij})$ are upper triangular matrices. Then

     (i) $A + B$ is upper triangular, with diag $(a_{11} + b_{11}, a_{22} + b_{22}, \ldots, a_{nn} + b_{nn})$.

     (ii) kA is upper triangular, with diag $(ka_{11}, ka_{22}, \ldots, ka_{nn})$.

     (iii) AB is upper triangular, with diag $(a_{11}b_{11}, a_{22}b_{22}, \ldots, a_{nn}b_{nn})$.

     (iv)   For any polynomial $f(x)$, the matrix $F(A)$ is upper triangular with

$$\text{diag}(f(a_{11}), f(a_{22}), \ldots, f(a_{nn})).$$

     (v) A is invertible if and only if each diagonal element $a_{ii} \neq 0$.

Analogously, a *lower triangular matrix* is a square matrix whose entries above the diagonal are all zero, and a theorem analogous to Theorem 4.3 holds for such matrices.

Symmetric Matrices

A real matrix A is *symmetric* if $A^T = A$. Equivalently, $A = (a_{ij})$ is symmetric if symmetric elements (mirror images in the diagonal) are equal, i.e., if each $a_{ij} = a_{ji}$. (Note that A must be square in order for $A^T = A$.)

A real matrix A is *skew-symmetric* if $A^T = -A$. Equivalently, $A = (a_{ij})$ is skew-symmetric if each $a_{ij} = -a_{ji}$. Clearly, the diagonal elements of a skew-symmetric matrix must be zero since $a_{ii} = -a_{ii}$ implies $a_{ii} = 0$.

Example 4.6.   Consider the following matrices:

$$
A = \begin{pmatrix} 2 & -3 & 5 \\ -3 & 6 & 7 \\ 5 & 7 & -8 \end{pmatrix}
\qquad
B = \begin{pmatrix} 0 & 3 & -4 \\ -3 & 0 & 5 \\ 4 & -5 & 0 \end{pmatrix}
\qquad
C = \begin{pmatrix} 1 & 0 & 0 \\ 0 & 0 & 1 \end{pmatrix}
$$

(a)   By inspection, the symmetric elements in A are equal, or $A^T = A$. Thus A is symmetric.

(b) By inspection, the diagonal elements of B are 0 and symmetric elements are negatives of each other. Thus B is skew-symmetric.

(c) Since C is not square, C is neither symmetric nor skew-symmetric.

If A and B are symmetric matrices, then $A + B$ and kA are symmetric. However, AB need not be symmetric. For example,

$$A = \begin{pmatrix} 1 & 2 \\ 2 & 3 \end{pmatrix} \text{ and } B = \begin{pmatrix} 4 & 5 \\ 5 & 6 \end{pmatrix} \text{ are symmetric, but } AB = \begin{pmatrix} 14 & 17 \\ 23 & 28 \end{pmatrix} \text{ is not symmetric.}$$

Thus the symmetric matrices do not form an algebra of matrices.

The following theorem is proved in Problem 4.29.

Theorem 4.4: If A is a square matrix, then (i) $A + A^T$ is symmetric; (ii) $A - A^T$ is skew-symmetric; (iii) $A = B + C$, for some symmetric matrix B and some skew-symmetric matrix C.

Orthogonal Matrices

A real matrix A is said to be *orthogonal* if $AA^T = A^TA = I$. Observe that an orthogonal matrix A is necessarily square and invertible, with inverse $A^{-1} = A^T$.

Example 4.7. Let $A = \begin{pmatrix} \frac{1}{9} & \frac{8}{9} & -\frac{4}{9} \\ \frac{4}{9} & -\frac{4}{9} & -\frac{7}{9} \\ \frac{8}{9} & \frac{1}{9} & \frac{4}{9} \end{pmatrix}$. Then

$$AA^T = \begin{pmatrix} \frac{1}{9} & \frac{8}{9} & -\frac{4}{9} \\ \frac{4}{9} & -\frac{4}{9} & -\frac{7}{9} \\ \frac{8}{9} & \frac{1}{9} & \frac{4}{9} \end{pmatrix}\begin{pmatrix} \frac{1}{9} & \frac{4}{9} & \frac{8}{9} \\ \frac{8}{9} & -\frac{4}{9} & \frac{1}{9} \\ -\frac{4}{9} & -\frac{7}{9} & \frac{4}{9} \end{pmatrix} = \frac{1}{81}\begin{pmatrix} 1+64+16 & 4-32+28 & 8+8-16 \\ 4-32+28 & 16+16+49 & 32-4-28 \\ 8+8-16 & 32-4-28 & 64+1+16 \end{pmatrix}$$

$$= \frac{1}{81}\begin{pmatrix} 81 & 0 & 0 \\ 0 & 81 & 0 \\ 0 & 0 & 81 \end{pmatrix} = \begin{pmatrix} 1 & 0 & 0 \\ 0 & 1 & 0 \\ 0 & 0 & 1 \end{pmatrix} = I$$

This means $A^TA = I$ and $A^T = A^{-1}$. Thus A is orthogonal.

Consider now an arbitrary 3×3 matrix

$$A = \begin{pmatrix} a_1 & a_2 & a_3 \\ b_1 & b_2 & b_3 \\ c_1 & c_2 & c_3 \end{pmatrix}$$

If A is orthogonal, then

$$AA^T = \begin{pmatrix} a_1 & a_2 & a_3 \\ b_1 & b_2 & b_3 \\ c_1 & c_2 & c_3 \end{pmatrix}\begin{pmatrix} a_1 & b_1 & c_1 \\ a_2 & b_2 & c_2 \\ a_3 & b_3 & c_3 \end{pmatrix} = \begin{pmatrix} 1 & 0 & 0 \\ 0 & 1 & 0 \\ 0 & 0 & 1 \end{pmatrix} = I$$

This yields

$$\begin{array}{lll} a_1^2 + a_2^2 + a_3^2 = 1 & a_1b_1 + a_2b_2 + a_3b_3 = 0 & a_1c_1 + a_2c_2 + a_3c_3 = 0 \\ b_1a_1 + b_2a_2 + b_3a_3 = 0 & b_1^2 + b_2^2 + b_3^2 = 1 & b_1c_1 + b_2c_2 + b_3c_3 = 0 \\ c_1a_1 + c_2a_2 + c_3a_3 = 0 & c_1b_1 + c_2b_2 + c_3b_3 = 0 & c_1^2 + c_2^2 + c_3^2 = 1 \end{array}$$

or, in other words,

$$\begin{array}{lll} u_1 \cdot u_1 = 1 & u_1 \cdot u_2 = 0 & u_1 \cdot u_3 = 0 \\ u_2 \cdot u_1 = 0 & u_2 \cdot u_2 = 1 & u_2 \cdot u_3 = 0 \\ u_3 \cdot u_1 = 0 & u_3 \cdot u_2 = 0 & u_3 \cdot u_3 = 1 \end{array}$$

where $u_1 = (a_1, a_2, a_3)$, $u_2 = (b_1, b_2, b_3)$, $u_3 = (c_1, c_2, c_3)$ are the rows of A. Thus the rows u_1, u_2, and u_3 are orthogonal to each other and have unit lengths, or, in other words, u_1, u_2, u_3 form an *orthonormal set of vectors*. The condition $A^T A = I$ similarly shows that the columns of A form an orthonormal set of vectors. Furthermore, since each step is reversible, the converse is true.

The above result for 3×3 matrices is true in general. That is,

Theorem 4.5:   Let A be a real matrix. Then the following are equivalent: (a) A is orthogonal; (b) the rows of A form an orthonormal set; (c) the columns of A form an orthonormal set.

For $n = 2$, we have the following result, proved in Problem 4.33.

Theorem 4.6:   Every 2×2 orthogonal matrix has the form $\begin{pmatrix} \cos\theta & \sin\theta \\ -\sin\theta & \cos\theta \end{pmatrix}$ or $\begin{pmatrix} \cos\theta & \sin\theta \\ \sin\theta & -\cos\theta \end{pmatrix}$ for some real number θ.

> **Remark:**   The condition that vectors $u_1, u_2, \ldots, u_m$ form an orthonormal set may be described simply by $u_i \cdot u_k = \delta_{ij}$, where δ_{ij} is the Kronecker delta [Example 4.3(a)].

Normal Matrices

A real matrix A is *normal* if A commutes with its transpose, that is, if $AA^T = A^T A$. Clearly, if A is symmetric, orthogonal or skew-symmetric, then A is normal. These, however, are not the only normal matrices.

Example 4.8.   Let $A = \begin{pmatrix} 6 & -3 \\ 3 & 6 \end{pmatrix}$. Then

$$AA^T = \begin{pmatrix} 6 & -3 \\ 3 & 6 \end{pmatrix}\begin{pmatrix} 6 & 3 \\ -3 & 6 \end{pmatrix} = \begin{pmatrix} 45 & 0 \\ 0 & 45 \end{pmatrix} \quad \text{and} \quad A^T A = \begin{pmatrix} 6 & 3 \\ -3 & 6 \end{pmatrix}\begin{pmatrix} 6 & -3 \\ 3 & 6 \end{pmatrix} = \begin{pmatrix} 45 & 0 \\ 0 & 45 \end{pmatrix}$$

Since $AA^T = A^T A$, the matrix A is normal.

The following theorem, proved in Problem 4.35, completely characterizes real 2×2 normal matrices.

Theorem 4.7:   Let A be a real 2×2 normal matrix. Then A is either symmetric or the sum of a diagonal matrix and a skew-symmetric matrix.

4.7   COMPLEX MATRICES

Let A be a complex matrix, i.e., a matrix whose entries are complex numbers. Recall (Section 2.9) that if $z = a + bi$ is a complex number, then $\bar{z} = a - bi$ is its conjugate. The conjugate of the complex matrix A, written $\bar{A}$, is the matrix obtained from A by taking the conjugate of each entry in A, that is, if $A = (a_{ij})$ then $\bar{A} = (b_{ij})$ where $b_{ij} = \overline{a_{ij}}$. [We denote this fact by writing $\bar{A} = (\overline{a_{ij}})$.]

The two operations of transpose and conjugation commute for any complex matrix A, that is, $(\bar{A})^T = \overline{(A^T)}$. In fact, the special notation A^H is used for the conjugate transpose of A. (Note that if A is real then $A^H = A^T$.)

Example 4.9.   Let $A = \begin{pmatrix} 2 + 8i & 5 - 3i & 4 - 7i \\ 6i & 1 - 4i & 3 + 2i \end{pmatrix}$. Then

$$A^H = \begin{pmatrix} \overline{2 + 8i} & \overline{6i} \\ \overline{5 - 3i} & \overline{1 - 4i} \\ \overline{4 - 7i} & \overline{3 + 2i} \end{pmatrix} = \begin{pmatrix} 2 - 8i & -6i \\ 5 + 3i & 1 + 4i \\ 4 + 7i & 3 - 2i \end{pmatrix}$$

Hermitian, Unitary, and Normal Complex Matrices

A square complex matrix A is said to be *Hermitian* or *skew-Hermitian* according as

$$A^H = A \qquad \text{or} \qquad A^H = -A$$

If $A = (a_{ij})$ is Hermitian, then $a_{ij} = \overline{a_{ji}}$ and hence each diagonal element a_{ii} must be real. Similarly, if A is skew-Hermitian then each diagonal element $a_{ii} = 0$.

A square complex matrix A is said to be *unitary* if

$$A^H = A^{-1}$$

A complex matrix A is unitary if and only if its rows (columns) form an orthonormal set of vectors relative to the inner product of complex vectors. (See Problem 4.39.)

Note that when A is real, Hermitian is the same as symmetric, and unitary is the same as orthogonal.

A square complex matrix A is said to be *normal* if

$$AA^H = A^H A$$

This definition reduces to the one for real matrices when A is real.

Example 4.10.   Consider the following matrices:

$$A = \frac{1}{2}\begin{pmatrix} 1 & -i & -1 + i \\ i & 1 & 1 + i \\ 1 + i & -1 + i & 0 \end{pmatrix} \qquad B = \begin{pmatrix} 3 & 1 - 2i & 4 + 7i \\ 1 + 2i & -4 & -2i \\ 4 - 7i & 2i & 2 \end{pmatrix} \qquad C = \begin{pmatrix} 2 + 3i & 1 \\ i & 1 + 2i \end{pmatrix}$$

(a) A is unitary if $A^H = A^{-1}$ or if $AA^H = A^H A = I$. As noted previously, we need only show that $AA^H = I$:

$$AA^H = A\bar{A}^T = \frac{1}{4}\begin{pmatrix} 1 & -i & -1 + i \\ i & 1 & 1 + i \\ 1 + i & -1 + i & 0 \end{pmatrix}\begin{pmatrix} 1 & -i & 1 - i \\ i & 1 & -1 - i \\ -1 - i & 1 - i & 0 \end{pmatrix}$$

$$= \frac{1}{4}\begin{pmatrix} 1 + 1 + 2 & -i - i + 2i & 1 - i + i - 1 + 0 \\ i + i - 2i & 1 + 1 + 2 & i + 1 - 1 - i \\ 1 + i - i - 1 + 0 & -i + 1 - 1 + i + 0 & 2 + 2 + 0 \end{pmatrix} = \begin{pmatrix} 1 & 0 & 0 \\ 0 & 1 & 0 \\ 0 & 0 & 1 \end{pmatrix} = I$$

Accordingly, A is unitary.

(b) B is Hermitian since its diagonal elements 3, -4, and 2 are real, and the symmetric elements, $1 - 2i$ and $1 + 2i$, $4 + 7i$ and $4 - 7i$, and $-2i$ and $2i$, are conjugates.

(c)   To show that C is normal, evaluate CC^H and $C^H C$:

$$CC^H = C\bar{C}^T = \begin{pmatrix} 2 + 3i & 1 \\ i & 1 + 2i \end{pmatrix}\begin{pmatrix} 2 - 3i & -i \\ 1 & 1 - 2i \end{pmatrix} = \begin{pmatrix} 14 & 4 - 4i \\ 4 + 4i & 6 \end{pmatrix}$$

$$C^H C = \bar{C}^T C = \begin{pmatrix} 2 - 3i & -i \\ 1 & 1 - 2i \end{pmatrix}\begin{pmatrix} 2 + 3i & 1 \\ i & 1 + 2i \end{pmatrix} = \begin{pmatrix} 14 & 4 - 4i \\ 4 + 4i & 6 \end{pmatrix}$$

Since $CC^H = C^H C$, the complex matrix C is normal.

4.8 SQUARE BLOCK MATRICES

A block matrix A is called a *square block matrix* if (i) A is a square matrix, (ii) the blocks form a square matrix, and (iii) the diagonal blocks are also square matrices. The latter two conditions will occur if and only if there are the same number of horizontal and vertical lines and they are placed symmetrically.

Consider the following two block matrices:

$$A = \begin{pmatrix} 1 & 2 & 3 & 4 & 5 \\ 1 & 1 & 1 & 1 & 1 \\ 9 & 8 & 7 & 6 & 5 \\ 4 & 4 & 4 & 4 & 4 \\ 3 & 5 & 3 & 5 & 3 \end{pmatrix} \qquad B = \begin{pmatrix} 1 & 2 & 3 & 4 & 5 \\ 1 & 1 & 1 & 1 & 1 \\ 9 & 8 & 7 & 6 & 5 \\ 4 & 4 & 4 & 4 & 4 \\ 3 & 5 & 3 & 5 & 3 \end{pmatrix}$$

Block matrix A is not a square block matrix since the second and third diagonal blocks are not square matrices. On the other hand, block matrix B is a square block matrix.

A *block diagonal matrix* M is a square block matrix where the nondiagonal blocks are all zero matrices. The importance of block diagonal matrices is that the algebra of the block matrix is frequently reduced to the algebra of the individual blocks. Specifically, suppose M is a block diagonal matrix and $f(x)$ is any polynomial. Then M and $f(M)$ have the following form:

$$M = \begin{pmatrix} A_{11} & & & \\ & A_{22} & & \\ & & \cdots & \\ & & & A_{rr} \end{pmatrix} \qquad f(M) = \begin{pmatrix} f(A_{11}) & & & \\ & f(A_{22}) & & \\ & & \cdots & \\ & & & f(A_{rr}) \end{pmatrix}$$

(As usual, we use blank spaces for patterns of zeros or zero blocks.)

Analogously, a square block matrix is called a *block upper triangular matrix* if the blocks below the diagonal are zero matrices, and a *block lower triangular matrix* if the blocks above the diagonal are zero matrices.

4.9 ELEMENTARY MATRICES AND APPLICATIONS

First recall (Section 1.8) the following operations on a matrix A, called *elementary row operations*:

[E_1] (Row-interchange) Interchange the ith row and the jth row:

$$R_i \leftrightarrow R_j$$

[E_2] (Row-scaling) Multiply the ith row by a nonzero scalar k:

$$kR_i \rightarrow R_i \quad k \neq 0$$

[E_3] (Row-addition) Replace the ith row by k times the jth row plus the ith row:

$$kR_j + R_i \rightarrow R_i$$

Each of the above operations has an inverse operation of the same type. Specifically (Problem 4.19):

(1) $R_j \rightarrow R_i$ is its own inverse.

(2) $kR_i \rightarrow R_i$ and $k^{-1}R_i \rightarrow R_i$ are inverses.

(3) $kR_j + R_i \rightarrow R_i$ and $-kR_j + R_i \rightarrow R_i$ are inverses.

Also recall (Section 1.8) that a matrix B is said to be *row equivalent* to a matrix A, written $A \sim B$, if B can be obtained from A by a finite sequence of elementary row operations. Since the elementary row operations are reversible, row equivalence is an equivalence relation; that is, (a) $A \sim A$; (b) if $A \sim B$, then

$B \sim A$; (c) if $A \sim B$ and $B \sim C$, then $A \sim C$. We also restate the following basic result on row equivalence:

Theorem 4.8:   Every matrix A is row equivalent to a unique matrix in row canonical form.

Elementary Matrices

Let e denote an elementary row operation and let $e(A)$ denote the result of applying the operation e on a matrix A. The matrix E obtained by applying e to the identity matrix,

$$E = e(I)$$

is called the *elementary matrix* corresponding to the elementary row operation e.

Example 4.11.   The 3-square elementary matrices corresponding to the elementary row operations $R_2 \leftrightarrow R_3$, $-6R_2 \rightarrow R_2$ and $-4R_1 + R_3 \rightarrow R_3$ are, respectively,

$$E_1 = \begin{pmatrix} 1 & 0 & 0 \\ 0 & 0 & 1 \\ 0 & 1 & 0 \end{pmatrix} \qquad E_2 = \begin{pmatrix} 1 & 0 & 0 \\ 0 & -6 & 0 \\ 0 & 0 & 1 \end{pmatrix} \qquad E_3 = \begin{pmatrix} 1 & 0 & 0 \\ 0 & 1 & 0 \\ -4 & 0 & 1 \end{pmatrix}$$

The following theorem, proved in Problem 4.18, shows the fundamental relationship between the elementary row operations and their corresponding elementary matrices.

Theorem 4.9:   Let e be an elementary row operation and E the corresponding m-square elementary matrix, i.e., $E = e(I_m)$. Then, for any $m \times n$ matrix A, $e(A) = EA$.

That is, the result of applying an elementary row operation e on a matrix A can be obtained by premultiplying A by the corresponding elementary matrix E.

Now suppose e' is the inverse of an elementary row operation e. Let E' and E be the corresponding matrices. We prove in Problem 4.19 that E is invertible and E' is its inverse. This means, in particular, that any product

$$P = E_k \cdots E_2 E_1$$

of elementary matrices is nonsingular.

Using Theorem 4.9, we are also able to prove (Problem 4.20) the following fundamental result on invertible matrices.

Theorem 4.10:   Let A be a square matrix. Then the following are equivalent:

   (i) A is invertible (nonsingular);

   (ii) A is row equivalent to the identity matrix I;

   (iii) A is a product of elementary matrices.

We also use Theorem 4.9 to prove the following theorems.

Theorem 4.11:   If $AB = I$, then $BA = I$ and hence $B = A^{-1}$.

Theorem 4.12: B is row equivalent to A if and only if there exists a nonsingular matrix P such that $B = PA$.

Application to Finding Inverses

Suppose a matrix A is invertible and, say, it is row reducible to the identity matrix I by the sequence of elementary operations $e_1, e_2, \ldots, e_n$. Let E_i be the elementary matrix corresponding to the operation e_i. Then, by Theorem 4.9,

$$E_n \cdots E_2 E_1 A = I \quad \text{and} \quad (E_n \cdots E_2 E_1 I)A = I \quad \text{so} \quad A^{-1} = E_n \cdots E_2 E_1 I$$

In other words, A^{-1} can be obtained by applying the elementary row operations $e_1, e_2, \ldots, e_n$ to the identity matrix I.

The above discussion leads us to the following (Gaussian elimination) algorithm which either finds the inverse of an n-square matrix A or determines that A is not invertible.

Algorithm: Inverse of a matrix A

Step 1. Form the $n \times 2n$ [block] matrix $M = (A \mid I)$; that is, A is in the left half of M and the identity matrix I is in the right half of M.

Step 2. Row reduce M to echelon form. If the process generates a zero row in the A-half of M, STOP (A is not invertible). Otherwise, the A-half will assume triangular form.

Step 3. Further row reduce M to the row canonical form $(I \mid B)$, where I has replaced A in the left half of the matrix.

Step 4. Set $A^{-1} = B$.

Example 4.12. Suppose we want to find the inverse of $A = \begin{pmatrix} 1 & 0 & 2 \\ 2 & -1 & 3 \\ 4 & 1 & 8 \end{pmatrix}$. First we form the block matrix $M = (A \mid I)$ and reduce M to echelon form:

$$M = \begin{pmatrix} 1 & 0 & 2 & \vdots & 1 & 0 & 0 \\ 2 & -1 & 3 & \vdots & 0 & 1 & 0 \\ 4 & 1 & 8 & \vdots & 0 & 0 & 1 \end{pmatrix} \sim \begin{pmatrix} 1 & 0 & 2 & \vdots & 1 & 0 & 0 \\ 0 & -1 & -1 & \vdots & -2 & 1 & 0 \\ 0 & 1 & 0 & \vdots & -4 & 0 & 1 \end{pmatrix} \sim \begin{pmatrix} 1 & 0 & 2 & \vdots & 1 & 0 & 0 \\ 0 & -1 & -1 & \vdots & -2 & 1 & 0 \\ 0 & 0 & -1 & \vdots & -6 & 1 & 1 \end{pmatrix}$$

In echelon form, the left half of M is in triangular form; hence A is invertible. Next we further reduce M to its row canonical form:

$$M \sim \begin{pmatrix} 1 & 0 & 0 & \vdots & -11 & 2 & 2 \\ 0 & -1 & 0 & \vdots & 4 & 0 & -1 \\ 0 & 0 & 1 & \vdots & 6 & -1 & -1 \end{pmatrix} \sim \begin{pmatrix} 1 & 0 & 0 & \vdots & -11 & 2 & 2 \\ 0 & 1 & 0 & \vdots & -4 & 0 & 1 \\ 0 & 0 & 1 & \vdots & 6 & -1 & -1 \end{pmatrix}$$

The identity matrix is in the left half of the final matrix; hence the right half is A^{-1}. In other words,

$$A^{-1} = \begin{pmatrix} -11 & 2 & 2 \\ -4 & 0 & 1 \\ 6 & -1 & -1 \end{pmatrix}$$

4.10 ELEMENTARY COLUMN OPERATIONS, MATRIX EQUIVALENCE

This section repeats some of the discussion of the preceding section using the columns of a matrix instead of the rows. (The choice of first using rows comes from the fact that the row operations are closely related to the operations with linear equations.) We also show the relationship between the row and column operations and their elementary matrices.

The elementary column operations which are analogous to the elementary row operations are as follows:

$[F_1]$ (Column-interchange)   Interchange the ith column and the jth column:

$$C_i \leftrightarrow C_j$$

$[F_2]$ (Column-scaling)   Multiply the ith column by a nonzero scalar k:

$$kC_i \rightarrow C_i \quad (k \neq 0)$$

$[F_3]$ (Column-addition)   Replace the ith column by k times the jth column plus the ith column:

$$kC_j + C_i \rightarrow C_i$$

Each of the above operations has an inverse operation of the same type just like the corresponding row operations.

Let f denote an elementary column operation. The matrix F, obtained by applying f to the identity matrix I, that is,

$$F = f(I)$$

is called the *elementary matrix* corresponding to the elementary column operation f.

Example 4.13.   The 3-square elementary matrices corresponding to the elementary column operations $C_3 \rightarrow C_1$, $-2C_3 \rightarrow C_3$, and $-5C_2 + C_3 \rightarrow C_3$ are, respectively,

$$F_1 = \begin{pmatrix} 0 & 0 & 1 \\ 0 & 1 & 0 \\ 1 & 0 & 0 \end{pmatrix} \qquad F_2 = \begin{pmatrix} 1 & 0 & 0 \\ 0 & 1 & 0 \\ 0 & 0 & -2 \end{pmatrix} \qquad F_3 = \begin{pmatrix} 0 & 0 & 1 \\ 0 & 1 & -5 \\ 0 & 0 & 1 \end{pmatrix}$$

Throughout the discussion below, e and f will denote, respectively, corresponding elementary row and column operations, and E and F will denote the corresponding elementary matrices.

Lemma 4.13:   Suppose A is any matrix. Then

$$f(A) = [e(A^T)]^T$$

that is, applying the column operation f to a matrix A gives the same result as applying the corresponding row operation e to A^T and then taking the transpose.

The proof of the lemma follows directly from the fact that the columns of A are the rows of A^T, and vice versa.

The above lemma shows that

$$F = f(I) = [e(I^T)]^T = [e(I)]^T = E^T$$

In other words,

Corollary 4.14: F is the transpose of E.

(Thus F is invertible since E is invertible.) Also, by the above lemma,

$$f(A) = [e(A^T)]^T = [EA^T]^T = (A^T)^T E^T = AF$$

This proves the following theorem (which is analogous to Theorem 4.9 for the elementary row operations):

Theorem 4.15:   For any matrix A, $f(A) = AF$.

That is, the result of applying an elementary column operation f on a matrix A can be obtained by postmultiplying A by the corresponding elementary matrix F.

A matrix B is said to be *column equivalent* to a matrix A if B can be obtained from A by a sequence of elementary column operations. Using the argument that is analogous to that for Theorem 4.12 yields:

Theorem 4.16: B is column equivalent to A if and only if there exists a nonsingular matrix Q such that $B = AQ$.

Matrix Equivalence

A matrix B is said to be *equivalent* to a matrix A if B can be obtained from A by a finite sequence of elementary row and column operations. Alternatively (Problem 4.23), B is equivalent to A if there exist nonsingular matrices P and Q such that $B = PAQ$. Just like row equivalence and column equivalence, equivalence of matrices is an equivalence relation.

The main result of this subsection, proved in Problem 4.25, is as follows:

Theorem 4.17: Every $m \times n$ matrix A is equivalent to a unique block matrix of the form

$$\left(\begin{array}{c|c} I_r & 0 \\ \hline 0 & 0 \end{array} \right)$$

where I_r is the $r \times r$ identity matrix. (The nonnegative integer r is called the *rank* of A.)

4.11 CONGRUENT SYMMETRIC MATRICES, LAW OF INERTIA

A matrix B is said to be *congruent* to a matrix A if there exists a nonsingular (invertible) matrix P such that

$$B = P^T A P$$

Suppose A is symmetric, i.e., $A^T = A$. Then

$$B^T = (P^T A P)^T = P^T A^T P^{TT} = P^T A P = B$$

and so B is symmetric. Since congruence is an equivalence relation (Problem 4.123), only symmetric matrices are congruent to each other and, in particular, only symmetric matrices are congruent to diagonal matrices.

The next theorem plays an important role in linear algebra.

Theorem 4.18 (Law of Inertia): Let A be a real symmetric matrix. Then there exists a nonsingular matrix P such that $B = P^T A P$ is diagonal. Moreover, every such diagonal matrix B has the same number **p** of positive entries and the same number **n** of negative entries.

The *rank* and *signature* of the above real symmetric matrix A are denoted and defined, respectively, by

$$\text{rank } A = \mathbf{p} + \mathbf{n} \qquad \text{and} \qquad \text{sig } A = \mathbf{p} - \mathbf{n}$$

These are uniquely defined by Theorem 4.18. [The notion of rank is actually defined for any matrix (Section 5.7), and the above definition agrees with the general definition.]

Diagonalization Algorithm

The following is an algorithm which diagonalizes (under congruence) a real symmetric matrix $A = (a_{ij})$.

Algorithm: Congruence diagonalization of a symmetric matrix

Step 1. Form the $n \times 2n$ [block] matrix $M = (A \mid I)$; that is, A is the left half of M and the identity matrix I is the right half of M.

Step 2. Examine the entry a_{11}.

Case I: $a_{11} \neq 0$. Apply the row operations $-a_{i1}R_1 + a_{11}R_i \to R_i$, $i = 2, \ldots, n$, and then the corresponding column operations $-a_{i1}C_1 + a_{11}C_i \to C_i$ to reduce the matrix M to the form

$$M = \begin{pmatrix} a_{11} & 0 & \vdots & * & * \\ 0 & B & \vdots & * & * \end{pmatrix} \tag{1}$$

Case II: $a_{11} = 0$ but $a_{ii} \neq 0$, for some $i > 1$. Apply the row operation $R_1 \leftrightarrow R_i$ and then the corresponding column operation $C_1 \leftrightarrow C_i$ to bring a_{ii} into the first diagonal position. This reduces the matrix to Case I.

Case III: All diagonal entries $a_{ii} = 0$. Choose i, j such that $a_{ij} \neq 0$ and apply the row operations $R_j + R_i \to R_i$ and the corresponding column operation $C_j \to C_i \to C_i$ to bring $2a_{ij} \neq 0$ into the ith diagonal position. This reduces the matrix to Case II.

In each of the cases, we finally reduce M to the form (1) where B is a symmetric matrix of order less than A.

Remark: The row operations will change both halves of M, but the column operations will only change the left half of M.

Step 3. Repeat Step 2 with each new matrix (neglecting the first row and column of the preceding matrix) until A is diagonalized, that is, until M is transformed into the form $M' = (D, Q)$ where D is diagonal.

Step 4. Set $P = Q^T$. Then $D = P^T A P$.

The justification of the above algorithm is as follows. Let $e_1, e_2, \ldots, e_k$ be all the elementary row operations in the algorithm and let $f_1, f_2, \ldots, f_k$ be the corresponding elementary column operations. Suppose E_i and F_i are the corresponding elementary matrices. By Corollary 4.14,

$$F_i = E_i^T$$

By the above algorithm,

$$Q = E_k \cdots E_2 E_1 I = E_k \cdots E_2 E_1$$

since the right half I of M is only changed by the row operations. On the other hand, the left half A of M is changed by both the row and column operations; therefore,

$$\begin{aligned} D &= E_k \cdots E_2 E_1 A F_1 F_2 \cdots F_k \\ &= (E_k \cdots E_2 E_1) A (E_k \cdots E_2 E_1)^T \\ &= QAQ^T = P^T A P \end{aligned}$$

where $P = Q^T$.

Example 4.14. Suppose $A = \begin{pmatrix} 1 & 2 & -3 \\ 2 & 5 & -4 \\ -3 & -4 & 8 \end{pmatrix}$, a symmetric matrix. To find a nonsingular matrix P such that

$B = P^T A P$ is diagonal, first form the block matrix $(A \mid I)$:

$$(A \mid I) = \begin{pmatrix} 1 & 2 & -3 & \vdots & 1 & 0 & 0 \\ 2 & 5 & -4 & \vdots & 0 & 1 & 0 \\ -3 & -4 & 8 & \vdots & 0 & 0 & 1 \end{pmatrix}$$

Apply the operations $-2R_1 + R_2 \to R_2$ and $3R_1 + R_3 \to R_3$ to $(A \mid I)$ and then the corresponding operations $-2C_1 + C_2 \to C_2$ and $3C_1 + C_3 \to C_3$ to A to obtain

$$\begin{pmatrix} 1 & 2 & -3 & \vdots & 1 & 0 & 0 \\ 0 & 1 & 2 & \vdots & -2 & 1 & 0 \\ 0 & 2 & -1 & \vdots & 3 & 0 & 1 \end{pmatrix} \quad \text{and then} \quad \begin{pmatrix} 1 & 0 & 0 & \vdots & 1 & 0 & 0 \\ 0 & 1 & 2 & \vdots & -2 & 1 & 0 \\ 0 & 2 & -1 & \vdots & 3 & 0 & 1 \end{pmatrix}$$

Next apply the operation $-2R_2 + R_3 \to R_3$ and then the corresponding operation $-2C_2 + C_3 \to C_3$ to obtain

$$\begin{pmatrix} 1 & 0 & 0 & \vdots & 1 & 0 & 0 \\ 0 & 1 & 2 & \vdots & -2 & 1 & 0 \\ 0 & 0 & -5 & \vdots & 7 & -2 & 1 \end{pmatrix} \quad \text{and then} \quad \begin{pmatrix} 1 & 0 & 0 & \vdots & 1 & 0 & 0 \\ 0 & 1 & 0 & \vdots & -2 & 1 & 0 \\ 0 & 0 & -5 & \vdots & 7 & -2 & 1 \end{pmatrix}$$

Now A has been diagonalized. Set

$$P = \begin{pmatrix} 1 & -2 & 7 \\ 0 & 1 & -2 \\ 0 & 0 & 1 \end{pmatrix} \quad \text{and then} \quad B = P^T A P = \begin{pmatrix} 1 & 0 & 0 \\ 0 & 1 & 0 \\ 0 & 0 & -5 \end{pmatrix}$$

Note that B has $\mathbf{p} = 2$ positive entries and $\mathbf{n} = 1$ negative entry.

4.12 QUADRATIC FORMS

A *quadratic form* q in variables $x_1, x_2, \ldots, x_n$ is a polynomial

$$q(x_1, x_2, \ldots, x_n) = \sum_{i < j} c_{ij} x_i x_j \tag{4.1}$$

(where each term has degree two). The quadratic form q is said to be *diagonalized* if

$$q(x_1, x_2, \ldots, x_n) = c_{11} x_1^2 + c_{22} x_2^2 + \cdots + c_{nn} x_n^2$$

that is, if q has no *cross product* terms $x_i x_j$ (where $i \neq j$).

The quadratic form (4.1) may be expressed uniquely in the matrix form

$$q(X) = X^T A X \tag{4.2}$$

where $X = (x_1, x_2, \ldots, x_n)^T$ and $A = (a_{ij})$ is a symmetric matrix. The entries of A can be obtained from (4.1) by setting

$$a_{ii} = c_{ii} \quad \text{and} \quad a_{ij} = a_{ji} = c_{ij}/2 \quad \text{(for } i \neq j)$$

that is, A has diagonal entry a_{ii} equal to the coefficient of x_i^2 and has entries a_{ij} and a_{ji} each equal to half the coefficient of $x_i x_j$. Thus

$$q(X) = (x_1, \ldots, x_n) \begin{pmatrix} a_{11} & a_{22} & \cdots & a_{1n} \\ a_{21} & a_{22} & \cdots & a_{2n} \\ \cdots\cdots\cdots\cdots\cdots\cdots \\ a_{n1} & a_{n2} & \cdots & a_{nn} \end{pmatrix} \begin{pmatrix} x_1 \\ x_2 \\ \vdots \\ x_n \end{pmatrix}$$

$$= \sum_{i, j} a_{ij} x_i x_j = a_{11} x_1^2 + a_{22} x_2^2 + \cdots + a_{nn} x_n^2 + 2 \sum_{i < j} a_{ij} x_i x_j$$

The above symmetric matrix A is called the *matrix representation* of the quadratic form q. Although many matrices A in (4.2) will yield the same quadratic form q, only one such matrix is symmetric.

Conversely, any symmetric matrix A defines a quadratic form q by (4.2). Thus there is a one-to-one correspondence between quadratic forms q and symmetric matrices A. Furthermore, a quadratic form q is diagonalized if and only if the corresponding symmetric matrix A is diagonal.

Example 4.15

(a) The quadratic form

$$q(x, y, z) = x^2 - 6xy + 8y^2 - 4xz + 5yz + 7z^2$$

may be expressed in the matrix form

$$q(x, y, z) = (x, y, z)\begin{pmatrix} 1 & -3 & -2 \\ -3 & 8 & \frac{5}{2} \\ -2 & \frac{5}{2} & 7 \end{pmatrix}\begin{pmatrix} x \\ y \\ z \end{pmatrix}$$

where the defining matrix is symmetric. The quadratic form may also be expressed in the matrix form

$$q(x, y, z) = (x, y, z)\begin{pmatrix} 1 & -6 & -4 \\ 0 & 8 & 5 \\ 0 & 0 & 7 \end{pmatrix}\begin{pmatrix} x \\ y \\ z \end{pmatrix}$$

where the defining matrix is upper triangular.

(b) The symmetric matrix $\begin{pmatrix} 2 & 3 \\ 3 & 5 \end{pmatrix}$ determines the quadratic form

$$q(x, y) = (x, y)\begin{pmatrix} 2 & 3 \\ 3 & 5 \end{pmatrix}\begin{pmatrix} x \\ y \end{pmatrix} = 2x^2 + 6xy + 5y^2$$

> **Remark:** For theoretical reasons, we will always assume a quadratic form q is represented by a symmetric matrix A. Since A is obtained from q by division by 2, we must also assume $1 + 1 \neq 0$ in our field K. This is always true when K is the real field **R** or the complex field **C**.

Change-of-Variable Matrix

Consider a change of variables, say from $x_1, x_2, \ldots, x_n$ to $y_1, y_2, \ldots, y_n$, by means of an invertible linear substitution of the form

$$x_i = p_{i1}y_1 + p_{i2}y_2 + \cdots + p_{in}y_n \qquad (i = 1, 2, \ldots, n)$$

(Here *invertible* means that one can solve for each of the y's uniquely in terms of the x's.) Such a linear substitution can be expressed in the matrix form

$$X = PY \qquad\qquad (4.3)$$

where

$$X = (x_1, x_2, \ldots, x_n)^T \qquad Y = (y_1, y_2, \ldots, y_n)^T \qquad \text{and} \qquad P = (p_{ij})$$

The matrix P is called the *change-of-variable matrix*; it is nonsingular since the linear substitution is invertible.

Conversely, any nonsingular matrix P defines an invertible linear substitution of variables, $X = PY$. Furthermore,

$$Y = P^{-1}X$$

yields the formula for the y's in terms of the x's.

There is a geometrical interpretation of the change-of-variable matrix P which is illustrated in the next example.

Example 4.16. Consider the cartesian plane $\mathbf{R}^2$ with the usual x and y axes, and consider the 2×2 nonsingular matrix

$$P = \begin{pmatrix} 2 & -1 \\ 1 & 1 \end{pmatrix}$$

The columns $u_1 = (2, 1)^T$ and $u_2 = (-1, 1)^T$ of P determine a new coordinate system of the plane, say with s and t axes, as shown in Fig. 4-1. That is

(1) The s axis is in the direction of u_1 and its unit length is the length of u_1.

(2) The t axis is in the direction of u_2 and its unit length is the length of u_2.

Any point Q in the plane will have coordinates relative to each coordinate system, say $Q(a, b)$ relative to the x and y axes and $Q(a', b')$ relative to the s and t axes. These coordinate vectors are related by the matrix P. Specifically,

$$\begin{pmatrix} a \\ b \end{pmatrix} = \begin{pmatrix} 2 & -1 \\ 1 & 1 \end{pmatrix} \begin{pmatrix} a' \\ b' \end{pmatrix} \quad \text{or} \quad X = PY$$

where $X = (a, b)^T$ and $Y = (a', b')^T$.

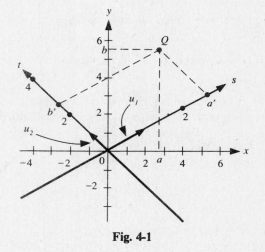

Fig. 4-1

Diagonalizing a Quadratic Form

Consider a quadratic form q in variables $x_1, x_2, \ldots, x_n$, say $q(X) = X^T A X$ (where A is a symmetric matrix). Suppose a change of variables is made in q using the linear substitution (*4.3*). Setting $X = PY$ in q yields the quadratic form

$$q(Y) = (PY)^T A (PY) = Y^T (P^T A P) Y$$

Thus $B = P^T A P$ is the matrix representation of the quadratic form in the new variables $y_1, y_2, \ldots, y_n$. Observe that the new matrix B is congruent to the original matrix A representing q.

The above linear substitution $Y = PX$ is said to *diagonalize* the quadratic form $q(X)$ if $q(Y)$ is diagonal, i.e., if $B = P^T A P$ is a diagonal matrix. Since B is congruent to A and A is a symmetric matrix, Theorem 4.18 may be restated as follows.

Theorem 4.19 (Law of Inertia): Let $q(X) = X^T A X$ be a real quadratic form. Then there exists an invertible linear substitution $Y = PX$ which diagonalizes q. Moreover, every other diagonal representation of q has the same number **p** of positive entries and the same number **n** of negative entries.

The *rank* and *signature* of a real quadratic form q are denoted and defined by

$$\text{rank } q = \mathbf{p} + \mathbf{n} \qquad \text{and} \qquad \text{sig } q = \mathbf{p} - \mathbf{n}$$

These are uniquely defined by Theorem 4.19.

Since diagonalizing a quadratic form is the same as diagonalizing under congruence a symmetric matrix, Algorithm 4.11 may be used here.

Example 4.17.   Consider the quadratic form

$$q(x, y, z) = x^2 + 4xy + 5y^2 - 6xz - 8yz + 8z^2 \qquad (1)$$

The (symmetric) matrix A which represents q is as follows:

$$A = \begin{pmatrix} 1 & 2 & -3 \\ 2 & 5 & -4 \\ -3 & -4 & 8 \end{pmatrix}$$

From Example 4.14, the following nonsingular matrix P diagonalizes the matrix A under congruence:

$$P = \begin{pmatrix} 1 & -2 & 7 \\ 0 & 1 & -2 \\ 0 & 0 & 1 \end{pmatrix} \qquad \text{and} \qquad B = P^T A P = \begin{pmatrix} 1 & 0 & 0 \\ 0 & 1 & 0 \\ 0 & 0 & -5 \end{pmatrix}$$

Accordingly, q may be diagonalized by the following linear substitution:

$$\begin{aligned} x &= r - 2s + 7t \\ y &= s - 2t \\ z &= t \end{aligned}$$

Specifically, substituting for x, y, and z in (1) yields the quadratic form

$$q(r, s, t) = r^2 + s^2 - 5t^2 \qquad (2)$$

Here $p = 2$ and $n = 1$; hence

$$\text{rank } q = 3 \qquad \text{and} \qquad \text{sig } q = 1$$

Remark: There is a geometrical interpretation of the Law of Inertia (Theorem 4.19) which we give here using the quadratic form q in Example 4.17. Consider the following surface S in $\mathbf{R}^3$:

$$q(x, y, z) = x^2 + 4xy + 5y^2 - 6xz - 8yz + 8z^2 = 25$$

Under the change of variables,

$$x = r - 2s + 7t \qquad y = s - 2t \qquad z = t$$

or, equivalently, relative to a new coordinate system with r, s, and t axes, the equation of S becomes

$$q(r, s, t) = r^2 + s^2 - 5t^2 = 25$$

Accordingly, S is a hyperboloid of one sheet, since there are two positive and one negative entry on the diagonal. Furthermore, S will always be a hyperboloid of one sheet regardless of the coordinate system. Thus any diagonal representation of the quadratic form $q(x, y, z)$ will contain two positive and one negative entries on the diagonal.

Positive Definite Symmetric Matrices and Quadratic Forms

A real symmetric matrix A is said to be *positive definite* if

$$X^T A X > 0$$

for every nonzero (column) vector X in $\mathbf{R}^n$. Analogously, a quadratic form q is said to be *positive definite* if $q(v) > 0$ for every nonzero vector in $\mathbf{R}^n$.

Alternatively, a real symmetric matrix A or its quadratic form q is *positive definite* if any diagonal representation has only positive diagonal entries. Such matrices and quadratic forms play a very important role in linear algebra. They are considered in Problems 4.54–4.60.

4.13 SIMILARITY

A function $f: \mathbf{R}^n \to \mathbf{R}^n$ may be viewed geometrically as "sending" or "mapping" each point Q into a point $f(Q)$ in the space $\mathbf{R}^n$. Suppose the function f can be represented in the form

$$f(Q) = AQ$$

where A is an $n \times n$ matrix and the coordinates of Q are written as a column vector. Furthermore, suppose P is a nonsingular matrix which may be viewed as introducing a new coordinate system in the space $\mathbf{R}^n$. (See Example 4.16.) Relative to this new coordinate system, we prove that f is represented by the matrix

$$B = P^{-1}AP$$

that is,

$$f(Q') = BQ'$$

where Q' denotes the column vector of the coordinates of Q relative to the new coordinate system.

Example 4.18.   Consider the function $f: \mathbf{R}^2 \to \mathbf{R}^2$ defined by

$$f(x, y) = (3x - 4y, 5x + 2y)$$

or, equivalently,

$$f\begin{pmatrix} x \\ y \end{pmatrix} = A\begin{pmatrix} x \\ y \end{pmatrix} \quad \text{where} \quad A = \begin{pmatrix} 3 & -4 \\ 5 & 2 \end{pmatrix}$$

Suppose a new coordinate system with s and t axes is introduced in $\mathbf{R}^2$ by means of the nonsingular matrix

$$P = \begin{pmatrix} 2 & -1 \\ 1 & 1 \end{pmatrix} \quad \text{and so} \quad P^{-1} = \begin{pmatrix} \frac{1}{3} & \frac{1}{3} \\ -\frac{1}{3} & \frac{2}{3} \end{pmatrix}$$

(See Fig. 4-1.) Relative to this new coordinate system of $\mathbf{R}^2$, the function f may be represented as

$$f\begin{pmatrix} s \\ t \end{pmatrix} = B\begin{pmatrix} s \\ t \end{pmatrix}$$

where

$$B = P^{-1}AP = \begin{pmatrix} \frac{1}{3} & \frac{1}{3} \\ -\frac{1}{3} & \frac{2}{3} \end{pmatrix}\begin{pmatrix} 3 & -4 \\ 5 & 2 \end{pmatrix}\begin{pmatrix} 2 & -1 \\ 1 & 1 \end{pmatrix} = \begin{pmatrix} \frac{14}{3} & -\frac{10}{3} \\ \frac{22}{3} & \frac{1}{3} \end{pmatrix}$$

In other words,

$$f(s, t) = (\tfrac{14}{3}s - \tfrac{10}{3}t, \tfrac{22}{3}s + \tfrac{1}{3}t)$$

The above discussion leads us to the following:

Definition: A matrix B is *similar* to a matrix A if there exists a nonsingular matrix P such that

$$B = P^{-1}AP$$

Similarity, like congruence, is an equivalence relation (Problem 4.125); hence we say that A and B are similar matrices when $B = P^{-1}AP$.

A matrix A is said to be *diagonalizable* if there exists a nonsingular matrix P such that $B = P^{-1}AP$ is a diagonal matrix. The question of whether or not a given matrix A is diagonalizable and of finding the matrix P when A is diagonalizable plays an important role in linear algebra. These questions will be addressed in Chapter 8.

4.14   LU FACTORIZATION

Suppose A is a nonsingular matrix which can be brought into (upper) triangular form U using only row-addition operations, that is, suppose A can be triangularized by the following algorithm which we write using computer algorithmic notation.

Algorithm: Triangularizing matrix $A = (a_{ij})$

Step 1.   Repeat for $i = 1, 2, \ldots, n - 1$;

Step 2.   Repeat for $j = i + 1, \ldots, n$
  (a)   Set $m_{ij} := a_{ij}/a_{ii}$
  (b)   Set $R_j := m_{ij}R_i + R_j$
 [End of Step 2 inner loop.]
 [End of Step 1 outer loop.]

Step 3.   Exit.

The numbers m_{ij} are called *multipliers*. Sometimes we keep track of these multipliers by means of the following lower triangular matrix L:

$$L = \begin{pmatrix} 1 & 0 & 0 & \cdots & 0 & 0 \\ -m_{21} & 1 & 0 & \cdots & 0 & 0 \\ -m_{31} & -m_{32} & 1 & \cdots & 0 & 0 \\ \cdots\cdots\cdots\cdots\cdots\cdots\cdots\cdots\cdots\cdots\cdots\cdots\cdots\cdots\cdots \\ -m_{n1} & -m_{n2} & -m_{n3} & \cdots & -m_{n, n-1} & 1 \end{pmatrix}$$

That is, L has 1s on the diagonal, 0s above the diagonal, and the negative of m_{ij} as its ij-entry below the diagonal.

The above lower triangular matrix L may be alternatively described as follows. Let $e_1, e_2, \ldots, e_k$ denote the sequence of elementary row operations in the above algorithm. The inverses of these operations are as follows. For $i = 1, 2, \ldots, n - 1$, we have

$$-m_{ij}R_i + R_j \to R_j \qquad (j = i + 1, \ldots, n)$$

Applying these inverse operations in reverse order to the identity matrix I yields the matrix L Thus

$$L = E_1^{-1}E_2^{-1} \cdots E_k^{-1}I$$

where $E_1, \ldots, E_k$ are the elementary matrices corresponding to the elementary operations $e_1, \ldots, e_k$.

On the other hand, the elementary operations $e_1, \ldots, e_k$ transform the original matrix A into the upper triangular matrix U. Thus $E_k \cdots E_2 E_1 A = U$. Accordingly,

$$A = (E_1^{-1}E_2^{-1} \cdots E_k^{-1})U = (E_1^{-1}E_2^{-1} \cdots E_k^{-1}I)U = LU$$

This gives us the classical LU factorization of such a matrix A. We formally state this result as a theorem.

Theorem 4.20:   Let A be a matrix as above. Then $A = LU$ where L is a lower triangular matrix with 1s on the diagonal and U is an upper triangular matrix with no 0s on the diagonal.

> **Remark:** We emphasize that the above theorem only applies to nonsingular matrices A which can be brought into triangular form without any row interchanges. Such matrices are said to be *LU-factorable* or to have an LU factorization.

Example 4.19.   Let $A = \begin{pmatrix} 1 & 2 & -3 \\ -3 & -4 & 13 \\ 2 & 1 & -5 \end{pmatrix}$. Then A may be reduced to triangular form by the operations $3R_1 + R_2 \to R_2$ and $-2R_1 + R_3 \to R_3$, and then $(\frac{3}{2})R_2 + R_3 \to R_3$:

$$A \sim \begin{pmatrix} 1 & 2 & -3 \\ 0 & 2 & 4 \\ 0 & -3 & 1 \end{pmatrix} \sim \begin{pmatrix} 1 & 2 & -3 \\ 0 & 2 & 4 \\ 0 & 0 & 7 \end{pmatrix}$$

This gives us the factorization $A = LU$ where

$$L = \begin{pmatrix} 1 & 0 & 0 \\ -3 & 1 & 0 \\ 2 & -\frac{3}{2} & 1 \end{pmatrix} \qquad U = \begin{pmatrix} 1 & 2 & -3 \\ 0 & 2 & 4 \\ 0 & 0 & 7 \end{pmatrix}$$

Note that the entries -3, 2, and $-\frac{3}{2}$ in L come from the above elementary row operations, and that U is the triangular form of A.

Applications to Linear Equations

Consider a computer algorithm M. Let $C(n)$ denote the running time of the algorithm as a function of the size n of the input data. [The function $C(n)$ is sometimes called the *time complexity* or simply the *complexity* of the algorithm M.] Frequently, $C(n)$ simply counts the number of multiplications and divisions executed by M, but does not count the number of additions and subtractions since they take much less time to execute.

Now consider a square system of linear equations

$$AX = B$$

where $A = (a_{ij})$ has an LU factorization and

$$X = (x_1, \ldots, x_n)^T \qquad \text{and} \qquad B = (b_1, \ldots, b_n)^T$$

Then the system may be brought into triangular form (in order to apply back-substitution) by applying the above algorithm to the augmented matrix $M = (A, b)$ of the system. The time complexity of the above algorithm and back-substitution are, respectively,

$$C(n) \approx n^3/2 \qquad \text{and} \qquad C(n) \approx n^2/2$$

where n is the number of equations.

On the other hand, suppose we already have the factorization $A = LU$. Then to triangularize the system we need only apply the row operations in the algorithm (retained by the matrix L) to the column vector B. In this case, the time complexity is

$$C(n) \approx n^2/2$$

Of course, to obtain the factorization $A = LU$ requires the original algorithm where $C(n) \approx n^3/2$. Thus nothing may be gained by first finding the LU factorization when a single system is involved. However, there are situations, illustrated below, where the LU factorization is useful.

Suppose that for a given matrix A we need to solve the system

$$AX = B$$

repeatedly for a sequence of different constant vectors, say $B_1, B_2, \ldots, B_k$. Also, suppose some of the B_i depend upon the solution of the system obtained while using preceding vectors B_j. In such a case, it is more efficient to first find the LU factorization of A, and then to use this factorization to solve the system for each new B.

Example 4.20. Consider the system

$$\begin{aligned} x - 2y - z &= k_1 \\ 2x - 5y - z &= k_2 \qquad \text{or} \qquad AX = B \\ -3x + 10y - 3z &= k_3 \end{aligned} \tag{1}$$

where $A = \begin{pmatrix} 1 & -2 & -1 \\ 2 & -5 & -1 \\ -3 & 10 & -3 \end{pmatrix}$ and $B = \begin{pmatrix} k_1 \\ k_2 \\ k_3 \end{pmatrix}$.

Suppose we want to solve the system for B_1, B_2, B_3, B_4, where $B_1 = (1, 2, 3)^T$ and

$$B_{j+1} = B_j + X_j \qquad \text{(for } j > 1)$$

where X_j is the solution of (1) obtained using B_j. Here it is more efficient to first obtain the LU factorization for A and then to use the LU factorization to solve the system for each of the B's. (See Problem 4.73.)

Solved Problems

ALGEBRA OF SQUARE MATRICES

4.1. Let $A = \begin{pmatrix} 1 & 3 & 6 \\ 2 & -5 & 8 \\ 4 & -2 & 7 \end{pmatrix}$. Find: (a) the diagonal and trace of A; (b) $A(u)$ where $u = (2, -3, 5)^T$;

(c) $A(v)$ where $v = (1, 7, -2)$.

(a) The diagonal consists of the elements from the upper left corner to the lower right corner of the matrix, that is, the elements a_{11}, a_{22}, a_{33}. Thus the diagonal of A consists of the scalars 1, -5, and 7. The trace of A is the sum of the diagonal elements; hence tr $A = 1 - 5 + 7 = 3$.

(b) $$A(u) = Au = \begin{pmatrix} 1 & 3 & 6 \\ 2 & -5 & 8 \\ 4 & -2 & 7 \end{pmatrix} \begin{pmatrix} 2 \\ -3 \\ 5 \end{pmatrix} = \begin{pmatrix} 2 - 9 + 30 \\ 4 + 15 + 40 \\ 8 + 6 + 35 \end{pmatrix} = \begin{pmatrix} 23 \\ 59 \\ 49 \end{pmatrix}$$

(c) By our convention, $A(v)$ is not defined for a row vector v.

4.2. Let $A = \begin{pmatrix} 1 & 2 \\ 4 & -3 \end{pmatrix}$. (a) Find A^2 and A^3. (b) Find $f(A)$, where $f(x) = 2x^3 - 4x + 5$.

(a)

$$A^2 = AA = \begin{pmatrix} 1 & 2 \\ 4 & -3 \end{pmatrix}\begin{pmatrix} 1 & 2 \\ 4 & -3 \end{pmatrix} = \begin{pmatrix} 1+8 & 2-6 \\ 4-12 & 8+9 \end{pmatrix} = \begin{pmatrix} 9 & -4 \\ -8 & 17 \end{pmatrix}$$

$$A^3 = AA^2 = \begin{pmatrix} 1 & 2 \\ 4 & -3 \end{pmatrix}\begin{pmatrix} 9 & -4 \\ -8 & 17 \end{pmatrix} = \begin{pmatrix} 9-16 & -4+34 \\ 36+24 & -16-51 \end{pmatrix} = \begin{pmatrix} -7 & 30 \\ 60 & -67 \end{pmatrix}$$

(b) To find $f(A)$, first substitute A for x and $5I$ for the constant 5 in the given polynomial $f(x) = 2x^3 - 4x + 5$:

$$f(A) = 2A^3 - 4A + 5I = 2\begin{pmatrix} -7 & 30 \\ 60 & -67 \end{pmatrix} - 4\begin{pmatrix} 1 & 2 \\ 4 & -3 \end{pmatrix} + 5\begin{pmatrix} 1 & 0 \\ 0 & 1 \end{pmatrix}$$

Then multiply each matrix by its respective scalar:

$$f(A) = \begin{pmatrix} -14 & 60 \\ 120 & -134 \end{pmatrix} + \begin{pmatrix} -4 & -8 \\ -16 & 12 \end{pmatrix} + \begin{pmatrix} 5 & 0 \\ 0 & 5 \end{pmatrix}$$

Lastly, add the corresponding elements in the matrices:

$$f(A) = \begin{pmatrix} -14-4+5 & 60-8+0 \\ 120-16+0 & -134+12+5 \end{pmatrix} = \begin{pmatrix} -13 & 52 \\ 104 & -117 \end{pmatrix}$$

4.3. Let $A = \begin{pmatrix} 2 & 2 \\ 3 & -1 \end{pmatrix}$. Find $g(A)$, where $g(x) = x^2 - x - 8$.

$$A^2 = \begin{pmatrix} 2 & 2 \\ 3 & -1 \end{pmatrix}\begin{pmatrix} 2 & 2 \\ 3 & -1 \end{pmatrix} = \begin{pmatrix} 4+6 & 4-2 \\ 6-3 & 6+1 \end{pmatrix} = \begin{pmatrix} 10 & 2 \\ 3 & 7 \end{pmatrix}$$

$$g(A) = A^2 - A - 8I = \begin{pmatrix} 10 & 2 \\ 3 & 7 \end{pmatrix} - \begin{pmatrix} 2 & 2 \\ 3 & -1 \end{pmatrix} - 8\begin{pmatrix} 1 & 0 \\ 0 & 1 \end{pmatrix}$$

$$= \begin{pmatrix} 10 & 2 \\ 3 & 7 \end{pmatrix} + \begin{pmatrix} -2 & -2 \\ -3 & 1 \end{pmatrix} + \begin{pmatrix} -8 & 0 \\ 0 & -8 \end{pmatrix} = \begin{pmatrix} 0 & 0 \\ 0 & 0 \end{pmatrix}$$

Thus A is a zero of $g(x)$.

4.4. Given $A = \begin{pmatrix} 1 & 3 \\ 4 & -3 \end{pmatrix}$. Find a *nonzero* column vector $u = \begin{pmatrix} x \\ y \end{pmatrix}$ such that $A(u) = 3u$.

First set up the matrix equation $Au = 3u$:

$$\begin{pmatrix} 1 & 3 \\ 4 & -3 \end{pmatrix}\begin{pmatrix} x \\ y \end{pmatrix} = 3\begin{pmatrix} x \\ y \end{pmatrix}$$

Write each side as a single matrix (column vector):

$$\begin{pmatrix} x+3y \\ 4x-3y \end{pmatrix} = \begin{pmatrix} 3x \\ 3y \end{pmatrix}$$

Set corresponding elements equal to each other to obtain the system of equations, and reduce it to echelon form:

$$\left.\begin{matrix} x+3y = 3x \\ 4x-3y = 3y \end{matrix}\right\} \rightarrow \left.\begin{matrix} 2x-3y = 0 \\ 4x-6y = 0 \end{matrix}\right\} \rightarrow \left.\begin{matrix} 2x-3y = 0 \\ 0 = 0 \end{matrix}\right\} \rightarrow 2x-3y = 0$$

The system reduces to one homogeneous equation in two unknowns, and so has an infinite number of solutions. To obtain a nonzero solution let, say, $y = 2$; then $x = 3$. That is, $u = (3, 2)^T$ has the desired property.

4.5. Let $A = \begin{pmatrix} 1 & 2 & -3 \\ 2 & 5 & -1 \\ 5 & 12 & -5 \end{pmatrix}$. Find all vectors $u = (x, y, z)^T$ such that $A(u) = 0$.

Set up the equation $Au = 0$ and then write each side as a single matrix:

$$\begin{pmatrix} 1 & 2 & -3 \\ 2 & 5 & -1 \\ 5 & 12 & -5 \end{pmatrix}\begin{pmatrix} x \\ y \\ z \end{pmatrix} = \begin{pmatrix} 0 \\ 0 \\ 0 \end{pmatrix} \quad \text{or} \quad \begin{pmatrix} x + 2y - 3z \\ 2x + 5y - z \\ 5x + 12y - 5z \end{pmatrix} = \begin{pmatrix} 0 \\ 0 \\ 0 \end{pmatrix}$$

Set corresponding elements equal to each other to obtain a homogeneous system, and reduce the system to echelon form:

$$\left.\begin{aligned} x + 2y - 3z &= 0 \\ 2x + 5y - z &= 0 \\ 5x + 12y - 5z &= 0 \end{aligned}\right\} \rightarrow \left\{\begin{aligned} x + 2y - 3z &= 0 \\ y + 5z &= 0 \\ 2y + 10z &= 0 \end{aligned}\right. \rightarrow \left\{\begin{aligned} x + 2y - 3z &= 0 \\ y + 5z &= 0 \end{aligned}\right.$$

In the echelon form, z is the free variable. To obtain the general solution, set $z = a$, where a is a parameter. Back-substitution yields $y = -5a$, and then $x = 13a$. Thus, $u = (13a, -5a, a)^T$ represents all vectors such that $Au = 0$.

4.6. Show that the collection $\mathbf{M}$ of all 2×2 matrices of the form $\begin{pmatrix} s & t \\ t & s \end{pmatrix}$ is a commutative algebra of matrices.

Clearly, $\mathbf{M}$ is nonempty. If $A = \begin{pmatrix} a & b \\ b & a \end{pmatrix}$ and $B = \begin{pmatrix} c & d \\ d & c \end{pmatrix}$ belong to $\mathbf{M}$, then

$$A + B = \begin{pmatrix} a + c & d + b \\ b + d & a + c \end{pmatrix} \qquad kA = \begin{pmatrix} ka & kb \\ kb & ka \end{pmatrix} \qquad AB = \begin{pmatrix} ac + bd & ad + bc \\ bc + ad & bd + ac \end{pmatrix}$$

also belong to $\mathbf{M}$. Thus $\mathbf{M}$ is an algebra of matrices. Furthermore,

$$BA = \begin{pmatrix} ca + db & cb + da \\ da + cb & db + ca \end{pmatrix}$$

Thus $BA = AB$ and so $\mathbf{M}$ is a commutative algebra of matrices.

4.7. Find all matrices $M = \begin{pmatrix} x & y \\ z & t \end{pmatrix}$ that commute with $A = \begin{pmatrix} 1 & 1 \\ 0 & 1 \end{pmatrix}$.

First find

$$AM = \begin{pmatrix} x + z & y + t \\ z & t \end{pmatrix} \quad \text{and} \quad MA = \begin{pmatrix} x & x + y \\ z & z + t \end{pmatrix}$$

Then set $AM = MA$ to obtain the four equations

$$x + z = x \qquad y + t = x + y \qquad z = z \qquad t = z + t$$

From the first or last equation, $z = 0$; from the second equation, $x = t$. Thus M is any matrix of the form $\begin{pmatrix} x & y \\ 0 & x \end{pmatrix}$.

4.8. Let $e_i = (0, \dots, 1, \dots, 0)^T$, where $i = 1, \dots, n$, be the (column) vector in $\mathbf{R}^n$ with 1 in the ith position and 0 elsewhere, and let A and B be $m \times n$ matrices.

(a) Show that Ae_i is the ith column of A.

    (b)   Suppose $Ae_i = Be_i$ for each i. Show that $A = B$.

    (c)   Suppose $Au = Bu$ for every vector u in $\mathbf{R}^n$. Show that $A = B$.

    (a)   Let $A = (a_{ij})$ and let $Ae_i = (b_1, \ldots, b_n)^T$. Then

$$b_k = R_k e_i = (a_{k1}, \ldots, a_{kn})(0, \ldots, 1, \ldots, 0)^T = a_{ki}$$

where R_k is the kth row of A. Thus

$$Ae_i = (a_{1i}, a_{2i}, \ldots, a_{ni})^T$$

the ith column of A.

    (b) $Ae_i = Be_i$ means A and B have the same ith column for each i. Thus $A = B$.

    (c)   If $Au = Bu$ for every vector u in $\mathbf{R}^n$, then $Ae_i = Be_i$ for each i. Thus $A = B$.

4.9.   Suppose A is an $m \times n$ matrix. Show that: (a) $I_m A = A$, (b) $AI_n = A$. (Thus $AI = IA = A$ when A is a square matrix.)

    We use the fact that $I = (\delta_{ij})$ where δ_{ij} is the Kronecker delta (Example 4.3).

    (a)   Suppose $I_m A = (f_{ij})$. Then

$$f_{ij} = \sum_{k=1}^{m} \delta_{ik} a_{kj} = \delta_{ii} a_{ij} = a_{ij}$$

Thus $I_m A = A$, since corresponding entries are equal.

    (b)   Suppose $AI_n = (g_{ij})$. Then

$$g_{ij} = \sum_{k=1}^{n} a_{ik} \delta_{kj} = a_{ij} \delta_{jj} = a_{ij}$$

Thus $AI_n = A$, since corresponding entries are equal.

4.10.   Prove Theorem 4.1.

    (i)   Let $A + B = (c_{ij})$. Then $c_{ij} = a_{ij} + b_{ij}$, so that

$$\operatorname{tr}(A + B) = \sum_{k=1}^{n} c_{kk} = \sum_{k=1}^{n} (a_{kk} + b_{kk}) = \sum_{k=1}^{n} a_{kk} + \sum_{k=1}^{n} b_{kk} = \operatorname{tr} A + \operatorname{tr} B$$

    (ii)   Let $kA = (c_{ij})$. Then $c_{ij} = ka_{ij}$, and

$$\operatorname{tr} kA = \sum^{n} ka_{jj} = k \sum^{n} a_{jj} = k \cdot \operatorname{tr} A$$

    (iii)   Let $AB = (c_{ij})$ and $BA = (d_{ij})$. Then $c_{ij} = \sum_{k=1}^{n} a_{ik} b_{kj}$ and $d_{ij} = \sum_{k=1}^{n} b_{ik} a_{kj}$, whence

$$\operatorname{tr} AB = \sum_{i=1}^{n} c_{ii} = \sum_{i=1}^{n} \sum_{k=1}^{n} a_{ik} b_{ki} = \sum_{k=1}^{n} \sum_{i=1}^{n} b_{ki} a_{ik} = \sum_{k=1}^{n} d_{kk} = \operatorname{tr} BA$$

4.11.   Prove Theorem 4.2.

    Suppose $f(x) = \sum_{i=1}^{r} a_i x^i$ and $g(x) = \sum_{j=1}^{s} b_j x^j$.

    (i)   We can assume $r = s = n$ by adding powers of x with 0 as their coefficients. Then

$$f(x) + g(x) = \sum_{i=1}^{n} (a_i + b_i) x^i$$

Hence

$$(f + g)(A) = \sum_{i=1}^{n} (a_i + b_i)A^i = \sum_{i=1}^{n} a_i A^i + \sum_{i=1}^{n} b_i A^i = f(A) + g(A)$$

(ii)   We have $f(x)g(x) = \sum_{i,j} a_i b_j x^{i+j}$. Then

$$f(A)g(A) = \left(\sum_i a_i A^i\right)\left(\sum_j b_j A^j\right) = \sum_{i,j} a_i b_j A^{i+j} = (fg)(A)$$

(iii)   Using $f(x)g(x) = g(x)f(x)$, we have

$$f(A)g(A) = (fg)(A) = (gf)(A) = g(A)f(A)$$

4.12.   Let $D_k = kI$, the scalar matrix belonging to the scalar k. Show that (a) $D_k A = kA$, (b) $BD_k = kB$, (c) $D_k + D_{k'} = D_{k+k'}$, and (d) $D_k D_{k'} = D_{kk'}$.

(a) $D_k A = (kI)A = k(IA) = kA$

(b) $BD_k = B(kI) = k(BI) = kB$

(c) $D_k + D_{k'} = kI + k'I = (k + k')I = D_{k+k'}$

(d) $D_k D_{k'} = (kI)(k'I) = kk'(II) = kk'I = D_{kk'}$

INVERTIBLE MATRICES, INVERSES

4.13.   Find the inverse of $\begin{pmatrix} 3 & 5 \\ 2 & 3 \end{pmatrix}$.

Method 1.   We seek scalars x, y, z, and w for which

$$\begin{pmatrix} 3 & 5 \\ 2 & 3 \end{pmatrix}\begin{pmatrix} x & y \\ z & w \end{pmatrix} = \begin{pmatrix} 1 & 0 \\ 0 & 1 \end{pmatrix} \quad \text{or} \quad \begin{pmatrix} 3x + 5z & 3y + 5w \\ 2x + 3z & 2y + 3w \end{pmatrix} = \begin{pmatrix} 1 & 0 \\ 0 & 1 \end{pmatrix}$$

or which satisfy

$$\begin{cases} 3x + 5z = 1 \\ 2x + 3z = 0 \end{cases} \quad \text{and} \quad \begin{cases} 3y + 5w = 0 \\ 2y + 3w = 1 \end{cases}$$

The solution of the first system is $x = -3$, $z = 2$, and of the second system is $y = 5$, $w = -3$. Thus the inverse of the given matrix is $\begin{pmatrix} -3 & 5 \\ 2 & -3 \end{pmatrix}$.

Method 2.   The general formula for the inverse of the 2×2 matrix $A = \begin{pmatrix} a & b \\ c & d \end{pmatrix}$ is

$$A^{-1} = \begin{pmatrix} d/|A| & -b/|A| \\ -c/|A| & a/|A| \end{pmatrix} = \frac{1}{|A|}\begin{pmatrix} d & -b \\ -c & a \end{pmatrix} \quad \text{where} \quad |A| = ad - bc$$

Thus if $A = \begin{pmatrix} 3 & 5 \\ 2 & 3 \end{pmatrix}$, then first find $|A| = (3)(3) - (5)(2) = -1 \neq 0$. Next interchange the diagonal elements, take the negatives of the other elements, and multiply by $1/|A|$:

$$A^{-1} = -1\begin{pmatrix} 3 & -5 \\ -2 & 3 \end{pmatrix} = \begin{pmatrix} -3 & 5 \\ 2 & -3 \end{pmatrix}$$

4.14.   Find the inverse of (a) $A = \begin{pmatrix} 1 & 2 & -4 \\ -1 & -1 & 5 \\ 2 & 7 & -3 \end{pmatrix}$ and (b) $B = \begin{pmatrix} 1 & 3 & -4 \\ 1 & 5 & -1 \\ 3 & 13 & -6 \end{pmatrix}$.

(a) Form the block matrix $M = (A \mid I)$ and row reduce M to echelon form:

$$M = \begin{pmatrix} 1 & 2 & -4 & \vdots & 1 & 0 & 0 \\ -1 & -1 & 5 & \vdots & 0 & 1 & 0 \\ 2 & 7 & -3 & \vdots & 0 & 0 & 1 \end{pmatrix} \sim \begin{pmatrix} 1 & 2 & -4 & \vdots & 1 & 0 & 0 \\ 0 & 1 & 1 & \vdots & 1 & 1 & 0 \\ 0 & 3 & 5 & \vdots & -2 & 0 & 1 \end{pmatrix}$$

$$\sim \begin{pmatrix} 1 & 2 & -4 & \vdots & 1 & 0 & 0 \\ 0 & 1 & 1 & \vdots & 1 & 1 & 0 \\ 0 & 0 & 1 & \vdots & -5 & -3 & 1 \end{pmatrix}$$

The left half of M is now in triangular form; hence A has an inverse. Further row reduce M to row canonical form:

$$M \sim \begin{pmatrix} 1 & 2 & 0 & \vdots & -9 & -6 & 2 \\ 0 & 1 & 0 & \vdots & \frac{7}{2} & \frac{5}{2} & -\frac{1}{2} \\ 0 & 0 & 1 & \vdots & -\frac{5}{2} & -\frac{3}{2} & \frac{1}{2} \end{pmatrix} \sim \begin{pmatrix} 1 & 0 & 0 & \vdots & -16 & -11 & 3 \\ 0 & 1 & 0 & \vdots & \frac{7}{2} & \frac{5}{2} & -\frac{1}{2} \\ 0 & 0 & 1 & \vdots & -\frac{5}{2} & -\frac{3}{2} & \frac{1}{2} \end{pmatrix} = (I \mid A^{-1})$$

Thus $A^{-1} = \begin{pmatrix} -16 & -11 & 3 \\ \frac{7}{2} & \frac{5}{2} & -\frac{1}{2} \\ -\frac{5}{2} & -\frac{3}{2} & \frac{1}{2} \end{pmatrix}$.

(b) Form the block matrix $M = (B \mid I)$ and row reduce to echelon form:

$$\begin{pmatrix} 1 & 3 & -4 & \vdots & 1 & 0 & 0 \\ 1 & 5 & -1 & \vdots & 0 & 1 & 0 \\ 3 & 13 & -6 & \vdots & 0 & 0 & 1 \end{pmatrix} \sim \begin{pmatrix} 1 & 3 & -4 & \vdots & 1 & 0 & 0 \\ 0 & 2 & 3 & \vdots & -1 & 1 & 0 \\ 0 & 4 & 6 & \vdots & -3 & 0 & 1 \end{pmatrix}$$

$$\sim \begin{pmatrix} 1 & 3 & -4 & \vdots & 1 & 0 & 0 \\ 0 & 2 & 3 & \vdots & -1 & 1 & 0 \\ 0 & 0 & 0 & \vdots & -1 & -2 & 1 \end{pmatrix}$$

In echelon form, M has a zero row in its left half; that is, B is not row reducible to triangular form. Accordingly, B is not invertible.

4.15. Prove the following:

(a) If A and B are invertible, then AB is invertible and $(AB)^{-1} = B^{-1}A^{-1}$.

(b) If $A_1, A_2, \ldots, A_n$ are invertible, then $(A_1 A_2 \cdots A_n)^{-1} = A_n^{-1} \cdots A_2^{-1} A_1^{-1}$.

(c) A is invertible if and only if A^T is invertible.

(d) The operations of inversion and transposing commute: $(A^T)^{-1} = (A^{-1})^T$.

(a) We have

$$(AB)(B^{-1}A^{-1}) = A(BB^{-1})A^{-1} = AIA^{-1} = AA^{-1} = I$$
$$(B^{-1}A^{-1})(AB) = B^{-1}(A^{-1}A)B = B^{-1}IB = B^{-1}B = I$$

Thus $B^{-1}A^{-1}$ is the inverse of AB.

(b) By induction on n and using Part (a), we have

$$(A_1 \cdots A_{n-1}A_n)^{-1} = [(A_1 \cdots A_{n-1})A_n]^{-1} = A_n^{-1}(A_1 \cdots A_{n-1})^{-1} = A_n^{-1} \cdots A_2^{-1}A_1^{-1}$$

(c) If A is invertible, then there exists a matrix B such that $AB = BA = I$. Then

$$(AB)^T = (BA)^T = I^T \quad \text{and so} \quad B^TA^T = A^TB^T = I$$

Hence A^T is invertible, with inverse B^T. The converse follows from the fact that $(A^T)^T = A$.

(d) By Part (c), B^T is the inverse of A^T; that is $B^T = (A^T)^{-1}$. But $B = A^{-1}$; hence $(A^{-1})^T = (A^T)^{-1}$.

4.16. Show that, if A has a zero row or a zero column, then A is not invertible.

By Problem 3.20, if A has a zero row, then AB would have a zero row. Thus if A were invertible, then $AA^{-1} = I$ would imply that I has a zero row. Therefore, A is not invertible. On the other hand, if A has a zero column, then A^T would have a zero row; and so A^T would not be invertible. Thus, again, A is not invertible.

ELEMENTARY MATRICES

4.17. Find the 3-square elementary matrices E_1, E_2, E_3 which correspond, respectively, to the row operations $R_1 \leftrightarrow R_2$, $-7R_3 \rightarrow R_3$ and $-3R_1 + R_2 \rightarrow R_2$.

Apply the operations to the identity matrix $I_3 = \begin{pmatrix} 1 & 0 & 0 \\ 0 & 1 & 0 \\ 0 & 0 & 1 \end{pmatrix}$ to obtain

$$E_1 = \begin{pmatrix} 0 & 1 & 0 \\ 1 & 0 & 0 \\ 0 & 0 & 1 \end{pmatrix} \qquad E_2 = \begin{pmatrix} 1 & 0 & 0 \\ 0 & 1 & 0 \\ 0 & 0 & -7 \end{pmatrix} \qquad E_3 = \begin{pmatrix} 1 & 0 & 0 \\ -3 & 1 & 0 \\ 0 & 0 & 1 \end{pmatrix}$$

4.18. Prove Theorem 4.9.

Let R_i be the ith row of A; we denote this by writing $A = (R_1, \ldots, R_m)$. If B is a matrix for which AB is defined, then it follows directly from the definition of matrix multiplication that $AB = (R_1 B, \ldots, R_m B)$. We also let

$$e_i = (0, \ldots, 0, \widehat{1}, 0, \ldots, 0), \qquad \frown = i$$

Here $\frown = i$ means that 1 is the ith component. By Problem 4.8, $e_i A = R_i$. We also remark that $I = (e_1, \ldots, e_m)$ is the identity matrix.

(i)   Let e be the elementary row operation $R_i \leftrightarrow R_j$. Then, for $\frown = i$ and $\widehat{\frown} = j$,

$$E = e(I) = (e_1, \ldots, \widehat{e_j}, \ldots, \widehat{\widehat{e_i}}, \ldots, e_m)$$

and

$$e(A) = (R_1, \ldots, \widehat{R_j}, \ldots, \widehat{\widehat{R_i}}, \ldots, R_m)$$

Thus

$$EA = (e_1 A, \ldots, \widehat{e_j A}, \ldots, \widehat{\widehat{e_i A}}, \ldots, e_m A) = (R_1, \ldots, \widehat{R_j}, \ldots, \widehat{\widehat{R_i}}, \ldots, R_m) = e(A)$$

(ii) Now let e be the elementary row operation $kR_i \rightarrow R_i$, $k \neq 0$. Then, for $\frown = i$,

$$E = e(I) = (e_1, \ldots, \widehat{ke_i}, \ldots, e_m) \qquad \text{and} \qquad e(A) = (R_1, \ldots, \widehat{kR_i}, \ldots, R_m)$$

Thus

$$EA = (e_1 A, \ldots, \widehat{ke_i A}, \ldots, e_m A) = (R_1, \ldots, \widehat{kR_i}, \ldots, R_m) = e(A)$$

(iii) Last, let e be the elementary row operation $kR_j + R_i \rightarrow R_i$. Then, for $\frown = i$,

$$E = e(I) = (e_1, \ldots, \widehat{ke_j + e_i}, \ldots, e_m) \qquad \text{and} \qquad e(A) = (R_1, \ldots, \widehat{kR_j + R_i}, \ldots, R_m)$$

Using $(ke_j + e_i)A = k(e_j A) + e_i A + kR_j + R_i$, we have

$$EA = (e_1 A, \ldots, \widehat{(ke_j + e_i)A}, \ldots, e_m A) = (R_1, \ldots, \widehat{kR_j + R_i}, \ldots, R_m) = e(A)$$

Thus we have proven the theorem.

4.19. Prove each of the following:

(a) Each of the following elementary row operations has an inverse operation of the same type.

[E_1]   Interchange the ith row and the jth row: $R_i \leftrightarrow R_j$.

$[E_2]$ Multiply the ith row by a nonzero scalar k: $kR_i \to R_i$, $k \neq 0$.

$[E_3]$ Replace the ith row by k times the jth row plus the ith row: $kR_j + R_i \to R_i$.

(b)   Every elementary matrix E is invertible, and its inverse is an elementary matrix.

(a)   Each operation is treated separately.

(1)   Interchanging the same two rows twice, we obtain the original matrix; that is, this operation is its own inverse.

(2)   Multiplying the ith row by k and then by k^{-1}, or by k^{-1} and then by k, we obtain the original matrix. In other words, the operations $kR_i \to R_i$ and $k^{-1}R_i \to R_i$ are inverses.

(3)   Applying the operation $kR_j + R_i \to R_i$ and then the operation $-kR_j + R_i \to R_i$, or applying the operation $-kR_j + R_i \to R_i$ and then the operation $kR_j + R_i \to R_i$, we obtain the original matrix. In other words, the operations $kR_j + R_i \to R_i$ and $-kR_j + R_i \to R_i$ are inverses.

(b)   Let E be the elementary matrix corresponding to the elementary row operation e: $e(I) = E$. Let e' be the inverse operation of e and let E' be its corresponding elementary matrix. Then

$$I = e'(e(I)) = e'(E) = E'E \qquad \text{and} \qquad I = e(e'(I)) = e(E') = EE'$$

Therefore E' is the inverse of E.

4.20.   Prove Theorem 4.10.

Suppose A is invertible and suppose A is row equivalent to a matrix B in row canonical form. Then there exist elementary matrices $E_1, E_2, \ldots, E_s$ such that $E_s \cdots E_2 E_1 A = B$. Since A is invertible and each elementary matrix E_i is invertible, B is invertible. But if $B \neq I$, then B has a zero row; hence B is not invertible. Thus $B = I$, and (a) implies (b).

If (b) holds, then there exist elementary matrices $E_1, E_2, \ldots, E_s$ such that $E_s \cdots E_2 E_1 A = I$, and so $A = (E_s \cdots E_2 E_1)^{-1} = E_1^{-1} E_2^{-1} \cdots E_s^{-1}$. But the E_i^{-1} are also elementary matrices. Thus (b) implies (c).

If (c) holds, then $A = E_1 E_2 \cdots E_s$. The E_i are invertible matrices; hence their product, A, is also invertible. Thus (c) implies (a). Accordingly, the theorem is proved.

4.21.   Prove Theorem 4.11.

Suppose A is not invertible. Then A is not row equivalent to the identity matrix I, and so A is row equivalent to a matrix with a zero row. In other words, there exist elementary matrices $E_1, \ldots, E_s$ such that $E_s \cdots E_2 E_1 A$ has a zero row. Hence $E_s \cdots E_2 E_1 AB = E_s \cdots E_2 E_1$, an invertible matrix, also has a zero row. But invertible matrices cannot have zero rows; hence A is invertible, with inverse A^{-1}. Then also,

$$B = IB = (A^{-1}A)B = A^{-1}(AB) = A^{-1}I = A^{-1}$$

4.22.   Prove Theorem 4.12.

If $B \sim A$, then $B = e_s(\ldots(e_2(e_1(A)))\ldots) = E_s \cdots E_2 E_1 A = PA$, where $P = E_s \cdots E_2 E_1$ is nonsingular. Conversely, suppose $B = PA$ where P is nonsingular. By Theorem 4.10, P is a product of elementary matrices and hence B can be obtained from A by a sequence of elementary row operations, i.e., $B \sim A$. Thus the theorem is proved.

4.23.   Show that B is equivalent to A if and only if there exist invertible matrices P and Q such that $B = PAQ$.

If B is equivalent to A, then $B = E_s \cdots E_2 E_1 A F_1 F_2 \cdots F_t \equiv PAQ$, where $P = E_s \cdots E_2 E_1$ and $Q = F_1 F_2 \cdots F_t$ are invertible. The converse follows from the fact that each step is reversible.

4.24.   Show that equivalence of matrices, written $\approx$, is an equivalence relation: (a) $A \approx A$, (b) If $A \approx B$, then $B \approx A$, (c) If $A \approx B$ and $B \approx C$, then $A \approx C$.

(a) $A = IAI$ where I is nonsingular; hence $A \approx A$.

(b) If $A \approx B$ then $A = PBQ$ where P and Q are nonsingular. Then $B = P^{-1}AQ^{-1}$ where P^{-1} and Q^{-1} are nonsingular. Hence $B \approx A$.

(c) If $A \approx B$ and $B \approx C$, then $A = PBQ$ and $B = P'CQ'$ where P, Q, P', Q' are nonsingular. Then

$$A = P(P'CQ')Q = (PP')C(QQ')$$

where PP' and QQ' are nonsingular. Hence $A \approx C$.

4.25. Prove Theorem 4.17.

The proof is constructive, in the form of an algorithm.

Step 1. Row reduce A to row canonical form, with leading nonzero entries $a_{11}, a_{2j_2}, \ldots, a_{rj_r}$.

Step 2. Interchange C_2 and C_{j_2}, interchange C_3 and $C_{j_3}, \ldots$, and interchange C_r and C_{j_r}. This gives a matrix in the form $\begin{pmatrix} I_r & B \\ \hline 0 & 0 \end{pmatrix}$, with leading nonzero entries $a_{11}, a_{22}, \ldots, a_{rr}$.

Step 3. Use column operations, with the a_{ii} as pivots, to replace each entry in B with a zero; i.e., for

$$i = 1, 2, \ldots, r \qquad \text{and} \qquad j = r + 1, r + 2, \ldots, n,$$

apply the operation $-b_{ij}C_i + C_j \to C_j$.

The final matrix has the desired form $\begin{pmatrix} I_r & 0 \\ \hline 0 & 0 \end{pmatrix}$.

SPECTRAL TYPES OF MATRICES

4.26. Find an upper triangular matrix A such that $A^3 = \begin{pmatrix} 8 & -57 \\ 0 & 27 \end{pmatrix}$.

Set $A = \begin{pmatrix} x & y \\ 0 & z \end{pmatrix}$. Then A^3 has the form $\begin{pmatrix} x^3 & * \\ 0 & z^3 \end{pmatrix}$. Thus $x^3 = 8$, so $x = 2$; $z^3 = 27$, so $z = 3$. Next calculate A^3 using $x = 2$ and $z = 3$:

$$A^2 = \begin{pmatrix} 2 & y \\ 0 & 3 \end{pmatrix}\begin{pmatrix} 2 & y \\ 0 & 3 \end{pmatrix} = \begin{pmatrix} 4 & 5y \\ 0 & 9 \end{pmatrix} \qquad \text{and} \qquad A^3 = \begin{pmatrix} 2 & y \\ 0 & 3 \end{pmatrix}\begin{pmatrix} 4 & 5y \\ 0 & 9 \end{pmatrix} = \begin{pmatrix} 8 & 19y \\ 0 & 27 \end{pmatrix}$$

Thus $19y = -57$, or $y = -3$. Accordingly, $A = \begin{pmatrix} 2 & -3 \\ 0 & 3 \end{pmatrix}$.

4.27. Prove Theorem 4.3(iii).

Let $AB = (c_{ij})$. Then

$$c_{ij} = \sum_{k=1}^{n} a_{ik}b_{kj} \qquad \text{and} \qquad c_{ii} = \sum_{k=1}^{n} a_{ik}b_{ki}$$

Suppose $i > j$. Then, for any k, either $i > k$ or $k > j$, so that either $a_{ik} = 0$ or $b_{kj} = 0$. Thus, $c_{ik} = 0$, and AB is upper triangular. Suppose $i = j$. Then, for $k < i$, $a_{ik} = 0$; and, for $k > i$, $b_{ki} = 0$. Hence $c_{ii} = a_{ii}b_{ii}$, as claimed.

4.28. What kinds of matrices are both upper triangular and lower triangular?

If A is both upper and lower triangular, then every entry off the main diagonal must be zero. Hence A is diagonal.

4.29. Prove Theorem 4.4.

    (i) $(A + A^T)^T = A^T + (A^T)^T = A^T + A = A + A^T$

    (ii) $(A - A^T)^T = A^T - (A^T)^T = A^T - A = -(A - A^T)$

    (iii) Choose $B \equiv \frac{1}{2}(A + A^T)$ and $C \equiv \frac{1}{2}(A - A^T)$, and appeal to (i) and (ii).

4.30. Write $A = \begin{pmatrix} 2 & 3 \\ 7 & 8 \end{pmatrix}$ as the sum of a symmetric matrix B and a skew-symmetric matrix C.

Calculate $A^T = \begin{pmatrix} 2 & 7 \\ 3 & 8 \end{pmatrix}$, $A + A^T = \begin{pmatrix} 4 & 10 \\ 10 & 16 \end{pmatrix}$, and $A - A^T = \begin{pmatrix} 0 & -4 \\ 4 & 0 \end{pmatrix}$. Then

$$B = \frac{1}{2}(A + A^T) = \begin{pmatrix} 2 & 5 \\ 5 & 8 \end{pmatrix} \qquad C = \frac{1}{2}(A - A^T) = \begin{pmatrix} 0 & -2 \\ 2 & 0 \end{pmatrix}$$

4.31. Find x, y, z, s, t if $A = \begin{pmatrix} x & \frac{2}{3} & \frac{2}{3} \\ \frac{2}{3} & \frac{1}{3} & y \\ z & s & t \end{pmatrix}$ is orthogonal.

Let R_1, R_2, R_3 denote the rows of A, and let C_1, C_2, C_3 denote the columns of A. Since R_1 is a unit vector, $x^2 + \frac{4}{9} + \frac{4}{9} = 1$, or $x = \pm\frac{1}{3}$. Since R_2 is a unit vector, $\frac{4}{9} + \frac{1}{9} + y^2 = 1$, or $y = \pm\frac{2}{3}$. Since $R_1 \cdot R_2 = 0$, we get $2x/3 + \frac{2}{9} + 2y/3 = 0$, or $3x + 3y = -1$. The only possibility is that $x = \frac{1}{3}$ and $y = -\frac{2}{3}$. Thus

$$A = \begin{pmatrix} \frac{1}{3} & \frac{2}{3} & \frac{2}{3} \\ \frac{2}{3} & \frac{1}{3} & -\frac{2}{3} \\ z & s & t \end{pmatrix}$$

Since the columns are unit vectors,

$$\frac{1}{9} + \frac{4}{9} + z^2 = 1 \qquad \frac{4}{9} + \frac{1}{9} + s^2 = 1 \qquad \frac{4}{9} + \frac{4}{9} + t^2 = 1$$

Thus $z = \pm\frac{2}{3}$, $s = \pm\frac{2}{3}$, and $t = \pm\frac{1}{3}$.

Case (i): $z = \frac{2}{3}$. Since C_1 and C_2 are orthogonal, $s = -\frac{2}{3}$; since C_1 and C_3 are orthogonal, $t = \frac{1}{3}$.

Case (ii): $z = -\frac{2}{3}$. Since C_1 and C_2 are orthogonal, $s = \frac{2}{3}$; since C_1 and C_3 are orthogonal, $t = -\frac{1}{3}$.

Hence there are exactly two possible solutions:

$$A = \begin{pmatrix} \frac{1}{3} & \frac{2}{3} & \frac{2}{3} \\ \frac{2}{3} & \frac{1}{3} & -\frac{2}{3} \\ \frac{2}{3} & -\frac{2}{3} & \frac{1}{3} \end{pmatrix} \quad \text{and} \quad \begin{pmatrix} \frac{1}{3} & \frac{2}{3} & \frac{2}{3} \\ \frac{2}{3} & \frac{1}{3} & -\frac{2}{3} \\ -\frac{2}{3} & \frac{2}{3} & -\frac{1}{3} \end{pmatrix}$$

4.32. Suppose $A = \begin{pmatrix} a & b \\ c & d \end{pmatrix}$ is orthogonal. Show that $a^2 + b^2 = 1$ and

$$A = \begin{pmatrix} a & b \\ b & -a \end{pmatrix} \quad \text{or} \quad A = \begin{pmatrix} a & b \\ -b & a \end{pmatrix}$$

Since A is orthogonal, the rows of A form an orthonormal set. Hence

$$a^2 + b^2 = 1 \qquad c^2 + d^2 = 1 \qquad ac + bd = 0$$

Similarly, the columns form an orthonormal set, so

$$a^2 + c^2 = 1 \qquad b^2 + d^2 = 1 \qquad ab + cd = 0$$

Therefore, $c^2 = 1 - a^2 = b^2$, whence $c = \pm b$.

Case (i): $c = +b$. Then $b(a + d) = 0$, or $d = -a$; the corresponding matrix is $\begin{pmatrix} a & b \\ b & -a \end{pmatrix}$.

Case (ii): $c = -b$. Then $b = (d - a) = 0$, or $d = a$; the corresponding matrix is $\begin{pmatrix} a & b \\ -b & a \end{pmatrix}$.

4.33. Prove Theorem 4.6.

Let a and b be any real numbers such that $a^2 + b^2 = 1$. Then there exists a real number θ such that $a = \cos \theta$ and $b = \sin \theta$. The result now follows from Problem 4.32.

4.34. Find a 3×3 orthogonal matrix P whose first row is a multiple of $u_1 = (1, 1, 1)$ and whose second row is a multiple of $u_2 = (0, -1, 1)$.

First find a vector u_3 orthogonal to u_1 and u_2, say (cross product) $u_3 = u_1 \times u_2 = (2, -1, -1)$. Let A be the matrix whose rows are u_1, u_2, u_3; and let P be the matrix obtained from A by normalizing the rows of A. Thus

$$A = \begin{pmatrix} 1 & 1 & 1 \\ 0 & -1 & 1 \\ 2 & -1 & -1 \end{pmatrix} \quad \text{and} \quad P = \begin{pmatrix} 1/\sqrt{3} & 1/\sqrt{3} & 1/\sqrt{3} \\ 0 & -1/\sqrt{2} & 1/\sqrt{2} \\ 2/\sqrt{6} & -1/\sqrt{6} & -1/\sqrt{6} \end{pmatrix}$$

4.35. Prove Theorem 4.7.

Suppose $A = \begin{pmatrix} a & b \\ c & d \end{pmatrix}$. Then

$$AA^T = \begin{pmatrix} a & b \\ c & d \end{pmatrix}\begin{pmatrix} a & c \\ b & d \end{pmatrix} = \begin{pmatrix} a^2 + b^2 & ac + bd \\ ac + bd & c^2 + d^2 \end{pmatrix}$$

$$A^T A = \begin{pmatrix} a & c \\ b & d \end{pmatrix}\begin{pmatrix} a & b \\ c & d \end{pmatrix} = \begin{pmatrix} a^2 + c^2 & ab + cd \\ ab + cd & b^2 + d^2 \end{pmatrix}$$

Since $AA^T = A^T A$, we get

$$a^2 + b^2 = a^2 + c^2 \qquad c^2 + d^2 = b^2 + d^2 \qquad ac + bd = ab + cd$$

The first equation yields $b^2 = c^2$; hence $b = c$ or $b = -c$.

Case (i): $b = c$ (which includes the case $b = -c = 0$). Then we obtain the symmetric matrix $A = \begin{pmatrix} a & b \\ b & d \end{pmatrix}$.

Case (ii): $b = -c \neq 0$. Then $ac + bd = b(d - a)$ and $ab + cd = b(a - d)$. Thus $b(d - a) = b(a - d)$, and so $2b(d - a) = 0$. Since $b \neq 0$, we get $a = d$. Thus A has the form

$$A = \begin{pmatrix} a & b \\ -b & a \end{pmatrix} = \begin{pmatrix} a & 0 \\ 0 & a \end{pmatrix} + \begin{pmatrix} 0 & b \\ -b & 0 \end{pmatrix}$$

which is the sum of a scalar matrix and a skew-symmetric matrix.

COMPLEX MATRICES

4.36. Find the conjugate of the matrix $A = \begin{pmatrix} 2 + i & 3 - 5i & 4 + 8i \\ 6 - i & 2 - 9i & 5 + 6i \end{pmatrix}$.

Take the conjugate of each element of the matrix (where $\overline{a + bi} = a - bi$):

$$\bar{A} = \overline{\begin{pmatrix} 2 + i & 3 - 5i & 4 + 8i \\ 6 - i & 2 - 9i & 5 + 6i \end{pmatrix}} = \begin{pmatrix} 2 - i & 3 + 5i & 4 - 8i \\ 6 + i & 2 + 9i & 5 - 6i \end{pmatrix}$$

4.37. Find A^H when $A = \begin{pmatrix} 2-3i & 5+8i \\ -4 & 3-7i \\ -6-i & 5i \end{pmatrix}$.

$A^H = \bar{A}^T$, the conjugate transpose of A. Hence,

$$A^H = \begin{pmatrix} \overline{2-3i} & \overline{-4} & \overline{-6-i} \\ \overline{5+8i} & \overline{3-7i} & \overline{5i} \end{pmatrix} = \begin{pmatrix} 2+3i & -4 & -6+i \\ 5-8i & 3+7i & -5i \end{pmatrix}$$

4.38. Write $A = \begin{pmatrix} 2+6i & 5+3i \\ 9-i & 4-2i \end{pmatrix}$ in the form $A = B + C$, where B is Hermitian and C is skew-Hermitian.

First find

$$A^H = \begin{pmatrix} 2-6i & 9+i \\ 5-3i & 4+2i \end{pmatrix} \qquad A + A^H = \begin{pmatrix} 4 & 14+4i \\ 14-4i & 8 \end{pmatrix} \qquad A - A^H = \begin{pmatrix} 12i & -4+2i \\ 4+2i & -4i \end{pmatrix}$$

Then the required matrices are

$$B = \frac{1}{2}(A + A^H) = \begin{pmatrix} 2 & 7+2i \\ 7-2i & 4 \end{pmatrix} \quad \text{and} \quad C = \frac{1}{2}(A - A^H) = \begin{pmatrix} 6i & -2+i \\ 2+i & -2i \end{pmatrix}$$

4.39. Define an orthonormal set of vectors in $\mathbf{C}^n$ and prove the following complex analogue of Theorem 4.5:

Theorem: Let A be a complex matrix. Then the following are equivalent: (*a*) A is unitary; (*b*) the rows of A form an orthonormal set; (*c*) the columns of A form an orthonormal set.

The vectors $u_1, u_2, \ldots, u_r$ in $\mathbf{C}^n$ form an orthonormal set if $u_i \cdot u_j = \delta_{ij}$ where the dot product in $\mathbf{C}^n$ is defined by

$$(a_1, a_2, \ldots, a_n) \cdot (b_1, b_2, \ldots, b_n) = a_1 \overline{b_1} + a_2 \overline{b_2} + \cdots + a_n \overline{b_n}$$

and δ_{ij} is the Kronecker delta [see Example 4.3(*a*)].

Let $R_1, \ldots, R_n$ denote the rows of A; then $\bar{R}_1^T, \ldots, \bar{R}_n^T$ are the columns of A^H. Let $AA^H = (c_{ij})$. By matrix multiplication, $c_{ij} = R_i \bar{R}_j^T = R_i \cdot R_j$. Then $AA^H = I$ iff $R_i \cdot R_j = \delta_{ij}$ iff $R_1, R_2, \ldots, R_n$ form an orthonormal set. Thus (*a*) and (*b*) are equivalent. Similarly, A is unitary iff A^H is unitary iff the rows of A^H are orthonormal iff the conjugates of the columns of A are orthonormal iff the columns of A are orthonormal. Thus (*a*) and (*c*) are equivalent, and the theorem is proved.

4.40. Show that $A = \begin{pmatrix} \frac{1}{2} - \frac{2}{3}i & \frac{2}{3}i \\ -\frac{2}{3}i & -\frac{1}{3} - \frac{2}{3}i \end{pmatrix}$ is unitary.

The rows form an orthonormal set:

$$(\tfrac{1}{3} - \tfrac{2}{3}i, \tfrac{2}{3}i) \cdot (\tfrac{1}{3} - \tfrac{2}{3}i, \tfrac{2}{3}i) = (\tfrac{1}{9} + \tfrac{4}{9}) + \tfrac{4}{9} = 1$$

$$(\tfrac{1}{3} - \tfrac{2}{3}i, \tfrac{2}{3}i) \cdot (-\tfrac{2}{3}i, -\tfrac{1}{3} - \tfrac{2}{3}i) = (\tfrac{2}{3}i + \tfrac{4}{9}) + (-\tfrac{2}{9}i - \tfrac{4}{9}) = 0$$

$$(-\tfrac{2}{3}i, -\tfrac{1}{3} - \tfrac{2}{3}i) \cdot (-\tfrac{2}{3}i, -\tfrac{1}{3} - \tfrac{2}{3}i) = \tfrac{4}{9} + (\tfrac{1}{9} + \tfrac{4}{9}) = 1$$

Thus A is unitary.

SQUARE BLOCK MATRICES

4.41. Determine which matrix is a square block matrix:

$$A = \begin{pmatrix} 1 & 2 & \vdots & 3 & \vdots & 4 & 5 \\ 1 & 1 & \vdots & 1 & \vdots & 1 & 1 \\ 9 & 8 & \vdots & 7 & 6 & 5 \\ 3 & 3 & \vdots & 3 & 3 & 3 \\ 1 & 3 & \vdots & 5 & \vdots & 7 & 9 \end{pmatrix} \qquad B = \begin{pmatrix} 1 & 2 & \vdots & 3 & \vdots & 4 & 5 \\ 1 & 1 & \vdots & 1 & \vdots & 1 & 1 \\ 9 & 8 & \vdots & 7 & 6 & 5 \\ 3 & 3 & \vdots & 3 & 3 & 3 \\ 1 & 3 & \vdots & 5 & \vdots & 7 & 9 \end{pmatrix}$$

Although A is a 5×5 square matrix and is a 3×3 block matrix, the second and third diagonal blocks are not square matrices. Thus A is not a square block matrix.

B is a square block matrix.

4.42. Complete the partitioning of $C = \begin{pmatrix} 1 & 2 & 3 & 4 & 5 \\ 1 & 1 & 1 & 1 & 1 \\ 9 & 8 & 7 & 6 & 5 \\ 3 & 3 & 3 & 3 & 3 \\ 1 & 3 & 5 & 7 & 9 \end{pmatrix}$ into a square block matrix.

One horizontal line is between the second and third rows; hence add a vertical line between the second and third columns. The other horizontal line is between the fourth and fifth rows; hence add a vertical line between the fourth and fifth columns. [The horizontal lines and the vertical lines must be symmetrically placed to obtain a square block matrix.] This yields the square block matrix

$$C = \begin{pmatrix} 1 & 2 & \vdots & 3 & \vdots & 4 & 5 \\ 1 & 1 & \vdots & 1 & \vdots & 1 & 1 \\ 9 & 8 & \vdots & 7 & \vdots & 6 & 5 \\ 3 & 3 & \vdots & 3 & \vdots & 3 & 3 \\ 1 & 3 & \vdots & 5 & \vdots & 7 & 9 \end{pmatrix}$$

4.43. Determine which of the following square block matrices are lower triangular, upper triangular, or diagonal:

$$A = \begin{pmatrix} 1 & 2 & \vdots & 0 \\ 3 & 4 & \vdots & 5 \\ 0 & 0 & \vdots & 6 \end{pmatrix} \quad B = \begin{pmatrix} 1 & \vdots & 0 & 0 & \vdots & 0 \\ 2 & \vdots & 3 & 4 & \vdots & 0 \\ 5 & \vdots & 0 & 6 & \vdots & 0 \\ 0 & \vdots & 7 & 8 & \vdots & 9 \end{pmatrix} \quad C = \begin{pmatrix} 1 & \vdots & 0 & 0 \\ 0 & \vdots & 2 & 3 \\ 0 & \vdots & 4 & 5 \end{pmatrix} \quad D = \begin{pmatrix} 1 & 2 & \vdots & 0 \\ 3 & 4 & \vdots & 5 \\ 0 & 6 & \vdots & 7 \end{pmatrix}$$

A is upper triangular since the block below the diagonal is a zero block.

B is lower triangular since all blocks above the diagonal are zero blocks.

C is diagonal since the blocks above and below the diagonal are zero blocks.

D is neither upper triangular nor lower triangular. Furthermore, no other partitioning of D will make it into either a block upper triangular matrix or a block lower triangular matrix.

4.44. Consider the following block diagonal matrices of which corresponding diagonal blocks have the same size:

$$M = \text{diag}\,(A_1, A_2, \ldots, A_r) \qquad \text{and} \qquad N = \text{diag}\,(B_1, B_2, \ldots, B_r)$$

Find: (a) $M + N$, (b) kM, (c) MN, (d) $f(M)$ for a given polynomial $f(x)$.

(a) Simply add the diagonal blocks: $M + N = \text{diag}\,(A_1 + B_1, A_2 + B_2, \ldots, A_r + B_r)$.

(b) Simply multiply the diagonal blocks by k: $kM = \text{diag}\,(kA_1, kA_2, \ldots, kA_r)$.

(c) Simply multiply corresponding diagonal blocks: $MN = \text{diag}(A_1 B_1, A_2 B_2, \ldots, A_r B_r)$.

(d) Find $f(A_i)$ for each diagonal block A_i. Then $f(M) = \text{diag}(f(A_1), f(A_2), \ldots, f(A_r))$.

4.45. Find M^2 where $M = \begin{pmatrix} 1 & 2 & & & \\ 3 & 4 & & & \\ & & 5 & & \\ & & & 1 & 3 \\ & & & 5 & 7 \end{pmatrix}$.

Since M is block diagonal, square each block:

$$\begin{pmatrix} 1 & 2 \\ 3 & 4 \end{pmatrix}\begin{pmatrix} 1 & 2 \\ 3 & 4 \end{pmatrix} = \begin{pmatrix} 7 & 10 \\ 15 & 22 \end{pmatrix}$$

$$(5)(5) = (25)$$

$$\begin{pmatrix} 1 & 3 \\ 5 & 7 \end{pmatrix}\begin{pmatrix} 1 & 3 \\ 5 & 7 \end{pmatrix} = \begin{pmatrix} 16 & 24 \\ 40 & 64 \end{pmatrix}$$

Then $M^2 = \begin{pmatrix} 7 & 10 & & & \\ 15 & 22 & & & \\ & & 25 & & \\ & & & 16 & 24 \\ & & & 40 & 64 \end{pmatrix}$.

CONGRUENT SYMMETRIC MATRICES, QUADRATIC FORMS

4.46. Let $A = \begin{pmatrix} 1 & -3 & 2 \\ -3 & 7 & -5 \\ 2 & -5 & 8 \end{pmatrix}$, a symmetric matrix. Find (a) a nonsingular matrix P such that $P^T A P$

is diagonal, i.e., the diagonal matrix $B = P^T A P$, and (b) the signature of A.

(a) First form the block matrix $(A \vdots I)$:

$$(A \vdots I) = \begin{pmatrix} 1 & -3 & 2 & \vdots & 1 & 0 & 0 \\ -3 & 7 & -5 & \vdots & 0 & 1 & 0 \\ 2 & -5 & 8 & \vdots & 0 & 0 & 1 \end{pmatrix}$$

Apply the row operations $3R_1 + R_2 \to R_2$ and $-2R_1 + R_3 \to R_3$ to $(A \vdots I)$ and then the corresponding column operations $3C_1 + C_2 \to C_2$ and $-2C_1 + C_3 \to C_3$ to A to obtain

$$\begin{pmatrix} 1 & -3 & 2 & \vdots & 1 & 0 & 0 \\ 0 & -2 & 1 & \vdots & 3 & 1 & 0 \\ 0 & 1 & 4 & \vdots & -2 & 0 & 1 \end{pmatrix} \quad \text{and then} \quad \begin{pmatrix} 1 & 0 & 0 & \vdots & 1 & 0 & 0 \\ 0 & -2 & 1 & \vdots & 3 & 1 & 0 \\ 0 & 1 & 4 & \vdots & -2 & 0 & 1 \end{pmatrix}$$

Next apply the row operation $R_2 + 2R_3 \to R_3$ and then the corresponding column operation $C_2 + 2C_3 \to C_3$ to obtain

$$\begin{pmatrix} 1 & 0 & 0 & \vdots & 1 & 0 & 0 \\ 0 & -2 & 1 & \vdots & 3 & 1 & 0 \\ 0 & 0 & 9 & \vdots & -1 & 1 & 2 \end{pmatrix} \quad \text{and then} \quad \begin{pmatrix} 1 & 0 & 0 & \vdots & 1 & 0 & 0 \\ 0 & -2 & 0 & \vdots & 3 & 1 & 0 \\ 0 & 0 & 18 & \vdots & -1 & 1 & 2 \end{pmatrix}$$

Now A has been diagonalized. Set

$$P = \begin{pmatrix} 1 & 3 & -1 \\ 0 & 1 & 1 \\ 0 & 0 & 2 \end{pmatrix} \quad \text{and then} \quad B = P^T A P = \begin{pmatrix} 1 & 0 & 0 \\ 0 & -2 & 0 \\ 0 & 0 & 18 \end{pmatrix}$$

(b) B has $\mathbf{p} = 2$ positive and $\mathbf{n} = 1$ negative diagonal element. Hence sig. $A = 2 - 1 = 1$.

QUADRATIC FORMS

4.47. Find the quadratic form $q(x, y)$ corresponding to the symmetric matrix $A = \begin{pmatrix} 5 & -3 \\ -3 & 8 \end{pmatrix}$.

$$q(x, y) = (x, y)\begin{pmatrix} 5 & -3 \\ -3 & 8 \end{pmatrix}\begin{pmatrix} x \\ y \end{pmatrix} = (5x - 3y, \ -3x + 8y)\begin{pmatrix} x \\ y \end{pmatrix}$$

$$= 5x^2 - 3xy - 3xy + 8y^2 = 5x^2 - 6xy + 8y^2$$

4.48. Find the symmetric matrix A which corresponds to the quadratic form

$$q(x, y, z) = 3x^2 + 4xy - y^2 + 8xz - 6yz + z^2$$

The symmetric matrix $A = (a_{ij})$ representing $q(x_1, \ldots, x_n)$ has the diagonal entry a_{ii} equal to the coefficient of x_i^2 and has the entries a_{ij} and a_{ji} each equal to half the coefficient of $x_i x_j$. Thus

$$A = \begin{pmatrix} 3 & 2 & 4 \\ 2 & -1 & -3 \\ 4 & -3 & 1 \end{pmatrix}$$

4.49. Find the symmetric matrix B which corresponds to the quadratic form:

$$(a)\ q(x, y) = 4x^2 + 5xy - 7y^2 \text{ and } (b)\ q(x, y, z) = 4xy + 5y^2.$$

(a) Here $B = \begin{pmatrix} 4 & \frac{5}{2} \\ \frac{5}{2} & -7 \end{pmatrix}$. (Division by 2 may introduce fractions even though the coefficients in q are integers.)

(b) Even though only x and y appears in the polynomial, the expression $q(x, y, z)$ indicates that there are three variables. In other words,

$$q(x, y, z) = 0x^2 + 4xy + 5y^2 + 0xz + 0yz + 0z^2$$

Thus

$$B = \begin{pmatrix} 0 & 2 & 0 \\ 2 & 5 & 0 \\ 0 & 0 & 0 \end{pmatrix}$$

4.50. Consider the quadratic form $q(x, y) = 3x^2 + 2xy - y^2$ and the linear substitution, $x = s - 3t$, $y = 2s + t$.

(a) Rewrite $q(x, y)$ in matrix notation, and find the matrix A representing the quadratic form.

(b) Rewrite the linear substitution using matrix notation, and find the matrix P corresponding to the substitution.

(c) Find $q(s, t)$ using direct substitution.

(d) Find $q(s, t)$ using matrix notation.

(a) Here $q(x, y) = (x, y)\begin{pmatrix} 3 & 1 \\ 1 & -1 \end{pmatrix}\begin{pmatrix} x \\ y \end{pmatrix}$. Hence $A = \begin{pmatrix} 3 & 1 \\ 1 & -1 \end{pmatrix}$ and $q(X) = X^T A X$ where $X = (x, y)^T$.

(b) We have $\begin{pmatrix} x \\ y \end{pmatrix} = \begin{pmatrix} 1 & -3 \\ 2 & 1 \end{pmatrix}\begin{pmatrix} s \\ t \end{pmatrix}$. Thus $P = \begin{pmatrix} 1 & -3 \\ 2 & 1 \end{pmatrix}$ and $X = PY$, where $X = (x, y)^T$ and $Y = (s, t)^T$.

(c)   Substitute for x and y in q to obtain

$$q(s, t) = 3(s - 3t)^2 + 2(s - 3t)(2s + t) - (2s + t)^2$$
$$= 3(s^2 - 6st + 9t^2) + 2(2s^2 - 5st - 3t^2) - (s^2 + 4st + t^2) = 3s^2 - 32st + 20t^2$$

(d)   Here $q(X) = X^T A X$ and $X = PY$. Thus $X^T = Y^T P^T$. Therefore,

$$q(s, t) = q(Y) = Y^T P^T A P Y = (s, t)\begin{pmatrix} 1 & 2 \\ -3 & 1 \end{pmatrix}\begin{pmatrix} 3 & 1 \\ 1 & -1 \end{pmatrix}\begin{pmatrix} 1 & -3 \\ 2 & 1 \end{pmatrix}\begin{pmatrix} s \\ t \end{pmatrix}$$

$$= (s, t)\begin{pmatrix} 3 & -16 \\ -16 & 20 \end{pmatrix}\begin{pmatrix} s \\ t \end{pmatrix} = 3s^2 - 32st + 20t^2$$

[As expected, the results in (c) and (d) are equal.]

4.51.   Let L be a linear substitution $X = PY$, as in Problem 4.50.

(a)   When is L nonsingular? orthogonal?

(b)   Describe one main advantage of an orthogonal substitution over a nonsingular substitution.

(c)   Is the linear substitution in Problem 4.50 nonsingular? orthogonal?

(a) L is said to be nonsingular or orthogonal according as the matrix P representing the substitution is nonsingular or orthogonal.

(b)   Recall that the columns of the matrix P representing the linear substitution introduces a new coordinate system. If P is orthogonal, then the new axes are perpendicular and have the same unit lengths as the original axes.

(c)   The matrix $P = \begin{pmatrix} 1 & -3 \\ 2 & 1 \end{pmatrix}$ is nonsingular, but not orthogonal; hence the linear substitution is non-singular, but not orthogonal.

4.52.   Let $q(x, y, z) = x^2 + 4xy + 3y^2 - 8xz - 12yz + 9z^2$. Find a nonsingular linear substitution expressing the variables x, y, z in terms of the variables r, s, t so that $q(r, s, t)$ is diagonal. Also find the signature of q.

Form the block matrix $(A \mid I)$ where A is the matrix which corresponds to the quadratic form:

$$(A \mid I) = \begin{pmatrix} 1 & 2 & -4 & \vdots & 1 & 0 & 0 \\ 2 & 3 & -6 & \vdots & 0 & 1 & 0 \\ -4 & -6 & 9 & \vdots & 0 & 0 & 1 \end{pmatrix}$$

Apply $-2R_1 + R_2 \to R_2$ and $4R_1 + R_4 \to R_3$ and the corresponding column operations, and then $2R_2 + R_3 \to R_3$ and the corresponding column operation to obtain

$$\begin{pmatrix} 1 & 0 & 0 & \vdots & 1 & 0 & 0 \\ 0 & -1 & 2 & \vdots & -2 & 1 & 0 \\ 0 & 2 & -7 & \vdots & 4 & 0 & 1 \end{pmatrix} \quad \text{and then} \quad \begin{pmatrix} 1 & 0 & 0 & \vdots & 1 & 0 & 0 \\ 0 & -1 & 0 & \vdots & -2 & 1 & 0 \\ 0 & 0 & -3 & \vdots & 0 & 2 & 1 \end{pmatrix}$$

Thus the linear substitution $x = r - 2s$, $y = s + 2t$, $z = t$ will yield the quadratic form

$$q(r, s, t) = r^2 - s^2 - 3t^2$$

Also, sig $q = 1 - 2 = -1$.

4.53.   Diagonalize the following quadratic form q by the method known as "completing the square":

$$q(x, y) = 2x^2 - 12xy + 5y^2$$

First factor out the coefficient of x^2 from the x^2 term and the xy term to get

$$q(x, y) = 2(x^2 - 6xy \qquad) + 5y^2$$

Next complete the square inside the parentheses by adding an appropriate multiple of y^2 and then subtract the corresponding amount outside the parentheses to get

$$q(x, y) = 2(x^2 - 6xy + 9y^2) + 5y^2 - 18y^2 = 2(x - 3y)^2 - 13y^2$$

(The -18 comes from the fact that the $9y^2$ inside the parentheses is multiplied by 2.) Let $s = x - 3y$, $t = y$. Then $x = s + 3t$, $y = t$. This linear substitution yields the quadratic form $q(s, t) = 2s^2 - 13t^2$.

POSITIVE DEFINITE QUADRATIC FORMS

4.54. Let $q(x, y, z) = x^2 + 2y^2 - 4xz - 4yz + 7z^2$. Is q positive definite?

Diagonalize (under congruence) the symmetric matrix A corresponding to q (by applying $2R_1 + R_3 \to R_3$ and $2C_1 + C_3 \to C_3$, and then $R_2 + R_3 \to R_3$ and $C_2 + C_3 \to C_3$):

$$A = \begin{pmatrix} 1 & 0 & -2 \\ 0 & 2 & -2 \\ -2 & -2 & 7 \end{pmatrix} \to \begin{pmatrix} 1 & 0 & 0 \\ 0 & 2 & -2 \\ 0 & -2 & 3 \end{pmatrix} \to \begin{pmatrix} 1 & 0 & 0 \\ 0 & 2 & 0 \\ 0 & 0 & 1 \end{pmatrix}$$

The diagonal representation of q only contains positive entries, 1, 2, and 1, on the diagonal; hence q is positive definite.

4.55. Let $q(x, y, z) = x^2 + y^2 + 2xz + 4yz + 3z^2$. Is q positive definite?

Diagonalize (under congruence) the symmetric matrix A corresponding to q:

$$A = \begin{pmatrix} 1 & 0 & 1 \\ 0 & 1 & 2 \\ 1 & 2 & 3 \end{pmatrix} \to \begin{pmatrix} 1 & 0 & 0 \\ 0 & 1 & 2 \\ 0 & 2 & 2 \end{pmatrix} \to \begin{pmatrix} 1 & 0 & 0 \\ 0 & 1 & 0 \\ 0 & 0 & -2 \end{pmatrix}$$

There is a negative entry -2 in the diagonal representation of q; hence q is not positive definite.

4.56. Show that $q(x, y) = ax^2 + bxy + cy^2$ is positive definite if and only if the discriminant $D = b^2 - 4ac < 0$.

Suppose $v = (x, y) \neq 0$, say $y \neq 0$. Let $t = x/y$. Then

$$q(v) = y^2[a(x/y)^2 + b(x/y) + c] = y^2(at^2 + bt + c)$$

However, $s = at^2 + bt + c$ lies above the t axis, i.e., is positive for every value of t if and only if $D = b^2 - 4ac < 0$. Thus q is positive definite if and only if $D < 0$.

4.57. Determine which quadratic form q is positive definite:

$$(a) \quad q(x, y) = x^2 - 4xy + 5y^2 \qquad (b) \quad q(x, y) = x^2 + 6xy + 3y^2$$

(a) **Method 1.**   Diagonalize by completing the square:

$$q(x, y) = x^2 - 4xy + 4y^2 + 5y^2 - 4y^2 = (x - 2y)^2 + y^2 = s^2 + t^2$$

where $s = x - 2y$, $t = y$. Thus q is positive definite.

Method 2.   Compute the discriminant $D = b^2 - 4ac = 16 - 20 = -4$. Since $D < 0$, q is positive definite.

(b) **Method 1.**   Diagonalize by completing the square:

$$q(x, y) = x^2 + 6xy + 9y^2 + 3y^2 - 9y^2 = (x + 3y)^2 - 6y^2 = s^2 - 6t^2$$

where $s = x + 3y$, $t = y$. Since -6 is negative, q is not positive definite.

Method 2.   Compute $D = b^2 - 4ac = 36 - 12 = 24$. Since $D > 0$, q is not positive definite.

4.58. Let B be any nonsingular matrix, and let $M = B^T B$. Show that (a) M is symmetric, and (b) M is positive definite.

(a) $M^T = (B^T B)^T = B^T B^{TT} = B^T B = M$; hence M is symmetric.

(b) Since B is nonsingular, $BX \neq 0$ for any nonzero $X \in \mathbf{R}^n$. Hence the dot product of BX with itself, $BX \cdot BX = (BX)^T(BX)$, is positive. Thus

$$q(X) = X^T M X = X^T(B^T B)X = (X^T B^T)(BX) = (BX)^T(BX) > 0$$

Thus M is positive definite.

4.59. Show that $q(X) = \| X \|^2$, the square of the norm of a vector X, is a positive definite quadratic form.

For $X = (x_1, x_2, \ldots, x_n)$, we have $q(X) = x_1^2 + x_2^2 + \cdots + x_n^2$. Now q is a polynomial with each term of degree two, and q is in diagonal form where all diagonal entries are positive. Thus q is a positive definite quadratic form.

4.60. Prove that the following two definitions of a positive definite quadratic form are equivalent:

(a)   The diagonal entries are all positive in any diagonal representation of q.

(b) $q(Y) > 0$, for any nonzero vector Y in $\mathbf{R}^n$.

Suppose $q(Y) = a_1 y_1^2 + a_2 y_2^2 + \cdots + a_n x_n^2$. If all the coefficients a_i are positive, then clearly $q(Y) > 0$ for any nonzero vector Y. Thus (a) implies (b). Conversely, suppose $a_k < 0$. Let $e_k = (0, \ldots, 1, \ldots, 0)$ be the vector whose entries are all 0 except 1 in the kth position. Then $q(e_k) = a_k < 0$ for $e_k \neq 0$. Thus (b) implies (a). Accordingly, (a) and (b) are equivalent.

SIMILARITY OF MATRICES

4.61. Consider the cartesian plane $\mathbf{R}^2$ with the usual x and y axes. The 2×2 nonsingular matrix

$$P = \begin{pmatrix} 1 & 3 \\ -1 & 2 \end{pmatrix}$$

determines a new coordinate system of the plane, say with s and t axes. (See Example 4.16.)

(a)   Plot the new s and t axis in the plane $\mathbf{R}^2$.

(b)   Find the coordinates of $Q(1, 5)$ in the new system.

(a)   Plot the s axis in the direction of the first column $u_1 = (1, -1)^T$ of P with unit length equal to the length u_1. Similarly, plot the t axis in the direction of the second column $u_2 = (3, 2)^T$ of P with unit length equal to the length of u_2. See Fig. 4-2.

(b)   Find $P^{-1} = \begin{pmatrix} \frac{2}{5} & -\frac{3}{5} \\ \frac{1}{5} & \frac{1}{5} \end{pmatrix}$, say by using the formula for the inverse of a 2×2 matrix. Then multiply the coordinate (column) vector of Q by P^{-1}:

$$P^{-1}Q = \begin{pmatrix} \frac{2}{5} & -\frac{3}{5} \\ \frac{1}{5} & \frac{1}{5} \end{pmatrix}\begin{pmatrix} 1 \\ 5 \end{pmatrix} = \begin{pmatrix} -\frac{13}{5} \\ \frac{6}{5} \end{pmatrix}$$

Thus $Q'(-\frac{13}{5}, \frac{6}{5})$ represents Q in the new system.

4.62. Let $f: \mathbf{R}^2 \to \mathbf{R}^2$ be defined by $f(x, y) = (2x - 5y, 3x + 4y)$.

(a)   Using $X = (x, y)^T$, write f in matrix notation, i.e., find the matrix A such that $f(X) = AX$.

(b)   Referring to the new coordinate s and t axes of $\mathbf{R}^2$ introduced in Problem 4.61, and using $Y = (s, t)^T$, find $f(s, t)$ by first finding the matrix B such that $f(Y) = BY$.

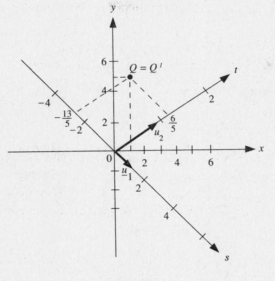

Fig. 4-2

(a) Here $f\begin{pmatrix} x \\ y \end{pmatrix} = \begin{pmatrix} 2 & -5 \\ 3 & 4 \end{pmatrix}\begin{pmatrix} x \\ y \end{pmatrix}$; hence $A = \begin{pmatrix} 2 & -5 \\ 3 & 4 \end{pmatrix}$.

(b) Find $B = P^{-1}AP = \begin{pmatrix} \frac{2}{5} & -\frac{3}{5} \\ \frac{1}{5} & \frac{1}{5} \end{pmatrix}\begin{pmatrix} 2 & -5 \\ 3 & 4 \end{pmatrix}\begin{pmatrix} 1 & 3 \\ -1 & 2 \end{pmatrix} = \begin{pmatrix} \frac{17}{5} & -\frac{59}{5} \\ \frac{6}{5} & \frac{13}{5} \end{pmatrix}$. Then

$$f\begin{pmatrix} s \\ t \end{pmatrix} = \begin{pmatrix} \frac{17}{5} & -\frac{59}{5} \\ \frac{6}{5} & \frac{13}{5} \end{pmatrix}\begin{pmatrix} s \\ t \end{pmatrix}$$

Thus $f(s, t) = (\frac{17}{5}s - \frac{59}{5}t, \frac{6}{5}s + \frac{13}{5}t)$.

4.63. Consider the space $\mathbf{R}^3$ with the usual x, y, z axes. The 3×3 nonsingular matrix

$$P = \begin{pmatrix} 1 & 3 & -2 \\ -2 & -5 & 2 \\ 1 & 2 & 1 \end{pmatrix}$$

determines a new coordinate system for $\mathbf{R}^3$, say with r, s, t axes. [Alternatively, P defines the linear substitution $X = PY$, where $X = (x, y, z)^T$ and $Y = (r, s, t)^T$.] Find the coordinates of the point $Q(1, 2, 3)$ in the new system.

First find P^{-1}. Form the block matrix $M = (P \mid I)$ and reduce M to row canonical form:

$$M = \begin{pmatrix} 1 & 3 & -2 & \vdots & 1 & 0 & 0 \\ -2 & -5 & 2 & \vdots & 0 & 1 & 0 \\ 1 & 2 & 1 & \vdots & 0 & 0 & 1 \end{pmatrix} \sim \begin{pmatrix} 1 & 3 & -2 & \vdots & 1 & 0 & 0 \\ 0 & 1 & -2 & \vdots & 2 & 1 & 0 \\ 0 & -1 & 3 & \vdots & -1 & 0 & 1 \end{pmatrix}$$

$$\sim \begin{pmatrix} 1 & 3 & -2 & \vdots & 1 & 0 & 0 \\ 0 & 1 & -2 & \vdots & 2 & 1 & 0 \\ 0 & 0 & 1 & \vdots & 1 & 1 & 1 \end{pmatrix} \sim \begin{pmatrix} 1 & 3 & 0 & \vdots & 3 & 2 & 2 \\ 0 & 1 & 0 & \vdots & 4 & 3 & 2 \\ 0 & 0 & 1 & \vdots & 1 & 1 & 1 \end{pmatrix}$$

$$\sim \begin{pmatrix} 1 & 0 & 0 & \vdots & -9 & -7 & -4 \\ 0 & 1 & 0 & \vdots & 4 & 3 & 2 \\ 0 & 0 & 1 & \vdots & 1 & 1 & 1 \end{pmatrix}$$

Accordingly,

$$P^{-1} = \begin{pmatrix} -9 & -7 & -4 \\ 4 & 3 & 2 \\ 1 & 1 & 1 \end{pmatrix} \quad \text{and} \quad P^{-1}Q = \begin{pmatrix} -9 & -7 & -4 \\ 4 & 3 & 2 \\ 1 & 1 & 1 \end{pmatrix}\begin{pmatrix} 1 \\ 2 \\ 3 \end{pmatrix} = \begin{pmatrix} -47 \\ 16 \\ 6 \end{pmatrix}$$

Thus $Q'(-47, 16, 6)$ represents Q in the new system.

4.64. Let $f : \mathbf{R}^3 \to \mathbf{R}^3$ be defined by

$$f(x, y, z) = (x + 2y - 3z, 2x + z, x - 3y + z)$$

and let P be the nonsingular change-of-variable matrix in Problem 4.63. [Thus, $X = PY$ where $X = (x, y, z)^T$ and $Y = (r, s, t)^T$.] Find: (a) the matrix A such that $f(X) = AX$, (b) the matrix B such that $f(Y) = BY$, and (c) $f(r, s, t)$.

(a) The coefficients of x, y and z give the matrix A:

$$f\begin{pmatrix} x \\ y \\ z \end{pmatrix} = \begin{pmatrix} 1 & 2 & -3 \\ 2 & 0 & 1 \\ 1 & -3 & 1 \end{pmatrix}\begin{pmatrix} x \\ y \\ z \end{pmatrix} \quad \text{and so} \quad A = \begin{pmatrix} 1 & 2 & -3 \\ 2 & 0 & 1 \\ 1 & -3 & 1 \end{pmatrix}$$

(b) Here B is similar to A with respect to P, that is,

$$B = P^{-1}AP = \begin{pmatrix} -9 & -7 & -4 \\ 4 & 3 & 2 \\ 1 & 1 & 1 \end{pmatrix}\begin{pmatrix} 1 & 2 & -3 \\ 2 & 0 & 1 \\ 1 & -3 & 1 \end{pmatrix}\begin{pmatrix} 1 & 3 & -2 \\ -2 & -5 & 2 \\ 1 & 2 & 1 \end{pmatrix} = \begin{pmatrix} 1 & -19 & 58 \\ 1 & 12 & -27 \\ 5 & 15 & -11 \end{pmatrix}$$

(c) Use the matrix B to obtain

$$f(r, s, t) = (r - 19s + 58t, r + 12s - 27t, 5r + 15s - 11t)$$

4.65. Suppose B is similar to A. Prove tr $B = $ tr A.

Since B is similar to A, there exists a nonsingular matrix P such that $B = P^{-1}AP$. Then, using Theorem 4.1,

$$\text{tr } B = \text{tr } P^{-1}AP = \text{tr } PP^{-1}A = \text{tr } A$$

LU FACTORIZATION

4.66. Find the *LU* factorization of $A = \begin{pmatrix} 1 & 3 & 2 \\ 2 & 5 & 6 \\ -3 & -2 & 7 \end{pmatrix}$.

Reduce A to triangular form by the operations $-2R_1 + R_2 \to R_2$ and $3R_1 + R_3 \to R_3$, and then $7R_2 + R_3 \to R_3$:

$$A \sim \begin{pmatrix} 1 & 3 & 2 \\ 0 & -1 & 2 \\ 0 & 7 & 13 \end{pmatrix} \sim \begin{pmatrix} 1 & 3 & 2 \\ 0 & -1 & 2 \\ 0 & 0 & 27 \end{pmatrix}$$

Use the negatives of the multipliers -2, 3, and 7 in the above row operations to form the matrix L, and use the triangular form of A to obtain the matrix U; that is,

$$L = \begin{pmatrix} 1 & 0 & 0 \\ 2 & 1 & 0 \\ -3 & -7 & 1 \end{pmatrix} \quad \text{and} \quad U = \begin{pmatrix} 1 & 3 & 2 \\ 0 & -1 & 2 \\ 0 & 0 & 27 \end{pmatrix}$$

(As a check, multiply L and U to verify that $A = LU$.)

4.67. Find the *LDU* factorization of the matrix *A* in Problem 4.66.

The $A = LDU$ factorization refers to the situation where *L* is a lower triangular matrix with 1s on the diagonal (as in the *LU* factorization of *A*), *D* is a diagonal matrix, and *U* is an upper triangular matrix with 1s on the diagonal. Thus simply factor out the diagonal entries in the matrix *U* in the above *LU* factorization of *A* to obtain the matrices *D* and *L*. Hence

$$L = \begin{pmatrix} 1 & 0 & 0 \\ 2 & 1 & 0 \\ -3 & -7 & 1 \end{pmatrix} \qquad D = \begin{pmatrix} 1 & 0 & 0 \\ 0 & -1 & 0 \\ 0 & 0 & \frac{1}{27} \end{pmatrix} \qquad U = \begin{pmatrix} 1 & 3 & 2 \\ 0 & 1 & -2 \\ 0 & 0 & 1 \end{pmatrix}$$

4.68. Find the *LU* factorization of $B = \begin{pmatrix} 1 & 4 & -3 \\ 2 & 8 & 1 \\ -5 & -9 & 7 \end{pmatrix}$.

Reduce *B* to triangular form by first applying the operations $-2R_1 + R_2 \to R_2$ and $5R_1 + R_3 \to R_3$:

$$B \sim \begin{pmatrix} 1 & 4 & -3 \\ 0 & 0 & 7 \\ 0 & 11 & -8 \end{pmatrix}$$

Observe that the second diagonal entry is 0. Thus *B* cannot be brought into triangular form without row interchange operations. In other words, *B* is not *LU*-factorable.

4.69. Find the *LU* factorization of $A = \begin{pmatrix} 1 & 2 & -3 & 4 \\ 2 & 3 & -8 & 5 \\ 1 & 3 & 1 & 3 \\ 3 & 8 & -1 & 13 \end{pmatrix}$ by a direct method.

First form the following matrices *L* and *U*:

$$L = \begin{pmatrix} 1 & 0 & 0 & 0 \\ l_{21} & 1 & 0 & 0 \\ l_{31} & l_{32} & 1 & 0 \\ l_{41} & l_{42} & l_{43} & 1 \end{pmatrix} \qquad \text{and} \qquad U = \begin{pmatrix} u_{11} & u_{12} & u_{13} & u_{14} \\ 0 & u_{22} & u_{23} & u_{24} \\ 0 & 0 & u_{33} & u_{34} \\ 0 & 0 & 0 & u_{44} \end{pmatrix}$$

That part of the product *LU* which determines the first row of *A* yields the four equations

$$u_{11} = 1 \qquad u_{12} = 2 \qquad u_{13} = -3 \qquad u_{14} = 4$$

and that part of the product *LU* which determines the first column of *A* yields the equations

$$l_{21}u_{11} = 2, \qquad l_{31}u_{11} = 1, \qquad l_{41}u_{11} = 3 \qquad \text{or} \qquad l_{21} = 2, \qquad l_{31} = 1, \qquad l_{41} = 3$$

Thus, at this point, the matrices *L* and *U* have the form

$$L = \begin{pmatrix} 1 & 0 & 0 & 0 \\ 2 & 0 & 0 & 0 \\ 1 & l_{32} & 1 & 0 \\ 3 & l_{42} & l_{43} & 1 \end{pmatrix} \qquad \text{and} \qquad U = \begin{pmatrix} 1 & 2 & -3 & 4 \\ 0 & u_{22} & u_{23} & u_{24} \\ 0 & 0 & u_{33} & u_{34} \\ 0 & 0 & 0 & u_{44} \end{pmatrix}$$

That part of the product *LU* which determines the remaining entries in the second row of *A* yields the equations

$$4 + u_{22} = 3 \qquad -6 + u_{23} = -8 \qquad 8 + u_{24} = 5$$

or $\qquad\qquad u_{22} = -1 \qquad\qquad u_{23} = -2 \qquad\qquad u_{24} = -3$

and that part of the product LU which determines the remaining entries in the second column of A yields the equations

$$2 + l_{32}u_{22} = 3, \qquad 6 + l_{42}u_{22} = 8 \quad \text{or} \quad l_{32} = -1, \qquad l_{42} = -2$$

Thus L and U now have the form

$$L = \begin{pmatrix} 1 & 0 & 0 & 0 \\ 2 & 1 & 0 & 0 \\ 1 & -1 & 1 & 0 \\ 3 & -2 & l_{43} & 1 \end{pmatrix} \quad \text{and} \quad U = \begin{pmatrix} 1 & 2 & -3 & 4 \\ 0 & -1 & -2 & -3 \\ 0 & 0 & u_{33} & u_{34} \\ 0 & 0 & 0 & u_{44} \end{pmatrix}$$

Continuing, using the third row, third column, and fourth row of A, we get

$$u_{33} = 2, \qquad u_{34} = -1, \quad \text{then} \quad l_{43} = 2, \quad \text{and lastly} \quad u_{44} = 3$$

Thus

$$L = \begin{pmatrix} 1 & 0 & 0 & 0 \\ 2 & 1 & 0 & 0 \\ 1 & -1 & 1 & 0 \\ 3 & -2 & 2 & 1 \end{pmatrix} \quad \text{and} \quad U = \begin{pmatrix} 1 & 2 & -3 & 4 \\ 0 & -1 & -2 & -3 \\ 0 & 0 & 2 & -1 \\ 0 & 0 & 0 & 3 \end{pmatrix}$$

4.70. Find the LDU factorization of matrix A in the Problem 4.69.

Here U should have 1s on the diagonal and D is a diagonal matrix. Thus, using the above LU factorization of A, factor out the diagonal entries in that U to obtain

$$D = \begin{pmatrix} 1 & & & \\ & -1 & & \\ & & 2 & \\ & & & 3 \end{pmatrix} \quad \text{and} \quad U = \begin{pmatrix} 1 & 2 & -3 & 4 \\ & 1 & 2 & 3 \\ & & 1 & -2 \\ & & & 1 \end{pmatrix}$$

The matrix L is the same as in Problem 4.69.

4.71. Given the factorization $A = LU$, where $L = (l_{ij})$ and $U = (u_{ij})$. Consider the system $AX = B$. Determine (a) the algorithm to find $L^{-1}B$, and (b) the algorithm that solves $UX = B$ by back-substitution.

(a) The entry l_{ij} in the matrix L corresponds to the elementary row operation $-l_{ij}R_i + R_j \rightarrow R_j$. Thus the algorithm which transform B into B' is as follows:

Algorithm P4.88A: Evaluating $L^{-1}B$

Step 1. Repeat for $j = 1$ to $n - 1$:

Step 2.   Repeat for $i = j + 1$ to n:
$$b_j := -l_{ij}b_i + b_j$$
[End of Step 2 inner loop.]
[End of Step 1 outer loop.]

Step 3. Exit.

[The complexity of this algorithm is $C(n) \approx n^2/2$.]

(b) The back-substitution algorithm follows:

Algorithm P4.88B: Back-substitution for system $UX = B$

Step 1. $x_n = b_n/u_{nn}$

Step 2. Repeat for $j = n - 1, n - 2, \ldots, 1$
$$x_j = (b_j - u_{j,j+1}x_{j+1} - \cdots - u_{jn}x_n)/u_{jj}$$

Step 3. Exit.

[The complexity here is also $C(n) \approx n^2/2$.]

4.72. Find the *LU* factorization of the matrix $A = \begin{pmatrix} 1 & 2 & 1 \\ 2 & 3 & 3 \\ -3 & -10 & 2 \end{pmatrix}$.

Reduce A to triangular form by the operations

$$(1) \quad -2R_1 + R_2 \rightarrow R_2, \qquad (2) \quad 3R_1 + R_3 \rightarrow R_3, \qquad (3) \quad -4R_2 + R_3 \rightarrow R_3$$

$$A \sim \begin{pmatrix} 1 & 2 & 1 \\ 0 & -1 & 1 \\ 0 & -4 & 5 \end{pmatrix} \sim \begin{pmatrix} 1 & 2 & 1 \\ 0 & -1 & 1 \\ 0 & 0 & 1 \end{pmatrix}$$

Thus $\qquad L = \begin{pmatrix} 1 & 0 & 0 \\ 2 & 1 & 0 \\ -3 & 4 & 1 \end{pmatrix}$ and $U = \begin{pmatrix} 1 & 2 & 1 \\ 0 & -1 & 1 \\ 0 & 0 & 1 \end{pmatrix}$

The entries 2, -3, and 4 in L are the negatives of the multipliers in the above row operations.

4.73. Solve the system $AX = B$ for B_1, B_2, B_3, where A is the matrix in Problem 4.72 and where $B_1 = (1, 1, 1)$, $B_2 = B_1 + X_1$, $B_3 = B_2 + X_2$ (here X_j is the solution when $B = B_j$).

(a) Find $L^{-1}B_1$ or, equivalently, apply the row operations (1), (2), and (3) to B_1 to yield

$$B_1 = \begin{pmatrix} 1 \\ 1 \\ 1 \end{pmatrix} \xrightarrow{\text{(1) and (2)}} \begin{pmatrix} 1 \\ -1 \\ 4 \end{pmatrix} \xrightarrow{\text{(3)}} \begin{pmatrix} 1 \\ -1 \\ 8 \end{pmatrix}$$

Solve $UX = B$ for $B = (1, -1, 8)$ by back-substitution to obtain $X_1 = (-25, 9, 8)$.

(b) Find $B_2 = B_1 + X_1 = (1, 1, 1) + (-25, 9, 8) = (-24, 10, 9)$. Apply the operations (1), (2), and (3) to B_2 to obtain $(-24, 58, -63)$, and then $B = (-24, 58, -295)$.
      Solve $UX = B$ by back-substitution to obtain $X_2 = (943, -353, -295)$.

(c) Find $B_3 = B_2 + X_2 = (-24, 10, 9) + (943, -353, -295) = (919, -343, -286)$. Apply the operations (1), (2), and (3) to B_3 to obtain $(919, -2181, 2671)$, and then $B = (919, -2181, 11395)$.
      Solve $UX = B$ by back-substitution to obtain $X_3 = (-37,628, 13,576, 11,395)$.

Supplementary Problems

ALGEBRA OF MATRICES

4.74. Let $A = \begin{pmatrix} 1 & 2 \\ 0 & 1 \end{pmatrix}$. Find A^n.

4.75. Suppose the 2×2 matrix B commutes with every 2×2 matrix A. Show that $B = \begin{pmatrix} k & 0 \\ 0 & k \end{pmatrix}$ for some scalar k, i.e., B is a scalar matrix.

4.76. Let $A = \begin{pmatrix} 5 & 2 \\ 0 & k \end{pmatrix}$. Find all numbers k for which A is a root of the polynomial

(a) $f(x) = x^2 - 7x + 10$,   (b) $g(x) = x^2 - 25$,   (c) $h(x) = x^2 - 4$

4.77. Let $B = \begin{pmatrix} 1 & 0 \\ 26 & 27 \end{pmatrix}$. Find a matrix A such that $A^3 = B$.

4.78. Let $A = \begin{pmatrix} 0 & 1 & 0 & 0 \\ 0 & 0 & 1 & 0 \\ 0 & 0 & 0 & 1 \\ 0 & 0 & 0 & 0 \end{pmatrix}$ and $B = \begin{pmatrix} 1 & 1 & 0 \\ 0 & 1 & 1 \\ 0 & 0 & 1 \end{pmatrix}$. Find: (a) A^n for all positive integers n, and (b) B^n for all positive integers n.

4.79. Find conditions on matrices A and B so that $A^2 - B^2 = (A + B)(A - B)$.

INVERTIBLE MATRICES, INVERSES, ELEMENTARY MATRICES

4.80. Find the inverse of each matrix: (a) $\begin{pmatrix} 1 & 3 & -2 \\ 2 & 8 & -3 \\ 1 & 7 & 1 \end{pmatrix}$, (b) $\begin{pmatrix} 2 & 1 & -1 \\ 5 & 2 & -3 \\ 0 & 2 & 1 \end{pmatrix}$, (c) $\begin{pmatrix} 1 & -2 & 0 \\ 2 & -3 & 1 \\ 1 & 1 & 5 \end{pmatrix}$.

4.81. Find the inverse of each matrix: (a) $\begin{pmatrix} 1 & 1 & 1 & 1 \\ 0 & 1 & 1 & 1 \\ 0 & 0 & 1 & 1 \\ 0 & 0 & 0 & 1 \end{pmatrix}$, (b) $\begin{pmatrix} 1 & 2 & 1 & 0 \\ 0 & 1 & -1 & 1 \\ 1 & 3 & 1 & -2 \\ 1 & 4 & -2 & 4 \end{pmatrix}$.

4.82. Express each matrix as a product of elementary matrices: (a) $\begin{pmatrix} 1 & 2 \\ 3 & 4 \end{pmatrix}$, (b) $\begin{pmatrix} 3 & -6 \\ -2 & 4 \end{pmatrix}$.

4.83. Express $A = \begin{pmatrix} 1 & 2 & 0 \\ 0 & 1 & 3 \\ 3 & 8 & 7 \end{pmatrix}$ as a product of elementary matrices.

4.84. Suppose A is invertible. Show that if $AB = AC$ then $B = C$. Give an example of a nonzero matrix A such that $AB = AC$ but $B \neq C$.

4.85. If A is invertible, show that kA is invertible when $k \neq 0$, with inverse $k^{-1}A^{-1}$.

4.86. Suppose A and B are invertible and $A + B \neq 0$. Show, by an example, that $A + B$ need not be invertible.

SPECIAL TYPES OF SQUARE MATRICES

4.87. Using only the elements 0 and 1, find all 3×3 nonsingular upper triangular matrices.

4.88. Using only the elements 0 and 1, find the number of: (a) 4×4 diagonal matrices, (b) 4×4 upper triangular matrices, (c) 4×4 nonsingular upper triangular matrices. Generalize to $n \times n$ matrices.

4.89. Find all real matrices A such that $A^2 = B$ where (a) $B = \begin{pmatrix} 4 & 21 \\ 0 & 25 \end{pmatrix}$, (b) $B = \begin{pmatrix} 1 & 4 \\ 0 & -9 \end{pmatrix}$.

4.90. Let $B = \begin{pmatrix} 1 & 8 & 5 \\ 0 & 9 & 5 \\ 0 & 0 & 4 \end{pmatrix}$. Find a matrix A with positive diagonal entries such that $A^2 = B$.

4.91. Suppose $AB = C$ where A and C are upper triangular.
   (a) Show, by an example, that B need not be upper triangular even when A and C are nonzero matrices.
   (b) Show that B is upper triangular when A is invertible.

4.92. Show that AB need not be symmetric, even though A and B are symmetric.

4.93. Let A and B be symmetric matrices. Show that AB is symmetric if and only if A and B commute.

4.94. Suppose A is a symmetric matrix. Show that (a) A^2 and, in general, A^n is symmetric; (b) $f(A)$ is symmetric for any polynomial $f(x)$; (c) $P^T A P$ is symmetric.

4.95. Find a 2×2 orthogonal matrix P whose first row is (a) $(2/\sqrt{29}, 5/\sqrt{29})$, (b) a multiple of $(3, 4)$.

4.96. Find a 3×3 orthogonal matrix P whose first two rows are multiples of (a) $(1, 2, 3)$ and $(0, -2, 3)$, respectively; (b) $(1, 3, 1)$ and $(1, 0, -1)$, respectively.

4.97. Suppose A and B are orthogonal. Show that A^T, A^{-1}, and AB are also orthogonal.

4.98. Which of the following matrices are normal?

$$A = \begin{pmatrix} 3 & -4 \\ 4 & 3 \end{pmatrix}, \quad B = \begin{pmatrix} 1 & -2 \\ 2 & 3 \end{pmatrix}, \quad C = \begin{pmatrix} 1 & 1 & 1 \\ 0 & 1 & 1 \\ 0 & 0 & 1 \end{pmatrix} \quad D = \begin{pmatrix} 2 & -1 & 3 \\ 1 & 2 & 1 \\ -3 & -1 & 2 \end{pmatrix}$$

4.99. Suppose A is a normal matrix. Show that: (a) A^T, (b) A^2 and, in general A^n, (c) $B = kI + A$ are also normal.

4.100. A matrix E is *idempotent* if $E^2 = E$. Show that $E = \begin{pmatrix} 2 & -2 & -4 \\ -1 & 3 & 4 \\ 1 & -2 & -3 \end{pmatrix}$ is idempotent.

4.101. Show that if $AB = A$ and $BA = B$, then A and B are idempotent.

4.102. A matrix A is *nilpotent of class p* if $A^p = 0$ but $A^{p-1} \neq 0$. Show that $A = \begin{pmatrix} 1 & 1 & 3 \\ 5 & 2 & 6 \\ -2 & -1 & -3 \end{pmatrix}$

is nilpotent of class 3.

4.103. Suppose A is nilpotent of class p. Show that $A^q = 0$ for $q > p$ but $A^q \neq 0$ for $q < p$.

4.104. A square matrix is *tridiagonal* if the nonzero entries occur only on the diagonal directly above the main diagonal (on the superdiagonal), or directly below the main diagonal (on the subdiagonal). Display the generic tridiagonal matrices of orders 4 and 5.

4.105. Show that the product of tridiagonal matrices need not be tridiagonal.

COMPLEX MATRICES

4.106. Find real numbers x, y, and z so that A is Hermitian, where

$$(a) \quad A = \begin{pmatrix} x + yi & 3 \\ 3 + zi & 0 \end{pmatrix}, \quad (b) \begin{pmatrix} 3 & x + 2i & yi \\ 3 - 2i & 0 & 1 + zi \\ yi & 1 - xi & -1 \end{pmatrix}$$

4.107. Suppose A is any complex matrix. Show that AA^H and $A^H A$ are both Hermitian.

4.108. Suppose A is any complex square matrix. Show that $A + A^H$ is Hermitian and $A - A^H$ is skew-Hermitian.

4.109. Which of the following matrices are unitary?

$$A = \begin{pmatrix} i/2 & -\sqrt{3}/2 \\ \sqrt{3}/2 & -i/2 \end{pmatrix}, \quad B = \frac{1}{2}\begin{pmatrix} 1+i & 1-i \\ 1-i & 1+i \end{pmatrix}, \quad C = \frac{1}{2}\begin{pmatrix} 1 & -i & -1+i \\ i & 1 & 1+i \\ 1+i & -1+i & 0 \end{pmatrix}$$

4.110. Suppose A and B are unitary matrices. Show that: (a) A^H is unitary, (b) A^{-1} is unitary, (c) AB is unitary.

4.111. Determine which of the following matrices are normal: $A = \begin{pmatrix} 3+4i & 1 \\ i & 2+3i \end{pmatrix}, B = \begin{pmatrix} 1 & 0 \\ 1-i & i \end{pmatrix}.$

4.112. Suppose A is a normal matrix and U is a unitary matrix. Show that $B = U^H A U$ is also normal.

4.113. Recall the following elementary row operations:

$$[E_1] \quad R_i \leftrightarrow R_j, \quad [E_2] \quad kR_i \to R_i, \quad k \neq 0, \quad [E_3] \quad kR_j + R_i \to R_i$$

For complex matrices, the respective corresponding Hermitian column operations are as follows:

$$[G_1] \quad C_i \leftrightarrow C_j, \quad [G_2] \quad \bar{k}C_i \to C_i, \quad k \neq 0, \quad [G_3] \quad \bar{k}C_j + C_i \to C_i$$

Show that the elementary matrix corresponding to $[G_i]$ is the conjugate transpose of the elementary matrix corresponding to $[E_i]$.

SQUARE BLOCK MATRICES

4.114. Using vertical lines, complete the partitioning of each matrix so that it is a square block matrix:

$$A = \begin{pmatrix} 1 & 2 & 3 & 4 & 5 \\ 1 & 1 & 1 & 1 & 1 \\ 9 & 8 & 7 & 6 & 5 \\ 2 & 2 & 2 & 2 & 2 \\ 3 & 3 & 3 & 3 & 3 \end{pmatrix}, \quad B = \begin{pmatrix} 1 & 2 & 3 & 4 & 5 \\ 1 & 1 & 1 & 1 & 1 \\ 9 & 8 & 7 & 6 & 5 \\ 2 & 2 & 2 & 2 & 2 \\ 3 & 3 & 3 & 3 & 3 \end{pmatrix}.$$

4.115. Partition each of the following matrices so that it becomes a block diagonal matrix with as many diagonal blocks as possible:

$$A = \begin{pmatrix} 1 & 0 & 0 \\ 0 & 0 & 2 \\ 0 & 0 & 3 \end{pmatrix}, \quad B = \begin{pmatrix} 1 & 2 & 0 & 0 & 0 \\ 3 & 0 & 0 & 0 & 0 \\ 0 & 0 & 4 & 0 & 0 \\ 0 & 0 & 5 & 0 & 0 \\ 0 & 0 & 0 & 0 & 6 \end{pmatrix}, \quad C = \begin{pmatrix} 0 & 1 & 0 \\ 0 & 0 & 0 \\ 0 & 2 & 0 \end{pmatrix}$$

4.116. Find M^2 and M^3 for each matrix M:

(a) $M = \begin{pmatrix} 2 & 0 & 0 & 0 \\ 0 & 1 & 4 & 0 \\ 0 & 2 & 1 & 0 \\ 0 & 0 & 0 & 3 \end{pmatrix}$, (b) $M = \begin{pmatrix} 1 & 1 & 0 & 0 \\ 2 & 3 & 0 & 0 \\ 0 & 0 & 1 & 2 \\ 0 & 0 & 4 & 5 \end{pmatrix}.$

4.117. Let $M = \text{diag}(A_1, \ldots, A_k)$ and $N = \text{diag}(B_1, \ldots, B_k)$ be block diagonal matrices where each pair of blocks A_i, B_i have the same size. Prove MN is block diagonal and

$$MN = \text{diag}(A_1 B_1, A_2 B_2, \ldots, A_k B_k)$$

REAL SYMMETRIC MATRICES AND QUADRATIC FORMS

4.118. Let $A = \begin{pmatrix} 1 & 1 & -2 & -3 \\ 1 & 2 & -5 & -1 \\ -2 & -5 & 6 & 9 \\ -3 & -1 & 9 & 11 \end{pmatrix}$. Find a nonsingular matrix P such that $B = P^T A P$ is diagonal. Also, find B and sig A.

4.119. For each quadratic form $q(x, y, z)$, find a nonsingular linear substitution expressing the variables x, y, z in terms of variables r, s, t such that $q(r, s, t)$ is diagonal.

    (a) $q(x, y, z) = x^2 + 6xy + 8y^2 - 4xz + 2yz - 9z^2$

    (b) $q(x, y, z) = 2x^2 - 3y^2 + 8xz + 12yz + 25z^2$

4.120. Find those values of k so that the given quadratic form is positive definite:

    (a) $q(x, y) = 2x^2 - 5xy + ky^2$

    (b) $q(x, y) = 3x^2 - kxy + 12y^2$

    (c) $q(x, y, z) = x^2 + 2xy + 2y^2 + 2xz + 6yz + kz^2$

4.121. Give an example of a quadratic form $q(x, y)$ such that $q(u) = 0$ and $q(v) = 0$ but $q(u + v) \neq 0$.

4.122. Show that any real symmetric matrix A is congruent to a diagonal matrix with only 1s, -1s, and 0s on the diagonal.

4.123. Show that congruence of matrices is an equivalence relation.

SIMILARITY OF MATRICES

4.124. Consider the space $\mathbf{R}^3$ with the usual x, y, z axes. The nonsingular matrix $P = \begin{pmatrix} 1 & -2 & -2 \\ 2 & -3 & -6 \\ 1 & 1 & -7 \end{pmatrix}$ determines a new coordinate system for $\mathbf{R}^3$, say with r, s, t axes. Find:

    (a)   The coordinates of the point $Q(1, 1, 1)$ in the new system,

    (b) $f(r, s, t)$ when $f(x, y, z) = (x + y, y + 2z, x - z)$,

    (c) $g(r, s, t)$ when $g(x, y, z) = (x + y - z, x - 3z, 2x + y)$.

4.125. Show that similarity of matrices is an equivalence relation.

LU FACTORIZATION

4.126. Find the *LU* and *LDU* factorization of each matrix:

    (a) $A = \begin{pmatrix} 1 & 3 & -1 \\ 2 & 5 & 1 \\ 3 & 4 & 2 \end{pmatrix}$,   (b) $B = \begin{pmatrix} 2 & 3 & 6 \\ 4 & 7 & 9 \\ 3 & 5 & 4 \end{pmatrix}$.

4.127. Let $A = \begin{pmatrix} 1 & -1 & -1 \\ 3 & -4 & -2 \\ 2 & -3 & -2 \end{pmatrix}$.

    (a)   Find the *LU* factorization of A.

(b) Let X_k denote the solution of $AX = B_k$. Find X_1, X_2, X_3, X_4 when $B_1 = (1, \ 1, \ 1)^T$ and $B_{k+1} = B_k + X_k$ for $k > 0$.

Answers to Supplementary Problems

4.74. $\begin{pmatrix} 1 & 2^n \\ 0 & 1 \end{pmatrix}$

4.76. (a) $k = 2$,   (b) $k = -5$,   (c) None

4.77. $\begin{pmatrix} 1 & 0 \\ 2 & 3 \end{pmatrix}$

4.78. (a) $A^2 = \begin{pmatrix} 0 & 0 & 1 & 0 \\ 0 & 0 & 0 & 1 \\ 0 & 0 & 0 & 0 \\ 0 & 0 & 0 & 0 \end{pmatrix}$, $A^3 = \begin{pmatrix} 0 & 0 & 0 & 1 \\ 0 & 0 & 0 & 0 \\ 0 & 0 & 0 & 0 \\ 0 & 0 & 0 & 0 \end{pmatrix}$, $A^k = 0$ for $k > 3$ (b) $B^n = \begin{pmatrix} 1 & n & n(n-1)/2 \\ 0 & 1 & n \\ 0 & 0 & 1 \end{pmatrix}$

4.79. $AB = BA$

4.80. (a) $\begin{pmatrix} \frac{29}{2} & -\frac{17}{2} & \frac{7}{2} \\ -\frac{5}{2} & \frac{3}{2} & -\frac{1}{2} \\ 3 & -2 & 1 \end{pmatrix}$,   (b) $\begin{pmatrix} 8 & -3 & -1 \\ -5 & 2 & 1 \\ 10 & -4 & -1 \end{pmatrix}$,   (c) $\begin{pmatrix} -8 & 5 & -1 \\ -\frac{9}{2} & \frac{5}{2} & -\frac{1}{2} \\ \frac{5}{2} & -\frac{3}{2} & \frac{1}{2} \end{pmatrix}$

4.81. (a) $\begin{pmatrix} 1 & -1 & 0 & 0 \\ 0 & 1 & -1 & 0 \\ 0 & 0 & 1 & -1 \\ 0 & 0 & 0 & 1 \end{pmatrix}$,   (b) $\begin{pmatrix} -10 & -20 & 4 & 7 \\ 3 & 6 & -1 & -2 \\ 5 & 8 & -2 & -3 \\ 2 & 3 & -1 & -1 \end{pmatrix}$

4.82. (a) $\begin{pmatrix} 1 & 0 \\ 3 & 1 \end{pmatrix}\begin{pmatrix} 1 & -1 \\ 0 & 1 \end{pmatrix}\begin{pmatrix} 1 & 0 \\ 0 & -2 \end{pmatrix}$ or $\begin{pmatrix} 1 & 0 \\ 3 & 0 \end{pmatrix}\begin{pmatrix} 1 & 0 \\ 0 & -2 \end{pmatrix}\begin{pmatrix} 1 & 2 \\ 0 & 1 \end{pmatrix}$,   (b) No product: matrix has no inverse.

4.83. $\begin{pmatrix} 1 & 0 & 0 \\ 0 & 1 & 0 \\ 3 & 0 & 1 \end{pmatrix}\begin{pmatrix} 1 & 0 & 0 \\ 0 & 1 & 0 \\ 0 & 2 & 1 \end{pmatrix}\begin{pmatrix} 1 & 0 & 0 \\ 0 & 1 & 3 \\ 0 & 0 & 1 \end{pmatrix}\begin{pmatrix} 1 & 2 & 0 \\ 0 & 1 & 0 \\ 0 & 0 & 1 \end{pmatrix}$

4.84. $A = \begin{pmatrix} 1 & 2 \\ 1 & 2 \end{pmatrix}$, $B = \begin{pmatrix} 0 & 0 \\ 1 & 1 \end{pmatrix}$, $C = \begin{pmatrix} 2 & 2 \\ 0 & 0 \end{pmatrix}$

4.86. $A = \begin{pmatrix} 1 & 2 \\ 0 & 3 \end{pmatrix}$, $B = \begin{pmatrix} 4 & 3 \\ 3 & 0 \end{pmatrix}$

4.87. All diagonal entries must be 1 to be nonsingular. There are eight possible choices for the entries above the diagonal:

$$\begin{pmatrix} 0 & 0 \\ * & 0 \end{pmatrix}, \begin{pmatrix} 0 & 0 \\ * & 1 \end{pmatrix}, \begin{pmatrix} 0 & 1 \\ * & 0 \end{pmatrix}, \begin{pmatrix} 0 & 1 \\ * & 1 \end{pmatrix}, \begin{pmatrix} 1 & 0 \\ * & 0 \end{pmatrix}, \begin{pmatrix} 1 & 0 \\ * & 1 \end{pmatrix}, \begin{pmatrix} 1 & 1 \\ * & 0 \end{pmatrix}, \begin{pmatrix} 1 & 1 \\ * & 1 \end{pmatrix}$$

4.88. (a) $2^4[2^n]$, (b) $2^{10}[2^{n(n+1)/2}]$, (c) $2^6[2^{n(n-1)/2}]$

4.89. (a) $\begin{pmatrix} 2 & 4 \\ 0 & 5 \end{pmatrix}, \begin{pmatrix} 2 & -7 \\ 0 & -5 \end{pmatrix}, \begin{pmatrix} -2 & 7 \\ 0 & 5 \end{pmatrix}, \begin{pmatrix} -2 & -4 \\ 0 & -5 \end{pmatrix}$, (b) None

4.90. $\begin{pmatrix} 1 & 2 & 1 \\ & 3 & 1 \\ & & 2 \end{pmatrix}$

4.91. (a) $A = \begin{pmatrix} 1 & 1 \\ 0 & 0 \end{pmatrix}, B = \begin{pmatrix} 1 & 2 \\ 3 & 4 \end{pmatrix}, C = \begin{pmatrix} 4 & 6 \\ 0 & 0 \end{pmatrix}$

4.92. $\begin{pmatrix} 1 & 2 \\ 2 & 2 \end{pmatrix}\begin{pmatrix} 3 & 3 \\ 3 & 1 \end{pmatrix} = \begin{pmatrix} 9 & 5 \\ 12 & 8 \end{pmatrix}$

4.95. (a) $\begin{pmatrix} 2/\sqrt{29} & 5/\sqrt{29} \\ -5/\sqrt{29} & 2/\sqrt{29} \end{pmatrix}$, (b) $\begin{pmatrix} \frac{3}{5} & \frac{4}{5} \\ -\frac{4}{5} & \frac{3}{5} \end{pmatrix}$

4.96. (a) $\begin{pmatrix} 1/\sqrt{14} & 2/\sqrt{14} & 3/\sqrt{14} \\ 0 & -2/\sqrt{13} & 3/\sqrt{13} \\ 12/\sqrt{157} & -3/\sqrt{157} & -2/\sqrt{157} \end{pmatrix}$, (b) $\begin{pmatrix} 1/\sqrt{11} & 3/\sqrt{11} & 1/\sqrt{11} \\ 1/\sqrt{2} & 0 & -1/\sqrt{2} \\ 3/\sqrt{22} & -2/\sqrt{22} & 3/\sqrt{22} \end{pmatrix}$

4.98. A, C

4.104. $\begin{pmatrix} a_{11} & a_{21} & & \\ a_{21} & a_{22} & a_{32} & \\ & a_{32} & a_{33} & a_{34} \\ & & a_{43} & a_{44} \end{pmatrix}$ $\begin{pmatrix} b_{11} & b_{21} & & & \\ b_{21} & b_{22} & b_{23} & & \\ & b_{32} & b_{33} & b_{34} & \\ & & b_{43} & b_{44} & b_{45} \\ & & & b_{54} & b_{55} \end{pmatrix}$

4.105. $\begin{pmatrix} 1 & 1 & 0 \\ 1 & 1 & 1 \\ 0 & 1 & 1 \end{pmatrix}\begin{pmatrix} 1 & 1 & 0 \\ 1 & 1 & 1 \\ 0 & 1 & 1 \end{pmatrix} = \begin{pmatrix} 2 & 2 & 1 \\ 2 & 3 & 2 \\ 1 & 2 & 2 \end{pmatrix}$

4.106. (a) $x = a$ (parameter), $y = 0, z = 0$; (b) $x = 3, y = 0, z = 3$

4.109. A, B, C

4.111. A

4.114. $A = \begin{pmatrix} 1 & 2 & 3 & 4 & 5 \\ 1 & 1 & 1 & 1 & 1 \\ 9 & 8 & 7 & 6 & 5 \\ 2 & 2 & 2 & 2 & 2 \\ 3 & 3 & 3 & 3 & 3 \end{pmatrix}, B = \begin{pmatrix} 1 & 2 & 3 & 4 & 5 \\ 1 & 1 & 1 & 1 & 1 \\ 9 & 8 & 7 & 6 & 5 \\ 2 & 2 & 2 & 2 & 2 \\ 3 & 3 & 3 & 3 & 3 \end{pmatrix}$

4.115. $A = \begin{pmatrix} 1 & 0 & 0 \\ 0 & 0 & 2 \\ 0 & 0 & 3 \end{pmatrix}, B = \begin{pmatrix} 1 & 2 & 0 & 0 & 0 \\ 3 & 0 & 0 & 0 & 0 \\ 0 & 0 & 4 & 0 & 0 \\ 0 & 0 & 5 & 0 & 0 \\ 0 & 0 & 0 & 0 & 6 \end{pmatrix}, C = \begin{pmatrix} 0 & 1 & 0 \\ 0 & 0 & 0 \\ 0 & 2 & 0 \end{pmatrix}$

(C, itself, is a block diagonal matrix; no further partitioning of C is possible.)

4.116. (a) $M^2 = \begin{pmatrix} 4 & & \\ & 9 & 8 \\ & 4 & 9 \\ & & & 9 \end{pmatrix}$, $M^3 = \begin{pmatrix} 8 & & \\ & 25 & 44 \\ & 22 & 25 \\ & & & 27 \end{pmatrix}$

(b) $M^2 = \begin{pmatrix} 3 & 4 & & \\ 8 & 11 & & \\ & & 9 & 12 \\ & & 24 & 33 \end{pmatrix}$, $M^3 = \begin{pmatrix} 11 & 15 & & \\ 30 & 41 & & \\ & & 57 & 78 \\ & & 156 & 213 \end{pmatrix}$

4.118. $P = \begin{pmatrix} 1 & -1 & -1 & 26 \\ 0 & 1 & 3 & 13 \\ 0 & 0 & 1 & 9 \\ 0 & 0 & 0 & 7 \end{pmatrix}$, $B = \begin{pmatrix} 1 & & & \\ & 1 & & \\ & & -7 & \\ & & & 469 \end{pmatrix}$, sig $A = 2$

4.119. (a) $x = r - 3s + 19t$, $y = s + 7t$, $z = t$,
$q(r, s, t) = r^2 - s^2 + 36t^2$, rank $q = 3$, sig $q = 1$

(b) $x = r - 2t$, $y = s + 2t$, $z = t$,
$q(r, s, t) = 2r^2 - 3s^2 + 29t^2$, rank $q = 3$, sig $q = 1$

(c) $x = r - 2s + 18t$, $y = s - 7t$, $z = t$,
$q(r, s, t) = r^2 + s^2 - 62t^2$, rank $q = 3$, sig $q = 1$

(d) $x = r - s - t$, $y = s - t$, $z = t$,
$q(x, y, z) = r^2 + 2s^2$, rank $q = 2$, sig $q = 2$

4.120. (a) $k > \frac{25}{8}$; (b) $k < -12$ or $k > 12$; (c) $k > 5$

4.121. $q(x, y) = x^2 - y^2$, $u = (1, 1)$, $v = (1, -1)$

4.122. Suppose A has been diagonalized to $P^T A P = \text{diag}(a_i)$. Let $Q = \text{diag}(b_i)$ be defined by
$b_i = \begin{cases} 1/\sqrt{|a_i|} & \text{if } a_i \neq 0 \\ 1 & \text{if } a_i = 0 \end{cases}$. Then $B = Q^T P^T A P Q = (PQ)^T A (PQ)$ has the required form.

4.124. (a) $Q(17, 5, 3)$, (b) $f(r, s, t) = (17r - 61s + 134t, 4r - 41s + 46t, 3r - 25s + 25t)$,
(c) $g(r, s, t) = (61r + s - 330t, 16r + 3s - 91t, 9r - 4s - 4t)$

4.126. (a) $A = \begin{pmatrix} 1 & & \\ 2 & 1 & \\ 3 & 5 & 1 \end{pmatrix} \begin{pmatrix} 1 & & \\ & -1 & \\ & & -10 \end{pmatrix} \begin{pmatrix} 1 & 3 & -1 \\ & 1 & -3 \\ & & 1 \end{pmatrix}$

(b) $B = \begin{pmatrix} 1 & & \\ 2 & 1 & \\ \frac{3}{2} & \frac{1}{2} & 1 \end{pmatrix} \begin{pmatrix} 2 & & \\ & 1 & \\ & & -\frac{7}{2} \end{pmatrix} \begin{pmatrix} 1 & \frac{3}{2} & 3 \\ & 1 & -3 \\ & & 1 \end{pmatrix}$

4.127. (a) $A = \begin{pmatrix} 1 & 0 & 0 \\ 3 & 1 & 0 \\ 2 & 1 & 1 \end{pmatrix} \begin{pmatrix} 1 & -1 & -1 \\ 0 & -1 & 1 \\ 0 & 0 & -1 \end{pmatrix}$

(b) $X_1 = \begin{pmatrix} 1 \\ 1 \\ -1 \end{pmatrix}$, $B_2 = \begin{pmatrix} 2 \\ 2 \\ 0 \end{pmatrix}$, $X_2 = \begin{pmatrix} 6 \\ 4 \\ 0 \end{pmatrix}$, $B_3 = \begin{pmatrix} 8 \\ 6 \\ 0 \end{pmatrix}$, $X_3 = \begin{pmatrix} 22 \\ 16 \\ -2 \end{pmatrix}$, $B_4 = \begin{pmatrix} 30 \\ 22 \\ -2 \end{pmatrix}$, $X_4 = \begin{pmatrix} 86 \\ 62 \\ -6 \end{pmatrix}$

Chapter 5

Vector Spaces

5.1 INTRODUCTION

This chapter introduces the underlying algebraic structure of linear algebra—that of a finite-dimensional vector space. The definition of a vector space involves an arbitrary field (see Appendix B) whose elements are called *scalars*. The following notation will be used (unless otherwise stated or implied):

$$K \qquad \text{the field of scalars}$$
$$a, b, c \text{ or } k \qquad \text{the elements of } K$$
$$V \qquad \text{the given vector space}$$
$$u, v, w \qquad \text{the elements of } V$$

Nothing essential is lost if the reader assumes that K is the real field **R** or the complex field **C**.

Length and orthogonality are not covered in this chapter since they are not considered as part of the fundamental structure of a vector space. They will be included as an additional structure in Chapter 6.

5.2 VECTOR SPACES

The following defines the notion of a vector space or linear space.

Definition: Let K be a given field and let V be a nonempty set with rules of addition and scalar multiplication which assigns to any $u, v \in V$ a sum $u + v \in V$ and to any $u \in V$, $k \in K$ a product $ku \in V$. Then V is called a *vector space over* K (and the elements of V are called *vectors*) if the following axioms hold (see Problem 5.3).

[A_1]   For any vectors $u, v, w \in V$, $(u + v) + w = u + (v + w)$.

[A_2]   There is a vector in V, denoted by 0 and called the *zero vector*, for which $u + 0 = u$ for any vector $u \in V$.

[A_3]   For each vector $u \in V$ there is a vector in V, denoted by $-u$, for which $u + (-u) = 0$.

[A_4]   For any vectors $u, v \in V$, $u + v = v + u$.

[M_1]   For any scalar $k \in K$ and any vectors $u, v \in V$, $k(u + v) = ku + kv$.

[M_2]   For any scalars $a, b \in K$ and any vector $u \in V$, $(a + b)u = au + bu$.

[M_3]   For any scalars $a, b \in K$ and any vector $u \in V$, $(ab)u = a(bu)$.

[M_4]   For the unit scalar $1 \in K$, $1u = u$ for any vector $u \in V$.

The above axioms naturally split into two sets. The first four are only concerned with the additive structure of V and can be summarized by saying that V is a *commutative group* (see Appendix B) under addition. It follows that any sum of vectors of the form

$$v_1 + v_2 + \cdots + v_m$$

141

requires no parentheses and does not depend upon the order of the summands, the zero vector 0 is unique, the *negative* $-u$ of u is unique, and the *cancellation law* holds; that is, for any vectors u, v, $w \in V$.

$$u + w = v + w \qquad \text{implies} \qquad u = v$$

Also, *subtraction* is defined by

$$u - v = u + (-v)$$

On the other hand, the remaining four axioms are concerned with the "action" of the field K on V. Observe that the labelling of the axioms reflects this splitting. Using these additional axioms we prove (Problem 5.1) the following simple properties of a vector space.

Theorem 5.1:   Let V be a vector space over a field K.

   (i)   For any scalar $k \in K$ and $0 \in V$, $k0 = 0$.

   (ii)   For $0 \in K$ and any vector $u \in V$, $0u = 0$.

   (iii)   If $ku = 0$, where $k \in K$ and $u \in V$, then $k = 0$ or $u = 0$.

   (iv)   For any $k \in K$ and any $u \in V$, $(-k)u = k(-u) = -ku$.

5.3   EXAMPLES OF VECTOR SPACES

This section lists a number of important examples of vector spaces which will be used throughout the text.

Space K^n

Let K be an arbitrary field. The notation K^n is frequently used to denote the set of all n-tuples of elements in K. Here K^n is viewed as a vector space over K where vector addition and scalar multiplication are defined by

$$(a_1, a_2, \ldots, a_n) + (b_1, b_2, \ldots, b_n) = (a_1 + b_1, a_2 + b_2, \ldots, a_n + b_n)$$

and

$$k(a_1, a_2, \ldots, a_n) = (ka_1, ka_2, \ldots, ka_n)$$

The zero vector in K^n is the n-tuple of zeros,

$$0 = (0, 0, \ldots, 0)$$

and the negative of a vector is defined by

$$-(a_1, a_2, \ldots, a_n) = (-a_1, -a_2, \ldots, -a_n)$$

The proof that K^n is a vector space is identical to the proof of Theorem 2.1, which we now regard as stating that $\mathbf{R}^n$ with the operations defined there is a vector space over $\mathbf{R}$.

Matrix Space $\mathbf{M}_{m, n}$

The notation $\mathbf{M}_{m, n}$, or simply $\mathbf{M}$, will be used to denote the set of all $m \times n$ matrices over an arbitrary field K. Then $\mathbf{M}_{m, n}$ is a vector space over K with respect to the usual operations of vector addition and scalar multiplication. (See Theorem 3.1.)

Polynomial Space P(t)

Let $\mathbf{P}(t)$ denote the set of all polynomials

$$a_0 + a_1 t + a_2 t^2 + \cdots + a_n t^n$$

with coefficients a_i in some field K. (See Appendix.) Then $\mathbf{P}(t)$ is vector space over K with respect to the usual operations of addition of polynomials and multiplication of a polynomial by a constant.

Function Space F(X)

Let X be any nonempty set and let K be an arbitrary field. Consider the set $F(X)$ of all functions from X into K. [Note that $F(X)$ is nonempty since X is nonempty.] The sum of two functions f, $g \in F(X)$ is the function $f + g \in F(X)$ defined by

$$(f + g)(x) = f(x) + g(x) \qquad \forall x \in X$$

and the product of a scalar $k \in K$ and a function $f \in F(X)$ is the function $kf \in F(X)$ defined by

$$(kf)(x) = kf(x) \qquad \forall x \in X$$

(The symbol $\forall$ means "for every.") Then $F(X)$ with the above operations is a vector space over K (Problem 5.5).

The zero vector in $F(X)$ is the zero function $\mathbf{0}$ which maps each $x \in X$ into $0 \in K$, that is,

$$\mathbf{0}(x) = 0 \qquad \forall x \in X$$

Also, for any function $f \in F(X)$, the function $-f$ defined by

$$(-f)(x) = -f(x) \qquad \forall x \in X$$

is the negative of the function f.

Fields and Subfields

Suppose E is a field which contains a subfield K. Then E may be viewed as a vector space over K as follows. Let the usual addition in E be the vector addition, and let the scalar product kv of $k \in K$ and $v \in E$ be the product of k and v as elements of the field E. Then E is a vector space over K, that is, the above eight axioms of a vector space are satisfied by E and K.

5.4 SUBSPACES

Let W be a subset of a vector space V over a field K. W is called a *subspace* of V if W is itself a vector space over K with respect to the operations of vector addition and scalar multiplication on V. Simple criteria for identifying subspaces follow (see Problem 5.4 for proof).

Theorem 5.2: Suppose W is a subset of a vector space V. Then W is a subspace of V if and only if the following hold:

(i) $0 \in W$ (or $W \neq 0$)

(ii) W is *closed under vector addition*, that is:
    For every $u, v \in W$, the sum $u + v \in W$.

(iii) W is *closed under scalar multiplication*, that is:
    For every $u \in W$, $k \in K$, the multiple $ku \in W$.

Conditions (ii) and (iii) may be combined into one condition as in (ii) below (see Problem 5.5 for proof).

Corollary 5.3: W is a subspace of V if and only if:

  (i) $0 \in W$ (or $W \neq 0$), and

  (ii) $au + bv \in W$ for every $u, v \in W$ and $a, b \in K$.

Example 5.1

(a) Let V be any vector space. Then the set $\{\mathbf{0}\}$ consisting of the zero vector alone, and also the entire space V are subspaces of V.

(b) Let W be the xy plane in $\mathbf{R}^3$ consisting of those vectors whose third component is 0; or, in other words $W = \{(a, b, 0) : a, b \in \mathbf{R}\}$. Note $0 = (0, 0, 0) \in W$ since the third component of 0 is 0. For any vectors $u = (a, b, 0)$ and $v = (c, d, 0)$ in W and any scalar $k \in \mathbf{R}$, we have

$$u + v = (a + c, b + d, 0) \qquad \text{and} \qquad ku = (ka, kb, 0)$$

belong to W. Thus W is a subspace of V.

(c) Let $V = \mathbf{M}_{n,n}$ the space of $n \times n$ matrices. Then the subset W_1 of (upper) triangular matrices and the subset W_2 of symmetric matrices are subspaces of V since they are nonempty and closed under matrix addition and scalar multiplication.

(d) Recall $\mathbf{P}(t)$ denotes the vector space of polynomials. Let $\mathbf{P}_n(t)$ denote the subset of $\mathbf{P}(t)$ and consist of those polynomials of degree $\leq n$, for a fixed n. Then $\mathbf{P}_n(t)$ is a subspace of $\mathbf{P}(t)$. This vector space $\mathbf{P}_n(t)$ will occur very often in our examples.

Example 5.2.   Let U and W be subspaces of a vector space V. We show that the intersection $U \cap W$ is also a subspace of V. Clearly $0 \in U$ and $0 \in W$ since U and W are subspaces; whence $0 \in U \cap W$. Now suppose $u, v \in U \cap W$. Then $u, v \in U$ and $u, v \in W$ and, since U and W are subspaces,

$$u + v, \qquad ku \in U \qquad \text{and} \qquad u + v, \qquad ku \in W$$

for any scalar k. Thus $u + v, ku \in U \cap W$ and hence $U \cap W$ is a subspace of V.

The result in the preceding example generalizes as follows.

Theorem 5.4:   The intersection of any number of subspaces of a vector space V is a subspace of V.

Recall that any solution u of a system $AX = B$ of linear equations in n unknowns may be viewed as a point in K^n; and thus the solution set of such a system is a subset of K^n. Suppose the system is homogeneous, i.e., suppose the system has the form $AX = 0$. Let W denote its solution set. Since $A0 = 0$, the zero vector $0 \in W$. Moreover, if u and v belong to W, i.e., if u and v are solutions of $AX = 0$, then $Au = 0$ and $Av = 0$. Therefore, for any scalars a and b in K, we have

$$A(au + bv) = aAu + bAv = a0 + b0 = 0 + 0 = 0$$

Thus $au + bv$ is also a solution of $AX = 0$ or, in other words, $au + bv \in W$. Accordingly, by the above Corollary 5.3, we have proved:

Theorem 5.5:   The solution set W of a homogenous system $AX = 0$ in n unknowns is a subspace of K^n.

We emphasize that the solution set of a nonhomogenous system $AX = B$ is not a subspace of K^n. In fact, the zero vector 0 does not belong to its solution set.

5.5 LINEAR COMBINATIONS, LINEAR SPANS

Let V be a vector space over a field K and let $v_1, v_2, \ldots, v_m \in V$. Any vector in V of the form

$$a_1 v_1 + a_2 v_2 + \cdots + a_m v_m$$

where the $a_i \in K$, is called a *linear combination* of $v_1, v_2, \ldots, v_m$. The set of all such linear combinations, denoted by

$$\text{span } (v_1, v_2, \ldots, v_m)$$

is called the *linear span* of $v_1, v_2, \ldots, v_m$.

Generally, for any subset S of V, span $S = \{0\}$ when S is empty and span S consists of all the linear combinations of vectors in S.

The following theorem is proved in Problem 5.16.

Theorem 5.6:   Let S be a subset of a vector space V.

(i)   Then span S is a subspace of V which contains S.

(ii)   If W is a subspace of V containing S, then span $S \subset W$.

On the other hand, given a vector space V, the vectors $u_1, u_2, \ldots, u_r$ are said to *span* or *generate* or to form a *spanning set* of V if

$$V = \text{span } (u_1, u_2, \ldots, u_r)$$

In other words, $u_1, u_2, \ldots, u_r$ span V if, for every $v \in V$, there exist scalars $a_1, a_2, \ldots, a_r$ such that

$$v = a_1 u_1 + a_2 u_2 + \cdots + a_r u_r$$

that is, if v is a linear combination of $u_1, u_2, \ldots, u_r$.

Example 5.3

(a)   Consider the vector space $\mathbf{R}^3$. The linear span of any nonzero vector $u \in \mathbf{R}^3$ consists of all scalar multiples of u; geometrically, span u is the line through the origin and the endpoint of u as shown in Fig. 5-1(a). Also, for any two vectors $u, v \in \mathbf{R}^3$ which are not multiples of each other, span (u, v) is the plane through the origin and the endpoints of u and v as shown in Fig. 5-1(b).

(b)   The vectors $e_1 = (1, 0, 0)$, $e_2 = (0, 1, 0)$ and $e_3 = (0, 0, 1)$ span the vector space $\mathbf{R}^3$. Specifically, for any vector $u = (a, b, c)$ in $\mathbf{R}^3$, we have

$$u = (a, b, c) = a(1, 0, 0) + b(0, 1, 0) + c(0, 0, 1) = ae_1 + be_2 + ce_3$$

That is, u is a linear combination of e_1, e_2, e_3.

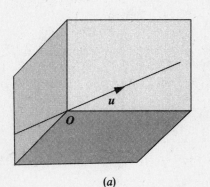

(a)   (b)

Fig. 5-1

(c) The polynomials $1, t, t^2, t^3, \ldots$ span the vector space $\mathbf{P}(t)$ of all polynomials, that is,

$$\mathbf{P}(t) = \text{span } (1, t, t^2, t^3, \ldots)$$

In other words, any polynomial is a linear combination of 1 and powers of t. Similarly, the polynomials $1, t, t^2, \ldots, t^n$ span the vector space $\mathbf{P}_n(t)$ of all polynomials of degree $\leq n$.

Row Space of a Matrix

Let A be an arbitrary $m \times n$ matrix over a field K:

$$A = \begin{pmatrix} a_{11} & a_{12} & \cdots & a_{1n} \\ a_{21} & a_{22} & \cdots & a_{2n} \\ \cdots\cdots\cdots\cdots\cdots\cdots \\ a_{m1} & a_{m2} & \cdots & a_{mn} \end{pmatrix}$$

The rows of A,

$$R_1 = (a_{11}, a_{12}, \ldots, a_{1n}), \ldots, R_m = (a_{m1}, a_{m2}, \ldots, a_{mn})$$

may be viewed as vectors in K^n and hence they span a subspace of K^n called the *row space* of A and denoted by rowsp A. That is,

$$\text{rowsp } A = \text{span } (R_1, R_2, \ldots, R_m)$$

Analogously, the columns of A may be viewed as vectors in K^m and hence span a subspace of K^m called the *column space* of A and denoted by colsp A. Alternatively, colsp A = rowsp A^T.

Now suppose we apply an elementary row operation on A,

(i) $R_i \leftrightarrow R_j$,   (ii) $kR_i \to R_i, k \neq 0$,   or   (iii) $kR_j + R_i \to R_i$

and obtain a matrix B. Then each row of B is clearly a row of A or a linear combination of rows of A. Hence the row space of B is contained in the row space of A. On the other hand, we can apply the inverse elementary row operation on B and obtain A; hence the row space of A is contained in the row space of B. Accordingly, A and B have the same row space. This leads us to the following theorem.

Theorem 5.7: Row equivalent matrices have the same row space.

In particular, we prove the following fundamental results about row equivalent matrices (proved in Problems 5.51 and 5.52, respectively).

Theorem 5.8: Row canonical matrices have the same row space if and only if they have the same nonzero rows.

Theorem 5.9: Every matrix is row equivalent to a unique matrix in row canonical form.

We apply the above results in the next example.

Example 5.4. Show that the subspace U of $\mathbf{R}^4$ spanned by the vectors

$$u_1 = (1, 2, -1, 3) \qquad u_2 = (2, 4, 1, -2) \qquad \text{and} \qquad u_3 = (3, 6, 3, -7)$$

and the subspace W of $\mathbf{R}^4$ spanned by the vectors

$$v_1 = (1, 2, -4, 11) \qquad \text{and} \qquad v_2 = (2, 4, -5, 14)$$

are equal; that is, $U = W$.

Method 1.   Show that each u_i is a linear combination of v_1 and v_2, and show that each v_i is a linear combination of u_1, u_2, and u_3. Observe that we have to show that six systems of linear equations are consistent.

Method 2.   Form the matrix A whose rows are the u_i, and row reduce A to row canonical form:

$$A = \begin{pmatrix} 1 & 2 & -1 & 3 \\ 2 & 4 & 1 & -2 \\ 3 & 6 & 3 & -7 \end{pmatrix} \sim \begin{pmatrix} 1 & 2 & -1 & 3 \\ 0 & 0 & 3 & -8 \\ 0 & 0 & 6 & -16 \end{pmatrix} \sim \begin{pmatrix} 1 & 2 & 0 & \frac{1}{3} \\ 0 & 0 & 1 & -\frac{8}{3} \\ 0 & 0 & 0 & 0 \end{pmatrix}$$

Now form the matrix B whose rows are v_1 and v_2, and row reduce B to row canonical form:

$$B = \begin{pmatrix} 1 & 2 & -4 & 11 \\ 2 & 4 & -5 & 14 \end{pmatrix} \sim \begin{pmatrix} 1 & 2 & -4 & 11 \\ 0 & 0 & 3 & -8 \end{pmatrix} \sim \begin{pmatrix} 1 & 2 & 0 & \frac{1}{3} \\ 0 & 0 & 1 & -\frac{8}{3} \end{pmatrix}$$

Since the nonzero rows of the reduced matrices are identical, the row spaces of A and B are equal and so $U = W$.

5.6 LINEAR DEPENDENCE AND INDEPENDENCE

The following defines the notion of linear dependence and independence. This concept plays an essential role in the theory of linear algebra and in mathematics in general.

Definition: Let V be a vector space over a field K. The vectors $v_1, \ldots, v_m \in V$ are said to be *linearly dependent over K*, or simply *dependent*, if there exist scalars $a_1, \ldots, a_m \in K$, not all of them 0, such that

$$a_1 v_1 + a_2 v_2 + \cdots + a_m v_m = 0 \qquad\qquad (*)$$

Otherwise, the vectors are said to be *linearly independent over K*, or simply *independent*.

Observe that the relation $(*)$ will always hold if the a's are all 0. If this relation holds only in this case, that is,

$$a_1 v_1 + a_2 v_2 + \cdots + a_m v_m = 0 \qquad \text{implies} \qquad a_1 = 0, \ldots, a_m = 0$$

then the vectors are linearly independent. On the other hand, if the relation $(*)$ also holds when one of the a's is not 0, then the vectors are linearly dependent.

A set $\{v_1, v_2, \ldots, v_m\}$ of vectors is said to be linearly dependent or independent according as the vectors $v_1, v_2, \ldots, v_m$ are linearly dependent or independent. An infinite set S of vectors is linearly dependent if there exist vectors $u_1, \ldots, u_k$ in S which are linearly dependent; otherwise S is linearly independent. The empty set $\varnothing$ is defined to be linearly independent.

The following remarks follow from the above definitions.

Remark 1:   If 0 is one of the vectors $v_1, \ldots, v_m$, say $v_1 = 0$, then the vectors must be linearly dependent; for

$$1v_1 + 0v_2 + \cdots + 0v_m = 1 \cdot 0 + 0 + \cdots + 0 = 0$$

and the coefficient of v_1 is not 0.

Remark 2:   Any nonzero vector v is, by itself, linearly independent; for

$$kv = 0, \quad v \neq 0 \qquad \text{implies} \qquad k = 0$$

Remark 3: If two of the vectors $v_1, v_2, \ldots, v_m$ are equal or one is a scalar multiple of the other, say $v_1 = kv_2$, then the vectors are linearly dependent. For

$$v_1 - kv_2 + 0v_3 + \cdots + 0v_m = 0$$

and the coefficient of v_1 is not 0.

Remark 4: Two vectors v_1 and v_2 are linearly dependent if and only if one of them is a multiple of the other.

Remark 5: If the set $\{v_1, \ldots, v_m\}$ is linearly independent, then any rearrangement of the vectors $\{v_{i_1}, v_{i_2}, \ldots, v_{i_m}\}$ is also linearly independent.

Remark 6: If a set S of vectors is linearly independent, then any subset of S is linearly independent. Alternatively, if S contains a linearly dependent subset, then S is linearly dependent.

Remark 7: In the real space $\mathbf{R}^3$, linear dependence of vectors can be described geometrically as follows: (a) Any two vectors u and v are linearly dependent if and only if they lie on the same line through the origin as shown in Fig. 5-2(a). (b) Any three vectors u, v, and w are linearly dependent if and only if they lie on the same plane through the origin as shown in Fig. 5-2(b).

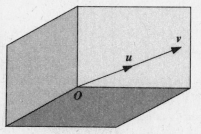
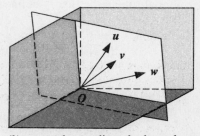

(a) u and v are linearly dependent   (b) u, v, and w are linearly dependent

Fig. 5-2

Other examples of linearly dependent and independent vectors follow.

Example 5.5

(a) The vectors $u = (1, -1, 0)$, $v = (1, 3, -1)$, and $w = (5, 3, -2)$ are linearly dependent since

$$3(1, -1, 0) + 2(1, 3, -1) - (5, 3, -2) = (0, 0, 0)$$

That is, $3u + 2v - w = 0$.

(b) We show that the vectors $u = (6, 2, 3, 4)$, $v = (0, 5, -3, 1)$, and $w = (0, 0, 7, -2)$ are linearly independent. For suppose $xu + yv + zw = 0$ where x, y and z are unknown scalars. Then

$$(0, 0, 0, 0) = x(6, 2, 3, 4) + y(0, 5, -3, 1) + z(0, 0, 7, -2)$$
$$= (6x, 2x + 5y, 3x - 3y + 7z, 4x + y - 2z)$$

and so, by the equality of the corresponding components,

$$\begin{aligned} 6x &= 0 \\ 2x + 5y &= 0 \\ 3x - 3y + 7z &= 0 \\ 4x + y - 2z &= 0 \end{aligned}$$

The first equation yields $x = 0$; the second equation with $x = 0$ yields $y = 0$; and the third equation with $x = 0$, $y = 0$ yields $z = 0$. Thus

$$xu + yv + zw = 0 \qquad \text{implies} \qquad x = 0, \ y = 0, \ z = 0$$

Accordingly u, v, and w are linearly independent.

Linear Combinations and Linear Dependence

The notions of linear combinations and linear dependence are closely related. Specifically, for more than one vector, we show that the vectors $v_1, v_2, \ldots, v_m$ are linearly dependent if and only if one of them is a linear combination of the others.

For suppose, say, v_i is a linear combination of the others:

$$v_i = a_1 v_1 + \cdots + a_{i-1} v_{i-1} + a_{i+1} v_{i+1} + \cdots + a_m v_m$$

Then by adding $-v_i$ to both sides, we obtain

$$a_1 v_1 + \cdots + a_{i-1} v_{i-1} - v_i + a_{i+1} v_{i+1} + \cdots + a_m v_m = 0$$

where the coefficient of v_i is not 0; hence the vectors are linearly dependent. Conversely, suppose the vectors are linearly dependent, say,

$$b_1 v_1 + \cdots + b_j v_j + \cdots + b_m v_m = 0 \qquad \text{where} \qquad b_j \neq 0$$

Then

$$v_j = -b_j^{-1} b_1 v_1 - \cdots - b_j^{-1} b_{j-1} v_{j-1} - b_j^{-1} b_{j+1} v_{j+1} - \cdots - b_j^{-1} b_m v_m$$

and so v_j is a linear combination of the other vectors.

We now formally state a slightly stronger statement than that above (see Problem 5.36 for the proof); this result has many important consequences.

Lemma 5.10: Suppose two or more nonzero vectors $v_1, v_2, \ldots, v_m$ are linearly dependent. Then one of the vectors is a linear combination of the preceding vectors, that is, there exists a $k > 1$ such that

$$v_k = c_1 v_1 + c_2 v_2 + \cdots + c_{k-1} v_{k-1}$$

Example 5.6. Consider the following matrix in echelon form:

$$A = \begin{pmatrix} 0 & 2 & 3 & 4 & 5 & 6 & 7 \\ 0 & 0 & 4 & -4 & 4 & -4 & 4 \\ 0 & 0 & 0 & 0 & 7 & 8 & 9 \\ 0 & 0 & 0 & 0 & 0 & 6 & -6 \\ 0 & 0 & 0 & 0 & 0 & 0 & 0 \end{pmatrix}$$

Observe that rows R_2, R_3, and R_4 have 0s in the second column (below the pivot element in R_1) and hence any linear combination of R_2, R_3, and R_4 must have a 0 as its second component. Thus R_1 cannot be a linear combination of the nonzero rows below it. Similarly, rows R_3 and R_4 have 0s in the third column below the pivot element in R_2; hence R_2 cannot be a linear combination of the nonzero rows below it. Finally, R_3 cannot be a multiple of R_4 since R_4 has a 0 in the fifth column below the pivot in R_3. Viewing the nonzero rows from the bottom up, R_4, R_3, R_2, R_1, no row is a linear combination of the previous rows. Thus the rows are linearly independent by Lemma 5.10.

The argument in the above example can be used for the nonzero rows of any echelon matrix. Thus we have the following very useful result (proved in Problem 5.37).

Theorem 5.11:   The nonzero rows of a matrix in echelon form are linearly independent.

5.7 BASIS AND DIMENSION

First we state two equivalent ways (Problem 5.30) to define a basis of a vector space V.

Definition A: A set $S = \{u_1, u_2, \ldots, u_n\}$ of vectors is a *basis* of V if the following two conditions hold:

    (1) $u_1, u_2, \ldots, u_n$ are linearly independent,

    (2) $u_1, u_2, \ldots, u_n$ span V.

Definition B: A set $S = \{u_1, u_2, \ldots, u_n\}$ of vectors is a *basis* of V if every vector $v \in V$ can be written uniquely as a linear combination of the basis vectors.

A vector space V is said to be of *finite dimension n* or to be *n-dimensional*, written

$$\dim V = n$$

if V has such a basis with n elements. This definition of dimension is well defined in view of the following theorem (proved in Problem 5.40).

Theorem 5.12:   Let V be a finite-dimensional vector space. Then every basis of V has the same number of elements.

The vector space $\{0\}$ is defined to have dimension 0. (In a certain sense this agrees with the above definition since, by definition, $\varnothing$ is independent and generates $\{0\}$.) When a vector space is not of finite dimension, it is said to be of *infinite dimension*.

Example 5.7

(a) Consider the vector space $M_{2,3}$ of all 2×3 matrices over a field K. Then the following six matrices form a basis of $M_{2,3}$:

$$\begin{pmatrix} 1 & 0 & 0 \\ 0 & 0 & 0 \end{pmatrix} \quad \begin{pmatrix} 0 & 1 & 0 \\ 0 & 0 & 0 \end{pmatrix} \quad \begin{pmatrix} 0 & 0 & 1 \\ 0 & 0 & 0 \end{pmatrix} \quad \begin{pmatrix} 0 & 0 & 0 \\ 1 & 0 & 0 \end{pmatrix} \quad \begin{pmatrix} 0 & 0 & 0 \\ 0 & 1 & 0 \end{pmatrix} \quad \begin{pmatrix} 0 & 0 & 0 \\ 0 & 0 & 1 \end{pmatrix}$$

More generally, in the vector space $M_{r,s}$ of $r \times s$ matrices let E_{ij} be the matrix with ij-entry 1 and 0 elsewhere. Then all such matrices E_{ij} form a basis of $M_{r,s}$, called the *usual basis* of $M_{r,s}$. Then $\dim M_{r,s} = rs$. In particular, $e_1 = (1, 0, \ldots, 0)$, $e_2 = (0, 1, 0, \ldots, 0)$, $\ldots$, $e_n = (0, 0, \ldots, 0, 1)$ form the usual basis for K^n.

(b) Consider the vector space $P_n(t)$ of polynomials of degree $\leq n$. The polynomials $1, t, t^2, \ldots, t^n$ form a basis of $P_n(t)$, and so $\dim P_n(t) = n + 1$.

The above fundamental theorem on dimension is a consequence of the following important "replacement lemma" (proved in Problem 5.39):

Lemma 5.13:   Suppose $\{v_1, v_2, \ldots, v_n\}$ spans V, and suppose $\{w_1, w_2, \ldots, w_m\}$ is linearly independent. Then $m \leq n$, and V is spanned by a set of the form

$$\{w_1, \ldots, w_m, v_{i_1}, \ldots, v_{i_{n-m}}\}$$

Thus, in particular, any $n + 1$ or more vectors in V are linearly dependent.

Observe in the above lemma that we have replaced m of the vectors in the spanning set by the m independent vectors and still retained a spanning set.

The following theorems (proved in Problems 5.41, 5.42, and 5.43, respectively) will be frequently used.

Theorem 5.14:   Let V be a vector space of finite dimension n.

  (i)   Any $n + 1$ or more vectors in V are linearly independent.

  (ii)   Any linearly independent set $S = \{u_1, u_2, \ldots, u_n\}$ with n elements is a basis of V.

  (iii)   Any spanning set $T = \{v_1, v_2, \ldots, v_n\}$ of V with n elements is a basis of V.

Theorem 5.15:   Suppose S spans a vector space V.

  (i)   Any maximum number of linearly independent vectors in S form a basis of V.

  (ii)   Suppose one deletes from S each vector which is a linear combination of preceding vectors in S. Then the remaining vectors form a basis of V.

Theorem 5.16:   Let V be a vector space of finite dimension and let $S = \{u_1, u_2, \ldots, u_r\}$ be a set of linearly independent vectors in V. Then S is part of a basis of V, that is, S may be extended to a basis of V.

Example 5.8

(a)   Consider the following four vectors in $\mathbf{R}^4$:

$$(1, 1, 1, 1) \qquad (0, 1, 1, 1) \qquad (0, 0, 1, 1) \qquad (0, 0, 0, 1)$$

Note that the vector will form a matrix in echelon form; hence the vectors are linearly independent. Furthermore, since dim $\mathbf{R}^4 = 4$, the vectors form a basis of $\mathbf{R}^4$.

(b)   Consider the following $n + 1$ polynomials in $\mathbf{P}_n(t)$:

$$1, t - 1, (t - 1)^2, \ldots, (t - 1)^n$$

The degree of $(t - 1)^k$ is k; hence no polynomial can be a linear combination of preceding polynomials. Thus the polynomials are linearly independent. Furthermore, they form a basis of $\mathbf{P}_n(t)$ since dim $\mathbf{P}_n(t) = n + 1$.

Dimension and Subspaces

The following theorem (proved in Problem 5.44) gives the basic relationship between the dimension of a vector space and the dimension of a subspace.

Theorem 5.17:   Let W be a subspace of an n-dimension vector space V. Then dim $W \leq n$. In particular if dim $W = n$, then $W = V$.

Example 5.9.   Let W be a subspace of the real space $\mathbf{R}^3$. Now dim $\mathbf{R}^3 = 3$; hence by Theorem 5.17 the dimension of W can only be 0, 1, 2, or 3. The following cases apply:

  (i)   dim $W = 0$, then $W = \{0\}$, a point;

  (ii)   dim $W = 1$, then W is a line through the origin;

  (iii)   dim $W = 2$, then W is a plane through the origin;

  (iv)   dim $W = 3$, then W is the entire space $\mathbf{R}^3$.

Rank of a Matrix

Let A be an arbitrary $m \times n$ matrix over a field K. Recall that the row space of A is the subspace of K^n spanned by its rows, and the column space of A is the subspace of K^m spanned by its columns.

The *row rank* of the matrix A is equal to the maximum number of linearly independent rows or, equivalently, to the dimension of the row space of A. Analogously, the *column rank* of A is equal to the maximum number of linearly independent columns or, equivalently, to the dimension of the column space of A.

Although rowsp A is a subspace of K^n, and colsp A is a subspace of K^m, where n may not equal m, we have the following important result (proved in Problem 5.53).

Theorem 5.18:   The row rank and the column rank of any matrix A are equal.

Definition: The *rank* of the matrix A, written rank A, is the common value of its row rank and column rank.

The rank of a matrix may be easily found by using row reduction as illustrated in the next example.

Example 5.10.   Suppose we want to find a basis and the dimension of the row space of

$$A = \begin{pmatrix} 1 & 2 & 0 & -1 \\ 2 & 6 & -3 & -3 \\ 3 & 10 & -6 & -5 \end{pmatrix}$$

We reduce A to echelon form using the elementary row operations:

$$A \sim \begin{pmatrix} 1 & 2 & 0 & -1 \\ 0 & 2 & -3 & -1 \\ 0 & 4 & -6 & -2 \end{pmatrix} \sim \begin{pmatrix} 1 & 2 & 0 & -1 \\ 0 & 2 & -3 & -1 \\ 0 & 0 & 0 & 0 \end{pmatrix}$$

Recall that row equivalent matrices have the same row space. Thus the nonzero rows of the echelon matrix, which are independent by Theorem 5.11 form a basis of the row space of A. Thus dim rowsp $A = 2$ and so rank $A = 2$.

5.8   LINEAR EQUATIONS AND VECTOR SPACES

Consider a system of m linear equations in n unknowns $x_1, \ldots, x_n$ over a field K:

$$\begin{aligned}
a_{11}x_1 + a_{12}x_2 + \cdots + a_{1n}x_n &= b_1 \\
a_{21}x_1 + a_{22}x_2 + \cdots + a_{2n}x_n &= b_2 \\
&\cdots\cdots\cdots\cdots\cdots\cdots\cdots\cdots\cdots\cdots\cdots \\
a_{m1}x_1 + a_{m2}x_2 + \cdots + a_{mn}x_n &= b_m
\end{aligned} \tag{5.1}$$

or the equivalent matrix equation

$$AX = B$$

where $A = (a_{ij})$ is the coefficient matrix, and $X = (x_i)$ and $B = (b_i)$ are the column vectors consisting of the unknowns and of the constants, respectively. Recall that the *augmented matrix* of the system is defined to be the matrix

$$(A, B) = \begin{pmatrix} a_{11} & a_{12} & \ldots & a_{1n} & b_1 \\ a_{21} & a_{22} & \ldots & a_{2n} & b_2 \\ \cdots & \cdots & \cdots & \cdots & \cdots \\ a_{m1} & a_{m2} & \ldots & a_{mn} & b_m \end{pmatrix}$$

Remark 1:   The linear equations (*5.1*) are said to be dependent or independent according as the corresponding vectors, i.e., the rows of the augmented matrix, are dependent or independent.

Remark 2: Two systems of linear equations are equivalent if and only if the corresponding augmented matrices are row equivalent, i.e., have the same row space.

Remark 3: We can always replace a system of equations by a system of independent equations, such as a system in echelon form. The number of independent equations will always be equal to the rank of the augmented matrix.

Observe that the system (5.1) is also equivalent to the vector equation

$$x_1 \begin{pmatrix} a_{11} \\ a_{21} \\ \cdots \\ a_{m1} \end{pmatrix} + x_2 \begin{pmatrix} a_{12} \\ a_{22} \\ \cdots \\ a_{m2} \end{pmatrix} + \cdots + x_n \begin{pmatrix} a_{1n} \\ a_{2n} \\ \cdots \\ a_{mn} \end{pmatrix} = \begin{pmatrix} b_1 \\ b_2 \\ \cdots \\ b_m \end{pmatrix}$$

The above comment gives us the following basic existence theorem.

Theorem 5.19: The following three statements are equivalent.

    (a) The system of linear equations $AX = B$ has a solution.

    (b) B is a linear combination of the columns of A.

    (c) The coefficient matrix A and the augmented matrix (A, B) have the same rank.

Recall that an $m \times n$ matrix A may be viewed as a function $A : K^n \to K^m$. Thus a vector B belongs to the image of A if and only if the equation $AX = B$ has a solution. This means that the image (range) of the function A, written Im A, is precisely the column space of A. Accordingly,

$$\dim (\text{Im } A) = \dim (\text{colsp } A) = \text{rank } A$$

We use this fact to prove (Problem 5.59) the following basic result on homogeneous systems of linear equations.

Theorem 5.20: The dimension of the solution space W of the homogeneous system of linear equations $AX = 0$ is $n - r$ where n is the number of unknowns and r is the rank of the coefficient matrix A.

In case the system $AX = 0$ is in echelon form, then it has precisely $n - r$ free variables, say, $x_{i_1}, x_{i_2}, \ldots, x_{i_{n-r}}$. Let v_j be the solution obtained by setting $x_{i_j} = 1$ (or any nonzero constant) and the remaining free variables equal to 0. Then the solutions $v_1, \ldots, v_{n-r}$ are linearly independent (Problem 5.58) and hence they form a basis for the solution space.

Example 5.11. Suppose we want to find the dimension and a basis of the solution space W of the following system:

$$\begin{aligned} x + 2y + 2z - s + 3t &= 0 \\ x + 2y + 3z + s + t &= 0 \\ 3x + 6y + 8z + s + 5t &= 0 \end{aligned}$$

First reduce the system to echelon form:

$$\begin{aligned} x + 2y + 2z - \ s + 3t &= 0 \\ z + 2s - 2t &= 0 \qquad \text{or} \qquad \begin{aligned} x + 2y + 2z - \ s + 3t &= 0 \\ z + 2s - 2t &= 0 \end{aligned} \\ 2z + 4s - 4t &= 0 \end{aligned}$$

The system in echelon form has 2 (nonzero) equations in 5 unknowns; and hence the system has $5 - 2 = 3$ free variables which are y, s and t. Thus $\dim W = 3$. To obtain a basis for W, set:

   (i) $y = 1, s = 0, t = 0$ to obtain the solution $v_1 = (-2, 1, 0, 0, 0)$,

(ii) $y = 0, s = 1, t = 0$ to obtain the solution $v_2 = (5, 0, -2, 1, 0)$,

(iii) $y = 0, s = 0, t = 1$ to obtain the solution $v_3 = (-7, 0, 2, 0, 1)$.

The set $\{v_1, v_2, v_3\}$ is a basis of the solution space W.

Two Basis-Finding Algorithms

Suppose we are given vectors $u_1, u_2, \ldots, u_r$ in K^n. Let

$$W = \text{span}\,(u_1, u_2, \ldots, u_r)$$

the subspace of K^n spanned by the given vectors. Each algorithm below finds a basis (and hence the dimension) of W.

Algorithm 5.8A (Row space algorithm)

Step 1. Form the matrix A whose *rows* are the given vectors.

Step 2. Row reduce A to an echelon form.

Step 3. Output the nonzero rows of the echelon matrix.

The above algorithm essentially already appeared in Example 5.10. The next algorithm is illustrated in Example 5.12 and uses the above result on nonhomogeneous systems of linear equations.

Algorithm 5.8B (Casting-Out algorithm)

Step 1. Form the matrix M whose *columns* are the given vectors.

Step 2. Row reduce M to echelon form.

Step 3. For each column C_k in the echelon matrix without a pivot, delete (cast-out) the vector v_k from the given vectors.

Step 4. Output the remaining vectors (which correspond to columns with pivots).

Example 5.12. Let W be the subspace of $\mathbf{R}^5$ spanned by the following vectors:

$$v_1 = (1, 2, 1, -2, 3) \qquad v_2 = (2, 5, -1, 3, -2) \qquad v_3 = (1, 3, -2, 5, -5)$$

$$v_4 = (3, 1, 2, -4, 1) \qquad v_5 = (5, 6, 1, -1, -1)$$

We use Algorithm *5.8B* to find the dimension and a basis of W.

First form the matrix M whose columns are the given vectors, and reduce the matrix to echelon form:

$$M = \begin{pmatrix} 1 & 2 & 1 & 3 & 5 \\ 2 & 5 & 3 & 1 & 6 \\ 1 & -1 & -2 & 2 & 1 \\ -2 & 3 & 5 & -4 & -1 \\ 3 & -2 & -5 & 1 & -1 \end{pmatrix} \sim \begin{pmatrix} 1 & 2 & 1 & 3 & 5 \\ 0 & 1 & 1 & -5 & -4 \\ 0 & -3 & -3 & -1 & -4 \\ 0 & 7 & 7 & 2 & 9 \\ 0 & -8 & -8 & -8 & -16 \end{pmatrix}$$

$$\sim \begin{pmatrix} 1 & 2 & 1 & 3 & 5 \\ 0 & 1 & 1 & 7 & -4 \\ 0 & 0 & 0 & -16 & -16 \\ 0 & 0 & 0 & 37 & 37 \\ 0 & 0 & 0 & -48 & -48 \end{pmatrix} \sim \begin{pmatrix} 1 & 2 & 1 & 3 & 5 \\ 0 & 1 & 1 & 7 & -4 \\ 0 & 0 & 0 & 1 & 1 \\ 0 & 0 & 0 & 0 & 0 \\ 0 & 0 & 0 & 0 & 0 \end{pmatrix}$$

Observe that the pivots in the echelon matrix appear in columns C_1, C_2, and C_4. The fact that column C_3 does not have a pivot means that the system $xv_1 + yv_2 = v_3$ has a solution and hence v_3 is a linear combination of v_1 and v_2. Similarly, the fact that column C_5 does not have a pivot means that v_5 is a linear combination of preceding vectors. Accordingly, the vectors v_1, v_2 and v_4, which correspond to the columns in the echelon matrix with pivots, form a basis of W and dim $W = 3$.

5.9 SUMS AND DIRECT SUMS

Let U and W be subsets of a vector space V. The sum of U and W, written $U + W$, consists of all sums $u + w$ where $u \in U$ and $w \in W$. That is,

$$U + W = \{u + w : u \in U, w \in W\}$$

Now suppose U and W are subspaces of a vector space V. Note that $0 = 0 + 0 \in U + W$, since $0 \in U$, $0 \in W$. Furthermore, suppose $u + w$ and $u' + w'$ belong to $U + W$, with $u, u' \in U$ and $w, w' \in W$. Then

$$(u + w) + (u' + w') = (u + u') + (w + w') \in U + W$$

and, for any scalar k,

$$k(u + w) = ku + kw \in U + W$$

Thus we have proven the following theorem.

Theorem 5.21:   The sum $U + W$ of the subspaces U and W of V is also a subspace of V.

Recall that the intersection $U \cap W$ is also a subspace of V. The following theorem, proved in Problem 5.69, relates the dimensions of these subspaces.

Theorem 5.22:   Let U and W be finite-dimensional subspaces of a vector space V. Then $U + W$ has finite dimension and

$$\dim (U + W) = \dim U + \dim W - \dim (U \cap W)$$

Example 5.13.   Suppose U and W are the xy and yz planes, respectively, in $\mathbf{R}^3$. That is,

$$U = \{(a, b, 0)\} \qquad \text{and} \qquad W = \{(0, b, c)\}$$

Note $\mathbf{R}^3 = U + W$; hence dim $(U + W) = 3$. Also dim $U = 2$ and dim $W = 2$. By Theorem 5.22,

$$3 = 2 + 2 - \dim (U \cap W) \qquad \text{or} \qquad \dim (U \cap W) = 1$$

This agrees with the facts that $U \cap W$ is the y axis (Fig. 5-3) and the y axis has dimension 1.

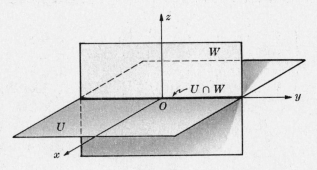

Fig. 5-3

Direct Sums

The vector space V is said to be the *direct sum* of its subspaces U and W, denoted by

$$V = U \oplus W$$

if every vector $v \in V$ can be written in one and only one way as $v = u + w$ where $u \in U$ and $w \in W$. The following theorem, proved in Problem 5.70, characterizes such a decomposition.

Theorem 5.23: The vector space V is the direct sum of its subspaces U and W if and only if (i) $V = U + W$ and (ii) $U \cap W = \{0\}$.

Example 5.14

(a) In the vector space $\mathbf{R}^3$, let U be the xy plane and let W be the yz plane:

$$U = \{(a, b, 0): a, b \in R\} \qquad \text{and} \qquad W = \{(0, b, c): b, c \in \mathbf{R}\}$$

Then $\mathbf{R}^3 = U + W$ since every vector in $\mathbf{R}^3$ is the sum of a vector in U and a vector in W. However, $\mathbf{R}^3$ is not the direct sum of U and W since such sums are not unique; for example,

$$(3, 5, 7) = (3, 1, 0) + (0, 4, 7) \qquad \text{and also} \qquad (3, 5, 7) = (3, -4, 0) + (0, 9, 7)$$

(b) In $\mathbf{R}^3$, let U be the xy plane and let W be the z axis:

$$U = \{(a, b, 0): a, b \in \mathbf{R}\} \qquad \text{and} \qquad W = \{(0, 0, c): c \in \mathbf{R}\}$$

Now any vector $(a, b, c) \in \mathbf{R}^3$ can be written as the sum of a vector in U and a vector in V in one and only one way:

$$(a, b, c) = (a, b, 0) + (0, 0, c)$$

Accordingly, $\mathbf{R}^3$ is the direct sum of U and W, that is, $\mathbf{R}^3 = U \oplus W$. Alternatively, $\mathbf{R}^3 = U \oplus W$ since $\mathbf{R}^3 = U + W$ and $U \cap W = \{0\}$.

General Direct Sums

The notion of a direct sum is extended to more than one factor in the obvious way. That is, V is the *direct sum* of subspaces $W_1, W_2, \ldots, W_r$, written

$$V = W_1 \oplus W_2 \oplus \cdots \oplus W_r$$

if every vector $v \in V$ can be written in one and only one way as

$$v = w_1 + w_2 + \cdots + w_r$$

where $w_1 \in W_1, w_2 \in W_2, \ldots, w_r \in W_r$.

The following theorems apply.

Theorem 5.24: Suppose $V = W_1 \oplus W_2 \oplus \cdots \oplus W_r$. Also, for each i, suppose S_i is a linearly independent subset of W_i. Then

(a) The union $S = \bigcup_i S_i$ is linearly independent in V.

(b) If S_i is a basis of W_i, then $S = \bigcup_i S_i$ is a basis of V.

(c) $\dim V = \dim W_1 + \dim W_2 + \cdots + \dim W_r$

Theorem 5.25:   Suppose $V = W_1 + W_2 + \cdots + W_r$ (where V has finite dimension) and suppose

$$\dim V = \dim W_1 + \dim W_2 + \cdots + \dim W_r$$

Then $V = W_1 \oplus W_2 \oplus \cdots \oplus W_r$.

5.10 COORDINATES

Let V be an n-dimensional vector space over a field K, and suppose

$$S = \{u_1, u_2, \ldots, u_n\}$$

is a basis of V. Then any vector $v \in V$ can be expressed uniquely as a linear combination of the basis vectors in S, say

$$v = a_1 u_1 + a_2 u_2 + \cdots + a_n u_n$$

These n scalars $a_1, a_2, \ldots, a_n$ are called the *coordinates* of v relative to the basis S; and they form the n-tuple $[a_1, a_2, \ldots, a_n]$ in K^n, called the *coordinate vector* of v relative to S. We denote this vector by $[v]_S$, or simply $[v]$, when S is understood. Thus

$$[v]_S = [a_1, a_2, \ldots, a_n]$$

Observe that brackets $[\ldots]$, not parentheses $(\ldots)$, are used to denote the coordinate vector.

Example 5.15

(a) Consider the vector space $\mathbf{P}_2(t)$ of polynomials of degree ≤ 2. The polynomials

$$p_1 = 1 \qquad p_2 = t - 1 \qquad p_3 = (t - 1)^2 = t^2 - 2t + 1$$

form a basis S of $\mathbf{P}_2(t)$. Let $v = 2t^2 - 5t + 6$. The coordinate vector of v relative to the basis S is obtained as follows.

Set $v = xp_1 + yp_2 + zp_3$ using unknown scalars x, y, z and simplify:

$$\begin{aligned}
2t^2 - 5t + 6 &= x(1) + y(t - 1) + z(t^2 - 2t + 1) \\
&= x + yt - y + zt^2 - 2zt + z \\
&= zt^2 + (y - 2z)t + (x - y + z)
\end{aligned}$$

Then set the coefficients of the same powers of t equal to each other:

$$\begin{aligned}
x - y + z &= 6 \\
y - 2z &= -5 \\
z &= 2
\end{aligned}$$

The solution of the above system is $x = 3, y = -1, z = 2$. Thus

$$v = 3p_1 - p_2 + 2p_3 \qquad \text{and so} \qquad [v] = [3, -1, 2]$$

(b) Consider real space $\mathbf{R}^3$. The vectors

$$u_1 = (1, -1, 0) \qquad u_2 = (1, 1, 0) \qquad u_3 = (0, 1, 1)$$

form a basis S of $\mathbf{R}^3$. Let $v = (5, 3, 4)$. The coordinates of v relative to the basis S are obtained as follows.

Set $v = xu_1 + yu_2 + zu_3$, that is, set v as a linear combination of the basis vectors using unknown scalars x, y, z:

$$\begin{aligned}
(5, 3, 4) &= x(1, -1, 0) + y(1, 1, 0) + z(0, 1, 1) \\
&= (x, -x, 0) + (y, y, 0) + (0, z, z) \\
&= (x + y, -x + y + z, z)
\end{aligned}$$

Then set the corresponding components equal to each other to obtain the equivalent system of linear equations

$$x + y = 5 \qquad -x + y + z = 3 \qquad z = 4$$

The solution of the system is $x = 3$, $y = 2$, $z = 4$. Thus

$$v = 3u_1 + 2u_2 + 4u_3 \qquad \text{and so} \qquad [v]_S = [3, 2, 4]$$

Remark: There is a geometrical interpretation of the coordinates of a vector v relative to a basis S for real space $\mathbf{R}^n$. We illustrate this by using the following basis of $\mathbf{R}^3$ appearing in Example 5.15.

$$S = \{u_1 = (1, -1, 0), \ u_2 = (1, 1, 0), \ u_3 = (0, 1, 1)\}$$

First consider the space $\mathbf{R}^3$ with the usual x, y, and z axes. Then the basis vectors determine a new coordinate system of $\mathbf{R}^3$, say with x', y', and z' axes, as shown in Fig. 5-4. That is:

(1)   The x' axis is in the direction of u_1.

(2)   The y' axis is in the direction of u_2.

(3)   The z' axis is in the direction of u_3.

Furthermore, the unit length in each of the axes will be equal, respectively, to the length of the corresponding basis vector. Then each vector $v = (a, b, c)$ or, equivalently, the point $P(a, b, c)$ in $\mathbf{R}^3$ will have new coordinates with respect to the new x', y', z' axes. These new coordinates are precisely the coordinates of v with respect to the basis S.

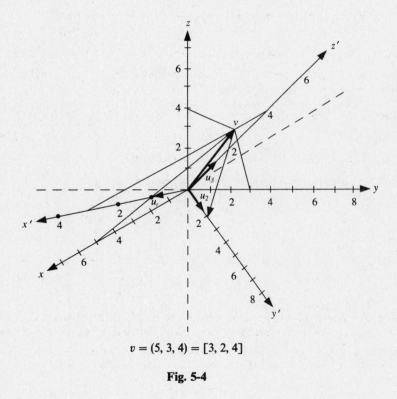

$$v = (5, 3, 4) = [3, 2, 4]$$

Fig. 5-4

Isomorphism of V with K^n

Consider a basis $S = \{u_1, u_2, \ldots, u_n\}$ of a vector space V over a field K. We have shown above that to each vector $v \in V$ there corresponds a unique n-tuple $[v]_S$ in K^n. On the other hand, for any n-tuple

$[c_1, c_2, \ldots, c_n] \in K^n$, there corresponds the vector $c_1 u_1 + c_2 u_2 + \cdots + c_n u_n$ in V. Thus the basis S induces a one-to-one correspondence between the vectors in V and the n-tuples in K^n. Furthermore, suppose

$$v = a_1 u_1 + a_2 u_2 + \cdots + a_n u_n \qquad \text{and} \qquad w = b_1 u_1 + b_2 u_2 + \cdots + b_n u_n$$

Then

$$v + w = (a_1 + b_1)u_1 + (a_2 + b_2)u_2 + \cdots + (a_n + b_n)u_n$$
$$kv = (ka_1)u_1 + (ka_2)u_2 + \cdots + (ka_n)u_n$$

where k is a scalar. Accordingly,

$$[v + w]_S = [a_1 + b_1, \ldots, a_n + b_n] = [a_1, \ldots, a_n] + [b_1, \ldots, b_n] = [v]_S + [w]_S$$

and

$$[kv]_S = [ka_1, ka_2, \ldots, ka_n] = k[a_1, a_2, \ldots, ka_n] = k[v]_S$$

Thus the above one-to-one correspondence between V and K^n preserves the vector space operations of vector and scalar multiplication; we then say that V and K^n are *isomorphic*, written $V \cong K^n$. We state this result formally.

Theorem 5.26:   Let V be an n-dimensional vector space over a field K. Then V and K^n are isomorphic.

The next example gives a practical application of the above result.

Example 5.16.   Suppose we want to determine whether or not the following matrices are linearly independent:

$$A = \begin{pmatrix} 1 & 2 & -3 \\ 4 & 0 & 1 \end{pmatrix} \qquad B = \begin{pmatrix} 1 & 3 & -4 \\ 6 & 5 & 4 \end{pmatrix} \qquad C = \begin{pmatrix} 3 & 8 & -11 \\ 16 & 10 & 9 \end{pmatrix}$$

The coordinate vectors of the above matrices relative to the usual basis [Example 5.7(a)] of $\mathbf{M}_{2,3}$ are as follows:

$$[A] = (1, 2, -3, 4, 0, 1) \qquad [B] = (1, 3, -4, 6, 5, 4) \qquad [C] = (3, 8, -11, 16, 10, 9)$$

Form the matrix M whose rows are the above coordinate vectors:

$$M = \begin{pmatrix} 1 & 2 & -3 & 4 & 0 & 1 \\ 1 & 3 & -4 & 6 & 5 & 4 \\ 3 & 8 & -11 & 16 & 10 & 9 \end{pmatrix}$$

Row reduce M to echelon form:

$$M \sim \begin{pmatrix} 1 & 2 & -3 & 4 & 0 & 1 \\ 0 & 1 & -1 & 2 & 5 & 3 \\ 0 & 2 & -2 & 4 & 10 & 6 \end{pmatrix} \sim \begin{pmatrix} 1 & 2 & -3 & 4 & 0 & 1 \\ 0 & 1 & -1 & 2 & 5 & 3 \\ 0 & 0 & 0 & 0 & 0 & 0 \end{pmatrix}$$

Since the echelon matrix has only two nonzero rows, the coordinate vectors $[A]$, $[B]$, and $[C]$ span a subspace of dimension 2 and so are linearly dependent. Accordingly, the original matrices A, B, and C are linearly dependent.

5.11   CHANGE OF BASIS

Section 5.10 showed that we can represent each vector in a vector space V by means of an n-tuple once we have selected a basis S of V. We ask the following natural question: How does our representation change if we select another basis? For the answer, we must first redefine some terms. Specifically, suppose $a_1, a_2, \ldots, a_n$ are the coordinates of a vector v relative to a basis S of V. Then we will represent

v by its *coordinate column vector*, denoted and defined by

$$[v]_S = \begin{pmatrix} a_1 \\ a_2 \\ \cdots \\ a_n \end{pmatrix} = (a_1, a_2, \ldots, a_n)^T$$

We emphasize that in this section $[v]_S$ is an $n \times 1$ matrix, not simply an element of K^n. (The meaning of $[v]_S$ will always be clear by its context.)

Suppose $S = \{u_1, u_2, \ldots, u_n\}$ is a basis of a vector space V and suppose $S' = \{v_1, v_2, \ldots, v_n\}$ is another basis. Since S is a basis, each vector in S' can be written uniquely as a linear combination of the elements in S. Say

$$v_1 = c_{11}u_1 + c_{12}u_2 + \cdots + c_{1n}u_n$$
$$v_2 = c_{21}u_1 + c_{22}u_2 + \cdots + c_{2n}u_n$$
$$\cdots\cdots\cdots\cdots\cdots\cdots\cdots\cdots\cdots\cdots\cdots$$
$$v_n = c_{n1}u_1 + c_{n2}u_2 + \cdots + c_{nn}u_n$$

Let P denote the transpose of the above matrix of coefficients:

$$P = \begin{pmatrix} c_{11} & c_{21} & \cdots & c_{n1} \\ c_{12} & c_{22} & \cdots & c_{n2} \\ \cdots\cdots\cdots\cdots\cdots\cdots\cdots \\ c_{1n} & c_{2n} & \cdots & c_{nn} \end{pmatrix}$$

That is, $P = (p_{ij})$ where $p_{ij} = c_{ji}$. Then P is called the *change-of-basis matrix* (or *transition matrix*) from the "old basis" S to the "new basis" S'.

> **Remark:** Since the vectors $v_1, v_2, \ldots, v_n$ in S' are linearly independent, the matrix P is invertible (Problem 5.84). In fact (Problem 5.80), its inverse P^{-1} is the change-of-basis matrix from the basis S' back to the basis S.

Example 5.17.   Consider the following two bases of $\mathbf{R}^2$:

$$S = \{u_1 = (1, 2), u_2 = (3, 5)\} \qquad E = \{e_1 = (1, 0), e_2 = (0, 1)\}$$

From $u_1 = e_1 + 2e_2$, $u_2 = 3e_1 + 5e_2$ it follows that

$$e_1 = -5u_1 + 2u_2$$
$$e_2 = 3u_1 - u_2$$

Writing the coefficients of u_1 and u_2 as columns gives us the change-of-basis matrix P from the basis S to the usual basis E:

$$P = \begin{pmatrix} -5 & 3 \\ 2 & -1 \end{pmatrix}$$

Furthermore, since E is the usual basis,

$$u_1 = (1, 2) = e_1 + 2e_2$$
$$u_2 = (3, 5) = 3e_1 + 5e_2$$

Writing the coefficients of e_1 and e_2 as columns gives us the change-of-basis matrix Q from the basis E back to the basis S:

$$Q = \begin{pmatrix} 1 & 3 \\ 2 & 5 \end{pmatrix}$$

Observe that P and Q are inverses:

$$PQ = \begin{pmatrix} -5 & 3 \\ 2 & -1 \end{pmatrix} \begin{pmatrix} 1 & 3 \\ 2 & 5 \end{pmatrix} = \begin{pmatrix} 1 & 0 \\ 0 & 1 \end{pmatrix} = I$$

The next theorem (proved in Problem 5.19) tells us how the coordinate (column) vectors are affected by a change of basis.

Theorem 5.27:   Let P be the change-of-basis matrix from a basis S to a basis S' in a vector space V. Then, for any vector $v \in V$, we have

$$P[v]_{S'} = [v]_S \quad \text{and hence} \quad P^{-1}[v]_S = [v]_{S'}$$

> **Remark:**   Although P is called the change-of-basis matrix from the old basis S to the new basis S', it is P^{-1} which transforms the coordinates of v relative to the original basis S into the coordinates of v relative to the new basis S'.

We illustrate the above theorem in the case dim $V = 3$. Suppose P is the change-of-basis matrix from the basis $S = \{u_1, u_2, u_3\}$ to the basis $S' = \{v_1, v_2, v_3\}$; say

$$v_1 = a_1 u_1 + a_2 u_2 + a_3 u_3$$
$$v_2 = b_1 u_1 + b_2 u_2 + b_3 u_3$$
$$v_3 = c_1 u_1 + c_2 u_2 + c_3 u_3$$

Hence

$$P = \begin{pmatrix} a_1 & b_1 & c_1 \\ a_2 & b_2 & c_2 \\ a_3 & b_3 & c_3 \end{pmatrix}$$

Now suppose $v \in V$ and, say $v = k_1 v_1 + k_2 v_2 + k_3 v_3$. Then, substituting for v_1, v_2, v_3 from above, we obtain

$$v = k_1(a_1 u_1 + a_2 u_2 + a_3 u_3) + k_2(b_1 u_1 + b_2 u_2 + b_3 u_3) + k_3(c_1 u_1 + c_2 u_2 + c_3 u_3)$$
$$= (a_1 k_1 + b_1 k_2 + c_1 k_3)u_1 + (a_2 k_1 + b_2 k_2 + c_2 k_3)u_2 + (a_3 k_1 + b_3 k_2 + c_3 k_3)u_3$$

Thus

$$[v]_{S'} = \begin{pmatrix} k_1 \\ k_2 \\ k_3 \end{pmatrix} \quad \text{and} \quad [v]_S = \begin{pmatrix} a_1 k_1 + b_1 k_2 + c_1 k_3 \\ a_2 k_1 + b_2 k_2 + c_2 k_3 \\ a_3 k_1 + b_3 k_2 + c_3 k_3 \end{pmatrix}$$

Accordingly,

$$P[v]_{S'} = \begin{pmatrix} a_1 & b_1 & c_1 \\ a_2 & b_2 & c_2 \\ a_3 & b_3 & c_3 \end{pmatrix} \begin{pmatrix} k_1 \\ k_2 \\ k_3 \end{pmatrix} = \begin{pmatrix} a_1 k_1 + b_1 k_2 + c_1 k_3 \\ a_2 k_1 + b_2 k_2 + c_2 k_3 \\ a_3 k_1 + b_3 k_2 + c_3 k_3 \end{pmatrix} = [v]_S$$

Also, multiplying the above equation by P^{-1}, we have

$$P^{-1}[v]_S = P^{-1}P[v]_{S'} = I[v]_{S'} = [v]_{S'}$$

> **Remark:**   Suppose $S = \{u_1, u_2, \ldots, u_n\}$ is a basis of a vector space V over a field K, and suppose $P = (p_{ij})$ is any nonsingular matrix over K. Then the n vectors
>
> $$v_i = p_{1i} u_1 + p_{2i} u_2 + \cdots + p_{ni} u_n \qquad i = 1, 2, \ldots, n$$

are linearly independent (Problem 5.84) and hence form another basis S' of V. More-over, P will be the change-of-basis matrix from S to the new basis S'.

Solved Problems

VECTOR SPACES

5.1. Prove Theorem 5.1.

(i) By axiom $[A_2]$, with $u = 0$, we have $0 + 0 = 0$. Hence by axiom $[M_1]$,

$$k0 = k(0 + 0) = k0 + k0$$

Adding $-k0$ to both sides gives the desired result.

(ii) By a property of K, $0 + 0 = 0$. Hence by axiom $[M_2]$, $0u = (0 + 0)u = 0u + 0u$. Adding $-0u$ to both sides yields the required result.

(iii) Suppose $ku = 0$ and $k \neq 0$. Then there exists a scalar k^{-1} such that $k^{-1}k = 1$; hence

$$u = 1u = (k^{-1}k)u = k^{-1}(ku) = k^{-1}0 = 0$$

(iv) Using $u + (-u) = 0$, we obtain $0 = k0 = k(u + (-u)) = ku + k(-u)$. Adding $-ku$ to both sides gives $-ku = k(-u)$.

Using $k + (-k) = 0$, we obtain $0 = 0u = (k + (-k))u = ku + (-k)u$. Adding $-ku$ to both sides yields $-ku = (-k)u$. Thus $(-k)u = k(-u) = -ku$.

5.2. Show that for any scalar k and any vectors u and v, $k(u - v) = ku - kv$.

Use the definition of subtraction, $u - v \equiv u + (-v)$, and the result $k(-v) = -kv$ to obtain:

$$k(u - v) = k(u + (-v)) = ku + k(-v) = ku + (-kv) = ku - kv$$

5.3. Let V be the set of all functions from a nonempty set X into a field K. For any functions $f, g \in V$ and any scalar $k \in K$, let $f + g$ and kf be the functions in V defined as follows:

$$(f + g)(x) = f(x) + g(x) \qquad \text{and} \qquad (kf)(x) = kf(x) \qquad \forall x \in X$$

Prove that V is a vector space over K.

Since X is nonempty, V is also nonempty. We now need to show that all the axioms of a vector space hold.

$[A_1]$ Let $f, g, h \in V$. To show that $(f + g) + h = f + (g + h)$, it is necessary to show that the function $(f + g) + h$ and the function $f + (g + h)$ both assign the same value to each $x \in X$. Now,

$$((f + g) + h)(x) = (f + g)(x) + h(x) = (f(x) + g(x)) + h(x) \qquad \forall x \in X$$
$$(f + (g + h))(x) = f(x) + (g + h)(x) = f(x) + (g(x) + h(x)) \qquad \forall x \in X$$

But $f(x)$, $g(x)$, and $h(x)$ are scalars in the field K where addition of scalars is associative; hence

$$(f(x) + g(x)) + h(x) = f(x) + (g(x) + h(x))$$

Accordingly, $(f + g) + h = f + (g + h)$.

$[A_2]$ Let $\mathbf{0}$ denote the zero function: $\mathbf{0}(x) = 0$, $\forall x \in X$. Then for any function $f \in V$,

$$(f + \mathbf{0})(x) = f(x) + \mathbf{0}(x) = f(x) + 0 = f(x) \qquad \forall x \in X$$

Thus $f + \mathbf{0} = f$, and $\mathbf{0}$ is the zero vector in V.

[A_3]   For any function $f \in V$, let $-f$ be the function defined by $(-f)(x) = -f(x)$. Then,

$$(f + (-f))(x) = f(x) + (-f)(x) = f(x) - f(x) = 0 = \mathbf{0}(x) \qquad \forall x \in X$$

Hence $f + (-f) = \mathbf{0}$.

[A_4]   Let $f, g \in V$. Then

$$(f + g)(x) = f(x) + g(x) = g(x) + f(x) = (g + f)(x) \qquad \forall x \in X$$

Hence $f + g = g + f$. [Note that $f(x) + g(x) = g(x) + f(x)$ follows from the fact that $f(x)$ and $g(x)$ are scalars in the field K where addition is commutative.]

[M_1]   Let $f, g \in K$. Then

$$(k(f + g))(x) = k((f + g)(x)) = k(f(x) + g(x)) = kf(x) + kg(x)$$
$$= (kf)(x) + (kg)(x) = (kf + kg)(x) \qquad \forall x \in X$$

Hence $k(f + g) = kf + kg$. (Note that $k(f(x) + g(x)) = kf(x) + kg(x)$ follows from the fact that $k, f(x)$ and $g(x)$ are scalars in the field K where multiplication is distributive over addition.)

[M_2]   Let $f \in V$ and $a, b \in K$. Then

$$((a + b)f)(x) = (a + b)f(x) = af(x) + bf(x) = (af)(x) + bf(x)$$
$$= (af + bf)(x), \qquad \forall x \in X$$

Hence $(a + b)f = af + bf$.

[M_3]   Let $f \in V$ and $a, b \in K$. Then,

$$((ab)f)(x) = (ab)f(x) = a(bf(x)) = a(bf)(x) = (a(bf))(x) \qquad \forall x \in X$$

Hence $(ab)f = a(bf)$.

[M_4]   Let $f \in V$. Then, for the unit $1 \in K$, $(1f)(x) = 1f(x)) = f(x)$, $\forall x \in X$. Hence $1f = f$.

Since all the axioms are satisfied, V is a vector space over K.

SUBSPACES

5.4.   Prove Theorem 5.2.

Suppose W satisfies (i), (ii), and (iii). By (i), W is nonempty; and by (ii) and (iii), the operations of vector addition and scalar multiplication are well defined for W. Moreover, the axioms [A_1], [A_4], [M_1], [M_2], [M_3] and [M_4] hold in W since the vectors in W belong to V. Hence we need only show that [A_2] and [A_3], also hold in W. By (i), W is nonempty, say $u \in W$. Then by (iii), $0u = 0 \in W$ and $v + 0 = v$ for every $v \in W$. Hence W satisfies [A_2]. Lastly, if $v \in W$ then $(-1)v = -v \in W$ and $v + (-v) = 0$; hence W is a subspace of V then clearly (i), (ii) and (iii) hold.

5.5.   Prove Corollary 5.3.

Suppose W satisfies (i) and (ii). Then, by (i), W is nonempty. Furthermore, if $v, w \in W$ then, by (ii), $v + w = 1v + 1w \in W$; and if $v \in W$ and $k \in K$ then, by (ii), $kv = kv + 0v \in W$. Thus by Theorem 5.2, W is a subspace of V.

Conversely, if W is a subspace of V then clearly (i) and (ii) hold in W.

5.6.   Show that W is a subspace of $\mathbf{R}^3$ where $W = \{(a, b, c): a + b + c = 0\}$, i.e., W consists of those vectors each with the property that the sum of its components is zero.

$0 = (0, 0, 0) \in W$ since $0 + 0 + 0 = 0$. Suppose $v = (a, b, c)$, $w = (a', b', c')$ belong to W, i.e., $a + b + c = 0$ and $a' + b' + c' = 0$. Then for any scalars k and k',

$$kv + k'w = k(a, b, c) + k'(a', b', c') = (ka, kb, kc) + (k'a', k'b', k'c') = (ka + k'a', kb + k'b', kc + k'c')$$

and furthermore

$$(ka + k'a') + (kb + k'b') + (kc + k'c') = k(a + b + c) + k'(a' + b' + c') = k0 + k'0 = 0$$

Thus $kv + k'w \in W$, and so W is a subspace of $\mathbf{R}^3$.

5.7. Let V be the vector space of all square $n \times n$ matrices over a field K. Show that W is a subspace of V where:

(a) W consists of the symmetric matrices, i.e., all matrices $A = (a_{ij})$ for which $a_{ji} = a_{ij}$;

(b) W consists of all matrices which commute with a given matrix T; that is

$$W = \{A \in V : AT = TA\}.$$

(a) $0 \in W$ since all entries of 0 are 0 and hence equal. Now suppose $A = (a_{ij})$ and $B = (b_{ij})$ belong to W, i.e., $a_{ji} = a_{ij}$ and $b_{ji} = b_{ij}$. For any scalars $a, b \in K$, $aA + bB$ is the matrix whose ij-entry is $aa_{ij} + bb_{ij}$. But $aa_{ji} + bb_{ji} = aa_{ij} + bb_{ij}$. Thus $aA + bB$ is also symmetric, and so W is a subspace of V.

(b) $0 \in W$ since $0T = 0 = T0$. Now suppose $A, B \in W$; that is, $AT = TA$ and $BT = TB$. For any scalars $a, b \in K$,

$$(aA + bB)T = (aA)T + (bB)T = a(AT) + b(BT) = a(TA) + b(TB)$$
$$= T(aA) + T(bB) = T(aA + bB)$$

Thus $aA + bB$ commutes with T, i.e., belongs to W; hence W is a subspace of V.

5.8. Let V be the vector space of all 2×2 matrices over the real field $\mathbf{R}$. Show that W is not a subspace of V where:

(a) W consists of all matrices with zero determinant;

(b) W consists of all matrices A for which $A^2 = A$.

(a) $\left[\text{Recall that } \det \begin{pmatrix} a & b \\ c & d \end{pmatrix} = ad - bc.\right]$ The matrices $A = \begin{pmatrix} 1 & 0 \\ 0 & 0 \end{pmatrix}$ and $B = \begin{pmatrix} 0 & 0 \\ 0 & 1 \end{pmatrix}$ belong to W since $\det(A) = 0$ and $\det(B) = 0$. But $A + B = \begin{pmatrix} 1 & 0 \\ 0 & 1 \end{pmatrix}$ does not belong to W since $\det(A + B) = 1$. Hence W is not a subspace of V.

(b) The unit matrix $I = \begin{pmatrix} 1 & 0 \\ 0 & 1 \end{pmatrix}$ belongs to W since

$$I^2 = \begin{pmatrix} 1 & 0 \\ 0 & 1 \end{pmatrix}\begin{pmatrix} 1 & 0 \\ 0 & 1 \end{pmatrix} = \begin{pmatrix} 1 & 0 \\ 0 & 1 \end{pmatrix} = I$$

But $2I = \begin{pmatrix} 2 & 0 \\ 0 & 2 \end{pmatrix}$ does not belong to W since

$$(2I)^2 = \begin{pmatrix} 2 & 0 \\ 0 & 2 \end{pmatrix}\begin{pmatrix} 2 & 0 \\ 0 & 2 \end{pmatrix} = \begin{pmatrix} 4 & 0 \\ 0 & 4 \end{pmatrix} \neq 2I$$

Hence W is not a subspace of V.

5.9. Let V be the vector space of all functions from the real field $\mathbf{R}$ into $\mathbf{R}$. Show that W is a subspace of V where W consists of the odd functions, i.e., those functions f for which $f(-x) = -f(x)$.

Let $\mathbf{0}$ denote the zero function: $\mathbf{0}(x) = 0$, for every $x \in \mathbf{R}$. $\mathbf{0} \in W$ since $\mathbf{0}(-x) = 0 = -0 = -\mathbf{0}(x)$. Suppose $f, g \in W$, i.e., $f(-x) = -f(x)$ and $g(-x) = -g(x)$. Then for any real numbers a and b,

$$(af + bg)(-x) = af(-x) + bg(-x) = -af(x) - bg(x) = -(af(x) + bg(x)) = -(af + bg)(x)$$

Hence $af + bg \in W$, and so W is a subspace of V.

5.10. Let V be the vector space of polynomials $a_0 + a_1 t + a_2 t^2 + \cdots + a_n t^n$ with real coefficients, i.e., $a_i \in \mathbf{R}$. Determine whether or not W is a subspace of V where:

(a) W consists of all polynomials with integral coefficients;

(b) W consists of all polynomials with degree ≤ 3;

(c) W consists of all polynomials $b_0 + b_1 t^2 + b_2 t^4 + \cdots + b_n t^{2n}$, i.e., polynomials with only even powers of t.

(a) No, since scalar multiples of vectors in W do not always belong to W. For example,

$$v = 3 + 5t + 7t^2 \in W \qquad \text{but} \qquad \tfrac{1}{2}v = \tfrac{3}{2} + \tfrac{5}{2}t + \tfrac{7}{2}t^2 \in W$$

(Observe that W is "closed" under vector addition, i.e., sums of elements in W belong to W.)

(b) and (c).   Yes. For, in each case, W is nonempty, the sum of elements in W belong to W, and the scalar multiples of any element in W belong to W.

5.11 Prove Theorem 5.4.

Let $\{W_i : i \in I\}$ be a collection of subspaces of V and let $W = \bigcap (W_i : i \in I)$. Since each W_i is a subspace, $0 \in W_i$ for every $i \in I$. Hence $0 \in W$. Suppose $u, v \in W$. Then $u, v \in W_i$ for every $i \in I$. Since each W_i is a subspace, $au + bv \in W_i$, for each $i \in I$. Hence $au + bv \in W$. Thus W is a subspace of V.

LINEAR COMBINATIONS, LINEAR SPANS

5.12. Express $v = (1, -2, 5)$ in $\mathbf{R}^3$ as a linear combination of the vectors u_1, u_2, u_3 where $u_1 = (1, -3, 2), u_2 = (2, -4, -1), u_3 = (1, -5, 7)$.

First set

$$(1, -2, 5) = x(1, -3, 2) + y(2, -4, -1) + z(1, -5, 7) = (x + 2y + z, -3x - 4y - 5z, 2x - y + 7z)$$

Form the equivalent system of equations and reduce to echelon form:

$$
\begin{array}{ccc}
\begin{array}{rcl}
x + 2y + z &=& 1 \\
-3x - 4y - 5z &=& -2 \\
2x - y + 7z &=& 5
\end{array}
\quad \text{or} \quad
\begin{array}{rcl}
x + 2y + z &=& 1 \\
2y - 2z &=& 1 \\
-5y + 5z &=& 3
\end{array}
\quad \text{or} \quad
\begin{array}{rcl}
x + 2y + z &=& 1 \\
2y - 2z &=& 1 \\
0 &=& 11
\end{array}
\end{array}
$$

The system does not have a solution. Thus v is not a linear combination of u_1, u_2, u_3.

5.13. Express the polynomial $v = t^2 + 4t - 3$ over $\mathbf{R}$ as a linear combination of the polynomials $p_1 = t^2 - 2t + 5, p_2 = 2t^2 - 3t, p_3 = t + 3$.

Set v as a linear combination of p_1, p_2, p_3 using unknowns x, y, z:

$$
\begin{aligned}
t^2 + 4t - 3 &= x(t^2 - 2t + 5) + y(2t^2 - 3t) + z(t + 3) \\
&= xt^2 - 2xt + 5x + 2yt^2 - 3yt + zt + 3z \\
&= (x + 2y)t^2 + (-2x - 3y + z)t + (5x + 3z)
\end{aligned}
$$

Set coefficients of the same powers of t equal to each other, and reduce the system to echelon form:

$$
\begin{array}{ccc}
\begin{array}{rcl}
x + 2y &=& 1 \\
-2x - 3y + z &=& 4 \\
5x + 3z &=& -3
\end{array}
\quad \text{or} \quad
\begin{array}{rcl}
x + 2y &=& 1 \\
y + z &=& 6 \\
-10y + 3z &=& -8
\end{array}
\quad \text{or} \quad
\begin{array}{rcl}
x + 2y &=& 1 \\
y + z &=& 6 \\
13z &=& 52
\end{array}
\end{array}
$$

The system is in triangular form and has a solution. Solving by back-substitution yields $x = -3$, $y = 2$, $z = 4$. Thus $v = -3p_1 + 2p_2 + 4p_3$.

5.14. Write the matrix $E = \begin{pmatrix} 3 & 1 \\ 1 & -1 \end{pmatrix}$ as a linear combination of the matrices

$$A = \begin{pmatrix} 1 & 1 \\ 1 & 0 \end{pmatrix} \qquad B = \begin{pmatrix} 0 & 0 \\ 1 & 1 \end{pmatrix} \quad \text{and} \quad C = \begin{pmatrix} 0 & 2 \\ 0 & -1 \end{pmatrix}$$

Set E as a linear combination of A, B, C using the unknowns x, y, z: $E = xA + yB + zC$.

$$\begin{pmatrix} 3 & 1 \\ 1 & -1 \end{pmatrix} = x\begin{pmatrix} 1 & 1 \\ 1 & 0 \end{pmatrix} + y\begin{pmatrix} 0 & 0 \\ 1 & 1 \end{pmatrix} + z\begin{pmatrix} 0 & 2 \\ 0 & -1 \end{pmatrix}$$
$$= \begin{pmatrix} x & x \\ x & 0 \end{pmatrix} + \begin{pmatrix} 0 & 0 \\ y & y \end{pmatrix} + \begin{pmatrix} 0 & 2z \\ 0 & -z \end{pmatrix} = \begin{pmatrix} x & x + 2z \\ x + y & y - z \end{pmatrix}$$

Form the equivalent system of equations by setting corresponding entries equal to each other:

$$x = 3 \qquad x + y = 1 \qquad x + 2z = 1 \qquad y - z = -1$$

Substitute $x = 3$ in the second and third equations to obtain $y = -2$ and $z = -1$. Since these values also satisfy the last equation, they form a solution of the system. Hence $E = 3A - 2B - C$.

5.15. Find a condition on a, b, c so that $w = (a, b, c)$ is a linear combination of $u = (1, -3, 2)$ and $v = (2, -1, 1)$, that is, so that w belongs to span (u, v).

Set $w = xu + yv$ using unknowns x and y:

$$(a, b, c) = x(1, -3, 2) + y(2, -1, 1) = (x + 2y, -3x - y, 2x + y)$$

Form the equivalent system and reduce to echelon form:

$$\begin{matrix} x + 2y = a \\ -3x - y = b \\ 2x + y = c \end{matrix} \quad \text{or} \quad \begin{matrix} x + 2y = a \\ 5y = 3a + b \\ -3y = -2a + c \end{matrix} \quad \text{or} \quad \begin{matrix} x + 2y = a \\ 5y = 3a + b \\ 0 = -a + 3b + 5c \end{matrix}$$

The system is consistent if and only if $a - 3b - 5c = 0$ and hence w is a linear combination of u and v when $a - 3b - 5c = 0$.

5.16. Prove Theorem 5.6.

Suppose S is empty. By definition span $S = \{0\}$. Hence span $S = \{0\}$ is a subspace and $S \subseteq$ span S. Suppose S is not empty, and $v \in S$. Then $1v = v \in$ span S; hence S is a subset of span S. Thus span S is not empty since S is not empty. Now suppose v, $w \in$ span S, say

$$v = a_1 v_1 + \cdots + a_m v_m \quad \text{and} \quad w = b_1 w_1 + \cdots + b_n w_n$$

where v_i, $w_j \in S$ and a_i, b_j are scalars. Then

$$v + w = a_1 v_1 + \cdots + a_m v_m + b_1 w_1 + \cdots + b_n w_n$$

and, for any scalar k,

$$kv = k(a_1 v_1 + \cdots + a_m v_m) = ka_1 v_1 + \cdots + ka_m v_m$$

belong to span S since each is a linear combination of vectors in S. Thus span S is a subspace of V.

Now suppose W is a subspace of V containing S and suppose $v_1, \ldots, v_m \in S \subseteq W$. Then all multiples $a_1 v_1, \ldots, a_m v_m \in W$, where $a_i \in K$, and hence the sum $a_1 v_1 + \cdots + a_m v_m \in W$. That is, W contains all linear combinations of elements of S. Consequently, span S is a subspace of W, as claimed.

LINEAR DEPENDENCE

5.17. Determine whether $u = 1 - 3t + 2t^2 - 3t^3$ and $v = -3 + 9t - 6t^2 + 9t^3$ are linearly dependent.

Two vectors are linearly dependent if and only if one is a multiple of the other. In this case, $v = -3u$.

5.18. Determine whether or not the following vectors in $\mathbf{R}^3$ are linearly dependent:

$$u = (1, -2, 1), \; v = (2, 1, -1), \; w = (7, -4, 1).$$

Method 1. Set a linear combination of the vectors equal to the zero vector using unknown scalars x, y, and z:

$$x(1, -2, 1) + y(2, 1, -1) + z(7, -4, 1) = (0, 0, 0)$$

Then

$$(x, -2x, x) + (2y, y, -y) + (7z, -4z, z) = (0, 0, 0)$$

or

$$(x + 2y + 7z, \; -2x + y - 4z, \; x - y + z) = (0, 0, 0)$$

Set corresponding components equal to each other to obtain the equivalent homogeneous system, and reduce to echelon form:

$$\begin{array}{lll} x + 2y + 7z = 0 & x + 2y + 7z = 0 & \\ -2x + y - 4z = 0 \quad \text{or} & 5y + 10z = 0 \quad \text{or} & x + 2y + 7z = 0 \\ x - y + z = 0 & -3y - 6z = 0 & y + 2z = 0 \end{array}$$

The system, in echelon form, has only two nonzero equations in the three unknowns; hence the system has a nonzero solution. Thus the original vectors are linearly dependent.

Method 2. Form the matrix whose rows are the given vectors, and reduce to echelon form using the elementary row operations:

$$\begin{pmatrix} 1 & -2 & 1 \\ 2 & 1 & -1 \\ 7 & -4 & 1 \end{pmatrix} \sim \begin{pmatrix} 1 & -2 & 1 \\ 0 & 5 & -3 \\ 0 & 10 & -6 \end{pmatrix} \sim \begin{pmatrix} 1 & -2 & 1 \\ 0 & 5 & -3 \\ 0 & 0 & 0 \end{pmatrix}$$

Since the echelon matrix has a zero row, the vectors are linearly dependent. (The three given vectors span a space of dimension 2.)

5.19. Consider the vector space $\mathbf{P}(t)$ of polynomials over $\mathbf{R}$. Determine whether the polynomials u, v, and w are linearly dependent where $u = t^3 + 4t^2 - 2t + 3$, $v = t^3 + 6t^2 - t + 4$, $w = 3t^3 + 8t^2 - 8t + 7$.

Set a linear combination of the polynomials u, v and w equal to the zero polynomial using unknown scalars x, y, and z; that is, set $xu + yv + zw = 0$. Thus

$$x(t^3 + 4t^2 - 2t + 3) + y(t^3 + 6t^2 - t + 4) + z(3t^3 + 8t^2 - 8t + 7) = 0$$

or

$$xt^3 + 4xt^2 - 2xt + 3x + yt^3 + 6yt^2 - yt + 4y + 3zt^3 + 8zt^2 - 8zt + 7z = 0$$

or

$$(x + y + 3z)t^3 + (4x + 6y + 8z)t^2 + (-2x - y - 8z)t + (3x + 4y + 7z) = 0$$

Set the coefficients of the powers of t each equal to 0 and reduce the system to echelon form:

$$\begin{array}{lll} x + y + 3z = 0 & x + y + 3z = 0 & \\ 4x + 6y + 8z = 0 & 2y - 4z = 0 & x + y + 3z = 0 \\ -2x - y - 8z = 0 \quad \text{or} & y - 2z = 0 \quad \text{or finally} & y - 2z = 0 \\ 3x + 4y + 7z = 0 & y - 2z = 0 & \end{array}$$

The system in echelon form has a free variable and hence a nonzero solution. We have shown that $xu + yv + zw = 0$ does not imply that $x = 0$, $y = 0$, $z = 0$; hence the polynomials are linearly dependent.

5.20. Let V be the vector space of functions from $\mathbf{R}$ into $\mathbf{R}$. Show that $f, g, h \in V$ are linearly independent, where $f(t) = \sin t$, $g(t) = \cos t$, $h(t) = t$.

Set a linear combination of the functions equal to the zero function 0 using unknown scalars x, y, and z: $xf + yg + zh = 0$; and then show that $x = 0$, $y = 0$, $z = 0$. We emphasize that $xf + yg + zh = 0$ means that, *for every value of* t, $xf(t) + yg(t) + zh(t) = 0$.

Thus, in the equation $x \sin t + y \cos t + zt = 0$, substitute

$t = 0$	to obtain	$x \cdot 0 + y \cdot 1 + z \cdot 0 = 0$	or	$y = 0$
$t = \pi/2$	to obtain	$x \cdot 1 + y \cdot 0 + z\pi/2 = 0$	or	$x + \pi z/2 = 0$
$t = \pi$	to obtain	$x \cdot 0 + y(-1) + z \cdot \pi = 0$	or	$-y + \pi z = 0$

Solve the system $\begin{cases} y = 0 \\ x + \pi z/2 = 0 \\ -y + \pi z = 0 \end{cases}$ to obtain only the zero solution: $x = 0$, $y = 0$, $z = 0$.

Hence f, g and h are linearly independent.

5.21. Let V be the vector space of 2×2 matrices over **R**. Determine whether the matrices A, B, $C \in V$ are linearly dependent, where:

$$A = \begin{pmatrix} 1 & 1 \\ 1 & 1 \end{pmatrix} \qquad B = \begin{pmatrix} 1 & 0 \\ 0 & 1 \end{pmatrix} \qquad C = \begin{pmatrix} 1 & 1 \\ 0 & 0 \end{pmatrix}$$

Set a linear combination of the matrices A, B, and C equal to the zero matrix using unknown scalars x, y, and z; that is, set $xA + yB + zC = 0$. Thus

$$x\begin{pmatrix} 1 & 1 \\ 1 & 1 \end{pmatrix} + y\begin{pmatrix} 1 & 0 \\ 0 & 1 \end{pmatrix} + z\begin{pmatrix} 1 & 1 \\ 0 & 0 \end{pmatrix} = \begin{pmatrix} 0 & 0 \\ 0 & 0 \end{pmatrix}$$

or

$$\begin{pmatrix} x + y + z & x + z \\ x & x + y \end{pmatrix} = \begin{pmatrix} 0 & 0 \\ 0 & 0 \end{pmatrix}$$

Set corresponding entries equal to each other to obtain the following equivalent system of linear equations:

$$x + y + z = 0 \qquad x + z = 0 \qquad x = 0 \qquad x + y = 0$$

Solving the above system we obtain only the zero solution, $x = 0$, $y = 0$, $z = 0$. We have shown that $xA + yB + zC$ implies $x = 0$, $y = 0$, $z = 0$; hence the matrices A, B, and C are linearly independent.

5.22. Suppose u, v, and w are linearly independent vectors. Show that $u + v$, $u - v$, and $u - 2v + w$ are also linearly independent.

Suppose $x(u + v) + y(u - v) + z(u - 2v + w) = 0$ where x, y, and z are scalars. Then

$$xu + xv + yu - yv + zu - 2zv + zw = 0$$

or

$$(x + y + z)u + (x - y - 2z)v + zw = 0$$

But u, v, and w are linearly independent; hence the coefficients in the above relation are each 0:

$$\begin{aligned} x + y + z &= 0 \\ x - y - 2z &= 0 \\ z &= 0 \end{aligned}$$

The only solution to the above system is $x = 0$, $y = 0$, $z = 0$. Hence $u + v$, $u - v$, and $u - 2v + w$ are linearly independent.

5.23. Show that the vectors $v = (1 + i, 2i)$ and $w = (1, 1 + i)$ in $\mathbf{C}^2$ are linearly dependent over the complex field **C** but are linearly independent over the real field **R**.

Recall that two vectors are linearly dependent (over a field K) if and only if one of them is a multiple of the other (by an element in K). Since

$$(1 + i)w = (1 + i)(1, 1 + i) = (1 + i, 2i) = v$$

v and w are linearly dependent over $\mathbf{C}$. On the other hand, v and w are linearly independent over $\mathbf{R}$ since no real multiple of w can equal v. Specifically, when k is real, the first component of $kw = (k, k + ki)$ is real and it can never equal the first component $1 + i$ of v which is complex.

BASIS AND DIMENSION

5.24. Determine whether $(1, 1, 1)$, $(1, 2, 3)$, and $(2, -1, 1)$ form a basis for the vector space $\mathbf{R}^3$.

The three vectors form a basis if and only if they are linearly independent. Thus form the matrix A whose rows are the given vectors, and row reduce to echelon form:

$$A = \begin{pmatrix} 1 & 1 & 1 \\ 1 & 2 & 3 \\ 2 & -1 & 1 \end{pmatrix} \sim \begin{pmatrix} 1 & 1 & 1 \\ 0 & 1 & 2 \\ 0 & -3 & -1 \end{pmatrix} \sim \begin{pmatrix} 1 & 1 & 1 \\ 0 & 1 & 2 \\ 0 & 0 & 5 \end{pmatrix}$$

The echelon matrix has no zero rows; hence the three vectors are linearly independent and so they form a basis for $\mathbf{R}^3$.

5.25. Determine whether $(1, 1, 1, 1)$, $(1, 2, 3, 2)$, $(2, 5, 6, 4)$, $(2, 6, 8, 5)$ form a basis of $\mathbf{R}^4$.

Form the matrix whose rows are the given vectors, and row reduce to echelon form:

$$B = \begin{pmatrix} 1 & 1 & 1 & 1 \\ 1 & 2 & 3 & 2 \\ 2 & 5 & 6 & 4 \\ 2 & 6 & 8 & 5 \end{pmatrix} \sim \begin{pmatrix} 1 & 1 & 1 & 1 \\ 0 & 1 & 2 & 1 \\ 0 & 3 & 4 & 2 \\ 0 & 4 & 6 & 3 \end{pmatrix} \sim \begin{pmatrix} 1 & 1 & 1 & 1 \\ 0 & 1 & 2 & 1 \\ 0 & 0 & -2 & -1 \\ 0 & 0 & -2 & -1 \end{pmatrix} \sim \begin{pmatrix} 1 & 1 & 1 & 1 \\ 0 & 1 & 2 & 1 \\ 0 & 0 & 2 & 1 \\ 0 & 0 & 0 & 0 \end{pmatrix}$$

The echelon matrix has a zero row; hence the four vectors are linearly dependent and do not form a basis of $\mathbf{R}^4$.

5.26. Consider the vector space $\mathbf{P}_n(t)$ of polynomials in t of degree $\leq n$. Determine whether or not $1 + t, t + t^2, t^2 + t^3, \ldots, t^{n-1} + t^n$ form a basis of $\mathbf{P}_n(t)$.

The polynomials are linearly independent since each one is of degree higher than the preceding ones. However, there are only n polynomials and $\dim \mathbf{P}_n(t) = n + 1$. Thus the polynomials do not form a basis of $\mathbf{P}_n(t)$.

5.27. Let V be the vector space of real 2×2 matrices. Determine whether

$$A = \begin{pmatrix} 1 & 1 \\ 0 & 0 \end{pmatrix} \qquad B = \begin{pmatrix} 0 & 1 \\ 1 & 0 \end{pmatrix} \qquad C = \begin{pmatrix} 0 & 0 \\ 1 & 1 \end{pmatrix} \qquad D = \begin{pmatrix} 0 & 0 \\ 0 & 1 \end{pmatrix}$$

form a basis for V.

The coordinate vectors (see Section 5.10) of the matrices relative to the usual basis are, respectively,

$$[A] = (1, 1, 0, 0) \qquad [B] = (0, 1, 1, 0) \qquad [C] = (0, 0, 1, 1) \qquad [D] = (0, 0, 0, 1)$$

The coordinate vectors form a matrix in echelon form and hence they are linearly independent. Thus the four corresponding matrices are linearly independent. Moreover, since $\dim V = 4$, they form a basis for V.

5.28. Let V be the vector space of 2×2 symmetric matrices over K. Show that $\dim V = 3$. [Recall that $A = (a_{ij})$ is symmetric iff $A = A^T$ or, equivalently, $a_{ij} = a_{ji}$.]

An arbitrary 2×2 symmetric matrix is of the form $A = \begin{pmatrix} a & b \\ b & c \end{pmatrix}$ where $a, b, c \in K$. (Note that there are three "variables.") Setting

(i) $a = 1, b = 0, c = 0$,   (ii) $a = 0, b = 1, c = 0$,   (iii) $a = 0, b = 0, c = 1$

we obtain the respective matrices

$$E_1 = \begin{pmatrix} 1 & 0 \\ 0 & 0 \end{pmatrix} \qquad E_2 = \begin{pmatrix} 0 & 1 \\ 1 & 0 \end{pmatrix} \qquad E_3 = \begin{pmatrix} 0 & 0 \\ 0 & 1 \end{pmatrix}$$

We show that $\{E_1, E_2, E_3\}$ is a basis of V, that is, that (a) it spans V, and (b) it is linearly independent.

(a)   For the above arbitrary matrix A in V, we have

$$A = \begin{pmatrix} a & b \\ b & c \end{pmatrix} = aE_1 + bE_2 + cE_3$$

Thus $\{E_1, E_2, E_3\}$ spans V.

(b)   Suppose $xE_1 + yE_2 + zE_3 = 0$, where x, y, z are unknown scalars. That is, suppose

$$x\begin{pmatrix} 1 & 0 \\ 0 & 0 \end{pmatrix} + y\begin{pmatrix} 0 & 1 \\ 1 & 0 \end{pmatrix} + z\begin{pmatrix} 0 & 0 \\ 0 & 1 \end{pmatrix} = \begin{pmatrix} 0 & 0 \\ 0 & 0 \end{pmatrix} \quad \text{or} \quad \begin{pmatrix} x & y \\ y & z \end{pmatrix} = \begin{pmatrix} 0 & 0 \\ 0 & 0 \end{pmatrix}$$

Setting corresponding entries equal to each other, we obtain $x = 0, y = 0, z = 0$. In other words,

$$xE_1 + yE_2 + zE_3 = 0 \qquad \text{implies} \qquad x = 0, y = 0, z = 0$$

Accordingly, $\{E_1, E_2, E_3\}$ is linearly independent.

Thus $\{E_1, E_2, E_3\}$ is a basis of V and so the dimension of V is 3.

5.29.   Consider the complex field $\mathbf{C}$ which contains the real field $\mathbf{R}$ which contains the rational field $\mathbf{Q}$. (Thus $\mathbf{C}$ is a vector space over $\mathbf{R}$, and $\mathbf{R}$ is a vector space over $\mathbf{Q}$.)

(a)   Show that $\mathbf{C}$ is a vector space of dimension 2 over $\mathbf{R}$.

(b)   Show that $\mathbf{R}$ is a vector space of infinite dimension over $\mathbf{Q}$.

(a)   We claim that $\{1, i\}$ is a basis of $\mathbf{C}$ over $\mathbf{R}$. For if $v \in \mathbf{C}$, then $v = a + bi = a \cdot 1 + b \cdot i$ where $a, b \in \mathbf{R}$; that is, $\{1, i\}$ spans $\mathbf{C}$ over $\mathbf{R}$. Furthermore, if $x \cdot 1 + y \cdot i = 0$ or $x + yi = 0$, where $x, y \in \mathbf{R}$, then $x = 0$ and $y = 0$; that is, $\{1, i\}$ is linearly independent over $\mathbf{R}$. Thus $\{1, i\}$ is a basis of $\mathbf{C}$ over $\mathbf{R}$, and so $\mathbf{C}$ is of dimension 2 over $\mathbf{R}$.

(b)   We claim that, for any n, $\{1, \pi, \pi^2, \ldots, \pi^n\}$ is linearly independent over $\mathbf{Q}$. For suppose $a_0 1 + a_1 \pi + a_2 \pi^2 + \cdots + a_n \pi^n = 0$, where the $a_i \in \mathbf{Q}$, and not all the a_i are 0. Then π is a root of the following nonzero polynomial over $\mathbf{Q}$: $a_0 + a_1 x + a_2 x^2 + \cdots + a_n x^n$. But it can be shown that π is a transcendental number, i.e., that π is not a root of any nonzero polynomial over $\mathbf{Q}$. Accordingly, the $n + 1$ real numbers $1, \pi, \pi^2, \ldots, \pi^n$ are linearly independent over $\mathbf{Q}$. Thus, for any finite n, $\mathbf{R}$ cannot be of dimension n over $\mathbf{Q}$; $\mathbf{R}$ is of infinite dimension over $\mathbf{Q}$.

5.30.   Let $S = \{u_1, u_2, \ldots, u_n\}$ be a subset of a vector space V. Show that the following two conditions are equivalent: (a) S is linearly independent and spans V, and (b) every vector $v \in V$ can be written uniquely as a linear combination of the vectors in S.

Suppose (a) holds. Since S spans V, the vector v is a linear combination of the u_i; say,

$$v = a_1 u_1 + a_2 u_2 + \cdots + a_n u_n$$

Suppose we also have

$$v = b_1 u_1 + b_2 u_2 + \cdots + b_n u_n$$

Subtracting, we get

$$0 = v - v = (a_1 - b_1)u_1 + (a_2 - b_2)u_2 + \cdots + (a_n - b_n)u_n$$

But the u_i are linearly independent; hence the coefficients in the above relation are each 0:

$$a_1 - b_1 = 0, \, a_2 - b_2 = 0, \, \ldots, \, a_n - b_n = 0$$

Therefore, $a_1 = b_1, \, a_2 = b_2, \, \ldots, \, a_n = b_n$; hence the representation of v as a linear combination of the u_i is unique. Thus (a) implies (b).

Suppose (b) holds. Then S spans V. Suppose

$$0 = c_1 u_1 + c_2 u_2 + \cdots + c_n u_n$$

However, we do have

$$0 = 0u_1 + 0u_2 + \cdots + 0u_n$$

By hypothesis, the representation of 0 as a linear combination of the u_i is unique. Hence each $c_i = 0$ and the u_i are linearly independent. Thus (b) implies (a).

DIMENSION AND SUBSPACES

5.31. Find a basis and the dimension of the subspace W of $\mathbf{R}^3$ where:

$$(a) \quad W = \{(a, b, c): a + b + c = 0\}, \qquad (b) \quad W = \{(a, b, c): a = b = c\},$$

$$(c) \quad W = (a, b, c): c = 3a\}$$

(a) Note $W \neq \mathbf{R}^3$ since, e.g., $(1, 2, 3) \notin W$. Thus dim $W < 3$. Note $u_1 = (1, 0, -1)$ and $u_2 = (0, 1, -1)$ are two independent vectors in W. Thus dim $W = 2$ and so u_1 and u_2 form a basis of W.

(b) The vector $u = (1, 1, 1) \in W$. Any vector $w \in W$ has the form $w = (k, k, k)$. Hence $w = ku$. Thus u spans W and dim $W = 1$.

(c) $W \neq \mathbf{R}^3$ since, e.g., $(1, 1, 1) \notin W$. Thus dim $W < 3$. The vectors $u_1 = (1, 0, 3)$ and $u_2 = (0, 1, 0)$ belong to W and are linearly independent. Thus dim $W = 2$ and u_1, u_2 form a basis of W.

5.32. Find a basis and the dimension of the subspace W of $\mathbf{R}^4$ spanned by

$$u_1 = (1, -4, -2, 1), \quad u_2 = (1, -3, -1, 2), \quad u_3 = (3, -8, -2, 7).$$

Apply the Row Space Algorithm 5.8A. Form a matrix where the rows are the given vectors, and row reduce it to echelon form:

$$\begin{pmatrix} 1 & -4 & -2 & 1 \\ 1 & -3 & -1 & 2 \\ 3 & -8 & -2 & 7 \end{pmatrix} \sim \begin{pmatrix} 1 & -4 & -2 & 1 \\ 0 & 1 & 1 & 1 \\ 0 & 4 & 4 & 4 \end{pmatrix} \sim \begin{pmatrix} 1 & -4 & -2 & 1 \\ 0 & 1 & 1 & 1 \\ 0 & 0 & 0 & 0 \end{pmatrix}$$

The nonzero rows in the echelon matrix form a basis of W and so dim $W = 2$. In particular, this means that the original three vectors are linearly dependent.

5.33. Let W be the subspace of $\mathbf{R}^4$ spanned by the vectors

$$u_1 = (1, -2, 5, -3), \quad u_2 = (2, 3, 1, -4) \quad u_3 = (3, 8, -3, -5).$$

(a) Find a basis and the dimension of W. (b) Extend the basis of W to a basis of the whole space $\mathbf{R}^4$.

(a) Form the matrix A whose rows are the given vectors, and row reduce it to echelon form:

$$A = \begin{pmatrix} 1 & -2 & 5 & -3 \\ 2 & 3 & 1 & -4 \\ 3 & 8 & -3 & -5 \end{pmatrix} \sim \begin{pmatrix} 1 & -2 & 5 & -3 \\ 0 & 7 & -9 & 2 \\ 0 & 14 & -18 & 4 \end{pmatrix} \sim \begin{pmatrix} 1 & -2 & 5 & -3 \\ 0 & 7 & -9 & 2 \\ 0 & 0 & 0 & 0 \end{pmatrix}$$

The nonzero rows $(1, -2, 5, -3)$ and $(0, 7, -9, 2)$ of the echelon matrix form a basis of the row space of A and hence of W. Thus, in particular, dim $W = 2$.

(b) We seek four linearly independent vectors which include the above two vectors. The four vectors $(1, -2, 5, -3)$, $(0, 7, -9, 2)$, $(0, 0, 1, 0)$, and $(0, 0, 0, 1)$ are linearly independent (since they form an echelon matrix), and so they form a basis of $\mathbf{R}^4$ which is an extension of the basis of W.

5.34. Let W be the subspace of $\mathbf{R}^5$ spanned by the vectors $u_1 = (1, 2, -1, 3, 4)$, $u_2 = (2, 4, -2, 6, 8)$, $u_3 = (1, 3, 2, 2, 6)$, $u_4 = (1, 4, 5, 1, 8)$, and $u_5 = (2, 7, 3, 3, 9)$. Find a subset of the vectors which form a basis of W.

Method 1. Here we use the Casting-Out Algorithm *5.8B*. Form the matrix whose columns are the given vectors and reduce it to echelon form:

$$\begin{pmatrix} 1 & 2 & 1 & 1 & 2 \\ 2 & 4 & 3 & 4 & 7 \\ -1 & -2 & 2 & 5 & 3 \\ 3 & 6 & 2 & 1 & 3 \\ 4 & 8 & 6 & 8 & 9 \end{pmatrix} \sim \begin{pmatrix} 1 & 2 & 1 & 1 & 2 \\ 0 & 0 & 1 & 2 & 3 \\ 0 & 0 & 3 & 6 & 5 \\ 0 & 0 & -1 & -2 & -3 \\ 0 & 0 & 2 & 4 & 1 \end{pmatrix} \sim \begin{pmatrix} 1 & 2 & 1 & 1 & 2 \\ 0 & 0 & 1 & 2 & 3 \\ 0 & 0 & 0 & 0 & -4 \\ 0 & 0 & 0 & 0 & 0 \\ 0 & 0 & 0 & 0 & -5 \end{pmatrix}$$

$$\sim \begin{pmatrix} 1 & 2 & 1 & 1 & 2 \\ 0 & 0 & 1 & 2 & 3 \\ 0 & 0 & 0 & 0 & -4 \\ 0 & 0 & 0 & 0 & 0 \\ 0 & 0 & 0 & 0 & 0 \end{pmatrix}$$

The pivot positions are in columns C_1, C_3, C_5. Hence the corresponding vectors u_1, u_2, u_3 form a basis of W and dim $W = 3$.

Method 2. Here we use a slight modification of the Row Reduction Algorithm *5.8A*. Form the matrix whose rows are the given vectors and reduce it to an "echelon" form but without interchanging any zero rows:

$$\begin{pmatrix} 1 & 2 & -1 & 3 & 4 \\ 2 & 4 & -2 & 6 & 8 \\ 1 & 3 & 2 & 2 & 6 \\ 1 & 4 & 5 & 1 & 8 \\ 2 & 7 & 3 & 3 & 9 \end{pmatrix} \sim \begin{pmatrix} 1 & 2 & -1 & 3 & 4 \\ 0 & 0 & 0 & 0 & 0 \\ 0 & 1 & 3 & -1 & 2 \\ 0 & 2 & 6 & -2 & 4 \\ 0 & 3 & 5 & -3 & 1 \end{pmatrix} \sim \begin{pmatrix} 1 & 2 & -1 & 3 & 4 \\ 0 & 0 & 0 & 0 & 0 \\ 0 & 1 & 3 & -1 & 2 \\ 0 & 0 & 0 & 0 & 0 \\ 0 & 0 & -4 & 0 & -5 \end{pmatrix}$$

The nonzero rows are the first, third, and fifth rows; hence u_1, u_3, u_5 form a basis of W. Thus, in particular, dim $W = 3$.

5.35. Let V be the vector space of real 2×2 matrices. Find the dimension and a basis of the subspace W of V spanned by

$$A = \begin{pmatrix} 1 & 2 \\ -1 & 3 \end{pmatrix} \qquad B = \begin{pmatrix} 2 & 5 \\ 1 & -1 \end{pmatrix} \qquad C = \begin{pmatrix} 5 & 12 \\ 1 & 1 \end{pmatrix} \qquad D = \begin{pmatrix} 3 & 4 \\ -2 & 5 \end{pmatrix}$$

The coordinate vectors (see Section 5.10) of the given matrices relative to the usual basis of V are as follows:

$$[A] = [1, 2, -1, 3] \qquad [B] = [2, 5, 1, -1] \qquad [C] = [5, 12, 1, 1] \qquad [D] = [3, 4, -2, 5]$$

Form a matrix whose rows are the coordinate vectors, and reduce it to echelon form:

$$\begin{pmatrix} 1 & 2 & -1 & 3 \\ 2 & 5 & 1 & -1 \\ 5 & 12 & 1 & 1 \\ 3 & 4 & -2 & 5 \end{pmatrix} \sim \begin{pmatrix} 1 & 2 & -1 & 3 \\ 0 & 1 & 3 & -7 \\ 0 & 2 & 6 & -14 \\ 0 & -2 & 1 & -4 \end{pmatrix} \sim \begin{pmatrix} 1 & 2 & -1 & 3 \\ 0 & 1 & 3 & -7 \\ 0 & 0 & 0 & 0 \\ 0 & 0 & 7 & -18 \end{pmatrix}$$

The nonzero rows are linearly independent, hence the corresponding matrices $\begin{pmatrix} 1 & 2 \\ -1 & 3 \end{pmatrix}$, $\begin{pmatrix} 0 & 1 \\ 3 & -7 \end{pmatrix}$, and

$\begin{pmatrix} 0 & 0 \\ 7 & -18 \end{pmatrix}$ form a basis of W and dim $W = 3$. (Note also that the matrices A, B, and D form a basis of W.)

THEOREMS ON LINEAR DEPENDENCE, BASIS, AND DIMENSION

5.36. Prove Lemma 5.10.

Since the v_i are linearly dependent, there exist scalars $a_1, \ldots, a_m$, not all 0, such that $a_1 v_1 + \cdots + a_m v_m = 0$. Let k be the largest integer such that $a_k \neq 0$. Then

$$a_1 v_1 + \cdots + a_k v_k + 0 v_{k+1} + \cdots + 0 v_m = 0 \qquad \text{or} \qquad a_1 v_1 + \cdots + a_k v_k = 0$$

Suppose $k = 1$; then $a_1 v_1 = 0$, $a_1 \neq 0$ and so $v_1 = 0$. But the v_i are nonzero vectors; hence $k > 1$ and

$$v_k = -a_k^{-1} a_1 v_1 - \cdots - a_k^{-1} a_{k-1} v_{k-1}$$

That is, v_k is a linear combination of the preceding vectors.

5.37. Prove Theorem 5.11.

Suppose $R_n, R_{n-1}, \ldots, R_1$ are linearly dependent. Then one of the rows, say R_m, is a linear combination of the preceding rows:

$$R_m = a_{m+1} R_{m+1} + a_{m+2} R_{m+2} + \cdots + a_n R_n \qquad (*)$$

Now suppose the kth component of R_m is its first nonzero entry. Then, since the matrix is in echelon form, the kth components of $R_{m+1}, \ldots, R_n$ are all 0, and so the kth component of $(*)$ is

$$a_{m+1} \cdot 0 + a_{m+2} \cdot 0 + \cdots + a_n \cdot 0 = 0$$

But this contradicts the assumption that the kth component of R_m is not 0. Thus $R_1, \ldots, R_n$ are linearly independent.

5.38. Suppose $\{v_1, \ldots, v_m\}$ spans a vector space V. Prove:

(a) If $w \in V$, then $\{w, v_1, \ldots, v_m\}$ is linearly dependent and spans V.

(b) If v_i is a linear combination of vectors $(v_1, v_2, \ldots, v_{i-1})$, then $\{v_1, \ldots, v_{i-1}, v_{i+1}, \ldots, v_m\}$ spans V.

(a) The vector w is a linear combination of the v_i since $\{v_i\}$ spans V. Accordingly, $\{w, v_1, \ldots, v_m\}$ is linearly dependent. Clearly, w with the v_i span V since the v_i by themselves span V. That is, $\{w, v_1, \ldots, v_m\}$ spans V.

(b) Suppose $v_i = k_1 v_1 + \cdots + k_{i-1} v_{i-1}$. Let $u \in V$. Since $\{v_i\}$ spans V, u is a linear combination of the v_i, say, $u = a_1 v_1 + \cdots + a_m v_m$. Substituting for v_i, we obtain

$$u = a_1 v_1 + \cdots + a_{i-1} v_{i-1} + a_i(k_1 v_1 + \cdots + k_{i-1} v_{i-1}) + a_{i+1} v_{i+1} + \cdots + a_m v_m$$
$$= (a_1 + a_i k_1) v_1 + \cdots + (a_{i-1} + a_i k_{i-1}) v_{i-1} + a_{i+1} v_{i+1} + \cdots + a_m v_m$$

Thus $\{v_1, \ldots, v_{i-1}, v_{i+1}, \ldots, v_m\}$ spans V. In other words, we can delete v_i from the spanning set and still retain a spanning set.

5.39. Prove Lemma 5.13.

It suffices to prove the theorem in the case that the v_i are all not 0. (Prove!) Since $\{v_i\}$ spans V, we have by Problem 5.38 that

$$\{w_1, v_1, \ldots, v_n\} \qquad (1)$$

is linearly dependent and also spans V. By Lemma 5.10, one of the vectors in (I) is a linear combination of the preceding vectors. This vector cannot be w_1, so it must be one of the v's, say v_j. Thus by Problem 5.38 we can delete v_j from the spanning set (I) and obtain the spanning set

$$\{w_1; v_1, \ldots, v_{j-1}, v_{j+1}, \ldots, v_n\} \tag{2}$$

Now we repeat the argument with the vector w_2. That is, since (2) spans V, the set

$$\{w_1, w_2, v_1, \ldots, v_{j-1}, v_{j+1}, \ldots, v_n\} \tag{3}$$

is linearly dependent and also spans V. Again by Lemma 5.10, one of the vectors in (3) is a linear combination of the preceding vectors. We emphasize that this vector cannot be w_1 or w_2 since $\{w_1, \ldots, w_m\}$ is independent; hence it must be one of the v's, say v_k. Thus by the preceding problem we can delete v_k from the spanning set (3) and obtain the spanning set

$$\{w_1, w_2, v_1, \ldots, v_{j-1}, v_{j+1}, \ldots, v_{k-1}, v_{k+1}, \ldots, v_n\}$$

We repeat the argument with w_3 and so forth. At each step we are able to add one of the w's and delete one of the v's in the spanning set. If $m \le n$, then we finally obtain a spanning set of the required form:

$$\{w_1, \ldots, w_m, v_{i_1}, \ldots, v_{i_{n-m}}\}$$

Last, we show that $m > n$ is not possible. Otherwise, after n of the above steps, we obtain the spanning set $\{w_1, \ldots, w_n\}$. This implies that w_{n+1} is a linear combination of $w_1, \ldots, w_n$ which contradicts the hypothesis that $\{w_i\}$ is linearly independent.

5.40. Prove Theorem 5.12.

Suppose $\{u_1, u_2, \ldots, u_n\}$ is a basis of V, and suppose $\{v_1, v_2, \ldots\}$ is another basis of V. Since $\{u_i\}$ spans V, the basis $\{v_1, v_2, \ldots\}$ must contain n or less vectors, or else it is linearly dependent by Problem 5.39 (Lemma 5.13). On the other hand, if the basis $\{v_1, v_2, \ldots\}$ contains less than n elements, then $\{u_1, u_2, \ldots, u_n\}$ is linearly dependent by Problem 5.39. Thus the basis $\{v_1, v_2, \ldots\}$ contains exactly n vectors, and so the theorem is true.

5.41. Prove Theorem 5.14.

Suppose $B = \{w_1, w_2, \ldots, w_n\}$ is a basis of V.

(i)   Since B spans V, any $n + 1$ or more vectors are linearly dependent by Lemma 5.13.

(ii) By Lemma 5.13, elements from B can be adjoined to S to form a spanning set of V with n elements. Since S already has n elements, S itself is a spanning set of V. Thus S is a basis of V.

(iii) Suppose T is linearly dependent. Then some v_i is a linear combination of the preceding vectors. By Problem 5.38, V is spanned by the vectors in T without v_i and there are $n - 1$ of them. By Lemma 5.13, the independent set B cannot have more than $n - 1$ elements. This contradicts the fact that B has n elements. Thus T is linearly independent and hence T is a basis of V.

5.42. Prove Theorem 5.15.

(i)   Suppose $\{v_1, \ldots, v_m\}$ is a maximal linearly independent subset of S, and suppose $w \in S$. Accordingly $\{v_1, \ldots, v_m, w\}$ is linearly dependent. No v_k can be a linear combination of preceding vectors; hence w is a linear combination of the v_i. Thus $w \in$ span v_i and hence $S \subseteq$ span v_i. This leads to

$$V = \text{span } S \subseteq \text{span } v_i \subseteq V$$

Thus $\{v_i\}$ spans V and, since it is linearly independent, it is a basis of V.

(ii) The remaining vectors form a maximal linearly independent subset of S and hence by part (i) it is a basis of V.

5.43. Prove Theorem 5.16.

Suppose $B = \{w_1, w_2, \ldots, w_n\}$ is a basis of V. Then B spans V and hence V is spanned by

$$S \cup B = \{u_1, u_2, \ldots, u_r, w_1, w_2, \ldots, w_n\}$$

By Theorem 5.15, we can delete from $S \cup B$ each vector which is a linear combination of preceding vectors to obtain a basis B' for V. Since S is linearly independent, no u_k is a linear combination of preceding vectors. Thus B' contains every vector in S. Thus S is part of the basis B' for V.

5.44. Prove Theorem 5.17.

Since V is of dimension n, any $n + 1$ or more vectors are linearly dependent. Furthermore, since a basis of W consists of linearly independent vectors, it cannot contain more than n elements. Accordingly, $\dim W \le n$.

In particular, if $\{w_1, \ldots, w_n\}$ is a basis of W, then, since it is an independent set with n elements, it is also a basis of V. Thus $W = V$ when $\dim W = n$.

ROWSPACE AND RANK OF A MATRIX

5.45. Determine whether the following matrices have the same row space:

$$A = \begin{pmatrix} 1 & 1 & 5 \\ 2 & 3 & 13 \end{pmatrix} \qquad B = \begin{pmatrix} 1 & -1 & -2 \\ 3 & -2 & -3 \end{pmatrix} \qquad C = \begin{pmatrix} 1 & -1 & -1 \\ 4 & -3 & -1 \\ 3 & -1 & 3 \end{pmatrix}$$

Matrices have the same row space if and only if their row canonical forms have the same nonzero rows; hence row reduce each matrix to row canonical form:

$$A = \begin{pmatrix} 1 & 1 & 5 \\ 2 & 3 & 13 \end{pmatrix} \sim \begin{pmatrix} 1 & 1 & 5 \\ 0 & 1 & 3 \end{pmatrix} \sim \begin{pmatrix} 1 & 0 & 2 \\ 0 & 1 & 3 \end{pmatrix}$$

$$B = \begin{pmatrix} 1 & -1 & -2 \\ 3 & -2 & -3 \end{pmatrix} \sim \begin{pmatrix} 1 & -1 & -2 \\ 0 & 1 & 3 \end{pmatrix} \sim \begin{pmatrix} 1 & 0 & 1 \\ 0 & 1 & 3 \end{pmatrix}$$

$$C = \begin{pmatrix} 1 & -1 & -1 \\ 4 & -3 & -1 \\ 3 & -1 & 3 \end{pmatrix} \sim \begin{pmatrix} 1 & -1 & -1 \\ 0 & 1 & 3 \\ 0 & 2 & 6 \end{pmatrix} \sim \begin{pmatrix} 1 & -1 & -1 \\ 0 & 1 & 3 \\ 0 & 0 & 0 \end{pmatrix} \quad \text{to} \quad \begin{pmatrix} 1 & 0 & 2 \\ 0 & 1 & 3 \\ 0 & 0 & 0 \end{pmatrix}$$

Since the nonzero rows of the reduced form of A and of the reduced form of C are the same, A and C have the same row space. On the other hand, the nonzero rows of the reduced form of B are not the same as the others, and so B has a different row space.

5.46. Show that $A = \begin{pmatrix} 1 & 3 & 5 \\ 1 & 4 & 3 \\ 1 & 1 & 9 \end{pmatrix}$ and $B = \begin{pmatrix} 1 & 2 & 3 \\ -2 & -3 & -4 \\ 7 & 12 & 17 \end{pmatrix}$ have the same column space.

Observe that A and B have the same column space if and only if the transposes A^T and B^T have the same row space. Thus reduce A^T and B^T to row canonical form:

$$A^T = \begin{pmatrix} 1 & 1 & 1 \\ 3 & 4 & 1 \\ 5 & 3 & 9 \end{pmatrix} \sim \begin{pmatrix} 1 & 1 & 1 \\ 0 & 1 & -2 \\ 0 & -2 & 4 \end{pmatrix} \sim \begin{pmatrix} 1 & 1 & 1 \\ 0 & 1 & -2 \\ 0 & 0 & 0 \end{pmatrix} \quad \text{to} \quad \begin{pmatrix} 1 & 0 & 3 \\ 0 & 1 & -2 \\ 0 & 0 & 0 \end{pmatrix}$$

$$B^T = \begin{pmatrix} 1 & -2 & 7 \\ 2 & -3 & 12 \\ 3 & -4 & 17 \end{pmatrix} \sim \begin{pmatrix} 1 & -2 & 7 \\ 0 & 1 & -2 \\ 0 & 2 & -4 \end{pmatrix} \sim \begin{pmatrix} 1 & -2 & 7 \\ 0 & 1 & -2 \\ 0 & 0 & 0 \end{pmatrix} \quad \text{to} \quad \begin{pmatrix} 1 & 0 & 3 \\ 0 & 1 & -2 \\ 0 & 0 & 0 \end{pmatrix}$$

Since A^T and B^T have the same row space, A and B have the same column space.

5.47. Consider the subspace $U = \text{span}\,(u_1, u_2, u_3)$ and $W = \text{span}\,(w_1, w_2, w_3)$ of $\mathbf{R}^3$ where:

$$u_1 = (1, 1, -1), \qquad u_2 = (2, 3, -1), \qquad u_3 = (3, 1, -5)$$
$$w_1 = (1, -1, -3), \qquad w_2 = (3, -2, -8), \qquad w_3 = (2, 1, -3)$$

Show that $U = W$.

Form the matrix A whose rows are the u_i, and row reduce A to row canonical form:

$$A = \begin{pmatrix} 1 & 1 & -1 \\ 2 & 3 & -1 \\ 3 & 1 & -5 \end{pmatrix} \sim \begin{pmatrix} 1 & 1 & -1 \\ 0 & 1 & 1 \\ 0 & -2 & -2 \end{pmatrix} \sim \begin{pmatrix} 1 & 0 & -2 \\ 0 & 1 & 1 \\ 0 & 0 & 0 \end{pmatrix}$$

Next form the matrix B whose rows are the w_i and row reduce B to row canonical form:

$$B = \begin{pmatrix} 1 & -1 & -3 \\ 3 & -2 & -8 \\ 2 & 1 & -3 \end{pmatrix} \sim \begin{pmatrix} 1 & -1 & -3 \\ 0 & 1 & 1 \\ 0 & 3 & 3 \end{pmatrix} \sim \begin{pmatrix} 1 & 0 & -2 \\ 0 & 1 & 1 \\ 0 & 0 & 0 \end{pmatrix}$$

Since A and B have the same row canonical form, the row spaces of A and B are equal and so $U = W$.

5.48. Find the rank of the matrix A where:

$$(a) \quad A = \begin{pmatrix} 1 & 2 & -3 \\ 2 & 1 & 0 \\ -2 & -1 & 3 \\ -1 & 4 & -2 \end{pmatrix}, \qquad (b) \quad A = \begin{pmatrix} 1 & 3 \\ 0 & -2 \\ 5 & -1 \\ -2 & 3 \end{pmatrix}$$

(a) Since row rank equals column rank, it is easier to form the transpose of A and then row reduce to echelon form:

$$A^t = \begin{pmatrix} 1 & 2 & -2 & -1 \\ 2 & 1 & -1 & 4 \\ -3 & 0 & 3 & -2 \end{pmatrix} \sim \begin{pmatrix} 1 & 2 & -2 & -1 \\ 0 & -3 & 3 & 6 \\ 0 & 6 & -3 & -5 \end{pmatrix} \sim \begin{pmatrix} 1 & 2 & -2 & -1 \\ 0 & -3 & 3 & 6 \\ 0 & 0 & 3 & 7 \end{pmatrix}$$

Thus rank $A = 3$.

(b) The two columns are linearly independent since one is not a multiple of the other. Thus rank $A = 2$.

5.49. Consider an arbitrary matrix $A = (a_{ij})$. Suppose $u = (b_1, \ldots, b_n)$ is a linear combination of the rows $R_1, \ldots, R_m$ of A; say $u = k_1 R_1 + \cdots + k_m R_m$. Show that

$$b_i = k_1 a_{1i} + k_2 a_{2i} + \cdots + k_m a_{mi} \qquad (i = 1, 2, \ldots, n)$$

where $a_{1i}, \ldots, a_{mi}$ are the entries of the ith column of A.

We are given $u = k_1 R_1 + \cdots + k_m R_m$; hence

$$(b_1, \ldots, b_n) = k_1(a_{11}, \ldots, a_{1n}) + \cdots + k_m(a_{m1}, \ldots, a_{mn})$$
$$= (k_1 a_{11} + \cdots + k_m a_{m1}, \ldots, k_1 a_{m1} + \cdots + k_m a_{mn})$$

Setting corresponding components equal to each other, we obtain the desired result.

5.50. Suppose $A = (a_{ij})$ and $B = (b_{ij})$ are echelon matrices with pivot entries:

$$a_{1j_1}, a_{2j_2}, \ldots, a_{rj_r} \qquad \text{and} \qquad b_{1k_1}, b_{2k_2}, \ldots, b_{sk_s}$$

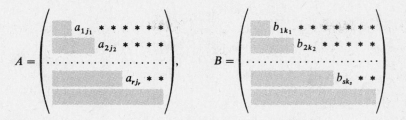

Suppose A and B have the same row space. Prove that the pivot entries of A and B are in the same positions, that is, prove that $j_1 = k_1, j_2 = k_2, \ldots, j_r = k_r$, and $r = s$.

Clearly $A = 0$ if and only if $B = 0$, and so we need only prove the theorem when $r \geq 1$ and $s \geq 1$. We first show that $j_1 = k_1$. Suppose $j_1 < k_1$. Then the j_1th column of B is zero. Since the first row of A is in the row space of B, we have by the preceding problem,

$$a_{1j_1} = c_1 0 + c_2 0 + \cdots + c_m 0 = 0$$

for scalars c_i. But this contradicts the fact that the pivot element $a_{1j_1} \neq 0$. Hence $j_1 \geq k_1$, and similarly $k_1 \geq j_1$. Thus $j_1 = k_1$.

Now let A' be the submatrix of A obtained by deleting the first row of A, and let B' be the submatrix of B obtained by deleting the first row of B. We prove that A' and B' have the same row space. The theorem will then follow by induction since A' and B' are also echelon matrices.

Let $R = (a_1, a_2, \ldots, a_n)$ be any row of A' and let $R_1, \ldots, R_m$ be the rows of B. Since R is in the row space of B, there exist scalars $d_1, \ldots, d_m$ such that $R = d_1 R_1 + d_2 R_2 + \cdots + d_m R_m$. Since A is in echelon form and R is not the first row of A, the j_1th entry of R is zero: $a_i = 0$ for $i = j_1 = k_1$. Furthermore, since B is in echelon form, all the entries in the k_1th column of B are 0 except the first: $b_{1k_1} \neq 0$, but $b_{2k_1} = 0, \ldots, b_{mk_1} = 0$. Thus

$$0 = a_{k_1} = d_1 b_{1k_1} + d_2 0 + \cdots + d_m 0 = d_1 b_{1k_1}$$

Now $b_{1k_1} \neq 0$ and so $d_1 = 0$. Thus R is a linear combination of $R_2, \ldots, R_m$ and so is in the row space of B'. Since R was any row of A', the row space of A' is contained in the row space of B'. Similarly, the row space of B' is contained in the row space of A'. Thus A' and B' have the same row space, and so the theorem is proved.

5.51. Prove Theorem 5.8.

Obviously, if A and B have the same nonzero rows then they have the same row space. Thus we only have to prove the converse.

Suppose A and B have the same row space, and suppose $R \neq 0$ is the ith row of A. Then there exist scalars $c_1, \ldots, c_s$ such that

$$R = c_1 R_1 + c_2 R_2 + \cdots + c_s R_s \tag{1}$$

where the R_i are the nonzero rows of B. The theorem is proved if we show that $R = R_i$, or $c_i = 1$ but $c_k = 0$ for $k \neq i$.

Let a_{ij_i} be the pivot entry in R, i.e., the first nonzero entry of R. By (1) and Problem 5.49,

$$a_{ij_i} = c_1 b_{1j_i} + c_2 b_{2j_i} + \cdots + c_s b_{sj_i} \tag{2}$$

But by the preceding problem b_{ij_i} is a pivot entry of B and, since B is row reduced, it is the only nonzero entry in the j_ith column of B. Thus from (2) we obtain $a_{ij_i} = c_i b_{ij_i}$. However, $a_{ij_i} = 1$ and $b_{ij_i} = 1$ since A and B are row reduced; hence $c_i = 1$.

Now suppose $k \neq i$, and b_{kj_k} is the pivot entry in R_k. By (1) and Problem 5.49,

$$a_{ij_k} = c_1 b_{1j_k} + c_2 b_{2j_k} + \cdots + c_s b_{sj_k} \tag{3}$$

Since B is row reduced, b_{kj_k} is the only nonzero entry in the j_kth column of B; hence by (3), $a_{ij_k} = c_k b_{kj_k}$. Furthermore, by the preceding problem a_{kj_k} is a pivot entry of A and, since A is row reduced, $a_{ij_k} = 0$. Thus $c_k b_{kj_k} = 0$ and, since $b_{kj_k} = 1$, $c_k = 0$. Accordingly $R = R_i$ and the theorem is proved.

5.52. Prove Theorem 5.9.

Suppose A is row equivalent to matrices A_1 and A_2 where A_1 and A_2 are in row canonical form. Then rowsp A = rowsp A_1 and rowsp A = rowsp A_2. Hence rowsp A_1 = rowsp A_2. Since A_1 and A_2 are in row canonical form, $A_1 = A_2$ by Theorem 5.8. Thus the theorem is proved.

5.53. Prove Theorem 5.18.

Let A be an arbitrary $m \times n$ matrix:

$$A = \begin{pmatrix} a_{11} & a_{12} & \cdots & a_{1n} \\ a_{21} & a_{22} & \cdots & a_{2n} \\ \cdots\cdots\cdots\cdots\cdots\cdots \\ a_{m1} & a_{m2} & \cdots & a_{mn} \end{pmatrix}$$

Let $R_1, R_2, \ldots, R_m$ denote its rows:

$$R_1 = (a_{11}, a_{12}, \ldots, a_{1n}), \ldots, R_m = (a_{m1}, a_{m2}, \ldots, a_{mn})$$

Suppose the row rank is r and that the following r vectors form a basis for the row space:

$$S_1 = (b_{11}, b_{12}, \ldots, b_{1n}), S_2 = (b_{21}, b_{22}, \ldots, b_{2n}), \ldots, S_r = (b_{r1}, b_{r2}, \ldots, b_{rn})$$

Then each of the row vectors is a linear combination of the S_i:

$$R_1 = k_{11}S_1 + k_{12}S_2 + \cdots + k_{1r}S_r$$
$$R_2 = k_{21}S_1 + k_{22}S_2 + \cdots + k_{2r}S_r$$
$$\cdots\cdots\cdots\cdots\cdots\cdots\cdots\cdots\cdots\cdots$$
$$R_m = k_{m1}S_1 + k_{m2}S_2 + \cdots + k_{mr}S_r$$

where the k_{ij} are scalars. Setting the ith components of each of the above vector equations equal to each other, we obtain the following system of equations, each valid for $i = 1, \ldots, n$:

$$a_{1i} = k_{11}b_{1i} + k_{12}b_{2i} + \cdots + k_{1r}b_{ri}$$
$$a_{2i} = k_{21}b_{1i} + k_{22}b_{2i} + \cdots + k_{2r}b_{ri}$$
$$\cdots\cdots\cdots\cdots\cdots\cdots\cdots\cdots\cdots\cdots$$
$$a_{mi} = k_{m1}b_{1i} + k_{m2}b_{2i} + \cdots + k_{mr}b_{ri}$$

Thus, for $i = 1, \ldots, n$,

$$\begin{pmatrix} a_{1i} \\ a_{2i} \\ \cdots \\ a_{mi} \end{pmatrix} = b_{1i}\begin{pmatrix} k_{11} \\ k_{21} \\ \cdots \\ k_{m1} \end{pmatrix} + b_{2i}\begin{pmatrix} k_{12} \\ k_{22} \\ \cdots \\ k_{m2} \end{pmatrix} + \cdots + b_{ri}\begin{pmatrix} k_{1r} \\ k_{2r} \\ \cdots \\ k_{mr} \end{pmatrix}$$

In other words, each of the columns of A is a linear combination of the r vectors

$$\begin{pmatrix} k_{11} \\ k_{21} \\ \cdots \\ k_{m1} \end{pmatrix}, \begin{pmatrix} k_{12} \\ k_{22} \\ \cdots \\ k_{m1} \end{pmatrix}, \ldots, \begin{pmatrix} k_{1r} \\ k_{2r} \\ \cdots \\ k_{mr} \end{pmatrix}$$

Thus the column space of the matrix A has dimension at most r, i.e., column rank $\leq r$. Hence, column rank $\leq$ row rank.

Similarly (or considering the transpose matrix A^t) we obtain row rank $\leq$ column rank. Thus the row rank and column rank are equal.

5.54. Suppose R is a row vector and A and B are matrices such that RB and AB are defined. Prove:

(a) RB is a linear combination of the rows of B.

(b) Row space of AB is contained in the row space of B.

(c)   Column space of AB is contained in the column space of A.

(d)   rank $AB \leq$ rank B and rank $AB \leq$ rank A.

(a)   Suppose $R = (a_1, a_2, \ldots, a_m)$ and $B = (b_{ij})$. Let $B_1, \ldots, B_m$ denote the rows of B and $B^1, \ldots, B^n$ its columns. Then

$$RB = (R \cdot B^1, R \cdot B^2, \ldots, R \cdot B^n)$$
$$= (a_1 b_{11} + a_2 b_{21} + \cdots + a_m b_{m1}, a_1 b_{12} + a_2 b_{22} + \cdots + a_m b_{m2}, \ldots, a_1 b_{1n} + a_2 b_{2n} + \cdots + a_m b_{mn})$$
$$= a_1(b_{11}, b_{12}, \ldots, b_{1n}) + a_2(b_{21}, b_{22}, \ldots, b_{2n}) + \cdots + a_m(b_{m1}, b_{m2}, \ldots, b_{mn})$$
$$= a_1 B_1 + a_2 B_2 + \cdots + a_m B_m$$

Thus RB is a linear combination of the rows of B, as claimed.

(b)   The rows of AB are $R_i B$ where R_i is the ith row of A. Thus, by part (a), each row of AB is in the row space of B. Thus rowsp $AB \subseteq$ rowsp B, as claimed.

(c)   Using part (b), we have:

$$\text{colsp } AB = \text{rowsp } (AB)^T = \text{rowsp } B^T A^T \subseteq \text{rowsp } A^T = \text{colsp } A$$

(d)   The row space of AB is contained in the row space of B; hence rank $AB \leq$ rank B. Furthermore, the column space of AB is contained in the column space of A; hence rank $AB \leq$ rank A.

5.55.   Let A be an n-square matrix. Show that A is invertible if and only if rank $A = n$.

Note that the rows of the n-square identity matrix I_n are linearly independent since I_n is in echelon form; hence rank $I_n = n$. Now if A is invertible then A is row equivalent to I_n; hence rank $A = n$. But if A is not invertible then A is row equivalent to a matrix with a zero row; hence rank $A < n$. That is, A is invertible if and only if rank $A = n$.

APPLICATIONS TO LINEAR EQUATIONS

5.56.   Find the dimension and a basis of the solution space W of the system

$$x + 2y + z - 3t = 0$$
$$2x + 4y + 4z - t = 0.$$
$$3x + 6y + 7z + t = 0$$

Reduce the system to echelon form:

$$x + 2y + z - 3t = 0 \qquad\qquad x + 2y + z - 3t = 0$$
$$2z + 5t = 0 \qquad \text{or} \qquad 2z + 5t = 0$$
$$4z + 10t = 0$$

The free variables are y and t, and dim $W = 2$. Set:

(i) $y = 1, z = 0$ to obtain the solution $u_1 = (-2, 1, 0, 0)$

(ii) $y = 0, t = 2$ to obtain the solution $u_2 = (11, 0, -5, 2)$

Then $\{u_1, u_2\}$ is a basis of W. [The choice $y = 0, t = 1$ in (ii), would introduce fractions in the solution.]

5.57.   Find a homogeneous system whose solution set W is spanned by

$$\{(1, -2, 0, 3), (1, -1, -1, 4), (1, 0, -2, 5)\}$$

Let $v = (x, y, z, t)$. Form the matrix M whose first rows are the given vectors and whose last row is v; and then row reduce to echelon form:

$$M = \begin{pmatrix} 1 & -2 & 0 & 3 \\ 1 & -1 & -1 & 4 \\ 1 & 0 & -2 & 5 \\ x & y & z & t \end{pmatrix} \text{ to } \begin{pmatrix} 1 & -2 & 0 & 3 \\ 0 & 1 & -1 & 1 \\ 0 & 2 & -2 & 2 \\ 0 & 2x+y & z & -3x+t \end{pmatrix} \text{ to } \begin{pmatrix} 1 & -2 & 0 & 3 \\ 0 & 1 & -1 & 1 \\ 0 & 0 & 2x+y+z & -5x-y+t \\ 0 & 0 & 0 & 0 \end{pmatrix}$$

The original first three rows show that W has dimension 2. Thus $v \in W$ if and only if the additional row does not increase the dimension of the row space. Hence we set the last two entries in the third row on the right equal to 0 to obtain the required homogeneous system

$$2x + y + z \quad = 0$$
$$5x + y \quad\quad -t = 0$$

5.58. Let $x_{i_1}, x_{i_2}, \ldots, x_{i_k}$ be the free variables of a homogeneous system of linear equations with n unknowns. Let v_j be the solution for which $x_{i_j} = 1$, and all other free variables $= 0$. Show that the solutions $v_1, v_2, \ldots, v_k$ are linearly independent.

Let A be the matrix whose rows are the v_i, respectively. We interchange column 1 and column i_1, then column 2 and column $i_2, \ldots$, and then column k and column i_k; and obtain the $k \times n$ matrix

$$B = (I, C) = \begin{pmatrix} 1 & 0 & 0 & \ldots & 0 & 0 & c_{1,k+1} & \cdots & c_{1n} \\ 0 & 1 & 0 & \ldots & 0 & 0 & c_{2,k+1} & \cdots & c_{2n} \\ \hdotsfor{9} \\ 0 & 0 & 0 & \ldots & 0 & 1 & c_{k,k+1} & \cdots & c_{kn} \end{pmatrix}$$

The above matrix B is in echelon form and so its rows are independent; hence rank $B = k$. Since A and B are column equivalent, they have the same rank, i.e., rank $A = k$. But A has k rows; hence these rows, i.e., the v_i, are linearly independent, as claimed.

5.59. Prove Theorem 5.20.

Suppose $u_1, u_2, \ldots, u_r$ form a basis for the column space of A (There are r such vectors since rank $A = r$.) By Theorem 5.19, each system $AX = u_i$ has a solution, say v_i. Hence

$$Av_1 = u_1, \ Av_2 = u_2, \ \ldots, \ Av_r = u_r \qquad (I)$$

Suppose dim $W = s$ and $w_1, w_2, \ldots, w_s$ form a basis of W. Let

$$B = \{v_1, v_2, \ldots, v_r, w_1, w_2, \ldots, w_s\}$$

We claim that B is a basis of K^n. Thus we need to prove that B spans K^n and that B is linearly independent.

(a) *Proof that B spans K^n.* Suppose $v \in K^n$ and $Av = u$. Then $u = Av$ belongs to the column space of A and hence Av is a linear combination of the u_i. Say

$$Av = k_1 u_1 + k_2 u_2 + \cdots + k_r u_r \qquad (2)$$

Let $v' = v - k_1 v_1 - k_2 v_2 - \cdots - k_r v_r$. Then, using (I) and (2),

$$A(v') = A(v - k_1 v_1 - k_2 v_2 - \cdots - k_r v_r)$$
$$= Av - k_1 Av_1 - k_2 Av_2 - \cdots - k_r Av_r$$
$$= Av - k_1 u_1 - k_2 u_2 - \cdots - k_r u_r = Av - Av = 0$$

Thus v' belongs to the solution W and hence v' is a linear combination of the w_i. Say $v' = c_1 w_1 + c_2 w_2 + \cdots + c_s w_s$. Then

$$v = v' + \sum_{i=1}^{r} k_i v_i = \sum_{i=1}^{r} k_i v_i + \sum_{j=1}^{s} c_j w_j$$

Thus v is a linear combination of the elements in B, and hence B spans K^n.

(b) *Proof that B is linearly independent.* Suppose

$$a_1 v_1 + a_2 v_2 + \cdots + a_r v_r + b_1 w_1 + b_2 w_2 + \cdots + b_s w_s = 0 \qquad (3)$$

Since $w_j \in W$, each $Aw_j = 0$. Using this fact and (I) and (3), we obtain

$$0 = A(0) = A\left(\sum_{i=1}^{r} a_i v_i + \sum_{j=1}^{s} b_j w_j\right) = \sum_{i=1}^{r} a_i Av_i + \sum_{j=1}^{s} b_j Aw_j$$
$$= \sum_{i=1}^{r} a_i u_i + \sum_{j=1}^{s} b_j 0 = a_1 u_1 + a_2 u_2 + \cdots + a_r u_r$$

Since $u_1, \ldots, u_r$ are linearly independent, each $a_i = 0$. Substituting this in (3) yields

$$b_1 w_1 + b_2 w_2 + \cdots + b_s w_s = 0$$

However, $w_1, \ldots, w_s$ are linearly independent. Thus each $b_j = 0$. Therefore, B is linearly independent.

Accordingly B is a basis of K^n. Since B has $r + s$ elements, we have $r + s = n$. Consequently, $\dim W = s = n - r$, as claimed.

SUMS, DIRECT SUMS, INTERSECTIONS

5.60. Let U and W be subspaces of a vector space V. Show that: (a) U and W are each contained in V; (b) $U + W$ is the smallest subspace of V containing U and W, that is, $U + W = \text{span}\,(U, W)$, the linear span of U and W; (c) $W + W = W$.

    (a) Let $u \in U$. By hypothesis W is a subspace of V and so $0 \in W$. Hence $u = u + 0 \in U + W$. Accordingly, U is contained in $U + W$. Similarly, W is contained in $U + W$.

    (b) Since $U + W$ is a subspace of V containing both U and W, it must also contain the linear span of U and W, that is, $\text{span}\,(U, W) \subseteq U + W$.

        On the other hand, if $v \in U + W$ then $v = u + w = 1u + 1w$ where $u \in U$ and $w \in W$; hence v is a linear combination of elements in $U \cup W$ and so belongs to span (U, W). Consequently $U + W \subseteq \text{span}\,(U, W)$.

    (c) Since W is a subspace of V, we have that W is closed under vector addition; hence $W + W \subseteq W$. By part (a), $W \subseteq W + W$. Hence $W + W = W$.

5.61. Give an example of a subset S of $\mathbf{R}^2$ such that: (a) $S + S \subset S$ (properly contained); (b) $S \subset S + S$ (properly contained); (c) $S + S = S$ but S is not a subspace of $\mathbf{R}^2$.

    (a) Let $S = \{(0, 5), (0, 6), (0, 7), \ldots\}$. Then $S + S \subset S$.

    (b) Let $S = \{(0, 0), (0, 1)\}$. Then $S \subset S + S$.

    (c) Let $S = \{(0, 0), (0, 1), (0, 2), (0, 3), \ldots\}$. Then $S + S = S$.

5.62. Suppose U and W are distinct 4-dimensional subspaces of a vector space V where $\dim V = 6$. Find the possible dimensions of $U \cap W$.

    Since U and W are distinct, $U + W$ properly contains U and W; consequently $\dim (U + W) > 4$. But $\dim (U + W)$ cannot be greater than 6, since $\dim V = 6$. Hence we have two possibilities: (i) $\dim (U + W) = 5$, or (ii) $\dim (U + W) = 6$. By Theorem 5.22,

$$\dim (U \cap W) = \dim U + \dim W - \dim (U + W) = 8 - \dim (U + W)$$

Thus (i) $\dim (U \cap W) = 3$, or (ii) $\dim (U \cap W) = 2$.

5.63. Consider the following subspaces of $\mathbf{R}^4$:

$$U = \text{span}\,\{(1, 1, 0, -1), (1, 2, 3, 0), (2, 3, 3, -1)\}$$
$$W = \text{span}\,\{(1, 2, 2, -2), (2, 3, 2, -3), (1, 3, 4, -3)\}$$

Find (a) $\dim (U + W)$ and (b) $\dim (U \cap W)$.

    (a) $U + W$ is the space spanned by all six vectors. Hence form the matrix whose rows are the given six vectors, and then row reduce to echelon form:

$$\begin{pmatrix} 1 & 1 & 0 & -1 \\ 1 & 2 & 3 & 0 \\ 2 & 3 & 3 & -1 \\ 1 & 2 & 2 & -2 \\ 2 & 3 & 2 & -3 \\ 1 & 3 & 4 & -3 \end{pmatrix} \sim \begin{pmatrix} 1 & 1 & 0 & -1 \\ 0 & 1 & 3 & 1 \\ 0 & 1 & 3 & 1 \\ 0 & 1 & 2 & -1 \\ 0 & 1 & 2 & -1 \\ 0 & 2 & 4 & -2 \end{pmatrix} \sim \begin{pmatrix} 1 & 1 & 0 & -1 \\ 0 & 1 & 3 & 1 \\ 0 & 1 & 2 & -1 \\ 0 & 0 & 0 & 0 \\ 0 & 0 & 0 & 0 \\ 0 & 0 & 0 & 0 \end{pmatrix}$$

$$\sim \begin{pmatrix} 1 & 1 & 0 & -1 \\ 0 & 1 & 3 & 1 \\ 0 & 0 & -1 & -2 \\ 0 & 0 & 0 & 0 \\ 0 & 0 & 0 & 0 \\ 0 & 0 & 0 & 0 \end{pmatrix}$$

Since the echelon matrix has three nonzero rows, $\dim (U + W) = 3$.

(b) First find $\dim U$ and $\dim W$. Form the two matrices whose rows are the generators of U and W, respectively, and then row reduce each to echelon form:

and

$$\begin{pmatrix} 1 & 1 & 0 & -1 \\ 1 & 2 & 3 & 0 \\ 2 & 3 & 3 & -1 \end{pmatrix} \sim \begin{pmatrix} 1 & 1 & 0 & -1 \\ 0 & 1 & 3 & 1 \\ 0 & 1 & 3 & 1 \end{pmatrix} \sim \begin{pmatrix} 1 & 1 & 0 & -1 \\ 0 & 1 & 3 & 1 \\ 0 & 0 & 0 & 0 \end{pmatrix}$$

$$\begin{pmatrix} 1 & 2 & 2 & -2 \\ 2 & 3 & 2 & -3 \\ 1 & 3 & 4 & -3 \end{pmatrix} \sim \begin{pmatrix} 1 & 2 & 2 & -2 \\ 0 & -1 & -2 & 1 \\ 0 & 1 & 2 & -1 \end{pmatrix} \sim \begin{pmatrix} 1 & 2 & 2 & -2 \\ 0 & -1 & -2 & 1 \\ 0 & 0 & 0 & 0 \end{pmatrix}$$

Since each of the echelon matrices has two nonzero rows, $\dim U = 2$ and $\dim W = 2$. Using Theorem 5.22, i.e., $\dim (U + W) = \dim U + \dim W - \dim (U \cap W)$, we have

$$3 = 2 + 2 - \dim (U \cap W) \qquad \text{or} \qquad \dim (U \cap W) = 1$$

5.64. Let U and W be the following subspaces of $\mathbf{R}^4$:

$$U = \{(a, b, c, d): b + c + d = 0\}, \qquad W = \{(a, b, c, d): a + b = 0, c = 2d\}$$

Find a basis and the dimension of: (a) U, (b) W, (c) $U \cap W$, (d) $U + W$.

(a) We seek a basis of the set of solutions (a, b, c, d) of the equation

$$b + c + d = 0 \qquad \text{or} \qquad 0 \cdot a + b + c + d = 0$$

The free variables are a, c, and d. Set:

    (1) $a = 1, c = 0, d = 0$ to obtain the solution $v_1 = (1, 0, 0, 0)$

    (2) $a = 0, c = 1, d = 0$ to obtain the solution $v_2 = (0, -1, 1, 0)$

    (3) $a = 0, c = 0, d = 1$ to obtain the solution $v_3 = (0, -1, 0, 1)$

The set $\{v_1, v_2, v_3\}$ is a basis of U, and $\dim U = 3$.

(b) We seek a basis of the set of solutions (a, b, c, d) of the system

$$\begin{aligned} a + b &= 0 \\ c &= 2d \end{aligned} \qquad \text{or} \qquad \begin{aligned} a + b &= 0 \\ c - 2d &= 0 \end{aligned}$$

The free variables are b and d. Set

    (1) $b = 1, d = 0$ to obtain the solution $v_1 = (-1, 1, 0, 0)$

    (2) $b = 0, d = 1$ to obtain the solution $v_2 = (0, 0, 2, 1)$

The set $\{v_1, v_2\}$ is a basis of W, and $\dim W = 2$.

(c) $U \cap W$ consists of those vectors (a, b, c, d) which satisfy the conditions defining U and the conditions defining W, i.e., the three equations

$$\begin{aligned} b + c + d &= 0 \\ a + b &= 0 \\ c &= 2d \end{aligned} \qquad \text{or} \qquad \begin{aligned} a + b &= 0 \\ b + c + d &= 0 \\ c - 2d &= 0 \end{aligned}$$

The free variable is d. Set $d = 1$ to obtain the solution $v = (3, -3, 2, 1)$. Thus $\{v\}$ is a basis of $U \cap W$, and $\dim (U \cap W) = 1$.

(d) By Theorem 5.22,

$$\dim (U + W) = \dim U + \dim W - \dim (U \cap W) = 3 + 2 - 1 = 4$$

Thus $U + W = \mathbf{R}^4$. Accordingly any basis of $\mathbf{R}^4$, say the usual basis, will be a basis of $U + W$.

5.65. Consider the following subspaces of $\mathbf{R}^5$:

$$U = \text{span } \{(1, 3, -2, 2, 3), (1, 4, -3, 4, 2), (2, 3, -1, -2, 9)\}$$
$$W = \text{span } \{(1, 3, 0, 2, 1), (1, 5, -6, 6, 3), (2, 5, 3, 2, 1)\}$$

Find a basis and dimension of (a) $U + W$, (b) $U \cap W$.

(a) $U + W$ is the space spanned by all six vectors. Hence form the matrix whose rows are the given six vectors, and then row reduce to echelon form:

$$
\begin{pmatrix}
1 & 3 & -2 & 2 & 3 \\
1 & 4 & -3 & 4 & 2 \\
2 & 3 & -1 & -2 & 9 \\
1 & 3 & 0 & 2 & 1 \\
1 & 5 & -6 & 6 & 3 \\
2 & 5 & 3 & 2 & 1
\end{pmatrix}
\sim
\begin{pmatrix}
1 & 3 & -2 & 2 & 3 \\
0 & 1 & -1 & 2 & -1 \\
0 & -3 & 3 & -6 & 3 \\
0 & 0 & 2 & 0 & -2 \\
0 & 2 & -4 & 4 & 0 \\
0 & -1 & 7 & -2 & -5
\end{pmatrix}
\sim
\begin{pmatrix}
1 & 3 & -2 & 2 & 3 \\
0 & 1 & -1 & 2 & -1 \\
0 & 0 & 0 & 0 & 0 \\
0 & 0 & 2 & 0 & -2 \\
0 & 0 & -2 & 0 & 2 \\
0 & 0 & 6 & 0 & -6
\end{pmatrix}
$$

$$
\sim
\begin{pmatrix}
1 & 3 & -2 & 2 & 3 \\
0 & 1 & -1 & 2 & -1 \\
0 & 0 & 2 & 0 & -2 \\
0 & 0 & 0 & 0 & 0 \\
0 & 0 & 0 & 0 & 0 \\
0 & 0 & 0 & 0 & 0
\end{pmatrix}
$$

The set of nonzero rows of the echelon matrix,

$$\{(1, 3, -2, 2, 3), (0, 1, -1, 2, -1), (0, 0, 2, 0, -2)\}$$

is a basis of $U + W$; thus $\dim (U + W) = 3$.

(b) First find homogeneous systems whose solution sets are U and W, respectively. Form the matrix whose first three rows span U and whose last row is (x, y, z, s, t) and then row reduce to an echelon form:

$$
\begin{pmatrix}
1 & 3 & -2 & 2 & 3 \\
1 & 4 & -3 & 4 & 2 \\
2 & 3 & -1 & -2 & 9 \\
x & y & z & s & t
\end{pmatrix}
\sim
\begin{pmatrix}
1 & 3 & -2 & 2 & 3 \\
0 & 1 & -1 & 2 & -1 \\
0 & -3 & 3 & -6 & 3 \\
0 & -3x+y & 2x+z & -2x+s & -3x+t
\end{pmatrix}
$$

$$
\sim
\begin{pmatrix}
1 & 3 & -2 & 2 & 3 \\
0 & 1 & -1 & 2 & -1 \\
0 & 0 & -x+y+z & 4x-2y+s & -6x+y+t \\
0 & 0 & 0 & 0 & 0
\end{pmatrix}
$$

Set the entries of the third row equal to 0 to obtain the homogeneous system whose solution space is U:

$$-x + y + z = 0 \qquad 4x - 2y + s = 0 \qquad -6x + y + t = 0$$

Now form the matrix whose first rows span W and whose last row is (x, y, z, s, t) and then row reduce to an echelon form:

$$\begin{pmatrix} 1 & 3 & 0 & 2 & 1 \\ 1 & 5 & -6 & 6 & 3 \\ 2 & 5 & 3 & 2 & 1 \\ x & y & z & s & t \end{pmatrix} \sim \begin{pmatrix} 1 & 3 & 0 & 2 & 1 \\ 0 & 2 & -6 & 4 & 2 \\ 0 & -1 & 3 & -2 & -1 \\ 0 & -3x+y & z & -2x+s & -x+t \end{pmatrix}$$

$$\sim \begin{pmatrix} 1 & 3 & 0 & 2 & 1 \\ 0 & 1 & -3 & 2 & 1 \\ 0 & 0 & -9x+3y+z & 4x-2y+s & 2x-y+t \\ 0 & 0 & 0 & 0 & 0 \end{pmatrix}$$

Set the entries of the third row equal to 0 to obtain the homogeneous system whose solution space is W:

$$-9x + 3y + z = 0 \qquad 4x - 2y + s = 0 \qquad 2x - y + t = 0$$

Combine both of the above systems to obtain a homogeneous system whose solution space is $U \cap W$, and then solve:

$$\begin{cases} -x + y + z & = 0 \\ 4x - 2y & + s & = 0 \\ -6x + y & + t = 0 \\ -9x + 3y + z & = 0 \\ 4x - 2y & + s & = 0 \\ 2x - y & + t = 0 \end{cases} \quad \text{or} \quad \begin{cases} -x + y + z & = 0 \\ 2y + 4z + s & = 0 \\ -5y - 6z & + t = 0 \\ -6y - 8z & = 0 \\ 2y + 4z + s & = 0 \\ y + 2z & + t = 0 \end{cases}$$

$$\begin{cases} -x + y + z & = 0 \\ 2y + 4z + s & = 0 \\ 8z + 5s + 2t = 0 \\ 4z + 3s & = 0 \\ s - 2t = 0 \end{cases} \quad \text{or} \quad \begin{cases} -x + y + z & = 0 \\ 2y + 4z + s & = 0 \\ 8z + 5s + 2t = 0 \\ s - 2t = 0 \end{cases}$$

There is one free variable, which is t; hence dim $(U \cap W) = 1$. Setting $t = 2$, we obtain the solution $x = 1, y = 4, z = -3, s = 4, t = 2$. Thus $\{(1, 4, -3, 4, 2)\}$ is a basis of $U \cap W$.

5.66. Let U and W be the subspaces of $\mathbf{R}^3$ defined by

$$U = \{(a, b, c): a = b = c\} \qquad \text{and} \qquad W = \{(0, b, c)\}$$

(Note that W is the yz plane.) Show that $\mathbf{R}^3 = U \oplus W$.

Note first that $U \cap W = \{0\}$, for $v = (a, b, c) \in U \cap W$ implies that

$$a = b = c \qquad \text{and} \qquad a = 0 \qquad \text{which implies} \qquad a = 0, b = 0, c = 0$$

We also claim that $\mathbf{R}^3 = U + W$. For if $v = (a, b, c) \in \mathbf{R}^3$, then

$$v = (a, a, a) + (0, b - a, c - a) \qquad \text{where} \qquad (a, a, a) \in U \qquad \text{and} \qquad (0, b - a, c - a) \in W$$

Both conditions, $U \cap W = \{0\}$ and $\mathbf{R}^3 = U + W$, imply $\mathbf{R}^3 = U \oplus W$.

5.67. Let V be the vector space of n-square matrices over a field K.

(a) Show that $V = U \oplus W$ where U and W are the subspaces of symmetric and antisymmetric matrices, respectively. (Recall M is symmetric iff $M = M^T$, and M is antisymmetric iff $M^T = -M$.)

(b) Show that $V \neq U \oplus W$ where U and W are the subspaces of upper and lower triangular matrices, respectively. (Note $V = U + W$.)

(a)   We first show that $V = U + W$. Let A be any arbitrary n-square matrix. Note that

$$A = \tfrac{1}{2}(A + A^T) + \tfrac{1}{2}(A - A^T)$$

We claim that $\tfrac{1}{2}(A + A^T) \in U$ and that $\tfrac{1}{2}(A - A^T) \in W$. For

$$(\tfrac{1}{2}(A + A^T))^T = \tfrac{1}{2}(A + A^T)^T = \tfrac{1}{2}(A^T + A^{TT}) = \tfrac{1}{2}(A + A^T)$$

that is, $\tfrac{1}{2}(A + A^T)$ is symmetric. Furthermore,

$$(\tfrac{1}{2}(A - A^T))^T = \tfrac{1}{2}(A - A^T)^T = \tfrac{1}{2}(A^T - A) = -\tfrac{1}{2}(A - A^T)$$

that is, $\tfrac{1}{2}(A - A^T)$ is antisymmetric.

  We next show that $U \cap W = \{0\}$. Suppose $M \in U \cap W$. Then $M = M^T$ and $M^T = -M$, which implies $M = -M$ and hence $M = 0$. Thus $U \cap W = \{0\}$. Accordingly, $V = U \oplus W$.

(b) $U \cap W \neq \{0\}$ since $U \cap W$ consists of all the diagonal matrices. Thus the sum cannot be direct.

5.68.   Suppose U and W are subspaces of a vector space V, and suppose that $S = \{u_i\}$ spans U and $T = \{w_j\}$ spans W. Show that $S \cup T$ spans $U + W$. (Accordingly, by induction, if S_i spans W_i for $i = 1, 2, \ldots, n$, then $S_1 \cup \cdots \cup S_n$ spans $W_1 + \cdots + W_n$.)

  Let $v \in U + W$. Then $v = u + w$ where $u \in U$ and $w \in W$. Since S spans U, u is a linear combination of the u_i's, and since T spans W, w is a linear combination of the w_j's:

$$u = a_1 u_{i_1} + a_2 u_{i_2} + \cdots + a_n u_{i_n} \qquad a_j \in K$$
$$w = b_1 w_{j_1} + b_2 w_{j_2} + \cdots + b_m w_{j_m} \qquad b_j \in K$$

Thus

$$v = u + w = a_1 u_{i_1} + a_2 u_{i_2} + \cdots + a_n u_{i_n} + b_1 w_{j_1} + b_2 w_{j_2} + \cdots + b_m w_{j_m}$$

Accordingly, $S \cup T = \{u_i, v_j\}$ spans $U + W$.

5.69.   Prove Theorem 5.22.

  Observe that $U \cap W$ is a subspace of both U and W. Suppose $\dim U = m$, $\dim W = n$, and $\dim (U \cap W) = r$. Suppose $\{v_1, \ldots, v_r\}$ is a basis of $U \cap W$. By Theorem 5.16, we can extend $\{v_i\}$ to a basis of U and to a basis of W; say,

$$\{v_1, \ldots, v_r, u_1, \ldots, u_{m-r}\} \qquad \text{and} \qquad \{v_1, \ldots, v_r, w_1, \ldots, w_{n-r}\}$$

are bases of U and W, respectively. Let

$$B = \{v_1, \ldots, v_r, u_1, \ldots, u_{m-r}, w_1, \ldots, w_{n-r}\}$$

Note that B has exactly $m + n - r$ elements. Thus the theorem is proved if we can show that B is a basis of $U + W$. Since $\{v_i, u_j\}$ spans U and $\{v_i, w_k\}$ spans W, the union $B = \{v_i, u_j, w_k\}$ spans $U + W$. Thus it suffices to show that B is independent.

  Suppose

$$a_1 v_1 + \cdots + a_r v_r + b_1 u_1 + \cdots + b_{m-r} u_{m-r} + c_1 w_1 + \cdots + c_{n-r} w_{n-r} = 0 \tag{1}$$

where a_i, b_j, c_k are scalars. Let

$$v = a_1 v_1 + \cdots + a_r v_r + b_1 u_1 + \cdots + b_{m-r} u_{m-r} \tag{2}$$

By (1), we also have that

$$v = -c_1 w_1 - \cdots - c_{n-r} w_{n-r} \tag{3}$$

Since $\{v_i, u_j\} \subset U$, $v \in U$ by (2); and since $\{w_k\} \subset W$, $v \in W$ by (3). Accordingly, $v \in U \cap W$. Now $\{v_i\}$ is a basis of $U \cap W$ and so there exist scalars $d_1, \ldots, d_r$ for which $v = d_1 v_1 + \cdots + d_r v_r$. Thus by (3) we have

$$d_1 v_1 + \cdots + d_r v_r + c_1 w_1 + \cdots + c_{n-r} w_{n-r} = 0$$

But $\{v_i, w_k\}$ is a basis of W and so is independent. Hence the above equation forces $c_1 = 0, \ldots, c_{n-r} = 0$. Substituting this into (1), we obtain

$$a_1 v_1 + \cdots + a_r v_r + b_1 u_1 + \cdots + b_{m-r} u_{m-r} = 0$$

But $\{v_i, u_j\}$ is a basis of U and so is independent. Hence the above equation forces $a_1 = 0, \ldots, a_r = 0$, $b_1 = 0, \ldots, b_{m-r} = 0$.

Since the equation (1) implies that the a_i, b_j and c_k are all 0, $B = \{v_i, u_j, w_k\}$ is independent and the theorem is proved.

5.70. Prove Theorem 5.23.

Suppose $V = U \oplus W$. Then any $v \in V$ can be uniquely written in the form $v = u + w$, where $u \in U$ and $w \in W$. Thus, in particular, $V = U + W$. Now suppose $v \in U \cap W$. Then:

(1) $v = v + 0$ where $v \in U, 0 \in W$; and (2) $v = 0 + v$ where $0 \in U, v \in W$

Since such a sum for v must be unique, $v = 0$. Accordingly, $U \cap W = \{0\}$.

On the other hand, suppose $V = U + W$ and $U \cap W = \{0\}$. Let $v \in V$. Since $V = U + W$, there exist $u \in U$ and $w \in W$ such that $v = u + w$. We need to show that such a sum is unique. Suppose also that $v = u' + w'$ where $u' \in U$ and $w' \in W$. Then

$$u + w = u' + w' \quad \text{and so} \quad u - u' = w' - w$$

But $u - u' \in U$ and $w' - w \in W$; hence, by $U \cap W = \{0\}$,

$$u - u' = 0, \ w' - w = 0 \quad \text{and so} \quad u = u', \ w = w'$$

Thus such a sum for $v \in V$ is unique and $V = U \oplus W$.

5.71. Prove Theorem 5.24 (for two factors): Suppose $V = U \oplus W$. Suppose $S = \{u_1, \ldots, u_m\}$ and $S' = \{w_1, \ldots, w_n\}$ are linearly independent subsets of U and W, respectively. Then: (a) $S \cup S'$ is linearly independent in V; (b) if S is a basis of U and S' is a basis of W, then $S \cup S'$ is a basis of V; and (c) dim $V =$ dim $U +$ dim W.

(a) Suppose $a_1 u_1 + \cdots + a_m u_m + b_1 w_1 + \cdots + b_n w_n = 0$, where a_i, b_j are scalars. Then

$$0 = (a_1 u_1 + \cdots + a_m u_m) + (b_1 w_1 + \cdots + b_n w_n) = 0 + 0$$

where $0, a_1 u_1 + \cdots + a_m u_m \in U$ and $0, b_1 w_1 + \cdots + b_n w_n \in W$. Since such a sum for 0 is unique, this leads to

$$a_1 u_1 + \cdots + a_m u_m = 0 \qquad b_1 w_1 + \cdots + b_n w_n = 0$$

Since S is linearly independent, each $a_i = 0$, and since S' is linearly independent, each $b_j = 0$. Thus $S \cup S'$ is linearly independent.

(b) By part (a), $S \cup S'$ linearly independent, and, by Problem 5.68, $S \cup S'$ span V. Thus $S \cup S'$ is a basis of V.

(c) Follows directly from part (b).

COORDINATE VECTORS

5.72. Let S be the basis of $\mathbf{R}^2$ consisting of $u_1 = (2, 1)$ and $u_2 = (1, -1)$. Find the coordinate vector $[v]$ of v relative to S where $v = (a, b)$.

Set $v = x u_1 + y u_2$ to obtain $(a, b) = (2x + y, x - y)$. Solve $2x + y = a$ and $x - y = b$ to obtain $x = (a + b)/3$, $y = (a - 2b)/3$. Thus $[v] = [(a + b)/3, (a - 2b)/3]$.

5.73. Consider the vector space $\mathbf{P}_3(t)$ of real polynomials in t of degree ≤ 3.

(a) Show that $S = \{1, 1 - t, (1 - t)^2, (1 - t)^3\}$ is a basis of $\mathbf{P}_3(t)$.

(b) Find the coordinate vector $[u]$ of $u = 2 - 3t + t^2 + 2t^3$ relative to S.

(a) The degree of $(1 - t)^k$ is k; hence no polynomial in S is a linear combination of preceding polynomials. Thus the polynomials are linearly independent and, since dim $\mathbf{P}_3(t) = 4$, they form a basis of $\mathbf{P}_3(t)$.

(b) Set u as a linear combination of the basis vectors using unknowns x, y, z, s:

$$\begin{aligned} u = 2 - 3t + t^2 + 2t^3 &= x(1) + y(1 - t) + z(1 - t)^2 + s(1 - t)^3 \\ &= x(1) + y(1 - t) + z(1 - 2t + t^2) + s(1 - 3t + 3t^2 - t^3) \\ &= x + y - yt + z - 2zt + zt^2 + s - 3st + 3st^2 - st^3 \\ &= (x + y + z + s) + (-y - 2z - 3s)t + (z + 3s)t^2 + (-s)t^3 \end{aligned}$$

Then set the coefficients of the same powers of t equal to each other:

$$x + y + z + s = 2 \qquad -y - 2z - 3s = -3 \qquad z + 3s = 1 \qquad -s = 2$$

Solving, $x = 2, y = -5, z = 7, s = -2$. Thus $[u] = [2, -5, 7, -2]$.

5.74. Consider the matrix $A = \begin{pmatrix} 2 & 3 \\ 4 & -7 \end{pmatrix}$ in the vector space V of 2×2 real matrices. Find the coordinate vector $[A]$ of the matrix A relative to $\left\{ \begin{pmatrix} 1 & 0 \\ 0 & 0 \end{pmatrix}, \begin{pmatrix} 0 & 1 \\ 0 & 0 \end{pmatrix}, \begin{pmatrix} 0 & 0 \\ 1 & 0 \end{pmatrix}, \begin{pmatrix} 0 & 0 \\ 0 & 1 \end{pmatrix} \right\}$, the usual basis of V.

We have

$$\begin{pmatrix} 2 & 3 \\ 4 & -7 \end{pmatrix} = x \begin{pmatrix} 1 & 0 \\ 0 & 0 \end{pmatrix} + y \begin{pmatrix} 0 & 1 \\ 0 & 0 \end{pmatrix} + z \begin{pmatrix} 0 & 0 \\ 1 & 0 \end{pmatrix} + t \begin{pmatrix} 0 & 0 \\ 0 & 1 \end{pmatrix} = \begin{pmatrix} x & y \\ z & t \end{pmatrix}$$

Thus $x = 2, y = 3, z = 4, t = -7$. Hence $[A] = [2, 3, 4, -7]$, whose components are the elements of A written row by row.

> **Remark:** The above result is true in general, that is, if A is any $m \times n$ matrix in the vector space V of $m \times n$ matrices over a field K, then the coordinate vector $[A]$ of A relative to the usual basis of V is the mn coordinate vector in K^{mn} whose components are the elements of A written row by row.

CHANGE OF BASIS

This section will represent a vector of $v \in V$ relative to a basis S of V by its coordinate column vector,

$$[v]_S = \begin{pmatrix} a_1 \\ a_2 \\ \cdots \\ a_n \end{pmatrix} = [a_1, a_2, \ldots, a_n]^T$$

(which is an $m \times 1$ matrix).

5.75. Consider the following bases of $\mathbf{R}^2$:

$$S_1 = \{u_1 = (1, -2), u_2 = (3, -4)\} \qquad \text{and} \qquad S_2 = \{v_1 = (1, 3), v_2 = (3, 8)\}$$

(a) Find the coordinates of an arbitrary vector $v = (a, b)$ in $\mathbf{R}^2$ relative to the basis $S_1 = \{u_1, u_2\}$.

(b) Find the change-of-basis matrix P from S_1 to S_2.

(c) Find the coordinates of an arbitrary vector $v = (a, b)$ in $\mathbf{R}^2$ relative to the basis $S_2 = \{v_1, v_2\}$.

(d) Find the change-of-basis matrix Q from S_2 back to S_1.

(e)   Verify that $Q = P^{-1}$.

(f)   Show that $P[v]_{S_2} = [v]_{S_1}$ for any vector $v = (a, b)$. (See Theorem 5.27.)

(g)   Show that $P^{-1}[v]_{S_1} = [v]_{S_2}$ for any vector $v = (a, b)$. (See Theorem 5.27.)

(a)   Let $v = xu_1 + yu_2$ for unknowns x and y:

$$\begin{pmatrix} a \\ b \end{pmatrix} = x\begin{pmatrix} 1 \\ -2 \end{pmatrix} + y\begin{pmatrix} 3 \\ -4 \end{pmatrix} \quad \text{or} \quad \begin{matrix} x + 3y = a \\ -2x - 4y = b \end{matrix} \quad \text{or} \quad \begin{matrix} x + 3y = a \\ 2y = 2a + b \end{matrix}$$

Solve for x and y in terms of a and b to get $x = -2a - \frac{3}{2}b$, $y = a + \frac{1}{2}b$. Thus

$$(a, b) = (-2a - \tfrac{3}{2}b)u_1 + (a + \tfrac{1}{2}b)u_2 \quad \text{or} \quad [(a, b)]_{S_1} = [-2a - \tfrac{3}{2}b, \, a + \tfrac{1}{2}b]^T$$

(b)   Use (a) to write each of the basis vectors v_1 and v_2 of S_2 as a linear combination of the basis vectors u_1 and u_2 of S_1:

$$v_1 = (1, 3) = (-2 - \tfrac{9}{2})u_1 + (1 + \tfrac{3}{2})u_2 = (-\tfrac{13}{2})u_1 + (\tfrac{5}{2})u_2$$

$$v_2 = (3, 8) = (-6 - 12)u_1 + (3 + 4)u_2 = -18u_1 + 7u_2$$

Then P is the matrix whose columns are the coordinates of v_1 and v_2 relative to the basis S_1, that is,

$$P = \begin{pmatrix} -\frac{13}{2} & -18 \\ \frac{5}{2} & 7 \end{pmatrix}$$

(c)   Let $v = xv_1 + yv_2$ for unknown scalars x and y:

$$\begin{pmatrix} a \\ b \end{pmatrix} = x\begin{pmatrix} 1 \\ 3 \end{pmatrix} + y\begin{pmatrix} 3 \\ 8 \end{pmatrix} \quad \text{or} \quad \begin{matrix} x + 3y = a \\ 3x + 8y = b \end{matrix} \quad \text{or} \quad \begin{matrix} x + 3y = a \\ -y = b - 3a \end{matrix}$$

Solve for x and y to get $x = -8a + 3b$, $y = 3a - b$. Thus

$$(a, b) = (-8a + 3b)v_1 + (3a - b)v_2 \quad \text{or} \quad [(a, b)]S_2 = [-8a + 3b, \, 3a - b]^T$$

(d)   Use (c) to express each of the basis vectors u_1 and u_2 of S_1 as a linear combination of the basis vectors v_1 and v_2 of S_2:

$$u_1 = (1, -2) = (-8 - 6)v_1 + (3 + 2)v_2 = -14v_1 + 5v_2$$

$$u_2 = (3, -4) = (-24 - 12)v_1 + (9 + 4)v_2 = -36v_1 + 13v_2$$

Write the coordinates of u_1 and u_2 relative to S_2 as columns to obtain $Q = \begin{pmatrix} -14 & -36 \\ 5 & 13 \end{pmatrix}$.

(e) $$QP = \begin{pmatrix} -14 & -36 \\ 5 & 13 \end{pmatrix}\begin{pmatrix} -\frac{13}{2} & -18 \\ \frac{5}{2} & 7 \end{pmatrix} = \begin{pmatrix} 1 & 0 \\ 0 & 1 \end{pmatrix} = I$$

(f)   Use (a), (b), and (c) to obtain

$$P[v]_{S_2} = \begin{pmatrix} -\frac{13}{2} & -18 \\ \frac{5}{2} & 7 \end{pmatrix}\begin{pmatrix} -8a + 3b \\ 3a - b \end{pmatrix} = \begin{pmatrix} -2a - \frac{3}{2}b \\ a + \frac{1}{2}b \end{pmatrix} = [v]_{S_1}$$

(g)   Use (a), (c), and (d) to obtain

$$P^{-1}[v]_{S_1} = Q[v]_{S_1} = \begin{pmatrix} -14 & -36 \\ 5 & 13 \end{pmatrix}\begin{pmatrix} -2a - \frac{3}{2}b \\ a + \frac{1}{2}b \end{pmatrix} = \begin{pmatrix} -8a + 3b \\ 3a - b \end{pmatrix} = [v]_{S_2}$$

5.76.   Suppose the following vectors form a basis S of K^n:

$$v_1 = (a_1, a_2, \ldots, a_n), \qquad v_2 = (b_1, b_2, \ldots, b_n), \qquad \ldots, \qquad v_n = (c_1, c_2, \ldots, c_n)$$

Show that the change-of-basis matrix from the usual basis $E = \{e_i\}$ of K^n to the basis S is the matrix P whose columns are the vectors $v_1, v_2, \ldots, v_n$, respectively.

Since $e_1, e_2, \ldots, e_n$ form the usual basis E of K^n, we have

$$v_1 = (a_1, a_2, \ldots, a_n) = a_1 e_1 + a_2 e_2 + \cdots + a_n e_n$$
$$v_2 = (b_1, b_2, \ldots, b_n) = b_1 e_1 + b_2 e_2 + \cdots + b_n e_n$$
$$\cdots\cdots\cdots\cdots\cdots\cdots\cdots\cdots\cdots\cdots\cdots\cdots$$
$$v_n = (c_1, c_2, \ldots, c_n) = c_1 e_1 + c_2 e_2 + \cdots + c_n e_n$$

Writing the coordinates as column we get

$$P = \begin{pmatrix} a_1 & b_1 & \ldots & c_1 \\ a_2 & b_2 & \ldots & c_2 \\ \cdots\cdots\cdots\cdots\cdots \\ a_n & b_n & \ldots & c_n \end{pmatrix}$$

as claimed.

5.77. Consider the basis $S = \{u_1 = (1, 2, 0),\ u_2 = (1, 3, 2),\ u_3 = (0, 1, 3)\}$ of $\mathbf{R}^3$. Find:

(a)   The change-of-basis matrix P from the usual basis $E = \{e_1, e_2, e_3\}$ of $\mathbf{R}^3$ to the basis S,

(b)   The change-of-basis matrix Q from the above basis S back to the usual basis E of $\mathbf{R}^3$.

(a)   Since E is the usual basis, simply write the basis vectors of S as columns:

$$P = \begin{pmatrix} 1 & 1 & 0 \\ 2 & 3 & 1 \\ 0 & 2 & 3 \end{pmatrix}$$

(b)   **Method 1.**   Express each basis vector of E as a linear combination of the basis vectors of S by first finding the coordinates of an arbitrary vector $v = (a, b, c)$ relative to the basis S. We have

$$\begin{pmatrix} a \\ b \\ c \end{pmatrix} = x \begin{pmatrix} 1 \\ 2 \\ 0 \end{pmatrix} + y \begin{pmatrix} 1 \\ 3 \\ 2 \end{pmatrix} + z \begin{pmatrix} 0 \\ 1 \\ 3 \end{pmatrix} \qquad \text{or} \qquad \begin{array}{rcl} x + y & = a \\ 2x + 3y + z & = b \\ 2y + 3z & = c \end{array}$$

Solve for x, y, z to get $x = 7a - 3b + c$, $y = -6a + 3b - c$, $z = 4a - 2b + c$. Thus

$$v = (a, b, c) = (7a - 3b + c)u_1 + (-6a + 3b - c)u_2 + (4a - 2b + c)u_3$$

or $$[v]_S = [(a, b, c)]_S = [7a - 3b + c,\ -6a + 3b - c,\ 4a - 2b + c]^T$$

Using the above formula for $[v]_S$ and then writing the coordinates of the e_i as columns yields

$$\begin{array}{l} e_1 = (1, 0, 0) = 7u_1 - 6u_2 + 4u_3 \\ e_2 = (0, 1, 0) = -3u_1 + 3u_2 - 2u_3 \\ e_3 = (0, 0, 1) = u_1 - u_2 + u_3 \end{array} \qquad \text{and} \qquad Q = \begin{pmatrix} 7 & -3 & 1 \\ -6 & 3 & -1 \\ 4 & -2 & 1 \end{pmatrix}$$

Method 2.   Find P^{-1} by row reducing $M = (P\ \vdots\ I)$ to the form $(I\ \vdots\ P^{-1})$:

$$M = \begin{pmatrix} 1 & 1 & 0 & \vdots & 1 & 0 & 0 \\ 2 & 3 & 1 & \vdots & 0 & 1 & 0 \\ 0 & 2 & 3 & \vdots & 0 & 0 & 1 \end{pmatrix} \sim \begin{pmatrix} 1 & 1 & 0 & \vdots & 1 & 0 & 0 \\ 0 & 1 & 1 & \vdots & -2 & 1 & 0 \\ 0 & 2 & 3 & \vdots & 0 & 0 & 1 \end{pmatrix}$$

$$\sim \begin{pmatrix} 1 & 1 & 0 & \vdots & 1 & 0 & 0 \\ 0 & 1 & 1 & \vdots & -2 & 1 & 0 \\ 0 & 0 & 1 & \vdots & 4 & -2 & 1 \end{pmatrix} \sim \begin{pmatrix} 1 & 1 & 0 & \vdots & 1 & 0 & 0 \\ 0 & 1 & 0 & \vdots & -6 & 3 & -1 \\ 0 & 0 & 1 & \vdots & 4 & -2 & 1 \end{pmatrix}$$

$$\sim \begin{pmatrix} 1 & 0 & 0 & \vdots & 7 & -3 & 1 \\ 0 & 1 & 0 & \vdots & -6 & 3 & 1 \\ 0 & 0 & 1 & \vdots & 4 & -2 & 1 \end{pmatrix}$$

$$\text{Thus } Q = P^{-1} = \begin{pmatrix} 7 & -3 & 1 \\ -6 & 3 & -1 \\ 4 & -2 & 1 \end{pmatrix}.$$

5.78. Suppose the x and y axes in the plane $\mathbf{R}^2$ are rotated counterclockwise $45°$ so that the new x' axis is along the line $y = x$, and the new y' axis is along the line $y = -x$. Find (a) the change-of-basis matrix P and (b) the new coordinates of the point $A(5, 6)$ under the given rotation.

    (a) The unit vectors in the direction of the new x' and y' axes are, respectively,

$$u_1 = (\sqrt{2}/2, \sqrt{2}/2) \quad \text{and} \quad u_2 = (-\sqrt{2}/2, \sqrt{2}/2)$$

    (The unit vectors in the direction of the original x and y axes are, respectively, usual basis vectors for $\mathbf{R}^2$.) Thus write the coordinates of u_1 and u_2 as columns to obtain

$$P = \begin{pmatrix} \sqrt{2}/2 & -\sqrt{2}/2 \\ \sqrt{2}/2 & \sqrt{2}/2 \end{pmatrix}$$

    (b) Multiply the coordinates of the point by P^{-1}:

$$\begin{pmatrix} \sqrt{2}/2 & \sqrt{2}/2 \\ -\sqrt{2}/2 & \sqrt{2}/2 \end{pmatrix}\begin{pmatrix} 5 \\ 6 \end{pmatrix} = \begin{pmatrix} 11\sqrt{2}/2 \\ \sqrt{2}/2 \end{pmatrix}$$

[Since P is orthogonal, P^{-1} is simply the transpose of P.]

5.79. Consider the bases $S = \{1, i\}$ and $S' = \{1 + i, 1 + 2i\}$ of the complex field $\mathbf{C}$ over the real field $\mathbf{R}$. Find (a) the change-of-basis matrix P from the S-basis to the S'-basis, and (b) find the change-of-basis matrix Q from the S'-basis back to the S-basis.

    (a) We have

$$\begin{aligned} 1 + i &= 1(1) + 1(i) \\ 1 + 2i &= 1(1) + 2(i) \end{aligned} \quad \text{and so} \quad P = \begin{pmatrix} 1 & 1 \\ 1 & 2 \end{pmatrix}$$

    (b) Use the formula for the inverse of a 2×2 matrix to obtain $Q = P^{-1} = \begin{pmatrix} 2 & -1 \\ -1 & 1 \end{pmatrix}$.

5.80. Suppose P is the change-of-basis matrix from a basis $\{u_i\}$ to a basis $\{w_i\}$, and suppose Q is the change-of-basis matrix from the basis $\{w_i\}$ back to the basis $\{u_i\}$. Prove that P is invertible and $Q = P^{-1}$.

    Suppose, for $i = 1, 2, \ldots, n$,

$$w_i = a_{i1}u_1 + a_{i2}u_2 + \cdots + a_{in}u_n = \sum_{j=1}^{n} a_{ij}u_j \tag{1}$$

and, for $j = 1, 2, \ldots, n$,

$$u_j = b_{j1}w_1 + b_{j2}w_2 + \cdots + b_{jn}w_n = \sum_{k=1}^{n} b_{jk}w_k \tag{2}$$

Let $A = (a_{ij})$ and $B = (b_{jk})$. Then $P = A^T$ and $Q = B^T$. Substituting (2) into (1) yields

$$w_1 = \sum_{j=1}^{n} a_{ij}\left(\sum_{k=1}^{n} b_{jk}w_k\right) = \sum_{k=1}^{n}\left(\sum_{j=1}^{n} a_{ij}b_{jk}\right)w_k$$

Since the $\{w_i\}$ is a basis $\sum a_{ij}b_{jk} = \delta_{ik}$ where δ_{ik} is the Kronecker delta, that is, $\delta_{ik} = 1$ if $i = k$ but $\delta_{ik} = 0$ if $i \neq k$. Suppose $AB = (c_{ik})$. Then $c_{ik} = \delta_{ik}$. Accordingly, $AB = I$, and so

$$QP = B^T A^T = (AB)^T = I^T = I$$

Thus $Q = P^{-1}$.

5.81.   Prove Theorem 5.27.

Suppose $S = \{u_1, \ldots, u_n\}$ and $S' = \{w_1, \ldots, w_n\}$, and suppose, for $i = 1, \ldots, n$,

$$w_i = a_{i1}u_1 + a_{i2}u_2 + \cdots + a_{in}u_n = \sum_{j=1}^{n} a_{ij}u_j$$

Then P is the n-square matrix whose jth row is

$$(a_{1j}, a_{2j}, \ldots, a_{nj}) \tag{1}$$

Also suppose $v = k_1 w_1 + k_2 w_2 + \cdots + k_n w_n = \sum_{i=1}^{n} k_i w_i$. Then

$$[v]_{S'} = [k_{1j}, k_{2j}, \ldots, k_{nj}]^T \tag{2}$$

Substituting for w_i in the equation for v, we obtain

$$v = \sum_{i=1}^{n} k_i w_i = \sum_{i=1}^{n} k_i \left(\sum_{j=1}^{n} a_{ij}u_j \right) = \sum_{j=1}^{n} \left(\sum_{i=1}^{n} a_{ij}k_i \right) u_j$$

$$= \sum_{j=1}^{n} (a_{1j}k_1 + a_{2j}k_2 + \cdots + a_{nj}k_n)u_j$$

Accordingly, $[v]_S$ is the column vector whose jth entry is

$$a_{1j}k_1 + a_{2j}k_2 + \cdots + a_{nj}k_n \tag{3}$$

On the other hand, the jth entry of $P[v]_{S'}$ is obtained by multiplying the jth row of P by $[v]_{S'}$, that is, (1) by (2). However, the product of (1) and (2) is (3); hence $P[v]_{S'}$ and $[v]_S$ have the same entries. Thus $P[v]_{S'} = [v]_{S'}$, as claimed.

Furthermore, multiplying the above by P^{-1} gives $P^{-1}[v]_S = P^{-1}P[v]_{S'} = [v]_{S'}$.

MISCELLANEOUS PROBLEMS

5.82.   Consider a finite sequence of vectors $S = \{v_1, v_2, \ldots, v_n\}$. Let T be the sequence of vectors obtained from S by one of the following "elementary operations": (i) interchange two vectors, (ii) multiply a vector by a nonzero scalar, (iii) add a multiple of one vector to another. Show that S and T span the same space W. Also show that T is independent if and only if S is independent.

Observe that, for each operation, the vectors in T are linear combinations of vectors in S. On the other hand, each operation has an inverse of the same type (Prove!); hence the vectors in S are linear combinations of the vectors in T. Thus S and T span the same space W. Also, T is independent if and only if dim $W = n$, and this is true iff S is also independent.

5.83.   Let $A = (a_{ij})$ and $B = (b_{ij})$ be row equivalent $m \times n$ matrices over a field K, and let $v_1, \ldots, v_n$ be any vectors in a vector space V over K. Let

$$u_1 = a_{11}v_1 + a_{12}v_2 + \cdots + a_{1n}v_n \qquad w_1 = b_{11}v_1 + b_{12}v_2 + \cdots + b_{1n}v_n$$
$$u_2 = a_{21}v_1 + a_{22}v_2 + \cdots + a_{2n}v_n \qquad w_2 = b_{21}v_1 + b_{22}v_2 + \cdots + b_{2n}v_n$$
$$\cdots\cdots\cdots\cdots\cdots\cdots\cdots\cdots\cdots \qquad \cdots\cdots\cdots\cdots\cdots\cdots\cdots\cdots\cdots$$
$$u_m = a_{m1}v_1 + a_{m2}v_2 + \cdots + a_{mn}v_n \qquad w_m = b_{m1}v_1 + b_{m2}v_2 + \cdots + b_{mn}v_n$$

Show that $\{u_i\}$ and $\{w_i\}$ span the same space.

Applying an "elementary operation" of Problem 5.82 to $\{u_i\}$ is equivalent to applying an elementary row operation to the matrix A. Since A and B are row equivalent, B can be obtained from A by a sequence of elementary row operations; hence $\{w_i\}$ can be obtained from $\{u_i\}$ by the corresponding sequence of operations. Accordingly, $\{u_i\}$ and $\{w_i\}$ span the same space.

5.84. Let $v_1, \ldots, v_n$ belong to a vector space V over a field K. Let

$$w_1 = a_{11}v_1 + a_{12}v_2 + \cdots + a_{1n}v_n$$
$$w_2 = a_{21}v_1 + a_{22}v_2 + \cdots + a_{2n}v_n$$
$$\cdots\cdots\cdots\cdots\cdots\cdots\cdots\cdots\cdots\cdots\cdots\cdots\cdots$$
$$w_n = a_{n1}v_1 + a_{n2}v_2 + \cdots + a_{nn}v_n$$

where $a_{ij} \in K$. Let P be the n-square matrix of coefficients, i.e., let $P = (a_{ij})$.

(a) Suppose P is invertible. Show that $\{w_i\}$ and $\{v_i\}$ span the same space; hence $\{w_i\}$ is independent if and only if $\{v_i\}$ is independent.

(b) Suppose P is not invertible. Show that $\{w_i\}$ is dependent.

(c) Suppose $\{w_i\}$ is independent. Show that P is invertible.

(a) Since P is invertible, it is row equivalent to the identity matrix I. Hence by the preceding problem $\{w_i\}$ and $\{v_i\}$ span the same space. Thus one is independent if and only if the other is.

(b) Since P is not invertible, it is row equivalent to a matrix with a zero row. This means that $\{w_i\}$ spans a space which has a spanning set of less than n elements. Thus $\{w_i\}$ is dependent.

(c) This is the contrapositive of the statement of (b) and so it follows from (b).

5.85. Suppose that $A_1, A_2, \ldots$ are linearly independent sets of vectors, and that $A_1 \subseteq A_2 \subseteq \cdots$. Show that the union $A = A_1 \cup A_2 \cup \cdots$ is also linearly independent.

Suppose A is linearly dependent. Then there exist vectors $v_1, \ldots, v_n \in A$ and scalars $a_1, \ldots, a_n \in K$, not all of them 0, such that

$$a_1v_1 + a_2v_2 + \cdots + a_nv_n = 0 \tag{1}$$

Since $A = \cup A_i$ and the $v_i \in A$, there exist sets $A_{i_1}, \ldots, A_{i_n}$ such that

$$v_1 \in A_{i_1}, v_2 \in A_{i_2}, \ldots, v_n \in A_{i_n}$$

Let k be the maximum index of the sets A_{i_j}: $k = \max(i_1, \ldots, i_n)$. It follows then, since $A_1 \subseteq A_2 \subseteq \cdots$, that each A_{i_j} is contained in A_k. Hence $v_1, v_2, \ldots, v_n \in A_k$ and so, by (1), A_k is linearly dependent, which contradicts our hypothesis. Thus A is linearly independent.

5.86. Let K be a subfield of a field L and L a subfield of a field E: that is, $K \subseteq L \subseteq E$. (Hence K is a subfield of E.) Suppose that E is of dimension n over L and L is of dimension m over K. Show that E is of dimension mn over K.

Suppose $\{v_1, \ldots, v_n\}$ is a basis of E over L and $\{a_1, \ldots, a_m\}$ is a basis of L over K. We claim that $\{a_i v_j : i = 1, \ldots, m, j = 1, \ldots, n\}$ is a basis of E over K. Note that $\{a_i v_j\}$ contains mn elements.

Let w be any arbitrary element in E. Since $\{v_1, \ldots, v_n\}$ spans E over L, w is a linear combination of the v_i with coefficients in L:

$$w = b_1v_1 + b_2v_2 + \cdots + b_nv_n \qquad b_i \in L \tag{1}$$

Since $\{a_1, \ldots, a_m\}$ spans L over K, each $b_i \in L$ is a linear combination of the a_j with coefficients in K:

$$b_1 = k_{11}a_1 + k_{12}a_2 + \cdots + k_{1m}a_m$$
$$b_2 = k_{21}a_1 + k_{22}a_2 + \cdots + k_{2m}a_m$$
$$\cdots\cdots\cdots\cdots\cdots\cdots\cdots\cdots\cdots\cdots\cdots\cdots\cdots$$
$$b_n = k_{n1}a_1 + k_{n2}a_2 + \cdots + k_{nm}a_m$$

where $k_{ij} \in K$. Substituting in (1), we obtain

$$
\begin{aligned}
w &= (k_{11}a_1 + \cdots + k_{1m}a_m)v_1 + (k_{21}a_1 + \cdots + k_{2m}a_m)v_2 + \cdots + (k_{n1}a_1 + \cdots + k_{nm}a_m)v_n \\
&= k_{11}a_1v_1 + \cdots + k_{1m}a_mv_1 + k_{21}a_1v_2 + \cdots + k_{2m}a_mv_2 + \cdots + k_{n1}a_1v_n + \cdots + k_{nm}a_mv_n \\
&= \sum_{i,j} k_{ji}(a_iv_j)
\end{aligned}
$$

where $k_{ji} \in K$. Thus w is a linear combination of the $a_i v_j$ with coefficients in K; hence $\{a_i v_j\}$ spans E over K.

The proof is complete if we show that $\{a_i v_j\}$ is linearly independent over K. Suppose, for scalars $x_{ji} \in K, \sum_{i,j} x_{ji}(a_i v_j) = 0$; that is,

$$(x_{11}a_1v_1 + x_{12}a_2v_1 + \cdots + x_{1m}a_mv_1) + \cdots + (x_{n1}a_1v_n + x_{n2}a_2v_n + \cdots + x_{nm}a_mv_n) = 0$$

or $\qquad (x_{11}a_1 + x_{12}a_2 + \cdots + x_{1m}a_m)v_1 + \cdots + (x_{n1}a_1 + x_{n2}a_2 + \cdots + x_{nm}a_m)v_n = 0$

Since $\{v_1, \ldots, v_n\}$ is linearly independent over L and since the above coefficients of the v_i belong to L, each coefficient must be 0:

$$x_{11}a_1 + x_{12}a_2 + \cdots + x_{1m}a_m = 0, \ldots, x_{n1}a_1 + x_{n2}a_2 + \cdots + x_{nm}a_m = 0$$

But $\{a_1, \ldots, a_m\}$ is linearly independent over K; hence, since the $x_{ji} \in K$,

$$x_{11} = 0, x_{12} = 0, \ldots, x_{1m} = 0, \ldots, x_{n1} = 0, x_{n2} = 0, \ldots, x_{nm} = 0$$

Accordingly, $\{a_i v_j\}$ is linearly independent over K and the theorem is proved.

Supplementary Problems

VECTOR SPACES

5.87. Let V be the set of ordered pairs (a, b) of real numbers with addition in V and scalar multiplication on V defined by

$$(a, b) + (c, d) = (a + c, b + d) \qquad \text{and} \qquad k(a, b) = (ka, 0)$$

Show that V satisfies all of the axioms of a vector space except $[M_4]$: $1u = u$. Hence $[M_4]$ is not a consequence of the other axioms.

5.88. Show that the following axiom $[A_4]$ can be derived from the other axioms of a vector space.
[A_4] For any vectors $u, v \in V, u + v = v + u$.

5.89. Let V be the set of infinite sequences $(a_1, a_2, \ldots)$ in a field K with addition in V and scalar multiplication on V defined by

$$(a_1, a_2, \ldots) + (b_1, b_2, \ldots) = (a_1 + b_1, a_2 + b_2, \ldots)$$

$$k(a_1, a_2, \ldots) = (ka_1, ka_2, \ldots)$$

where $a_i, b_j, k \in K$. Show that V is a vector space over K.

SUBSPACES

5.90. Determine whether or not W is a subspace of $\mathbf{R}^3$ where W consists of those vectors $(a, b, c) \in \mathbf{R}^3$ for which (a) $a = 2b$; (b) $a \le b \le c$; (c) $ab = 0$; (d) $a = b = c$; (e) $a = b^2$.

5.91. Let V be the vector space of n-square matrices over a field K. Show that W is a subspace of V if W consists of all matrices which are (a) antisymmetric ($A^t = -A$), (b) (upper) triangular, (c) diagonal, (d) scalar.

5.92. Let $AX = B$ be a nonhomogeneous system of linear equations in n unknowns over a field K. Show that the solution set of the system is not a subspace of K^n.

5.93. Discuss whether or not $\mathbf{R}^2$ is a subspace of $\mathbf{R}^3$.

5.94. Suppose U and W are subspaces of V for which $U \cup W$ is also a subspace. Show that either $U \subseteq W$ or $W \subseteq U$.

5.95. Let V be the vector space of all functions from the real field $\mathbf{R}$ into $\mathbf{R}$. Show that W is a subspace of V in each of the following cases.

  (a) W consists of all bounded functions. [Here $f: \mathbf{R} \to \mathbf{R}$ is bounded if there exists $M \in \mathbf{R}$ such that $|f(x)| \leq M, \forall x \in \mathbf{R}$.]
  (b) W consists of all even functions. [Here $f: \mathbf{R} \to \mathbf{R}$ is even if $f(-x) = f(x), \forall x \in \mathbf{R}$.]
  (c) W consists of all continuous functions.
  (d) W consists of all differentiable functions.
  (e) W consists of all integrable functions in, say, the interval $0 \leq x \leq 1$.
  (The last three cases require some knowledge of analysis.)

5.96. Let V be the vector space (Problem 5.106) of infinite sequences $(a_1, a_2, \ldots)$ in a field K. Show that W is a subspace of V where (a) W consists of all sequences with 0 as the first component, and (b) W consists of all sequences with only a finite number of nonzero components.

LINEAR COMBINATIONS, LINEAR SPANS

5.97. Show that the complex numbers $w = 2 + 3i$ and $z = 1 - 2i$ span the complex field $\mathbf{C}$ as a vector space over the real field $\mathbf{R}$.

5.98. Show that the polynomials $(1 - t)^3, (1 - t)^2, 1 - t$, and 1 span the space $P_3^{(t)}$ of polynomials of degree ≤ 3.

5.99. Find one vector in $\mathbf{R}^3$ which spans the intersection of U and W where U is the xy plane: $U = \{(a, b, 0)\}$, and W is the space spanned by the vectors $(1, 2, 3)$ and $(1, -1, 1)$.

5.100. Prove that span S is the intersection of all the subspaces of V containing S.

5.101. Show that span $S = $ span $(S \cup \{0\})$. That is, by joining or deleting the zero vector from a set, we do not change the space spanned by the set.

5.102. Show that if $S \subseteq T$, then span $S \subseteq$ span T.

5.103. Show that span (span S) = span S.

5.104. Let $W_1, W_2, \ldots$ be subspaces of a vector space V for which $W_1 \subseteq W_2 \subseteq \cdots$. Let $W = W_1 \cup W_2 \cup \cdots$. (a) Show that W is a subspace of V. (b) Suppose S_i spans W_i for $i = 1, 2, \ldots$. Show that $S = S_1 \cup S_2 \cup \cdots$ spans W.

LINEAR DEPENDENCE AND INDEPENDENCE

5.105. Determine whether the following vectors in $\mathbf{R}^4$ are linearly dependent or independent:

  (a) $(1, 3, -1, 4), (3, 8, -5, 7), (2, 9, 4, 23)$ (b) $(1, -2, 4, 1), (2, 1, 0, -3), (3, -6, 1, 4)$

5.106. Let V be the vector space of polynomials of degree ≤ 3 over $\mathbf{R}$. Determine whether $u, v, w \in V$ are linearly dependent or independent where:

  (a) $u = t^3 - 4t^2 + 2t + 3, v = t^3 + 2t^2 + 4t - 1, w = 2t^3 - t^2 - 3t + 5$
  (b) $u = t^3 - 5t^2 - 2t + 3, v = t^3 - 4t^2 - 3t + 4, w = 2t^3 - 7t^2 - 7t + 9$

5.107. Show that (a) the vectors $(1 - i, i)$ and $(2, -1 + i)$ in $\mathbf{C}^2$ are linearly dependent over the complex field $\mathbf{C}$ but are linearly independent over the real field $\mathbf{R}$; (b) the vectors $(3 + \sqrt{2}, 1 + \sqrt{2})$ and $(7, 1 + 2\sqrt{2})$ in $\mathbf{R}^2$ are linearly dependent over the real field $\mathbf{R}$ but are linearly independent over the rational field $\mathbf{Q}$.

5.108. Suppose $\{u_1, \ldots, u_r, w_1, \ldots, w_s\}$ is a linearly independent subset of V. Prove that span $u_i \cap$ span $w_j = \{0\}$. (Recall that span u_i is the subspace of V spanned by the u_i.)

5.109. Suppose $v_1, v_2, \ldots, v_n$ are linearly independent vectors. Prove the following:

(a) $\{a_1v_1, a_2v_2, \ldots, a_nv_n\}$ is linearly independent where each $a_i \neq 0$.

(b) $\{v_1, \ldots, v_{i-1}, w, v_{i+1}, \ldots, v_n\}$ is linearly independent where $w = b_1v_1 + \cdots + b_iv_i + \cdots + b_nv_n$ and $b_i \neq 0$.

5.110. Suppose $(a_{11}, \ldots, a_{1n}), \ldots, (a_{m1}, \ldots, a_{mn})$ are linearly independent vectors in K^n, and suppose $v_1, \ldots, v_n$ are linearly independent vectors in a vector space V over K. Show that the vectors

$$w_1 = a_{11}v_1 + \cdots + a_{1n}v_n, \ldots, w_m = a_{m1}v_1 + \cdots + a_{mn}v_n$$

are also linearly independent.

5.111. Suppose A is any n-square matrix and suppose $u_1, u_2, \ldots, u_r$ are $n \times 1$ column vectors. Show that, if $Au_1, Au_2, \ldots, Au_r$ are linearly independent (column) vectors, then $u_1, u_2, \ldots, u_r$ are linearly independent.

BASIS AND DIMENSION

5.112. Find a subset of u_1, u_2, u_3, u_4 which gives a basis for $W = $ span (u_1, u_2, u_3, u_4) of $\mathbf{R}^5$ where:

(a) $u_1 = (1, 1, 1, 2, 3), u_2 = (1, 2, -1, -2, 1), u_3 = (3, 5, -1, -2, 5), u_4 = (1, 2, 1, -1, 4)$

(b) $u_1 = (1, -2, 1, 3, -1), u_2 = (-2, 4, -2, -6, 2), u_3 = (1, -3, 1, 2, 1), u_4 = (3, -7, 3, 8, -1)$

(c) $u_1 = (1, 0, 1, 0, 1), u_2 = (1, 1, 2, 1, 0), u_3 = (1, 2, 3, 1, 1), u_4 = (1, 2, 1, 1, 1)$

(d) $u_1 = (1, 0, 1, 1, 1), u_2 = (2, 1, 2, 0, 1), u_3 = (1, 1, 2, 3, 4), u_4 = (4, 2, 5, 4, 6)$

5.113. Let U and W be the following subspaces of $\mathbf{R}^4$:

$$U = \{(a, b, c, d): b - 2c + d = 0\} \qquad W = \{(a, b, c, d): a = d, b = 2c\}$$

Find a basis and the dimension of (a) U, (b) W, (c) $U \cap W$.

5.114. Find a basis and the dimension of the solution space W of each homogeneous system:

$$\begin{array}{ll}
x + 2y - 2z + 2s - t = 0 & x + 2y - z + 3s - 4t = 0 \\
x + 2y - z + 3s - 2t = 0 & 2x + 4y - 2z - s + 5t = 0 \\
2x + 4y - 7z + s + t = 0 & 2x + 4y - 2z + 4s - 2t = 0 \\
\qquad\qquad (a) & \qquad\qquad (b)
\end{array}$$

5.115. Find a homogeneous system whose solution set W is spanned by the three vectors:

$$(1, -2, 0, 3, -1) \qquad (2, -3, 2, 5, -3) \qquad (1, -2, 1, 2, -2)$$

5.116. Let V be the vector space of polynomials in t of degree $\leq n$. Determine whether or not each of the following is a basis of V:

(a) $\{1, 1 + t, 1 + t + t^2, 1 + t + t^2 + t^3, \ldots, 1 + t + t^2 + \cdots + t^{n-1} + t^n\}$

(b) $\{1 + t, t + t^2, t^2 + t^3, \ldots, t^{n-2} + t^{n-1}, t^{n-1} + t^n\}$

5.117. Find a basis and the dimension of the subspace W of $\mathbf{P}(t)$ spanned by the polynomials

(a) $u = t^3 + 2t^2 - 2t + 1, v = t^3 + 3t^2 - t + 4$, and $w = 2t^3 + t^2 - 7t - 7$

(b) $u = t^3 + t^2 - 3t + 2, v = 2t^3 + t^2 + t - 4$, and $w = 4t^3 + 3t^2 - 5t + 2$

5.118. Let V be the space of 2×2 matrices over $\mathbf{R}$. Find a basis and the dimension of the subspace W of V spanned by the matrices

$$\begin{pmatrix} 1 & -5 \\ -4 & 2 \end{pmatrix} \qquad \begin{pmatrix} 1 & 1 \\ -1 & 5 \end{pmatrix} \qquad \begin{pmatrix} 2 & -4 \\ -5 & 7 \end{pmatrix} \quad \text{and} \quad \begin{pmatrix} 1 & -7 \\ -5 & 1 \end{pmatrix}$$

ROWSPACE AND RANK OF A MATRIX

5.119. Consider the following subspaces of $\mathbf{R}^3$:

$$U_1 = \text{span } [(1, 1, -1), (2, 3, -1), (3, 1, -5)]$$
$$U_2 = \text{span } [(1, -1, -3), (3, -2, -8), (2, 1, -3)]$$
$$U_3 = \text{span } [(1, 1, 1), (1, -1, 3), (3, -1, 7)]$$

Determine which of the subspaces are identical.

5.120. Find the rank of each matrix:

$$\begin{pmatrix} 1 & 3 & -2 & 5 & 4 \\ 1 & 4 & 1 & 3 & 5 \\ 1 & 4 & 2 & 4 & 3 \\ 2 & 7 & -3 & 6 & 13 \end{pmatrix} \qquad \begin{pmatrix} 1 & 2 & -3 & -2 & -3 \\ 1 & 3 & -2 & 0 & -4 \\ 3 & 8 & -7 & -2 & -11 \\ 2 & 1 & -9 & -10 & -3 \end{pmatrix} \qquad \begin{pmatrix} 1 & 1 & 2 \\ 4 & 5 & 5 \\ 5 & 8 & 1 \\ -1 & -2 & 2 \end{pmatrix} \qquad \begin{pmatrix} 2 & 1 \\ 3 & -7 \\ -6 & 1 \\ 5 & -8 \end{pmatrix}$$

$$\qquad\qquad (a) \qquad\qquad\qquad\qquad\qquad (b) \qquad\qquad\qquad\qquad\qquad (c) \qquad\qquad\qquad (d)$$

5.121. Show that if any row is deleted from a matrix in echelon (row canonical) form then the resulting matrix is still in echelon (row canonical) form.

5.122. Let A and B be arbitrary $m \times n$ matrices. Show that rank $(A + B) \le$ rank $A +$ rank B.

5.123. Give examples of 2×2 matrices A and B such that:

(a) rank $(A + B) <$ rank A, rank B (c) rank $(A + B) >$ rank A, rank B

(b) rank $(A + B) =$ rank $A =$ rank B

SUMS, DIRECT SUMS, INTERSECTIONS

5.124. Suppose U and W are 2-dimensional subspaces of $\mathbf{R}^3$. Show that $U \cap W \ne \{0\}$.

5.125. Suppose U and W are subspaces of V and that dim $U = 4$, dim $W = 5$, and dim $V = 7$. Find the possible dimensions of $U \cap W$.

5.126. Let U and W be subspaces of $\mathbf{R}^3$ for which dim $U = 1$, dim $W = 2$, and $U \nsubseteq W$. Show that $\mathbf{R}^3 = U \oplus W$.

5.127. Let U be the subspace of $\mathbf{R}^5$ spanned by

$$\{(1, 3, -3, -1, -4) \qquad (1, 4, -1, -2, -2) \qquad (2, 9, 0, -5, -2)\}$$

and let W be the subspace spanned by

$$\{(1, 6, 2, -2, 3) \qquad (2, 8, -1, -6, -5) \qquad (1, 3, -1, -5, -6)\}$$

Find (a) dim $(U + W)$, (b) dim $(U \cap W)$.

5.128. Let V be the vector space of polynomials over $\mathbf{R}$. Find (a) dim $(U + W)$, (b) dim $(U \cap W)$, where

$$U = \text{span}\,(t^3 + 4t^2 - t + 3,\ t^3 + 5t^2 + 5,\ 3t^3 + 10t^2 - 5t + 5)$$
$$W = \text{span}\,(t^3 + 4t^2 + 6,\ t^3 + 2t^2 - t + 5,\ 2t^3 + 2t^2 - 3t + 9)$$

5.129. Let U be the subspace of $\mathbf{R}^5$ spanned by

$$(1, -1, -1, -2, 0) \qquad (1, -2, -2, 0, -3) \qquad \text{and} \qquad (1, -1, -2, -2, 1)$$

and let W be the subspace spanned by

$$(1, -2, -3, 0, -2) \qquad (1, -1, -3, 2, -4) \qquad \text{and} \qquad (1, -1, -2, 2, -5)$$

(a) Find two homogeneous systems whose solution spaces are U and W, respectively.

(b) Find a basis and the dimension of $U \cap W$.

5.130. Let U_1, U_2, and U_3 be the following subspaces of $\mathbf{R}^3$:

$$U_1 = \{(a, b, c): a + b + c = 0\} \qquad U_2 = \{(a, b, c): a = c\} \qquad U_3 = \{(0, 0, c): c \in \mathbf{R}\}$$

Show that: $(a)\ \mathbf{R}^3 = U_1 + U_2$, $(b)\ \mathbf{R}^3 = U_2 + U_3$, $(c)\ \mathbf{R}^3 = U_1 + U_3$. When is the sum direct?

5.131. Suppose U, V, and W are subspaces of a vector space. Prove that

$$(U \cap V) + (U \cap W) \subseteq U \cap (V + W)$$

Find subspaces of $\mathbf{R}^2$ for which equality does not hold.

5.132. The sum of arbitrary nonempty subsets (not necessarily subspaces) S and T of a vector space V is defined by $S + T = \{s + t : s \in S, t \in T\}$. Show that this operation satisfies:

(a) Commutative law: $S + T = T + S$ (c) $S + \{0\} = \{0\} + S = S$

(b) Associative law: $(S_1 + S_2) + S_3 = S_1 + (S_2 + S_3)$ (d) $S + V = V + S = V$

5.133. Suppose $W_1, W_2, \ldots, W_r$ are subspaces of a vector space V. Show that:

(a) span $(W_1, W_2, \ldots, W_r) = W_1 + W_2 + \cdots + W_r$

(b) If S_i spans W_i for $i = 1, \ldots, r$, then $S_1 \cup S_2 \cup \cdots \cup S_r$ spans $W_1 + W_2 + \cdots + W_r$.

5.134. Prove Theorem 5.24.

5.135. Prove Theorem 5.25.

5.136. Let U and W be vector spaces over a field K. Let V be the set of ordered pairs (u, w) where u belongs to U and w to W: $V = \{(u, w): u \in U, w \in W\}$. Show that V is a vector space over K with addition in V and scalar multiplication on V defined by

$$(u, w) + (u', w') = (u + u', w + w') \qquad \text{and} \qquad k(u, w) = (ku, kw)$$

where $u, u' \in U$, $w, w' \in W$, and $k \in K$. (This space V is called the *external direct sum* of U and W.)

5.137. Let V be the external direct sum of the vector spaces U and W over a field K. (See Problem 5.136.) Let $\hat{U} = \{(u, 0): u \in U\}$, $\qquad \hat{W} = \{(0, w): w \in W\}$. Show that

(a) $\hat{U}$ and $\hat{W}$ are subspaces of V and that $V = \hat{U} \oplus \hat{W}$;

(b) U is isomorphic to $\hat{U}$ under the correspondence $u \leftrightarrow (u, 0)$, and that W is isomorphic to $\hat{W}$ under the correspondence $w \leftrightarrow (0, w)$;

(c) dim $V =$ dim $U +$ dim W.

5.138. Suppose $V = U \oplus W$. Let $\hat{V}$ be the external direct product of U and W. Show that V is isomorphic to $\hat{V}$ under the correspondence $v = u + w \leftrightarrow (u, w)$.

COORDINATE VECTORS

5.139. Consider the basis $S = \{u_1 = (1, -2), u_2 = (4, -7)\}$ of $\mathbf{R}^2$. Find the coordinate vector $[v]$ of v relative to S where (a) $v = (3, 5)$, (b) $v = (1, 1)$, and (c) $v = (a, b)$.

5.140. Consider the vector space $\mathbf{P}_3(t)$ of polynomials of degree ≤ 3 and the basis $S = \{1, t + 1, t^2 + t, t^3 + t^2\}$ of $\mathbf{P}_3(t)$. Find the coordinate vector of v relative to S where (a) $v = 2 - 3t + t^2 + 2t^3$; (b) $v = 3 - 2t - t^2$; and (c) $v = a + bt + ct^2 + dt^3$.

5.141. Let S be the following basis of the vector space W of 2×2 real symmetric matrices:

$$\left\{ \begin{pmatrix} 1 & -1 \\ -1 & 2 \end{pmatrix}, \begin{pmatrix} 4 & 1 \\ 1 & 0 \end{pmatrix}, \begin{pmatrix} 3 & -2 \\ -2 & 1 \end{pmatrix} \right\}$$

Find the coordinate vector of the matrix $A \in W$ relative to the above basis S where (a) $A = \begin{pmatrix} 1 & -5 \\ -5 & 5 \end{pmatrix}$ and (b) $A = \begin{pmatrix} 1 & 2 \\ 2 & 4 \end{pmatrix}$.

CHANGE OF BASIS

5.142. Find the change-of-basis matrix P from the usual basis $E = \{(1, 0), (0, 1)\}$ of $\mathbf{R}^2$ to the basis S, the change-of-basis matrix Q from S back to E, and the coordinate vector of $v = (a, b)$ relative to S where

(a) $S = \{(1, 2), (3, 5)\}$ (c) $S = \{(2, 5), (3, 7)\}$

(b) $S = \{(1, -3), (3, -8)\}$ (d) $S = \{(2, 3), (4, 5)\}$

5.143. Consider the following bases of $\mathbf{R}^2$: $S = \{u_1 = (1, 2), u_2 = (2, 3)\}$ and $S' = \{v_1 = (1, 3), v_2 = (1, 4)\}$. Find: (a) the change-of-basis matrix P from S to S', and (b) the change-of-basis matrix Q from S' back to S.

5.144. Suppose that the x and y axes in the plane $\mathbf{R}^2$ are rotated counterclockwise $30°$ to yield new x' and y' axes for the plane. Find: (a) the unit vectors in the direction of the new x' and y' axes, (b) the change-of-basis matrix P for the new coordinate system, and (c) the new coordinates of each of the following points under the new coordinate system: $A(1, 3)$, $B(2, -5)$, $C(a, b)$.

5.145. Find the change-of-basis matrix P from the usual basis E of $\mathbf{R}^3$ to the basis S, the change-of-basis matrix Q from S back to E, and the coordinate vector of $v = (a, b, c)$ relative to S where S consists of the vectors:

(a) $u_1 = (1, 1, 0), u_2 = (0, 1, 2), u_3 = (0, 1, 1)$ (c) $u_1 = (1, 2, 1), u_2 = (1, 3, 4), u_3 = (2, 5, 6)$

(b) $u_1 = (1, 0, 1), u_2 = (1, 1, 2), u_3 = (1, 2, 4)$

5.146. Suppose $S_1, S_2,$ and S_3 are bases of a vector space V, and suppose P is the change-of-basis matrix from S_1 to S_2 and Q is the change-of-basis matrix from S_2 to S_3. Prove that the product PQ is the change-of-basis matrix from S_1 to S_3.

MISCELLANEOUS PROBLEMS

5.147. Determine the dimension of the vector space W of n-square: (a) symmetric matrices over a field K, (b) antisymmetric matrices over K.

5.148. Let V be a vector space of dimension n over a field K, and let K be a vector space of dimension m over a subfield F. (Hence V may also be viewed as a vector space over the subfield F.) Prove that the dimension of V over F is mn.

5.149. Let $t_1, t_2, \ldots, t_n$ be symbols, and let K be any field. Let V be the set of expressions

$$a_1 t_1 + a_2 t_2 + \cdots + a_n t_n \qquad \text{where} \qquad a_i \in K$$

Define addition in V by

$$(a_1 t_1 + a_2 t_2 + \cdots + a_n t_n) + (b_1 t_1 + b_2 t_2 + \cdots + b_n t_n) = (a_1 + b_1)t_1 + (a_2 + b_2)t_2 + \cdots + (a_n + b_n)t_n$$

Define scalar multiplication on V by

$$k(a_1 t_1 + a_2 t_2 + \cdots + a_n t_n) = ka_1 t_1 + ka_2 t_2 + \cdots + ka_n t_n$$

Show that V is a vector space over K with the above operations. Also show that $\{t_1, \ldots, t_n\}$ is a basis of V where, for $i = 1, \ldots, n$,

$$t_i = 0t_1 + \cdots + 0t_{i-1} + 1t_i + 0t_{i+1} + \cdots + 0t_n$$

Answers to Supplementary Problems

5.90. (a) Yes.

(b) No; e.g., $(1, 2, 3) \in W$ but $-2(1, 2, 3) \notin W$.

(c) No; e.g., $(1, 0, 0), (0, 1, 0) \in W$, but not their sum.

(d) Yes.

(e) No; e.g., $(9, 3, 0) \in W$ but $2(9, 3, 0) \notin W$.

5.92. $X = 0$ is not a solution of $AX = B$.

5.93. No. Although one may "identify" the vector $(a, b) \in \mathbf{R}^2$ with, say, $(a, b, 0)$ in the xy plane in $\mathbf{R}^3$, they are distinct elements belonging to distinct, disjoint sets.

5.95. (a) Let $f, g \in W$ with M_f and M_g bounds for f and g, respectively. Then for any scalars $a, b \in \mathbf{R}$,

$$|(af + bg)(x)| = |af(x) + bg(x)| \le |af(x)| + |bg(x)| = |a||f(x)| + |b||g(x)| \le |a|M_f + |b|M_g$$

That is, $|a|M_f + |b|M_g$ is a bound for the function $af + bg$.

(b) $(af + bg)(-x) = af(-x) + bg(-x) = af(x) + bg(x) = (af + bg)(x)$.

5.99. $(2, -5, 0)$.

5.105. (a) Dependent, (b) Independent.

5.106. (a) Independent, (b) Dependent.

5.107. (a) $(2, -1 + i) = (1 + i)(1 - i, i)$; (b) $(7, 1 + 2\sqrt{2}) = (3 - \sqrt{2})(3 + \sqrt{2}, 1 + \sqrt{2})$.

5.112. (a) u_1, u_2, u_4; (b) u_1, u_3; (c) u_1, u_2, u_3, u_4; (d) u_1, u_2, u_3.

5.113. (a) Basis, $\{(1, 0, 0, 0), (0, 2, 1, 0), (0, -1, 0, 1)\}$; dim $U = 3$.

(b) Basis, $\{(1, 0, 0, 1), (0, 2, 1, 0)\}$; dim $W = 2$.

(c) Basis, $\{(0, 2, 1, 0)\}$; dim $(U \cap W) = 1$. *Hint.* $U \cap W$ must satisfy all three conditions on a, b, c and d.

5.114. (a) Basis, $\{(2, -1, 0, 0, 0), (4, 0, 1, -1, 0), (3, 0, 1, 0, 1)\}$; dim $W = 3$.

(b) Basis, $\{(2, -1, 0, 0, 0), (1, 0, 1, 0, 0)\}$; dim $W = 2$.

5.115. $\begin{cases} 5x + y - z - s = 0 \\ x + y - z - t = 0 \end{cases}$

5.116. (a) Yes,   (b) No. For dim $V = n + 1$, but the set contains only n elements.

5.117. (a) dim $W = 2$,   (b) dim $W = 3$

5.118. dim $W = 2$

5.119. U_1 and U_2.

5.120. (a) 3,   (b) 2,   (c) 3,   (d) 2

5.123. (a) $A = \begin{pmatrix} 1 & 1 \\ 0 & 0 \end{pmatrix}$, $B = \begin{pmatrix} -1 & -1 \\ 0 & 0 \end{pmatrix}$ (c) $A = \begin{pmatrix} 1 & 0 \\ 0 & 0 \end{pmatrix}$, $B = \begin{pmatrix} 0 & 0 \\ 0 & 1 \end{pmatrix}$

       (b) $A = \begin{pmatrix} 1 & 0 \\ 0 & 0 \end{pmatrix}$, $B = \begin{pmatrix} 0 & 2 \\ 0 & 0 \end{pmatrix}$

5.125. dim $(U \cap W) = 2, 3$ or 4.

5.127. (a) dim $(U + W) = 3$,   (b) dim $(U \cap W) = 2$.

5.128. (a) dim $(U + W) = 3$,   dim $(U \cap W) = 1$.

5.129. (a) $\begin{cases} 3x + 4y - z & - t = 0 \\ 4x + 2y & + s = 0 \end{cases}$, $\begin{cases} 4x + 2y & - s = 0 \\ 9x + 2y + z & + t = 0 \end{cases}$

       (b) $\{(1, -2, -5, 0, 0), (0, 0, 1, 0, -1)\}$. dim $(U \cap W) = 2$.

5.130. The sum is direct in (b) and (c).

5.131. In $\mathbf{R}^2$, let U, V, and W be, respectively, the line $y = x$, the x axis, and the y axis.

5.139. (a) $[-41, 11]$,   (b) $[-11, 3]$,   (c) $[-7a - 4b, 2a + b]$

5.140. (a) $[4, -2, -1, 2]$,   (b) $[4, -1, -1, 0]$,   (c) $[a - b + c - d, b - c + d, c - d, d]$

5.141. (a) $[2, -1, 1]$,   (b) $[3, 1, -2]$

5.142. (a) $P = \begin{pmatrix} 1 & 3 \\ 2 & 5 \end{pmatrix}$, $Q = \begin{pmatrix} -5 & 3 \\ 2 & -1 \end{pmatrix}$, $[v] = \begin{pmatrix} -5a + b \\ 2a - b \end{pmatrix}$

       (b) $P = \begin{pmatrix} 1 & 3 \\ -3 & -8 \end{pmatrix}$, $Q = \begin{pmatrix} -8 & -3 \\ 3 & 1 \end{pmatrix}$, $[v] = \begin{pmatrix} -8a - 3b \\ 3a - b \end{pmatrix}$

       (c) $P = \begin{pmatrix} 2 & 3 \\ 5 & 7 \end{pmatrix}$, $Q = \begin{pmatrix} -7 & 3 \\ 5 & -2 \end{pmatrix}$, $[v] = \begin{pmatrix} -7a + 3b \\ 5a - 2b \end{pmatrix}$

       (d) $P = \begin{pmatrix} 2 & 4 \\ 3 & 5 \end{pmatrix}$, $Q = \begin{pmatrix} -\frac{5}{2} & 2 \\ \frac{3}{2} & -1 \end{pmatrix}$, $[v] = \begin{pmatrix} (-\frac{5}{2})a + 2b \\ (\frac{3}{2})a - b \end{pmatrix}$

5.143. (a) $P = \begin{pmatrix} 3 & 5 \\ -1 & -1 \end{pmatrix}$,   (b) $Q = \begin{pmatrix} 2 & 5 \\ -1 & -3 \end{pmatrix}$

5.144. (a) $(\sqrt{3}/2, \frac{1}{2}), (-\frac{1}{2}, \sqrt{3}/2)$

       (b) $P = \begin{pmatrix} \sqrt{3}/2 & -\frac{1}{2} \\ \frac{1}{2} & \sqrt{3}/2 \end{pmatrix}$

(c) $[A] = [(\sqrt{3} - 3)/2, (1 + 3\sqrt{3})/2],$

$[B] = [(2\sqrt{3} + 5)/2, (2 - 5\sqrt{3})/2],$

$[C] = [(\sqrt{3}a - b)/2, (a + \sqrt{3}b)/2]$

5.145. Since E is the usual basis, simply let P be the matrix whose columns are u_1, u_2, u_3. Then $Q = P^{-1}$ and $[v] = P^{-1}v = Qv$.

(a) $P = \begin{pmatrix} 1 & 0 & 0 \\ 1 & 1 & 1 \\ 0 & 2 & 1 \end{pmatrix}$, $Q = \begin{pmatrix} 1 & 0 & 0 \\ 1 & -1 & 1 \\ -2 & 2 & -1 \end{pmatrix}$, $[v] = \begin{pmatrix} a \\ a - b + c \\ -2a + 2b - c \end{pmatrix}$

(b) $P = \begin{pmatrix} 1 & 1 & 1 \\ 0 & 1 & 2 \\ 1 & 2 & 4 \end{pmatrix}$, $Q = \begin{pmatrix} 0 & -2 & 1 \\ 2 & 3 & -2 \\ -1 & -1 & 1 \end{pmatrix}$, $[v] = \begin{pmatrix} -2b + c \\ 2a + 3b - 2c \\ -a - b + c \end{pmatrix}$

(c) $P = \begin{pmatrix} 1 & 1 & 2 \\ 2 & 3 & 5 \\ 1 & 4 & 6 \end{pmatrix}$, $Q = \begin{pmatrix} -2 & 2 & -1 \\ -7 & 4 & -1 \\ 5 & -3 & 1 \end{pmatrix}$, $[v] = \begin{pmatrix} -2a + 2b - c \\ -7a + 4b - c \\ 5a - 3b + c \end{pmatrix}$

5.147. (a) $n(n + 1)/2$,   (b) $n(n - 1)/2$

5.148. *Hint:* The proof is almost identical to that given in Problem 5.86 for the special case when V is an extension field of K.

Chapter 6

Inner Product Spaces, Orthogonality

6.1 INTRODUCTION

The definition of a vector space V involves an arbitrary field K. In this chapter we restrict K to be either the real field $\mathbf{R}$ or the complex field $\mathbf{C}$. Specifically, we first assume, unless otherwise stated or implied, that $K = \mathbf{R}$, in which case V is called a *real vector space*, and in the last sections we extend our results to the case that $K = \mathbf{C}$, in which case V is called a *complex vector space*.

Recall that the concepts of "length" and "orthogonality" did not appear in the investigation of arbitrary vector spaces (although they did appear in Chapter 2 on the spaces $\mathbf{R}^n$ and $\mathbf{C}^n$). In this chapter we place an additional structure on a vector space V to obtain an inner product space, and in this context these concepts are defined.

As in Chapter 5, we adopt the following notation (unless otherwise stated or implied):

$$
\begin{array}{ll}
V & \text{the given vector space} \\
u, v, w & \text{vectors in } V \\
K & \text{the field of scalars} \\
a, b, c, \text{ or } k & \text{scalars in } K
\end{array}
$$

We emphasize that V shall denote a vector space of finite dimension unless otherwise stated or implied. In fact, many of the theorems in this chapter are not valid for spaces of infinite dimension. This is illustrated by some of the examples and problems.

6.2 INNER PRODUCT SPACES

We begin with a definition.

Definition: Let V be a real vector space. Suppose to each pair of vectors $u, v \in V$ there is assigned a real number, denoted by $\langle u, v \rangle$. This function is called a *(real) inner product* on V if it satisfies the following axioms:

$[I_1]$ (Linear Property) $\langle au_1 + bu_2, v \rangle = a\langle u_1, v \rangle + b\langle u_2, v \rangle$

$[I_2]$ (Symmetric Property) $\langle u, v \rangle = \langle v, u \rangle$

$[I_3]$ (Positive Definite Property) $\langle u, u \rangle \geq 0$; and $\langle u, u \rangle = 0$ if and only if $u = 0$.

The vector space V with an inner product is called a *(real)* inner product space.

Axiom $[I_1]$ is equivalent to the following two conditions:

(a) $\quad \langle u_1 + u_2, v \rangle = \langle u_1, v \rangle + \langle u_2, v \rangle \qquad$ and $\qquad (b) \quad \langle ku, v \rangle = k\langle u, w \rangle$

Using $[I_1]$ and the symmetry axiom $[I_2]$, we obtain

$$\langle u, cv_1 + dv_2 \rangle = \langle cv_1 + dv_2, u \rangle = c\langle v_1, u \rangle + d\langle v_2, u \rangle = c\langle u, v_1 \rangle + d\langle u, v_2 \rangle$$

or, equivalently, the two conditions

$$(a) \quad \langle u, v_1 + v_2 \rangle = \langle u, v_1 \rangle + \langle u, v_2 \rangle \quad \text{and} \quad (b) \quad \langle u, kv \rangle = k \langle u, v \rangle$$

That is, the inner product function is also linear in its second position (variable). By induction, we obtain

$$\langle a_1 u_1 + \cdots + a_r u_r, v \rangle = a_1 \langle u_1, v \rangle + a_2 \langle u_2, v \rangle + \cdots + a_r \langle u_r, v \rangle$$

and

$$\langle u, b_1 v_1 + b_2 v_2 + \cdots + b_s v_s \rangle = b_1 \langle u, v_1 \rangle + b_2 \langle u, v_2 \rangle + \cdots + b_s \langle u, v_s \rangle$$

Combining these two properties yields the following general formula:

$$\left\langle \sum_{i=1}^{r} a_i u_i, \sum_{j=1}^{s} b_j v_j \right\rangle = \sum_{i=1}^{r} \sum_{j=1}^{s} a_i b_j \langle u_i, v_j \rangle$$

The following remarks are in order.

Remark 1: Axiom $[I_1]$ by itself implies that

$$\langle 0, 0 \rangle = \langle 0v, 0 \rangle = 0 \langle v, 0 \rangle = 0$$

Accordingly, $[I_1]$, $[I_2]$, and $[I_3]$ are equivalent to $[I_1]$, $[I_2]$, and the following axiom:

$$[I_3'] \quad \text{If } u \neq 0, \text{ then } \langle u, u \rangle > 0$$

That is, a function satisfying $[I_1]$, $[I_2]$, and $[I_3']$ is an inner product.

Remark 2: By $[I_3]$, $\langle u, u \rangle$ is nonnegative and hence its positive real square root exists. We use the notation

$$\| u \| = \sqrt{\langle u, u \rangle}$$

This nonnegative real number $\| u \|$ is called the norm or length of u. This function does satisfy the axioms of a norm for a vector space. (See Theorem 6.25 and Section 6.9.) We note that the relation $\| u \|^2 = \langle u, u \rangle$ will be frequently used.

Example 6.1.   Consider the vector space $\mathbf{R}^n$. The *dot product* (or *scalar product*) in $\mathbf{R}^n$ is defined by

$$u \cdot v = a_1 b_1 + a_2 b_2 + \cdots + a_n b_n$$

where $u = (a_i)$ and $v = (b_i)$. This function defines an inner product on $\mathbf{R}^n$. The norm $\| u \|$ of the vector $u = (a_i)$ in this space follows:

$$\| u \| = \sqrt{u \cdot u} = \sqrt{a_1^2 + a_2^2 + \cdots + a_n^2}$$

On the other hand, by the Pythagorean Theorem, the distance from the origin O in $\mathbf{R}^3$ to the point $P(a, b, c)$, shown in Fig. 6-1, is given by $\sqrt{a^2 + b^2 + c^2}$. This is precisely the same as the above defined norm of the vector $v = (a, b, c)$ in $\mathbf{R}^3$. Since the Pythagorean Theorem is a consequence of the axioms of Euclidean geometry, the vector space $\mathbf{R}^n$ with the above inner product and norm is called *Euclidean n-space*. Although there are many ways to define an inner product on $\mathbf{R}^n$, we shall assume this as the inner product on $\mathbf{R}^n$, unless otherwise stated or implied; it is called the *usual inner product* on $\mathbf{R}^n$.

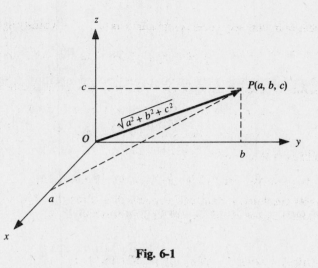

Fig. 6-1

Remark: Frequently, the vectors in $\mathbf{R}^n$ are represented by $n \times 1$ column matrices. In such a case, the usual inner product on $\mathbf{R}^n$ (Example 6.1) may be defined by

$$\langle u, v \rangle = u^T v$$

Example 6.2

(a)   Let V be the vector space of real continuous functions on the interval $a \le t \le b$. Then the following is an inner product on V:

$$\langle f, g \rangle = \int_a^b f(t) g(t) \, dt$$

where $f(t)$ and $g(t)$ are now any continuous functions on $[a, b]$.

(b)   Let V be again the vector space of continuous functions on the interval $a \le t \le b$. Let $w(t)$ be a given continuous function which is positive on the interval $a \le t \le b$. Then the following is also an inner product on V:

$$\langle f, g \rangle = \int_a^b w(t) f(t) g(t) \, dt$$

In this case, $w(t)$ is called a *weight function* for the inner product.

Example 6.3

(a)   Let V denote the vector space of $m \times n$ matrices over $\mathbf{R}$. The following is an inner product in V:

$$\langle A, B \rangle = \operatorname{tr} (B^T A)$$

where tr stands for trace, the sum of the diagonal elements. If $A = (a_{ij})$ and $B = (b_{ij})$, then

$$\langle A, B \rangle = \operatorname{tr} (B^T A) = \sum_{i=1}^m \sum_{j=1}^n a_{ij} b_{ij}$$

the sum of the products of corresponding entries. In particular,

$$\| A \|^2 = \langle A, A \rangle = \sum_{i=1}^m \sum_{j=1}^n a_{ij}^2$$

the sum of the squares of all the elements of A.

(b) Let V be the vector space of infinite sequences of real numbers $(a_1, a_2, \ldots)$ satisfying

$$\sum_{i=1}^{\infty} a_i^2 = a_1^2 + a_2^2 + \cdots < \infty$$

i.e., the sum converges. Addition and scalar multiplication are defined componentwise:

$$(a_1, a_2, \ldots) + (b_1, b_2, \ldots) = (a_1 + b_1, a_2 + b_2, \ldots)$$

$$k(a_1, a_2, \ldots) = (ka_1, ka_2, \ldots)$$

An inner product is defined in V by

$$\langle (a_1, a_2, \ldots), (b_1, b_2, \ldots) \rangle = a_1 b_1 + a_2 b_2 + \cdots$$

The above sum converges absolutely for any pair of points in V (Problem 6.12); hence the inner product is well defined. This inner product space is called l_2-*space* (or *Hilbert space*).

6.3 CAUCHY–SCHWARZ INEQUALITY, APPLICATIONS

The following formula (proved in Problem 6.10) is called the Cauchy–Schwarz inequality; it is used in many branches of mathematics.

Theorem 6.1 (Cauchy–Schwarz): For any vectors $u, v \in V$,

$$\langle u, v \rangle^2 \leq \langle u, u \rangle \langle v, v \rangle \qquad \text{or, equivalently,} \qquad |\langle u, v \rangle| \leq \| u \| \, \| v \|$$

Next we examine this inequality in specific cases.

Example 6.4

(a) Consider any real $a_1, \ldots, a_n, b_1, \ldots, b_n$. Then by the Cauchy–Schwarz inequality,

$$(a_1 b_1 + a_2 b_2 + \cdots + a_n b_n)^2 \leq (a_1^2 + \cdots + a_n^2)(b_1^2 + \cdots + b_n^2)$$

that is, $(u \cdot v)^2 \leq \| u \|^2 \, \| v \|^2$ where $u = (a_i)$ and $v = (b_i)$.

(b) Let f and g be any real continuous functions defined on the unit interval $0 \leq t \leq 1$. Then by the Cauchy–Schwarz inequality,

$$(\langle f, g \rangle)^2 = \left(\int_0^1 f(t) g(t) \, dt \right)^2 \leq \int_0^1 f^2(t) \, dt \int_0^1 g^2(t) \, dt = \| f \|^2 \, \| g \|^2$$

Here V is the inner product space of Example 6.2(a).

The next theorem (proved in Problem 6.11) gives basic properties of a norm; the proof of the third property requires the Cauchy–Schwarz inequality.

Theorem 6.2: Let V be an inner product space. Then the norm in V satisfies the following properties:

$[N_1]$ $\| v \| \geq 0$; and $\| v \| = 0$ if and only if $v = 0$.

$[N_2]$ $\| kv \| = |k| \, \| v \|$.

$[N_3]$ $\| u + v \| \leq \| u \| + \| v \|$.

The above properties $[N_1]$, $[N_2]$, and $[N_3]$ are those that have been chosen as the axioms of an abstract norm in a vector space (see Section 6.9). Thus the above theorem says that the norm defined by an inner product is an actual norm. The property $[N_3]$ is frequently called the *triangle inequality* because if we view $u + v$ as the side of the triangle formed with u and v (as shown in Fig. 6-2), then $[N_3]$ states that the length of one side of a triangle is less than or equal to the sum of the lengths of the other two sides.

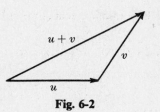

Fig. 6-2

The following remarks are in order.

Remark 1: If $\| u \| = 1$, or, equivalently, if $\langle u, u \rangle = 1$, then u is called a *unit vector* and is said to be *normalized*. Every nonzero vector $v \in V$ can be multiplied by the reciprocal of its length to obtain the unit vector

$$u = \frac{1}{\| v \|} v$$

which is a positive multiple of v. This process is called *normalizing v*.

Remark 2: The nonnegative real number $d(u, v) = \| u - v \|$ is called the distance between u and v; this function does satisfy the axioms of a metric space (see Theorem 6.19).

Remark 3: For any nonzero vectors $u, v \in V$, the angle between u and v is defined to be the angle θ such that $0 \le \theta \le \pi$ and

$$\cos \theta = \frac{\langle u, v \rangle}{\| u \| \, \| v \|}$$

By the Cauchy–Schwarz inequality, $-1 \le \cos \theta \le 1$ and so the angle θ always exists and is unique.

6.4 ORTHOGONALITY

Let V be an inner product space. The vectors $u, v \in V$ are said to be *orthogonal* and u is said to be *orthogonal* to v if

$$\langle u, v \rangle = 0$$

The relation is clearly symmetric; that is, if u is orthogonal to v, then $\langle v, u \rangle = 0$ and so v is orthogonal to u. We note that $0 \in V$ is orthogonal to every $v \in V$ for

$$\langle 0, v \rangle = \langle 0v, v \rangle = 0 \langle v, v \rangle = 0$$

Conversely, if u is orthogonal to every $v \in V$, then $\langle u, u \rangle = 0$ and hence $u = 0$ by $[I_3]$. Observe that u and v are orthogonal if and only if $\cos \theta = 0$ where θ is the angle between u and v, and this is true if and only if u and v are "perpendicular," i.e., $\theta = \pi/2$ (or $\theta = 90°$).

Example 6.5

(a) Consider an arbitrary vector $u = (a_1, a_2, \ldots, a_n)$ in $\mathbf{R}^n$. Then a vector $v = (x_1, x_2, \ldots, x_n)$ is orthogonal to u if

$$\langle u, v \rangle = a_1 x_1 + a_2 x_2 + \cdots + a_n x_n = 0$$

In other words, v is orthogonal to u if v satisfies a homogeneous equation whose coefficients are the elements of u.

(b) Suppose we want a nonzero vector which is orthogonal to $v_1 = (1, 3, 5)$ and $v_2 = (0, 1, 4)$ in $\mathbf{R}^3$. Let $w = (x, y, z)$. We want

$$0 = \langle v_1, w \rangle = x + 3y + 5z \qquad \text{and} \qquad 0 = \langle v_2, w \rangle = y + 4z$$

Thus we obtain the homogeneous system

$$x + 3y + 5z = 0 \qquad y + 4z = 0$$

Set $z = 1$ to obtain $y = -4$ and $x = 7$; then $w = (7, -4, 1)$ is orthogonal to v_1 and v_2. Normalizing w, we obtain

$$\hat{w} = w/\| w \| = (7/\sqrt{66},\ -4/\sqrt{66},\ 1/\sqrt{66})$$

which is a unit vector orthogonal to v_1 and v_2.

Orthogonal Complements

Let S be a subset of an inner product space V. The orthogonal complement of S, denoted by $S^{\perp}$ (read "S perp") consists of those vectors in V which are orthogonal to every vector $u \in S$:

$$S^{\perp} = \{v \in V : \langle v, u \rangle = 0 \text{ for every } u \in S\}$$

In particular, for a given vector u in V, we have

$$u^{\perp} = \{v \in V : \langle v, u \rangle = 0\}$$

That is, $u^{\perp}$ consists of all vectors in V which are orthogonal to the given vector u.

We show that $S^{\perp}$ is a subspace of V. Clearly $0 \in S^{\perp}$ since 0 is orthogonal to every vector in V. Now suppose $v, w \in S^{\perp}$. Then, for any scalars a and b and any vector $u \in S$, we have

$$\langle av + bw, u \rangle = a\langle v, u \rangle + b\langle w, u \rangle = a \cdot 0 + b \cdot 0 = 0$$

Thus $av + bw \in S^{\perp}$ and therefore $S^{\perp}$ is a subspace of V.

We state this result formally.

Proposition 6.3:   Let S be a subset of an inner product space V. Then $S^{\perp}$ is a subspace of V.

Remark 1:   Suppose u is a nonzero vector in $\mathbf{R}^3$. Then there is a geometrical description of $u^{\perp}$. Specifically, $u^{\perp}$ is the plane in $\mathbf{R}^3$ through the origin 0 and perpendicular to the vector u, as shown in Fig. 6-3.

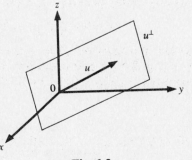

Fig. 6-3

Remark 2: Consider a homogeneous system of linear equations over $\mathbf{R}$:

$$a_{11}x_1 + a_{12}x_2 + \cdots + a_{1n}x_n = 0$$
$$a_{21}x_1 + a_{22}x_2 + \cdots + a_{2n}x_n = 0$$
$$\cdots\cdots\cdots\cdots\cdots\cdots\cdots\cdots\cdots\cdots\cdots\cdots$$
$$a_{m1}x_1 + a_{m2}x_2 + \cdots + a_{mn}x_n = 0$$

Recall that the solution space W may be viewed as the solution of the equivalent matrix equation $AX = 0$ where $A = (a_{ij})$ and $X = (x_i)$. This gives another interpretation of W using the notion of orthogonality. Specifically, each solution vector $v = (x_1, x_2, \ldots, x_n)$ is orthogonal to each row of A; and consequently W is the orthogonal complement of the row space of A.

Example 6.6. Suppose we want to find a basis for the subspace $u^\perp$ in $\mathbf{R}^3$ where $u = (1, 3, -4)$. Note $u^\perp$ consists of all vectors (x, y, z) such that

$$\langle (x, y, z), (1, 3, -4) \rangle = 0 \qquad \text{or} \qquad x + 3y - 4z = 0$$

The free variables are y and z. Set

(1) $y = -1, z = 0$ to obtain the solution $w_1 = (3, -1, 0)$

(2) $y = 0, z = 1$ to obtain the solution $w_2 = (4, 0, 1)$

The vectors w_1 and w_2 form a basis for the solution space of the equation and hence a basis for $u^\perp$.

Suppose W is a subspace of V. Then both W and $W^\perp$ are subspaces of V. The next theorem, whose proof (Problem 6.35) requires results of later sections, is a basic result in linear algebra.

Theorem 6.4: Let W be a subspace of V. Then V is the direct sum of W and $W^\perp$, that is, $V = W \oplus W^\perp$.

Example 6.7. Let W be the z axis in $\mathbf{R}^3$, i.e., $W = \{(0, 0, c): c \in \mathbf{R}\}$. Then $W^\perp$ is the xy plane, or, in other words, $W^\perp = \{(a, b, 0): a, b \in \mathbf{R}\}$ as shown in Fig. 6-4. As noted previously, $\mathbf{R}^3 = W \oplus W^\perp$.

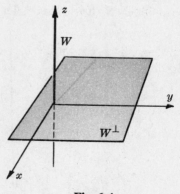

Fig. 6-4

6.5 ORTHOGONAL SETS AND BASES, PROJECTIONS

A set S of vectors in V is called *orthogonal* if each pair of vectors in S are orthogonal, and S is called *orthonormal* if S is orthogonal and each vector in S has unit length. In other words, $S = \{u_1, u_2, \ldots, u_r\}$

is *orthogonal* if

$$\langle u_i, u_j \rangle = 0 \qquad \text{for} \quad i \neq j$$

and S is *orthonormal* if

$$\langle u_i, u_j \rangle = \delta_{ij} = \begin{cases} 0 & \text{for } i \neq j \\ 1 & \text{for } i = j \end{cases}$$

Normalizing an orthogonal set S refers to the process of multiplying each vector in S by the reciprocal of its length in order to transform S into an orthonormal set of vectors.

A basis S of a vector space V is called an *orthogonal basis* or an *orthonormal basis* according as S is an orthogonal set or an orthonormal set of vectors.

The following theorems, proved in Problems 6.20 and 6.21, respectively, apply.

Theorem 6.5: Suppose S is an orthogonal set of nonzero vectors. Then S is linearly independent.

Theorem 6.6 (Pythagoras): Suppose $\{u_1, u_2, \ldots, u_r\}$ is an orthogonal set of vectors. Then

$$\| u_1 + u_2 + \cdots + u_r \|^2 = \| u_1 \|^2 + \| u_2 \|^2 + \cdots + \| u_r \|^2$$

Here we prove the above Pythagorean Theorem in the special and familiar case for two vectors. Specifically, suppose $\langle u, v \rangle = 0$. Then

$$\| u + v \|^2 = \langle u + v, u + v \rangle = \langle u, u \rangle + 2\langle u, v \rangle + \langle v, v \rangle = \langle u, u \rangle + \langle v, v \rangle = \| u \|^2 + \| v \|^2$$

which gives our result.

Example 6.8

(a) Consider the usual basis E of Euclidean 3-space $\mathbf{R}^3$:

$$E = \{e_1 = (1, 0, 0), \, e_2 = (0, 1, 0), \, e_3 = (0, 0, 1)\}$$

It is clear that

$$\langle e_1, e_2 \rangle = \langle e_1, e_3 \rangle = \langle e_2, e_3 \rangle = 0 \qquad \text{and} \qquad \langle e_1, e_1 \rangle = \langle e_2, e_2 \rangle = \langle e_3, e_3 \rangle = 1$$

Thus E is an orthonormal basis of $\mathbf{R}^3$. More generally, the usual basis of $\mathbf{R}^n$ is orthonormal for every n.

(b) Let V be the vector space of real continuous functions on the interval $-\pi \leq t \leq \pi$ with inner product defined by $\langle f, g \rangle = \int_{-\pi}^{\pi} f(t) g(t) \, dt$. The following is a classical example of an orthogonal subset of V:

$$\{1, \cos t, \cos 2t, \ldots, \sin t, \sin 2t, \ldots\}$$

The above orthogonal set plays a fundamental role in the theory of Fourier series.

(c) Consider the following set S of vectors in $\mathbf{R}^4$:

$$S = \{u = (1, 2, -3, 4), \, v = (3, 4, 1, -2), \, w = (3, -2, 1, 1)\}$$

Note that

$$\langle u, v \rangle = 3 + 8 - 3 + 8 = 0 \qquad \langle u, w \rangle = 3 - 4 - 3 + 4 = 0 \qquad \langle v, w \rangle = 9 - 8 + 1 - 2 = 0$$

Thus S is orthogonal. We normalize S to obtain an orthonormal set by first finding

$$\| u \|^2 = 1 + 4 + 9 + 16 = 30 \qquad \| v \|^2 = 9 + 16 + 1 + 4 = 30 \qquad \| w \|^2 = 9 + 4 + 1 + 1 = 15$$

Then the following form the desired orthonormal set of vectors:

$$\hat{u} = (1/\sqrt{30}, \, 2/\sqrt{30}, \, -3/\sqrt{30}, \, 4/\sqrt{30})$$
$$\hat{v} = (3/\sqrt{30}, \, 4/\sqrt{30}, \, 1/\sqrt{30}, \, -2/\sqrt{30})$$
$$\hat{w} = (3/\sqrt{15}, \, -2/\sqrt{30}, \, 1/\sqrt{15}, \, 1/\sqrt{15})$$

We also have $u + v + w = (7, 4, -1, 3)$ and $\| u + v + w \|^2 = 49 + 16 + 1 + 9 = 75$. Thus

$$\| u \|^2 + \| v \|^2 + \| w \|^2 = 30 + 30 + 15 = 75 = \| u + v + w \|^2$$

which verifies the Pythagorean Theorem for the orthogonal set S.

Example 6.9.   Consider the vector $u = (1, 1, 1, 1)$ in $\mathbf{R}^4$. Suppose we want to find an orthogonal basis of $u^\perp$, the orthogonal complement of **u**. Note that $u^\perp$ is the solution space of the linear equation

$$x + y + z + t = 0 \qquad\qquad (1)$$

Find a nonzero solution v_1 of (1), say $v_1 = (0, 0, 1, -1)$. We want our second basis vector v_2 to be a solution to (1) and also orthogonal to v_1, i.e., to be a solution of the system

$$x + y + z + t = 0 \qquad z - t = 0 \qquad\qquad (2)$$

Find a nonzero solution v_2 of (2), say $v_2 = (0, 2, -1, -1)$. We want our third basis vector to be a solution of (1) and also orthogonal to v_1 and v_2, i.e., to be a solution of the system

$$x + y + z + t = 0 \qquad 2y - z - t = 0 \qquad z - 1 = 0 \qquad\qquad (3)$$

Find a nonzero solution of (3), say $v_3 = (-3, 1, 1, 1)$. Then $\{v_1, v_2, v_3\}$ is an orthogonal basis of $u^\perp$. (Observe that we chose the intermediate solutions v_1 and v_2 in such a way that each new system is already in echelon form. This makes the calculations simpler.) We can find an orthonormal basis for $u^\perp$ by normalizing the above orthogonal basis for $u^\perp$. We have

$$\|v_1\|^2 = 0 + 0 + 1 + 1 = 2 \qquad \|v_2\|^2 = 0 + 4 + 1 + 1 = 6 \qquad \|v_3\|^2 = 9 + 1 + 1 + 1 = 12$$

Thus the following is an orthonormal basis for $u^\perp$.

$$v_1 = (0, 0, 1/\sqrt{2}, -1/\sqrt{2}) \qquad v_2 = (0, 2/\sqrt{6}, -1/\sqrt{6}, -1/\sqrt{6}) \qquad v_3 = (-3/\sqrt{12}, 1/\sqrt{12}, 1/\sqrt{12}, 1/\sqrt{12})$$

Example 6.10.   Let S consist of the following three vectors in $\mathbf{R}^3$:

$$u_1 = (1, 2, 1) \qquad u_2 = (2, 1, -4) \qquad u_3 = (3, -2, 1)$$

Then S is orthogonal since $u_1, u_2,$ and u_3 are orthogonal to each other:

$$\langle u_1, u_2 \rangle = 2 + 2 - 4 = 0 \qquad \langle u_1, u_3 \rangle = 3 - 4 + 1 = 0 \qquad \langle u_2, u_3 \rangle = 6 - 2 - 4 = 0$$

Thus S is linearly independent and, since S has 3 elements, S is an orthogonal basis for $\mathbf{R}^3$.

Suppose we want to write $v = (4, 1, 18)$ as a linear combination of u_1, u_2, u_3. First set v as a linear combination of u_1, u_2, u_3 using unknowns x, y, z as follows:

$$(4, 1, 18) = x(1, 1, 1) + y(2, 1, -4) + z(3, -2, 1) \qquad\qquad (1)$$

Method 1.   Expand (1) to obtain

$$x + 2y + 3z = 4 \qquad 2x + y - 2z = 1 \qquad x - 4y + z = 18$$

from which $x = 4, y = -3, z = 2$. Thus $v = 4u_1 - 3u_2 + 2u_3$.

Method 2.   (This method uses the fact that the basis vectors are orthogonal, and the arithmetic is much simpler.) Take the inner product of (1) with u_1 to get

$$(4, 1, 18) \cdot (1, 2, 1) = x(1, 2, 1) \cdot (1, 2, 1) \qquad \text{or} \qquad 24 = 6x \qquad \text{or} \qquad x = 4$$

(The two last terms drop out since u_1 is orthogonal to u_2 and to u_3.) Take the inner product of (1) with u_2 to get

$$(4, 1, 18) \cdot (2, 1, -4) = y(2, 1, -4) \cdot (2, 1, -4) \qquad \text{or} \qquad -63 = 21y \qquad \text{or} \qquad y = -3$$

Finally, take the inner product of (1) with u_3 to get

$$(4, 1, 18) \cdot (3, -2, 1) = z(3, -2, 1) \cdot (3, -2, 1) \qquad \text{or} \qquad 28 = 14z \qquad \text{or} \qquad z = 2$$

Thus $v = 4u_1 - 3u_2 + 2u_3$.

The procedure in Method 2 in Example 6.10 is true in general; that is,

Theorem 6.7:   Suppose $\{u_1, u_2, \ldots, u_n\}$ is an orthogonal basis for V. Then, for any $v \in V$,

$$v = \frac{\langle v, u_1 \rangle}{\langle u_1, u_1 \rangle} u_1 + \frac{\langle v, u_2 \rangle}{\langle u_2, u_2 \rangle} u_2 + \cdots + \frac{\langle v, u_n \rangle}{\langle u_n, u_n \rangle} u_n$$

(See Problem 6.5 for the proof.)

Remark: The above scalar,

$$k_i \equiv \frac{\langle v, u_i \rangle}{\langle u_i, u_i \rangle} = \frac{\langle v, u_i \rangle}{\| u_i \|^2}$$

is called the *Fourier coefficient* of v with respect to u_i since it is analogous to a coefficient in the Fourier series of a function. This scalar also has a geometric interpretation which is discussed below.

Projections

Consider a nonzero vector w in an inner product space V. For any $v \in V$, we show (Problem 6.24) that

$$c = \frac{\langle v, w \rangle}{\langle w, w \rangle} = \frac{\langle v, w \rangle}{\| w \|^2}$$

is the unique scalar such that $v' = v - cw$ is orthogonal to w. The projection of v along w, as indicated by Fig. 6-5, is denoted and defined by

$$\text{proj}(v, w) = cw = \frac{\langle v, w \rangle}{\langle w, w \rangle} w = \frac{\langle v, w \rangle}{\| w \|^2} w$$

The scalar c is also called the Fourier coefficient of v with respect to w or the component of v along w.

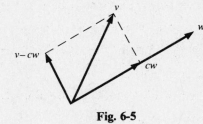

Fig. 6-5

Example 6.11

(a) We find the component c and the projection cw of $v = (1, 2, 3, 4)$ along $w = (1, -3, 4, -2)$ in $\mathbf{R}^4$. First we compute

$$\langle v, w \rangle = 1 - 6 + 12 - 8 = -1 \qquad \text{and} \qquad \| w \|^2 = 1 + 9 + 16 + 4 = 30$$

Then $c = -\frac{1}{30}$ and $\text{proj}(v, w) = cw = (-\frac{1}{30}, \frac{1}{10}, -\frac{2}{15}, \frac{1}{30})$.

(b) Let V be the vector space of polynomials with inner product $\langle f, g \rangle = \int_0^1 f(t) g(t)\, dt$. We find the component (Fourier coefficient) c and the projection cg of $f(t) = 2t - 1$ along $g(t) = t^2$. First we compute

$$\langle f, g \rangle = \int_0^1 (2t^3 - t^2)\, dt = \left[\frac{t^4}{2} - \frac{t^3}{3} \right]_0^1 = \frac{1}{6} \qquad \langle g, g \rangle = \int_0^1 t^4\, dt = \left[\frac{t^5}{5} \right]_0^1 = \frac{1}{5}$$

Then $c = \frac{5}{6}$ and $\text{proj}(f, g) = cg = 5t^2/6$.

The above notion may be generalized as follows.

Theorem 6.8: Suppose $w_1, w_2, \ldots, w_r$ form an orthogonal set of nonzero vectors in V. Let v be any vector in V. Define $v' = v - c_1 w_1 - c_2 w_2 - \cdots - c_r w_r$ where

$$c_1 = \frac{\langle v, w_1 \rangle}{\| w_1 \|^2}, \qquad c_2 = \frac{\langle v, w_2 \rangle}{\| w_2 \|^2}, \qquad \ldots, \qquad c_r = \frac{\langle v, w_r \rangle}{\| w_r \|^2}$$

Then v' is orthogonal to $w_1, w_2, \ldots, w_r$.

Note that the c_i in the above theorem are, respectively, the components (Fourier coefficients) of v along the w_i, Furthermore, the following theorem (proved in Problem 6.31) shows that $c_1 w_1 + \cdots + c_r w_r$ is the closest approximation to v as a linear combination of $w_1, \ldots, w_r$.

Theorem 6.9: Suppose $w_1, w_2, \ldots, w_r$ form an orthogonal set of nonzero vectors in V. Let V be any vector in V and let c_i be the component of v along w_i. Then, for any scalars $a_1, \ldots, a_r$,

$$\left\| v - \sum_{k=1}^{r} c_k w_k \right\| \leq \left\| v - \sum_{k=1}^{r} a_k w_k \right\|$$

The next theorem (proved in Problem 6.32) is known as the *Bessel inequality*.

Theorem 6.10: Suppose $\{e_1, e_2, \ldots, e_r\}$ is an orthonormal set of vectors in V. Let v be any vector in V and let c_i be the Fourier coefficient of v with respect to u_i. Then

$$\sum_{k=1}^{r} c_k^2 \leq \| v \|^2$$

Remark: The notion of projection includes that of a vector along a subspace as follows. Suppose W is a subspace of V and $v \in V$. By Theorem 6.4, $V = W \oplus W^\perp$; hence v can be expressed uniquely in the form

$$v = w + w' \qquad \text{where} \quad w \in W, \quad w' \in W^\perp.$$

We call w the *projection of v along W* and denote it by proj (v, W). (See Fig. 6-6.) In particular, if $W = \text{span}(w_1, \ldots, w_r)$ where the w_i forms an orthogonal set, then

$$\text{proj}(v, W) = c_1 w_1 + c_2 w_2 + \cdots + c_r w_r$$

where c_i is the component of v along w_i, as above.

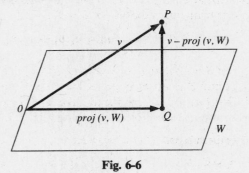

Fig. 6-6

6.6 GRAM–SCHMIDT ORTHOGONALIZATION PROCESS

Suppose $\{v_1, v_2, \ldots, v_n\}$ is a basis for an inner product space V. An orthogonal basis $\{w_1, w_2, \ldots, w_n\}$ for V as follows. Set

$$w_1 = v_1$$

$$w_2 = v_2 - \frac{\langle v_2, w_1 \rangle}{\| w_1 \|^2} w_1$$

$$w_3 = v_3 - \frac{\langle v_3, w_1 \rangle}{\| w_1 \|^2} w_1 - \frac{\langle v_3, w_2 \rangle}{\| w_2 \|^2} w_2$$

$$\cdots \cdots \cdots \cdots \cdots \cdots \cdots \cdots \cdots \cdots \cdots \cdots$$

$$w_n = v_n - \frac{\langle v_n, w_1 \rangle}{\| w_1 \|^2} w_1 - \frac{\langle v_n, w_2 \rangle}{\| w_2 \|^2} w_2 - \cdots - \frac{\langle v_n, w_n \rangle}{\| w_n \|^2} w_n$$

In other words, for $k = 2, 3, \ldots, n$, we define

$$w_k = v_k - c_{k1}w_1 - c_{k2}w_2 - \cdots - c_{k,k-1}w_{k-1}$$

where $c_{ki} = \langle v_k, w_i \rangle / \| w_i \|^2$ is the component of v_k along w_i. By Theorem 6.8, each w_k is orthogonal to the preceding w's. Thus $w_1, w_2, \ldots, w_n$ form an orthogonal basis for V as claimed. Normalizing each w_k will then yield an orthonormal basis for V.

The above construction is known as the *Gram–Schmidt orthogonalization process.* The following remarks are in order.

Remark 1: Each vector w_k is a linear combination of v_k and preceding w's; hence by induction each w_k is a linear combination of $v_1, v_2, \ldots, v_k$.

Remark 2: Suppose $w_1, w_2, \ldots, w_r$ are linearly independent. Then they form a basis for $U = \operatorname{span} w_i$. Applying the Gram–Schmidt orthogonalization process to the w_i yields an orthogonal basis for U.

Remark 3: In hand calculations, it may be simpler to clear the fraction in any new w_k by multiplying w_k by an appropriate scalar as this does not affect the orthogonality (Problem 6.76).

The following theorems, proved in Problems 6.32 and 6.33, respectively, use the above algorithm and remarks.

Theorem 6.11: Let $\{v_1, v_2, \ldots, v_n\}$ be any basis of an inner product space V. Then there exists an orthonormal basis $\{u_1, u_2, \ldots, u_n\}$ of V such that the change-of-basis matrix from $\{v_i\}$ to $\{u_i\}$ is triangular; that is, for $k = 1, \ldots, n$,

$$u_k = a_{k1}v_1 + a_{k2}v_2 + \cdots + a_{kk}v_k$$

Theorem 6.12: Suppose $S = \{w_1, w_2, \ldots, w_r\}$ is an orthogonal basis for a subspace W of V. Then one may extend S to an orthogonal basis for V, that is, one may find vectors $w_{r+1}, \ldots, w_n$, such that $\{w_1, w_2, \ldots, w_n\}$ is an orthogonal basis for V.

Example 6.12.   Consider the subspace U of $\mathbf{R}^4$ spanned by

$$v_1 = (1, 1, 1, 1) \qquad v_2 = (1, 2, 4, 5) \qquad v_3 = (1, -3, -4, -2)$$

We find an orthonormal basis for U by first finding an orthogonal basis of U using the Gram–Schmidt algorithm. First set $w_1 = u_1 = (1, 1, 1, 1)$. Next find

$$v_2 - \frac{\langle v_2, w_1 \rangle}{\| w_1 \|^2} w_1 = (1, 2, 4, 5) - \frac{12}{4}(1, 1, 1, 1) = (-2, -1, 1, 2)$$

Set $w_2 = (-2, -1, 1, 2)$. Then find

$$v_3 - \frac{\langle v_3, w_1 \rangle}{\| w_1 \|^2} w_1 - \frac{\langle v_3, w_2 \rangle}{\| w_2 \|^2} w_2 = (1, -3, -4, -2) - \frac{-8}{4}(1, 1, 1, 1) - \frac{-7}{10}(-2, -1, 1, 2)$$

$$= \left(\frac{8}{5}, -\frac{17}{10}, -\frac{13}{10}, \frac{7}{5} \right)$$

Clear fractions to obtain $w_3 = (16, -17, -13, 14)$. (In hand calculations, it is usually simpler to clear fractions as this does not affect the orthogonality.) Last, normalize the orthogonal basis

$$w_1 = (1, 1, 1, 1) \qquad w_2 = (-2, -1, 1, 2) \qquad w_3 = (16, -17, -13, 14)$$

Since $\| w_1 \|^2 = 4$, $\| w_2 \|^2 = 10$, $\| w_3 \|^2 = 910$, the following vectors form an orthonormal basis of U:

$$u_1 = \frac{1}{2}(1, 1, 1, 1) \qquad u_2 = \frac{1}{\sqrt{10}}(-2, -1, 1, 2) \qquad u_3 = \frac{1}{\sqrt{910}}(16, -17, -13, 14)$$

Example 6.13.   Let V be the vector space of polynomials $f(t)$ with inner product $\langle f, g \rangle = \int_{-1}^{1} f(t) g(t) \, dt$. We apply the Gram–Schmidt algorithm to the set $\{1, t, t^2, t^3\}$ to obtain an orthogonal basis $\{f_0, f_1, f_2, f_3\}$ with integer coefficients for the subspace U of polynomials of degree ≤ 3. Here we use the fact that if $r + s = n$ then

$$\langle t^r, t^s \rangle = \int_{-1}^{1} t^n \, dt = \left[\frac{t^{n+1}}{n+1} \right]_{-1}^{1} = \begin{cases} 2/(n+1) & \text{if } n \text{ is even} \\ 0 & \text{if } n \text{ is odd} \end{cases},$$

First set $f_0 = 1$. Then find

$$t - \frac{\langle t, 1 \rangle}{\langle 1, 1 \rangle} \cdot 1 = t - \frac{0}{2} \cdot 1 = t$$

Let $f_1 = t$. Then find

$$t^2 - \frac{\langle t^2, 1 \rangle}{\langle 1, 1 \rangle} \cdot 1 - \frac{\langle t^2, t \rangle}{\langle t, t \rangle} t = t^2 - \frac{\frac{2}{3}}{2} \cdot 1 - \frac{0}{\frac{2}{3}} t = t^2 - \frac{1}{3}$$

Multiply by 3 to obtain $f_2 = 3t^2 - 1$. Next find

$$t^3 - \frac{\langle t^3, 1 \rangle}{\langle 1, 1 \rangle} \cdot 1 - \frac{\langle t^3, t \rangle}{\langle t, t \rangle} t - \frac{\langle t^3, 3t^2 - 1 \rangle}{\langle 3t^2 - 1, 3t^2 - 1 \rangle} (3t^2 - 1) = t^3 - 0.1 - \frac{\frac{2}{5}}{\frac{2}{3}} t - 0(3t^2 - 1) = t^3 - \frac{3}{5} t$$

Multiply by 5 to obtain $f_3 = 5t^3 - 3t$. That is, $\{1, t, 3t^2 - 1, 5t^3 - 3t\}$ is the required orthogonal basis for U.

> **Remark:**   Normalizing the polynomials in Example 6.13 so that $p(1) = 1$ for each polynomial $p(t)$ yields the polynomials
>
> $$1, \quad t, \quad \tfrac{1}{2}(3t^2 - 1), \quad \tfrac{1}{2}(5t^3 - 3t)$$
>
> These are the first four *Legendre* polynomials (which are important in the study of differential equations).

6.7   INNER PRODUCTS AND MATRICES

This section investigates two types of matrices which play a special role in the theory of real inner product spaces V: positive definite matrices and orthogonal matrices. In this context, vectors in $\mathbf{R}^n$ will be represented by column vectors. (Thus $\langle u, v \rangle = u^T v$ denotes the usual inner product in $\mathbf{R}^n$.)

Positive Definite Matrices

Let A be a real symmetric matrix. Recall (Section 4.11) that A is congruent to a diagonal matrix B, i.e., that there exists a nonsingular matrix P such that $B = P^T A P$ is diagonal, and that the number of positive entries in B is an invariant of A (Theorem 4.18, Law of Inertia). The matrix A is said to be positive definite if all the diagonal entries of B are positive. Alternatively, A is said to be positive definite if $X^T A X > 0$ for every nonzero vector X in $\mathbf{R}^n$.

Example 6.14.   Let $A = \begin{pmatrix} 1 & 0 & -1 \\ 0 & 1 & -2 \\ -1 & -2 & 8 \end{pmatrix}$. We reduce A to a (congruent) diagonal matrix (see Section 4.11) by applying $R_1 + R_3 \to R_3$ and the corresponding column operation $C_1 + C_3 \to C_3$, and then applying $2R_2 + R_3 \to R_3$ and $2C_2 + C_3 \to C_3$:

$$A \sim \begin{pmatrix} 1 & 0 & 0 \\ 0 & 1 & -2 \\ 0 & -2 & 7 \end{pmatrix} \sim \begin{pmatrix} 1 & 0 & 0 \\ 0 & 1 & 0 \\ 0 & 0 & 3 \end{pmatrix}$$

Since the diagonal matrix has only positive diagonal entries A is a positive definite matrix.

Remark: A 2×2 symmetric matrix $A = \begin{pmatrix} a & b \\ c & d \end{pmatrix} = \begin{pmatrix} a & b \\ b & d \end{pmatrix}$ is positive definite if and only if the diagonal entries a and d are positive and the determinant $\det(A) = ad - bc = ad - b^2$ is positive.

The following theorem, proved in Problem 6.40, applies.

Theorem 6.13: Let A be a real positive definite matrix. Then the function $\langle u, v \rangle = u^T A v$ is an inner product on $\mathbf{R}^n$.

Theorem 6.13 says that every positive definite matrix A determines an inner product. The following discussion and Theorem 6.15 may be viewed as the converse of this result.

Let V be any inner product space and let $S = \{u_1, u_2, \ldots, u_n\}$ be any basis for V. The following matrix A is called the matrix representation of the inner product on V relative to the basis S:

$$A = \begin{pmatrix} \langle u_1, u_1 \rangle & \langle u_1, u_2 \rangle & \cdots & \langle u_1, u_n \rangle \\ \langle u_2, u_1 \rangle & \langle u_2, u_2 \rangle & \cdots & \langle u_2, u_n \rangle \\ \cdots & \cdots & \cdots & \cdots \\ \langle u_n, u_1 \rangle & \langle u_n, u_2 \rangle & \cdots & \langle u_n, u_n \rangle \end{pmatrix}$$

That is, $A = (a_{ij})$ where $a_{ij} = \langle u_i, u_j \rangle$.

Observe that A is symmetric since the inner product is symmetric, that is, $\langle u_i, u_j \rangle = \langle u_j, u_i \rangle$. Also, A depends on both the inner product on V and the basis S for V. Moreover, if S is an orthogonal basis then A is diagonal, and if S is an orthonormal matrix then A is the identity matrix.

Example 6.15. The following three vectors form a basis S for Euclidean space $\mathbf{R}^3$:

$$u_1 = (1, 1, 0) \qquad u_2 = (1, 2, 3) \qquad u_3 = (1, 3, 5)$$

Computing each $\langle u_i, u_j \rangle = \langle u_j, u_i \rangle$ yields:

$$\langle u_1, u_1 \rangle = 1 + 1 + 0 = 2 \qquad \langle u_1, u_2 \rangle = 1 + 2 + 0 = 3 \qquad \langle u_1, u_3 \rangle = 1 + 3 + 0 = 4$$

$$\langle u_2, u_2 \rangle = 1 + 4 + 9 = 14 \qquad \langle u_2, u_3 \rangle = 1 + 6 + 15 = 22 \qquad \langle u_3, u_3 \rangle = 1 + 9 + 25 = 35$$

Thus
$$A = \begin{pmatrix} 2 & 3 & 4 \\ 3 & 14 & 22 \\ 4 & 22 & 35 \end{pmatrix}$$

is the matrix representation of the usual inner product on $\mathbf{R}^3$ relative to the basis S.

The following theorems, proved in Problems 6.41 and 6.42, respectively, apply.

Theorem 6.14: Let A be the matrix representation of an inner product relative to a basis S for V. Then, for any vectors $u, v \in V$, we have

$$\langle u, v \rangle = [u]^T A[v]$$

where $[u]$ and $[v]$ denote the (column) coordinate vectors relative to the basis S.

Theorem 6.15: Let A be the matrix representation of any inner product on V. Then A is a positive definite matrix.

Orthogonal Matrices

Recall (Section 4.6) that a real matrix P is orthogonal if P is nonsingular and $P^{-1} = P^T$, that is, if $PP^T = P^TP = I$. This subsection further investigates these matrices. First we recall (Theorem 4.5) an important characterization of such matrices.

Theorem 6.16: Let P be a real matrix. Then the following three properties are equivalent:

(i) P is orthogonal, that is, $P^T = P^{-1}$.

(ii) The rows of P form an orthonormal set of vectors.

(iii) The columns of P form an orthonormal set of vectors.

(The above theorem is true only using the usual inner product on $\mathbf{R}^n$. It is not true if $\mathbf{R}^n$ is given any other inner product.)

Remark: Every 2×2 orthogonal matrix has the form $\begin{pmatrix} \cos\theta & \sin\theta \\ -\sin\theta & \cos\theta \end{pmatrix}$ or

$\begin{pmatrix} \cos\theta & \sin\theta \\ \sin\theta & -\cos\theta \end{pmatrix}$ for some real number θ (Theorem 4.6).

Example 6.16

Let $P = \begin{pmatrix} 1/\sqrt{3} & 1/\sqrt{3} & 1/\sqrt{3} \\ 0 & 1/\sqrt{2} & -1/\sqrt{2} \\ 2/\sqrt{6} & -1/\sqrt{6} & -1/\sqrt{6} \end{pmatrix}$. The rows are orthogonal to each other and are unit vectors, that is, the rows

form an orthonormal set of vectors. Thus P is orthogonal.

The following two theorems, proved in Problems 6.48 and 6.49, respectively, show important relationships between orthogonal matrices and orthonormal bases of an inner product space V.

Theorem 6.17: Suppose $E = \{e_i\}$ and $E' = \{e_i'\}$ are orthonormal bases of V. Let P be the change-of-basis matrix from the basis E to the basis E'. Then P is orthogonal.

Theorem 6.18: Let $\{e_1, \ldots, e_n\}$ be an orthonormal basis of an inner product space V. Let $P = (a_{ij})$ be an orthogonal matrix. Then the following n vectors form an orthonormal basis for V:

$$e_i' = a_{1i}e_1 + a_{2i}e_2 + \cdots + a_{ni}e_n \qquad (i = 1, 2, \ldots, n)$$

6.8 COMPLEX INNER PRODUCT SPACES

This section considers vector spaces V over the complex field $\mathbf{C}$. First we recall some properties of complex numbers (Section 2.9). Suppose $z \in \mathbf{C}$, say, $z = a + bi$ where $a, b \in \mathbf{R}$. Then

$$\bar{z} = a - bi \qquad z\bar{z} = a^2 + b^2 \qquad \text{and} \qquad |z| = \sqrt{a^2 + b^2}$$

Also, for any $z, z_1, z_2 \in \mathbf{C}$,

$$\overline{z_1 + z_2} = \overline{z_1} + \overline{z_2} \qquad \overline{z_1 z_2} = \overline{z_1} \cdot \overline{z_2} \qquad \bar{\bar{z}} = z$$

and z is real if and only if $\bar{z} = z$.

Definition: Let V be a vector space over $\mathbf{C}$. Suppose to each pair of vectors $u, v \in V$ there is assigned a complex number, denoted by $\langle u, v \rangle$. This function is called a *(complex) inner product* on V if it satisfies the following axioms:

[I_1^*] (Linear Property) $\langle au_1 + bu_2, v \rangle = a\langle u_1, v \rangle + b\langle u_2, v \rangle$

[I_2^*] (Conjugate Symmetric Property) $\langle u, v \rangle = \overline{\langle v, u \rangle}$

[I_3^*] (Positive Definite Property) $\langle u, u \rangle \geq 0$; and $\langle u, u \rangle = 0$ if and only if $u = 0$.

The vector space V over $\mathbf{C}$ with an inner product is called a *(complex) inner product space*.

Observe that a complex inner product differs only slightly from a real inner product space (only [I_2^*] differs from [I_2]). In fact, many of the definitions and properties of a complex inner product space are the same as that of a real inner product space. However, some of the proofs must be adapted to the complex case.

Axiom [I_1^*] is also equivalent to the following two conditions:

> (*a*) $\langle u_1 + u_2, v \rangle = \langle u_1, v \rangle + \langle u_2, v \rangle$ and   (*b*) $\langle ku, v \rangle = k\langle u, v \rangle$

On the other hand,

$$\langle u, kv \rangle = \overline{\langle kv, u \rangle} = \overline{k\langle v, u \rangle} = \bar{k}\overline{\langle v, u \rangle} = \bar{k}\overline{\langle u, v \rangle} = \bar{k}\langle u, v \rangle$$

(In other words, we must take the conjugate of a complex scalar when it is taken out of the second position of the inner product.) In fact, we show (Problem 6.50) that the inner product is *conjugate linear* in the second position, that is,

$$\langle u, av_1 + bv_2 \rangle = \bar{a}\langle u, v_1 \rangle = \bar{b}\langle u, v_2 \rangle$$

One can analogously prove (Problem 6.96)

$$\langle a_1u_1 + a_2u_2, b_1v_1 + b_2v_2 \rangle = a_1\bar{b}_1\langle u_1, v_1 \rangle + a_1\bar{b}_2\langle u_1, v_2 \rangle + a_2\bar{b}_1\langle u_2, v_1 \rangle + a_2\bar{b}_2\langle u_2, v_2 \rangle$$

and, by induction,

$$\left\langle \sum_{i=1}^{m} a_iu_i, \sum_{j=1}^{n} b_jv_j \right\rangle = \sum_{i,j} a_i\bar{b}_j\langle u_i, v_j \rangle$$

The following similar remarks are in order:

> **Remark 1:** Axiom [I_1^*] by itself implies that $\langle 0, 0 \rangle = \langle 0v, 0 \rangle = 0\langle v, 0 \rangle = 0$. Accordingly, [$I_1^*$], [$I_2^*$], and [$I_3^*$] are equivalent to [$I_1^*$], [$I_2^*$] and the following axiom:
>
> [$I_3^{*\prime}$] If $u \neq 0$,   then $\langle u, u \rangle > 0$
>
> That is, a function satisfying [I_1^*], [I_2^*], and [$I_3^{*\prime}$] is a (complex) inner product on V.

> **Remark 2:** By [I_2^*], $\langle u, u \rangle = \overline{\langle u, u \rangle}$. Thus $\langle u, u \rangle$ must be real. By [I_3^*], $\langle u, u \rangle$ must be nonnegative, and hence its positive real square root exists. As with real inner product spaces, we define $\| u \| = \sqrt{\langle u, u \rangle}$ to be the norm or length of u.

> **Remark 3:** Besides the norm, we define the notions of orthogonality, orthogonal complement, orthogonal and orthonormal sets as before. In fact, the definitions of distance and Fourier coefficient and projection are the same as with the real case.

Example 6.17. Let $u = (z_i)$ and $v = (w_i)$ be vectors in $\mathbb{C}^n$. Then

$$\langle u, v \rangle = \sum_{k=1}^{n} z_k \bar{w}_k = z_1 \bar{w}_1 + z_2 \bar{w}_2 + \cdots + z_n \bar{w}_n$$

is an inner product on $\mathbb{C}^n$ called the usual or standard inner product on $\mathbb{C}^n$. (We assume this inner product on $\mathbb{C}^n$ unless otherwise stated or implied.) In the case that u and v are real, we have $\bar{w}_i = w_i$ and

$$\langle u, v \rangle = z_1 \bar{w}_1 + z_2 \bar{w}_2 + \cdots + z_n \bar{w}_n = z_1 w_1 + z_2 w_2 + \cdots + z_n w_n$$

In other words, this inner product reduces to the analogous one on $\mathbb{R}^n$ when the entries are real.

> **Remark:** Assuming u and v are column vectors, then the above inner product may be defined by $\langle u, v \rangle = u^T \bar{v}$.

Example 6.18

(a) Let V be the vector space of complex continuous functions on the (real) interval $a \le t \le b$. Then the following is the usual inner product on V:

$$\langle f, g \rangle = \int_a^b f(t) \overline{g(t)} \, dt$$

(b) Let U be the vector space of $m \times n$ matrices over $\mathbb{C}$. Suppose $A = (z_{ij})$ and $B = (w_{ij})$ are elements of U. Then the following is the usual inner product on U:

$$\langle A, B \rangle = \operatorname{tr} B^H A = \sum_{i=1}^{m} \sum_{j=1}^{n} \bar{w}_{ij} z_{ij}$$

As usual, $B^H = \bar{B}^T$, that is, B^H is the conjugate transpose of B.

The following is a list of theorems for complex inner product spaces which are analogous to the ones for the real case (Theorem 6.19 is proved in Problem 6.53).

Theorem 6.19 (Cauchy–Schwarz): Let V be a complex inner product space. Then

$$|\langle u, v \rangle| \le \|u\| \, \|v\|$$

Theorem 6.20: Let W be a subspace of a complex inner product space V. Then $V = W \oplus W$.

Theorem 6.21: Suppose $\{u_1, u_2, \ldots, u_n\}$ is an orthogonal basis for a complex vector space V. Then, for any $v \in V$,

$$v = \frac{\langle v, u_1 \rangle}{\|u_1\|^2} u_1 + \frac{\langle v, u_2 \rangle}{\|u_2\|^2} u_2 + \cdots + \frac{\langle v, u_n \rangle}{\|u_n\|^2} u_n$$

Theorem 6.22: Suppose $\{u_1, u_2, \ldots, u_n\}$ is a basis for a complex inner product space V. Let $A = (a_{ij})$ be the complex matrix defined by $a_{ij} = \langle u_i, u_j \rangle$. Then, for any $u, v \in V$,

$$\langle u, v \rangle = [u]^T A \overline{[v]}$$

where $[u]$ and $[v]$ are the coordinate column vectors in the given basis $\{u_i\}$. (*Remark:* This matrix A is said to represent the inner product on V.)

Theorem 6.23: Let A be a Hermitian matrix (i.e., $A^H = \bar{A}^T = A$) such that $X^T A \bar{X}$ is real and positive for every nonzero vector $X \in \mathbb{C}^n$. Then $\langle u, v \rangle = u^T A \bar{v}$ is an inner product on $\mathbb{C}^n$.

Theorem 6.24: Let A be the matrix which represents an inner product on V. Then A is Hermitian, and $X^T A X$ is real and positive for any nonzero vector in $\mathbb{C}^n$.

6.9 NORMED VECTOR SPACES

We begin with a definition.

Definition: Let V be a real or complex vector space. Suppose to each $v \in V$ there is assigned a real number, denoted by $\| v \|$. This function $\| \cdot \|$ is called a *norm* on V if it satisfies the following axioms:

[N_1] $\| v \| \geq 0$; and $\| v \| = 0$ if and only if $v = 0$.

[N_2] $\| kv \| = |k| \, \| v \|$.

[N_3] $\| u + v \| \leq \| u \| + \| v \|$.

The vector space V with a norm is called a normed vector space.

The following remarks are in order.

      Remark 1: Axiom [N_2] by itself implies that $\| 0 \| = \| 0v \| = 0 \| v \| = 0$. Accordingly, [$N_1$], [$N_2$], and [$N_3$] are equivalent to [$N_2$], [$N_3$], and the following axiom:

$$[N_1'] \quad \text{If } v \neq 0, \quad \text{then} \quad \| v \| > 0$$

That is, a function $\| \cdot \|$ satisfying [N_1'], [N_2], and [N_3] is a norm on a vector space V.

      Remark 2: Suppose V is an inner product space. The norm on V defined by $\| v \| = \sqrt{\langle v, v \rangle}$ does satisify [N_1], [N_2], and [N_3]. Thus every inner product space V is a normed vector space. On the other hand, there may be norms on a vector space V which do not come from an inner product on V.

      Remark 3: Let V be a normed vector space. The distance between vectors u, $v \in V$ is denoted and defined by $d(u, v) = \| u - v \|$.

The following theorem is the main reason why $d(u, v)$ is called the distance between u and v.

Theorem 6.25: Let V be a normed vector space. Then the function $d(u, v) = \| u - v \|$ satisfies the following three axioms of a metric space:

[M_1] $d(u, v) \geq 0$; and $d(u, v) = 0$ if and only if $u = v$.

[M_2] $d(u, v) = d(v, u)$.

[M_3] $d(u, v) \leq d(u, w) + d(w, v)$.

Norms on $\mathbf{R}^n$ and $\mathbf{C}^n$

The following define three important norms on $\mathbf{R}^n$ and $\mathbf{C}^n$:

$$\| (a_1, \ldots, a_n) \|_\infty = \max \, (|a_i|)$$

$$\| (a_1, \ldots, a_n) \|_1 = |a_1| + |a_2| + \cdots + |a_n|$$

$$\| (a_1, \ldots, a_n) \|_2 = \sqrt{|a_1|^2 + |a_2|^2 + \cdots + |a_n|^2}$$

(Note that subscripts are used to distinguish between the three norms.) The norms $\| \cdot \|_\infty$, $\| \cdot \|_1$, and $\| \cdot \|_2$ are called the *infinity-norm*, *one-norm*, and *two-norm*, respectively. Observe that $\| \cdot \|_2$ is the norm on $\mathbf{R}^n(\mathbf{C}^n)$ induced by the usual inner product on $\mathbf{R}^n(\mathbf{C}^n)$. (We will let d_∞, d_1, and d_2 denote the corresponding distance functions.)

Example 6.19. Consider the vectors $u = (1, -5, 3)$ and $v = (4, 2, -3)$ in $\mathbf{R}^3$.

(a) The infinity-norm chooses the maximum of the absolute values of the components. Hence

$$\|u\|_\infty = 5 \quad\text{and}\quad \|v\|_\infty = 4$$

(b) The one-norm adds the absolute values of the components. Thus

$$\|u\|_1 = 1 + 5 + 3 = 9 \quad\text{and}\quad \|v\|_1 = 4 + 2 + 3 = 9$$

(c) The two-norm is equal to the square root of the sum of the square of the components (i.e., the norm induced by the usual inner product on $\mathbf{R}^3$). Thus

$$\|u\|_2 = \sqrt{1 + 25 + 9} = \sqrt{35} \quad\text{and}\quad \|v\|_2 = \sqrt{16 + 4 + 9} = \sqrt{29}$$

(d) Since $u - v = (1 - 4, -5 - 2, 3 + 3) = (-3, -7, 6)$, we have

$$d_\infty(u, v) = 7 \qquad d_1(u, v) = 3 + 7 + 6 = 16 \qquad d_2(u, v) = \sqrt{9 + 49 + 36} = \sqrt{94}$$

Example 6.20. Consider the cartesian plane $\mathbf{R}^2$ shown in Fig. 6-7.

(a) Let D_1 be the set of points $u = (x, y)$ in $\mathbf{R}^2$ such that $\|u\|_2 = 1$. Then D_1 consists of the points (x, y) such that $\|u\|_2^2 = x^2 + y^2 = 1$. Thus D_1 is the unit circle as shown in Fig. 6-7.

(b) Let D_2 be the set of points $u = (x, y)$ in $\mathbf{R}^2$ such that $\|u\|_1 = 1$. Then D_2 consists of the points (x, y) such that $\|u\|_1 = |x| + |y| = 1$. Thus D_2 is the diamond inside the unit circle as shown in Fig. 6-7.

(c) Let D_3 be the set of points $u = (x, y)$ in $\mathbf{R}^2$ such that $\|u\|_\infty = 1$. Then D_3 consists of the points (x, y) such that $\|u\|_\infty = \max(|x|, |y|) = 1$. Thus D_3 is the square circumscribing the unit circle as shown in Fig. 6-7.

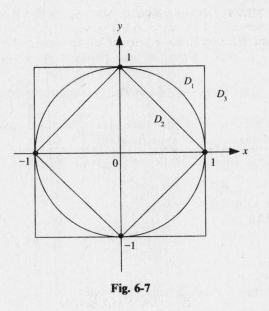

Fig. 6-7

Norms on $C[a, b]$

Consider the vector space $V = C[a, b]$ of continuous functions on the interval $a \le t \le b$. Recall that the following defines an inner product on V:

$$\langle f, g \rangle = \int_a^b f(t) g(t) \, dt$$

Accordingly, the above inner product defines the following norm on $V = C[a, b]$ (which is analogous to the $\| \cdot \|_2$ norm on $\mathbf{R}^n$):

$$\| f \|_2 = \int_a^b [f(t)]^2 \, dt$$

The following example defines two other norms on $V = C[a, b]$.

Example 6.21

(a) Let $\| f \|_1 = \int_a^b |f(t)| \, dt$. (This norm is analogous to the $\| \cdot \|_1$ norm on $\mathbf{R}^n$.) There is a geometrical description of $\| f \|_1$ and the distance $d_1(f, g)$. As shown in Fig. 6-8, $\| f \|_1$ is the area between the function $|f|$ and the t axis, and $d_1(f, g)$ is the area between the functions f and g.

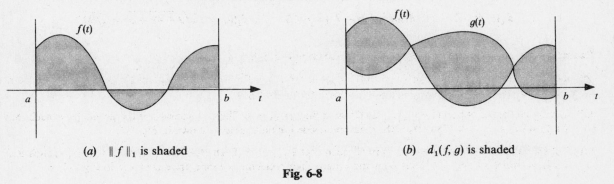

(a) $\| f \|_1$ is shaded   (b) $d_1(f, g)$ is shaded

Fig. 6-8

(b) Let $\| f \|_\infty = \max (|f(t)|)$. (This norm is analogous to the $\| \cdot \|_\infty$ on $\mathbf{R}^n$.) There is a geometrical description of $\| f \|_\infty$ and the distance function $d_\infty(f, g)$. As shown in Fig. 6-9, $\| f \|_\infty$ is the maximum distance between f and the t axis, and $d_\infty(f, g)$ is the maximum distance between f and g.

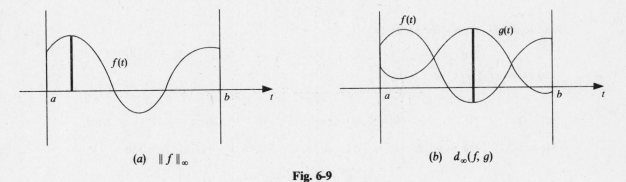

(a) $\| f \|_\infty$ (b) $d_\infty(f, g)$

Fig. 6-9

Solved Problems

INNER PRODUCTS

6.1.   Expand $\langle 5u_1 + 8u_2, 6v_1 - 7v_2 \rangle$.

Use the linearity in both positions to get

$$\langle 5u_1 + 8u_2, 6v_1 - 7v_2 \rangle = \langle 5u_1, 6v_1 \rangle + \langle 5u_1, -7v_2 \rangle + \langle 8u_2, 6v_1 \rangle + \langle 8u_2, -7v_2 \rangle$$
$$= 30\langle u_1, v_1 \rangle - 35\langle u_1, v_2 \rangle + 48\langle u_2, v_1 \rangle - 56\langle u_2, v_2 \rangle$$

[*Remark:* Observe the similarity between the above expansion and the expansion of $(5a + 8b)(6c - 7d)$ in ordinary algebra.]

6.2. Consider the following vectors in $\mathbf{R}^3$: $u = (1, 2, 4)$, $v = (2, -3, 5)$, $w = (4, 2, -3)$. Find: (a) $u \cdot v$, (b) $u \cdot w$, (c) $v \cdot w$, (d) $(u + v) \cdot W$, (e) $\| u \|$, (f) $\| v \|$, (g) $\| u + v \|$.

   (a) Multiply corresponding components and add to get $u \cdot v = 2 - 6 + 20 = 16$.

   (b) $u \cdot w = 4 + 4 - 12 = -4$.

   (c) $v \cdot w = 8 - 6 - 15 = -13$.

   (d) First find $u + v = (3, -1, 9)$. Then $(u + v) \cdot w = 12 - 2 - 27 = -17$. Alternatively, using $[I_1]$, $(u + v) \cdot w = u \cdot w + v \cdot w = -4 - 13 = -17$.

   (e) First find $\| u \|^2$ by squaring the components of u and adding:

$$\| u \|^2 = 1^2 + 2^2 + 4^2 = 1 + 4 + 16 = 21 \qquad \text{and so} \qquad \| u \| = \sqrt{21}$$

   (f) $\| v \|^2 = 4 + 9 + 25 = 38$ and so $\| v \| = \sqrt{38}$.

   (g) From (d), $u + v = (3, -1, 9)$. Hence $\| u + v \|^2 = 9 + 1 + 81 = 91$ and $\| u + v \| = \sqrt{91}$.

6.3. Verify that the following is an inner product in $\mathbf{R}^2$:

$$\langle u, v \rangle = x_1 y_1 - x_1 y_2 - x_2 y_1 + 3x_2 y_2 \qquad \text{where} \qquad u = (x_1, x_2), v = (y_1, y_2)$$

Method 1.   We verify the three axioms of an inner product. Letting $w = (z_1, z_2)$, we find

$$au + bw = a(x_1, x_2) + b(z_1, z_2) = (ax_1 + bz_1, ax_2 + bz_2)$$

Thus

$$\begin{aligned}
\langle au + bw, v \rangle &= \langle (ax_1, ax_2 + bz_2), (y_1, y_2) \rangle \\
&= (ax_1 + bz_1)y_1 - (ax_1 + bz_1)y_2 - (ax_2 + bz_2)y_1 + 3(ax_2 + bz_2)y_2 \\
&= a(x_1 y_1 - x_1 y_2 - x_2 y_1 + 3x_2 y_2) + b(z_1 y_1 - z_1 y_2 - z_2 y_1 + 3z_2 y_2) \\
&= a\langle u, v \rangle + b\langle w, v \rangle
\end{aligned}$$

and so axiom $[I_1]$ is satisfied. Also,

$$\langle v, u \rangle = y_1 x_1 - y_1 x_2 - y_2 x_1 + 3y_2 x_2 = x_1 y_1 - x_1 y_2 - x_2 y_1 + 3x_2 y_2 = \langle u, v \rangle$$

and axiom $[I_2]$ is satisfied. Finally,

$$\langle u, u \rangle = x_1^2 - 2x_1 x_2 + 3x_2^2 = x_1^2 - 2x_1 x_2 + x_2^2 + 2x_2^2 = (x_1 - x_2)^2 + 2x_2^2 \geq 0$$

Also, $\langle u, u \rangle = 0$ if and only if $x_1 = 0$, $x_2 = 0$, i.e., $u = 0$. Hence the last axiom $[I_3]$ is satisfied.

Method 2.   We argue via matrices. That is, we can write $\langle u, v \rangle$ in matrix notation:

$$\langle u, v \rangle = u^t A v = (x_1, x_2) \begin{pmatrix} 1 & -1 \\ -1 & 3 \end{pmatrix} \begin{pmatrix} y_1 \\ y_2 \end{pmatrix}$$

Since A is real and symmetric, we need only show that A is positive definite. Applying the elementary row operation $R_1 + R_2 \rightarrow R_2$ and then the corresponding elementary column operation $C_1 + C_2 \rightarrow C_2$, we transform A into diagonal form $\begin{pmatrix} 1 & 0 \\ 0 & 2 \end{pmatrix}$. Thus A is positive definite. Accordingly, $\langle u, v \rangle$ is an inner product.

6.4. Consider the vectors $u = (1, 5)$ and $v = (3, 4)$ in $\mathbf{R}^2$. Find:

   (a) $\langle u, v \rangle$ with respect to the usual inner product in $\mathbf{R}^2$,

   (b) $\langle u, v \rangle$ with respect to the inner product in $\mathbf{R}^2$ in Problem 6.3,

   (c) $\| v \|$ using the usual inner product in $\mathbf{R}^2$,

   (d) $\| v \|$ using the inner product in $\mathbf{R}^2$ in Problem 6.3.

(a) $\langle u, v \rangle = 3 + 20 = 23$.

(b) $\langle u, v \rangle = 1 \cdot 3 - 1 \cdot 4 - 5 \cdot 3 + 3 \cdot 5 \cdot 4 = 3 - 4 - 15 + 60 = 44$.

(c) $\| v \|^2 = \langle v, v \rangle = \langle (3, 4), (3, 4) \rangle = 9 + 16 = 25$; hence $\| v \| = 5$.

(d) $\| v \|^2 = \langle v, v \rangle = \langle (3, 4), (3, 4) \rangle = 9 - 12 - 12 + 48 = 33$; hence $\| v \| = \sqrt{33}$.

6.5. Consider the vector space V of polynomials with inner product defined by $\int_0^1 f(t) g(t) \, dt$ and the polynomials $f(t) = t + 2$, $g(t) = 3t - 2$, and $h(t) = t^2 - 2t - 3$. Find: (a) $\langle f, g \rangle$ and $\langle f, h \rangle$ and (b) $\| f \|$ and $\| g \|$. (c) Normalize f and g.

(a) Integrate as follows:

$$\langle f, g \rangle = \int_0^1 (t + 2)(3t - 2) \, dt = \int_0^1 (3t^2 + 4t - 4) \, dt = [t^3 + 2t^2 - 4t]_0^1 = -1$$

$$\langle f, h \rangle = \int_0^1 (t + 2)(t^2 - 2t - 3) \, dt = \left[\frac{t^4}{4} - \frac{7t^2}{2} - 6t \right]_0^1 = -\frac{37}{4}$$

(b) $$\langle f, f \rangle = \int_0^1 (t + 2)(t + 2) \, dt = \frac{19}{3} \text{ and } \| f \| = \sqrt{\langle f, f \rangle} + \sqrt{19/3} = \frac{\sqrt{57}}{3}.$$

$$\langle g, g \rangle = \int_0^1 (3t - 2)(3t - 2) = 1; \text{ hence } \| g \| = \sqrt{1} - 1.$$

(c) Since $\| f \| = \dfrac{\sqrt{57}}{3}$, $\hat{f} = \dfrac{1}{\| f \|} f = \dfrac{3}{\sqrt{57}} (t + 2)$. Note g is already a unit vector since $\| g \| = 1$; hence $\hat{g} = g = 3t - 2$.

6.6. Let V be the vector space of 2×3 real matrices with inner product $\langle A, B \rangle = \text{tr } B^T A$ and consider the matrices

$$A = \begin{pmatrix} 9 & 8 & 7 \\ 6 & 5 & 4 \end{pmatrix} \qquad B = \begin{pmatrix} 1 & 2 & 3 \\ 4 & 5 & 6 \end{pmatrix} \qquad C = \begin{pmatrix} 3 & -5 & 2 \\ 1 & 0 & -4 \end{pmatrix}$$

Find: (a) $\langle A, B \rangle$, $\langle A, C \rangle$, and $\langle B, C \rangle$; (b) $\langle 2A + 3B, 4C \rangle$; and (c) $\| A \|$ and $\| B \|$. (d) Normalize A and B.

(a) $\left[\text{Use } \langle A, B \rangle = \text{tr } B^T A = \displaystyle\sum_{i=1}^{m} \sum_{j=1}^{n} a_{ij} b_{ij}, \text{ the sum of the products of corresponding entries.} \right]$

$$\langle A, B \rangle = 9 + 16 + 21 + 24 + 25 + 24 = 119$$

$$\langle A, C \rangle = 27 - 40 + 14 + 6 + 0 - 16 = -9$$

$$\langle B, C \rangle = 3 - 10 + 6 + 4 + 0 - 24 = -21$$

(b) Find $2A + 3B = \begin{pmatrix} 21 & 22 & 23 \\ 24 & 25 & 26 \end{pmatrix}$ and $4C = \begin{pmatrix} 12 & -20 & 8 \\ 4 & 0 & -16 \end{pmatrix}$. Then

$$\langle 2A + 3B, 4C \rangle = 252 - 440 + 184 + 96 + 0 - 416 = -324$$

Alternatively, using the linear property of inner products,

$$\langle 2A + 3B, 4C \rangle = 8\langle A, C \rangle + 12\langle B, C \rangle = 8(-9) + 12(-21) = -324$$

(c) $\left[\text{Use } \| A \|^2 = \langle A, A \rangle = \displaystyle\sum_{i=1}^{m} \sum_{j=1}^{n} a_{ij}^2, \text{ the sum of the squares of all the elements of } A. \right]$

$$\| A \|^2 = \langle A, A \rangle = 9^2 + 8^2 + 7^2 + 6^2 + 5^2 + 4^2 = 271 \quad \text{and so} \quad \| A \| = \sqrt{271}$$

$$\| B \|^2 = \langle B, B \rangle = 1^2 + 2^2 + 3^2 + 4^2 + 5^2 + 6^2 = 91 \quad \text{and so} \quad \| B \| = \sqrt{91}$$

(d)

$$\hat{A} = \frac{1}{\|A\|} A = \frac{1}{\sqrt{271}} A = \begin{pmatrix} 9/\sqrt{271} & 8/\sqrt{271} & 7/\sqrt{271} \\ 6/\sqrt{271} & 5/\sqrt{271} & 4/\sqrt{271} \end{pmatrix}$$

$$\hat{B} = \frac{1}{\|B\|} B = \frac{1}{\sqrt{91}} B = \begin{pmatrix} 1/\sqrt{91} & 2/\sqrt{91} & 3/\sqrt{91} \\ 4/\sqrt{91} & 5/\sqrt{91} & 6/\sqrt{91} \end{pmatrix}$$

6.7. Find the distance $d(u, v)$ between the vectors:

(a) $u = (1, 3, 5, 7)$ and $v = (4, -2, 8, 1)$ in $\mathbf{R}^4$;

(b) $u = t + 2$ and $v = 3t - 2$ where $\langle u, v \rangle = \int_0^1 u(t) v(t) \, dt$.

Use $d(u, v) = \| u - v \|$.

(a) $u - v = (-3, 5, -3, 6)$. Thus

$$\| u - v \|^2 = 9 + 25 + 9 + 36 = 79 \qquad \text{and so} \qquad d(u, v) = \| u - v \| = \sqrt{79}$$

(b) $u - v = -2t + 4$. Thus

$$\| u - v \|^2 = \langle u - v, u - v \rangle = \int_0^1 (-2t + 4)(-2t + 4) \, dt$$

$$= \int_0^1 (4t^2 - 16t + 16) \, dt = \left[\frac{4}{3} t^3 - 8t^2 + 16t \right]_0^1 = \frac{28}{3}$$

Hence $d(u, v) = \sqrt{\frac{28}{3}} = \frac{2}{3}\sqrt{21}$.

6.8. Find $\cos \theta$ where θ is the angle between:

(a) $u = (1, -3, 2)$ and $v = (2, 1, 5)$ in $\mathbf{R}^3$;

(b) $u = (1, 3, -5, 4)$ and $v = (2, -3, 4, 1)$ in $\mathbf{R}^4$;

(c) $f(t) = 2t - 1$ and $g(t) = t^2$, where $\langle f, g \rangle = \int_0^1 f(t) g(t) \, dt$;

(d) $A = \begin{pmatrix} 2 & 1 \\ 3 & -1 \end{pmatrix}$ and $B = \begin{pmatrix} 0 & -1 \\ 2 & 3 \end{pmatrix}$, where $\langle A, B \rangle = \operatorname{tr} B^T A$.

Use $\cos \theta = \dfrac{\langle u, v \rangle}{\| u \| \| v \|}$.

(a) Compute $\langle u, v \rangle = 2 - 3 + 10 = 9$, $\| u \|^2 = 1 + 9 + 4 = 14$, $\| v \|^2 = 4 + 1 + 25 = 30$. Thus,

$$\cos \theta = \frac{9}{\sqrt{14}\sqrt{30}} = \frac{9}{\sqrt{105}}$$

(b) Here $\langle u, v \rangle = 2 - 9 - 20 + 4 = -23$, $\| u \|^2 = 1 + 9 + 25 + 16 = 51$, $\| v \|^2 = 4 + 9 + 16 + 1 = 30$. Thus

$$\cos \theta = \frac{-23}{\sqrt{51}\sqrt{30}} = \frac{-23}{3\sqrt{170}}$$

(c) Compute

$$\langle f, g \rangle = \int_0^1 (2t^3 - t^2) \, dt = \left[\frac{t^4}{2} - \frac{t^3}{3} \right]_0^1 = \frac{1}{2} - \frac{1}{3} = \frac{1}{6}$$

$$\| f \|^2 = \langle f, f \rangle = \int_0^1 (4t^2 - 4t + 1) \, dt = \frac{1}{3} \qquad \text{and} \qquad \| g \|^2 = \langle g, g \rangle = \int_0^1 t^4 \, dt = \frac{1}{3}$$

Thus $\cos \theta = \dfrac{\frac{1}{6}}{(1/\sqrt{3})(1/\sqrt{5})} = \dfrac{\sqrt{15}}{6}$.

(d) Compute $\langle A,\ B\rangle = 0 - 1 + 6 - 3 = 2$, $\|A\|^2 = 4 + 1 + 9 + 1 = 15$, $\|B\|^2 = 0 + 1 + 4 + 9 = 14$.
     Thus

$$\cos\theta = \frac{2}{\sqrt{15}\sqrt{14}} = \frac{2}{\sqrt{210}}$$

6.9. Verify each of the following:

(a) Parallelogram Law (Fig. 6-10): $\|u + v\|^2 + \|u - v\|^2 = 2\|u\|^2 + 2\|v\|^2$.

(b) Polar form for $\langle u, v\rangle$ (which shows that the inner product can be obtained from the norm function): $\langle u, v\rangle = \frac{1}{4}(\|u + v\|^2 - \|u - v\|^2)$.

Expand each of the following to obtain:

$$\|u + v\|^2 = \langle u + v, u + v\rangle = \|u\|^2 + 2\langle u, v\rangle + \|v\|^2 \qquad (1)$$

$$\|u - v\|^2 = \langle u - v, u - v\rangle = \|u\|^2 - 2\langle u, v\rangle + \|v\|^2 \qquad (2)$$

Add (1) and (2) to get the Parallelogram Law (a). Subtract (2) from (1) to obtain

$$\|u + v\|^2 - \|u - v\|^2 = 4\langle u, v\rangle$$

Divide by 4 to obtain the (real) polar form (b). (The polar form for the complex case is different.)

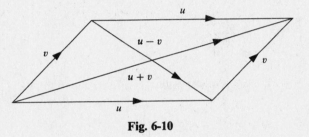

Fig. 6-10

6.10. Prove Theorem 6.1 (Cauchy–Schwarz).

For any real number t,

$$\langle tu + v, tu + c\rangle = t^2\langle u, u\rangle + 2t\langle u, v\rangle + \langle v, v\rangle = t^2\|u\|^2 + 2t\langle u, v\rangle + \|v\|^2$$

Let $a = \|u\|^2$, $b = 2\langle u, v\rangle$, and $c = \|v\|^2$. Since $\|tu + v\|^2 \geq 0$, we have

$$at^2 + bt + c \geq 0$$

for every value of t. This means that the quadratic polynomial cannot have two real roots. This implies that $b^2 - 4ac \leq 0$ or $b^2 \leq 4ac$. Thus

$$4\langle u, v\rangle^2 \leq 4\|u\|^2\|v\|^2$$

Dividing by 4 gives us our result. (*Remark:* The Cauchy–Schwarz inequality for complex inner product spaces appears in Problem 6.53.)

6.11. Prove Theorem 6.2.

If $v \neq 0$, then $\langle v, v\rangle > 0$ and hence $\|v\| = \sqrt{\langle v, v\rangle} > 0$. If $v = 0$ then $\langle 0, 0\rangle = 0$. Consequently $\|0\| = \sqrt{0} = 0$. Thus $[N_1]$ is true.

We have $\|kv\|^2 = \langle kv, kv\rangle = k^2\langle v, v\rangle = k^2\|v\|^2$. Taking the square root of both sides gives $[N_2]$.

Using the Cauchy–Schwarz inequality, we obtain

$$\|u + v\|^2 = \langle u + v, u + v\rangle = \langle u, u\rangle + \langle u, v\rangle + \langle u, v\rangle + \langle v, v\rangle$$

$$\leq \|u\|^2 + 2\|u\|\|v\| + \|v\|^2 = (\|u\| + \|v\|)^2$$

Taking the square root of both sides yields $[N_3]$.

6.12. Let $(a_1, a_2, \ldots)$ and $(b_1, b_2, \ldots)$ be any pair of points in l_2-space of Example 6.3(b). Show that the inner product is well defined, i.e., show that the sum $\sum\limits_{i=1}^{\infty} a_i b_i = a_1 b_1 + a_2 b_2 + \cdots$ converges absolutely.

By Example 6.4(a) (Cauchy–Schwarz inequality),

$$|a_1 b_1| + \cdots + |a_n b_n| \le \sqrt{\sum_{i=1}^{n} a_i^2} \sqrt{\sum_{i=1}^{n} b_i^2} \le \sqrt{\sum_{i=1}^{\infty} a_i^2} \sqrt{\sum_{i=1}^{\infty} b_i^2}$$

which holds for every n. Thus the (monotonic) sequence of sums $S_n = |a_1 b_1| + \cdots + |a_n b_n|$ is bounded, and therefore converges. Hence the infinite sum converges absolutely.

ORTHOGONALITY, ORTHOGONAL COMPLEMENTS, ORTHOGONAL SETS

6.13. Find k so that the following pairs are orthogonal:

(a) $u = (1, 2, k, 3)$ and $v = (3, k, 7, -5)$ in $\mathbf{R}^4$;

(b) $f(t) = t + k$ and $g(t) = t^2$ where $\langle f, g \rangle = \int_0^1 f(t) g(t)\, dt$.

(a) First find $\langle u, v \rangle = (1, \ 2, \ k, \ 3) \cdot (3, \ k, \ 7, \ -5) = 3 + 2k + 7k - 15 = 9k - 12$. Then set $\langle u, v \rangle = 9k - 12 = 0$ to find $k = \frac{4}{3}$.

(b) First find

$$\langle f, g \rangle = \int_0^1 (t + k) t^2\, dt = \int_0^1 (t^3 + kt^2)\, dt = \left[\frac{t^4}{4} + \frac{kt^3}{3} \right]_0^1 = \frac{1}{4} + \frac{k}{3}$$

Set $\langle f, g \rangle = \dfrac{1}{4} + \dfrac{k}{3} = 0$ to obtain $k = -\dfrac{3}{4}$.

6.14. Consider $u = (0, 1, -2, 5)$ in $\mathbf{R}^4$. Find a basis for the orthogonal complement $u^\perp$ of u.

We seek all vectors (x, y, z, t) in $\mathbf{R}^4$ such that

$$\langle (x, y, z, t), (0, 1, -2, 5) \rangle = 0 \qquad \text{or} \qquad 0x + y - 2z + 5t = 0$$

The free variables are x, z, and t. Accordingly,

(1) Set $x = 1$, $z = 0$, $t = 0$ to obtain the solution $w_1 = (1, 0, 0, 0)$.

(2) Set $x = 0$, $z = 1$, $t = 0$ to obtain the solution $w_2 = (0, 2, 1, 0)$.

(3) Set $x = 0$, $z = 0$, $t = 1$ to obtain the solution $w_3 = (0, -5, 0, 1)$.

The vector w_1, w_2, w_3 form a basis of the solution space of the equation and hence a basis for $u^\perp$.

6.15. Let W be the subspace of $\mathbf{R}^5$ spanned by $u = (1, 2, 3, -1, 2)$ and $v = (2, 4, 7, 2, -1)$. Find a basis of the orthogonal complement $W^\perp$ of W.

We seek all vectors $w = (x, y, z, s, t)$ such that

$$\langle w, u \rangle = \ x + 2y + 3z - \ s + 2t = 0$$

$$\langle w, v \rangle = 2x + 4y + 7z + 2s - \ t = 0$$

Eliminating x from the second equation, we find the equivalent system

$$x + 2y + 3z - \ s + 2t = 0$$

$$z + 4s - 5t = 0$$

The free variables are y, s, and t. Therefore,

(1)　Set $y = -1$, $s = 0$, $t = 0$ to obtain the solution $w_1 = (2, -1, 0, 0, 0)$.

(2)　Set $y = 0$, $s = 1$, $t = 0$ to find the solution $w_2 = (13, 0, -4, 1, 0)$.

(3)　Set $y = 0$, $s = 0$, $t = 1$ to obtain the solution $w_3 = (-17, 0, 5, 0, 1)$.

The set $\{w_1, w_2, w_3\}$ is a basis of $W^\perp$.

6.16.　Let $w = (1, 2, 3, 1)$ be a vector in $\mathbf{R}^4$. Find an orthogonal basis for $w^\perp$.

Find a nonzero solution of $x + 2y + 3z + t = 0$; say $v_1 = (0, 0, 1, -3)$. Now find a nonzero solution of the system

$$x + 2y + 3z + t = 0 \qquad z - 3t = 0$$

say $v_2 = (0, -5, 3, 1)$. Lastly, find a nonzero solution of the system

$$x + 2y + 3z + t = 0 \qquad -5y + 3z + t = 0 \qquad z - 3t = 0$$

say $v_3 = (-14, 2, 3, 1)$. Thus v_1, v_2, v_3 form an orthogonal basis for $w^\perp$. (Compare with Problem 6.14, where the basis need not be orthogonal.)

6.17.　Let S consist of the following vectors in $\mathbf{R}^3$:

$$u_1 = (1, 1, 1) \qquad u_2 = (1, 2, -3) \qquad u_3 = (5, -4, -1)$$

(a)　Show that S is orthogonal and S is a basis for $\mathbf{R}^3$.

(b)　Write $v = (1, 5, -7)$ as a linear combination of u_1, u_2, u_3.

(a)　Compute

$$\langle u_1, u_2 \rangle = 1 + 2 - 3 = 0 \qquad \langle u_1, u_3 \rangle = 5 - 4 - 1 = 0 \qquad \langle u_2, u_3 \rangle = 5 - 8 + 3 = 0$$

Since each inner product equals 0, S is orthogonal and hence S is linearly independent. Thus S is a basis for $\mathbf{R}^3$ since any three linearly independent vectors form a basis for $\mathbf{R}^3$.

(b)　Let $v = xu_1 + yu_2 + zu_3$ for unknown scalars x, y, z, i.e.,

$$(1, 5, -7) = x(1, 1, 1) + y(1, 2, -3) + z(5, -4, -1) \tag{1}$$

Method 1.　Expand (1) obtaining the system

$$x + y + 5z = 1 \qquad x + 2y - 4z = 5 \qquad x - 3y - z = -7$$

Solve the system to obtain $x = -\frac{1}{3}$, $y = \frac{16}{7}$, $z = -\frac{4}{21}$.

Method 2.　(This method uses the fact that the basis vectors are orthogonal, and the arithmetic is simpler.) Take the inner product of (1) with u_1 to get

$$(1, 5, -7) \cdot (1, 1, 1) = x(1, 1, 1) \cdot (1, 1, 1) \qquad \text{or} \qquad -1 = 3x \qquad \text{or} \qquad x = -\tfrac{1}{3}$$

(The two last terms drop out since u_1 is orthogonal to u_2 and to u_3.) Take the inner product of (1) with u_2 to get

$$(1, 5, -7) \cdot (1, 2, -3) = y(1, 2, -3) \cdot (1, 2, -3) \qquad \text{or} \qquad 32 = 14y \qquad \text{or} \qquad y = \tfrac{16}{7}$$

Take the inner product of (1) with u_3 to get

$$(1, 5, -7) \cdot (5, -4, -1) = z(5, -4, -1) \cdot (5, -4, -1) \qquad \text{or} \qquad -8 = 42z \qquad \text{or} \qquad z = -\tfrac{4}{21}$$

In either case, we get $v = (-\tfrac{1}{3})u_1 + (\tfrac{16}{7})u_2 - (\tfrac{4}{21})u_3$.

6.18.　Let S consist of the following vectors in $\mathbf{R}^4$:

$$u_1 = (1, 1, 0, -1) \qquad u_2 = (1, 2, 1, 3) \qquad u_3 = (1, 1, -9, 2) \qquad u_4 = (16, -13, 1, 3)$$

(a) Show that S is orthogonal and a basis of $\mathbf{R}^4$.

(b) Find the coordinates of an arbitrary vector $v = (a, b, c, d)$ in $\mathbf{R}^4$ relative to the basis S.

(a) Compute

$$u_1 \cdot u_2 = 1 + 2 + 0 - 3 = 0 \qquad u_1 \cdot u_3 = 1 + 1 + 0 - 2 = 0 \qquad u_1 \cdot u_4 = 16 - 13 + 0 - 3 = 0$$

$$u_2 \cdot u_3 = 1 + 2 - 9 + 6 = 0 \qquad u_2 \cdot u_4 = 16 - 26 + 1 + 9 = 0 \qquad u_3 \cdot u_4 = 16 - 13 - 9 + 6 = 0$$

Thus S is orthogonal and hence S is linearly independent. Accordingly, S is a basis for $\mathbf{R}^4$ since any four linearly independent vectors form a basis of $\mathbf{R}^4$.

(b) Since S is orthogonal, we need only find the Fourier coefficients of v with respect to the basis vectors, as in Theorem 6.7. Thus

$$k_1 = \frac{\langle v, u_1 \rangle}{\langle u_1, u_1 \rangle} = \frac{a + b - d}{3} \qquad\qquad k_3 = \frac{\langle v, u_3 \rangle}{\langle u_3, u_3 \rangle} = \frac{a + b - 9c + 2d}{87}$$

$$k_2 = \frac{\langle v, u_2 \rangle}{\langle u_2, u_2 \rangle} = \frac{a + 2b + c + 3d}{15} \qquad\qquad k_4 = \frac{\langle v, u_4 \rangle}{\langle u_4, u_4 \rangle} = \frac{16a - 13b + c + 3d}{435}$$

are the coordinates of v with respect to the basis S.

6.19. Suppose S, S_1, and S_2 are subsets of V. Prove the following:

(a) $S \subseteq S^{\perp\perp}$ (b) If $S_1 \subseteq S_2$, then $S_2^{\perp} \subseteq S_1^{\perp}$ (c) $S^{\perp} = \operatorname{span} S^{\perp}$

(a) Let $w \in S$. Then $\langle w, v \rangle = 0$ for every $v \in S^{\perp}$; hence $w \in S^{\perp\perp}$. Accordingly, $S \subseteq S^{\perp\perp}$.

(b) Let $w \in S_2^{\perp}$. Then $\langle w, v \rangle = 0$ for every $v \in S_2$. Since $S_1 \subseteq S_2$, $\langle w, v \rangle = 0$ for every $v \in S_1$. Thus $w \in S_1^{\perp}$, and hence $S_2^{\perp} \subseteq S_1^{\perp}$.

(c) Since $S \subseteq \operatorname{span} S$, we have $\operatorname{span} S^{\perp} \subseteq S^{\perp}$. Suppose $u \in S^{\perp}$ and suppose $v \in \operatorname{span} S^{\perp}$. Then there exist $w_1, w_2, \ldots, w_k$ in S such that

$$v = a_1 w_1 + a_2 w_2 + \cdots + a_k w_k$$

Then, using $u \in S^{\perp}$, we have

$$\langle u, v \rangle = \langle u, a_1 w_1 + a_2 w_2 + \cdots + a_k w_k \rangle = a_1 \langle u, w_1 \rangle + a_2 \langle u, w_2 \rangle + \cdots + a_k \langle u, w_k \rangle$$
$$= a_1 \cdot 0 + a_2 \cdot 0 + \cdots + a_k \cdot 0 = 0$$

Thus $u \in \operatorname{span} S^{\perp}$. Accordingly, $S^{\perp} \subseteq \operatorname{span} S^{\perp}$. Both inclusions give $S^{\perp} = \operatorname{span} S^{\perp}$.

6.20. Prove Theorem 6.5.

Suppose $S = \{u_1, u_2, \ldots, u_r\}$ and suppose

$$a_1 u_1 + a_2 u_2 + \cdots + a_r u_r = 0 \tag{1}$$

Taking the inner product of (1) with u_1, we get

$$0 = \langle 0, u_1 \rangle = \langle a_1 u_1 + a_2 u_2 + \cdots + a_r u_r, u_1 \rangle$$
$$= a_1 \langle u_1, u_1 \rangle + a_2 \langle u_2, u_1 \rangle + \cdots + a_r \langle u_r, u_1 \rangle$$
$$= a_1 \langle u_1, u_1 \rangle + a_2 \cdot 0 + \cdots + a_r \cdot 0 = a_1 \langle u_1, u_1 \rangle$$

Since $u_1 \neq 0$, we have $\langle u_1, u_1 \rangle \neq 0$. Thus $a_1 = 0$. Similarly, for $i = 2, \ldots, r$, taking the inner product for (1) with u_i,

$$0 = \langle 0, u_i \rangle = \langle a_1 u_1 + \cdots + a_r u_r, u_i \rangle$$
$$= a_1 \langle u_1, u_i \rangle + \cdots + a_i \langle u_i, u_i \rangle + \cdots + a_r \langle u_r, u_i \rangle = a_i \langle u_i, u_i \rangle$$

But $\langle u_i, u_i \rangle \neq 0$ and hence $a_i = 0$. Thus S is linearly independent.

6.21. Prove Theorem 6.6 (Pythagoras).

Expanding the inner product, we have

$$\| u_1 + u_2 + \cdots + u_r \|^2 = \langle u_1 + u_2 + \cdots + u_r, u_1 + u_2 + \cdots + u_r \rangle$$

$$= \langle u_1, u_1 \rangle + \langle u_2, u_2 \rangle + \cdots + \langle u_r, u_r \rangle + \sum_{i \neq j} \langle u_i, u_j \rangle$$

The theorem follows from the fact that $\langle u_i, u_i \rangle = \| u_i \|^2$ and $\langle u_i, u_j \rangle = 0$ for $i \neq j$.

6.22. Prove Theorem 6.7.

Suppose $v = k_1 u_1 + k_2 u_2 + \cdots + k_n u_n$. Taking the inner product of both sides with u_1 yields

$$\langle v, u_1 \rangle = \langle k_1 u_1 + k_2 u_2 + \cdots + k_n u_n, u_1 \rangle$$

$$= k_1 \langle u_1, u_1 \rangle + k_2 \langle u_2, u_1 \rangle + \cdots + k_n \langle u_n, u_1 \rangle$$

$$= k_1 \langle u_1, u_1 \rangle + k_2 \cdot 0 + \cdots + k_n \cdot 0 = k_1 \langle u_1, u_1 \rangle$$

Thus $k_1 = \langle v, u_1 \rangle / \langle u_1, u_1 \rangle$. Similarly, for $i = 2, \ldots, n$,

$$\langle v, u_i \rangle = \langle k_1 u_1 + k_2 u_2 + \cdots + k_n u_n, u_i \rangle$$

$$= k_1 \langle u_1, u_i \rangle + k_2 \langle u_2, u_i \rangle + \cdots + k_n \langle u_n, u_i \rangle$$

$$= k_1 \cdot 0 + \cdots + k_i \langle u_i, u_i \rangle + \cdots + k_n \cdot 0 = k_i \langle u_i, u_i \rangle$$

Thus $k_i = \langle v, u_i \rangle / \langle u_i, u_i \rangle$. Substituting for k_i in the equation $u = k_1 u_1 + \cdots + k_n u_n$, we obtain the desired result.

6.23. Suppose $E = \{e_1, e_2, \ldots, e_n\}$ is an orthonormal basis of V. Prove:

(a) For any $u \in V$, we have $u = \langle u, e_1 \rangle e_1 + \langle u, e_2 \rangle e_2 + \cdots + \langle u, e_n \rangle e_n$.

(b) $\langle a_1 e_1 + \cdots + a_n e_n, b_1 e_1 + \cdots + b_n e_n \rangle = a_1 b_1 + a_2 b_2 + \cdots + a_n b_n$.

(c) For any $u, v \in V$, we have $\langle u, v \rangle = \langle u, e_1 \rangle \langle v, e_1 \rangle + \cdots + \langle u, e_n \rangle \langle v, e_n \rangle$.

(a) Suppose $u = k_1 e_1 + k_2 k_2 + \cdots + k_n e_n$. Taking the inner product of u with e_1,

$$\langle u, e_1 \rangle = \langle k_1 e_1 + k_2 e_2 + \cdots + k_n e_n, e_1 \rangle$$

$$= k_1 \langle e_1, e_1 \rangle + k_2 \langle e_2, e_1 \rangle + \cdots + k_n \langle e_n, e_1 \rangle$$

$$= k_1 \cdot 1 + k_2 \cdot 0 + \cdots + k_n \cdot 0 = k_1$$

Similarly, for $i = 2, \ldots, n$

$$\langle u, e_i \rangle = \langle k_1 e_1 + \cdots + k_i e_i + \cdots + k_n e_n, e_i \rangle$$

$$= k_1 \langle e_1, e_i \rangle + \cdots + k_i \langle e_i, e_i \rangle + \cdots + k_n \langle e_n, e_i \rangle$$

$$= k_1 \cdot 0 + \cdots + k_i \cdot 1 + \cdots + k_n \cdot 0 = k_i$$

Substituting $\langle u, e_i \rangle$ for k_i in the equation $u = k_1 e_1 + \cdots + k_n e_n$, we obtain the desired result.

(b) We have

$$\left\langle \sum_{i=1}^{n} a_i e_i, \sum_{j=1}^{n} b_j e_j \right\rangle = \sum_{i,j=1}^{n} a_j b_j \langle e_i, e_j \rangle = \sum_{i=1}^{n} a_i b_i \langle e_i, e_i \rangle + \sum_{i \neq 1} a_i b_j \langle e_i, e_j \rangle$$

But $\langle e_i, e_j \rangle = 0$ for $i \neq j$, and $\langle e_i, e_j \rangle = 1$ for $i = j$; hence, as required,

$$\left\langle \sum_{i=1}^{n} a_i e_i, \sum_{j=1}^{n} b_j e_j \right\rangle = \sum_{i=1}^{n} a_i b_i = a_1 b_1 + a_2 b_2 + \cdots + a_n b_n$$

(c) By (a), we have

$$u = \langle u, e_1 \rangle e_1 + \cdots + \langle u, e_n \rangle e_n \quad \text{and} \quad v = \langle v, e_1 \rangle e_1 + \cdots + \langle v, e_n \rangle e_n$$

Thus, by (b),

$$\langle u, v \rangle = \langle u, e_1 \rangle \langle v, e_1 \rangle + \langle u, e_2 \rangle \langle v, e_2 \rangle + \cdots + \langle u, e_n \rangle \langle v, e_n \rangle$$

PROJECTIONS, GRAM–SCHMIDT ALGORITHM, APPLICATIONS

6.24. Suppose $w \neq 0$. Let v be any vector in V. Show that

$$c = \frac{\langle v, w \rangle}{\langle w, w \rangle} = \frac{\langle v, w \rangle}{\| w \|^2}$$

is the unique scalar such that $v' = v - cw$ is orthogonal to w.

In order for v' to be orthogonal to w we must have

$$\langle v - cw, w \rangle = 0 \qquad \text{or} \qquad \langle v, w \rangle - c\langle w, w \rangle = 0 \qquad \text{or} \qquad \langle v, w \rangle = c\langle w, w \rangle$$

Thus $c = \langle v, w \rangle / \langle w, w \rangle$. Conversely, suppose $c = \langle v, w \rangle / \langle w, w \rangle$. Then

$$\langle v - cw, w \rangle = \langle v, w \rangle - c\langle w, w \rangle = \langle v, w \rangle - \frac{\langle v, w \rangle}{\langle w, w \rangle} \langle w, w \rangle = 0$$

6.25. Find the Fourier coefficient c and the projection of $v = (1, -2, 3, -4)$ along $w = (1, 2, 1, 2)$ in $\mathbf{R}^4$.

Compute $\langle v, w \rangle = 1 - 4 + 3 - 8 = -8$ and $\| w \|^2 = 1 + 4 + 1 + 4 = 10$. Then $c = -\frac{8}{10} = -\frac{4}{5}$ and proj $(v, w) = cw = (-\frac{4}{5}, -\frac{8}{5}, -\frac{4}{5}, -\frac{8}{5})$.

6.26. Find an orthonormal basis for the subspace U of $\mathbf{R}^4$ spanned by the vectors

$$v_1 = (1, 1, 1, 1) \qquad v_2 = (1, 1, 2, 4) \qquad v_3 = (1, 2, -4, -3)$$

First find an orthogonal basis of U using the Gram–Schmidt algorithm. Begin the algorithm by setting $w_1 = u_1 = (1, 1, 1, 1)$. Next find

$$v_2 - \frac{\langle v_2, w_1 \rangle}{\| w_1 \|^2} w_1 = (1, 1, 2, 4) - \frac{8}{4}(1, 1, 1, 1) = (-1, -1, 0, 2)$$

Set $w_2 = (-1, -1, 0, 2)$. Then find

$$v_3 - \frac{\langle v_3, w_1 \rangle}{\| w_1 \|^2} w_1 - \frac{\langle v_3, w_2 \rangle}{\| w_2 \|^2} w_2 = (1, 2, -4, -3) - \frac{-4}{4}(1, 1, 1, 1) - \frac{-9}{6}(-1, -1, 0, 2)$$

$$= (\tfrac{1}{2}, \tfrac{3}{2}, -3, 1)$$

Clear fractions to obtain $w_3 = (1, 3, -6, 2)$. Last, normalize the orthogonal basis consisting of w_1, w_2, w_3. Since $\| w_1 \|^2 = 4$, $\| w_2 \|^2 = 6$, and $\| w_3 \|^2 = 50$, the following vectors form an orthonormal basis of U:

$$u_1 = \frac{1}{2}(1, 1, 1, 1) \qquad u_2 = \frac{1}{\sqrt{6}}(-1, -1, 0, 2) \qquad u_3 = \frac{1}{5\sqrt{2}}(1, 3, -6, 2)$$

6.27. Let V be the vector space of polynomials $f(t)$ with inner product $\langle f, g \rangle = \int_0^1 f(t) g(t) \, dt$. Apply the Gram–Schmidt algorithm to the set $\{1, t, t^2\}$ to obtain an orthogonal set $\{f_0, f_1, f_2\}$ with integer coefficients.

First set $f_0 = 1$. Then find

$$t - \frac{\langle t, 1 \rangle}{\langle 1, 1 \rangle} \cdot 1 = t - \frac{\frac{1}{2}}{1} \cdot 1 = t - \frac{1}{2}$$

Clear fractions to obtain $f_1 = 2t - 1$. Then find

$$t^2 - \frac{\langle t^2, 1 \rangle}{\langle 1, 1 \rangle} \cdot 1 - \frac{\langle t^2, 2t - 1 \rangle}{\langle 2t - 1, 2t - 1 \rangle} \cdot (2t - 1) = t^2 - \frac{\frac{1}{3}}{1} \cdot 1 - \frac{\frac{1}{6}}{\frac{1}{3}} \cdot (2t - 1) = t^2 - t + \frac{1}{6}$$

Clear fractions to obtain $f_2 = 6t^2 - 6t + 1$. Thus $\{1, 2t - 1, 6t^2 - 6t + 1\}$ is the required orthogonal set.

6.28. Suppose $v = (1, 3, 5, 7)$. Find the projection of v onto W (or, find $w \in W$ which minimizes $\| v - w \|$) where W is the subspace of $\mathbf{R}^4$ spanned by:

    (a) $u_1 = (1, 1, 1, 1)$ and $u_2 = (1, -3, 4, -2)$; and

    (b) $v_1 = (1, 1, 1, 1)$ and $v_2 = (1, 2, 3, 2)$.

    (a) Since u_1 and u_2 are orthogonal, we need only compute the Fourier coefficients:

$$c_1 = \frac{\langle v, u_1 \rangle}{\| u_1 \|^2} = \frac{1 + 3 + 5 + 7}{1 + 1 + 1 + 1} = \frac{16}{4} = 4$$

$$c_2 = \frac{\langle v, u_2 \rangle}{\| u_2 \|^2} = \frac{1 - 9 + 20 - 14}{1 + 9 + 16 + 4} = \frac{-2}{30} = \frac{-1}{15}$$

Then

$$w = \text{proj}\,(v, W) = c_1 u_1 + c_2 u_2 = 4(1, 1, 1, 1) - \tfrac{1}{15}(1, -3, 4, -2) = (\tfrac{59}{15}, \tfrac{63}{5}, \tfrac{56}{15}, \tfrac{62}{15})$$

    (b) Since v_1 and v_2 are not orthogonal, first apply the Gram–Schmidt algorithm to find an orthogonal basis for W. Set $w_1 = v_1 = (1, 1, 1, 1)$. Then find

$$v_2 - \frac{\langle v_2, w_1 \rangle}{w_1^2} w_1 = (1, 2, 3, 2) - \frac{8}{4}(1, 1, 1, 1) = (-1, 0, 1, 0)$$

Set $w_2 = (-1, 0, 1, 0)$. Now compute

$$c_1 = \frac{\langle v, w_1 \rangle}{\| w_1 \|^2} = \frac{1 + 3 + 5 + 7}{1 + 1 + 1 + 1} = \frac{16}{4} = 4 \quad \text{and} \quad c_2 = \frac{\langle v, w_2 \rangle}{\| w_2 \|^2} = \frac{-1 + 0 + 5 + 0}{1 + 0 + 1 + 0} = \frac{-6}{2} = -3$$

Then $w = \text{proj}\,(v, W) = c_1 w_1 + c_2 w_2 = 4(1, 1, 1, 1) - 3(-1, 0, 1, 0) = (7, 4, 1, 4)$.

6.29. Suppose w_1 and w_2 are nonzero orthogonal vectors. Let v be any vector in V. Find c_1 and c_2 so that v' is orthogonal to w_1 and w_2 where $v' = c - c_1 w_1 - c_2 w_2$.

    If v' is orthogonal to w_1, then

$$0 = \langle v - c_1 w_1 - c_2 w_2, w_1 \rangle = \langle v, w_1 \rangle - c_1 \langle w_1, w_1 \rangle - c_2 \langle w_2, w_1 \rangle$$
$$= \langle v, w_1 \rangle - c_1 \langle w_1, w_1 \rangle - c_2 0 = \langle v, w_1 \rangle - c_1 \langle w_1, w_1 \rangle$$

Thus $c_1 = \langle v, w_1 \rangle / \langle w_1, w_1 \rangle$. (That is, c_1 is the component of v along w_1.) Similarly, if v' is orthogonal to w_2, then

$$0 = \langle v - c_1 w_1 - c_2 w_2, w_2 \rangle = \langle v, w_2 \rangle - c_2 \langle w_2, w_2 \rangle$$

Thus $c_2 = \langle v, w_2 \rangle / \langle w_2, w_2 \rangle$. (That is, c_2 is the component of v along w_2.)

6.30. Prove Theorem 6.8.

    For $i = 1, 2, \ldots, r$ and using $\langle w_i, w_j \rangle = 0$ for $i \neq j$, we have

$$\langle v - c_1 w_1 - c_2 w_2 - \cdots - c_r w_r, w_i \rangle = \langle v, w_i \rangle - c_1 \langle w_1, w_i \rangle - \cdots - c_i \langle w_i w_i \rangle - \cdots - c_r \langle w_r, w_i \rangle$$
$$= \langle v, w_i \rangle - c_1 \cdot 0 - \cdots - c_i \langle w_i, w_i \rangle - \cdots - c_r \cdot 0$$
$$= \langle v, w_i \rangle = c_i \langle w_i, w_i \rangle = \langle v, w_i \rangle - \frac{\langle v, w_i \rangle}{\langle w_i, w_i \rangle} \langle w_i, w_i \rangle$$
$$= 0$$

Thus the theorem is proved.

6.31. Prove Theorem 6.9.

By Theorem 6.8, $v - \sum c_k w_k$ is orthogonal to every w_i and hence orthogonal to any linear combination of $w_1, w_2, \ldots, w_r$. Therefore, using the Pythagorean Theorem and summing from $k = 1$ to r,

$$\left\| v - \sum a_k w_k \right\|^2 = \left\| v - \sum c_k w_k + \sum (c_k - a_k)w_k \right\|^2 = \left\| v - \sum c_k w_k \right\|^2 + \left\| \sum (c_k - a_k)w_k \right\|^2$$

$$\geq \left\| v - \sum c_k w_k \right\|^2$$

The square root of both sides gives us our theorem.

6.32. Prove Theorem 6.10 (Bessel inequality).

Note that $c_i = \langle v, e_i \rangle$ since $\|e_i\| = 1$. Then, using $\langle e_i, e_j \rangle = 0$ for $i \neq j$ and summing from $k = 1$ to r, we get

$$0 \leq \left\langle v - \sum c_k e_k, v - \sum c_k, e_k \right\rangle = \langle v, v \rangle - 2\left\langle v, \sum c_k e_k \right\rangle + \sum c_k^2$$

$$= \langle v, v \rangle - \sum 2c_k \langle v, e_k \rangle + \sum c_k^2 = \langle v, v \rangle - \sum 2c_k^2 + \sum c_k^2$$

$$= \langle v, v \rangle - \sum c_k^2$$

This gives us our inequality.

6.33. Prove Theorem 6.11.

The proof uses the Gram–Schmidt algorithm and following Remarks 1 and 2 (Section 6.6). That is, apply the algorithm to $\{v_i\}$ to obtain an orthogonal basis $\{w_1, \ldots, w_n\}$, and then normalize $\{w_i\}$ to obtain an orthonormal basis $\{u_i\}$ of V. The specific algorithm guarantees that each w_k is a linear combination of $v_1, \ldots, v_k$, and hence each u_k is a linear combination of $v_1, \ldots, v_k$.

6.34. Prove Theorem 6.12.

Extend S to a basis $S' = \{w_1, \ldots, w_r, v_{r+1}, \ldots, v_n\}$ for V. Applying the Gram–Schmidt algorithm to S', we first obtain $w_1, w_2, \ldots, w_r$ since S is orthogonal and then we obtain vectors $w_{r+1}, \ldots, w_n$ where $\{w_1, w_2, \ldots, w_n\}$ is an orthogonal basis for V. Thus the theorem is proved.

6.35. Prove Theorem 6.4.

By Theorem 6.11, there exists an orthogonal basis $\{u_1, \ldots, u_r\}$ of W; and, by Theorem 6.12, we can extend it to an orthogonal basis $\{u_1, u_2, \ldots, u_n\}$ of V. Hence $u_{r+1}, \ldots, u_n \in W^\perp$. If $v \in V$, then

$$v = a_1 u_1 + \cdots + a_n u_n \quad \text{where} \quad a_1 u_1 + \cdots + a_r u_r \in W, \quad a_{r+1}u_{r+1} + \cdots + a_n u_n \in W^\perp$$

Accordingly, $V = W + W^\perp$.

On the other hand, if $w \in W \cap W^\perp$, then $\langle w, w \rangle = 0$. This yields $w = 0$; hence $W \cap W^\perp = \{0\}$.

The two conditions, $V = W + W^\perp$ and $W \cap W^\perp = \{0\}$, give the desired result $V = W \oplus W^\perp$.

Note that we have proved the theorem only for the case that V has finite dimension; we remark that the theorem also holds for spaces of arbitrary dimension.

6.36. Suppose W is a subspace of a finite-dimensional space V. Show that $W = W^{\perp\perp}$.

By Theorem 6.4, $V = W \oplus W^{-1}$ and, also, $V = W^\perp \oplus W^{\perp\perp}$. Hence

$$\dim W = \dim V - \dim W^\perp \quad \text{and} \quad \dim W^{\perp\perp} = \dim V - \dim W^\perp$$

This yields $\dim W = \dim W^{\perp\perp}$. But $W \subset W^{\perp\perp}$ [Problem 6.19(a)], hence $W = W^{\perp\perp}$, as required.

INNER PRODUCTS AND POSITIVE DEFINITE MATRICES

6.37. Determine whether or not A is positive definite where $A = \begin{pmatrix} 1 & 0 & 1 \\ 0 & 1 & 2 \\ 1 & 2 & 3 \end{pmatrix}$.

Use the Diagonalization Algorithm of Section 4.11 to transform A into a (congruent) diagonal matrix. Apply $-R_1 + R_3 \to R_3$ and $-C_1 + C_3 \to C_3$ and then $-2R_2 + R_3 \to R_3$ and $-2C_2 + C_3 \to C_3$ to obtain

$$A \sim \begin{pmatrix} 1 & 0 & 0 \\ 0 & 1 & 2 \\ 0 & 2 & 2 \end{pmatrix} \sim \begin{pmatrix} 1 & 0 & 0 \\ 0 & 1 & 0 \\ 0 & 0 & -2 \end{pmatrix}$$

There is a negative entry -2 in the diagonal matrix; hence A is not positive definite.

6.38. Find the matrix A which represents the usual inner product on $\mathbf{R}^2$ relative to each of the following bases of $\mathbf{R}^2$:

(a) $\{v_1 = (1, 3),\ v_2 = (2, 5)\}$,   (b) $\{w_1 = (1, 2),\ w_2 = (4, -2)\}$

(a) Compute $\langle v_1, v_1 \rangle = 1 + 9 = 10$, $\langle v_1, v_2 \rangle = 2 + 15 = 17$, $\langle v_2, v_2 \rangle = 4 + 25 = 29$.   Thus
$A = \begin{pmatrix} 10 & 17 \\ 17 & 29 \end{pmatrix}$.

(b) Compute $\langle w_1, w_1 \rangle = 1 + 4 = 5$, $\langle w_1, w_2 \rangle = 4 - 4 = 0$, $\langle w_2, w_2 \rangle = 16 + 4 = 20$. Thus $A = \begin{pmatrix} 5 & 0 \\ 0 & 20 \end{pmatrix}$.

(Since the basis vectors are orthogonal, the matrix A is diagonal.)

6.39. Consider the vector space V of polynomials $f(t)$ of degree ≤ 2 with inner product $\langle f, g \rangle = \int_{-1}^{1} f(t) g(t)\, dt$.

(a) Find $\langle f, g \rangle$ where $f(t) = t + 2$ and $g(t) = t^2 - 3t + 4$.

(b) Find the matrix A of the inner product with respect to the basis $\{1, t, t^2\}$ of V.

(c) Verify Theorem 6.14 by showing that $\langle f, g \rangle = [f]^T A [g]$ with respect to the basis $\{1, t, t^2\}$.

(a) $\langle f, g \rangle = \int_{-1}^{1} (t + 2)(t^2 - 3t + 4)\, dt = \int_{-1}^{1} (t^3 - t^2 - 2t + 8)\, dt = \left[\dfrac{t^4}{4} - \dfrac{t^3}{3} - t^2 + 8t \right]_{-1}^{1} = \dfrac{46}{3}$

(b) Here we use the fact that, if $r + s = n$,

$$\langle t^r, t^s \rangle = \int_{-1}^{1} t^n\, dt = \left[\frac{t^{n+1}}{n+1} \right]_{-1}^{1} = \begin{cases} 2/(n+1) & \text{if } n \text{ is even} \\ 0 & \text{if } n \text{ is odd} \end{cases}$$

Then $\langle 1, 1 \rangle = 2$, $\langle 1, t \rangle = 0$, $\langle 1, t^2 \rangle = \frac{2}{3}$, $\langle t, t \rangle = \frac{2}{3}$, $\langle t, t^2 \rangle = 0$, $\langle t^2, t^2 \rangle = \frac{2}{5}$. Thus

$$A = \begin{pmatrix} 2 & 0 & \frac{2}{3} \\ 0 & \frac{2}{3} & 0 \\ \frac{2}{3} & 0 & \frac{2}{5} \end{pmatrix}$$

(c) We have $[f]^T = (2, 1, 0)$ and $[g]^T = (4, -3, 1)$ relative to the given basis. Then

$$[f]^T A [g] = (2, 1, 0) \begin{pmatrix} 2 & 0 & \frac{2}{3} \\ 0 & \frac{2}{3} & 0 \\ \frac{2}{3} & 0 & \frac{2}{5} \end{pmatrix} \begin{pmatrix} 4 \\ -3 \\ 1 \end{pmatrix} = (4, \tfrac{2}{3}, \tfrac{4}{3}) \begin{pmatrix} 4 \\ -3 \\ 1 \end{pmatrix} = \tfrac{46}{3} = \langle f, g \rangle$$

6.40. Prove Theorem 6.13.

For any vectors u_1, u_2, and v,

$$\langle v_1 + u_2, v \rangle = (u_1 + u_2)^T Av = (u_1^T + u_2^T)Av = u_1^T Av + u_2^T Av = \langle u_1, v \rangle + \langle u_2, v \rangle$$

and, for any scalar k and vectors u, v,

$$\langle ku, v \rangle = (ku)^T Av = ku^T Av = k\langle u, v \rangle$$

Thus $[I_1]$ is satisfied.

Since $u^T Av$ is a scalar, $(u^T Av)^T = u^T Av$. Also, $A^T = A$ since A is symmetric. Therefore,

$$\langle u, v \rangle = u^T Av = (u^T Av)^T = v^T A^t t^{TT} = v^T Au = \langle v, u \rangle$$

Thus $[I_2]$ is satisfied.

Lastly, since A is positive definite, $X^T AX > 0$ for any nonzero $X \in \mathbf{R}^n$. Thus, for any nonzero vector v, $\langle v, v \rangle = v^T Av > 0$. Also, $\langle 0, 0 \rangle = 0^T A0 = 0$. Thus $[I_3]$ is satisfied. Accordingly, the function $\langle u, v \rangle = u^T Av$ is an inner product.

6.41. Prove Theorem 6.14.

Suppose $S = \{w_1, w_2, \ldots, w_n\}$ and $A = (k_{ij})$. Hence $k_{ij} = \langle w_i, w_j \rangle$. Suppose

$$u = a_1 w_1 + a_2 w_2 + \cdots + a_n w_n \qquad \text{and} \qquad v = b_1 w_1 + b_2 w_2 + \cdots + b_n w_n$$

Then

$$\langle u, v \rangle = \sum_{i=1}^{n} \sum_{j=1}^{n} a_i b_j \langle w_i, w_j \rangle \tag{1}$$

On the other hand,

$$[u]^T A[v] = (a_1, a_2, \ldots, a_n)\begin{pmatrix} k_{11} & k_{22} & \ldots & k_{1n} \\ k_{21} & k_{22} & \ldots & k_{2n} \\ \hline & \ldots\ldots\ldots & & \\ k_{n1} & k_{n2} & \ldots & k_{nn} \end{pmatrix}\begin{pmatrix} b_1 \\ b_2 \\ \vdots \\ b_n \end{pmatrix}$$

$$= \left(\sum_{i=1}^{n} a_i k_{i1}, \sum_{i=1}^{n} a_i k_{i2}, \ldots, \sum_{i=1}^{n} a_i k_{in} \right)\begin{pmatrix} b_1 \\ b_2 \\ \vdots \\ b_n \end{pmatrix} = \sum_{j=1}^{n} \sum_{i=1}^{n} a_i b_j k_{ij} \tag{2}$$

Since $k_{ij} = \langle w_i, w_j \rangle$, the final sums in (1) and (2) are equal. Thus $\langle u, v \rangle = [u]^T A[v]$.

6.42. Prove Theorem 6.15.

Since $\langle w_i, w_j \rangle = \langle w_j, w_i \rangle$ for any basis vectors w_i and w_j, the matrix A is symmetric. Let X be any nonzero vector in $\mathbf{R}^n$. Then $[u] = X$ for some nonzero vector $u \in V$. Using Theorem 6.14, we have $X^T AX = [u]^T A[u] = \langle u, u \rangle > 0$. Thus A is positive definite.

INNER PRODUCTS AND ORTHOGONAL MATRICES

6.43. Find an orthogonal matrix P whose first row is $u_1 = (\frac{1}{3}, \frac{2}{3}, \frac{2}{3})$.

First find a nonzero vector $w_2 = (x, y, z)$ which is orthogonal to u_1, i.e., for which

$$0 = \langle u_1, w_2 \rangle = \frac{x}{3} + \frac{2y}{3} + \frac{2z}{3} = 0 \qquad \text{or} \qquad x + 2y + 2z = 0$$

One such solution is $w_2 = (0, 1, -1)$. Normalize w_2 to obtain the second row of P, i.e., $u_2 = (0, 1/\sqrt{2}, -1/\sqrt{2})$.

Next find a nonzero vector $w_3 = (x, y, z)$ which is orthogonal to both u_1 and u_2, i.e., for which

$$0 = \langle u_1, w_3 \rangle = \frac{x}{3} + \frac{2y}{3} + \frac{2z}{3} = 0 \qquad \text{or} \qquad x + 2y + 2z = 0$$

$$0 = \langle u_2, w_3 \rangle = \frac{y}{\sqrt{2}} - \frac{y}{\sqrt{2}} = 0 \qquad \text{or} \qquad y - z = 0$$

Set $z = -1$ and find the solution $w_3 = (4, -1, -1)$. Normalize w_3 and obtain the third row of P, i.e., $u_3 = (4/\sqrt{18}, -1/\sqrt{18}, -1/\sqrt{18})$. Thus,

$$P = \begin{pmatrix} \frac{1}{3} & \frac{2}{3} & \frac{2}{3} \\ 0 & 1/\sqrt{2} & -1/\sqrt{2} \\ 4/3\sqrt{2} & -1/3\sqrt{2} & -1/3\sqrt{2} \end{pmatrix}$$

We emphasize that the above matrix P is not unique.

6.44. Let $A = \begin{pmatrix} 1 & 1 & -1 \\ 1 & 3 & 4 \\ 7 & -5 & 2 \end{pmatrix}$. Determine whether or not (a) the rows of A are orthogonal, (b) A is an orthogonal matrix, and (c) the columns of A are orthogonal.

(a)   Yes, since

$$(1, 1, -1) \cdot (1, 3, 4) = 1 + 3 - 4 = 0 \qquad (1, 1, -1) \cdot (7, -5, 2) = 7 - 5 - 2 = 0$$

$$(1, 3, 4) \cdot (7, -5, 2) = 7 - 15 + 8 = 0$$

(b)   No, since the rows of A are not unit vectors, e.g., $(1, 1, -1)^2 = 1 + 1 + 1 = 3$.

(c)   No, e.g., $(1, 1, 7) \cdot (1, 3, -5) = 1 + 3 - 35 = -31 \neq 0$.

6.45. Let B be the matrix obtained by normalizing each row of A in Problem 6.44. (a) Find B. (b) Is B an orthogonal matrix? (c) Are the columns of B orthogonal?

(a)   We have

$$\| (1, 1, -1) \|^2 = 1 + 1 + 1 = 3 \qquad \| (1, 3, 4) \|^2 = 1 + 9 + 16 = 26$$

$$\| (7, -5, 2) \|^2 = 49 + 25 + 4 = 78$$

Thus

$$B = \begin{pmatrix} 1/\sqrt{3} & 1/\sqrt{3} & -1/\sqrt{3} \\ 1/\sqrt{26} & 3/\sqrt{26} & 4/\sqrt{26} \\ 7/\sqrt{78} & -5/\sqrt{78} & 2/\sqrt{78} \end{pmatrix}$$

(b)   Yes, since the rows of B are still orthogonal and are now unit vectors.

(c)   Yes, since the rows of B form an orthonormal set of vectors, then, by Theorem 6.15, the columns of B must automatically form an orthonormal set.

6.46. Prove each of the following:

(a) P is orthogonal if and only if P^T is orthogonal.

(b)   If P is orthogonal, then P^{-1} is orthogonal.

(c)   If P and Q are orthogonal, then PQ is orthogonal.

(a)   We have $(P^T)^T = P$. Thus P is orthogonal iff $PP^T = I$ iff $P^{TT}P^T = I$ iff P^T is orthogonal.

(b)   We have $P^T = P^{-1}$ since P is orthogonal. Thus, by (a), P^{-1} is orthogonal.

(c) We have $P^T = P^T$ and $Q^T = Q^{-1}$. Thus

$$(PQ)(PQ)^T = PQQ^TP^T = PQQ^{-1}P^{-1} = I$$

Thus $(PQ)^T = (PQ)^{-1}$, and so PQ is orthogonal.

6.47. Suppose P is an orthogonal matrix. Show that:

    (a) $\langle Pu, Pv \rangle = \langle u, v \rangle$ for any $u, v \in V$;   (b) $\| Pu \| = \| u \|$ for every $u \in V$.

    (a)   Using $P^TP = I$, we have

$$\langle Pu, Pv \rangle = (Pu)^T(Pv) = u^TP^TPv = u^Tv = \langle u, v \rangle$$

    (b)   Using $P^TP = I$, we have

$$\| Pu \|^2 = \langle Pu, Pu \rangle = u^TP^TPu = u^Tu = \langle u, u \rangle = \| u \|^2$$

Taking the square root of both sides gives us our result.

6.48. Prove Theorem 6.17.

    Suppose

$$e_i' = b_{i1}e_1 + b_{i2}e_2 + \cdots + b_{in}e_n \qquad i = 1, \ldots, n \qquad (1)$$

Using Problem 6.23(b) and the fact that E' is orthonormal, we get

$$\delta_{ij} = \langle e_i', e_j' \rangle = b_{i1}b_{j1} + b_{i2}b_{j2} + \cdots + b_{in}b_{jn} \qquad (2)$$

Let $B = (b_{ij})$ be the matrix of coefficients in (1). (Then $P = B^T$.) Suppose $BB^T = (c_{ij})$. Then

$$c_{ij} = b_{i1}b_{j1} + b_{i2}b_{j2} + \cdots + b_{in}b_{jn} \qquad (3)$$

By (2) and (3), we have $c_{ij} = \delta_{ij}$. Thus $BB^T = I$. Accordingly, B is orthogonal, and hence $P = B^T$ is orthogonal.

6.49. Prove Theorem 6.18.

    Since $\{e_i\}$ is orthonormal, we get, by Problem 6.23(b),

$$\langle e_i', e_j' \rangle = a_{1i}a_{1j} + a_{2i}a_{2j} + \cdots + a_{ni}a_{nj} = \langle C_i, C_j \rangle$$

where C_i denotes the ith column of the orthogonal matrix $P = (a_{ij})$. Since P is orthogonal, its columns form an orthonormal set. This implies $\langle e_i', e_j' \rangle = \langle C_i, C_j \rangle = \delta_{ij}$. Thus $\{e_i'\}$ is an orthonormal basis.

COMPLEX INNER PRODUCT SPACES

6.50. Let V be a complex inner product space. Verify the relation

$$\langle u, av_1 + bv_2 \rangle = \bar{a}\langle u, v_1 \rangle + \bar{b}\langle u, v_2 \rangle$$

Using $[I_2^*]$, $[I_1^*]$, and then $[I_2^*]$, we find

$$\langle u, av_1 + bv_2 \rangle = \overline{\langle av_1 + bv_2, u \rangle} = \overline{a\langle v_1, u \rangle + b\langle v_2, u \rangle} = \bar{a}\overline{\langle v_1, u \rangle} + \bar{b}\overline{\langle v_2, u \rangle} + \bar{a}\langle u, v_1 \rangle + \bar{b}\langle u, v_2 \rangle$$

6.51. Suppose $\langle u, v \rangle = 3 + 2i$ in a complex inner product space V. Find:

    (a) $\langle (2 - 4i)u, v \rangle$;   (b) $\langle u, (4 + 3i)v \rangle$;   (c) $\langle (3 - 6i)u, (5 - 2i)v \rangle$

    (a) $\langle (2 - 4i)u, v \rangle = (2 - 4i)\langle u, v \rangle = (2 - 4i)(3 + 2i) = 14 - 18i$

    (b) $\langle u, (4 + 3i)v \rangle = \overline{(4 + 3i)}\langle u, v \rangle = (4 - 3i)(3 + 2i) = 18 - i$

    (c) $\langle (3 - 6i)u, (5 - 2i)v \rangle = (3 - 6i)\overline{(5 - 2i)}\langle u, v \rangle = (3 - 6i)(5 + 2i)(3 + 2i) = 137 - 30i$

6.52. Find the Fourier coefficient (component) c and the projection cw of $v = (3 + 4i, 2 - 3i)$ along $w = (5 + i, 2i)$ in $\mathbf{C}^2$.

Recall $c = \langle v, w \rangle / \langle w, w \rangle$. Compute

$$\langle v, w \rangle = (3 + 4i)(\overline{5 + i}) + (2 - 3i)(\overline{2i}) = (3 + 4i)(5 - i) + (2 - 3i)(-2i)$$
$$= 19 + 17i - 6 - 4i = 13 + 13i$$
$$\langle w, w \rangle = 25 + 1 + 4 = 30$$

Thus $c = (13 + 13i)/30 = \frac{13}{30} + 13i/30$. Accordingly,

$$\text{proj}\,(v, w) = cw = (\tfrac{26}{15} + 39i/15, \ -\tfrac{13}{15} + i/15)$$

6.53. Prove Theorem 6.19 (Cauchy–Schwarz).

If $v = 0$, the inequality reduces to $0 \le 0$ and hence is valid. Now suppose $v \ne 0$. Using $z\bar{z} = |z|^2$ (for any complex number z) and $\langle v, u \rangle = \overline{\langle u, v \rangle}$, we expand $\| u - \langle u, v \rangle tv \|^2 \ge 0$ where t is any real value:

$$0 \le \| u - \langle u, v \rangle tv \|^2 = \langle u - \langle u, v \rangle tv, \ u - \langle u, v \rangle tv \rangle$$
$$= \langle u, u \rangle - \overline{\langle u, v \rangle} t \langle u, v \rangle - \langle u, v \rangle t \langle v, u \rangle + \langle u, v \rangle \overline{\langle u, v \rangle} t^2 \langle v, v \rangle$$
$$= \| u \|^2 - 2t |\langle u, v \rangle|^2 + |\langle u, v \rangle|^2 t^2 \| v \|^2$$

Set $t = 1/\| v \|^2$ to find $0 \le \| u \|^2 - \dfrac{|\langle u, v \rangle|^2}{\| v \|^2}$, from which $|\langle u, v \rangle|^2 \le \| v \|^2 \| v \|^2$. Taking the square root of both sides, we obtain the required inequality.

6.54. Find an orthogonal basis for $u^\perp$ in $\mathbf{C}^3$ where $u = (1, i, 1 + i)$.

Here $u^\perp$ consists of all vectors $w = (x, y, z)$ such that

$$\langle w, u \rangle = x - iy + (1 - i)z = 0$$

Find one such solution, say $w_1 = (0, 1 - i, i)$. Then find a solution of the system

$$x - iy + (1 - i)z = 0 \qquad (1 + i)y - iz = 0$$

Here z is a free variable. Set $z = 1$ to obtain $y = i/(1 + i) = (1 + i)/2$ and $x = (3i - 3)2$. Multiplying by 2 yields the solution $w_2 = (3i - 3, 1 + i, 2)$. The vectors w_1 and w_2 form an orthogonal basis for $u^\perp$.

6.55. Find an orthonormal basis of the subspace W of $\mathbf{C}^3$ spanned by

$$v_1 = (1, i, 0) \quad \text{and} \quad v_2 = (1, 2, 1 - i).$$

Apply the Gram–Schmidt algorithm. Set $w_1 = v_1 = (1, i, 0)$. Compute

$$v_2 - \frac{\langle v_2, w_1 \rangle}{\| w_1 \|^2} w_1 = (1, 2, 1 - i) - \frac{1 - 2i}{2}(1, i, 0) = \left(\frac{1}{2} + i, \ 1 - \frac{1}{2}i, \ 1 - i \right)$$

Multiply by 2 to clear fractions obtaining $w_2 = (1 + 2i, 2 - i, 2 - 2i)$. Next find $\| w_1 \| = \sqrt{2}$ and then $\| w_2 \| = \sqrt{18}$. Normalizing $\{w_1, w_2\}$ we obtain the following required orthonormal basis of W:

$$\left\{ u_1 = \left(\frac{1}{\sqrt{2}}, \frac{i}{\sqrt{2}}, 0 \right), \qquad u_2 = \left(\frac{1 + 2i}{\sqrt{18}}, \frac{2 - i}{\sqrt{18}}, \frac{2 - 2i}{\sqrt{18}} \right) \right\}$$

6.56. Find the matrix P which represents the usual inner product on $\mathbf{C}^3$ relative to the basis $\{1, i, 1 - i\}$.

Compute

$$\langle 1, 1 \rangle = 1 \qquad\qquad \langle 1, i \rangle = \bar{i} = -i \qquad\qquad \langle 1, 1 - i \rangle = \overline{1 - i} = 1 + i$$

$$\langle i, i \rangle = j\bar{i} = 1 \qquad \langle i, 1 - i \rangle = i(\overline{1 - i}) = -1 + i \qquad \langle 1 - i, 1 - i \rangle = 2$$

Then, using $\langle u, v \rangle = \overline{\langle v, u \rangle}$, we have

$$P = \begin{pmatrix} 1 & -i & 1+i \\ i & 1 & -1+i \\ 1-i & -1-i & 2 \end{pmatrix}$$

(As expected, P is Hermitian, that is, $P^H = P$.)

NORMED VECTOR SPACES

6.57. Consider vectors $u = (1, 3, -6, 4)$ and $v = (3, -5, 1, -2)$ in $\mathbf{R}^4$. Find:

(a) $\|u\|_\infty$ and $\|v\|_\infty$ (c) $\|u\|_2$ and $\|v\|_2$

(b) $\|u\|_1$ and $\|v\|_1$ (d) $d_\infty(u, v)$, $d_1(u, v)$ and $d_2(u, v)$

(a) The infinity-norm chooses the maximum of the absolute values of the components. Hence

$$\|u\|_\infty = 6 \quad \text{and} \quad \|v\|_\infty = 5$$

(b) The one-norm adds the absolute values of the components. Thus

$$\|u\|_1 = 1 + 3 + 6 + 4 = 14 \qquad \|v\|_1 = 3 + 5 + 1 + 2 = 11$$

(c) The two-norm is equal to the square root of the sum of the squares of the components (i.e., the norm induced by the usual inner product on $\mathbf{R}^3$). Thus

$$\|u\|_2 = \sqrt{1 + 9 + 36 + 16} = \sqrt{62} \quad \text{and} \quad \|v\|_2 = \sqrt{9 + 25 + 1 + 4} = \sqrt{39}$$

(d) First find $u - v = (-2, 8, -7, 6)$. Then

$$d_\infty(u, v) = \|u - v\|_\infty = 8$$

$$d_1(u, v) = \|u - v\|_1 = 2 + 8 + 7 + 6 = 23$$

$$d_2(u, v) = \|u - v\|_2 = \sqrt{4 + 64 + 49 + 36} = \sqrt{153}$$

6.58. Consider the function $f(t) = t^2 - 4t$ in $C[0, 3]$. (a) Find $\|f\|_\infty$. (b) Plot $f(t)$ in the plane $\mathbf{R}^2$. (c) Find $\|f\|_1$. (d) Find $\|f\|_2$.

(a) We seek $\|f\|_\infty = \max(|f(t)|)$. Since $f(t)$ is differentiable on $[0, 3]$, $|f(t)|$ has a maximum at a critical point of $f(t)$, i.e., when the derivative $f'(t) = 0$, or at an endpoint of $[0, 3]$. Since $f'(t) = 2t - 4$, we set $2t - 4 = 0$ and obtain $t = 2$ as a critical point. Compute

$$f(2) = 4 - 8 = -4 \qquad f(0) = 0 - 0 = 0 \qquad f(3) = 9 - 12 = -3$$

Thus $\|f\|_\infty = |f(2)| = |-4| = 4$.

(b) Compute $f(t)$ for various values of t in $[0, 3]$, e.g.,

t	0	1	2	3
$f(t)$	0	-3	-4	-3

Plot the points in $\mathbf{R}^2$ and then draw a continuous curve through the points as shown in Fig. 6-11.

(c) We seek $\|f\|_1 = \int_0^3 |f(t)|\, dt$. As indicated by Fig. 6-11, $f(t)$ is negative in $[0, 3]$; hence $|f(t)| = -(t^2 - 4t) = 4t - t^2$. Thus

$$\|f\|_1 = \int_0^3 (4t - t^2)\, dt = \left[2t^2 - \frac{t^3}{3}\right]_0^3 = 18 - 9 = 9$$

(d) $$\|f\|_2^2 = \int_0^3 [f(t)]^2\, dt = \int_0^3 (t^4 - 8t^3 + 16t^2)\, dt = \left[\frac{t^5}{5} - 2t^4 + \frac{16t^3}{3}\right]_0^3 = \frac{153}{5}$$

Thus $\|f\|_2 = \sqrt{\frac{153}{5}}$.

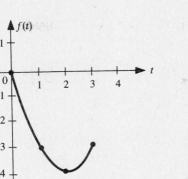

Fig. 6-11

6.59. Prove Theorem 6.25.

If $u \neq v$, then $u - v \neq 0$, and hence $d(u, v) = \| u - v \| > 0$. Also, $d(u, u) = \| u - u \| = \| 0 \| = 0$. Thus $[M_1]$ is satisfied. We also have

$$d(u, v) = \| u - v \| = \| -1(v - u) \| = |-1| \, \| v - u \| = \| v - u \| = d(v, u)$$

and

$$d(u, v) = \| u - v \| = \| (u - w) + (w - v) \| \leq \| u - w \| + \| w - v \| = d(u, w) + d(w, v)$$

Thus $[M_2]$ and $[M_3]$ are satisfied.

Supplementary Problems

INNER PRODUCTS

6.60. Verify that the following is an inner product on $\mathbf{R}^2$ where $u = (x_1, x_2)$ and $v = (y_1, y_2)$:

$$f(u, v) = x_1 y_1 - 2x_1 y_2 - 2x_2 y_1 + 5x_2 y_2$$

6.61. Find the values of k so that the following is an inner product on $\mathbf{R}^2$ where $u = (x_1, x_2)$ and $v = (y_1, y_2)$:

$$f(u, v) = x_1 y_1 = 3x_1 y_2 - 3x_2 y_1 + kx_2 y_2$$

6.62. Consider the vectors $u = (1, -3)$ and $v = (2, 5)$ in $\mathbf{R}^2$. Find:

(a) $\langle u, v \rangle$ with respect to the usual inner product in $\mathbf{R}^2$.

(b) $\langle u, v \rangle$ with respect to the inner product in $\mathbf{R}^2$ in Problem 6.60.

(c) $\| v \|$ using the usual inner product in $\mathbf{R}^2$.

(d) $\| v \|$ using the inner product in $\mathbf{R}^2$ in Problem 6.60.

6.63. Show that each of the following is not an inner product on $\mathbf{R}^3$ where $u = (x_1, x_2, x_3)$ and $v = (y_1, y_2, y_3)$:

(a) $\langle u, v \rangle = x_1 y_1 + x_2 y_2$ and   (b) $\langle u, v \rangle = x_1 y_2 x_3 + y_1 x_2 y_3$

6.64. Let V be the vector space of $m \times n$ matrices over $\mathbf{R}$. Show that $\langle A, B \rangle = \operatorname{tr} B^t A$ defines an inner product in V.

6.65. Let V be the vector space of polynomials over **R**. Show that $\langle f, g \rangle = \int_0^1 f(t) g(t)\, dt$ defines an inner product in V.

6.66. Suppose $|\langle u, v \rangle| = \| u \| \, \| v \|$. (That is, the Cauchy–Schwarz inequality reduces to an equality.) Show that u and v are linearly dependent.

6.67. Suppose $f(u, v)$ and $g(u, v)$ are inner products on a vector space V. Prove:

(a) The sum $f + g$ is an inner product on v where $(f + g)(u, v) = f(u, v) + g(u, v)$.

(b) The scalar product kf, for $k > 0$, is an inner product on V where $(kf)(u, v) = kf(u, v)$.

ORTHOGONALITY, ORTHOGONAL COMPLEMENTS, ORTHOGONAL SETS

6.68. Let V be the vector space of polynomials over **R** of degree ≤ 2 with inner product $\langle f, g \rangle = \int_0^1 f(t) g(t)\, dt$. Find a basis of the subspace W orthogonal to $h(t) = 2t + 1$.

6.69. Find a basis of the subspace W of $\mathbf{R}^4$ orthogonal to $u_1 = (1, -2, 3, 4)$ and $u_2 = (3, -5, 7, 8)$.

6.70. Find a basis for the subspace W of $\mathbf{R}^5$ orthogonal to the vectors $u_1 = (1, 1, 3, 4, 1)$ and $u_2 = (1, 2, 1, 2, 1)$.

6.71. Let $w = (1, -2, -1, 3)$ be a vector in $\mathbf{R}^4$. Find (a) an orthogonal and (b) an orthonormal basis for $w^\perp$.

6.72. Let W be the subspace of $\mathbf{R}^4$ orthogonal to $u_1 = (1, 1, 2, 2)$ and $u_2 = (0, 1, 2, -1)$. Find (a) an orthogonal and (b) an orthonormal basis for W. (Compare with Problem 6.69.)

6.73. Let S consist of the following vectors in $\mathbf{R}^4$:

$$u_1 = (1, 1, 1, 1) \qquad u_2 = (1, 1, -1, -1) \qquad u_3 = (1, -1, 1, -1) \qquad u_4 = (1, -1, -1, 1)$$

(a) Show that S is orthogonal and a basis of $\mathbf{R}^4$.

(b) Write $v = (1, 3, -5, 6)$ as a linear combination of u_1, u_2, u_3, u_4.

(c) Find the coordinates of an arbitrary vector $v = (a, b, c, d)$ in $\mathbf{R}^4$ relative to the basis S.

(d) Normalize S to obtain an orthonormal basis of $\mathbf{R}^4$.

6.74. Let V be the vector space of 2×2 matrices over **R** with inner product $\langle A, B \rangle = \text{tr } B^T A$. Show that the following is an orthonormal basis of V:

$$\left\{ \begin{pmatrix} 1 & 0 \\ 0 & 0 \end{pmatrix}, \begin{pmatrix} 0 & 1 \\ 0 & 0 \end{pmatrix}, \begin{pmatrix} 0 & 0 \\ 1 & 0 \end{pmatrix}, \begin{pmatrix} 0 & 0 \\ 0 & 1 \end{pmatrix} \right\}$$

6.75. Let V be the vector space of 2×2 matrices over **R** with inner product $\langle A, B \rangle = \text{tr } B^T A$. Find an orthogonal basis for the orthogonal complement of (a) the diagonal and (b) the symmetric matrices.

6.76. Suppose $\{u_1, u_2, \ldots, u_r\}$ is an orthogonal set of vectors. Show that $\{k_1 u_1, k_2 u_2, \ldots, k_r u_r\}$ is orthogonal for any scalars $k_1, k_2, \ldots, k_r$.

6.77. Let U and W be subspaces of a finite-dimensional inner product space V. Show that: (a) $(U + W)^\perp = U^\perp \cap W^\perp$; and (b) $(U \cap W)^\perp = U^\perp + W^\perp$.

PROJECTIONS, GRAM–SCHMIDT ALGORITHM, APPLICATIONS

6.78. Find an orthogonal and an orthonormal basis for the subspace U of $\mathbf{R}^4$ spanned by the vectors $v_1 = (1, 1, 1, 1)$, $v_2 = (1, -1, 2, 2)$, $v_3 = (1, 2, -3, -4)$.

6.79. Let V be the vector space of polynomials $f(t)$ with inner product $\langle f, g \rangle = \int_0^2 f(t)g(t)\,dt$. Apply the Gram–Schmidt algorithm to the set $\{1, t, t^2\}$ to obtain an orthogonal set $\{f_0, f_1, f_2\}$ with integer coefficients.

6.80. Suppose $v = (1, 2, 3, 4, 6)$. Find the projection of v onto W (or find $w \in W$ which minimizes $\|v - w\|$) where W is the subspace of $\mathbf{R}^5$ spanned by:

    (a) $u_1 = (1, 2, 1, 2, 1)$ and $u_2 = (1, -1, 2, -1, 1)$;   (b) $v_1 = (1, 2, 1, 2, 1)$ and $v_2 = (1, 0, 1, 5, -1)$

6.81. Let $V = C[-1, 1]$ with inner product $\langle f, g \rangle = \int_{-1}^1 f(t)g(t)\,dt$. Let W be the subspace of V of polynomials of degree ≤ 3. Find the projection of $f(t) = t^5$ onto W. [*Hint*: Use the (Legendre) polynomials $1, t, 3t^2 - 1$, $5t^3 - 3t$ in Example 6.13.]

6.82. Let $V = C[0, 1]$ with inner product $\langle f, g \rangle = \int_0^1 f(t)g(t)\,dt$. Let W be the subspace of V of polynomials of degree ≤ 2. Find the projection of $f(t) = t^3$ onto W. (*Hint*: Use the polynomials $1, 2t - 1, 6t^2 - 6t + 1$ in Problem 6.27.)

6.83. Let U be the subspace of $\mathbf{R}^4$ spanned by the vectors

$$v_1 = (1, 1, 1, 1) \qquad v_2 = (1, -1, 2, 2) \qquad v_3 = (1, 2, -3, -4)$$

Find the projection of $v = (1, 2, -3, 4)$ onto U. (*Hint*: Use Problem 6.78.)

INNER PRODUCTS AND POSITIVE DEFINITE MATRICES, ORTHOGONAL MATRICES

6.84. Find the matrix A which represents the usual inner product on $\mathbf{R}^2$ relative to each of the following bases of $\mathbf{R}^2$: (a) $\{v_1 = (1, 4), v_2 = (2, -3)\}$ and (b) $\{w_1 = (1, -3), w_2 = (6, 2)\}$.

6.85. Consider the following inner product on $\mathbf{R}^2$:

$$f(u, v) = x_1 y_1 - 2x_1 y_2 - 2x_2 y_1 + 5x_2 y_2 \qquad \text{where} \quad u = (x_1, x_2) \quad \text{and} \quad v = (y_1, y_2)$$

Find the matrix B which represents this inner product on $\mathbf{R}^2$ relative to each basis in Problem 6.84.

6.86. Find the matrix C which represents the usual basis on $\mathbf{R}^3$ relative to the basis S of $\mathbf{R}^3$ consisting of the vectors: $u_1 = (1, 1, 1)$, $u_2 = (1, 2, 1)$, $u_3 = (1, -1, 3)$.

6.87. Consider the vector space V of polynomials $f(t)$ of degree ≤ 2 with inner product $\langle f, g \rangle = \int_0^1 f(t)g(t)\,dt$.

    (a) Find $\langle f, g \rangle$ where $f(t) = t + 2$ and $g(t) = t^2 - 3t + 4$.

    (b) Find the matrix A of the inner product with respect to the basis $\{1, t, t^2\}$ of V.

    (c) Verify Theorem 6.14 that $\langle f, g \rangle = [f]^T A [g]$ with respect to the basis $\{1, t, t^2\}$.

6.88. Determine which of the following matrices are positive definite:

    (a) $\begin{pmatrix} 1 & 3 \\ 3 & 5 \end{pmatrix}$,   (b) $\begin{pmatrix} 3 & 4 \\ 4 & 7 \end{pmatrix}$,   (c) $\begin{pmatrix} 4 & 2 \\ 2 & 1 \end{pmatrix}$,   (d) $\begin{pmatrix} 6 & -7 \\ -7 & 9 \end{pmatrix}$

6.89. Determine whether or not A is positive definite where

    (a) $A = \begin{pmatrix} 2 & -2 & 1 \\ -2 & 3 & -2 \\ 1 & -2 & 2 \end{pmatrix}$ (b) $A = \begin{pmatrix} 1 & 1 & 2 \\ 1 & 2 & 6 \\ 2 & 6 & 9 \end{pmatrix}$

6.90. Suppose A and B are positive definite matrices. Show that: (a) $A + B$ is positive definite; (b) kA is positive definite for $k > 0$.

6.91. Suppose B is a real nonsingular matrix. Show that: (a) $B^T B$ is symmetric, and (b) $B^T B$ is positive definite.

6.92. Find the number and exhibit all 2×2 orthogonal matrices of the form $\begin{pmatrix} \frac{1}{3} & x \\ y & z \end{pmatrix}$.

6.93. Find a 3×3 orthogonal matrix P whose first two rows are multiples of $u = (1, 1, 1)$ and $v = (1, -2, 3)$, respectively.

6.94. Find a symmetric orthogonal matrix P whose first row is $(\frac{1}{3}, \frac{2}{3}, \frac{2}{3})$. (Compare with Problem 6.43.)

6.95. Real matrices A and B are said to be *orthogonally equivalent* if there exists an orthogonal matrix P such that $B = P^T A P$. Show that this relation is an equivalence relation.

COMPLEX INNER PRODUCT SPACES

6.96. Verify that

$$\langle a_1 u_1 + a_2 u_2, b_1 v_1 + b_2 v_2 \rangle = a_1 \bar{b}_1 \langle u_1, v_1 \rangle + a_1 \bar{b}_2 \langle u_1, v_2 \rangle + a_2 \bar{b}_1 \langle u_2, v_1 \rangle + a_2 \bar{b}_2 \langle u_2, v_2 \rangle$$

More generally, prove that $\left\langle \displaystyle\sum_{i=1}^m a_i u_i, \sum_{j=1}^n b_j v_j \right\rangle = \displaystyle\sum_{i,j} a_i \bar{b}_j \langle u_i, v_i \rangle$.

6.97. Consider $u = (1 + i, 3, 4 - i)$ and $v = (3 - 4i, 1 + i, 2i)$ in $\mathbf{C}^3$. Find:

(a) $\langle u, v \rangle$,   (b) $\langle v, u \rangle$,   (c) $\| u \|$,   (d) $\| v \|$,   (e) $d(u, v)$.

6.98. Find the Fourier coefficient c and the projection cw of

(a) $u = (3 + i, 5 - 2i)$ along $w = (5 + i, 1 + i)$ in $\mathbf{C}^2$;

(b) $u = (1 - i, 3i, 1 + i)$ along $w = (1, 2 - i, 3 + 2i)$ in $\mathbf{C}^3$.

6.99. Let $u = (z_1, z_2)$ and $v = (w_1, w_2)$ belong to $\mathbf{C}^2$. Verify that the following is an inner product on $\mathbf{C}^2$:

$$f(u, v) = z_1 \bar{w}_1 + (1 + i)z_1 \bar{w}_2 + (1 - i)z_2 \bar{w}_1 + 3z_2 \bar{w}_2$$

6.100. Find an orthogonal basis and an orthonormal basis for the subspace W of $\mathbf{C}^3$ spanned by $u_1 = (1, i, 1)$ and $u_2 = (1 + i, 0, 2)$.

6.101. Let $u = (z_1, z_2)$ and $v = (w_1, w_2)$ belong to $\mathbf{C}^2$. For what values of $a, b, c, d \in \mathbf{C}$ is the following an inner product on $\mathbf{C}^2$?

$$f(u, v) = az_1 \bar{w}_1 + bz_1 \bar{w}_2 + cz_2 \bar{w}_1 + dz_2 \bar{w}_2$$

6.102. Prove the following polar form for an inner product in a complex space V:

$$\langle u, v \rangle = \tfrac{1}{4}\| u + v \|^2 - \tfrac{1}{4}\| u - v \|^2 + \tfrac{1}{4}\| u + iv \|^2 - \tfrac{1}{4}\| u - iv \|^2$$

[Compare with Problem 6.9(b).]

6.103. Let V be a real inner product space. Show that:

(i) $\| u \| = \| v \|$ if and only if $\langle u + v, u - v \rangle = 0$;

(ii) $\| u + v \|^2 = \| u \|^2 + \| v \|^2$ if and only if $\langle u, v \rangle = 0$.

Show by counterexamples that the above statements are not true for, say, $\mathbf{C}^2$.

6.104. Find the matrix P which represents the usual inner product on $\mathbf{C}^3$ relative to the basis $\{1, 1 + i, 1 - 2i\}$.

6.105. A complex matrix A is *unitary* if it is invertible and $A^{-1} = A^H$. Alternatively, A is unitary if its rows (columns) form an orthonormal set of vectors (relative to the usual inner product of $\mathbf{C}^n$). Find a unitary matrix whose first row is (a) a multiple of $(1, 1 - i)$; and (b) $(\frac{1}{2}, \frac{1}{2}i, \frac{1}{2} - \frac{1}{2}i)$.

NORMED VECTOR SPACES

6.106. Consider vectors $u = (1, -3, 4, 1, -2)$ and $v = (3, 1, -2, -3, 1)$ in $\mathbf{R}^5$. Find:

    (a) $\|u\|_\infty$ and $\|v\|_\infty$ (c) $\|u\|_2$ and $\|v\|_2$

    (b) $\|u\|_1$ and $\|v\|_1$ (d) $d_\infty(u, v)$, $d_1(u, v)$ and $d_2(u, v)$

6.107. Consider vectors $u = (1 + i, 2 - 4i)$ and $v = (1 - i, 2 + 3i)$ in $\mathbf{C}^2$. Find:

    (a) $\|u\|_\infty$ and $\|v\|_\infty$ (c) $\|u\|_2$ and $\|v\|_2$

    (b) $\|u\|_1$ and $\|v\|_1$ (d) $d_1(u, v)$, $d_\infty(u, v)$ and $d_2(u, v)$

6.108. Consider the functions $f(t) = 5t - t^2$ and $g(t) = 3t - t^2$ in $C[0, 4]$. Find: (a) $d_\infty(f, g)$, (b) $d_1(f, g)$, and (c) $d_2(f, g)$.

6.109. Prove that: (a) $\|\cdot\|_1$ is a norm on $\mathbf{R}^n$; (b) $\|\cdot\|_\infty$ is a norm on $\mathbf{R}^n$.

6.110. Prove that: (a) $\|\cdot\|_1$ is a norm on $C[a, b]$; (b) $\|\cdot\|_\infty$ is a norm on $C[a, b]$.

Answers to Supplementary Problems

6.61. $k > 9$

6.62. (a) -13,   (b) -71,   (c) $\sqrt{29}$,   (d) $\sqrt{89}$

6.63. Let $u = (0, 0, 1)$. Then $\langle u, u \rangle = 0$ in both cases.

6.68. $\{7t^2 - 5t, 12t^2 - 5\}$

6.69. $\{(1, 2, 1, 0), (4, 4, 0, 1)\}$

6.70. $(-1, 0, 0, 0, 1), (-6, 2, 0, 1, 0), (-5, 2, 1, 0, 0)$

6.71. (a) $(0, 0, 3, 1), (0, 3, -3, 1), (2, 10, -9, 3)$;   (b) $(0, 0, 3, 1)/\sqrt{10}, (0, 3, -3, 1)/\sqrt{19}, (2, 10, -9, 3)/\sqrt{194}$

6.72. (a) $(0, 2, -1, 0), (-15, 1, 2, 5)$;   (b) $(0, 2, -1, 0)/\sqrt{5}, (-15, 1, 2, 5)/\sqrt{255}$

6.73. (b) $v = (5u_1 + 3u_2 - 13u_3 + 9u_4)/4$

    (c) $[v] = [a + b + c + d, a + b - c - d, a - b + c - d, a - b - c + d]/4$

6.75. (a) $\begin{pmatrix} 0 & 1 \\ 0 & 0 \end{pmatrix}, \begin{pmatrix} 0 & 0 \\ 1 & 0 \end{pmatrix}$ and   (b) $\begin{pmatrix} 0 & -1 \\ 1 & 0 \end{pmatrix}$

6.78. $w_1 = (1, 1, 1, 1), w_2 = (0, -2, 1, 1), w_3 = (12, -4, -1, -7)$.

6.79. $f_0 = 1, f_1 = t - 1, f_3 = 3t^2 - 6t + 2$

6.80. (a) proj $(v, W) = (21, 27, 26, 27, 21)/8$

  (b) First find orthogonal basis for W: $w_1 = (1, 2, 1, 2, 1)$, $w_2 = (0, 2, 0, -3, 2)$. Then proj $(v, W) = (34, 76, 34, 56, 42)/17$

6.81. proj $(f, W) = 10t^3/9 - 5t/21$

6.82. proj $(f, W) = 3t^2/2 - 3t/5 + \frac{1}{20}$

6.83. proj $(v, U) = (-14, 158, 47, 89)/70$

6.84. (a) $\begin{pmatrix} 17 & -10 \\ -10 & 13 \end{pmatrix}$, (b) $\begin{pmatrix} 10 & 0 \\ 0 & 40 \end{pmatrix}$

6.85. (a) $\begin{pmatrix} 65 & -68 \\ -68 & 73 \end{pmatrix}$, (b) $\begin{pmatrix} 58 & 23 \\ 23 & 8 \end{pmatrix}$

6.86. $\begin{pmatrix} 3 & 4 & 3 \\ 4 & 6 & 2 \\ 3 & 2 & 11 \end{pmatrix}$

6.87. (a) $\frac{83}{12}$, (b) $\begin{pmatrix} 1 & \frac{1}{2} & \frac{1}{3} \\ \frac{1}{2} & \frac{1}{3} & \frac{1}{4} \\ \frac{1}{3} & \frac{1}{4} & \frac{1}{5} \end{pmatrix}$

6.88. (a) No, (b) yes, (c) no, (d) yes

6.89. (a) Yes, (b) no

6.92. Four: $\begin{pmatrix} \frac{1}{3} & \sqrt{8}/3 \\ \sqrt{8}/3 & -\frac{1}{3} \end{pmatrix}, \begin{pmatrix} \frac{1}{3} & \sqrt{8}/3 \\ -\sqrt{8}/3 & -\frac{1}{3} \end{pmatrix}, \begin{pmatrix} \frac{1}{3} & -\sqrt{8}/3 \\ \sqrt{8}/3 & \frac{1}{3} \end{pmatrix}, \begin{pmatrix} \frac{1}{3} & -\sqrt{8}/3 \\ -\sqrt{8}/3 & -\frac{1}{3} \end{pmatrix}$

6.93. $P = \begin{pmatrix} 1/\sqrt{3} & 1/\sqrt{3} & 1/\sqrt{3} \\ 1/\sqrt{14} & -2/\sqrt{14} & 3/\sqrt{14} \\ 5/\sqrt{38} & -2/\sqrt{38} & -3/\sqrt{38} \end{pmatrix}$

6.94. $\begin{pmatrix} \frac{1}{3} & \frac{2}{3} & \frac{1}{3} \\ \frac{2}{3} & -\frac{2}{3} & \frac{1}{3} \\ \frac{2}{3} & \frac{1}{3} & -\frac{2}{3} \end{pmatrix}$

6.97. (a) $-4i$, (b) $4i$, (c) $\sqrt{28}$, (d) $\sqrt{31}$, (e) $\sqrt{59}$

6.98. (a) $c = (19 - 5i)/28$, (b) $c = (3 + 6i)/19$

6.100. $\{v_1 = (1, i, 1)/\sqrt{3}, v_2 = (2i, 1 - 3i, 3 - i)/\sqrt{24}\}$

6.101. a and d real and positive, $c = \bar{b}$ and $ad - bc$ positive.

6.103. $u = (1, 2), v = (i, 2i)$

6.104. $P = \begin{pmatrix} 1 & 1 - i & 1 + 2i \\ 1 + i & 2 & -2 + 3i \\ 1 - 2i & -2 - 3i & 5 \end{pmatrix}$

6.105. (a) $\begin{pmatrix} 1/\sqrt{3} & (1-i)/\sqrt{3} \\ (1+i)/\sqrt{3} & -1/\sqrt{3} \end{pmatrix}$,   (b) $\begin{pmatrix} \frac{1}{2} & \frac{1}{2}i & \frac{1}{2}-\frac{1}{2}i \\ i/\sqrt{2} & -1/\sqrt{2} & 0 \\ \frac{1}{2} & -\frac{1}{2}i & -\frac{1}{2}+\frac{1}{2}i \end{pmatrix}$

6.106. (a) 4 and 3,   (b) 11 and 13,   (c) $\sqrt{31}$ and $\sqrt{24}$,   (d) 6, 19, and 9

6.107. (a) $\sqrt{20}$ and $\sqrt{13}$,   (b) $\sqrt{2}+\sqrt{20}$ and $\sqrt{2}+\sqrt{13}$,   (c) $\sqrt{22}$ and $\sqrt{15}$,   (d) 7, 9, and $\sqrt{53}$

6.108. (a) 8,   (b) 16,   (c) $\frac{256}{3}$

Chapter 7

Determinants

7.1 INTRODUCTION

Each n-square matrix $A = (a_{ij})$ is assigned a special scalar called the *determinant of A*, denoted by $\det (A)$ or $|A|$ or

$$\begin{vmatrix} a_{11} & a_{12} & \cdots & a_{1n} \\ a_{21} & a_{22} & \cdots & a_{2n} \\ \vdots & & & \\ a_{n1} & a_{n2} & \cdots & a_{nn} \end{vmatrix}$$

We emphasize that an $n \times n$ array of scalars enclosed by straight lines, called a *determinant of order n*, is not a matrix but denotes the determinant of the enclosed array of scalars, i.e., the enclosed matrix.

The determinant function was first discovered in the investigation of systems of linear equations. We shall see that the determinant is an indispensable tool in investigating and obtaining properties of square matrices.

The definition of the determinant and most of its properties also apply in the case where the entries of a matrix come from a ring.

We begin with the special case of determinants of orders one, two, and three. Then we define a determinant of arbitrary order. This general definition is preceded with a discussion of permutations, which is necessary for our general definition of the determinant.

7.2 DETERMINANTS OF ORDERS ONE AND TWO

The determinants of orders one and two are defined as follows:

$$|a_{11}| = a_{11}$$

$$\begin{vmatrix} a_{11} & a_{12} \\ a_{21} & a_{22} \end{vmatrix} = a_{11}a_{22} - a_{12}a_{21}$$

Thus the determinant of a 1×1 matrix $A = (a_{11})$ is the scalar a_{11} itself, that is, $\det (A) = |a_{11}| = a_{11}$. The determinant of order two may easily be remembered by using the following diagram:

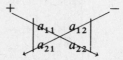

That is, the determinant is equal to the product of the elements along the plus-labeled arrow minus the product of the elements along the minus-labeled arrow. (There is an analogous diagram for determinants of order three but not for higher-order determinants.)

Example 7.1

(a) Since the determinant of order one is the scalar itself, we have $\det (24) = 24$, $\det (-6) = -6$, and $\det (t + 2) = t + 2$.

246

(b) $\begin{vmatrix} 5 & 4 \\ 2 & 3 \end{vmatrix} = (5)(3) - (4)(2) = 15 - 8 = 7$ and $\begin{vmatrix} 2 & 1 \\ -4 & 6 \end{vmatrix} = (2)(6) - (1)(-4) = 12 + 4 = 16.$

Consider two linear equations in two unknowns:

$$a_1 x + b_1 y = c_1$$
$$a_2 x + b_2 y = c_2$$

Recall (Problem 1.60) that the system has a unique solution if and only if $D \equiv a_1 b_2 - a_2 b_1 \neq 0$; and that solution is

$$x = \frac{b_2 c_1 - b_1 c_2}{a_1 b_2 - a_2 b_1} \qquad y = \frac{a_1 c_1 - a_2 c_1}{a_1 b_2 - a_2 b_1}$$

The solution may be expressed completely in terms of determinants:

$$x = \frac{N_x}{D} = \frac{b_2 c_1 - b_1 c_2}{a_1 b_2 - a_2 b_1} = \frac{\begin{vmatrix} c_1 & b_1 \\ c_2 & b_2 \end{vmatrix}}{\begin{vmatrix} a_1 & b_1 \\ a_2 & b_2 \end{vmatrix}} \qquad y = \frac{N_y}{D} = \frac{a_1 c_2 - a_2 c_1}{a_1 b_2 - a_2 b_1} = \frac{\begin{vmatrix} a_1 & c_1 \\ a_2 & c_2 \end{vmatrix}}{\begin{vmatrix} a_1 & b_1 \\ a_2 & b_2 \end{vmatrix}}$$

Here D, the determinant of the matrix of coefficients, appears in the denominator of both quotients. The numerators N_x and N_y of the quotients for x and y, respectively, can be obtained by substituting the column of constant terms in place of the column of coefficients of the given unknown in the matrix of coefficients.

Example 7.2. Solve by determinants: $\begin{cases} 2x - 3y = 7 \\ 3x + 5y = 1 \end{cases}$.

The determinant D of the matrix of coefficients is

$$D = \begin{vmatrix} 2 & -3 \\ 3 & 5 \end{vmatrix} = (2)(5) - (3)(-3) = 10 + 9 = 19$$

Since $D \neq 0$, the system has a unique solution. To obtain the numerator N_x replace, in the matrix of coefficients, the coefficients of x by the constant terms:

$$N_x = \begin{vmatrix} 7 & -3 \\ 1 & 5 \end{vmatrix} = (7)(5) - (1)(-3) = 35 + 3 = 38$$

To obtain the numerator N_y replace, in the matrix of coefficients, the coefficients of y by the constant terms:

$$N_y = \begin{vmatrix} 2 & 7 \\ 3 & 1 \end{vmatrix} = (2)(1) - (3)(7) = 2 - 21 = -19$$

Thus the unique solution of the system is

$$x = \frac{N_x}{D} = \frac{38}{19} = 2 \qquad \text{and} \qquad y = \frac{N_x}{D} = \frac{-19}{19} = -1$$

Remark: The result in Example 7.2 actually holds for any system of n linear equations in n unknowns, and this general result will be discussed in Section 7.5. We emphasize that this result is important for theoretical reasons, that is, in practice, Gaussian elimination, not determinants, is usually used to find the solution of a system of linear equations.

7.3 DETERMINANTS OF ORDER THREE

Consider an arbitrary 3×3 matrix $A = (a_{ij})$. The determinant of A is defined as follows:

$$\det (A) = \begin{vmatrix} a_{11} & a_{12} & a_{13} \\ a_{21} & a_{22} & a_{23} \\ a_{31} & a_{32} & a_{33} \end{vmatrix} = \begin{aligned} & a_{11}a_{22}a_{33} + a_{12}a_{23}a_{31} + a_{13}a_{21}a_{32} \\ & - a_{13}a_{22}a_{31} - a_{12}a_{21}a_{33} - a_{11}a_{23}a_{32} \end{aligned}$$

Observe that there are six products, each product consisting of three elements of the original matrix. Three of the products are plus-labeled (keep their sign) and three of the products are minus-labeled (change their sign).

The diagrams in Fig. 7-1 may help to remember the above six products in det (A). That is, the determinant is equal to the sum of the products of the elements along the three plus-labeled arrows in Fig. 7-1 plus the sum of the negatives of the products of the elements along the three minus-labeled arrows. We emphasize that there are no such diagrammatic devices to remember determinants of higher order.

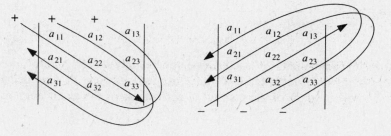

Fig. 7-1

Example 7.3

$$\begin{vmatrix} 2 & 1 & 1 \\ 0 & 5 & -2 \\ 1 & -3 & 4 \end{vmatrix} = (2)(5)(4) + (1)(-2)(1) + (1)(-3)(0) - (1)(5)(1) - (-3)(-2)(2) - (4)(1)(0)$$

$$= 40 - 2 + 0 - 5 - 12 - 0 = 21$$

The determinant of the 3×3 matrix $A = (a_{ij})$ may be rewritten as:

$$\det (A) = a_{11}(a_{22}a_{33} - a_{23}a_{32}) - a_{12}(a_{21}a_{33} - a_{23}a_{31}) + a_{13}(a_{21}a_{32} - a_{22}a_{31})$$

$$= a_{11} \begin{vmatrix} a_{22} & a_{23} \\ a_{32} & a_{33} \end{vmatrix} - a_{12} \begin{vmatrix} a_{21} & a_{23} \\ a_{31} & a_{33} \end{vmatrix} + a_{13} \begin{vmatrix} a_{21} & a_{22} \\ a_{31} & a_{32} \end{vmatrix}$$

which is a linear combination of three determinants of order two whose coefficients (with alternating signs) form the first row of the given matrix. This linear combination may be indicated in the form

$$a_{11} \begin{vmatrix} a_{11} & a_{12} & a_{13} \\ a_{21} & a_{22} & a_{23} \\ a_{31} & a_{32} & a_{33} \end{vmatrix} - a_{12} \begin{vmatrix} a_{11} & a_{12} & a_{13} \\ a_{21} & a_{22} & a_{23} \\ a_{31} & a_{32} & a_{33} \end{vmatrix} + a_{13} \begin{vmatrix} a_{11} & a_{12} & a_{13} \\ a_{21} & a_{22} & a_{23} \\ a_{31} & a_{32} & a_{33} \end{vmatrix}$$

Note that each 2×2 matrix can be obtained by deleting, in the original matrix, the row and column containing its coefficient.

Example 7.4

$$\begin{vmatrix} 1 & 2 & 3 \\ 4 & -2 & 3 \\ 0 & 5 & -1 \end{vmatrix} = 1 \begin{vmatrix} 1 & 2 & 3 \\ 4 & -2 & 3 \\ 0 & 5 & -1 \end{vmatrix} - 2 \begin{vmatrix} 1 & 2 & 3 \\ 4 & -2 & 3 \\ 0 & 5 & -1 \end{vmatrix} + 3 \begin{vmatrix} 1 & 2 & 3 \\ 4 & -2 & 3 \\ 0 & 5 & -1 \end{vmatrix}$$

$$= 1 \begin{vmatrix} -2 & 3 \\ 5 & -1 \end{vmatrix} - 2 \begin{vmatrix} 4 & 3 \\ 0 & -1 \end{vmatrix} + 3 \begin{vmatrix} 4 & -2 \\ 0 & 5 \end{vmatrix}$$

$$= 1(2 - 15) - 2(-4 + 0) + 3(20 + 0) = -13 + 8 + 60 = 55$$

7.4 PERMUTATIONS

A permutation σ of the set $\{1, 2, \ldots, n\}$ is a one-to-one mapping of the set onto itself or, equivalently, a rearrangement of the numbers $1, 2, \ldots, n$. Such a permutation σ is denoted by

$$\sigma = \begin{pmatrix} 1 & 2 & \cdots & n \\ j_1 & j_2 & \cdots & j_n \end{pmatrix} \qquad \text{or} \qquad \sigma = j_1 j_2 \cdots j_n \qquad \text{where } j_i = \sigma(i)$$

The set of all such permutations is denoted by S_n, and the number of such permutations is $n!$. If $\sigma \in S_n$, then the inverse mapping $\sigma^{-1} \in S_n$; and if $\sigma, \tau \in S_n$, then the composition mapping $\sigma \circ \tau \in S_n$. Also, the identity mapping $\varepsilon = \sigma \circ \sigma^{-1}$ belongs to S_n. (In fact, $\varepsilon = 12 \ldots n$.)

Example 7.5

(a) There are $2! = 2 \cdot 1 = 2$ permutations in S_2: the permutations 12 and 21.

(b) There are $3! = 3 \cdot 2 \cdot 1 = 6$ permutations in S_3: the permutations 123, 132, 213, 231, 312, 321.

Consider an arbitrary permutation σ in S_n; say $\sigma = j_1 j_2 \cdots j_n$. We say σ is an even or odd permutation according as to whether there is an even or odd number of inversions in σ. By an *inversion* in σ, we mean a pair of integers (i, k) such that $i > k$ but i precedes k in σ. We then define the sign or parity of σ, written sgn σ, by

$$\text{sgn } \sigma = \begin{cases} 1 & \text{if } \sigma \text{ is even} \\ -1 & \text{if } \sigma \text{ is odd} \end{cases}$$

Example 7.6

(a) Consider the permutation $\sigma = 35142$ in S_5. For each element, count the number of elements smaller than it and to the right of it. Thus

$$\text{3 produces the inversions } (3, 1) \text{ and } (3, 2);$$
$$\text{5 produces the inversions } (5, 1), (5, 4), (5, 2);$$
$$\text{4 produces the inversion } (4, 2);$$

(Note that 1 and 2 produce no inversions.) Since there are, in all, six inversions, σ is even and sgn $\sigma = 1$.

(b) The identity permutation $\varepsilon = 123 \ldots n$ is even because there are no inversions in ε.

(c) In S_2, the permutation 12 is even and 21 is odd.

In S_3, the permutations 123, 231 and 312 are even, and the permutations 132, 213 and 321 are odd.

(d) Let τ be the permutation which interchanges two numbers i and j and leaves the other numbers fixed, that is,

$$\tau(i) = j \qquad \tau(j) = i \qquad \tau(k) = k \qquad k \neq i, j$$

We call τ a transposition. If $i < j$, then

$$\tau = 12 \ldots (i - 1)j(i + 1) \ldots (j - 1)i(j + 1) \ldots n$$

There are $2(j - i - 1) + 1$ inversions in τ as follows:

$$(j, i), (j, x), (x, i) \qquad \text{where } x = i + 1, \ldots, j - 1$$

Thus the transposition τ is odd.

7.5 DETERMINANTS OF ARBITRARY ORDER

Let $A = (a_{ij})$ be an n-square matrix over a field K:

$$A = \begin{pmatrix} a_{11} & a_{12} & \cdots & a_{1n} \\ a_{21} & a_{22} & \cdots & a_{2n} \\ \cdots\cdots\cdots\cdots\cdots\cdots\cdots \\ a_{n1} & a_{n2} & \cdots & a_{nn} \end{pmatrix}$$

Consider a product of n elements of A such that one and only one element comes from each row and one and only one element comes from each column. Such a product can be written in the form

$$a_{1j_1} a_{2j_2} \cdots a_{nj_n}$$

that is, where the factors come from successive rows and so the first subscripts are in the natural order $1, 2, \ldots, n$. Now since the factors come from different columns, the sequence of second subscripts form a permutation $\sigma = j_1 j_2 \cdots j_n$ in S_n. Conversely, each permutation in S_n determines a product of the above form. Thus the matrix A contains $n!$ such products.

Definition: The determinant of $A = (a_{ij})$, denoted by det (A) or $|A|$, is the sum of all the above $n!$ products where each such product is multiplied by sgn σ. That is,

$$|A| = \sum_\sigma (\text{sgn } \sigma) a_{1j_1} a_{2j_2} \cdots a_{nj_n}$$

or

$$|A| = \sum_{\sigma \in S_n} (\text{sgn } \sigma) a_{1\sigma(1)} a_{2\sigma(2)} \cdots a_{n\sigma(n)}$$

The determinant of the n-square matrix A is said to be of order n.

The next example shows that the above definition agrees with the previous definition of determinants of orders one, two, and three.

Example 7.7

(a) Let $A = (a_{11})$ be a 1×1 matrix. Since S_1 has only one permutation which is even, det $(A) = a_{11}$, the number itself.

(b) Let $A = (a_{ij})$ be a 2×2 matrix. In S_2, the permutation 12 is even and the permutation 21 is odd. Hence

$$\det (A) = \begin{vmatrix} a_{11} & a_{12} \\ a_{21} & a_{22} \end{vmatrix} = a_{11}a_{22} - a_{12}a_{21}$$

(c) Let $A = (a_{ij})$ be a 3×3 matrix. In S_3, the permutations 123, 231 and 312 are even, and the permutations 321, 213 and 132 are odd. Hence

$$\det (A) = \begin{vmatrix} a_{11} & a_{12} & a_{13} \\ a_{21} & a_{22} & a_{23} \\ a_{31} & a_{32} & a_{33} \end{vmatrix} = a_{11}a_{22}a_{33} + a_{12}a_{23}a_{31} + a_{13}a_{21}a_{32} - a_{13}a_{22}a_{31} - a_{12}a_{21}a_{33} - a_{11}a_{23}a_{32}$$

As n increases, the number of terms in the determinant becomes astronomical. Accordingly, we use indirect methods to evaluate determinants rather than its definition. In fact, we prove a number of properties about determinants which will permit us to shorten the computation considerably. In particular, we show that a determinant of order n is equal to a linear combination of determinants of order $n - 1$ as in case $n = 3$ above.

7.6   PROPERTIES OF DETERMINANTS

We now list basic properties of the determinant.

Theorem 7.1:   The determinant of a matrix A and its transpose A^T are equal; that is, $|A| = |A^T|$.

By this theorem, proved in Problem 7.21, any theorem about the determinant of a matrix A which concerns the rows of A will have an analogous theorem concerning the columns of A.

The next theorem, proved in Problem 7.23, gives certain cases for which the determinant can be obtained immediately.

Theorem 7.2:   Let A be a square matrix.

    (*a*)   If A has a row (column) of zeros, then $|A| = 0$.

    (*b*)   If A has two identical rows (columns), then $|A| = 0$.

    (*c*)   If A is triangular, i.e., A has zeros above or below the diagonal, then $|A| =$ product of diagonal elements. Thus in particular, $|I| = 1$ where I is the identity matrix.

The next theorem, proved in Problem 7.22, shows how the determinant of a matrix is affected by the elementary row and column operations.

Theorem 7.3:   Suppose B is obtained from A by an elementary row (column) operation.

    (*a*)   If two rows (columns) of A were interchanged, then $|B| = -|A|$.

    (*b*)   If a row (column) of A was multiplied by a scalar k, then $|B| = k|A|$.

    (*c*)   If a multiple of a row (column) was added to another row (column), then $|B| = |A|$.

We now state two of the most important and useful theorems on determinants.

Theorem 7.4:   Let A be any n-square matrix. Then the following are equivalent:

    (i) A is invertible, i.e., A has an inverse A^{-1}.

    (ii) $AX = 0$ has only the zero solution.

    (iii) The determinant of A is not zero: $|A| \neq 0$.

        Remark: Depending on the author and text, a nonsingular matrix A is defined to be an invertible matrix A, or a matrix A for which $|A| \neq 0$, or a matrix A for which $AX = 0$ has only the zero solution. The above theorem shows that all such definitions are equivalent.

Theorem 7.5:   The determinant is a multiplicative function. That is, the determinant of a product of two matrices A and B is equal to the product of their determinants: $|AB| = |A||B|$.

We shall prove the above two theorems (Problems 7.27 and 7.28, respectively) using the theory of elementary matrices and the following lemma (proved in Problem 7.25).

Lemma 7.6:   Let E be an elementary matrix. Then, for any matrix A, $|EA| = |E||A|$.

We comment that one can also prove the preceding two theorems directly without resorting to the theory of elementary matrices.

Recall that matrices A and B are similar if there exists a nonsingular matrix P such that $B = P^{-1}AP$. Using the multiplicative property of the determinant (Theorem 7.5), we are able to prove (Problem 7.30):

Theorem 7.7:   Suppose A and B are similar matrices. Then $|A| = |B|$.

7.7 MINORS AND COFACTORS

Consider an n-square matrix $A = (a_{ij})$. Let M_{ij} denote the $(n-1)$-square submatrix of A obtained by deleting its ith row and jth column. The determinant $|M_{ij}|$ is called the minor of the element a_{ij} of A, and we define the cofactor of a_{ij}, denoted by A_{ij}, to be the "signed" minor:

$$A_{ij} = (-1)^{i+j}|M_{ij}|$$

Note that the "signs" $(-1)^{i+j}$ accompanying the minors form a chessboard pattern with $+$'s on the main diagonal:

$$\begin{pmatrix} + & - & + & - & \cdots \\ - & + & - & + & \cdots \\ + & - & + & - & \cdots \\ \hdotsfor{5} \end{pmatrix}$$

We emphasize that M_{ij} denotes a matrix whereas A_{ij} denotes a scalar.

> **Remark:**   The above sign $(-1)^{i+j}$ of the cofactor A_{ij} is frequently obtained using the checkerboard pattern. Specifically, beginning with " $+$ " and alternating signs, i.e., $+, -, +, -, \ldots$, count from the main diagonal to the appropriate square.

Example 7.8.   Consider the matrix $A = \begin{pmatrix} 2 & 3 & 4 \\ 5 & 6 & 7 \\ 8 & 9 & 1 \end{pmatrix}$.

$$|M_{23}| = \begin{vmatrix} 2 & 3 & 4 \\ 5 & 6 & 7 \\ 8 & 9 & 1 \end{vmatrix} = \begin{vmatrix} 2 & 3 \\ 8 & 9 \end{vmatrix} = 18 - 24 = -6 \text{ and so } A_{23} = (-1)^{2+3}|M_{23}| = (-1)\cdot(-6) = 6.$$

The following theorem, proved in Problem 7.31, applies.

Theorem 7.8:   The determinant of the matrix $A = (a_{ij})$ is equal to the sum of the products obtained by multiplying the elements of any row (column) by their respective cofactors:

$$|A| = a_{i1}A_{i1} + a_{i2}A_{i2} + \cdots + a_{in}A_{in} = \sum_{j=1}^{n} a_{ij}A_{ij}$$

and $$|A| = a_{1j}A_{1j} + a_{2j}A_{2j} + \cdots + a_{nj}A_{nj} = \sum_{i=1}^{n} a_{ij}A_{ij}$$

The above formulas for $|A|$ are called the *Laplace expansions* of the determinant of A by the ith row and the jth column, respectively. Together with the elementary row (column) operations, they offer a method of simplifying the computation of $|A|$, as described below.

Evelution of Determinants

The following algorithm reduces the evaluation of a determinant of order n to the evaluation of a determinant of order $n-1$.

Algorithm 7.7 (Reduction of the order of a determinant)

Here $A = (a_{ij})$ is a nonzero n-square matrix with $n > 1$.

Step 1. Choose an element $a_{ij} = 1$ or, if lacking, $a_{ij} \neq 0$.

Step 2. Using a_{ij} as a pivot, apply elementary row [column] operations to put 0s in all the other positions in the column [row] containing a_{ij}.

Step 3. Expand the determinant by the column [row] containing a_{ij}.

The following remarks are in order.

> **Remark 1:** Algorithm 7.7 is usually used for determinants of order four or more. With determinants of order less than four, one uses the specific formulas for the determinant.

> **Remark 2:** Gaussian elimination or, equivalently, repeated use of Algorithm 7.7 together with row interchanges, can be used to transform a matrix A into an upper triangular matrix whose determinant is the product of its diagonal entries. However, one must keep track of the number of row interchanges since each row interchange changes the sign of the determinant. (See Problem 7.11.)

Example 7.9. Compute the determinant of $A = \begin{pmatrix} 5 & 4 & 2 & 1 \\ 2 & 3 & 1 & -2 \\ -5 & -7 & -3 & 9 \\ 1 & -2 & -1 & 4 \end{pmatrix}$ by Algorithm 7.7.

Use $a_{23} = 1$ as a pivot to put 0s in the other positions of the third column, that is, apply the row operations $-2R_2 + R_1 \to R_1$, $3R_2 + R_3 \to R_3$, and $R_2 + R_4 \to R_4$. By Theorem 7.3(c), the value of the determinant does not change by these operations; that is,

$$|A| = \begin{vmatrix} 5 & 4 & 2 & 1 \\ 2 & 3 & 1 & -2 \\ -5 & -7 & -3 & 9 \\ 1 & -2 & -1 & 4 \end{vmatrix} = \begin{vmatrix} 1 & -2 & 0 & 5 \\ 2 & 3 & 1 & -2 \\ 1 & 2 & 0 & 3 \\ 3 & 1 & 0 & 2 \end{vmatrix}$$

Now if we expand by the third column, we may neglect all terms which contain 0. Thus

$$|A| = (-1)^{2+3} \begin{vmatrix} 1 & -2 & 0 & 5 \\ 2 & 3 & 1 & -2 \\ 1 & 2 & 0 & 3 \\ 3 & 1 & 0 & 2 \end{vmatrix} = - \begin{vmatrix} 1 & -2 & 5 \\ 1 & 2 & 3 \\ 3 & 1 & 2 \end{vmatrix} = -(4 - 18 + 5 - 30 - 3 + 4) = -(-38) = 38$$

7.8 CLASSICAL ADJOINT

Consider an n-square matrix $A = (a_{ij})$ over a field K. The classical adjoint (traditionally, just "adjoint") of A, denoted adj A, is the transpose of the matrix of cofactors of A:

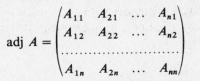

$$\text{adj } A = \begin{pmatrix} A_{11} & A_{21} & \cdots & A_{n1} \\ A_{12} & A_{22} & \cdots & A_{n2} \\ \cdots\cdots\cdots\cdots\cdots\cdots\cdots \\ A_{1n} & A_{2n} & \cdots & A_{nn} \end{pmatrix}$$

We say "classical adjoint" instead of simply "adjoint" because the term adjoint is currently used for an entirely different concept.

Example 7.10. Let $A = \begin{pmatrix} 2 & 3 & -4 \\ 0 & -4 & 2 \\ 1 & -1 & 5 \end{pmatrix}$. The cofactors of the nine elements of A are

$$A_{11} = + \begin{vmatrix} -4 & 2 \\ -1 & 5 \end{vmatrix} = -18 \qquad A_{12} = - \begin{vmatrix} 0 & 2 \\ 1 & 5 \end{vmatrix} = 2 \qquad A_{13} = + \begin{vmatrix} 0 & -4 \\ 1 & -1 \end{vmatrix} = 4$$

$$A_{21} = - \begin{vmatrix} 3 & -4 \\ -1 & 5 \end{vmatrix} = -11 \qquad A_{22} = + \begin{vmatrix} 2 & -4 \\ 1 & 5 \end{vmatrix} = 14 \qquad A_{23} = - \begin{vmatrix} 2 & 3 \\ 1 & -1 \end{vmatrix} = 5$$

$$A_{31} = + \begin{vmatrix} 3 & -4 \\ -4 & 2 \end{vmatrix} = -10 \qquad A_{32} = - \begin{vmatrix} 2 & -4 \\ 0 & 2 \end{vmatrix} = -4 \qquad A_{33} = + \begin{vmatrix} 2 & 3 \\ 0 & -4 \end{vmatrix} = -8$$

The transpose of the above matrix of cofactors yields the classical adjoint of A:

$$\text{adj } A = \begin{pmatrix} -18 & -11 & -10 \\ 2 & 14 & -4 \\ 4 & 5 & -8 \end{pmatrix}$$

The following theorem, proved in Problem 7.33, applies.

Theorem 7.9: For any square matrix A,

$$A \cdot (\text{adj } A) = (\text{adj } A) \cdot A = |A| I$$

where I is the identity matrix. Thus, if $|A| \neq 0$,

$$A^{-1} = \frac{1}{|A|} (\text{adj } A)$$

Observe that the above theorem gives us another method of obtaining the inverse of a given matrix.

Example 7.11. Consider the matrix A of Example 7.10 for which $|A| = -46$. We have

$$A(\text{adj } A) = \begin{pmatrix} 2 & 3 & -4 \\ 0 & -4 & 2 \\ 1 & -1 & 5 \end{pmatrix} \begin{pmatrix} -18 & -11 & -10 \\ 2 & 14 & -4 \\ 4 & 5 & -8 \end{pmatrix} = \begin{pmatrix} -46 & 0 & 0 \\ 0 & -46 & 0 \\ 0 & 0 & -46 \end{pmatrix} = -46 \begin{pmatrix} 1 & 0 & 0 \\ 0 & 1 & 0 \\ 0 & 0 & 1 \end{pmatrix}$$

$$= -46I = |A| I$$

Furthermore, by Theorem 7.9,

$$A^{-1} = \frac{1}{|A|} (\text{adj } A) = \begin{pmatrix} -18/(-46) & -11/(-46) & -10/(-46) \\ 2/(-46) & 14/(-46) & -4/(-46) \\ 4/(-46) & 5/(-46) & -8/(-46) \end{pmatrix} = \begin{pmatrix} \frac{9}{23} & \frac{11}{46} & \frac{5}{23} \\ -\frac{1}{23} & -\frac{7}{23} & \frac{2}{23} \\ -\frac{2}{23} & -\frac{5}{46} & \frac{4}{23} \end{pmatrix}$$

7.9 APPLICATIONS TO LINEAR EQUATIONS, CRAMER'S RULE

Consider a system of n linear equations in n unknowns:

$$a_{11}x_1 + a_{12}x_2 + \cdots + a_{1n}x_n = b_1$$
$$a_{21}x_1 + a_{22}x_2 + \cdots + a_{2n}x_n = b_2$$
$$\cdots\cdots\cdots\cdots\cdots\cdots\cdots\cdots\cdots\cdots\cdots\cdots\cdots\cdots\cdots$$
$$a_{n1}x_1 + a_{n2}x_2 + \cdots + a_{nn}x_n = b_n$$

The above system can be written in the form $AX = B$ where $A = (a_{ij})$ is the (square) matrix of coefficients and $B = (b_i)$ is the column vector of constants. Let A_i be the matrix obtained from A by replacing the ith column of A by the column vector B. Let $D = \det(A)$ and let $N_i = \det(A_i)$ for $i = 1, 2, \ldots, n$. The fundamental relationship between determinants and the solution of the above system follows.

Theorem 7.10:   The above system has a unique solution if and only if $D \neq 0$. In this case the unique solution is given by

$$x_1 = N_1/D, \qquad x_2 = N_2/D, \qquad \ldots, \qquad x_n = N_n/D$$

The above theorem, proved in Problem 7.34, is known as "Cramer's rule" for solving systems of linear equations. We emphasize that the theorem only refers to a system with the same number of equations as unknowns, and that it only gives the solution when $D \neq 0$. In fact, if $D = 0$ the theorem does not tell whether or not the system has a solution. However, in the case of a homogeneous system we have the following useful result (to be proved in Problem 7.65).

Theorem 7.11:   The homogeneous system $Ax = 0$ has a nonzero solution if and only if $D = |A| = 0$.

Example 7.12.   Solve, using determinants: $\begin{cases} 2x + y - z = 3 \\ x + y + z = 1 \\ x - 2y - 3z = 4 \end{cases}$.

First compute the determinant D of the matrix of coefficients:

$$D = \begin{vmatrix} 2 & 1 & -1 \\ 1 & 1 & 1 \\ 1 & -2 & -3 \end{vmatrix} = -6 + 1 + 2 + 1 + 4 + 3 = 5$$

Since $D \neq 0$, the system has a unique solution. To compute N_x, N_y, and N_z, replace the coefficients of x, y, and z in the matrix of coefficients by the constant terms:

$$N_x = \begin{vmatrix} 3 & 1 & -1 \\ 1 & 1 & 1 \\ 4 & -2 & -3 \end{vmatrix} = -9 + 4 + 2 + 4 + 6 + 3 = 10$$

$$N_y = \begin{vmatrix} 2 & 3 & -1 \\ 1 & 1 & 1 \\ 1 & 4 & -3 \end{vmatrix} = -6 + 3 - 4 + 1 - 8 + 9 = -5$$

$$N_z = \begin{vmatrix} 2 & 1 & 3 \\ 1 & 1 & 1 \\ 1 & -2 & 4 \end{vmatrix} = 8 + 1 - 6 - 3 + 4 - 4 = 0$$

Thus the unique solution is $x = N_x/D = 2$, $y = N_y/D = -1$, $z = N_z/D = 0$.

7.10 SUBMATRICES, GENERAL MINORS, PRINCIPAL MINORS

Let $A = (a_{ij})$ be an n-square matrix. Each ordered set $i_1, i_2, \ldots, i_r$ of r-row indices, and each ordered set $j_1, j_2, \ldots, j_r$ or r-column indices, defines the following submatrix of A of order r:

$$A_{i_1, i_2, \ldots, i_r}^{j_1, j_2, \ldots, j_r} = \begin{pmatrix} a_{i_1, j_1} & a_{i_1, j_2} & \cdots & a_{i_1, j_r} \\ a_{i_2, j_1} & a_{i_2, j_2} & \cdots & a_{i_2, j_2} \\ \hdotsfor{4} \\ a_{i_r, j_1} & a_{i_r, j_2} & \cdots & a_{i_r, j_r} \end{pmatrix}$$

The determinant $|A_{i_1, i_2, \ldots, i_r}^{j_1, j_2, \ldots, j_r}|$ is called a minor of A of order r and

$$(-1)^{i_1 + i_2 + \cdots + i_r + j_1 + j_2 + \cdots + j_r} |A_{i_1, i_2, \ldots, i_r}^{j_1, j_2, \ldots, j_r}|$$

is the corresponding signed minor. (Note that a minor of order $n - 1$ is a minor in the sense of Section 7.7 and the corresponding signed minor of order $n - 1$ is a cofactor.) Furthermore, if i'_k and j'_k denote, respectively, the remaining row and column indices, then

$$|A^{j'_1, \ldots, j'_{n-r}}_{i'_1, \ldots, i'_{n-r}}|$$

is the *complementary minor*.

Example 7.13. Suppose $A = (a_{ij})$ is a 5-square matrix. The row subscripts 3 and 5 and the column subscripts 1 and 4 define the submatrix

$$A^{1, 4}_{3, 5} = \begin{pmatrix} a_{31} & a_{34} \\ a_{51} & a_{54} \end{pmatrix}$$

and the minor and signed minor are, respectively,

$$|A^{1, 4}_{3, 5}| = \begin{vmatrix} a_{31} & a_{34} \\ a_{51} & a_{54} \end{vmatrix} = a_{31}a_{54} - a_{34}a_{51} \quad \text{and} \quad (-1)^{3+5+1+4}|A^{1, 4}_{3, 5}| = -|A^{1, 4}_{3, 5}|$$

The remaining row subscripts are 1, 2, and 4 and the remaining column subscripts are 2, 3, and 5; hence the complementary minor of $|A^{1, 4}_{3, 5}|$ is

$$|A^{2, 3, 5}_{1, 2, 4}| = \begin{vmatrix} a_{12} & a_{13} & a_{15} \\ a_{22} & a_{23} & a_{35} \\ a_{42} & a_{43} & a_{45} \end{vmatrix}$$

Principal Minors

A minor is principal if the row and column indices are the same or, in other words, if the diagonal elements of the minor come from the diagonal of the matrix.

Example 7.14. Consider the following minors of a 5-square matrix $A = (a_{ij})$:

$$M_1 = \begin{vmatrix} a_{22} & a_{24} & a_{25} \\ a_{42} & a_{44} & a_{45} \\ a_{52} & a_{54} & a_{55} \end{vmatrix} \qquad M_2 = \begin{vmatrix} a_{11} & a_{13} & a_{15} \\ a_{21} & a_{23} & a_{25} \\ a_{51} & a_{53} & a_{55} \end{vmatrix} \qquad M_3 = \begin{vmatrix} a_{22} & a_{25} \\ a_{52} & a_{55} \end{vmatrix}$$

Here M_1 and M_3 are principal minors since all their diagonal elements belong to the diagonal of A. On the other hand, M_2 is not principal since a_{23} belongs to the diagonal of M_2 but not to that of A.

The following remarks are in order.

> **Remark 1:** The sign of a principal minor is always $+1$ since the sum of the row and identical column subscripts is even.

> **Remark 2:** A minor is principal if and only if its complementary minor is also principal.

> **Remark 3:** A real symmetric matrix A is positive definite if and only if all the principal minors of A are positive.

7.11 BLOCK MATRICES AND DETERMINANTS

The following is the main result of this section.

Theorem 7.12: Suppose M is an upper (lower) triangular block matrix with the diagonal blocks $A_1, A_2, \ldots, A_n$. Then

$$\det(M) = \det(A_1)\det(A_2)\cdots\det(A_n)$$

The proof appears in Problem 7.35.

Example 7.15. Find $|M|$ where $M = \begin{pmatrix} 2 & 3 & 4 & 7 & 8 \\ -1 & 5 & 3 & 2 & 1 \\ 0 & 0 & 2 & 1 & 5 \\ 0 & 0 & 3 & -1 & 4 \\ 0 & 0 & 5 & 2 & 6 \end{pmatrix}$.

Note that M is an upper triangular block matrix. Evaluate the determinant of each diagonal block:

$$\begin{vmatrix} 2 & 3 \\ -1 & 5 \end{vmatrix} = 10 + 3 = 13 \qquad \begin{vmatrix} 2 & 1 & 5 \\ 3 & -1 & 4 \\ 5 & 2 & 6 \end{vmatrix} = -12 + 20 + 30 + 25 - 16 - 18 = 29$$

Then $|M| = (13)(29) = 377$.

Remark: Suppose $M = \begin{pmatrix} A & B \\ C & D \end{pmatrix}$ where A, B, C, D are square matrices. Then it is not generally true that $|M| = |A||D| - |B||C|$. (See Problem 7.77.)

7.12 DETERMINANTS AND VOLUME

Determinants are related to the notions of area and volume as follows. Let $u_1, u_2, \ldots, u_n$ be vectors in $\mathbf{R}^n$. Let S be the parallelepiped determined by the vectors; that is,

$$S = \{a_1 u_1 + a_2 u_2 + \cdots + a_n u_n : 0 \le a_i \le 1 \text{ for } i = 1, \ldots, n\}$$

(When $n = 2$, S is a parallelogram.) Let $V(S)$ denote the volume of S (or area of S when $n = 2$). Then

$$V(S) = \text{absolute value of det } (A)$$

where A is the matrix with rows $u_1, u_2, \ldots, u_n$. In general, $V(S) = 0$ if and only if the vectors $u_1, \ldots, u_n$ do not form a coordinate system for $\mathbf{R}^n$, i.e., if and only if the vectors are linearly dependent.

7.13 MULTILINEARITY AND DETERMINANTS

Let V be a vector space over a field K. Let $\mathscr{A} = V^n$, that is, $\mathscr{A}$ consists of all the n-tuples

$$A = (A_1, A_2, \ldots, A_n)$$

where the A_i are vectors in V. The following definitions apply:

Definition: A function $D: \mathscr{A} \to K$ is said to be multilinear if it is linear in each of the components; that is:

(i) If $A_i = B + C$, then

$$D(A) = D(\ldots, B + C, \ldots) = D(\ldots, B, \ldots) + D(\ldots, C, \ldots)$$

(ii) If $A_i = kB$ where $k \in K$, then

$$D(A) = D(\ldots, kB, \ldots) = kD(\ldots, B, \ldots)$$

We also say n-linear for multilinear if there are n components.

Definition: A function $D: \mathscr{A} \to K$ is said to be alternating if $D(A) = 0$ whenever A has two identical elements; that is,

$$D(A_1, A_2, \ldots, A_n) = 0 \qquad \text{whenever} \quad A_i = A_j, \ i \neq j$$

Now let **M** denote the set of all n-square matrices A over a field K. We may view A as an n-tuple consisting of its row vectors $A_1, A_2, \ldots, A_n$; that is, we may view A in the form $A = (A_1, A_2, \ldots, A_n)$. The following basic result (proved in Problem 7.36) applies (where I denotes the identity matrix):

Theorem 7.13:   There exists a unique function $D: \mathbf{M} \to K$ such that

(i) D is multilinear,   (ii) D is alternating,   (iii) $D(I) = 1$

This function D is none other than the determinant function; that is, for any matrix $A \in \mathbf{M}$, $D(A) = |A|$.

Solved Problems

COMPUTATION OF DETERMINANTS OF ORDERS TWO AND THREE

7.1.   Evaluate the determinant of each of the following matrix:

$$\begin{pmatrix} 6 & 5 \\ 2 & 3 \end{pmatrix}, \qquad \begin{pmatrix} 3 & -2 \\ 4 & 5 \end{pmatrix}, \qquad \begin{pmatrix} 4 & -5 \\ -1 & -2 \end{pmatrix}$$

$$\begin{vmatrix} 6 & 5 \\ 2 & 3 \end{vmatrix} = (6)(3) - (5)(2) = 18 - 10 = 8, \qquad \begin{vmatrix} 3 & -2 \\ 4 & 5 \end{vmatrix} = 15 + 8 = 23, \qquad \begin{vmatrix} 4 & -5 \\ -1 & -2 \end{vmatrix} = -8 - 5 = -13$$

7.2.   Find the determinant of $\begin{pmatrix} t-5 & 7 \\ -1 & t+3 \end{pmatrix}$.

$$\begin{vmatrix} t-5 & 7 \\ -1 & t+3 \end{vmatrix} = (t-5)(t+3) + 7 = t^2 - 2t - 15 + 7 = t^2 - 2t - 8$$

7.3.   Find those values of k for which $\begin{vmatrix} k & k \\ 4 & 2k \end{vmatrix} = 0.$

Set $\begin{vmatrix} k & k \\ 4 & 2k \end{vmatrix} = 2k^2 - 4k = 0$, or $2k(k-2) = 0$. Hence $k = 0$; and $k = 2$. That is, if $k = 0$ or $k = 2$, the determinant is zero.

7.4. Evaluate the determinants of the following matrices:

(a) $\begin{pmatrix} 1 & -2 & 3 \\ 2 & 4 & -1 \\ 1 & 5 & -2 \end{pmatrix}$,   (b) $\begin{pmatrix} a_1 & b_1 & c_1 \\ a_2 & b_2 & c_2 \\ a_3 & b_3 & c_3 \end{pmatrix}$

Use the diagrams in Fig. 7-1.

(a) $\begin{vmatrix} 1 & -2 & 3 \\ 2 & 4 & -1 \\ 1 & 5 & -2 \end{vmatrix} = (1)(4)(-2) + (-2)(-1)(1) + (3)(5)(2) - (1)(4)(3) - (5)(-1)(1) - (-2)(-2)(2)$
$= -8 + 2 + 30 - 12 + 5 - 8 = 9$

(b) $\begin{vmatrix} a_1 & b_1 & c_1 \\ a_2 & b_2 & c_2 \\ a_3 & b_3 & c_3 \end{vmatrix} = a_1 b_2 c_3 + b_1 c_2 a_3 + c_1 b_3 a_2 - a_3 b_2 c_1 - b_3 c_2 a_1 - c_3 b_1 a_2$

7.5. Compute the determinant of $\begin{pmatrix} 2 & 3 & 4 \\ 5 & 6 & 7 \\ 8 & 9 & 1 \end{pmatrix}$.

First simplify the entries by subtracting twice the first row from the second row, that is, by applying $-2R_1 + R_2 \to R_2$:

$$\begin{vmatrix} 2 & 3 & 4 \\ 5 & 6 & 7 \\ 8 & 9 & 1 \end{vmatrix} = \begin{vmatrix} 2 & 3 & 4 \\ 1 & 0 & -1 \\ 8 & 9 & 1 \end{vmatrix} = 0 - 24 + 36 - 0 + 18 - 3 = 27$$

7.6. Find the determinant of A where:

(a) $A = \begin{pmatrix} \frac{1}{2} & -1 & -\frac{1}{3} \\ \frac{3}{4} & \frac{1}{2} & -1 \\ 1 & -4 & 1 \end{pmatrix}$,   (b) $A = \begin{pmatrix} t+3 & -1 & 1 \\ 5 & t-3 & 1 \\ 6 & -6 & t+4 \end{pmatrix}$

(a) First multiply the first row by 6 and the second row by 4. Then

$$6 \cdot 4 |A| = 24|A| = \begin{vmatrix} 3 & -6 & -2 \\ 3 & 2 & -4 \\ 1 & -4 & 1 \end{vmatrix} = 6 + 24 + 24 + 4 - 48 + 18 = 28$$

Hence $|A| = \frac{28}{24} = \frac{7}{6}$. (Observe that the original multiplications eliminated the fractions, so the arithmetic is simpler.)

(b) Add the second column to the first column, and then add the third column to the second column to produce 0s; that is, apply $C_2 + C_1 \to C_1$ and $C_3 + C_2 \to C_2$:

$$|A| = \begin{vmatrix} t+2 & 0 & 1 \\ t+2 & t-2 & 1 \\ 0 & t-2 & t+4 \end{vmatrix}$$

Now factor $t+2$ from the first column and $t-2$ from the second column to get

$$|A| = (t+2)(t-2)\begin{vmatrix} 1 & 0 & 1 \\ 1 & 1 & 1 \\ 0 & 1 & t+4 \end{vmatrix}$$

Finally subtract the first column from the third column to obtain

$$|A| = (t+2)(t-2)\begin{vmatrix} 1 & 0 & 0 \\ 1 & 1 & 0 \\ 0 & 1 & t+4 \end{vmatrix} = (t+2)(t-2)(t+4)$$

COMPUTATION OF DETERMINANTS OF ARBITRARY ORDERS

7.7. Compute the determinant of $A = \begin{pmatrix} 2 & 5 & -3 & -2 \\ -2 & -3 & 2 & -5 \\ 1 & 3 & -2 & 2 \\ -1 & -6 & 4 & 3 \end{pmatrix}$.

Use $a_{31} = 1$ as a pivot and apply the row operations $-2R_3 + R_1 \to R_1$, $2R_3 + R_1 \to R_2$, and $R_3 + R_4 \to R_4$:

$$|A| = \begin{vmatrix} 2 & 5 & -3 & -2 \\ -2 & -3 & 2 & -5 \\ 1 & 3 & -2 & 2 \\ -1 & -6 & 4 & 3 \end{vmatrix} = \begin{vmatrix} 0 & -1 & 1 & -6 \\ 0 & 3 & -2 & -1 \\ 1 & 3 & -2 & 2 \\ 0 & -3 & 2 & 5 \end{vmatrix} = + \begin{vmatrix} -1 & 1 & -6 \\ 3 & -2 & -1 \\ -3 & 2 & 5 \end{vmatrix}$$

$$= 10 + 3 - 36 + 36 - 2 - 15 = -4$$

7.8. Evaluate the determinant of $A = \begin{pmatrix} 1 & 2 & 2 & 3 \\ 1 & 0 & -2 & 0 \\ 3 & -1 & 1 & -2 \\ 4 & -3 & 0 & 2 \end{pmatrix}$.

Use $a_{21} = 1$ as a pivot, and apply $2C_1 + C_3 \to C_3$:

$$|A| = \begin{vmatrix} 1 & 2 & 4 & 3 \\ 1 & 0 & 0 & 0 \\ 3 & -1 & 7 & -2 \\ 4 & -3 & 6 & 2 \end{vmatrix} = - \begin{vmatrix} 2 & 4 & 3 \\ -1 & 7 & -2 \\ -3 & 8 & 2 \end{vmatrix} = -(28 + 24 - 24 + 63 + 32 + 8) = -131$$

7.9. Find the determinant of $C = \begin{pmatrix} 6 & 2 & 1 & 0 & 5 \\ 2 & 1 & 1 & -2 & 1 \\ 1 & 1 & 2 & -2 & 3 \\ 3 & 0 & 2 & 3 & -1 \\ -1 & -1 & -3 & 4 & 2 \end{pmatrix}$.

First reduce $|C|$ to a determinant of order four, and then to a determinant of order three. Use $c_{22} = 1$ as a pivot and apply $-2R_2 + R_1 \to R_1$, $-R_2 + R_3 \to R_3$, and $R_2 + R_5 \to R_5$:

$$|C| = \begin{vmatrix} 2 & 0 & -1 & 4 & 3 \\ 2 & 1 & 1 & -2 & 1 \\ -1 & 0 & 1 & 0 & 2 \\ 3 & 0 & 2 & 3 & -1 \\ 1 & 0 & -2 & 2 & 6 \end{vmatrix} = \begin{vmatrix} 2 & -1 & 4 & 3 \\ -1 & 1 & 0 & 2 \\ 3 & 2 & 3 & -1 \\ 1 & -2 & 2 & 3 \end{vmatrix} = \begin{vmatrix} 1 & 1 & 4 & -1 \\ 0 & 1 & 0 & 0 \\ 5 & 2 & 3 & -5 \\ -1 & -2 & 2 & 7 \end{vmatrix}$$

$$= \begin{vmatrix} 1 & 4 & -1 \\ 5 & 3 & -5 \\ -1 & 2 & 7 \end{vmatrix} = 21 + 20 - 10 - 3 + 10 - 140 = -102$$

7.10. Find the determinant of each of the following matrices:

(a) $A = \begin{pmatrix} 5 & 6 & 7 & 8 \\ 0 & 0 & 0 & 0 \\ 1 & -3 & 5 & -7 \\ 8 & 4 & 2 & 6 \end{pmatrix}$,   (b) $B = \begin{pmatrix} 5 & 6 & 7 & 6 \\ 1 & -3 & 5 & -3 \\ 4 & 9 & -3 & 9 \\ 2 & 7 & 8 & 7 \end{pmatrix}$,

(c) $C = \begin{pmatrix} 2 & 3 & 4 & 5 \\ 0 & -3 & 7 & -8 \\ 0 & 0 & 5 & 6 \\ 0 & 0 & 0 & 4 \end{pmatrix}$

(a) Since A has a row of zeros, det $(A) = 0$.

(b) Since the second and fourth columns of B are equal, det $(B) = 0$.

(c) Since C is triangular, det (C) is equal to the product of the diagonal entries. Hence det $(C) = -120$.

7.11. Describe the Gaussian elimination algorithm for calculating the determinant of an n-square matrix $A = (a_{ij})$.

The algorithm uses Gaussian elimination to transform A into an upper triangular matrix (whose determinant is the product of its diagonal entries). Since the algorithm involves exchanging rows, which changes the sign of the determinant, one must keep track of such changes using some variable, say SIGN. The algorithm will also use "pivoting"; that is, the element with the greatest absolute value will be used as the pivot. The algorithm follows.

Step 1. Set SIGN = 0. [This initializes the variable SIGN.]

Step 2. Find the entry a_{i1} in the first column with greatest absolute value.
   (a) If $a_{i1} = 0$, then set det $(A) = 0$ and EXIT.
   (b) If $i \neq 1$, then interchange the first and ith rows and set SIGN = SIGN + 1.

Step 3. Use a_{11} as a pivot and elementary row operations of the form $kR_q + R_p \to R_p$ to put 0s below a_{11}.

Step 4. Repeat Steps 2 and 3 with the submatrix obtained by omitting the first row and the first column.

Step 5. Continue the above process until A is an upper triangular matrix.

Step 6. Set det $(A) = (-1)^{\text{SIGN}} a_{11} a_{22} \cdots a_{nn}$, and EXIT.

Note that the elementary row operation of the form $kR_p \to R_p$ (which multiplies a row by a scalar), which is permitted in the Gaussian algorithm for a system of linear equations, is barred here, as it changes the value of the determinant.

7.12. Use the Gaussian elimination algorithm in Problem 7.11 to find the determinant of

$$A = \begin{pmatrix} 3 & 8 & 6 \\ -2 & -3 & 1 \\ 5 & 10 & 15 \end{pmatrix}$$

First row reduce the matrix to an upper triangular form keeping track of the number of row interchanges:

$$A \sim \begin{pmatrix} 5 & 10 & 15 \\ -2 & -3 & 1 \\ 3 & 8 & 6 \end{pmatrix} \sim \begin{pmatrix} 5 & 10 & 15 \\ 0 & 1 & 7 \\ 0 & 2 & -3 \end{pmatrix} \sim \begin{pmatrix} 5 & 10 & 15 \\ 0 & 2 & -3 \\ 0 & 1 & 7 \end{pmatrix} \sim \begin{pmatrix} 5 & 10 & 15 \\ 0 & 2 & -3 \\ 0 & 0 & \frac{17}{2} \end{pmatrix}$$

A is now in triangular form and SIGN = 2 since there were two interchanges of rows. Hence $A = (-1)^{\text{SIGN}} \cdot 5 \cdot 2 \cdot (\frac{17}{2}) = 85$.

7.13. Evaluate $|B| = \begin{vmatrix} 0.921 & 0.185 & 0.476 & 0.614 \\ 0.782 & 0.157 & 0.527 & 0.138 \\ 0.872 & 0.484 & 0.637 & 0.799 \\ 0.312 & 0.555 & 0.841 & 0.448 \end{vmatrix}$.

Multiply the row containing the pivot a_{ij} by $1/a_{ij}$ so that the pivot is equal to 1:

$$|B| = 0.921 \begin{vmatrix} 1 & 0.201 & 0.517 & 0.667 \\ 0.782 & 0.157 & 0.527 & 0.138 \\ 0.872 & 0.484 & 0.637 & 0.799 \\ 0.312 & 0.555 & 0.841 & 0.448 \end{vmatrix} = 0.921 \begin{vmatrix} 1 & 0.201 & 0.571 & 0.667 \\ 0 & 0 & 0.123 & -0.384 \\ 0 & 0.309 & 0.196 & 0.217 \\ 0 & 0.492 & 0.680 & 0.240 \end{vmatrix}$$

$$= 0.921 \begin{vmatrix} 0 & 0.123 & -0.384 \\ 0.309 & 0.196 & 0.217 \\ 0.492 & 0.680 & 0.240 \end{vmatrix} = 0.921(-0.384) \begin{vmatrix} 0 & -0.320 & 1 \\ 0.309 & 0.196 & 0.217 \\ 0.492 & 0.680 & 0.240 \end{vmatrix}$$

$$= 0.921(-0.384) \begin{vmatrix} 0 & 0 & 1 \\ 0.309 & 0.265 & 0.217 \\ 0.492 & 0.757 & 0.240 \end{vmatrix} = 0.921(-0.384) \begin{vmatrix} 0.309 & 0.265 \\ 0.492 & 0.757 \end{vmatrix}$$

$$= 0.921(-0.384)(0.104) = -0.037$$

COFACTORS, CLASSICAL ADJOINTS

7.14. Consider the matrix $A = \begin{pmatrix} 2 & 1 & -3 & 4 \\ 5 & -4 & 7 & -2 \\ 4 & 0 & 6 & -3 \\ 3 & -2 & 5 & 2 \end{pmatrix}$. Find the cofactor of the 7 in A, that is, A_{23}.

We have

$$A_{23} = (-1)^{2+3} \begin{vmatrix} 2 & 1 & -3 & 4 \\ 5 & -4 & 7 & -2 \\ 4 & 0 & 6 & -3 \\ 3 & -2 & 5 & 2 \end{vmatrix} = -\begin{vmatrix} 2 & 1 & 4 \\ 4 & 0 & -3 \\ 3 & -2 & 2 \end{vmatrix} = -(0 - 9 - 32 - 0 - 12 - 8) = -(-61) = 61$$

The exponent $2 + 3$ comes from the fact that 7 appears in the second row, third column.

7.15. Consider the matrix $B = \begin{pmatrix} 1 & 1 & 1 \\ 2 & 3 & 4 \\ 5 & 8 & 9 \end{pmatrix}$. Find: (a) $|B|$, (b) adj B, and (c) B^{-1} using adj B.

(a) $\quad |B| = 27 + 20 + 16 - 15 - 32 - 18 = -2$

(b) Take the transpose of the matrix of cofactors:

$$\text{adj } B = \begin{pmatrix} \begin{vmatrix} 3 & 4 \\ 8 & 9 \end{vmatrix} & -\begin{vmatrix} 2 & 4 \\ 5 & 9 \end{vmatrix} & \begin{vmatrix} 2 & 3 \\ 5 & 8 \end{vmatrix} \\ -\begin{vmatrix} 1 & 1 \\ 8 & 9 \end{vmatrix} & \begin{vmatrix} 1 & 1 \\ 5 & 9 \end{vmatrix} & -\begin{vmatrix} 1 & 1 \\ 5 & 8 \end{vmatrix} \\ \begin{vmatrix} 1 & 1 \\ 3 & 4 \end{vmatrix} & -\begin{vmatrix} 1 & 1 \\ 2 & 4 \end{vmatrix} & \begin{vmatrix} 1 & 1 \\ 2 & 3 \end{vmatrix} \end{pmatrix}^T = \begin{pmatrix} -5 & 2 & 1 \\ -1 & 4 & -3 \\ 1 & -2 & 1 \end{pmatrix}^T = \begin{pmatrix} -5 & -1 & 1 \\ 2 & 4 & -2 \\ 1 & -3 & 1 \end{pmatrix}$$

(c) Since $|B| \neq 0$,

$$B^{-1} = \frac{1}{|B|}(\text{adj } B) = \frac{1}{-2}\begin{pmatrix} -5 & -1 & 1 \\ 2 & 4 & -2 \\ 1 & -3 & 1 \end{pmatrix} = \begin{pmatrix} \frac{5}{2} & \frac{1}{2} & -\frac{1}{2} \\ -1 & -2 & 1 \\ -\frac{1}{2} & \frac{3}{2} & -\frac{1}{2} \end{pmatrix}$$

7.16. Consider an arbitrary 2-square matrix $A = \begin{pmatrix} a & b \\ c & d \end{pmatrix}$. (a) Find adj A. (b) Show that adj (adj A) = A.

(a) $\text{adj } A = \begin{pmatrix} +|d| & -|c| \\ -|b| & +|a| \end{pmatrix}^T = \begin{pmatrix} d & -c \\ -b & a \end{pmatrix}^T = \begin{pmatrix} d & -b \\ -c & a \end{pmatrix}$

(b) $\text{adj (adj } A) = \text{adj}\begin{pmatrix} d & -b \\ -c & a \end{pmatrix} = \begin{pmatrix} +|a| & -|-c| \\ -|-b| & +|d| \end{pmatrix}^T = \begin{pmatrix} a & c \\ b & d \end{pmatrix}^T = \begin{pmatrix} a & b \\ c & d \end{pmatrix} = A$

DETERMINANTS AND LINEAR EQUATIONS, CRAMER'S RULE

7.17. Solve the following systems by using determinants.

$$\begin{cases} ax - 2by = c \\ 3ax - 5by = 2c \end{cases} \quad \text{where } ab \neq 0$$

First find $D = \begin{vmatrix} a & -2b \\ 3a & -5b \end{vmatrix} = -5ab + 6ab = ab$. Since $D = ab \neq 0$, the system has a unique solution.

Next find

$$N_x = \begin{vmatrix} c & -2b \\ 2c & -5b \end{vmatrix} = -5bc + 4bc = -bc \quad \text{and} \quad N_y = \begin{vmatrix} a & c \\ 3a & 2c \end{vmatrix} = 2ac - 3ac = -ac$$

Then $x = N_x/D = -bc/ab = -c/a$ and $y = N_y/D = -ac/ab = -c/b$.

7.18. Solve, using determinants: $\begin{cases} 3y + 2x = z + 1 \\ 3x + 2z = 8 - 5y \\ 3z - 1 = x - 2y \end{cases}$

First arrange the equations in standard form:

$$\begin{aligned} 2x + 3y - z &= 1 \\ 3x + 5y + 2z &= 8 \\ x - 2y - 3z &= -1 \end{aligned}$$

Compute the determinant D of the matrix of coefficients:

$$D = \begin{vmatrix} 2 & 3 & -1 \\ 3 & 5 & 2 \\ 1 & -2 & -3 \end{vmatrix} = -30 + 6 + 6 + 5 + 8 + 27 = 22$$

Since $D \neq 0$, the system has a unique solution. To compute N_x, N_y and N_z, replace the corresponding coefficients of the unknown in the matrix of coefficients by the constant terms:

$$N_x = \begin{vmatrix} 1 & 3 & -1 \\ 8 & 5 & 2 \\ -1 & -2 & -3 \end{vmatrix} = -15 - 6 + 16 - 5 + 4 + 72 = 66$$

$$N_y = \begin{vmatrix} 2 & 1 & -1 \\ 3 & 8 & 2 \\ 1 & -1 & -3 \end{vmatrix} = -48 + 2 + 3 + 8 + 4 + 9 = -22$$

$$N_z = \begin{vmatrix} 2 & 3 & 1 \\ 3 & 5 & 8 \\ 1 & -2 & -1 \end{vmatrix} = -10 + 24 - 6 - 5 + 32 + 9 = 44$$

Hence

$$x = \frac{N_x}{D} = \frac{66}{22} = 3 \qquad y = \frac{N_y}{D} = \frac{-22}{22} = -1 \qquad z = \frac{N_z}{D} = \frac{44}{22} = 2$$

7.19. Solve the following system by using Cramer's rule:

$$\begin{aligned} 2x_1 + x_2 + 5x_3 + x_4 &= 5 \\ x_1 + x_2 - 3x_3 - 4x_4 &= -1 \\ 3x_1 + 6x_2 - 2x_3 + x_4 &= 8 \\ 2x_1 + 2x_2 + 2x_3 - 3x_4 &= 2 \end{aligned}$$

Compute

$$D = \begin{vmatrix} 2 & 1 & 5 & 1 \\ 1 & 1 & -3 & -4 \\ 3 & 6 & -2 & 1 \\ 2 & 2 & 2 & -3 \end{vmatrix} = -120 \qquad N_1 = \begin{vmatrix} 5 & 1 & 5 & 1 \\ -1 & 1 & -3 & -4 \\ 8 & 6 & -2 & 1 \\ 2 & 2 & 2 & -3 \end{vmatrix} = -240$$

$$N_2 = \begin{vmatrix} 2 & 5 & 5 & 1 \\ 1 & -1 & -3 & -4 \\ 3 & 8 & -2 & 1 \\ 2 & 2 & 2 & -3 \end{vmatrix} = -24 \qquad N_3 = \begin{vmatrix} 2 & 1 & 5 & 1 \\ 1 & 1 & -1 & -4 \\ 3 & 6 & 8 & 1 \\ 2 & 2 & 2 & -3 \end{vmatrix} = 0$$

$$N_4 = \begin{vmatrix} 2 & 1 & 5 & 5 \\ 1 & 1 & -3 & -1 \\ 3 & 6 & -2 & 8 \\ 2 & 2 & 2 & 2 \end{vmatrix} = -96$$

Then $x_1 = N_1/D = 2$, $x_2 = N_2/D = \frac{1}{5}$, $x_3 = N_3/D = 0$, $x_4 = N_4/D = \frac{4}{5}$.

7.20. Use determinants to find those values of k for which the following system has a unique solution:

$$\begin{aligned} kx + y + z &= 1 \\ x + ky + z &= 1 \\ x + y + kz &= 1 \end{aligned}$$

The system has a unique solution when $D \neq 0$, where D is the determinant of the matrix of coefficients. Compute

$$D = \begin{vmatrix} k & 1 & 1 \\ 1 & k & 1 \\ 1 & 1 & k \end{vmatrix} = k^3 + 1 + 1 - k - k - k = k^3 - 3k + 2 = (k-1)^2(k+2)$$

Thus the system has a unique solution when $(k-1)^2(k+2) \neq 0$, that is, when $k \neq 1$ and $k \neq -2$. (Gaussian elimination indicates that the system has no solution when $k = -2$, and the system has an infinite number of solutions when $k = 1$.)

PROOF OF THEOREMS

7.21. Prove Theorem 7.1.

If $A = (a_{ij})$, then $A^T = (b_{ij})$, with $b_{ij} = a_{ji}$. Hence

$$|A^T| = \sum_{\sigma \in S_n} (\text{sgn } \sigma) b_{1\sigma(2)} b_{2\sigma(2)} \cdots b_{n\sigma(n)} = \sum_{\sigma \in S_n} (\text{sgn } \sigma) a_{\sigma(1),\,1} a_{\sigma(2),\,2} \cdots a_{\sigma(n),\,n}$$

Let $\tau = \sigma^{-1}$. By Problem 7.43, sgn τ = sgn σ, and $a_{\sigma(1),\,1} a_{\sigma(2),\,2} \cdots a_{\sigma(n),\,n} = a_{1\tau(1)} a_{2\tau(2)} \cdots a_{n\tau(n)}$. Hence

$$|A^T| = \sum_{\sigma \in S_n} (\text{sgn } \tau) a_{1\tau(1)} a_{2\tau(2)} \cdots a_{n\tau(n)}$$

However, as σ runs through all the elements of S_n, $\tau = \sigma^{-1}$ also runs through all the elements of S_n. Thus $|A^T| = |A|$.

7.22. Prove Theorem 7.3(a).

We prove the theorem for the case that two columns are interchanged. Let τ be the transposition which interchanges the two numbers corresponding to the two columns of A that are interchanged. If $A = (a_{ij})$ and $B = (b_{ij})$, then $b_{ij} = a_{i\tau(j)}$. Hence, for any permutation σ,

$$b_{1\sigma(1)} b_{2\sigma(2)} \cdots b_{n\sigma(n)} = a_{1\tau\sigma(1)} a_{2\tau\sigma(2)} \cdots a_{n\tau\sigma(n)}$$

Thus

$$|B| = \sum_{\sigma \in S_n} (\text{sgn } \sigma) b_{1\sigma(1)} b_{2\sigma(2)} \cdots b_{n\sigma(n)} = \sum_{\sigma \in S_n} (\text{sgn } \sigma) a_{1\tau\sigma(1)} a_{2\tau\sigma(2)} \cdots a_{n\tau\sigma(n)}$$

Since the transposition τ is an odd permutation, sgn $\tau\sigma$ = sgn $\tau \cdot$ sgn $\sigma = -$sgn σ. Thus sgn $\sigma = -$sgn $\tau\sigma$, and so

$$|B| = -\sum_{\sigma \in S_n} (\text{sgn } \tau\sigma) a_{1\tau\sigma(1)} a_{2\tau\sigma(2)} \cdots a_{n\tau\sigma(n)}$$

But as σ runs through all the elements of S_n, $\tau\sigma$ also runs through all the elements of S_n; hence $|B| = -|A|$.

7.23. Prove Theorem 7.2.

(a) Each term in $|A|$ contains a factor from every row and so from the row of zeros. Thus each term of $|A|$ is zero and so $|A| = 0$.

(b) Suppose $1 + 1 \neq 0$ in K. If we interchange the two identical rows of A, we still obtain the matrix A. Hence, by Problem 7.22, $|A| = -|A|$ and so $|A| = 0$.

Now suppose $1 + 1 = 0$ in K. Then sgn $\sigma = 1$ for every $\sigma \in S_n$. Since A has two identical rows, we can arrange the terms of A into pairs of equal terms. Since each pair is 0, the determinant of A is zero.

(c) Suppose $A = (a_{ij})$ is lower triangular, that is, the entries above the diagonal are all zero: $a_{ij} = 0$ whenever $i < j$. Consider a term t of the determinant of A:

$$t = (\text{sgn } \sigma) a_{1i_1} a_{2i_2} \cdots a_{ni_n} \qquad \text{where } \sigma = i_1 i_2 \cdots i_n$$

Suppose $i_1 \neq 1$. Then $1 < i_1$ and so $a_{1i_1} = 0$; hence $t = 0$. That is, each term for which $i_1 \neq 1$ is zero.

Now suppose $i_1 = 1$ but $i_2 \neq 2$. Then $2 < i_2$ and so $a_{2i_2} = 0$; hence $t = 0$. Thus each term for which $i_1 \neq 1$ or $i_2 \neq 2$ is zero.

Similarly we obtain that each term for which $i_1 \neq 1$ or $i_2 \neq 2$ or $\cdots$ or $i_n \neq n$ is zero. Accordingly, $|A| = a_{11} a_{22} \cdots a_{nn}$ = product of diagonal elements.

7.24. Prove Theorem 7.3.

(a) Proved in Problem 7.22.

(b) If the jth row of A is multiplied by k, then every term in $|A|$ is multiplied by k and so $|B| = k|A|$. That is,

$$|B| = \sum_{\sigma} (\text{sgn } \sigma) a_{1i_1} a_{2i_2} \cdots (k a_{ji_j}) \cdots a_{ni_n} = k \sum_{\sigma} (\text{sgn } \sigma) a_{1i_1} a_{2i_2} \cdots a_{ni_n} = k|A|$$

(c) Suppose c times the kth row is added to the jth row of A. Using the symbol $\wedge$ to denote the jth position in a determinant term, we have

$$|B| = \sum_{\sigma} (\text{sgn } \sigma) a_{1i_1} a_{2i_2} \cdots \widehat{(c a_{ki_k} + a_{ji_j})} \cdots a_{ni_n}$$

$$= c \sum_{\sigma} (\text{sgn } \sigma) a_{1i_1} a_{2i_2} \cdots \widehat{a_{ki_k}} \cdots a_{ni_n} + \sum_{\sigma} (\text{sgn } \sigma) a_{1i_1} a_{2i_2} \cdots \widehat{a_{ji_j}} \cdots s_{ni_n}$$

The first sum is the determinant of a matrix whose kth and jth rows are identical; hence, by Theorem 7.2(b), the sum is zero. The second sum is the determinant of A. Thus $|B| = c \cdot 0 + |A| = |A|$.

7.25. Prove Lemma 7.6.

Consider the following elementary row operations: (i) multiply a row by a constant $k \neq 0$; (ii) interchange two rows; (iii) add a multiple of one row to another. Let E_1, E_2, and E_3 be the corresponding elementary matrices. That is, E_1, E_2, and E_3 are obtained by applying the above operations, respectively, to the identity matrix I. By Problem 7.24,

$$|E_1| = k|I| = k \qquad |E_2| = -|I| = -1 \qquad |E_3| = |I| = 1$$

Recall (Theorem 4.12) that $E_i A$ is identical to the matrix obtained by applying the corresponding operation to A. Thus, by Theorem 7.3,

$$|E_1 A| = k|A| = |E_1||A| \qquad |E_2 A| = -|A| = |E_2||A| \qquad |E_3 A| = |A| = 1|A| = |E_3||A|$$

and the lemma is proved.

7.26. Suppose B is row equivalent to a square matrix A. Show that $|B| = 0$ if and only if $|A| = 0$.

By Theorem 7.3, the effect of an elementary row operation is to change the sign of the determinant or to multiply the determinant by a nonzero scalar. Therefore $|B| = 0$ if and only if $|A| = 0$.

7.27. Prove Theorem 7.4.

The proof is by the Gaussian algorithm. If A is invertible, it is row equivalent to I. But $|I| \neq 0$; hence, by Problem 7.26, $|A| \neq 0$. If A is not invertible, it is row equivalent to a matrix with a zero row; hence, det $A = 0$. Thus (i) and (iii) are equivalent.

If $AX = 0$ has only the solution $X = 0$, then A is row equivalent to I and A is invertible. Conversely, if A is invertible with inverse A^{-1}, then

$$X = IX = (A^{-1}A)X = A^{-1}(AX) = A^{-1}0 = 0$$

is the only solution of $AX = 0$. Thus (i) and (ii) are equivalent.

7.28. Prove Theorem 7.5.

If A is singular, then AB is also singular and so $|AB| = 0 = |A||B|$. On the other hand, if A is nonsingular, then $A = E_n \cdots E_2 E_1$, a product of elementary matrices. Then, using Lemma 7.6 and induction, we obtain

$$|AB| = |E_n \cdots E_2 E_1 B| = |E_n| \cdots |E_2||E_1||B| = |A||B|$$

7.29. Suppose P is invertible. Show that $|P^{-1}| = |P|^{-1}$.

$P^{-1}P = I$. Hence $1 = |I| = |P^{-1}P| = |P^{-1}||P|$, and so $|P^{-1}| = |P|^{-1}$.

7.30. Prove Theorem 7.7.

Since A and B are similar, there exists an invertible matrix P such that $B = P^{-1}AP$. Then by the preceding problem, $|B| = |P^{-1}AP| = |P^{-1}||A||P| = |A||P^{-1}||P| = |A|$.

We remark that although the matrices P^{-1} and A may not commute, their determinants $|P^{-1}|$ and $|A|$ do commute since they are scalars in the field K.

7.31. If $A = (a_{ij})$, prove that $|A| = a_{i1}A_{i1} + a_{i2}A_{i2} + \cdots + a_{in}A_{in}$, where A_{ij} is the cofactor of a_{ij}.

Each term in $|A|$ contains one and only one entry of the ith row $(a_{i1}, a_{i2}, \ldots, a_{in})$ of A. Hence we can write $|A|$ in the form

$$|A| = a_{i1}A_{i1}^* + a_{i2}A_{i2}^* + \cdots + a_{in}A_{in}^*$$

(Note A_{ij}^* is a sum of terms involving no entry of the ith row of A.) Thus the theorem is proved if we can show that

$$A_{ij}^* = A_{ij} = (-1)^{i+j}|M_{ij}|$$

where M_{ij} is the matrix obtained by deleting the row and column containing the entry a_{ij}. (Historically, the expression A_{ij}^* was defined as the cofactor of a_{ij}, and so the theorem reduces to showing that the two definitions of the cofactor are equivalent.)

First we consider the case that $i = n, j = n$. Then the sum of terms in $|A|$ containing a_{nn} is

$$a_{nn}A_{nn}^* = a_{nn}\sum_\sigma (\text{sgn } \sigma)a_{1\sigma(1)}a_{2\sigma(2)} \cdots a_{n-1, \sigma(n-1)}$$

where we sum over all permutations $\sigma \in S_n$ for which $\sigma(n) = n$. However, this is equivalent (prove!) to summing over all permutations of $\{1, \ldots, n-1\}$. Thus $A_{nn}^* = |M_{nn}| = (-1)^{n+n}|M_{nn}|$.

Now we consider any i and j. We interchange the ith row with each succeeding row until it is last, and we interchange the jth column with each succeeding column until it is last. Note that the determinant $|M_{ij}|$ is not affected since the relative positions of the other rows and columns are not affected by these interchanges. However, the "sign" of $|A|$ and of A_{ij}^* is changed $n - i$ and then $n - j$ times. Accordingly,

$$A_{ij}^* = (-1)^{n-i+n-j}|M_{ij}| = (-1)^{i+j}|M_{ij}|$$

7.32. Let $A = (a_{ij})$ and let B be the matrix obtained from A by replacing the ith row of A by the row vector $(b_{i1}, \ldots, b_{in})$. Show that

$$|B| = b_{i1}A_{i1} + b_{i2}A_{i2} + \cdots + b_{in}A_{in}$$

Furthermore, show that, for $j \neq i$,

$$a_{j1}A_{i1} + a_{j2}A_{i2} + \cdots + a_{jn}A_{in} = 0 \qquad \text{and} \qquad a_{1j}A_{1i} + a_{2j}A_{2i} + \cdots + a_{nj}A_{ni} = 0$$

Let $B = (b_{ij})$. By Theorem 7.8,

$$|B| = b_{i1}B_{i1} + b_{i2}B_{i2} + \cdots + b_{in}B_{in}$$

Since B_{ij} does not depend upon the ith row of B, $B_{ij} = A_{ij}$ for $j = 1, \ldots, n$. Hence

$$|B| = b_{i1}A_{i1} + b_{i2}A_{i2} + \cdots + b_{in}A_{in}$$

Now let A' be obtained from A by replacing the ith row of A by the jth row of A. Since A' has two identical rows, $|A'| = 0$. Thus, by the above result,

$$|A'| = a_{j1}A_{i1} + a_{j2}A_{i2} + \cdots + a_{jn}A_{in} = 0$$

Using $|A^t| = |A|$, we also obtain that $a_{1j}A_{1i} + a_{2j}A_{2i} + \cdots + a_{nj}A_{ni} = 0$.

7.33.   Prove Theorem 7.9.

Let $A = (a_{ij})$ and let $A \cdot (\text{adj } A) = (b_{ij})$. The ith row of A is

$$(a_{i1}, a_{i2}, \ldots, a_{in}) \tag{1}$$

Since adj A is the transpose of the matrix of cofactors, the jth column of adj A is the transpose of the cofactors of the jth row of A: That is,

$$(A_{j1}, A_{j2}, \ldots, A_{jn})^T \tag{2}$$

Now b_{ij}, the ij-entry in $A \cdot (\text{adj } A)$, is obtained by multiplying (*1*) and (*2*):

$$b_{ij} = a_{i1}A_{j1} + a_{i2}A_{j2} + \cdots + a_{in}A_{jn}$$

By Theorem 7.8 and Problem 7.32,

$$b_{ij} = \begin{cases} |A| & \text{if } i = j \\ 0 & \text{if } i \neq j \end{cases}$$

Accordingly, $A \cdot (\text{adj } A)$ is the diagonal matrix with each diagonal element $|A|$. In other words, $A \cdot (\text{adj } A) = |A| I$. Similarly, $(\text{adj } A) \cdot A = |A| I$.

7.34.   Prove Theorem 7.10.

By previous results, $AX = B$ has a unique solution if and only if A is invertible, and A is invertible if and only if $D = |A| \neq 0$.

Now suppose $D \neq 0$. By Theorem 7.9, $A^{-1} = (1/D)(\text{adj } A)$. Multiplying $AX = B$ by A^{-1} we obtain

$$X = A^{-1}AX = (1/D)(\text{adj } A)B \tag{1}$$

Note that the ith row of $(1/D)(\text{adj } A)$ is $(1/D)(A_{1i}, A_{2i}, \ldots, A_{ni})$. If $B = (b_1, b_2, \ldots b_n)^T$ then, by (*1*),

$$x_i = (1/D)(b_1 A_{1i} + b_2 A_{2i} + \cdots + b_n A_{ni})$$

However, as in Problem 7.32, $b_1 A_{1i} + b_2 A_{2i} + \cdots + b_n A_{ni} = N_i$, the determinant of the matrix obtained by replacing the ith column of A by the column vector B. Thus $x_i = (1/D)N_i$, as required.

7.35.   Prove Theorem 7.12.

We need only prove the theorem for $n = 2$, that is, when M is a square block matrix of the form $M = \begin{pmatrix} A & C \\ 0 & B \end{pmatrix}$. The proof of the general theorem follows easily by induction.

Suppose $A = (a_{ij})$ is r-square, $B = (b_{ij})$ is s-square, and $M = (m_{ij})$ is n-square, where $n = r + s$. By definition,

$$\det M = \sum_{\sigma \in S_n} (\text{sgn } \sigma) m_{1\sigma(1)} m_{2\sigma(2)} \cdots m_{n\sigma(n)}$$

If $i > r$ and $j \leq r$, then $m_{ij} = 0$. Thus we need only consider those permutations σ such that

$$\sigma\{r + 1, r + 2, \ldots, r + s\} = \{r + 1, r + 2, \ldots, r + s\} \qquad \text{and} \qquad \sigma\{1, 2, \ldots, r\} = \{1, 2, \ldots, r\}$$

Let $\sigma_1(k) = \sigma(k)$ for $k \leq r$, and let $\sigma_2(k) = \sigma(r + k) - r$ for $k \leq s$. Then

$$(\text{sgn } \sigma) m_{1\sigma(1)} m_{2\sigma(2)} \cdots m_{n\sigma(n)} = (\text{sgn } \sigma_1) a_{1\sigma_1(1)} a_{2\sigma_1(2)} \cdots a_{r\sigma_2(r)} (\text{sgn } \sigma_2) b_{1\sigma_2(1)} b_{2\sigma_2(2)} \cdots b_{s\sigma_2(s)}$$

which implies $\det M = (\det A)(\det B)$.

7.36.   Prove Theorem 7.13.

Let D be the determinant function: $D(A) = |A|$. We must show that D satisfies (i), (ii), and (iii), and that D is the only function satisfying (i), (ii), and (iii).

By Theorem 7.2, D satisfies (ii) and (iii); hence we need show that it is multilinear. Suppose the ith row of $A = (a_{ij})$ has the form $(b_{i1} + c_{i1}, b_{i2} + c_{i2}, \ldots, b_{in} + c_{in})$. Then

$$
\begin{aligned}
D(A) &= D(A_1, \ldots, B_i + C_i, \ldots, A_n) \\
&= \sum_{S_n} (\text{sgn } \sigma) a_{1\sigma(1)} \cdots a_{i-1,\,\sigma(i-1)} (b_{i\sigma(i)} + c_{i\sigma(i)}) \cdots a_{n\sigma(n)} \\
&= \sum_{S_n} (\text{sgn } \sigma) a_{i\sigma(1)} \cdots b_{i\sigma(i)} \cdots a_{n\sigma(n)} + \sum_{S_n} (\text{sgn } \sigma) a_{1\sigma(1)} \cdots c_{i\sigma(i)} \cdots a_{n\sigma(n)} \\
&= D(A_1, \ldots, B_i, \ldots, A_n) + D(A_1, \ldots, C_i, \ldots, A_n)
\end{aligned}
$$

Also, by Theorem 7.3(b),

$$
D(A_1, \ldots, kA_i, \ldots, A_n) = kD(A_1, \ldots, A_i, \ldots, A_n)
$$

Thus D is multilinear, i.e., D satisfies (i).

We next must prove the uniqueness of D. Suppose D satisfies (i), (ii), and (iii). If $\{e_1, \ldots, e_n\}$ is the usual basis of K^n, then by (iii), $D(e_1, e_2, \ldots, e_n) = D(I) = 1$. Using (ii) we also have that

$$
D(e_{i_1}, e_{i_2}, \ldots, e_{i_n}) = \text{sgn } \sigma \qquad \text{where } \sigma = i_1 i_2 \cdots i_n \tag{1}
$$

Now suppose $A = (a_{ij})$. Observe that the kth row A_k of A is

$$
A_k = (a_{k1}, a_{k2}, \ldots, a_{kn}) = a_{k1}e_1 + a_{k2}e_2 + \cdots + a_{kn}e_n
$$

Thus

$$
D(A) = D(a_{11}e_1 + \cdots + a_{1n}e_n, a_{21}e_1 + \cdots + a_{2n}e_n, \ldots, a_{n1}e_1 + \cdots + a_{nn}e_n)
$$

Using the multilinearity of D, we can write $D(A)$ as a sum of terms of the form

$$
\begin{aligned}
D(A) &= \sum D(a_{1i_1}e_{i_1}, a_{2i_2}e_{i_2}, \ldots, a_{ni_n}e_{i_n}) \\
&= \sum (a_{1i_1}a_{2i_2} \cdots a_{ni_n}) D(e_{i_1}, e_{i_2}, \ldots, e_{i_n})
\end{aligned} \tag{2}
$$

where the sum is summed over all sequences $i_1 i_2 \cdots i_n$ where $i_k \in \{1, \ldots, n\}$. If two of the indices are equal, say $i_j = i_k$ but $j \neq k$, then by (ii),

$$
D(e_{i_1}, e_{i_2}, \ldots, e_{i_n}) = 0
$$

Accordingly, the sum in (2) need only be summed over all permutations $\sigma = i_1 i_2 \cdots i_n$. Using (1), we finally have that

$$
\begin{aligned}
D(A) &= \sum_{\sigma} (a_{1i_1}a_{2i_2} \cdots a_{ni_n}) D(e_{i_1}, e_{i_2}, \ldots, e_{i_n}) \\
&= \sum_{\sigma} (\text{sgn } \sigma) a_{1i_1}a_{2i_2} \cdots a_{ni_n} \qquad \text{where } \sigma = i_1 i_2 \cdots i_n
\end{aligned}
$$

Hence D is the determinant function and so the theorem is proved.

PERMUTATIONS

7.37. Determine the parity of $\sigma = 542163$.

Method 1.   We need to obtain the number of pairs (i, j) for which $i > j$ and i precedes j in σ. There are:

> 3 numbers (5, 4 and 2) greater than and preceding 1,
>
> 2 numbers (5 and 4) greater than and preceding 2,
>
> 3 numbers (5, 4 and 6) greater than and preceding 3,
>
> 1 number (5) greater than and preceding 4,
>
> 0 numbers greater than and preceding 5,
>
> 0 numbers greater than and preceding 6.

Since $3 + 2 + 3 + 1 + 0 + 0 = 9$ is odd, σ is an odd permutation and so sgn $\sigma = -1$.

Method 2.   Bring 1 to the first position as follows:

$$5 \quad 4 \quad 2 \quad ① \quad 6 \quad 3 \qquad \text{to} \qquad 1 \quad 5 \quad 4 \quad 2 \quad 6 \quad 3$$

Bring 2 to the second position:

$$1 \quad 5 \quad 4 \quad ② \quad 6 \quad 3 \qquad \text{to} \qquad 1 \quad 2 \quad 5 \quad 4 \quad 6 \quad 3$$

Bring 3 to the third position:

$$1 \quad 2 \quad 5 \quad 4 \quad 6 \quad ③ \qquad \text{to} \qquad 1 \quad 2 \quad 3 \quad 5 \quad 4 \quad 6$$

Bring 4 to the fourth position:

$$1 \quad 2 \quad 3 \quad 5 \quad ④ \quad 6 \qquad \text{to} \qquad 1 \quad 2 \quad 3 \quad 4 \quad 5 \quad 6$$

Note that 5 and 6 are in the "correct" positions. Count the number of numbers "jumped": $3 + 2 + 3 + 1 = 9$. Since 9 is odd, σ is an odd permutation. (*Remark:* This method is essentially the same as the preceding method.)

Method 3.   An interchange of two numbers in a permutation is equivalent to multiplying the permutation by a transposition. Hence transform σ to the identity permutation using transpositions; such as,

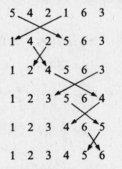

Since an odd number, 5, of transpositions was used (and since odd × odd = odd), σ is an odd permutation.

7.38.   Let $\sigma = 24513$ and $\tau = 41352$ be permutations in S_5. Find:
(*a*) the composition permutation $\tau \circ \sigma$, (*b*) $\sigma \circ \tau$, (*c*) σ^{-1}.

Recall that $\sigma = 24513$ and $\tau = 41352$ are short ways of writing

$$\sigma = \begin{pmatrix} 1 & 2 & 3 & 4 & 5 \\ 2 & 4 & 5 & 1 & 3 \end{pmatrix} \quad \text{and} \quad \tau = \begin{pmatrix} 1 & 2 & 3 & 4 & 5 \\ 4 & 1 & 3 & 5 & 2 \end{pmatrix}$$

which means

$$\sigma(1) = 2 \qquad \sigma(2) = 4 \qquad \sigma(3) = 5 \qquad \sigma(4) = 1 \quad \text{and} \quad \sigma(5) = 3$$

and

$$\tau(1) = 4 \qquad \tau(2) = 1 \qquad \tau(3) = 3 \qquad \tau(4) = 5 \quad \text{and} \quad \tau(5) = 2$$

(*a*)   The effect of σ and then τ on $1, 2, \ldots, 5$ is as follows:

$$\begin{array}{ccccccc}
 & 1 & 2 & 3 & 4 & 5 \\
\sigma & \downarrow & \downarrow & \downarrow & \downarrow & \downarrow \\
 & 2 & 4 & 5 & 1 & 3 \\
\tau & \downarrow & \downarrow & \downarrow & \downarrow & \downarrow \\
 & 1 & 5 & 2 & 4 & 3
\end{array}$$

Thus $\tau \circ \sigma = \begin{pmatrix} 1 & 2 & 3 & 4 & 5 \\ 1 & 5 & 2 & 4 & 3 \end{pmatrix}$, or $\tau \circ \sigma = 15243$.

(b) The effect of τ and then σ on 1, 2, ..., 5 is as follows:

$$
\begin{array}{ccccc}
 & 1 & 2 & 3 & 4 & 5 \\
\tau & \downarrow & \downarrow & \downarrow & \downarrow & \downarrow \\
 & 4 & 1 & 3 & 5 & 2 \\
\sigma & \downarrow & \downarrow & \downarrow & \downarrow & \downarrow \\
 & 1 & 2 & 5 & 3 & 4
\end{array}
$$

Thus $\sigma \circ \tau = 12534$.

(c) By definition, $\sigma^{-1}(j) = k$ if and only if $\sigma(k) = j$; hence

$$
\sigma^{-1} = \begin{pmatrix} 2 & 4 & 5 & 1 & 3 \\ 1 & 2 & 3 & 4 & 5 \end{pmatrix} = \begin{pmatrix} 1 & 2 & 3 & 4 & 5 \\ 4 & 1 & 5 & 2 & 3 \end{pmatrix} \qquad \text{or} \qquad \sigma^{-1} = 41523
$$

7.39. Consider any permutation $\sigma = j_1 j_2 \cdots j_n$. Show that, for each inversion (i, k) in σ, there is a pair (i^*, k^*) such that

$$
i^* < k^* \qquad \text{and} \qquad \sigma(i^*) > \sigma(k^*) \tag{1}
$$

and vice versa. Thus σ is even or odd according as to whether there is an even or odd number of pairs satisfying (1).

   Choose i^* and k^* so that $\sigma(i^*) = i$ and $\sigma(k^*) = k$. Then $i > k$ if and only if $\sigma(i^*) > \sigma(k^*)$, and i precedes k in σ if and only if $i^* < k^*$.

7.40. Consider the polynomial $g = g(x_1, \ldots, x_n) = \prod_{i<j} (x_i - x_j)$. Write out explicitly the polynomial $g = g(x_1, x_2, x_3, x_4)$.

   The symbol $\prod$ is used for a product of terms in the same way that the symbol $\sum$ is used for a sum of terms. That is, $\prod_{i<j} (x_i - x_j)$ means the product of all terms $(x_i - x_j)$ for which $i < j$. Hence

$$
g = g(x_1, \ldots, x_4) = (x_1 - x_2)(x_1 - x_3)(x_1 - x_4)(x_2 - x_3)(x_2 - x_4)(x_3 - x_4)
$$

7.41. Let σ be an arbitrary permutation. For the above polynomial g in Problem 7.40, define $\sigma(g) = \prod_{i<j} (x_{\sigma(i)} - x_{\sigma(j)})$. Show that

$$
\sigma(g) = \begin{cases} g & \text{if } \sigma \text{ is even} \\ -g & \text{if } \sigma \text{ is odd} \end{cases}
$$

Accordingly, $\sigma(g) = (\text{sgn } \sigma)g$.

   Since σ is one-to-one and onto,

$$
\sigma(g) = \prod_{i<j} (x_{\sigma(i)} - x_{\sigma(j)}) = \prod_{i<j \text{ or } i>j} (x_i - x_j)
$$

Thus $\sigma(g) = g$ or $\sigma(g) = -g$ according as to whether there is an even or an odd number of terms of the form $(x_i - x_j)$ where $i > j$. Note that for each pair (i, j) for which

$$
i < j \qquad \text{and} \qquad \sigma(i) > \sigma(j) \tag{1}
$$

there is a term $(x_{\sigma(i)} - x_{\sigma(j)})$ in $\sigma(g)$ for which $\sigma(i) > \sigma(j)$. Since σ is even if and only if there is an even number of pairs satisfying (1), we have $\sigma(g) = g$ if and only if σ is even; hence $\sigma(g) = -g$ if and only if σ is odd.

7.42. Let $\sigma, \tau \in S_n$. Show that sgn $(\tau \circ \sigma) = (\text{sgn } \tau)(\text{sgn } \sigma)$. Thus the product of two even or two odd permutations is even, and the product of an odd and an even permutation is odd.

Using Problem 7.41, we have

$$\text{sgn } (\tau \circ \sigma)g = (\tau \circ \sigma)(g) = \tau(\sigma(g)) = \tau((\text{sgn } \sigma)g) = (\text{sgn } \tau)(\text{sgn } \sigma)g$$

Accordingly, $\text{sgn } (\tau \circ \sigma) = (\text{sgn } \tau)(\text{sgn } \sigma)$.

7.43. Consider the permutation $\sigma = j_1 j_2 \cdots j_n$. Show that $\text{sgn } \sigma^{-1} = \text{sgn } \sigma$ and, for scalars a_{ij}, show that

$$a_{j_1 1} +_{j_2 2} \cdots a_{j_n n} = a_{1 k_1} + a_{2 k_2} \cdots a_{n k_n}$$

where $\sigma^{-1} = k_1 k_2 \cdots k_n$.

We have $\sigma^{-1} \circ \sigma = \varepsilon$, the identity permutation. Since ε is even, σ^{-1} and σ are both even or both odd. Hence $\text{sgn } \sigma^{-1} = \text{sgn } \sigma$.

Since $\sigma = j_1 j_2 \cdots j_n$ is a permutation, $a_{j_1 1} a_{j_2 2} \cdots a_{j_n n} = a_{1 k_1} a_{2 k_2} \cdots a_{n k_n}$. Then $k_1, k_2, \ldots, k_n$ have the property that

$$\sigma(k_1) = 1, \ \sigma(k_2) = 2, \ \ldots, \ \sigma(k_n) = n$$

Let $\tau = k_1 k_2 \cdots k_n$. Then for $i = 1, \ldots, n$,

$$(\sigma \circ \tau)(i) = \sigma(\tau(i)) = \sigma(k_i) = i$$

Thus $\sigma \circ \tau = \varepsilon$, the identity permutation; hence $\tau = \sigma^{-1}$.

MISCELLANEOUS PROBLEMS

7.44. Without expanding the determinant, show that $\begin{vmatrix} 1 & a & b+c \\ 1 & b & c+a \\ 1 & c & a+b \end{vmatrix} = 0$.

Add the second column to the third column, and remove the common factor from the third column; this yields

$$\begin{vmatrix} 1 & a & b+c \\ 1 & b & c+a \\ 1 & c & a+b \end{vmatrix} = \begin{vmatrix} 1 & a & a+b+c \\ 1 & b & a+b+c \\ 1 & c & a+b+c \end{vmatrix} = (a+b+c)\begin{vmatrix} 1 & a & 1 \\ 1 & b & 1 \\ 1 & c & 1 \end{vmatrix} = (a+b+c)(0) = 0$$

(We use the fact that a determinant with two identical columns is zero.)

7.45. Show that the difference product $g(x_1, \ldots, x_n)$ of Problem 7.40 can be represented by means of the Vandermonde determinant of $x_1, x_2, \ldots, x_{n-1}, x$ defined by:

$$V_{n-1}(x) \equiv \begin{vmatrix} 1 & 1 & \cdots & 1 & 1 \\ x_1 & x_2 & \cdots & x_{n-1} & x \\ x_1^2 & x_2^2 & \cdots & x_{n-1}^2 & x^2 \\ \cdots\cdots\cdots\cdots\cdots\cdots\cdots\cdots\cdots \\ x_1^{n-1} & x_2^{n-1} & \cdots & x_{n-1}^{n-1} & x^{n-1} \end{vmatrix}$$

This is a polynomial in x of degree $n-1$, of which the roots are $x_1, x_2, \ldots, x_{n-1}$; moreover, the leading coefficient (the cofactor of x^{n-1}) is equal to $V_{n-2}(x_{n-1})$. Thus, from algebra,

$$V_{n-1}(x) = (x - x_1)(x - x_2) \cdots (x - x_{n-1}) V_{n-2}(x_{n-1})$$

so that, by recursion,

$$V_{n-1}(x) = [(x - x_1) \cdots (x - x_{n-1})][(x_{n-1} - x_1) \cdots (x_{n-1} - x_{n-2})] V_{n-3}(x_{n-2})$$
$$= \cdots$$
$$= [(x - x_1) \cdots (x - x_{n-1})][(x_{n-1} - x_1) \cdots (x_{n-1} - x_{n-2})] \cdots [(x_2 - x_1)]$$

It follows that

$$V_{n-1}(x_n) = \prod_{n \geq i > j \geq 1} (x_i - x_j) = (-1)^{n(n-1)/2} \prod_{1 \leq i < j \leq n} (x_i - x_j)$$

Thus $g(x_1, \ldots, x_n) = (-1)^{n(n-1)/2} V_{n-1}(x_n)$.

7.46. Find the volume $V(S)$ of the parallelepiped S in $\mathbf{R}^3$ determined by the vectors $u_1 = (1, 2, 4)$, $u_2 = (2, 1, -3)$, and $u_3 = (5, 7, 9)$.

Evaluate the determinant $\begin{vmatrix} 1 & 2 & 4 \\ 2 & 1 & -3 \\ 5 & 7 & 9 \end{vmatrix} = 9 - 30 + 56 - 20 + 21 - 36 = 0$. Thus $V(S) = 0$ or, in

other words, u_1, u_2 and u_3 lie in a plane.

7.47. Find the volume $V(S)$ of the parallelepiped S in $\mathbf{R}^4$ determined by the vectors $u_1 = (2, -1, 4, -3)$, $u_2 = (-1, 1, 0, 2)$, $u_3 = (3, 2, 3, -1)$, and $u_4 = (1, -2, 2, 3)$.

Evaluate the following determinant, using u_{22} as a pivot and applying $C_2 + C_1 \rightarrow C_1$ and $-2C_2 + C_4 \rightarrow C_4$:

$$\begin{vmatrix} 2 & -1 & 4 & -3 \\ -1 & 1 & 0 & 2 \\ 3 & 2 & 3 & -1 \\ 1 & -2 & 2 & 3 \end{vmatrix} = \begin{vmatrix} 1 & -1 & 4 & -1 \\ 0 & 1 & 0 & 0 \\ 5 & 2 & 3 & -5 \\ -1 & -2 & 2 & 7 \end{vmatrix} = \begin{vmatrix} 1 & 4 & -1 \\ 5 & 3 & -5 \\ -1 & 2 & 7 \end{vmatrix}$$

$$= 21 + 20 - 10 - 3 + 10 - 140 = -102$$

Hence $V(S) = 102$.

7.48. Find det (M) where $M = \begin{pmatrix} 3 & 4 & 0 & 0 & 0 \\ 2 & 5 & 0 & 0 & 0 \\ 0 & 9 & 2 & 0 & 0 \\ 0 & 5 & 0 & 6 & 7 \\ 0 & 0 & 4 & 3 & 4 \end{pmatrix}$.

Partition M into a (lower) triangular block matrix as follows:

$$M = \begin{pmatrix} 3 & 4 & 0 & 0 & 0 \\ 2 & 5 & 0 & 0 & 0 \\ \hline 0 & 9 & 2 & 0 & 0 \\ \hline 0 & 5 & 0 & 6 & 7 \\ 0 & 0 & 4 & 3 & 4 \end{pmatrix}$$

Evaluate the determinant of each diagonal block:

$$\begin{vmatrix} 3 & 4 \\ 2 & 5 \end{vmatrix} = 15 - 8 = 7 \qquad |2| = 2 \qquad \begin{vmatrix} 6 & 7 \\ 3 & 4 \end{vmatrix} = 24 - 21 = 3$$

Hence $|M| = 7 \cdot 2 \cdot 3 = 42$.

7.49. Find the minor, signed minor, and complementary minor of $A_{1, 3}^{2, 3}$ where

$$A = \begin{pmatrix} 1 & 0 & 4 & -1 \\ 3 & 2 & -2 & 5 \\ 0 & 1 & -3 & 7 \\ 6 & -4 & -5 & -1 \end{pmatrix}$$

The row subscripts are 1 and 3 and the column subscripts are 2 and 3; hence the minor is

$$|A_{1,3}^{2,3}| = \begin{vmatrix} a_{12} & a_{13} \\ a_{32} & a_{33} \end{vmatrix} = \begin{vmatrix} 0 & 4 \\ 1 & -3 \end{vmatrix} = 0 - 4 = -4$$

and the signed minor is

$$(-1)^{1+3+2+3}|A_{1,3}^{2,3}| = -(-4) = 4$$

The missing row subscripts are 2 and 4 and the missing column subscripts are 1 and 4. Hence the complementary minor is

$$|A_{2,4}^{1,4}| = \begin{vmatrix} a_{21} & a_{24} \\ a_{41} & a_{44} \end{vmatrix} = \begin{vmatrix} 3 & 5 \\ 6 & -1 \end{vmatrix} = -3 - 30 = -33$$

7.50. Let $A = (a_{ij})$ be a 3-square matrix. Describe the sum S_k of the principal minors of orders (a) $k = 1$, (b) $k = 2$, (c) $k = 3$.

(a) The principal minors of order one are the diagonal elements. Thus $S_1 = a_{11} + a_{22} + a_{33} = \text{tr } A$, the trace of A.

(b) The principal minors of order two are the cofactors of the diagonal elements. Thus $S_2 = A_{11} + A_{22} + A_{33}$ where A_{ii} is the cofactor of a_{ii}.

(c) There is only one principal minor of order three, the determinant of A. Thus $S_3 = \det(A)$.

7.51. Find the number N_k and the sum S_k of all principal minors of order (a) $k = 1$, (b) $k = 2$, (c) $k = 3$, and (d) $k = 4$ of the matrix

$$A = \begin{pmatrix} 1 & 3 & 0 & -1 \\ -4 & 2 & 5 & 1 \\ 1 & 0 & 3 & -2 \\ 3 & -2 & 1 & 4 \end{pmatrix}$$

Each (nonempty) subset of the diagonal (or, equivalently, each nonempty subset of $\{1, 2, 3, 4\}$) determines a principal minor of A, and $N_k = \binom{n}{k} = \dfrac{n!}{(n-k)!}$ of them are of order k.

(a) $N_1 = \binom{4}{1} = 4$ and

$$S_1 = |1| + |2| + |3| + |4| = 1 + 2 + 3 + 4 = 10$$

(b) $N_2 = \binom{4}{2} = 6$ and

$$S_2 = \begin{vmatrix} 1 & 3 \\ -4 & 2 \end{vmatrix} + \begin{vmatrix} 1 & 0 \\ 1 & 3 \end{vmatrix} + \begin{vmatrix} 1 & -1 \\ 3 & 4 \end{vmatrix} + \begin{vmatrix} 2 & 5 \\ 0 & 3 \end{vmatrix} + \begin{vmatrix} 2 & 1 \\ -2 & 4 \end{vmatrix} + \begin{vmatrix} 3 & -2 \\ 1 & 4 \end{vmatrix}$$

$$= 14 + 3 + 7 + 6 + 10 + 14 = 54$$

(c) $N_3 = \binom{4}{3} = 4$ and

$$S_3 = \begin{vmatrix} 1 & 3 & 0 \\ -4 & 2 & 5 \\ 1 & 0 & 3 \end{vmatrix} + \begin{vmatrix} 1 & 3 & -1 \\ -4 & 2 & 1 \\ 3 & -2 & 4 \end{vmatrix} + \begin{vmatrix} 1 & 0 & -1 \\ 1 & 3 & -2 \\ 3 & 1 & 4 \end{vmatrix} + \begin{vmatrix} 2 & 5 & 1 \\ 0 & 3 & -2 \\ -2 & 1 & 4 \end{vmatrix}$$

$$= 57 + 65 + 22 + 54 = 198$$

(d) $N_4 = 1$ and $S_4 = \det(A) = 378$.

7.52. Let V be the vector space of 2 by 2 matrices $M = \begin{pmatrix} a & b \\ c & d \end{pmatrix}$ over $\mathbf{R}$. Determine whether or not $D: V \to \mathbf{R}$ is 2-linear (with respect to the rows) where: (a) $D(M) = a + d$, (b) $D(M) = ad$.

(a) No. For example, suppose $A = (1, 1)$ and $B = (3, 3)$. Then

$$D(A, B) = D\begin{pmatrix} 1 & 1 \\ 3 & 3 \end{pmatrix} = 4 \quad \text{and} \quad D(2A, B) = D\begin{pmatrix} 2 & 2 \\ 3 & 3 \end{pmatrix} = 5 \neq 2D(A, B)$$

(b) Yes. Let $A = (a_1, a_2)$, $B = (b_1, b_2)$, and $C = (c_1, c_2)$; then

$$D(A, C) = D\begin{pmatrix} a_1 & a_2 \\ c_1 & c_2 \end{pmatrix} = a_1 c_2 \quad \text{and} \quad D(B, C) = D\begin{pmatrix} b_1 & b_2 \\ c_1 & c_2 \end{pmatrix} = b_1 c_2$$

Hence for any scalars $s, t \in \mathbf{R}$,

$$D(sA + tB, C) = D\begin{pmatrix} sa_1 + tb_1 & sa_2 + tb_2 \\ c_1 & c_2 \end{pmatrix} = (sa_1 + tb_1)c_2$$
$$= s(a_1 c_2) + t(b_1 c_2) = sD(A, C) + tD(B, C)$$

That is, D is linear with respect to the first row.
Furthermore,

$$D(C, A) = D\begin{pmatrix} c_1 & c_2 \\ a_1 & a_2 \end{pmatrix} = c_1 a_2 \quad \text{and} \quad D(C, B) = D\begin{pmatrix} c_1 & c_2 \\ b_1 & b_2 \end{pmatrix} = c_1 b_2$$

Hence for any scalars $s, t \in \mathbf{R}$,

$$D(C, aA + tB) = D\begin{pmatrix} c_1 & c_2 \\ sa_1 + tb_1 & sa_2 + tb_2 \end{pmatrix} = c_1(sa_2 + tb_2)$$
$$= s(c_1 a_2) + t(c_1 b_2) = sD(C, A) + tD(C, B)$$

That is, D is linear with respect to the second row.
Both linearity conditions imply that D is 2-linear.

7.53. Let D be a 2-linear, alternating function. Show that $D(A, B) = -D(B, A)$. More generally, show that if D is multilinear and alternating, then

$$D(\ldots, A, \ldots, B, \ldots) = -D(\ldots, B, \ldots, A, \ldots)$$

that is, the sign is changed whenever two components are interchanged.

Since D is alternating, $D(A + B, A + B) = 0$. Furthermore, since D is multilinear,

$$0 = D(A + B, A + B) = D(A, A + B) + D(B, A + B)$$
$$= D(A, A) + D(A, B) + D(B, A) + D(B, B)$$

But $D(A, A) = 0$ and $D(B, B) = 0$. Hence

$$0 = D(A, B) + D(B, A) \quad \text{or} \quad D(A, B) = -D(B, A)$$

Similarly,

$$0 = D(\ldots, A + B, \ldots, A + B, \ldots)$$
$$= D(\ldots, A, \ldots, A, \ldots) + D(\ldots, A, \ldots, B, \ldots) + D(\ldots, B, \ldots, A, \ldots) + D(\ldots, B, \ldots, B, \ldots)$$
$$= D(\ldots, A, \ldots, B, \ldots) + D(\ldots, B, \ldots, A, \ldots)$$

and thus $D(\ldots, A, \ldots, B, \ldots) = -D(\ldots, B, \ldots, A, \ldots)$.

Supplementary Problems

COMPUTATION OF DETERMINANTS

7.54. Compute the determinant of each matrix:

$$(a)\ \begin{pmatrix} 2 & 1 & 1 \\ 0 & 5 & -2 \\ 1 & -3 & 4 \end{pmatrix}, \quad (b)\ \begin{pmatrix} 3 & -2 & -4 \\ 2 & 5 & -1 \\ 0 & 6 & 1 \end{pmatrix}, \quad (c)\ \begin{pmatrix} -2 & -1 & 4 \\ 6 & -3 & -2 \\ 4 & 1 & 2 \end{pmatrix}, \quad (d)\ \begin{pmatrix} 7 & 6 & 5 \\ 1 & 2 & 1 \\ 3 & -2 & 1 \end{pmatrix}$$

7.55. Evaluate the determinant of each matrix:

$$(a)\ \begin{pmatrix} t-2 & 4 & 3 \\ 1 & t+1 & -2 \\ 0 & 0 & t-4 \end{pmatrix}, \quad (b)\ \begin{pmatrix} t-1 & 3 & -3 \\ -3 & t+5 & -3 \\ -6 & 6 & y-4 \end{pmatrix}, \quad (c)\ \begin{pmatrix} t+3 & -1 & 1 \\ 7 & t-5 & 1 \\ 6 & -6 & t+2 \end{pmatrix}$$

7.56. For each matrix in Problem 7.54, determine those values of t for which the determinant is zero.

7.57. Evaluate the determinant of each matrix: $(a)\ \begin{pmatrix} 1 & 2 & 2 & 3 \\ 1 & 0 & -2 & 0 \\ 3 & -1 & 1 & -2 \\ 4 & -3 & 0 & 2 \end{pmatrix}, \quad (b)\ \begin{pmatrix} 2 & 1 & 3 & 2 \\ 3 & 0 & 1 & -2 \\ 1 & -1 & 4 & 3 \\ 2 & 2 & -1 & 1 \end{pmatrix}.$

7.58. Evaluate each determinant:

$$(a)\ \begin{vmatrix} 1 & 2 & -1 & 3 & 1 \\ 2 & -1 & 1 & -2 & 3 \\ 3 & 1 & 0 & 2 & -1 \\ 5 & 1 & 2 & -3 & 4 \\ -2 & 3 & -1 & 1 & -2 \end{vmatrix}, \quad (b)\ \begin{vmatrix} 1 & 3 & 5 & 7 & 9 \\ 2 & 4 & 2 & 4 & 2 \\ 0 & 0 & 1 & 2 & 3 \\ 0 & 0 & 5 & 6 & 2 \\ 0 & 0 & 2 & 3 & 1 \end{vmatrix}, \quad (c)\ \begin{vmatrix} 1 & 2 & 3 & 4 & 5 \\ 5 & 4 & 3 & 2 & 1 \\ 0 & 0 & 6 & 5 & 1 \\ 0 & 0 & 0 & 7 & 4 \\ 0 & 0 & 0 & 2 & 3 \end{vmatrix}$$

COFACTORS, CLASSICAL ADJOINTS, INVERSES

7.59. Let $A = \begin{pmatrix} 1 & 2 & 2 \\ 3 & 1 & 0 \\ 1 & 1 & 1 \end{pmatrix}$. Find (a) adj A, and (b) A^{-1}.

7.60. Find the classical adjoint of each matrix in Problem 7.57.

7.61. Determine the general 2 by 2 matrix A for which $A = $ adj A.

7.62. Suppose A is diagonal and B is triangular; say,

$$A = \begin{pmatrix} a_1 & 0 & \dots & 0 \\ 0 & a_2 & \dots & 0 \\ \multicolumn{4}{c}{\dotfill} \\ 0 & 0 & \dots & a_n \end{pmatrix} \quad \text{and} \quad B = \begin{pmatrix} b_1 & c_{12} & \dots & c_{1n} \\ 0 & b_2 & \dots & c_{2n} \\ \multicolumn{4}{c}{\dotfill} \\ 0 & 0 & \dots & b_n \end{pmatrix}$$

(a) Show that adj A is diagonal and adj B is triangular.

(b) Show that B is invertible iff all $b_i \neq 0$; hence A is invertible iff all $a_i \neq 0$.

(c) Show that the inverses of A and B (if either exists) are of the form

$$A^{-1} = \begin{pmatrix} a_1^{-1} & 0 & \cdots & 0 \\ 0 & a_2^{-1} & \cdots & 0 \\ \cdots\cdots\cdots\cdots\cdots\cdots \\ 0 & 0 & \cdots & a_n^{-1} \end{pmatrix}, \qquad B^{-1} = \begin{pmatrix} b_1^{-1} & d_{12} & \cdots & d_{1n} \\ 0 & b_2^{-1} & \cdots & d_{2n} \\ \cdots\cdots\cdots\cdots\cdots\cdots \\ 0 & 0 & \cdots & b_n^{-1} \end{pmatrix}$$

That is, the diagonal elements of A^{-1} and B^{-1} are the inverses of the corresponding diagonal elements of A and B.

DETERMINANTS AND LINEAR EQUATIONS

7.63. Solve by determinants: (a) $\begin{cases} 3x + 5y = 8 \\ 4x - 2y = 1 \end{cases}$, (b) $\begin{cases} 2x - 3y = -1 \\ 4x + 7y = -1 \end{cases}$.

7.64. Solve by determinants: (a) $\begin{cases} 2x - 5y + 2z = 7 \\ x + 2y - 4z = 3, \\ 3x - 4y - 6z = 5 \end{cases}$ (b) $\begin{cases} 2z + 3 = y + 3x \\ x - 3z = 2y + 1. \\ 3y + z = 2 - 2x \end{cases}$

7.65. Prove Theorem 7.11.

PERMUTATIONS

7.66. Determine the parity of these permutations in S_5: (a) $\sigma = 32154$, (b) $\tau = 13524$, (c) $\pi = 42531$.

7.67. For the permutations σ, τ and π in Problem 7.66, find (a) $\tau \circ \sigma$, (b) $\pi \circ \sigma$, (c) σ^{-1}, (d) τ^{-1}.

7.68. Let $\tau \in S_n$. Show that $\tau \circ \sigma$ runs through S_n as σ runs through S_n; that is, $S_n = \{\tau \circ \sigma : \sigma \in S_n\}$.

7.69. Let $\sigma \in S_n$ have the property that $\sigma(n) = n$. Let $\sigma^* \in S_{n-1}$ be defined by $\sigma^*(x) = \sigma(x)$. (a) Show that sgn $\sigma^* = $ sgn σ. (b) Show that as σ runs through S_n, where $\sigma(n) = n$, σ^* runs through S_{n-1}; that is, $S_{n-1} = \{\sigma^* : \sigma \in S_n, \sigma(n) = n\}$.

7.70. Consider a permutation $\sigma = j_1 j_2 \cdots j_n$. Let $\{e_i\}$ be the usual basis of K^n, and let A be the matrix whose ith row is e_{j_i}, i.e., $A = (e_{j_1}, e_{j_2}, \ldots, e_{j_n})$. Show that $|A| = $ sgn σ.

MISCELLANEOUS PROBLEMS

7.71. Find the volume $V(S)$ of the parallelepiped S in $\mathbf{R}^3$ determined by the vectors:
(a) $u_1 = (1, 2, -3)$, $u_2 = (3, 4, -1)$, $u_3 = (2, -1, 5)$, (b) $u_1 = (1, 1, 3)$, $u_2 = (1, -2, -4)$, $u_3 = (4, 1, 2)$

7.72. Find the volume $V(S)$ of the parallelepiped S in $\mathbf{R}^4$ determined by the vectors:

$$u_1 = (1, -2, 5, -1) \qquad u_2 = (2, 1, -2, 1) \qquad u_3 = (3, 0, 1, -2) \qquad u_4 = (1, -1, 4, -1)$$

7.73. Find the minor M_1, the signed minor M_2, and the complementary minor M_3 of $A_{1,4}^{3,4}$ where:

(a) $A = \begin{pmatrix} 1 & 2 & 3 & 2 \\ 1 & 0 & -2 & 3 \\ 3 & -1 & 2 & 5 \\ 4 & -3 & 0 & -1 \end{pmatrix}$, (b) $A = \begin{pmatrix} 1 & 3 & -1 & 5 \\ 2 & -3 & 1 & 4 \\ 0 & -5 & 2 & 1 \\ 3 & 0 & 5 & -2 \end{pmatrix}$

7.74. For $k = 1, 2, 3$, find the sum S_k of all principal minors of order k for:

$$(a) \quad A = \begin{pmatrix} 1 & 3 & 2 \\ 2 & -4 & 3 \\ 5 & -2 & 1 \end{pmatrix}, \qquad (b) \quad B = \begin{pmatrix} 1 & 5 & -4 \\ 2 & 6 & 1 \\ 3 & -2 & 0 \end{pmatrix}, \qquad (c) \quad C = \begin{pmatrix} 1 & -4 & 3 \\ 2 & 1 & 5 \\ 4 & -7 & 11 \end{pmatrix}.$$

7.75. For $k = 1, 2, 3, 4$, find the sum S_k of all principal minors of order k for

$$A = \begin{pmatrix} 1 & 2 & 3 & -1 \\ 1 & -2 & 0 & 5 \\ 0 & 1 & -2 & 2 \\ 4 & 0 & -1 & -3 \end{pmatrix}$$

7.76. Let A be an n-square matrix. Prove $|kA| = k^n |A|$.

7.77. Let A, B, C and D be commuting n-square matrices. Consider the $2n$-square block matrix $M = \begin{pmatrix} A & B \\ C & D \end{pmatrix}$. Prove that $|M| = |A||D| - |B||C|$. Show that the result may not be true if the matrices do not commute.

7.78. Suppose A is orthogonal, that is, $A^T A = I$. Show that $\det(A) = \pm 1$.

7.79. Let V be the space of 2×2 matrices $M = \begin{pmatrix} a & b \\ c & d \end{pmatrix}$ over $\mathbf{R}$. Determine whether or not $D: V \to \mathbf{R}$ is 2-linear (with respect to the rows) where: (a) $D(M) = ac - bd$, (b) $D(M) = ab - cd$, (c) $D(M) = 0$, and (d) $D(M) = 1$.

7.80. Let V be the space of m-square matrices viewed as m-tuples of row vectors. Suppose $D: V \to K$ is m-linear and alternating. Show that if $A_1, A_2, \ldots, A_m$ are linearly dependent, then $D(A_1, \ldots, A_m) = 0$.

7.81. Let V be the space of m-square matrices (as above), and suppose $D: V \to K$. Show that the following weaker statement is equivalent to D being alternating:

$$D(A_1, A_2, \ldots, A_n) = 0 \qquad \text{whenever } A_i = A_{i+1} \text{ for some } i$$

7.82. Let V be the space of n-square matrices over K. Suppose $B \in V$ is invertible and so $\det(B) \neq 0$. Define $D: V \to K$ by $D(A) = \det(AB)/\det(B)$ where $A \in V$. Hence

$$D(A_1, A_2, \ldots, A_n) = \det(A_1 B, A_2 B, \ldots, A_n B)/\det(B)$$

where A_i is the ith row of A and so $A_i B$ is the ith row of AB. Show that D is multilinear and alternating, and that $D(I) = 1$. (This method is used by some texts to prove $|AB| = |A||B|$.)

7.83. Let A be an n-square matrix. The determinantal rank of A is the order of the largest submatrix of A (obtained by deleting rows and columns of A) whose determinant is not zero. Show that the determinantal rank of A is equal to its rank, i.e., the maximum number of linearly independent rows (or columns).

Answers to Supplementary Problems

7.54. (a) 21, (b) -11, (c) 100, (d) 0

7.55. (a) $(t + 2)(t - 3)(t - 4)$, (b) $(t + 2)^2(t - 4)$, (c) $(t + 2)^2(t - 4)$

7.56. (a) $3, 4, -2$; (b) $4, -2$; (c) $4, -2$

7.57. (*a*) -131,   (*b*) -55

7.58. (*a*) -12,   (*b*) -42,   (*c*) -468

7.59. adj $A = \begin{pmatrix} 1 & 0 & -2 \\ -3 & -1 & 6 \\ 2 & 1 & -5 \end{pmatrix}$, $A^{-1} = \begin{pmatrix} -1 & 0 & 2 \\ 3 & 1 & -6 \\ -2 & -1 & 5 \end{pmatrix}$

7.60. (*a*) $\begin{pmatrix} -16 & -29 & -26 & -2 \\ -30 & -38 & -16 & 29 \\ -8 & 51 & -13 & -1 \\ -13 & 1 & 28 & -18 \end{pmatrix}$,   (*b*) $\begin{pmatrix} 21 & -14 & -17 & -19 \\ -44 & 11 & 33 & 11 \\ -29 & 1 & 13 & 21 \\ 17 & 7 & -19 & -18 \end{pmatrix}$

7.61. $A = \begin{pmatrix} k & 0 \\ 0 & k \end{pmatrix}$

7.63. (*a*) $x = \frac{21}{26}, y = \frac{29}{26}$;   (*b*) $x = -\frac{5}{13}, y = \frac{1}{13}$

7.64. (*a*) $x = 5, y = 1, z = 1$.   (*b*) Since $D = 0$, the system cannot be solved by determinants.

7.66. sgn $\sigma = 1$, sgn $\tau = -1$, sgn $\pi = -1$

7.67. (*a*) $\tau \circ \sigma = 53142$,   (*b*) $\pi \circ \sigma = 52413$,   (*c*) $\sigma^{-1} = 32154$,   (*d*) $\tau^{-1} = 14253$

7.71. (*a*) 30,   (*b*) 0

7.72. 17

7.73. (*a*) $-3, -3, -1$;   (*b*) $-23, -23, -10$

7.74. (*a*) $-2, -17, 73$;   (*b*) $7, 10, 105$;   (*c*) $13, 54, 0$

7.75. $S_1 = -6, S_2 = 13, S_3 = 62, S_4 = -219$

7.79. (*a*) Yes,   (*b*) No,   (*c*) Yes,   (*d*) No

Chapter 8

Eigenvalues and Eigenvectors, Diagonalization

8.1 INTRODUCTION

Consider an n-square matrix A over a field K. Recall (Section 4.13) that A induces a function $f: K^n \to K^n$ defined by

$$f(X) = AX$$

where X is any point (column vector) in K^n. (We then view A as the matrix which represents the function f relative to the usual basis E for K^n.)

Suppose a new basis is chosen for K^n, say

$$S = \{u_1, u_2, \ldots, u_n\}$$

(Geometrically, S determines a new coordinate system for K^n.) Let P be the matrix whose columns are the vectors $u_1, u_2, \ldots, u_n$. Then (Section 5.11) P is the change-of-basis matrix from the usual basis E to S. Also, by Theorem 5.27,

$$X' = P^{-1}X$$

gives the coordinates of X in the new basis S. Furthermore, the matrix

$$B = P^{-1}AP$$

represents the function f in the new system S; that is, $f(X') = BX'$.

The following two questions are addressed in this chapter:

(1) Given a matrix A, can we find a nonsingular matrix P (which represents a new coordinate system S), so that

$$B = P^{-1}AP$$

is a diagonal matrix? If the answer is yes, then we say that A is *diagonalizable*.

(2) Given a real matrix A, can we find an orthogonal matrix P (which represents a new orthonormal system S) so that

$$B = P^{-1}AP$$

is a diagonal matrix? If the answer is yes, then we say that A is *orthogonally diagonalizable*.

Recall that matrices A and B are said to be similar (orthogonally similar) if there exists a nonsingular (orthogonal) matrix P such that $B = P^{-1}AP$. Thus the above questions ask, in other words, whether or not a given matrix A is similar (orthogonally similar) to a diagonal matrix.

The answers are closely related to the roots of certain polynomials associated with A. The particular underlying field K also plays an important part in this theory since the existence of roots of the polynomials depends on K.

8.2 POLYNOMIALS AND MATRICES

Consider a polynomial $f(t)$ over a field K; say

$$f(t) = a_n t^n + \cdots + a_1 t + a_0$$

Recall that if A is a square matrix over K, then we define

$$f(A) = a_n A^n + \cdots + a_1 A + a_0 I$$

where I is the identity matrix. In particular, we say that A is a root or zero of the polynomial $f(t)$ if $f(A) = 0$.

Example 8.1. Let $A = \begin{pmatrix} 1 & 2 \\ 3 & 4 \end{pmatrix}$, and let $f(t) = 2t^2 - 3t + 7$, $g(t) = t^2 - 5t - 2$. Then

$$f(A) = 2\begin{pmatrix} 1 & 2 \\ 3 & 4 \end{pmatrix}^2 - 3\begin{pmatrix} 1 & 2 \\ 3 & 4 \end{pmatrix} + 7\begin{pmatrix} 1 & 0 \\ 0 & 1 \end{pmatrix} = \begin{pmatrix} 18 & 14 \\ 21 & 39 \end{pmatrix}$$

and

$$g(A) = \begin{pmatrix} 1 & 2 \\ 3 & 4 \end{pmatrix}^2 - 5\begin{pmatrix} 1 & 2 \\ 3 & 4 \end{pmatrix} - 2\begin{pmatrix} 1 & 0 \\ 0 & 1 \end{pmatrix} = \begin{pmatrix} 0 & 0 \\ 0 & 0 \end{pmatrix}$$

Thus A is a zero of $g(t)$.

The following theorem, proved in Problem 8.26, applies.

Theorem 8.1: Let f and g be polynomials over K, and let A be an n-square matrix over K. Then

   (i) $(f + g)(A) = f(A) + g(A)$

   (ii) $(fg)(A) = f(A)g(A)$

   and, for any scalar $k \in K$,

   (iii) $(kf)(A) = kf(A)$

Furthermore, since $f(t)g(t) = g(t)f(t)$ for any polynomials $f(t)$ and $g(t)$,

$$f(A)g(A) = g(A)f(A)$$

That is, any two polynomials in the matrix A commute.

8.3 CHARACTERISTIC POLYNOMIAL, CAYLEY–HAMILTON THEOREM

Consider an n-square matrix A over a field K:

$$A = \begin{pmatrix} a_{11} & a_{12} & \cdots & a_{1n} \\ a_{21} & a_{22} & \cdots & a_{2n} \\ \cdots\cdots\cdots\cdots\cdots\cdots\cdots \\ a_{n1} & a_{n2} & \cdots & a_{nn} \end{pmatrix}$$

The matrix $tI_n - A$, where I_n is the n-square identity matrix and t is an indeterminant, is called the characteristic matrix of A:

$$tI_n - A = \begin{pmatrix} t - a_{11} & -a_{12} & \cdots & -a_{1n} \\ -a_{21} & t - a_{22} & \cdots & -a_{2n} \\ \cdots\cdots\cdots\cdots\cdots\cdots\cdots\cdots \\ -a_{n1} & -a_{n2} & \cdots & t - a_{nn} \end{pmatrix}$$

Its determinant

$$\Delta_A(t) = \det(tI_n - A)$$

which is a polynomial in t, is called the characteristic polynomial of A. We also call

$$\Delta_A(t) = \det(tI_n - A) = 0$$

the characteristic equation of A.

Now each term in the determinant contains one and only one entry from each row and from each column; hence the above characteristic polynomial is of the form

$$\Delta_A(t) = (t - a_{11})(t - a_{22}) \cdots (t - a_{nn})$$
$$+ \text{ terms with at most } n - 2 \text{ factors of the form } t - a_{ii}$$

Accordingly,

$$\Delta_A(t) = t^n - (a_{11} + a_{22} + \cdots + a_{nn})t^{n-1} + \text{ terms of lower degree}$$

Recall that the trace of A is the sum of its diagonal elements. Thus the characteristic polynomial $\Delta_A(t) = \det(tI_n - A)$ of A is a monic polynomial of degree n, and the coefficient of t^{n-1} is the negative of the trace of A. (A polynomial is monic if its leading coefficient is 1.)

Furthermore, if we set $t = 0$ in $\Delta_A(t)$, we obtain

$$\Delta_A(0) = |-A| = (-1)^n |A|$$

But $\Delta_A(0)$ is the constant term of the polynomial $\Delta_A(t)$. Thus the constant term of the characteristic polynomial of the matrix A is $(-1)^n |A|$ where n is the order of A.

We now state one of the most important theorems in linear algebra (proved in Problem 8.27):

Cayley–Hamilton Theorem 8.2:   Every matrix is a zero of its characteristic polynomial.

Example 8.2.   Let $B = \begin{pmatrix} 1 & 2 \\ 3 & 2 \end{pmatrix}$. Its characteristic polynomial is

$$\Delta(t) = |tI - B| = \begin{vmatrix} t - 1 & -2 \\ -3 & t - 2 \end{vmatrix} = (t - 1)(t - 2) - 6 = t^2 - 3t - 4$$

As expected from the Cayley–Hamilton Theorem, B is a zero of $\Delta(t)$:

$$\Delta(B) = B^2 - 3B - 4I = \begin{pmatrix} 7 & 6 \\ 9 & 10 \end{pmatrix} + \begin{pmatrix} -3 & -6 \\ -9 & -6 \end{pmatrix} + \begin{pmatrix} -4 & 0 \\ 0 & -4 \end{pmatrix} = \begin{pmatrix} 0 & 0 \\ 0 & 0 \end{pmatrix}$$

Now suppose A and B are similar matrices, say $B = P^{-1}AP$ where P is invertible. We show that A and B have the same characteristic polynomial. Using $tI = P^{-1}tIP$,

$$|tI - B| = |tI - P^{-1}AP| = |P^{-1}tIP - P^{-1}AP|$$
$$= |P^{-1}(tI - A)P| = |P^{-1}||tI - A||P|$$

Since determinants are scalars and commute, and since $|P^{-1}||P| = 1$, we finally obtain

$$|tI - B| = |tI - A|$$

Thus we have proved

Theorem 8.3:   Similar matrices have the same characteristic polynomial.

Characteristic Polynomials of Degree Two and Three

Let A be a matrix of order two or three. Then there is an easy formula for its characteristic polynomial $\Delta(t)$. Specifically:

(1)   Suppose $A = \begin{pmatrix} a_{11} & a_{12} \\ a_{21} & a_{22} \end{pmatrix}$. Then

$$\Delta(t) = t^2 - (a_{11} + a_{22})t + \begin{vmatrix} a_{11} & a_{12} \\ a_{21} & a_{22} \end{vmatrix} = t^2 - (\text{tr } A)t + \det(A)$$

(Here tr A denotes the trace of A, that is, the sum of the diagonal elements of A.)

(2) Suppose $A = \begin{pmatrix} a_{11} & a_{12} & a_{13} \\ a_{21} & a_{22} & a_{23} \\ a_{31} & a_{32} & a_{33} \end{pmatrix}$. Then

$$\Delta(t) = t^3 - (a_{11} + a_{22} + a_{33})t^2 + \left(\begin{vmatrix} a_{22} & a_{23} \\ a_{32} & a_{33} \end{vmatrix} + \begin{vmatrix} a_{11} & a_{13} \\ a_{31} & a_{33} \end{vmatrix} + \begin{vmatrix} a_{11} & a_{12} \\ a_{21} & a_{22} \end{vmatrix}\right)t - \begin{vmatrix} a_{11} & a_{12} & a_{13} \\ a_{21} & a_{22} & a_{23} \\ a_{31} & a_{32} & a_{33} \end{vmatrix}$$

$$= t^3 - (\text{tr } A)t^2 + (A_{11} + A_{22} + A_{33})t - \det(A)$$

(Here A_{11}, A_{22}, A_{33} denote, respectively, the cofactors of the diagonal elements a_{11}, a_{22}, a_{33}.)

Consider again a 3-square matrix $A = (a_{ij})$. As noted above,

$$S_1 = \text{tr } A \qquad S_2 = A_{11} + A_{22} + A_{33} \qquad S_3 = \det(A)$$

are the coefficients of its characteristic polynomial with alternating signs. On the other hand, each S_k is the sum of all the principal minors of A of order k. The next theorem, whose proof lies beyond the scope of this text, tells us that this result is true in general.

Theorem 8.4:   Let A be an n-square matrix. Then its characteristic polynomial is

$$\Delta(t) = t^n - S_1 t^{n-1} + S_2 t^{n-2} - \cdots + (-1)^n S_n$$

where S_k is the sum of the principal minors of order k.

Characteristic Polynomial and Block Triangular Matrices

Suppose M is a block triangular matrix, say $M = \begin{pmatrix} A_1 & B \\ 0 & A_2 \end{pmatrix}$ where A_1 and A_2 are square matrices. Then the characteristic matrix of M,

$$tI - M = \begin{pmatrix} tI - A_1 & -B \\ 0 & tI - A_2 \end{pmatrix}$$

is also a block triangular matrix with diagonal blocks $tI - A_1$ and $tI - A_2$. Thus, by Theorem 7.12,

$$|tI - M| = \begin{vmatrix} tI - A_1 & -B \\ 0 & tI - A_2 \end{vmatrix} = |tI - A_1||tI - A_2|$$

That is, the characteristic polynomial of M is the product of the characteristic polynomials of the diagonal blocks A_1 and A_2.

By induction, we obtain the following useful result.

Theorem 8.5:   Suppose M is a block triangular matrix with diagonal blocks $A_1, A_2, \ldots, A_r$. Then the characteristic polynomial of M is the product of the characteristic polynomials of the diagonal blocks A_i, that is,

$$\Delta_M(t) = \Delta_{A_1}(t)\Delta_{A_2}(t) \cdots \Delta_{A_r}(t)$$

Example 8.3.   Consider the matrix

$$M = \begin{pmatrix} 9 & -1 & 5 & 7 \\ 8 & 3 & 2 & -4 \\ 0 & 0 & 3 & 6 \\ 0 & 0 & -1 & 8 \end{pmatrix}$$

Then M is a block triangular matrix with diagonal blocks $A = \begin{pmatrix} 9 & -1 \\ 8 & 3 \end{pmatrix}$ and $B = \begin{pmatrix} 3 & 6 \\ -1 & 8 \end{pmatrix}$. Here

$$\text{tr } A = 9 + 3 = 12 \qquad \det (A) = 27 + 8 = 35 \qquad \text{and so} \qquad \Delta_A(t) = t^2 - 12t + 35 = (t - 5)(t - 7)$$

$$\text{tr } B = 3 + 8 = 11 \qquad \det (B) = 24 + 6 = 30 \qquad \text{and so} \qquad \Delta_B(t) = t^2 - 11t + 30 = (t - 5)(t - 6)$$

Accordingly, the characteristic polynomial of M is the product

$$\Delta_M(t) = \Delta_A(t)\Delta_B(t) = (t - 5)^2(t - 6)(t - 7)$$

8.4 EIGENVALUES AND EIGENVECTORS

Let A be an n-square matrix over a field K. A scalar $\lambda \in K$ is called an *eigenvalue* of A if there exists a nonzero (column) vector $v \in K^n$ for which

$$Av = \lambda v$$

Every vector satisfying this relation is then called an *eigenvector* of A belonging to the eigenvalue λ. Note that each scalar multiple kv is such an eigenvector since

$$A(kv) = k(v) = k(\lambda v) = \lambda(kv)$$

The set E_λ of all eigenvectors belonging to λ is a subspace of K^n (Problem 8.16), called the *eigenspace* of λ. (If dim $E_\lambda = 1$, then E_λ is called an *eigenline* and λ is called a *scaling factor*.)

The terms characteristic value and characteristic vector (or proper value and proper vector) are frequently used instead of eigenvalue and eigenvector.

Example 8.4. Let $A = \begin{pmatrix} 1 & 2 \\ 3 & 2 \end{pmatrix}$ and let $v_1 = (2, 3)^T$ and $v_2 = (1, -1)^T$. Then

$$Av_1 = \begin{pmatrix} 1 & 2 \\ 3 & 2 \end{pmatrix}\begin{pmatrix} 2 \\ 3 \end{pmatrix} = \begin{pmatrix} 8 \\ 12 \end{pmatrix} = 4\begin{pmatrix} 2 \\ 3 \end{pmatrix} = 4v_1$$

and

$$Av_2 = \begin{pmatrix} 1 & 2 \\ 3 & 2 \end{pmatrix}\begin{pmatrix} 1 \\ -1 \end{pmatrix} = \begin{pmatrix} -1 \\ 1 \end{pmatrix} = (-1)v_2$$

Thus v_1 and v_2 are eigenvectors of A belonging, respectively, to the eigenvalues $\lambda_1 = 4$ and $\lambda_2 = -1$ of A.

The following theorem, proved in Problem 8.28, is the main tool for computing eigenvalues and eigenvectors (Section 8.5).

Theorem 8.6: Let A be an n-square matrix over a field K. Then the following are equivalent:

    (i)   A scalar $\lambda \in K$ is an eigenvalue of A.

    (ii) The matrix $M = \lambda I - A$ is singular.

    (iii) The scalar λ is a root of the characteristic polynomial $\Delta(t)$ of A.

The eigenspace E_λ of λ is then the solution space of the homogeneous system $(\lambda I - A)X = 0$.

Sometimes it is more convenient to solve the homogeneous system $(A - \lambda I)X = 0$ rather than $(\lambda I - A)X = 0$ when computing eigenvectors. Both systems, of course, yield the same solution space.

Some matrices may have no eigenvalues and hence no eigenvectors. However, using the Fundamental Theorem of Algebra (every polynomial over **C** has a root) and Theorem 8.6, we obtain the following result.

Theorem 8.7:   Let A be an n-square matrix over the complex field $\mathbf{C}$. Then A has at least one eigenvalue.

Now suppose λ is an eigenvalue of a matrix A. The *algebraic multiplicity* of λ is defined to be the multiplicity of λ as a root of the characteristic polynomial of A. The *geometric multiplicity* of λ is defined to be the dimension of its eigenspace.

The following theorem, proved in Chapter 10, applies.

Theorem 8.8:   Let λ be an eigenvalue of a matrix A. Then the geometric multiplicity of λ does not exceed its algebraic multiplicity.

Diagonalizable Matrices

A matrix A is said to be *diagonalizable* (under similarity) if there exists a nonsingular matrix P such that $D = P^{-1}AP$ is a diagonal matrix, i.e., if A is similar to a diagonal matrix D. The following theorem, proved in Problem 8.29, characterizes such matrices.

Theorem 8.9:   An n-square matrix A is similar to a diagonal matrix D if and only if A has n linearly independent eigenvectors. In this case, the diagonal elements of D are the corresponding eigenvalues and $D = P^{-1}AP$ where P is the matrix whose columns are the eigenvectors.

Suppose a matrix A can be diagonalized as above, say $P^{-1}AP = D$ where D is diagonal. Then A has the extremely usful *diagonal factorization*

$$A = PDP^{-1}$$

Using this factorization, the algebra of A reduces to the algebra of the diagonal matrix D which can be easily calculated. Specifically, suppose $D = \operatorname{diag}(k_1, k_2, \ldots, k_n)$. Then

$$A^m = (PDP^{-1})^m = PD^mP^{-1} = P \operatorname{diag}(k_1^m, \ldots, k_n^m)P^{-1}$$

and, more generally, for any polynomial $f(t)$,

$$f(A) = f(PDP^{-1}) = Pf(D)P^{-1} = P \operatorname{diag}(f(k_1), \ldots, f(k_n))P^{-1}$$

Furthermore, if the diagonal entries of D are nonnegative, then the following matrix B is a "square root" of A:

$$B = P \operatorname{diag}(\sqrt{k_1}, \ldots, \sqrt{k_n})P^{-1}$$

that is, $B^2 = A$.

Example 8.5.   Consider the matrix $A = \begin{pmatrix} 1 & 2 \\ 3 & 2 \end{pmatrix}$. By Example 8.5, A has two linearly independent eigenvectors $\begin{pmatrix} 2 \\ 3 \end{pmatrix}$ and $\begin{pmatrix} 1 \\ -1 \end{pmatrix}$. Set $P = \begin{pmatrix} 2 & 1 \\ 3 & -1 \end{pmatrix}$, and so $P^{-1} = \begin{pmatrix} \frac{1}{5} & \frac{1}{5} \\ \frac{3}{5} & -\frac{2}{5} \end{pmatrix}$. Then A is similar to the diagonal matrix

$$B = P^{-1}AP = \begin{pmatrix} \frac{1}{5} & \frac{1}{5} \\ \frac{3}{5} & -\frac{2}{5} \end{pmatrix}\begin{pmatrix} 1 & 2 \\ 3 & 2 \end{pmatrix}\begin{pmatrix} 2 & 1 \\ 3 & -1 \end{pmatrix} = \begin{pmatrix} 4 & 0 \\ 0 & -1 \end{pmatrix}$$

As expected, the diagonal elements 4 and -1 of the diagonal matrix B are the eigenvalues corresponding to the given eigenvectors. In particular, A has the factorization

$$A = PDP^{-1} = \begin{pmatrix} 2 & 1 \\ 3 & -1 \end{pmatrix}\begin{pmatrix} 4 & 0 \\ 0 & -1 \end{pmatrix}\begin{pmatrix} \frac{1}{5} & \frac{1}{5} \\ \frac{3}{5} & -\frac{2}{5} \end{pmatrix}$$

Accordingly,

$$A^4 = PD^4P^{-1} = \begin{pmatrix} 2 & 1 \\ 3 & -1 \end{pmatrix}\begin{pmatrix} 256 & 0 \\ 0 & 1 \end{pmatrix}\begin{pmatrix} \frac{1}{5} & \frac{1}{5} \\ \frac{3}{5} & -\frac{2}{5} \end{pmatrix} = \begin{pmatrix} 103 & 102 \\ 153 & 154 \end{pmatrix}$$

Furthermore, if $f(t) = t^3 - 7t^2 + 9t - 2$, then

$$f(A) = Pf(D)P^{-1} = \begin{pmatrix} 2 & 1 \\ 3 & -1 \end{pmatrix}\begin{pmatrix} -14 & 0 \\ 0 & -19 \end{pmatrix}\begin{pmatrix} \frac{1}{5} & \frac{1}{5} \\ \frac{3}{5} & -\frac{2}{5} \end{pmatrix} = \begin{pmatrix} -17 & 2 \\ 3 & -16 \end{pmatrix}$$

Remark: Throughout this chapter, we use the fact that the inverse of the matrix

$$P = \begin{pmatrix} a & b \\ c & d \end{pmatrix} \quad \text{is the matrix} \quad P^{-1} = \begin{pmatrix} d/|P| & -b/|P| \\ -c/|P| & a/|P| \end{pmatrix}$$

That is, P^{-1} is obtained by interchanging the diagonal elements a and d of P, taking the negatives of the nondiagonal elements b and c, and dividing each element by the determinant $|P|$.

The following two theorems, proved in Problems 8.30 and 8.31, respectively, will be subsequently used.

Theorem 8.10: Let $v_1, \ldots, v_n$ be nonzero eigenvectors of a matrix A belonging to distinct eigenvalues $\lambda_1, \ldots, \lambda_n$. Then $v_1, \ldots, v_n$ are linearly independent.

Theorem 8.11: Suppose the characteristic polynomial $\Delta(t)$ of an n-square matrix A is a product of n distinct factors, say, $\Delta(t) = (t - a_1)(t - a_2) \cdots (t - a_n)$. Then A is similar to a diagonal matrix whose diagonal elements are the a_i.

8.5 COMPUTING EIGENVALUES AND EIGENVECTORS, DIAGONALIZING MATRICES

This section computes the eigenvalues and eigenvectors for a given square matrix A and determines whether or not a nonsingular matrix P exists such that $P^{-1}AP$ is diagonal. Specifically, the following algorithm will be applied to the matrix A.

Diagonalization Algorithm 8.5:

The input is an n-square matrix A.

Step 1. Find the characteristic polynomial $\Delta(t)$ of A.

Step 2. Find the roots of $\Delta(t)$ to obtain the eigenvalues of A.

Step 3. Repeat (a) and (b) for each eigenvalue λ of A:

(a) Form $M = A - \lambda I$ by subtracting λ down the diagonal of A, or form $M' = \lambda I - A$ by substituting $t = \lambda$ in $tI - A$.

(b) Find a basis for the solution space of the homogeneous system $MX = 0$. (These basis vectors are linearly independent eigenvectors of A belonging to λ.)

Step 4. Consider the collection $S = \{v_1, v_2, \ldots, v_m\}$ of all eigenvectors obtained in Step 3:

(a) If $m \neq n$, then A is not diagonalizable.

(b) If $m = n$, let P be the matrix whose columns are the eigenvectors $v_1, v_2, \ldots, v_n$. Then

$$D = P^{-1}AP = \begin{pmatrix} \lambda_1 & & & \\ & \lambda_2 & & \\ & & \cdots & \\ & & & \lambda_n \end{pmatrix}$$

where λ_i is the eigenvalue corresponding to the eigenvector v_i.

Example 8.6. The Diagonalization Algorithm is applied to $A = \begin{pmatrix} 4 & 2 \\ 3 & -1 \end{pmatrix}$.

1. The characteristic polynomial $\Delta(t)$ of A is the determinant

$$\Delta(t) = |tI - A| = \begin{vmatrix} t-4 & -1 \\ -3 & t+1 \end{vmatrix} = t^2 - 3t - 10 = (t - 5)(t + 2)$$

Alternatively, tr $A = 4 - 1 = 3$ and $|A| = -4 - 6 = -10$; so $\Delta(t) = t^2 - 3t - 10$.

2. Set $\Delta(t) = (t - 5)(t + 2) = 0$. The roots $\lambda_1 = 5$ and $\lambda_2 = -2$ are the eigenvalues of A.

3. (i) We find an eigenvector v_1 of A belonging to the eigenvalue $\lambda_1 = 5$.

Subtract $\lambda_1 = 5$ down the diagonal of A to obtain the matrix $M = \begin{pmatrix} -1 & 2 \\ 3 & -6 \end{pmatrix}$. The eigenvectors belonging to $\lambda_1 = 5$ form the solution of the homogeneous system $MX = 0$, that is,

$$\begin{pmatrix} -1 & 2 \\ 3 & -6 \end{pmatrix}\begin{pmatrix} x \\ y \end{pmatrix} = \begin{pmatrix} 0 \\ 0 \end{pmatrix} \quad \text{or} \quad \begin{cases} -x + 2y = 0 \\ 3x - 6y = 0 \end{cases} \quad \text{or} \quad -x + 2y = 0$$

The system has only one independent solution; for example, $x = 2$, $y = 1$. Thus $v_1 = (2, 1)$ is an eigenvector which spans the eigenspace of $\lambda_1 = 5$.

(ii) We find an eigenvector v_2 of A belonging to the eigenvalue $\lambda_2 = -2$.

Subtract -2 (or add 2) down the diagonal of A to obtain $M = \begin{pmatrix} 6 & 2 \\ 3 & 1 \end{pmatrix}$ which yields the homogeneous system

$$\begin{cases} 6x + 2y = 0 \\ 3x + y = 0 \end{cases} \quad \text{or} \quad 3x + y = 0$$

The system has only one independent solution; for example, $x = -1$, $y = 3$. Thus $v_2 = (-1, 3)$ is an eigenvector which spans the eigenspace of $\lambda_2 = -2$.

4. Let P be the matrix whose columns are the above eigenvectors: $P = \begin{pmatrix} 2 & -1 \\ 1 & 3 \end{pmatrix}$. Then $P^{-1} = \begin{pmatrix} \frac{3}{7} & \frac{1}{7} \\ -\frac{1}{7} & \frac{2}{7} \end{pmatrix}$ and $D = P^{-1}AP$ is the diagonal matrix whose diagonal entries are the respective eigenvalues:

$$D = P^{-1}AP = \begin{pmatrix} \frac{3}{7} & \frac{1}{7} \\ -\frac{1}{7} & \frac{2}{7} \end{pmatrix}\begin{pmatrix} 4 & 2 \\ 3 & -1 \end{pmatrix}\begin{pmatrix} 2 & -1 \\ 1 & 3 \end{pmatrix} = \begin{pmatrix} 5 & 0 \\ 0 & -2 \end{pmatrix}$$

Accordingly, A has the "diagonal factorization"

$$A = PDP^{-1} = \begin{pmatrix} 2 & -1 \\ 1 & 3 \end{pmatrix}\begin{pmatrix} 5 & 0 \\ 0 & -2 \end{pmatrix}\begin{pmatrix} \frac{3}{7} & \frac{1}{7} \\ -\frac{1}{7} & \frac{2}{7} \end{pmatrix}$$

If $f(t) = t^4 - 4t^3 - 3t^2 + 5$, then we can calculate

$$f(A) = Pf(D)P^{-1} = \begin{pmatrix} 2 & -1 \\ 1 & 3 \end{pmatrix}\begin{pmatrix} 55 & 0 \\ 0 & 41 \end{pmatrix}\begin{pmatrix} \frac{3}{7} & \frac{1}{7} \\ -\frac{1}{7} & \frac{2}{7} \end{pmatrix} = \begin{pmatrix} 53 & 4 \\ 6 & 43 \end{pmatrix}$$

Example 8.7. Consider the matrix $B = \begin{pmatrix} 5 & 1 \\ -4 & 1 \end{pmatrix}$. Here tr $B = 5 + 1 = 6$ and $|B| = 5 + 4 = 9$. Hence $\Delta(t) = t^2 - t + 9 = (t - 3)^2$ is the characteristic polynomial of B. Accordingly, $\lambda = 3$ is the only eigenvalue of B.

Subtract $\lambda = 3$ down the diagonal of B to obtain the matrix $M = \begin{pmatrix} 2 & 1 \\ -4 & -2 \end{pmatrix}$ which corresponds to the homogeneous system

$$\begin{cases} 2x + y = 0 \\ -4x - 2y = 0 \end{cases} \quad \text{or} \quad 2x + y = 0$$

The system has only one independent solution; for example, $x = 1$, $y = -2$. Thus $v = (1, -2)$ is the only independent eigenvector of the matrix B. Accordingly, B is not diagonalizable since there does not exist a basis consisting of eigenvectors of B.

Example 8.8. Consider the matrix $A = \begin{pmatrix} 2 & -5 \\ 1 & -2 \end{pmatrix}$. Here tr $A = 2 - 2 = 0$ and $|A| = -4 + 5 = 1$. Thus $\Delta(t) = t^2 + 1$ is the characteristic polynomial of A. We consider two cases:

(a) A is a matrix over the real field **R**. Then $\Delta(t)$ has no (real) roots. Thus A has no eigenvalues and no eigenvectors, and so A is not diagonizable.

(b) A is a matrix over the complex field **C**. Then $\Delta(t) = (t - i)(t + i)$ has two roots, i and $-i$. Thus A has two distinct eigenvalues i and $-i$, and hence A has two independent eigenvectors. Accordingly, there exists a nonsingular matrix P over the complex field **C** for which

$$P^{-1}AP = \begin{pmatrix} i & 0 \\ 0 & -i \end{pmatrix}$$

Therefore, A is diagonalizable (over **C**).

8.6 DIAGONALIZING REAL SYMMETRIC MATRICES

There are many real matrices A which are not diagonalizable. In fact, some such matrices may not have any (real) eigenvalues. However, if A is a real symmetric matrix, then these problems do not exist. Namely:

Theorem 8.12: Let A be a real square matrix. Then each root λ of its characteristic polynomial is real.

Theorem 8.13: Let A be a real square matrix. Suppose u and v are nonzero eigenvectors of A belonging to distinct eigenvalues λ_1 and λ_2. Then u and v are orthogonal, i.e., $\langle u, v \rangle = 0$.

The above two theorems gives us the following fundamental result:

Theorem 8.14: Let A be a real symmetric matrix. Then there exists an orthogonal matrix P such that $D = P^{-1}AP$ is diagonal.

We can choose the columns of the above matrix P to be normalized orthogonal eigenvectors of A; then the diagonal entries of D are the corresponding eigenvalues.

Example 8.9. Let $A = \begin{pmatrix} 2 & -2 \\ -2 & 5 \end{pmatrix}$. We find an orthogonal matrix P such that $P^{-1}AP$ is diagonal. Here tr $A = 2 + 5 = 7$ and $|A| = 10 - 4 = 6$. Hence $\Delta(t) = t^2 - 7t + 6 = (t - 6)(t - 1)$ is the characteristic polynomial of A. The eigenvalues of A are 6 and 1. Substitute $\lambda = 6$ into the matrix $\lambda I - A$ to obtain the corresponding homogeneous system of linear equations

$$4x + 2y = 0 \qquad 2x + y = 0$$

A nonzero solution is $v_1 = (1, -2)$. Next substitute $\lambda = 1$ into the matrix $\lambda I - A$ to find the corresponding homogeneous system

$$-x + 2y = 0 \qquad 2x - 4y = 0$$

A nonzero solution is $(2, 1)$. As expected by Theorem 8.13, v_1 and v_2 are orthogonal. Normalize v_1 and v_2 to obtain the orthonormal vectors

$$u_1 = (1/\sqrt{5}, -2/\sqrt{5}) \qquad u_2 = (2/\sqrt{5}, 1/\sqrt{5})$$

Finally let P be the matrix whose columns are u_1 and u_2, respectively. Then

$$P = \begin{pmatrix} 1/\sqrt{5} & 2/\sqrt{5} \\ -2/\sqrt{5} & 1/\sqrt{5} \end{pmatrix} \quad \text{and} \quad P^{-1}AP = \begin{pmatrix} 6 & 0 \\ 0 & 1 \end{pmatrix}$$

As expected, the diagonal entries of $P^{-1}AP$ are the eigenvalues corresponding to the columns of P.

Application to Quadratic Forms

Recall (Section 4.12) that a real quadratic form $q(x_1, x_2, \ldots, x_n)$ can be expressed in the matrix form

$$q(X) = X^T A X$$

where $X = (x_1, \ldots, x_n)^T$ and A is a real symmetric matrix, and recall that under a change of variables $X = PY$, where $Y = (y_1, \ldots, y_n)$ and P is a nonsingular matrix, the quadratic form has the form

$$q(Y) = Y^T B Y$$

where $B = P^T A P$. (Thus B is congruent to A.)

Now if P is an orthogonal matrix, then $P^T = P^{-1}$. In such a case, $B = P^T A P = P^{-1} A P$ and so B is orthogonally similar to A. Accordingly, the above method for diagonalizing a real symmetric matrix A can be used to diagonalize a quadratic form q under an orthogonal change of coordinates as follows,

Orthogonal Diagonalization Algorithm 8.6:

The input is a quadratic form $q(X)$.

Step 1. Find the symmetric matrix A which represents q and find its characteristic polynomial $\Delta(t)$.

Step 2. Find the eigenvalues of A which are the roots of $\Delta(t)$.

Step 3. For each eigenvalue λ of A in Step 2, find an orthogonal basis of its eigenspace.

Step 4. Normalize all eigenvectors in Step 3 which then forms an orthonormal basis of $\mathbf{R}^n$.

Step 5. Let P be the matrix whose columns are the normalized eigenvectors in Step 4.

Then $X = PY$ is the required orthogonal change of coordinates, and the diagonal entries of $P^T A P$ will be the eigenvalues $\lambda_1, \ldots, \lambda_n$ which correspond to the columns of P.

8.7 MINIMUM POLYNOMIAL

Let A be an n-square matrix over a field K and let $J(A)$ denote the collection of all polynomials $f(t)$ for which $f(A) = 0$. [Note $J(A)$ is not empty since the characteristic polynomial $\Delta_A(t)$ of A belongs to $J(A)$.] Let $m(t)$ be the monic polynomial of minimal degree in $J(A)$. Then $m(t)$ is called the *minimum polynomial* of A. [Such a polynomial $m(t)$ exists and is unique (Problem 8.25).]

Theorem 8.15: The minimum polynomial $m(t)$ of A divides every polynomial which has A as a zero. In particular, $m(t)$ divides the characteristic polynomial $\Delta(t)$ of A.

(The proof is given in Problem 8.32.) There is an even stronger relationship between $m(t)$ and $\Delta(t)$.

Theorem 8.16:   The characteristic and minimum polynomials of a matrix A have the same irreducible factors.

This theorem, proved in Problem 8.33(b), does not say that $m(t) = \Delta(t)$; only that any irreducible factor of one must divide the other. In particular, since a linear factor is irreducible, $m(t)$ and $\Delta(t)$ have the same linear factors; hence they have the same roots. Thus we have:

Theorem 8.17:   A scalar λ is an eigenvalue for a matrix A if and only if λ is a root of the minimum polynomial of A.

Example 8.10.   Find the minimum polynomial $m(t)$ of $A = \begin{pmatrix} 2 & 2 & -5 \\ 3 & 7 & -15 \\ 1 & 2 & -4 \end{pmatrix}$.

First find the characteristic polynomial $\Delta(t)$ of A:

$$\Delta(t) = |tI - A| = \begin{vmatrix} t-2 & -2 & 5 \\ -3 & t-7 & 15 \\ -1 & -2 & t+4 \end{vmatrix} = t^3 - 5t^2 + 7t - 3 = (t-1)^2(t-3)$$

Alternatively, $\Delta(t) = t^3 - (\text{tr } A)t^2 + (A_{11} + A_{22} + A_{33})t - |A| = t^3 - 5t^2 + 7t - 3 = (t-1)^2(t-3)$ (where A_{ii} is the cofactor of a_{ii} in A).

The minimum polynomial $m(t)$ must divide $\Delta(t)$. Also, each irreducible factor of $\Delta(t)$, that is, $t-1$ and $t-3$, must also be a factor of $m(t)$. Thus $m(t)$ is exactly only of the following:

$$f(t) = (t-3)(t-1) \qquad \text{or} \qquad g(t) = (t-3)(t-1)^2$$

We know, by the Cayley–Hamilton Theorem, that $g(A) = \Delta(A) = 0$; hence we need only test $f(t)$. We have

$$f(A) = (A - I)(A - 3I) = \begin{pmatrix} 1 & 2 & -5 \\ 3 & 6 & -15 \\ 1 & 2 & -5 \end{pmatrix}\begin{pmatrix} -1 & 2 & -5 \\ 3 & 4 & -15 \\ 1 & 2 & -7 \end{pmatrix} = \begin{pmatrix} 0 & 0 & 0 \\ 0 & 0 & 0 \\ 0 & 0 & 0 \end{pmatrix}$$

Thus $f(t) = m(t) = (t-1)(t-3) = t^2 - 4t + 3$ is the minimum polynomial of A.

Example 8.11.   Consider the following n-square matrix where $a \neq 0$:

$$M = \begin{pmatrix} \lambda & a & 0 & \ldots & 0 & 0 \\ 0 & \lambda & a & \ldots & 0 & 0 \\ \ldots\ldots\ldots\ldots\ldots\ldots\ldots \\ 0 & 0 & 0 & \ldots & \lambda & a \\ 0 & 0 & 0 & \ldots & 0 & \lambda \end{pmatrix}$$

Note that M has λ's on the diagonal, a's on the superdiagonal, and 0s elsewhere. This matrix, especially when $a = 1$, is important in linear algebra. One can show that

$$f(t) = (t - \lambda)^n$$

is both the characteristic and minimum polynomial of M.

Example 8.12.   Consider an arbitrary monic polynomial $f(t) = t^n + a_{n-1}t^{n-1} + \cdots + a_1 t + a_0$. Let A be the n-square matrix with 1s on the subdiagonal, the negative of the coefficients in the last column and 0s elsewhere as follows:

$$A = \begin{pmatrix} 0 & 0 & \ldots & 0 & -a_0 \\ 1 & 0 & \ldots & 0 & -a_1 \\ 0 & 1 & \ldots & 0 & -a_2 \\ \ldots\ldots\ldots\ldots\ldots\ldots\ldots\ldots \\ 0 & 0 & \ldots & 1 & -a_{n-1} \end{pmatrix}$$

Then A is called the *companion matrix* of the polynomial $f(t)$. Moreover, the minimum polynomial $m(t)$ and the characteristic polynomial $\Delta(t)$ of the above companion matrix A are both equal to $f(t)$.

Minimum Polynomial and Block Diagonal Matrices

The following theorem, proved in Problem 8.34, applies.

Theorem 8.18: Suppose M is a block diagonal matrix with diagonal blocks $A_1, A_2, \ldots, A_r$. Then the minimum polynomial of M is equal to the least common multiple (LCM) of the minimum polynomials of the diagonal blocks A_i.

> **Remark:** We emphasize that this theorem applies to block diagonal matrices, whereas the analogous Theorem 8.5 on characteristic polynomials applies to block triangular matrices.

Example 8.13. Find the characteristic polynomial $\Delta(t)$ and the minimum polynomial $m(t)$ of the matrix

$$A = \begin{pmatrix} 2 & 5 & 0 & 0 & 0 \\ 0 & 2 & 0 & 0 & 0 \\ 0 & 0 & 4 & 2 & 0 \\ 0 & 0 & 3 & 5 & 0 \\ 0 & 0 & 0 & 0 & 7 \end{pmatrix}$$

Note A is a block diagonal matrix with diagonal blocks

$$A_1 = \begin{pmatrix} 2 & 5 \\ 0 & 2 \end{pmatrix} \qquad A_2 = \begin{pmatrix} 4 & 2 \\ 3 & 5 \end{pmatrix} \qquad A_3 = (7)$$

Then $\Delta(t)$ is the product of the characteristic polynomials $\Delta_1(t)$, $\Delta_2(t)$, and $\Delta_3(t)$ of A_1, A_2, and A_3, respectively. Since A_1 and A_3 are triangular, $\Delta_1(t) = (t-2)^2$ and $\Delta_3(t) = (t-7)$. Also,

$$\Delta_2(t) = t^2 - (\operatorname{tr} A_2)t + |A_2| = t^2 - 9t + 14 = (t-2)(t-7)$$

Thus $\Delta(t) = (t-2)^3(t-7)^2$. [As expected, deg $\Delta(t) = 5$.]

The minimum polynomials $m_1(t)$, $m_2(t)$, and $m_3(t)$ of the diagonal blocks A_1, A_2, and A_3, respectively, are equal to the characteristic polynomials; that is,

$$m_1(t) = (t-2)^2 \qquad m_2(t) = (t-2)(t-7) \qquad m_3(t) = t-7$$

But $m(t)$ is equal to the least common multiple of $m_1(t)$, $m_2(t)$, $M_3(t)$. Thus $m(t) = (t-2)^2(t-7)$.

Solved Problems

POLYNOMIALS AND MATRICES, CHARACTERISTIC POLYNOMIAL

8.1. Let $A = \begin{pmatrix} 1 & -2 \\ 4 & 5 \end{pmatrix}$. Find $f(A)$ where: (a) $f(t) = t^2 - 3t + 7$, and (b) $f(t) = t^2 - 6t + 13$.

(a) $f(A) = A^2 - 3A + 7I = \begin{pmatrix} 1 & -2 \\ 4 & 5 \end{pmatrix}^2 - 3\begin{pmatrix} 1 & -2 \\ 4 & 5 \end{pmatrix} + 7\begin{pmatrix} 1 & 0 \\ 0 & 1 \end{pmatrix}$

$= \begin{pmatrix} -7 & -12 \\ 24 & 17 \end{pmatrix} + \begin{pmatrix} -3 & 6 \\ -12 & -15 \end{pmatrix} + \begin{pmatrix} 7 & 0 \\ 0 & 7 \end{pmatrix} = \begin{pmatrix} -3 & -6 \\ 12 & 9 \end{pmatrix}$

(b) $f(A) = A^2 - 6A + 13I = \begin{pmatrix} -7 & -12 \\ 24 & 17 \end{pmatrix} + \begin{pmatrix} -6 & 12 \\ -24 & -30 \end{pmatrix} + \begin{pmatrix} 13 & 0 \\ 0 & 13 \end{pmatrix} = \begin{pmatrix} 0 & 0 \\ 0 & 0 \end{pmatrix}$

[Thus A is a root of $f(t)$.]

8.2. Find the characteristic polynomial $\Delta(t)$ of the matrix $A = \begin{pmatrix} 2 & -3 \\ 5 & 1 \end{pmatrix}$.

Form the characteristic matrix $tI - A$:

$$tI - A = \begin{pmatrix} t & 0 \\ 0 & t \end{pmatrix} + \begin{pmatrix} -2 & 3 \\ -5 & -1 \end{pmatrix} = \begin{pmatrix} t-2 & 3 \\ -5 & t-1 \end{pmatrix}$$

The characteristic polynomial $\Delta(t)$ of A is its determinant:

$$\Delta(t) = |tI - A| = \begin{vmatrix} t-2 & 3 \\ -5 & t-1 \end{vmatrix} = (t-2)(t-1) + 15 = t^2 - 3t + 17$$

Alternatively, $\operatorname{tr} A = 2 + 1 = 3$ and $|A| = 2 + 15 = 17$; hence $\Delta(t) = t^2 - 3t + 17$.

8.3. Find the characteristic polynomial $\Delta(t)$ of the matrix $A = \begin{pmatrix} 1 & 6 & -2 \\ -3 & 2 & 0 \\ 0 & 3 & -4 \end{pmatrix}$.

$$\Delta(t) = |tI - A| = \begin{vmatrix} t-1 & -6 & 2 \\ 3 & t-2 & 0 \\ 0 & -3 & t+4 \end{vmatrix} = (t-1)(t-2)(t+4) - 18 + 18(t+4) = t^3 + t^2 - 8t + 62$$

Alternatively, $\operatorname{tr} A = 1 + 2 - 4 = -1$, $A_{11} = \begin{vmatrix} 2 & 0 \\ 3 & -4 \end{vmatrix} = -8$, $A_{22} = \begin{vmatrix} 1 & -2 \\ 0 & -4 \end{vmatrix} = -4$, $A_{33} = \begin{vmatrix} 1 & 6 \\ -3 & 2 \end{vmatrix} = 2 + 18 = 20$, $A_{11} + A_{22} + A_{33} = -8 - 4 + 20 = 8$, and $|A| = -8 + 18 - 72 = -62$. Thus

$$\Delta(t) = t^3 - (\operatorname{tr} A)t^2 + (A_{11} + A_{22} + A_{33})t - |A| = t^3 + t^2 - 8t + 62$$

8.4. Find the characteristic polynomial $\Delta(t)$ of the following matrices:

$$(a) \quad R = \begin{pmatrix} 1 & 2 & 3 & 4 \\ 0 & 2 & 8 & -6 \\ 0 & 0 & 3 & -5 \\ 0 & 0 & 0 & 4 \end{pmatrix}, \qquad (b) \quad S = \begin{pmatrix} 2 & 5 & 7 & -9 \\ 1 & 4 & -6 & 4 \\ 0 & 0 & 6 & -5 \\ 0 & 0 & 2 & 3 \end{pmatrix}$$

(a) Since R is triangular, $\Delta(t) = (t-1)(t-2)(t-3)(t-4)$.

(b) Note S is block triangular with diagonal blocks $A_1 = \begin{pmatrix} 2 & 5 \\ 1 & 4 \end{pmatrix}$ and $A_2 = \begin{pmatrix} 6 & -5 \\ 2 & 3 \end{pmatrix}$. Thus

$$\Delta(t) = \Delta_{A_1}(t)\Delta_{A_2}(t) = (t^2 - 6t + 3)(t^2 - 9t + 28)$$

EIGENVALUES AND EIGENVECTORS

8.5. Let $A = \begin{pmatrix} 1 & 4 \\ 2 & 3 \end{pmatrix}$. Find: (a) all eigenvalues of A and the corresponding eigenvectors, (b) an invertible matrix P such that $D = P^{-1}AP$ is diagonal, and (c) A^5 and $f(A)$ where $f(t) = t^4 - 3t^3 - 7t^2 + 6t - 15$.

(a) Form the characteristic matrix $tI - A$ of A:

$$tI - A = \begin{pmatrix} t & 0 \\ 0 & t \end{pmatrix} - \begin{pmatrix} 1 & 4 \\ 2 & 3 \end{pmatrix} = \begin{pmatrix} t-1 & -4 \\ -2 & t-3 \end{pmatrix} \tag{1}$$

The characteristic polynomial $\Delta(t)$ of A is its determinant:

$$\Delta(t) = |tI - A| = \begin{vmatrix} t-1 & -4 \\ -2 & t-3 \end{vmatrix} = t^2 - 4t - 5 = (t-5)(t+1)$$

Alternatively, tr $A = 1 + 3 = 4$ and $|A| = 3 - 8 = -5$, so $\Delta(t) = t^2 - 4t - 5$. The roots $\lambda_1 = 5$ and $\lambda_2 = -1$ of the characteristic polynomial $\Delta(t)$ are the eigenvalues of A.

We obtain the eigenvectors of A belonging to the eigenvalue $\lambda_1 = 5$. Substitute $t = 5$ in the characteristic matrix (1) to obtain the matrix $M = \begin{pmatrix} 4 & -4 \\ -2 & 2 \end{pmatrix}$. The eigenvectors belonging to $\lambda_1 = 5$ form the solution of the homogeneous system $MX = 0$, that is,

$$\begin{pmatrix} 4 & -4 \\ -2 & 2 \end{pmatrix}\begin{pmatrix} x \\ y \end{pmatrix} = \begin{pmatrix} 0 \\ 0 \end{pmatrix} \quad \text{or} \quad \begin{cases} 4x - 4y = 0 \\ -2x + 2y = 0 \end{cases} \quad \text{or} \quad x - y = 0$$

The system has only one independent solution; for example, $x = 1$, $y = 1$. Thus $v_1 = (1, 1)$ is an eigenvector which spans the eigenspace of $\lambda_1 = 5$.

We obtain the eigenvectors of A belonging to the eigenvalue $\lambda_2 = -1$. Substitute $t = -1$ into $tI - A$ to obtain $M = \begin{pmatrix} -2 & -4 \\ -2 & -4 \end{pmatrix}$ which yields the homogeneous system

$$\begin{cases} -2x - 4y = 0 \\ -2x - 4y = 0 \end{cases} \quad \text{or} \quad x + 2y = 0$$

The system has only one independent solution; for example, $x = 2$, $y = -1$. Thus $v_2 = (2, -1)$ is an eigenvector which spans the eigenspace of $\lambda_2 = -1$.

(b) Let P be the matrix whose columns are the above eigenvectors: $P = \begin{pmatrix} 1 & 2 \\ 1 & -1 \end{pmatrix}$. Then $D = P^{-1}AP$ is the diagonal matrix whose diagonal entries are the respective eigenvalues:

$$D = P^{-1}AP = \begin{pmatrix} \frac{1}{3} & \frac{2}{3} \\ \frac{1}{3} & -\frac{1}{3} \end{pmatrix}\begin{pmatrix} 1 & 4 \\ 2 & 3 \end{pmatrix}\begin{pmatrix} 1 & 2 \\ 1 & -1 \end{pmatrix} = \begin{pmatrix} 5 & 0 \\ 0 & -1 \end{pmatrix}$$

[*Remark:* Here P is the change-of-base matrix from the usual basis E of $\mathbf{R}^2$ to the basis $S = \{v_1, v_2\}$. Hence B is the matrix representation of the function determined by A in this new basis.]

(c) Use the diagonal factorization of A,

$$A = PDP^{-1} = \begin{pmatrix} 1 & 2 \\ 1 & -1 \end{pmatrix}\begin{pmatrix} 5 & 0 \\ 0 & -1 \end{pmatrix}\begin{pmatrix} \frac{1}{3} & \frac{2}{3} \\ \frac{1}{3} & -\frac{1}{3} \end{pmatrix}$$

to obtain

$$A^5 = PD^5P^{-1} = \begin{pmatrix} 1 & 2 \\ 1 & -1 \end{pmatrix}\begin{pmatrix} 3125 & 0 \\ 0 & -1 \end{pmatrix}\begin{pmatrix} \frac{1}{3} & \frac{2}{3} \\ \frac{1}{3} & -\frac{1}{3} \end{pmatrix} = \begin{pmatrix} 1041 & 2084 \\ 1042 & 2083 \end{pmatrix}$$

and

$$f(A) = Pf(D)P^{-1} = \begin{pmatrix} 1 & 2 \\ 1 & -1 \end{pmatrix}\begin{pmatrix} 90 & 0 \\ 0 & -24 \end{pmatrix}\begin{pmatrix} \frac{1}{3} & \frac{2}{3} \\ \frac{1}{3} & -\frac{1}{3} \end{pmatrix} = \begin{pmatrix} 14 & 76 \\ 38 & 52 \end{pmatrix}$$

8.6. Find all eigenvalues and a maximum set S of linearly independent eigenvectors for the following matrices:

$$(a) \quad A = \begin{pmatrix} 5 & 6 \\ 3 & -2 \end{pmatrix} \qquad (b) \quad C = \begin{pmatrix} 5 & -1 \\ 1 & 3 \end{pmatrix}$$

Which of the matrices can be diagonalized? If so, find the required nonsingular matrix P.

(a) Find the characteristic polynomial $\Delta(t) = t^2 - 3t - 28 = (t-7)(t+4)$. Thus the eigenvalues of A are $\lambda_1 = 7$ and $\lambda_2 = -4$.

(i)   Subtract $\lambda_1 = 7$ down the diagonal of A to obtain $M = \begin{pmatrix} -2 & 6 \\ 3 & -9 \end{pmatrix}$ which corresponds to the system

$$\begin{cases} -2x + 6y = 0 \\ 3x - 9y = 0 \end{cases} \quad \text{or} \quad x - 3y = 0$$

Here $v_1 = (3, 1)$ is a nonzero solution (spanning the solution space) and so v_1 is the eigenvector of $\lambda_1 = 7$.

(ii) Subtract $\lambda_2 = -4$ (or add 4) down the diagonal of A to obtain $M = \begin{pmatrix} 9 & 6 \\ 3 & 2 \end{pmatrix}$ which corresponds to the system $3x + 2y = 0$. Here $v_2 = (2, -3)$ is a solution and hence an eigenvector of $\lambda_2 = -4$.

Then $S = \{v_1 = (3, 1), v_2 = (2, -3)\}$ is a maximum set of linearly independent eigenvectors of A. Since S is a basis for $\mathbf{R}^2$, A is diagonalizable. Let P be the matrix whose columns are v_1 and v_2. Then

$$P = \begin{pmatrix} 3 & 2 \\ 1 & -3 \end{pmatrix} \quad \text{and} \quad P^{-1}AP = \begin{pmatrix} 7 & 0 \\ 0 & -4 \end{pmatrix}$$

(b) Find $\Delta(t) = t^2 - 8t + 16 = (t - 4)^2$. Thus $\lambda = 4$ is the only eigenvalue. Subtract $\lambda = 4$ down the diagonal of C to obtain $M = \begin{pmatrix} 1 & -1 \\ 1 & -1 \end{pmatrix}$ which corresponds to the homogeneous system $x + y = 0$. Here $v = (1, 1)$ is a nonzero solution of the system and hence v is an eigenvector of C belonging to $\lambda = 4$. Since there are no other eigenvalues, the singleton set $S = \{v = (1, 1)\}$ is a maximum set of linearly independent eigenvectors. Furthermore, C is not diagonalizable since the number of linearly independent eigenvectors is not equal to the dimension of the vector space $\mathbf{R}^2$. In particular, no such nonsingular matrix P exists.

8.7.   Let $A = \begin{pmatrix} 2 & 2 \\ 1 & 3 \end{pmatrix}$. Find: (a) all eigenvalues of A and the corresponding eigenvectors; (b) an invertible matrix P such that $D = P^{-1}AP$ is diagonal; (c) A^6; and (d) a "square root" of A, i.e., a matrix B such that $B^2 = A$.

(a) Here $\Delta(t) = t^2 - \operatorname{tr} A + |A| = t^2 - 5t + 4 = (t - 1)(t - 4)$. Hence $\lambda_1 = 1$ and $\lambda_2 = 4$ are eigenvalues of A. We find corresponding eigenvectors:

(i)   Subtract $\lambda_1 = 1$ down the diagonal of A to obtain $M = \begin{pmatrix} 1 & 2 \\ 1 & 2 \end{pmatrix}$ which corresponds to the homogeneous system $x + 2y = 0$. Here $v_1 = (2, -1)$ is a nonzero solution of the system and so an eigenvector of A belonging to $\lambda_1 = 1$.

(ii) Subtract $\lambda_2 = 4$ down the diagonal of A to obtain $M = \begin{pmatrix} -2 & 2 \\ 1 & -1 \end{pmatrix}$ which corresponds to the homogeneous system $x - y = 0$. Here $v_2 = (1, 1)$ is a nonzero solution and so an eigenvector of A belonging to $\lambda_2 = 4$.

(b) Let P be the matrix whose columns are v_1 and v_2. Then

$$P = \begin{pmatrix} 2 & 1 \\ -1 & 1 \end{pmatrix} \quad \text{and} \quad D = P^{-1}AP = \begin{pmatrix} 1 & 0 \\ 0 & 4 \end{pmatrix}$$

(c) Use the diagonal factorization of A,

$$A = PDP^{-1} = \begin{pmatrix} 2 & 1 \\ -1 & 1 \end{pmatrix} \begin{pmatrix} 1 & 0 \\ 0 & 4 \end{pmatrix} \begin{pmatrix} \frac{1}{3} & -\frac{1}{3} \\ \frac{1}{3} & \frac{2}{3} \end{pmatrix}$$

to obtain

$$A^6 = PD^6P^{-1} = \begin{pmatrix} 2 & 1 \\ -1 & 1 \end{pmatrix} \begin{pmatrix} 1 & 0 \\ 0 & 4096 \end{pmatrix} \begin{pmatrix} \frac{1}{3} & -\frac{1}{3} \\ \frac{1}{3} & \frac{2}{3} \end{pmatrix} = \begin{pmatrix} 1366 & 2730 \\ 1365 & 2731 \end{pmatrix}$$

(d) Here $\begin{pmatrix} \pm 1 & 0 \\ 0 & \pm 2 \end{pmatrix}$ are square roots of D. Hence

$$B = P\sqrt{D}\,P^{-1} = \begin{pmatrix} 2 & 1 \\ -1 & 1 \end{pmatrix}\begin{pmatrix} 1 & 0 \\ 0 & 2 \end{pmatrix}\begin{pmatrix} \frac{1}{3} & -\frac{1}{3} \\ \frac{1}{3} & \frac{2}{3} \end{pmatrix} = \begin{pmatrix} \frac{4}{3} & \frac{2}{3} \\ \frac{1}{3} & \frac{5}{3} \end{pmatrix}$$

is a square root of A (which is similar to a diagonal matrix with nonnegative elements).

8.8. Suppose $A = \begin{pmatrix} 4 & 1 & -1 \\ 2 & 5 & -2 \\ 1 & 1 & 2 \end{pmatrix}$. Find: (a) the characteristic polynomial $\Delta(t)$ of A, (b) the eigen-

values of A, and (c) a maximum set of linearly independent eigenvectors of A. (d) Is A diago-nalizable? If yes, find P such that $P^{-1}AP$ is diagonal.

(a) We have

$$\Delta(t) = |tI - A| = \begin{vmatrix} t - 4 & -1 & 1 \\ -2 & t - 5 & 2 \\ -1 & -1 & t - 2 \end{vmatrix} = t^3 - 11t^2 + 39t - 45$$

Alternatively, $\Delta(t) = t^3 - (\text{tr } A)t^2 + (A_{11} + A_{22} + A_{33})t - |A| = t^3 - 11t^2 + 39t - 45$. (Here A_{ii} is the cofactor of a_{ii} in the matrix A.)

(b) Assuming $\Delta(t)$ has a rational root, it must be among $\pm 1, \pm 3, \pm 5, \pm 9, \pm 15, \pm 45$. Testing, we get

$$\begin{array}{r} 3 \underline{}\,1 - 11 + 39 - 45 \\ 3 - 24 + 45 \\ \hline 1 - 8 + 15 + 0 \end{array}$$

Thus $t = 3$ is a root of $\Delta(t)$ and

$$\Delta(t) = (t - 3)(t^2 - 8t + 15) = (t - 3)(t - 5)(t - 3) = (t - 3)^2(t - 5)$$

Accordingly, $\lambda_1 = 3$ and $\lambda_2 = 5$ are the eigenvalues of A.

(c) Find independent eigenvectors for each eigenvalue of A.

(i) Subtract $\lambda_1 = 3$ down the diagonal of A to obtain the matrix $M = \begin{pmatrix} 1 & 1 & -1 \\ 2 & 2 & -2 \\ 1 & 1 & -1 \end{pmatrix}$ which

corresponds to the homogeneous system $x + y - z = 0$. Here $u = (1, -1, 0)$ and $v = (1, 0, 1)$ are two independent solutions.

(ii) Subtract $\lambda_2 = 5$ down the diagonal of A to obtain $M = \begin{pmatrix} -1 & 1 & -1 \\ 2 & 0 & -2 \\ 1 & 1 & -3 \end{pmatrix}$ which corresponds to

the homogeneous system

$$\begin{cases} -x + y - z = 0 \\ 2x - 2z = 0 \\ x + y - 3z = 0 \end{cases} \quad \text{or} \quad \begin{cases} x -   z = 0 \\ y - 2z = 0 \end{cases}$$

Only z is a free variable. Here $w = (1, 2, 1)$ is a solution.

Thus $\{u = (1, -1, 0),\ v = (1, 0, 1),\ w = (1, 2, 1)\}$ is a maximum set of linearly independent eigenvectors of A.

> **Remark:** The vectors u and v were chosen so they were independent solution of the homogeneous system $x + y - z = 0$. On the other hand, w is automatically inde-pendent of u and v since w belongs to a different eigenvalue of A. Thus the three vectors are linearly independent.

(d) *A* is diagonalizable since it has three linearly independent eigenvectors. Let *P* be the matrix with column *u*, *v*, *w*. Then

$$P = \begin{pmatrix} 1 & 1 & 1 \\ -1 & 0 & 2 \\ 0 & 1 & 1 \end{pmatrix} \quad \text{and} \quad P^{-1}AP = \begin{pmatrix} 3 & & \\ & 3 & \\ & & 5 \end{pmatrix}$$

8.9. Suppose $B = \begin{pmatrix} -3 & 1 & -1 \\ -7 & 5 & -1 \\ -6 & 6 & -2 \end{pmatrix}$. Find: (*a*) the characteristic polynomial $\Delta(t)$ and eigenvalues

of *B*; and (*b*) a maximum set *S* of linearly independent eigenvectors of *B*. (*c*) Is *B* diagonalizable? If yes, find *P* such that $P^{-1}BP$ is diagonal.

(*a*) We have

$$\Delta(t) = |tI - B| = \begin{vmatrix} t+3 & -1 & 1 \\ 7 & t-5 & 1 \\ 6 & -6 & t+2 \end{vmatrix} = t^3 - 12t - 16$$

Therefore, $\Delta(t) = (t+2)^2(t-4)$. Thus $\lambda_1 = -2$ and $\lambda_2 = 4$ are the eigenvalues of *B*.

(*b*) Find a basis for the eigenspace of each eigenvalue.

(i) Substitute $t = -2$ into $tI - B$ to obtain the homogeneous system

$$\begin{pmatrix} 1 & -1 & 1 \\ 7 & -7 & 1 \\ 6 & -6 & 0 \end{pmatrix}\begin{pmatrix} x \\ y \\ z \end{pmatrix} = \begin{pmatrix} 0 \\ 0 \\ 0 \end{pmatrix} \quad \text{or} \quad \begin{cases} x - y + z = 0 \\ 7x - 7y + z = 0 \\ 6x - 6y = 0 \end{cases} \quad \text{or} \quad \begin{cases} x - y + z = 0 \\ x - y = 0 \end{cases}$$

The system has only one independent solution, e.g., $x = 1$, $y = 1$, $z = 0$. Thus $u = (1, 1, 0)$ forms a basis for the eigenspace of $\lambda_1 = -2$.

(ii) Substitute $t = 4$ into $tI - B$ to obtain the homogeneous system

$$\begin{pmatrix} 7 & -1 & 1 \\ 7 & -1 & 1 \\ 6 & -6 & 6 \end{pmatrix}\begin{pmatrix} x \\ y \\ z \end{pmatrix} = \begin{pmatrix} 0 \\ 0 \\ 0 \end{pmatrix} \quad \text{or} \quad \begin{cases} 7x - y + z = 0 \\ 7x - y + z = 0 \\ 6x - 6y + 6z = 0 \end{cases} \quad \text{or} \quad \begin{cases} 7x - y + z = 0 \\ x = 0 \end{cases}$$

The system has only one independent solution, e.g., $x = 0$, $y = 1$, $z = 1$. Thus $v = (0, 1, 1)$ forms a basis of the eigenspace of $\lambda_2 = 4$.

Thus $S = \{u, v\}$ is a maximum set of linearly independent eigenvectors of *B*.

(*c*) Since *B* has a maximum of two independent eigenvectors, *B* is not similar to a diagonal matrix, i.e., *B* is not diagonalizable.

8.10. Find the algebraic and geometric multiplicities of the eigenvalue $\lambda_1 = -2$ for matrix *B* in Problem 8.9.

The algebraic multiplicity of λ_1 is two. However, the geometric multiplicity of λ_1 is one, since $\dim E_{\lambda_1} = 1$.

8.11. Let $A = \begin{pmatrix} 1 & -1 \\ 2 & -1 \end{pmatrix}$. Find all eigenvalues and corresponding eigenvectors of *A* assuming *A* is a real matrix. Is *A* diagonalizable? If yes, find *P* such that $P^{-1}AP$ is diagonal.

The characteristic polynomial of *A* is $\Delta(t) = t^2 + 1$ which has no root in **R**. Thus *A*, viewed as a real matrix, has no eigenvalues and no eigenvectors, and hence *A* is not diagonalizable over **R**.

8.12. Repeat Problem 8.11 assuming now that A is a matrix over the complex field $\mathbf{C}$.

The characteristic polynomial of A is still $\Delta(t) = t^2 + 1$. (It does not depend on the field K.) Over $\mathbf{C}$, $\Delta(t)$ does factor; specifically, $\Delta(t) = t^2 + 1 = (t - i)(t + i)$. Thus $\lambda_1 = i$ and $\lambda_2 = -i$ are eigenvalues of A.

(i) Substitute $t = i$ in $tI - A$ to obtain the homogeneous system

$$\begin{pmatrix} i - 1 & 1 \\ -2 & i + 1 \end{pmatrix}\begin{pmatrix} x \\ y \end{pmatrix} = \begin{pmatrix} 0 \\ 0 \end{pmatrix} \quad \text{or} \quad \begin{cases} (i - 1)x + y = 0 \\ -2x + (i + 1)y = 0 \end{cases} \quad \text{or} \quad (i - 1)x + y = 0$$

The system has only one independent solution, e.g., $x = 1$, $y = 1 - i$. Thus $v_1 = (1, 1 - i)$ is an eigenvector which spans the eigenspace of $\lambda_1 = i$.

(ii) Substitute $t = -i$ into $tI - A$ to obtain the homogeneous system

$$\begin{pmatrix} -i - 1 & 1 \\ -2 & -i - 1 \end{pmatrix}\begin{pmatrix} x \\ y \end{pmatrix} = \begin{pmatrix} 0 \\ 0 \end{pmatrix} \quad \text{or} \quad \begin{cases} (-i - 1)x + y = 0 \\ -2x + (-i - 1)y = 0 \end{cases} \quad \text{or} \quad (-i - 1)x + y = 0$$

The system has only one independent solution, e.g., $x = 1$, $y = 1 + i$. Thus $v_2 = (1, 1 + i)$ is an eigenvector of A which spans the eigenspace of $\lambda_2 = -i$.

As a complex matrix, A is diagonalizable. Let P be the matrix whose columns are v_1 and v_2. Then

$$P = \begin{pmatrix} 1 & 1 \\ 1 - i & 1 + i \end{pmatrix} \quad \text{and} \quad P^{-1}AP = \begin{pmatrix} i & 0 \\ 0 & -i \end{pmatrix}$$

8.13. Let $B = \begin{pmatrix} 2 & 4 \\ 3 & 1 \end{pmatrix}$. Find: (a) all eigenvalues of B and the corresponding eigenvectors; (b) an invertible matrix P such that $D = P^{-1}BP$ is diagonal; and (c) B^6.

(a) Here $\Delta(t) = t^2 - \text{tr } B + |B| = t^2 - 3t - 10 = (t - 5)(t + 2)$. Thus $\lambda_1 = 5$ and $\lambda_2 = -2$ are the eigenvalues of B.

(i) Subtract $\lambda_1 = 5$ down the diagonal of B to obtain $M = \begin{pmatrix} -3 & 4 \\ 3 & -4 \end{pmatrix}$ which corresponds to the homogeneous system $3x - 4y = 0$. Here $v_1 = (4, 3)$ is a nonzero solution.

(ii) Subtract $\lambda_2 = -2$ (or add 2) down the diagonal of B to obtain $M = \begin{pmatrix} 4 & 4 \\ 3 & 3 \end{pmatrix}$ which corresponds to the system $x + y = 0$ which has a nonzero solution $v_2 = (1, -1)$.

(Since B has two independent eigenvectors, B is diagonalizable.)

(b) Let P be the matrix whose columns are v_1 and v_2. Then

$$P = \begin{pmatrix} 4 & 1 \\ 3 & -1 \end{pmatrix} \quad \text{and} \quad D = P^{-1}BP = \begin{pmatrix} 5 & 0 \\ 0 & -2 \end{pmatrix}$$

(c) Use the diagonal factorization of B,

$$B = PDP^{-1} = \begin{pmatrix} 4 & 1 \\ 3 & -1 \end{pmatrix}\begin{pmatrix} 5 & 0 \\ 0 & -2 \end{pmatrix}\begin{pmatrix} \frac{1}{7} & \frac{1}{7} \\ \frac{3}{7} & -\frac{4}{7} \end{pmatrix}$$

to obtain

$$B^6 = PD^6P^{-1} = \begin{pmatrix} 4 & 1 \\ 3 & -1 \end{pmatrix}\begin{pmatrix} 15{,}625 & 0 \\ 0 & 64 \end{pmatrix}\begin{pmatrix} \frac{1}{7} & \frac{1}{7} \\ \frac{3}{7} & -\frac{4}{7} \end{pmatrix} = \begin{pmatrix} 8956 & 8892 \\ 6669 & 6733 \end{pmatrix}$$

8.14. Determine whether or not A is diagonalizable where $A = \begin{pmatrix} 1 & 2 & 3 \\ 0 & 2 & 3 \\ 0 & 0 & 3 \end{pmatrix}$.

Since A is triangular, the eigenvalues of A are the diagonal elements 1, 2, and 3. Since they are distinct, A has three independent eigenvectors and thus A is similar to a diagonal matrix (Theorem 8.11). (We emphasize that here we do not need to compute eigenvectors to tell that A is diagonalizable. We will have to compute eigenvectors if we want to find P such that $P^{-1}AP$ is diagonal.)

8.15.   Suppose A and B are n-square matrices.

     (a)   Show that 0 is an eigenvalue of A if and only if A is singular.

     (b)   Show that AB and BA have the same eigenvalues.

     (c)   Suppose A is nonsingular (invertible) and λ is an eigenvalue of A. Show that λ^{-1} is an eigenvalue of A^{-1}.

     (d)   Show that A and its transpose A^T have the same characteristic polynomial.

     (a)   We have that 0 is an eigenvalue of A if and only if there exists a nonzero vector v such that $A(v) = 0v = 0$, i.e., that A is singular.

     (b)   By part (a) and the fact that the product of nonsingular matrices is nonsingular, the following statements are equivalent: (i) 0 is an eigenvalue of AB, (ii) AB is singular, (iii) A or B is singular, (iv) BA is singular, (v) 0 is an eigenvalue of BA.

         Now suppose λ is a nonzero eigenvalue of AB. Then there exists a nonzero vector v such that $ABv = \lambda v$. Set $w = Bv$. Since $\lambda \neq 0$ and $v \neq 0$,

$$Aw = ABv = \lambda v \neq 0 \qquad \text{and so} \qquad w \neq 0$$

But w is an eigenvector of BA belonging to the eigenvalue λ since

$$BAw = BABv = B\lambda v = \lambda Bv = \lambda w$$

Hence λ is an eigenvalue of BA. Similarly, any nonzero eigenvalue of BA is also an eigenvalue of AB. Thus AB and BA have the same eigenvalues.

     (c)   By part (a) $\lambda \neq 0$. By definition of an eigenvalue, there exists a nonzero vector v for which $A(v) = \lambda v$. Applying A^{-1} to both sides, we obtain $v = A^{-1}(\lambda v) = \lambda A^{-1}(v)$. Hence $A^{-1}(v) = \lambda^{-1}v$; that is, λ^{-1} is an eigenvalue of A^{-1}.

     (d)   By the transpose operation, $(tI - A)^T = tI^T - A^T$. Since a matrix and its transpose have the same determinant, $|tI - A| = |(tI - A)^T| = |tI - A^T|$. Thus A and A^T have the same characteristic polynomial.

8.16.   Let λ be an eigenvalue of an n-square matrix A over K. Let E_λ be the eigenspace of λ, i.e., the set of all eigenvectors of A belonging to λ. Show that E_λ is a subspace of K^n, that is, show that: (a) if $v \in E_\lambda$, then $kv \in E_\lambda$ for any scalar $k \in K$; and (b) if $u, v \in E_\lambda$, then $u + v \in E_\lambda$.

     (a)   Since $v \in E_\lambda$, we have $A(v) = \lambda v$. Then

$$A(kv) = kA(v) = k(\lambda v) = \lambda(kv)$$

        Thus $kv \in E_2$.

     (b)   Since $u, v \in E_\lambda$, we have $A(u) = \lambda v$ and $A(v) = \lambda v$. Then

$$A(u + v) = A(u) + A(v) = \lambda u + \lambda v = \lambda(u + v)$$

        Thus $u + v \in E_\lambda$.

DIAGONALIZING REAL SYMMETRIC MATRICES AND REAL QUADRATIC FORMS

8.17.   Let $A = \begin{pmatrix} 3 & 2 \\ 2 & 3 \end{pmatrix}$. Find a (real) orthogonal matrix P for which P^TAP is diagonal.

     The characteristic polynomial $\Delta(t)$ of A is

$$\Delta(t) = |tI - A| = \begin{vmatrix} t - 3 & -2 \\ -2 & t - 3 \end{vmatrix} = t^2 - 6t + 5 = (t - 5)(t - 1)$$

and thus the eigenvalues of A are 5 and 1.

Substitute $t = 5$ into the matrix $tI - A$ to obtain the corresponding homogeneous system of linear equations

$$2x - 2y = 0 \qquad -2x - 2y = 0$$

A nonzero solution is $v_2 = (1, -1)$. Normalize v_1 to find the unit solution $u_1 = (1/\sqrt{2}, 1/\sqrt{2})$.

Next substitute $t = 1$ into the matrix $tI - A$ to obtain the corresponding homogeneous system of linear equations

$$-2x - 2y = 0 \qquad -2x - 2y = 0$$

A nonzero solution is $v_2 = (1, -1)$. Normalize v_2 to find the unit solution $u_2 = (1/\sqrt{2}, -1/\sqrt{2})$.

Finally let P be the matrix whose columns are u_1 and u_2, respectively; then

$$P = \begin{pmatrix} 1/\sqrt{2} & 1/\sqrt{2} \\ 1/\sqrt{2} & -1/\sqrt{2} \end{pmatrix} \quad \text{and} \quad P^T A P = \begin{pmatrix} 5 & 0 \\ 0 & 1 \end{pmatrix}$$

As expected, the diagonal entries of $P^T A P$ are the eigenvalues of A.

8.18. Suppose $C = \begin{pmatrix} 11 & -8 & 4 \\ -8 & -1 & -2 \\ 4 & -2 & -4 \end{pmatrix}$. Find: (a) the characteristic polynomial $\Delta(t)$ of C; (b) the eigenvalues of C or, in other words, the roots of $\Delta(t)$; (c) a maximum set S of nonzero orthogonal eigenvectors of C; and (d) an orthogonal matrix P such that $P^{-1}CP$ is diagonal.

(a) We have

$$\Delta(t) = t^3 - (\text{tr } C)t^2 + (C_{11} + C_{22} + C_{33})t - |C| = t^3 - 6t^2 - 135t - 400$$

[Here C_{ii} is the cofactor of c_{ii} in $C = (c_{ij})$.]

(b) If $\Delta(t)$ has a rational root, it must divide 400. Testing $t = -5$, we get

$$-5 \begin{array}{|rrrr} 1 - & 6 - & 135 - & 400 \\ & -5 + & 55 + & 400 \\ \hline 1 - & 11 - & 80 + & 0 \end{array}$$

Thus $t + 5$ is a factor of $\Delta(t)$ and

$$\Delta(t) = (t + 5)(t^2 - 11t - 80) = (t + 5)^2(t - 16)$$

Accordingly, the eigenvalues of C are $\lambda = -5$ (with multiplicity two) and $\lambda = 16$ (with multiplicity one).

(c) Find an orthogonal basis for each eigenspace.

Subtract $\lambda = -5$ down the diagonal of C to obtain the homogeneous system

$$16x - 8y + 4z = 0 \qquad -8x + 4y - 2z = 0 \qquad 4x - 2y + z = \bar{0}$$

That is, $4x - 2y + z = 0$. The system has two independent solutions. One solution is $v_1 = (0, 1, 2)$. We seek a second solution $v_2 = (a, b, c)$ which is orthogonal to v_1; i.e., such that

$$4a - 2b + c = 0 \quad \text{and also} \quad b - 2c = 0$$

One such solution is $v_2 = (-5, -8, 4)$.

Subtract $\lambda = 16$ down the diagonal of C to obtain the homogeneous system

$$-5x - 8y + 4z = 0 \qquad -8x - 17y - 2z = 0 \qquad 4x - 2y - 20z = 0$$

This system yields a nonzero solution $v_3 = (4, -2, 1)$. (As expected from Theorem 8.13, the eigenvector v_3 is orthogonal to v_1 and v_2.)

Then v_1, v_2, v_3 form a maximum set of nonzero orthogonal eigenvectors of C.

(d)   Normalize v_1, v_2, v_3 to obtain the orthonormal basis

$$u_1 = (0, 1/\sqrt{5}, 2/\sqrt{5}) \qquad u_2 = (-5/\sqrt{105}, -8/\sqrt{105}, 4/\sqrt{105}) \qquad u_3 = (4/\sqrt{21}, -2/\sqrt{21}, 1/\sqrt{21})$$

Then P is the matrix whose columns are u_1, u_2, u_3. Thus

$$P = \begin{pmatrix} 0 & -5/\sqrt{105} & 4/\sqrt{21} \\ 1/\sqrt{5} & -8/\sqrt{105} & -2/\sqrt{21} \\ 2/\sqrt{5} & 4/\sqrt{105} & 1/\sqrt{21} \end{pmatrix} \quad \text{and} \quad P^T C P = \begin{pmatrix} -5 & & \\ & -5 & \\ & & 16 \end{pmatrix}$$

8.19.   Let $q(x, y) = 3x^2 - 6xy + 11y^2$. Find an orthogonal change of coordinates which diagonalizes q.

Find the symmetric matrix A representing q and its characteristic polynomial $\Delta(t)$:

$$A = \begin{pmatrix} 3 & -3 \\ -3 & 11 \end{pmatrix} \quad \text{and} \quad \Delta(t) = \begin{vmatrix} t-3 & 3 \\ 3 & t-11 \end{vmatrix} = t^2 - 14t + 24 = (t-2)(t-12)$$

The eigenvalues are 2 and 12; hence a diagonal form of q is

$$q(x', y') = 2x'^2 + 12y'^2$$

The corresponding change of coordinates is obtained by finding a corresponding set of eigenvectors of A. Set $t = 2$ into the matrix $tI - A$ to obtain the homogeneous system

$$-x + 3y = 0 \qquad 3x - 9y = 0$$

A nonzero solution is $v_1 = (3, 1)$. Next substitute $t = 12$ into the matrix $tI - A$ to obtain the homogeneous system

$$9x + 3y = 0 \qquad 3x + y = 0$$

A nonzero solution is $v_2 = (-1, 3)$. Normalize v_1 and v_2 to obtain the orthonormal basis

$$u_1 = (3/\sqrt{10}, 1/\sqrt{10}) \qquad u_2 = (-1/\sqrt{10}, 3/\sqrt{10})$$

The change-of-basis matrix P and the required change of coordinates follow:

$$P = \begin{pmatrix} 3/\sqrt{10} & -1/\sqrt{10} \\ 1/\sqrt{10} & 3/\sqrt{10} \end{pmatrix} \quad \text{and} \quad \begin{pmatrix} x \\ y \end{pmatrix} = P \begin{pmatrix} x' \\ y' \end{pmatrix} \quad \text{or} \quad \begin{cases} x = (3x' - y')/\sqrt{10} \\ y = (x' + 3y')/\sqrt{10} \end{cases}$$

One can also express x' and y' in terms of x and y by using $P^{-1} = P^T$, that is,

$$x' = (3x + y)/\sqrt{10} \qquad y' = (-x + 3y)/\sqrt{10}$$

8.20.   Consider the quadratic form $q(x, y, z) = 3x^2 + 2xy + 3y^2 + 2xz + 2yz + 3z^2$. Find:

(a)   The symmetric matrix A which represents q and its characteristic polynomial $\Delta(t)$,

(b)   The eigenvalues of A or, in other words, the roots of $\Delta(t)$,

(c)   A maximum set S of nonzero orthogonal eigenvectors of A,

(d)   An orthogonal change of coordinates which diagonalizes q.

(a)   Recall $A = (a_{ij})$ is the symmetric matrix where a_{ii} is the coefficient of x_i^2 and $a_{ij} = a_{ji}$ is one-half the coefficient of $x_i x_j$. Thus

$$A = \begin{pmatrix} 3 & 1 & 1 \\ 1 & 3 & 1 \\ 1 & 1 & 3 \end{pmatrix} \quad \text{and} \quad \Delta(t) = \begin{vmatrix} t-3 & -1 & -1 \\ -1 & t-3 & -1 \\ -1 & -1 & t-3 \end{vmatrix} = t^3 - 9t^2 + 24t - 20$$

(b) If $\Delta(t)$ has a rational root, it must divide the constant 20, or, in other words, it must be among $\pm 1, \pm 2, \pm 4, \pm 5, \pm 10, \pm 20$. Testing $t = 2$, we get

$$
\begin{array}{r|rrrr}
2 & 1 & -9 & +24 & -20 \\
 &   & 2 & -14 & +20 \\
\hline
 & 1 & -7 & +10 & +\ 0
\end{array}
$$

Thus

$$\Delta(t) = (t - 2)(t^2 - 7t + 10) = (t - 2)^2(t - 5)$$

Hence the eigenvalues of A are 2 (with multiplicity two) and 5 (with multiplicity one).

(c) Find an orthogonal basis for each eigenspace.

Subtract $\lambda = 2$ down the diagonal of A to obtain the corresponding homogeneous system

$$x + y + z = 0 \qquad x + y + z = 0 \qquad x + y + z = 0$$

That is, $x + y + z = 0$. The system has two independent solutions. One such solution is $v_1 = (0, 1, -1)$. We seek a second solution $v_2 = (a, b, c)$ which is orthogonal to v_1; that is, such that

$$a + b + c = 0 \qquad \text{and also} \qquad b - c = 0$$

For example, $v_2 = (2, -1, -1)$. Thus $v_1 = (0, 1, -1)$, $v_2 = (2, -1, -1)$ form an orthogonal basis for the eigenspace of $\lambda = 2$.

Subtract $\lambda = 5$ down the diagonal of A to obtain the corresponding homogeneous system

$$-2x + y + z = 0 \qquad x - 2y + z = 0 \qquad x + y - 2z = 0$$

This system yields a nonzero solution $v_3 = (1, 1, 1)$. (As expected from Theorem 8.13, the eigenvector v_3 is orthogonal to v_1 and v_2.)

Then v_1, v_2, v_3 form a maximum set of nonzero orthogonal eigenvectors of A.

(d) Normalize v_1, v_2, v_3 to obtain the orthonormal basis

$$u_1 = (0, 1/\sqrt{2}, -1/\sqrt{2}) \qquad u_2 = (2/\sqrt{6}, -1/\sqrt{6}, -1/\sqrt{6}) \qquad u_3 = (1/\sqrt{3}, 1/\sqrt{3}, 1/\sqrt{3})$$

Let P be the matrix whose columns are u_1, u_2, u_3. Then

$$
P = \begin{pmatrix} 0 & 2/\sqrt{6} & 1/\sqrt{3} \\ 1/\sqrt{2} & -1/\sqrt{6} & 1/\sqrt{3} \\ -1/\sqrt{2} & -1/\sqrt{6} & 1/\sqrt{3} \end{pmatrix} \qquad \text{and} \qquad P^T A P = \begin{pmatrix} 2 & & \\ & 2 & \\ & & 5 \end{pmatrix}
$$

Thus the required orthogonal change of coordinates is

$$x = \frac{2y'}{\sqrt{6}} + \frac{z'}{\sqrt{3}}$$

$$y = \frac{x'}{\sqrt{2}} - \frac{y'}{\sqrt{6}} + \frac{z'}{\sqrt{3}}$$

$$z = -\frac{x'}{\sqrt{2}} - \frac{y'}{\sqrt{6}} + \frac{z'}{\sqrt{3}}$$

Under this change of coordinates, q is transformed into the diagonal form

$$q(x', y', z') = 2x'^2 + 2y'^2 + 5z'^2$$

MINIMUM POLYNOMIAL

8.21. Find the minimum polynomial $m(t)$ of the matrix $A = \begin{pmatrix} 2 & -1 & 1 \\ 6 & -3 & 4 \\ 3 & -2 & 3 \end{pmatrix}$.

First find the characteristic polynomial $\Delta(t)$ of A:

$$\Delta(t) = |tI - A| = \begin{vmatrix} t-2 & 1 & -1 \\ -6 & t+3 & -4 \\ -3 & 2 & t-3 \end{vmatrix} = t^3 - 4t^2 + 5t - 2 = (t-2)(t-1)^2$$

Alternatively, $\Delta(t) = t^3 - (\text{tr } A)t^2 + (A_{11} + A_{22} + A_{33})t - |A| = t^3 - 4t^2 + 5t - 2 = (t-2)(t-1)^2$. (Here A_{ii} is the cofactor of a_{ii} in A.)

The minimum polynomial $m(t)$ must divide $\Delta(t)$. Also, each irreducible factor of $\Delta(t)$, that is, $t - 2$ and $t - 1$, must also be a factor of $m(t)$. Thus $m(t)$ is exactly only of the following:

$$f(t) = (t-2)(t-1) \qquad \text{or} \qquad g(t) = (t-2)(t-1)^2$$

We know, by the Caley–Hamilton Theorem, that $g(A) = \Delta(A) = 0$; hence we need only test $f(t)$. We have

$$f(A) = (A - 2I)(A - I) = \begin{pmatrix} 2 & -2 & 2 \\ 6 & -5 & 4 \\ 3 & -2 & 1 \end{pmatrix}\begin{pmatrix} 3 & -2 & 2 \\ 6 & -4 & 4 \\ 3 & -2 & 2 \end{pmatrix} = \begin{pmatrix} 0 & 0 & 0 \\ 0 & 0 & 0 \\ 0 & 0 & 0 \end{pmatrix}$$

Thus $f(t) = m(t) = (t-2)(t-1) = t^2 - 3t + 2$ is the minimum polynomial of A.

8.22. Find the minimum polynomial $m(t)$ of the matrix (where $a \neq 0$): $B = \begin{pmatrix} \lambda & a & 0 \\ 0 & \lambda & a \\ 0 & 0 & \lambda \end{pmatrix}$.

The characteristic polynomial of B is $\Delta(t) = (t - \lambda)^3$. [Note $m(t)$ is exactly one of $t - \lambda$, $(t - \lambda)^2$, or $(t - \lambda)^3$.] We find $(B - \lambda I)^2 \neq 0$; thus $m(t) = \Delta(t) = (t - \lambda)^3$.

(Remark: This matrix is a special case of Example 8.11 and Problem 8.61.)

8.23. Find the minimum polynomial $m(t)$ of the following matrix: $M' = \begin{pmatrix} 4 & 1 & 0 & 0 & 0 \\ 0 & 4 & 1 & 0 & 0 \\ 0 & 0 & 4 & 0 & 0 \\ 0 & 0 & 0 & 4 & 1 \\ 0 & 0 & 0 & 0 & 4 \end{pmatrix}$.

Here M' is block diagonal with diagonal blocks

$$A' = \begin{pmatrix} 4 & 1 & 0 \\ 0 & 4 & 1 \\ 0 & 0 & 4 \end{pmatrix} \qquad \text{and} \qquad B' = \begin{pmatrix} 4 & 1 \\ 0 & 4 \end{pmatrix}$$

The characteristic and minimum polynomial of A' is $f(t) = (t-4)^3$, and the characteristic and minimum polynomial of B' is $g(t) = (t-4)^2$. Thus $\Delta(t) = f(t)g(t) = (t-4)^5$ is the characteristic polynomial of M', but $m(t) = \text{LCM}\,[f(t), g(t)] = (t-4)^3$ (which is the size of the largest block) is the minimum polynomial of M'.

8.24. Find a matrix A whose minimum polynomial is:

(a) $f(t) = t^3 - 8t^2 + 5t + 7$,   (b) $f(t) = t^4 - 3t^3 - 4t^2 + 5t + 6$

Let A be the companion matrix (see Example 8.12) of $f(t)$. Then

$$(a) \quad A = \begin{pmatrix} 0 & 0 & -7 \\ 1 & 0 & -5 \\ 0 & 1 & 8 \end{pmatrix}, \qquad (b) \quad A = \begin{pmatrix} 0 & 0 & 0 & -6 \\ 1 & 0 & 0 & -5 \\ 0 & 1 & 0 & 4 \\ 0 & 0 & 1 & 3 \end{pmatrix}$$

(Remark: The polynomial $f(t)$ is also the characteristic polynomial of A.)

8.25. Show that the minimum polynomial of a matrix A exists and is unique.

By the Cayley–Hamilton Theorem, A is a zero of some nonzero polynomial (see also Problem 8.37). Let n be the lowest degree for which a polynomial $f(t)$ exists such that $f(A) = 0$. Dividing $f(t)$ by its leading coefficient, we obtain a monic polynomial $m(t)$ of degree n which has A as a zero. Suppose $m'(t)$ is another monic polynomial of degree n for which $m'(A) = 0$. Then the difference $m(t) - m'(t)$ is a nonzero polynomial of degree less than n which has A as a zero. This contradicts the original assumption on n; hence $m(t)$ is a unique minimum polynomial.

PROOFS OF THEOREMS

8.26. Prove Theorem 8.1.

Suppose $f = a_n t^n + \cdots + a_1 t + a_0$ and $g = b_m t^m + \cdots + b_1 t + b_0$. Then by definition,

$$f(A) = a_n A^n + \cdots + a_1 A + a_0 I \quad \text{and} \quad g(A) = b_m A^m + \cdots + b_1 A + b_0 I$$

(i) Suppose $m \le n$ and let $b_i = 0$ if $i > m$. Then

$$f + g = (a_n + b_n)t^n + \cdots + (a_1 + b_1)t + (a_0 + b_0)$$

Hence

$$(f + g)(A) = (a_n + b_n)A^n + \cdots + (a_1 + b_1)A + (a_0 + b_0)I$$
$$= a_n A^n + b_n A^n + \cdots + a_1 A + b_1 A + a_0 I + b_0 I = f(A) + g(A)$$

(ii) By definition, $fg = c_{n+m}t^{n+m} + \cdots + c_1 t + c_0 = \sum_{k=0}^{n+m} c_k t^k$ where

$$c_k = a_0 b_k + a_1 b_{k-1} + \cdots + a_k b_0 = \sum_{i=0}^{k} a_i b_{k-i}$$

Hence $(fg)(A) = \sum_{k=0}^{n+m} c_k A^k$ and

$$f(A)g(A) = \left(\sum_{i=0}^{n} a_i A^i\right)\left(\sum_{j=0}^{m} b_j A^j\right) = \sum_{i=0}^{n}\sum_{j=0}^{m} a_i b_j A^{i+j} = \sum_{k=0}^{n+m} c_k A^k = (fg)(A)$$

(iii) By definition, $kf = ka_n t^n + \cdots + ka_1 t + ka_0$, and so

$$(kf)(A) = ka_n A^n + \cdots + ka_1 A + ka_0 I = k(a_n A^n + \cdots + a_1 A + a_0 I) = kf(A)$$

8.27. Prove the Cayley–Hamilton Theorem 8.2.

Let A be an arbitrary n-square matrix and let $\Delta(t)$ be its characteristic polynomial; say,

$$\Delta(t) = |tI - A| = t^n + a_{n-1}t^{n-1} + \cdots + a_1 t + a_0$$

Now let $B(t)$ denote the classical adjoint of the matrix $tI - A$. The elements of $B(t)$ are cofactors of the matrix $tI - A$ and hence are polynomials in t of degree not exceeding $n - 1$. Thus

$$B(t) = B_{n-1}t^{n-1} + \cdots + B_1 t + B_0$$

where the B_i are n-square matrices over K which are independent of t. By the fundamental property of the classical adjoint (Theorem 7.9)

$$(tI - A)B(t) = |tI - A|I$$

or

$$(tI - A)(B_{n-1}t^{n-1} + \cdots + B_1 t + B_0) = (t^n + a_{n-1}t^{n-1} + \cdots + a_1 t + a_0)I$$

Removing parentheses and equating the coefficients of corresponding powers of t,

$$B_{n-1} = I$$
$$B_{n-2} - AB_{n-1} = a_{n-1}I$$
$$B_{n-3} - AB_{n-2} = a_{n-2}I$$
$$\dots\dots\dots\dots\dots\dots\dots$$
$$B_0 - AB_1 = a_1 I$$
$$-AB_0 = a_0 I$$

Multiplying the above matrix equations by A^n, A^{n-1}, ..., A, I, respectively,

$$A^n B_{n-1} = A^n$$
$$A^{n-1}B_{n-2} - A^n B_{n-1} = a_{n-1}A^{n-1}$$
$$A^{n-2}B_{n-3} - A^{n-1}B_{n-2} = a_{n-2}A^{n-2}$$
$$\dots\dots\dots\dots\dots\dots\dots\dots$$
$$AB_0 - A^2 B_1 = a_1 A$$
$$-AB_0 = a_0 I$$

Adding the above matrix equations,

$$0 = A^n + a_{n-1}A^{n-1} + \cdots + a_1 A + a_0 I$$

In other words, $\Delta(A) = 0$. That is, A is a zero of its characteristic polynomial.

8.28.   Prove Theorem 8.6.

The scalar λ is an eigenvalue of A if and only if there exists a nonzero vector v such that

$$Av = \lambda v \qquad \text{or} \qquad (\lambda I)v - Av = 0 \qquad \text{or} \qquad (\lambda I - A)v = 0$$

or $M = \lambda I - A$ is singular. In such a case λ is a root of $\Delta(t) = |tI - A|$. Also, v is in the eigenspace E_λ of λ if and only if the above relations hold; hence v is a solution of $(\lambda I - A)X = 0$.

8.29.   Prove Theorem 8.9.

Suppose A has n linearly independent eigenvectors v_1, v_2, ..., v_n with corresponding eigenvalues λ_1, λ_2, ..., λ_n. Let P be the matrix whose columns are v_1, ..., v_n. Then P is nonsingular. Also, the columns of AP are Av_1, ..., Av_n. But $Av_k = \lambda v_k$. Hence the columns of AP are $\lambda_1 v_1$, ..., $\lambda_n v_n$. On the other hand, let $D = \text{diag}(\lambda_1, \lambda_2, ..., \lambda_n)$, that is, the diagonal matrix with diagonal entries λ_k. Then PD is also a matrix with columns $\lambda_k v_k$. Accordingly,

$$AP = PD \qquad \text{and hence} \qquad D = P^{-1}AP$$

as required.

Conversely, suppose there exists a nonsingular matrix P for which

$$P^{-1}AP = \text{diag}(\lambda_1, \lambda_2, ..., \lambda_n) = D \qquad \text{and so} \qquad AP = PD$$

Let v_1, v_2, ..., v_n be the column vectors of P. Then the columns of AP are Av_k and the columns of PD are $\lambda_k v_k$. Accordingly, since $AP = PD$, we have

$$Av_1 = \lambda_1 v_1, \; Av_2 = \lambda_2 v_2, \; ..., \; Av_n = \lambda_n v_n$$

Furthermore, since P is nonsingular, v_1, v_2, ..., v_n are nonzero and hence, they are eigenvectors of A belonging to the eigenvalues that are the diagonal elements of D. Moreover, they are linearly independent. Thus the theorem is proved.

8.30. Prove Theorem 8.10.

The proof is by induction on n. If $n = 1$, then v_1 is linearly independent since $v_1 \neq 0$. Assume $n > 1$. Suppose

$$a_1 v_1 + a_2 v_2 + \cdots + a_n v_n = 0 \tag{1}$$

where the a_i are scalars. Multiply (1) by A and obtain

$$a_1 A v_1 + a_2 A v_2 + \cdots + a_n A v_n = A0 = 0$$

By hypothesis, $A v_i = \lambda_i v_i$. Thus on substitution we obtain

$$a_1 \lambda_1 v_1 + a_2 \lambda_2 v_2 + \cdots + a_n \lambda_n v_n = 0 \tag{2}$$

On the other hand, multiplying (1) by λ_n, we get

$$a_1 \lambda_n v_1 + a_2 \lambda_n v_2 + \cdots + a_n \lambda_n v_n = 0 \tag{3}$$

Subtracting (3) from (2) yields

$$a_1 (\lambda_1 - \lambda_n) v_1 + a_2 (\lambda_2 - \lambda_n) v_2 + \cdots + a_{n-1} (\lambda_{n-1} - \lambda_n) v_{n-1} = 0$$

By induction, $v_1, v_2, \ldots, v_{n-1}$ are linearly independent; hence each of the above coefficients is 0. Since the λ_i are distinct, $\lambda_i - \lambda_n \neq 0$ for $i \neq n$. Hence $a_1 = \cdots = a_{n-1} = 0$. Substituting this into (1), we get $a_n v_n = 0$, and hence $a_n = 0$. Thus the v_i are linearly independent.

8.31. Prove Theorem 8.11.

By Theorem 8.6, the a_i are eigenvalues of A. Let v_i be corresponding eigenvectors. By Theorem 8.10, the v_i are linearly independent and hence form a basis of K^n. Thus A is diagonalizable by Theorem 8.9.

8.32. Prove Theorem 8.15.

Suppose $f(t)$ is a polynomial for which $f(A) = 0$. By the division algorithm, there exist polynomials $q(t)$ and $r(t)$ for which $f(t) = m(t)q(t) + r(t)$ and $r(t) = 0$ or $\deg r(t) < \deg m(t)$. Substituting $t = A$ in this equation, and using that $f(A) = 0$ and $m(A) = 0$, we obtain $r(A) = 0$. If $r(t) \neq 0$, then $r(t)$ is a polynomial of degree less than $m(t)$ which has A as a zero; this contradicts the definition of the minimum polynomial. Thus $r(t) = 0$ and so $f(t) = m(t)q(t)$, i.e., $m(t)$ divides $f(t)$.

8.33. Let $m(t)$ be the minimum polynomial of an n-square matrix A.

(a) Show that the characteristic polynomial of A divides $(m(t))^n$.

(b) Prove Theorem 8.16.

(a) Suppose $m(t) = t^r + c_1 t^{r-1} + \cdots + c_{r-1} t + c_r$. Consider the following matrices:

$$B_0 = I$$
$$B_1 = A + c_1 I$$
$$B_2 = A^2 + c_1 A + c_2 I$$
$$\cdots\cdots\cdots\cdots\cdots\cdots\cdots\cdots\cdots$$
$$B_{r-1} = A^{r-1} + c_1 A^{r-2} + \cdots + c_{r-1} I$$

Then
$$B_0 = I$$
$$B_1 - AB_0 = c_1 I$$
$$B_2 - AB_1 = c_2 I$$
$$\cdots\cdots\cdots\cdots\cdots\cdots$$
$$B_{r-1} - AB_{r-2} = c_{r-1} I$$

Also,
$$-AB_{r-1} = c_r I - (A^r + c_1 A^{r-1} + \cdots + c_{r-1} A + c_r I)$$
$$= c_r I - m(A)$$
$$= c_r I$$

Set
$$B(t) = t^{r-1}B_0 + t^{r-2}B_1 + \cdots + tB_{r-2} + B_{r-1}$$

Then
$$
\begin{aligned}
(tI - A) \cdot B(t) &= (t^r B_0 + t^{r-1}B_1 + \cdots + tB_{r-1}) - (t^{r-1}AB_0 + t^{r-2}AB_1 + \cdots + AB_{r-1}) \\
&= t^r B_0 + t^{r-1}(B_1 - AB_0) + t^{r-2}(B_2 - AB_1) + \cdots + t(B_{r-1} - AB_{r-2}) - AB_{r-1} \\
&= t^r I + c_1 t^{r-1}I + c_2 t^{r-2}I + \cdots + c_{r-1}tI + c_r I \\
&= m(t)I
\end{aligned}
$$

The determinant of both sides gives $|tI - A||B(t)| = |m(t)I| = (m(t))^n$. Since $|B(t)|$ is a polynomial, $|tI - A|$ divides $(m(t))^n$; that is, the characteristic polynomial of A divides $(m(t))^n$.

(b) Suppose $f(t)$ is an irreducible polynomial. If $f(t)$ divides $m(t)$ then, since $m(t)$ divides $\Delta(t)$, $f(t)$ divides $\Delta(t)$. On the other hand, if $f(t)$ divides $\Delta(t)$ then, by part (a), $f(t)$ divides $(m(t))^n$. But $f(t)$ is irreducible; hence $f(t)$ also divides $m(t)$. Thus $m(t)$ and $\Delta(t)$ have the same irreducible factors.

8.34. Prove Theorem 8.18.

We prove the theorem for the case $r = 2$. The general theorem follows easily by induction. Suppose $M = \begin{pmatrix} A & 0 \\ 0 & B \end{pmatrix}$ where A and B are square matrices. We need to show that the minimum polynomial $m(t)$ of M is the least common multiple of the minimum polynomials $g(t)$ and $h(t)$ of A and B, respectively.

Since $m(t)$ is the minimum polynomial of M, $m(M) = \begin{pmatrix} m(A) & 0 \\ 0 & m(B) \end{pmatrix} = 0$ and hence $m(A) = 0$ and $m(B) = 0$. Since $g(t)$ is the minimum polynomial of A, $g(t)$ divides $m(t)$. Similarly, $h(t)$ divides $m(t)$. Thus $m(t)$ is a multiple of $g(t)$ and $h(t)$.

Now let $f(t)$ be another multiple of $g(t)$ and $h(t)$; then $f(M) = \begin{pmatrix} f(A) & 0 \\ 0 & f(B) \end{pmatrix} = \begin{pmatrix} 0 & 0 \\ 0 & 0 \end{pmatrix} = 0$. But $m(t)$ is the minimum polynomial of M; hence $m(t)$ divides $f(t)$. Thus $m(t)$ is the least common multiple of $g(t)$ and $h(t)$.

8.35. Suppose A is a real symmetric matrix viewed as a matrix over $\mathbf{C}$.

(a) Prove that $\langle Au, v \rangle = \langle u, Av \rangle$ for the inner product in $\mathbf{C}^n$.

(b) Prove Theorems 8.12 and 8.13 for the matrix A.

(a) We use the fact that the inner product in $\mathbf{C}^n$ is defined by $\langle u, v \rangle = u^T \bar{v}$. Since A is real symmetric, $A = A^T = \bar{A}$. Thus
$$\langle Au, v \rangle = (Au)^T \bar{v} = u^T A^T \bar{v} = u^T \bar{A}\bar{v} = u^T \overline{Av} = \langle u, Av \rangle$$

(b) We use the fact that in $\mathbf{C}^n$, $\langle ku, v \rangle = k\langle u, v \rangle$ but $\langle u, kv \rangle = \bar{k}\langle u, v \rangle$.

(1) There exists $v \neq 0$ such that $Av = \lambda v$. Then
$$\lambda\langle v, v \rangle = \langle \lambda v, v \rangle = \langle Av, v \rangle = \langle v, Av \rangle = \langle v, \lambda v \rangle = \bar{\lambda}\langle v, v \rangle$$

But $\langle v, v \rangle \neq 0$ since $v \neq 0$. Thus $\lambda = \bar{\lambda}$ and so λ is real.

(2) Here $Au = \lambda_1 u$ and $Av = \lambda_2 v$ and, by (1), λ_2 is real. Then
$$\lambda_1\langle u, v \rangle = \langle \lambda_1 u, v \rangle = \langle Au, v \rangle = \langle u, Av \rangle = \langle u, \lambda_2 v \rangle = \bar{\lambda}_2\langle u, v \rangle = \lambda_2\langle u, v \rangle$$

Since $\lambda_1 \neq \lambda_2$, we have $\langle u, v \rangle = 0$.

MISCELLANEOUS PROBLEMS

8.36. Suppose A be a 2×2 symmetric matrix with eigenvalues 1 and 9 and suppose $u = (1, 3)^T$ is an eigenvector belonging to the eigenvalue 1. Find: (a) an eigenvector v belonging to the eigenvalue 9, (b) the matrix A, and (c) a square root of A, i.e., a matrix B such that $B^2 = A$.

(a)   Since A is symmetric, v must be orthogonal to u. Set $v = (-3, 1)^T$.

(b)   Let P be the matrix whose columns are the eigenvectors u and v. Then, by the diagonal factorization of A, we have

$$A = PDP^{-1} = \begin{pmatrix} 1 & -3 \\ 3 & 1 \end{pmatrix} \begin{pmatrix} 1 & 0 \\ 0 & 9 \end{pmatrix} \begin{pmatrix} \frac{1}{10} & \frac{3}{10} \\ -\frac{3}{10} & \frac{1}{10} \end{pmatrix} = \begin{pmatrix} \frac{41}{5} & -\frac{12}{5} \\ -\frac{12}{5} & \frac{9}{5} \end{pmatrix}$$

(Alternatively, A is the matrix for which $Au = u$ and $Av = 9v$.)

(c)   Use the diagonal factorization of A to obtain

$$B = P\sqrt{D}\, P^{-1} = \begin{pmatrix} 1 & -3 \\ 3 & 1 \end{pmatrix} \begin{pmatrix} 1 & 0 \\ 0 & 3 \end{pmatrix} \begin{pmatrix} \frac{1}{10} & \frac{3}{10} \\ -\frac{3}{10} & \frac{1}{10} \end{pmatrix} = \begin{pmatrix} \frac{14}{5} & -\frac{3}{5} \\ -\frac{3}{5} & \frac{6}{5} \end{pmatrix}$$

8.37. Let A be an n-square matrix. Without using the Caley–Hamilton theorem, show that A is a root of a nonzero polynomial.

Let $N = n^2$. Consider the following $N + 1$ matrices

$$I, A, A^2, \ldots, A^N$$

Recall that the vector space V of $n \times n$ matrices has dimension $N = n^2$. Thus the above $N + 1$ matrices are linearly dependent. Thus there exist scalars $a_0, a_1, a_2, \ldots, a_N$, not all zero, for which

$$a_N A^N + \cdots + a_1 A + a_0 I = 0$$

Thus A is a root of the polynomial $f(t) = a_N t^N + \cdots + a_1 t + a_0$.

8.38. Suppose A is an n-square matrix. Prove the following:

(a) A is nonsingular if and only if the constant term of the minimum polynomial of A is not zero.

(b)   If A is nonsingular, then A^{-1} is equal to a polynomial in A of degree not exceeding n.

(a)   Suppose $f(t) = t^r + a_{r-1}t^{r-1} + \cdots + a_1 t + a_0$ is the minimum (characteristic) polynomial of A. Then the following are equivalent: (i) A is nonsingular, (ii) 0 is not a root of $f(t)$, and (iii) the constant term a_0 is not zero. Thus the statement is true.

(b)   Let $m(t)$ be the minimum polynomial of A. Then $m(t) = t^r + a_{r-1}t^{r-1} + \cdots + a_1 t + a_0$, where $r \le n$. Since A is nonsingular, $a_0 \ne 0$ by part (a). We have

$$m(A) = A^r + a_{r-1}A^{r-1} + \cdots + a_1 A + a_0 I = 0$$

Thus

$$-\frac{1}{a_0}(A^{r-1} + a_{r-1}A^{r-2} + \cdots + a_1 I)A = I$$

Accordingly,

$$A^{-1} = -\frac{1}{a_0}(A^{r-1} + a_{r-1}A^{r-2} + \cdots + a_1 I)$$

8.39. Let F be an extension of a field K. Let A be an n-square matrix over K. Note that A may also be viewed as a matrix $\hat{A}$ over F. Clearly $|tI - A| = |tI - \hat{A}|$, that is, A and $\hat{A}$ have the same characteristic polynomial. Show that A and $\hat{A}$ also have the same minimum polynomial.

Let $m(t)$ and $m'(t)$ be the minimum polynomials of A and $\hat{A}$, respectively. Now $m'(t)$ divides every polynomial over F which has A as a zero. Since $m(t)$ has A as a zero and since $m(t)$ may be viewed as a polynomial over F, $m'(t)$ divides $m(t)$. We show now that $m(t)$ divides $m'(t)$.

Since $m'(t)$ is a polynomial over F which is an extension of K, we may write

$$m'(t) = f_1(t)b_1 + f_2(t)b_2 + \cdots + f_n(t)b_n$$

where $f_i(t)$ are polynomials over K, and $b_1, \ldots, b_n$ belong to F and are linearly independent over K. We have

$$m'(A) = f_1(A)b_1 + f_2(A)b_2 + \cdots + f_n(A)b_n = 0 \tag{1}$$

Let $a_{ij}^{(k)}$ denote the ij-entry of $f_k(A)$. The above matrix equation implies that, for each pair (i, j),

$$a_{ij}^{(1)}b_1 + a_{ij}^{(2)}b_2 + \cdots + a_{ij}^{(n)}b_n = 0$$

Since the b_i are linearly independent over K and since the $a_{ij}^{(k)} \in K$, every $a_{ij}^{(k)} = 0$. Then

$$f_1(A) = 0, \ f_2(A) = 0, \ldots, f_n(A) = 0$$

Since the $f_i(t)$ are polynomials over K which have A as a zero and since $m(t)$ is the minimum polynomial of A as a matrix over K, $m(t)$ divides each of the $f_i(t)$. Accordingly, by (1), $m(t)$ must also divide $m'(t)$. But monic polynomials which divide each other are necessarily equal. That is, $m(t) = m'(t)$, as required.

Supplementary Problems

POLYNOMIALS AND MATRICES

8.40. Let $f(t) = 2t^2 - 5t + 6$ and $g(t) = t^3 - 2t^2 + t + 3$. Find $f(A)$, $g(A)$, $f(B)$, and $g(B)$ where $A = \begin{pmatrix} 2 & -3 \\ 5 & 1 \end{pmatrix}$ and $B = \begin{pmatrix} 1 & 2 \\ 0 & 3 \end{pmatrix}$.

8.41. Let $A = \begin{pmatrix} 1 & 1 \\ 0 & 1 \end{pmatrix}$. Find A^2, A^3, A^n.

8.42. Let $B = \begin{pmatrix} 8 & 12 & 0 \\ 0 & 8 & 12 \\ 0 & 0 & 8 \end{pmatrix}$. Find a real matrix A such that $B = A^3$.

8.43. Show that, for any square matrix A, $(P^{-1}AP)^n = P^{-1}A^nP$ where P is invertible. More generally, show that $f(P^{-1}AP) = P^{-1}f(A)P$ for any polynomial $f(t)$.

8.44. Let $f(t)$ be any polynomial. Show that $(a) f(A^T) = (f(A))^T$, and (b) if A is symmetric, then $f(A)$ is symmetric.

EIGENVALUES AND EIGENVECTORS

8.45. Let $A = \begin{pmatrix} 5 & 6 \\ -2 & -2 \end{pmatrix}$. Find: (a) all eigenvalues and linearly independent eigenvectors; (b) P such that $D = P^{-1}AP$ is diagonal; (c) A^{10} and $f(A)$ where $f(t) = t^4 - 5t^3 + 7t^2 - 2t + 5$; and (d) B such that $B^2 = A$.

8.46. For each of the following matrices, find all eigenvalues and a basis for each eigenspace:

$$(a) \quad A = \begin{pmatrix} 3 & 1 & 1 \\ 2 & 4 & 2 \\ 1 & 1 & 3 \end{pmatrix}, \qquad (b) \quad B = \begin{pmatrix} 1 & 2 & 2 \\ 1 & 2 & -1 \\ -1 & 1 & 4 \end{pmatrix}, \qquad (c) \quad C = \begin{pmatrix} 1 & 1 & 0 \\ 0 & 1 & 0 \\ 0 & 0 & 1 \end{pmatrix}$$

When possible, find invertible matrices P_1, P_2, and P_3 such that $P_1^{-1}AP_1$, $P_2^{-1}BP_2$, and $P_3^{-1}CP_3$ are diagonal.

8.47. Consider the matrices $A = \begin{pmatrix} 2 & -1 \\ 1 & 4 \end{pmatrix}$ and $B = \begin{pmatrix} 3 & -1 \\ 13 & -3 \end{pmatrix}$. Find all eigenvalues and linearly independent eigenvectors assuming (a) A and B are matrices over the real field $\mathbf{R}$, and (b) A and B are matrices over the complex field $\mathbf{C}$.

8.48. Suppose v is a nonzero eigenvector of matrices A and B. Show that v is also an eigenvector of the matrix $kA + k'B$ where k and k' are any scalars.

8.49. Suppose v is a nonzero eigenvector of a matrix A belonging to the eigenvalue λ. Show that for $n > 0$, v is also an eigenvector of A^n belonging to λ^n.

8.50. Suppose λ is an eigenvalue of a matrix A. Show that $f(\lambda)$ is an eigenvalue of $f(A)$ for any polynomial $f(t)$.

8.51. Show that similar matrices have the same eigenvalues.

8.52. Show that matrices A and A^T have the same eigenvalues. Give an example where A and A^T have different eigenvectors.

CHARACTERISTIC AND MINIMUM POLYNOMIALS

8.53. Find the characteristic and minimum polynomials of each of the following matrices:

$$A = \begin{pmatrix} 2 & 5 & 0 & 0 & 0 \\ 0 & 2 & 0 & 0 & 0 \\ 0 & 0 & 4 & 2 & 0 \\ 0 & 0 & 3 & 5 & 0 \\ 0 & 0 & 0 & 0 & 7 \end{pmatrix} \qquad B = \begin{pmatrix} 3 & 1 & 0 & 0 & 0 \\ 0 & 3 & 0 & 0 & 0 \\ 0 & 0 & 3 & 1 & 0 \\ 0 & 0 & 0 & 3 & 1 \\ 0 & 0 & 0 & 0 & 3 \end{pmatrix} \qquad C = \begin{pmatrix} \lambda & 0 & 0 & 0 & 0 \\ 0 & \lambda & 0 & 0 & 0 \\ 0 & 0 & \lambda & 0 & 0 \\ 0 & 0 & 0 & \lambda & 0 \\ 0 & 0 & 0 & 0 & \lambda \end{pmatrix}$$

8.54. Let $A = \begin{pmatrix} 1 & 1 & 0 \\ 0 & 2 & 0 \\ 0 & 0 & 1 \end{pmatrix}$ and $B = \begin{pmatrix} 2 & 0 & 0 \\ 0 & 2 & 2 \\ 0 & 0 & 1 \end{pmatrix}$. Show that A and B have different characteristic polynomials (and so are not similar), but have the same minimum polynomial. Thus nonsimilar matrices may have the same minimum polynomial.

8.55. Consider a square block matrix $M = \begin{pmatrix} A & B \\ C & D \end{pmatrix}$. Show that $tI - M = \begin{pmatrix} tI - A & -B \\ -C & tI - D \end{pmatrix}$ is the characteristic matrix of M.

8.56. Let A be an n-square matrix for which $A^k = 0$ for some $k > n$. Show that $A^n = 0$.

8.57. Show that a matrix A and its transpose A^T have the same minimum polynomial.

8.58. Suppose $f(t)$ is an irreducible monic polynomial for which $f(A) = 0$ for a matrix A. Show that $f(t)$ is the minimum polynomial of A.

8.59. Show that A is a scalar matrix kI if and only if the minimum polynomial of A is $m(t) = t - k$.

8.60. Find a matrix A whose minimum polynomial is (a) $t^3 - 5t^2 + 6t + 8$, (b) $t^4 - 5t^3 - 2t + 7t + 4$.

8.61. Consider the following n-square matrices (where $a \neq 0$):

$$N = \begin{pmatrix} 0 & 1 & 0 & \dots & 0 & 0 \\ 0 & 0 & 1 & \dots & 0 & 0 \\ \dots & \dots & \dots & \dots & \dots \\ 0 & 0 & 0 & \dots & 0 & 1 \\ 0 & 0 & 0 & \dots & 0 & 0 \end{pmatrix} \qquad M = \begin{pmatrix} \lambda & a & 0 & \dots & 0 & 0 \\ 0 & \lambda & a & \dots & 0 & 0 \\ \dots & \dots & \dots & \dots & \dots \\ 0 & 0 & 0 & \dots & \lambda & a \\ 0 & 0 & 0 & \dots & 0 & \lambda \end{pmatrix}$$

Here N has 1s on the first diagonal above the main diagonal and 0s elsewhere, and M has λ's on the main diagonal, a's on the first diagonal above the main diagonal and 0s elsewhere.

(a) Show that, for $k < n$, N^k has 1s on the kth diagonal above the main diagonal and 0s elsewhere, and show that $N^n = 0$.

(b) Show that the characteristic polynomial and minimal polynomial of N is $f(t) = t^n$.

(c) Show that the characteristic and minimum polynomial of M is $g(t) = (t - \lambda)^n$. (Hint: Note that $M = \lambda I + aN$.)

DIAGONALIZATION

8.62. Let $A = \begin{pmatrix} a & b \\ c & d \end{pmatrix}$ be a matrix over the real field $\mathbf{R}$. Find necessary and sufficient conditions on a, b, c, and d so that A is diagonalizable, i.e., has two linearly independent eigenvectors.

8.63. Repeat Problem 8.62 for the case that A is a matrix over the complex field $\mathbf{C}$.

8.64. Show that any matrix A is diagonalizable if and only if its minimum polynomial is a product of distinct linear factors.

8.65. Suppose E is a matrix such that $E^2 = E$.

(a) Find the minimum polynomial $m(t)$ of E.

(b) Show that E is diagonalizable and, moreover, E is similar to the diagonal matrix $A = \begin{pmatrix} I_r & 0 \\ 0 & 0 \end{pmatrix}$ where r is the rank of E.

DIAGONALIZATION OF REAL SYMMETRIC MATRICES AND QUADRATIC FORMS

8.66. For each of the following symmetric matrices A, find an orthogonal matrix P for which $P^{-1}AP$ is diagonal:

$$(a) \quad A = \begin{pmatrix} 1 & 2 \\ 2 & -2 \end{pmatrix}, \qquad (b) \quad A = \begin{pmatrix} 5 & 4 \\ 4 & -1 \end{pmatrix}, \qquad (c) \quad A = \begin{pmatrix} 7 & 3 \\ 3 & -1 \end{pmatrix}$$

8.67. Find an orthogonal transformation of coordinates which diagonalizes each quadratic form:

$$(a) \quad q(x, y) = 2x^2 - 6xy + 10y^2, \qquad (b) \quad q(x, y) = x^2 + 8xy - 5y^2$$

8.68. Find an orthogonal transformation of coordinates which diagonalizes the following quadratic form $q(x, y, z) = 2xy + 2xz + 2yz$.

8.69. Let A be a 2×2 real symmetric matrix with eigenvalues 2 and 3, and let $u = (1, 2)$ be an eigenvector belonging to 2. Find an eigenvector v belonging to 3 and find A.

Answers to Supplementary Problems

8.40. $f(A) = \begin{pmatrix} -26 & -3 \\ 5 & -27 \end{pmatrix}$, $\quad g(A) = \begin{pmatrix} -40 & 39 \\ -65 & -27 \end{pmatrix}$, $\quad f(B) = \begin{pmatrix} 3 & 6 \\ 0 & 9 \end{pmatrix}$, $\quad g(B) = \begin{pmatrix} 3 & 12 \\ 0 & 15 \end{pmatrix}$

8.41. $A^2 = \begin{pmatrix} 1 & 2 \\ 0 & 1 \end{pmatrix}$, $\quad A^3 = \begin{pmatrix} 1 & 3 \\ 0 & 1 \end{pmatrix}$, $\quad A^n = \begin{pmatrix} 1 & n \\ 0 & 1 \end{pmatrix}$

8.42. *Hint:* Let $A = \begin{pmatrix} 2 & a & b \\ 0 & 2 & c \\ 0 & 0 & 2 \end{pmatrix}$. Set $B = A^3$ and then obtain conditions on a, b, and c.

8.45. (a) $\lambda_1 = 1, u = (3, -2); \lambda_2 = 2, v = (2, -1)$ (b) $P = \begin{pmatrix} -1 & -2 \\ 2 & 3 \end{pmatrix}$

(c) $A^{10} = \begin{pmatrix} 4093 & 6138 \\ -2046 & -3066 \end{pmatrix}, f(A) = \begin{pmatrix} 2 & -6 \\ 2 & 9 \end{pmatrix}$ (d) $B = \begin{pmatrix} -3 + 4\sqrt{2} & -6 + 6\sqrt{2} \\ 2 - 2\sqrt{2} & 4 - 3\sqrt{2} \end{pmatrix}$

8.46. (a) $\lambda_1 = 2, u = (1, -1, 0), v = (1, 0, -1); \lambda_2 = 6, w = (1, 2, 1)$

(b) $\lambda_1 = 3, u = (1, 1, 0), v = (1, 0, 1); \lambda_2 = 1, w = (2, -1, 1)$

(c) $\lambda = 1, u = (1, 0, 0), v = (0, 0, 1)$

Let $P_1 = \begin{pmatrix} 1 & 1 & 1 \\ -1 & 0 & 2 \\ 0 & -1 & 1 \end{pmatrix}$ and $P_2 = \begin{pmatrix} 1 & 1 & 2 \\ 1 & 0 & -1 \\ 0 & 1 & 1 \end{pmatrix}$. P_3 does not exist since C has at most two linearly

independent eigenvectors, and so cannot be diagonalized.

8.47. (a) For A, $\lambda = 3$, $u = (1, -1)$; B has no eigenvalues (in **R**);

(b) For A, $\lambda = 3$, $u = (1, -1)$; for B, $\lambda_1 = 2i$, $u = (1, 3 - 2i)$; $\lambda_2 = -2i$, $v = (1, 3 + 2i)$.

8.52. Let $A = \begin{pmatrix} 1 & 1 \\ 0 & 1 \end{pmatrix}$. Then $\lambda = 1$ is the only eigenvalue and $v = (1, 0)$ spans the eigenspace of $\lambda = 1$. On the

other hand, for $A^t = \begin{pmatrix} 1 & 0 \\ 1 & 1 \end{pmatrix}$, $\lambda = 1$ is still the only eigenvalue, but $w = (0, 1)$ spans the eigenspace of $\lambda = 1$.

8.53. (a) $\Delta(t) = (t - 2)^3 (t - 7)^2; m(t) = (t - 2)^2 (t - 7)$

(b) $\Delta(t) = (t - 3)^5; m(t) = (t - 3)^3$

(c) $\Delta(t) = (t - \lambda)^5; m(t) = t - \lambda$

8.60. (a) $A = \begin{pmatrix} 0 & 0 & -8 \\ 1 & 0 & -6 \\ 0 & 1 & 5 \end{pmatrix}$, (b) $A = \begin{pmatrix} 0 & 0 & 0 & -4 \\ 1 & 0 & 0 & -7 \\ 0 & 1 & 0 & 2 \\ 0 & 0 & 1 & 5 \end{pmatrix}$

8.65. (a) If $E = I$, $m(t) = (t - 1)$; if $E = 0$, $m(t) = t$; otherwise $m(t) = t(t - 1)$.

(b) *Hint:* Use (a)

8.66. (a) $P = \begin{pmatrix} 2/\sqrt{5} & -1/\sqrt{5} \\ -1/\sqrt{5} & 2/\sqrt{5} \end{pmatrix}$, (b) $P = \begin{pmatrix} 2/\sqrt{5} & -1/\sqrt{5} \\ -1/\sqrt{5} & 2/\sqrt{5} \end{pmatrix}$, (c) $P = \begin{pmatrix} 3/\sqrt{10} & -1/\sqrt{10} \\ -1/\sqrt{10} & 3/\sqrt{10} \end{pmatrix}$

8.67. (a) $x = (3x' - y')/\sqrt{10}, y = (x' + 3y')/\sqrt{10}$, (b) $x = (2x' - y')/\sqrt{5}, y = (x' + 2y')/\sqrt{5}$

8.68. $x = x'/\sqrt{3} + y'/\sqrt{2} + z'/\sqrt{6}, y = x'/\sqrt{3} - y'/\sqrt{2} + z'/\sqrt{6}, z = x'/\sqrt{3} - 2z'/\sqrt{6}$

8.69. $v = (2, -1), A = \begin{pmatrix} \frac{14}{5} & -\frac{2}{5} \\ -\frac{2}{5} & \frac{11}{5} \end{pmatrix}$

Chapter 9

Linear Mappings

9.1 INTRODUCTION

The main subject matter of linear algebra is finite-dimensional vector spaces and linear mappings between such spaces. Vector spaces were introduced in Chapter 5. This chapter introduces us to the linear mappings. First, however, we begin with a discussion of mappings in general.

9.2 MAPPINGS

Let A and B be arbitrary nonempty sets. Suppose to each element of A there is assigned a unique element of B; the collection of such assignments is called a mapping (or map) from A into B. The set A is called the *domain* of the mapping and B is called the *codomain*. A mapping f from A into B is denoted by

$$f : A \to B$$

We write $f(a)$, read "f of a," for the element of B that f assigns to $a \in A$; it is called the *value* of f at a or the *image* of a under f.

> **Remark:** The term function is used synonymously with the word mapping, although some texts reserve the word function for a real-valued or complex-valued mapping, i.e., which maps a set into $\mathbf{R}$ or $\mathbf{C}$.

Consider a mapping $f : A \to B$. If A' is any subset of A, then $f(A')$ denotes the set of images of elements of A'; and if B' is any subset of B, then $f^{-1}(B')$ denotes the set of elements of A each of whose image lies in B':

$$f(A') = \{f(a) : a \in A'\} \qquad \text{and} \qquad f^{-1}(B') = \{a \in A : f(a) \in B'\}$$

We call $f(A')$ the image of A' and $f^{-1}(B')$ the inverse image or preimage of B'. In particular, the set of all images, i.e., $f(A)$, is called the image (or range) of f.

To each mapping $f : A \to B$ there corresponds the subset of $A \times B$ given by $\{(a, f(a)) : a \in A\}$. We call this set the graph of f. Two mappings $f : A \to B$ and $g : A \to B$ are defined to be equal, written $f = g$, if $f(a) = g(a)$ for every $a \in A$, that is, if they have the same graph. Thus we do not distinguish between a function and its graph. The negation of $f = g$ is written $f \neq g$ and is the statement:

There exists an $a \in A$ for which $f(a) \neq g(a)$.

Sometimes the "barred" arrow $\mapsto$ is used to denote the image of an arbitrary element $x \in A$ under a mapping $f : A \to B$ by writing

$$x \mapsto f(x)$$

This is illustrated in the following example.

Example 9.1

(a) Let $f : \mathbf{R} \to \mathbf{R}$ be the mapping which assigns to each real number x its square x^2:

$$x \mapsto x^2 \qquad \text{or} \qquad f(x) = x^2$$

Here the image of -3 is 9 so we may write $f(-3) = 9$.

(b) Consider the 2×3 matrix $A = \begin{pmatrix} 1 & -3 & 5 \\ 2 & 4 & -1 \end{pmatrix}$. If we write the vectors in $\mathbf{R}^3$ and $\mathbf{R}^2$ as column vectors, then A determines the mapping $F : \mathbf{R}^3 \to \mathbf{R}^2$ defined by

$$v \mapsto Av \qquad \text{that is} \qquad F(v) = Av \qquad v \in \mathbf{R}^3$$

Thus if $v = \begin{pmatrix} 3 \\ 1 \\ -2 \end{pmatrix}$, then $F(v) = Av = \begin{pmatrix} 1 & -3 & 5 \\ 2 & 4 & -1 \end{pmatrix} \begin{pmatrix} 3 \\ 1 \\ -2 \end{pmatrix} = \begin{pmatrix} -10 \\ 12 \end{pmatrix}$.

(c) Let V be the vector space of polynomials in the variable t over the real field $\mathbf{R}$. Then the derivative defines a mapping $\mathbf{D} : V \to V$ where, for any polynomial $f \in V$, we let $\mathbf{D}(f) = df/dt$. For example, $\mathbf{D}(3t^2 - 5t + 2) = 6t - 5$.

(d) Let V be the vector space of polynomials in t over $\mathbf{R}$ [as in (c)]. Then the integral from, say 0 to 1 defines a mapping $\mathbf{J} : V \to \mathbf{R}$ where, for any polynomial $f \in V$, we let $\mathbf{J}(f) = \int_0^1 f(t) \, dt$. For example,

$$\mathbf{J}(3t^2 - 5t + 2) = \int_0^1 (3t^2 - 5t + 2) \, dt = \tfrac{1}{2}$$

Note that this map is from the vector space V into the scalar field $\mathbf{R}$ whereas the map in (c) is from V into itself.

Remark: Every $m \times n$ matrix A over a field K determines the mapping $F : K^n \to K^m$ defined by

$$v \mapsto Av$$

where the vectors in K^n and K^m are written as column vectors. For convenience, we shall usually denote the above mapping by A, the same symbol used for the matrix.

Composition of Mapping

Consider two mappings $f : A \to B$ and $g : B \to C$ illustrated below:

$$\textcircled{A} \xrightarrow{\ f\ } \textcircled{B} \xrightarrow{\ g\ } \textcircled{C}$$

Let $a \in A$; then $f(a) \in B$, the domain of g. Hence we can obtain the image of $f(a)$ under the mapping g, that is, $g(f(a))$. This map

$$a \mapsto g(f(a))$$

from A into C is called the composition or product of f and g, and is denoted by $g \circ f$. In other words, $(g \circ f) : A \to C$ is the mapping defined by

$$(g \circ f)(a) = g(f(a))$$

Our first theorem tells us that composition of mappings satisfies the associative law.

Theorem 9.1: Let $f : A \to B$, $g : B \to C$, and $h : C \to D$. Then $h \circ (g \circ f) = (h \circ g) \circ f$.

We prove this theorem now. If $a \in A$, then

$$(h \circ (g \circ f))(a) = h((g \circ f)(a)) = h(g(f(a)))$$

and

$$((h \circ g) \circ f)(a) = (h \circ g)(f(a)) = h(g(f(a)))$$

Thus $(h \circ (g \circ f))(a) = ((h \circ g) \circ f)(a)$ for every $a \in A$, and so $h \circ (g \circ f) = (h \circ g) \circ f$.

Remark: Let $F : A \to B$. Some texts write aF instead of $F(a)$ for the image of $a \in A$ under F. With this notation, the composition of functions $F : A \to B$ and $G : B \to C$ is denoted by $F \circ G$ and not by $G \circ F$ as used in this text.

Injective (One-to-One) and Surjective (Onto) Mappings

We formally introduce some special types of mappings.

Definition: A mapping $f : A \to B$ is said to be *one-to-one* (or *one-one* or 1-1) or *injective* if different elements of A have distinct images; that is,

$$\text{if } a \neq a' \quad \text{implies} \quad f(a) \neq f(a')$$

or, equivalently, $\qquad \text{if } f(a) = f(a') \quad \text{implies} \quad a = a'$

Definition: A mapping $f : A \to B$ is said to be *onto* (or f maps A *onto* B) or *surjective* if every $b \in B$ is the image of at least one $a \in A$.

A mapping $f : A \to B$ which is both one-to-one and onto is said to be a *one-to-one correspondence* between A and B or *bijective*.

Example 9.2

(a) Let $f : \mathbf{R} \to \mathbf{R}$, $g : \mathbf{R} \to \mathbf{R}$, and $h : \mathbf{R} \to \mathbf{R}$ be defined by $f(x) = 2^x$, $g(x) = x^3 - x$ and $h(x) = x^2$. The graphs of these mappings are shown in Fig. 9.1. The mapping f is one-to-one; geometrically, this means that each horizontal line does not contain more than one point of f. The mapping g is onto; geometrically, this means that each horizontal line contains at least one point of g. The mapping h is neither one-to-one nor onto; for example, 2 and -2 have the same image 4, and -16 is not the image of any element of $\mathbf{R}$.

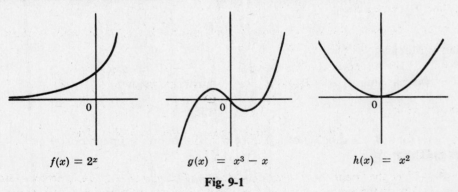

$$f(x) = 2^x \qquad\qquad g(x) = x^3 - x \qquad\qquad h(x) = x^2$$

Fig. 9-1

(b) Let A be any set. The mapping $f : A \to A$ defined by $f(a) = a$, i.e., which assigns to each element in A itself, is called the identity mapping on A and is denoted by 1_A or 1 or I.

(c) Let $f : A \to B$. We call $g : B \to A$ the inverse of f, written f^{-1}, if

$$f \circ g = 1_B \quad \text{and} \quad g \circ f = 1_A$$

We emphasize that f has an inverse if and only if f is both one-to-one and onto (Problem 9.11). Also, if $b \in B$ then $f^{-1}(b) = a$ where a is the unique element of A for which $f(a) = b$.

9.3 LINEAR MAPPINGS

Let V and U be vector spaces over the same field K. A mapping $F : V \to U$ is called a linear mapping (or linear transformation or vector space homomorphism) if it satisfies the following two conditions:

(1)   For any $v, w \in V$, $F(v + w) = F(v) + F(w)$.

(2)   For any $k \in K$ and any $v \in V$, $F(kv) = kF(v)$.

In other words, $F : V \to U$ is linear if it " preserves " the two basic operations of a vector space, that of vector addition and that of scalar multiplication.

Substituting $k = 0$ into (2) we obtain $F(0) = 0$. That is, every linear mapping takes the zero vector into the zero vector.

Now for any scalars $a, b \in K$ and any vectors $v, w \in V$ we obtain, by applying both conditions of linearity,

$$F(av + bw) = F(av) + F(bw) = aF(v) + bF(w)$$

More generally, for any scalars $a_i \in K$ and any vectors $v_i \in V$ we obtain the basic property of linear mappings:

$$F(a_1 v_1 + a_2 v_2 + \cdots + a_n v_n) = a_1 F(v_1) + a_2 F(v_2) + \cdots + a_n F(v_n)$$

> **Remark:** The condition $F(av + bw) = aF(v) + bF(w)$ completely characterizes linear mappings and is sometimes used as its definition.

Example 9.3

(a)   Let A be any $m \times n$ matrix over a field K. As noted previously, A determines a mapping $F : K^n \to K^m$ by the assignment $v \mapsto Av$. (Here the vectors in K^n and K^m are written as columns.) We claim that F is linear. For, by properties of matrices,

$$F(v + w) = A(v + w) = Av + Aw = F(v) + F(w)$$

and
$$F(kv) = A(kv) = kAv = kF(v)$$

where $v, w \in K^n$ and $k \in K$.

(b)   Let $F : \mathbf{R}^3 \to \mathbf{R}^3$ be the "projection" mapping into the xy plane: $F(x, y, z) = (x, y, 0)$. We show that F is linear. Let $v = (a, b, c)$ and $w = (a', b', c')$. Then

$$F(v + w) = F(a + a', b + b', c + c') = (a + a', b + b', 0)$$
$$= (a, b, 0) + (a', b', 0) = F(v) + F(w)$$

and, for any $k \in \mathbf{R}$,

$$F(kv) = F(ka, kb, kc) = (ka, kb, 0) = k(a, b, 0) = kF(v)$$

That is, F is linear.

(c)   Let $F : \mathbf{R}^2 \to \mathbf{R}^2$ be the "translation" mapping defined by $F(x, y) = (x + 1, y + 2)$. Observe that $F(0) = F(0, 0) = (1, 2) \neq 0$. That is, the zero vector is not mapped onto the zero vector. Hence F is not linear.

(d)   Let $F : V \to U$ be the mapping which assigns $0 \in U$ to every $v \in V$. Then, for any $v, w \in V$ and any $k \in K$, we have

$$F(v + w) = 0 = 0 + 0 = F(v) + F(w) \qquad \text{and} \qquad F(kv) = 0 = k0 = kF(v)$$

Thus F is linear. We call F the *zero mapping* and shall usually denote it by 0.

(e)   Consider the identity mapping $I : V \to V$ which maps each $v \in V$ into itself. Then, for any $v, w \in V$ and any $a, b \in K$, we have

$$I(av + bw) = av + bw = aI(v) + bI(w)$$

Thus I is linear.

(f)   Let V be the vector space of polynomials in the variable t over the real field $\mathbf{R}$. Then the differential mapping $\mathbf{D} : V \to V$ and the integral mapping $\mathbf{J} : V \to \mathbf{R}$, defined in Example 9.1(c) and (d), are linear. For it is proven in calculus that for any $u, v \in V$ and $k \in \mathbf{R}$,

$$\frac{d(u + v)}{dt} = \frac{du}{dt} + \frac{dv}{dt} \quad \text{and} \quad \frac{d(ku)}{dt} = k\frac{du}{dt}$$

that is, $\mathbf{D}(u + v) = \mathbf{D}(u) + \mathbf{D}(v)$ and $\mathbf{D}(ku) = k\,\mathbf{D}(u)$; and also,

$$\int_0^1 [u(t) + v(t)] \; dt = \int_0^1 u(t) \; dt + \int_0^1 v(t) \; dt$$

and

$$\int_0^1 ku(t) \; dt = k \int_0^1 u(t) \; dt$$

that is, $\mathbf{J}(u + v) = \mathbf{J}(u) + \mathbf{J}(v)$ and $\mathbf{J}(ku) = k\mathbf{J}(u)$.

(g) Let $F : V \to U$ be a linear mapping which is both one-to-one and onto. Then an inverse mapping $F^{-1} : U \to V$ exists. We will show (Problem 9.15) that this inverse mapping is also linear.

Our next theorem (proved in Problem 9.12) gives us an abundance of examples of linear mappings; in particular, it tells us that a linear mapping is completely determined by its values on the elements of a basis.

Theorem 9.2: Let V and U be vector spaces over a field K. Let $\{v_1, v_2, \ldots, v_n\}$ be a basis of V and let $u_1, u_2, \ldots, u_n$ be any vectors in U. Then there exists a unique linear mapping $F : V \to U$ such that $F(v_1) = u_1, F(v_2) = u_2, \ldots, F(v_n) = u_n$.

We emphasize that the vectors $u_1, \ldots, u_n$ in Theorem 9.2 are completely arbitrary; they may be linearly dependent or they may be equal to each other.

Vector Space Isomorphism

The notion of two vector spaces being isomorphic was defined in Chapter 5 when we investigated the coordinates of a vector relative to a basis. We now redefine this concept.

Definition: Two vector spaces V and U over K are said to be *isomorphic* if there exists a bijective linear mapping $F : V \to U$. The mapping F is then called an *isomorphism* between V and U.

Example 9.4. Let V be a vector space over K of dimension n and let S be a basis of V. Then as noted previously the mapping $v \mapsto [v]_S$, which maps each $v \in V$ into its coordinate vector relative to the basis S, is an isomorphism between V and K^n.

9.4 KERNEL AND IMAGE OF A LINEAR MAPPING

We begin by defining two concepts.

Definition: Let $F : V \to U$ be a linear mapping. The image of F, written Im F, is the set of image points in U:

$$\text{Im } F = \{u \in U : F(v) = u \text{ for some } v \in V\}$$

The kernel of F, written Ker F, is the set of elements in V which map into $0 \in U$:

$$\text{Ker } F = \{v \in V : F(v) = 0\}$$

The following theorem is easily proven (Problem 9.22):

Theorem 9.3:　Let $F : V \to U$ be a linear mapping. Then the image of F is a subspace of U and the kernel of F is a subspace of V.

Now suppose that the vectors $v_1, \ldots, v_n$ span V and that $F : V \to U$ is linear. We show that the vectors $F(v_1), \ldots, F(v_n) \in U$ span Im F. For suppose $u \in$ Im F; then $F(v) = u$ for some vector $v \in V$. Since the v_i span V and since $v \in V$, there exist scalars $a_1, \ldots, a_n$ for which

$$v = a_1 v_1 + a_2 v_2 + \cdots + a_n v_n.$$

Accordingly,

$$u = F(v) = F(a_1 v_1 + a_2 v_2 + \cdots + a_n v_n) = a_1 F(v_1) + a_2 F(v_2) + \cdots + a_n F(v_n)$$

and hence the vectors $F(v_1), \ldots, F(v_n)$ span Im F.

We formally state the above useful result.

Proposition 9.4:　Suppose $v_1, v_2, \ldots, v_n$ span a vector space V and $F : V \to U$ is linear. Then $F(v_1), F(v_2), \ldots, F(v_n)$ span Im F.

Example 9.5

(a)　Let $F : \mathbf{R}^3 \to \mathbf{R}^3$ be the projection mapping into the xy plane. That is,

$$F(x, y, z) = (x, y, 0)$$

(See Fig. 9-2.) Clearly the image of F is the entire xy plane. That is,

$$\text{Im } F = \{(a, b, 0) : a, b \in \mathbf{R}\}$$

Note that the kernel of F is the z axis. That is,

$$\text{Ker } F = \{(0, 0, c) : c \in \mathbf{R}\}$$

since these points and only these points map into the zero vector $0 = (0, 0, 0)$.

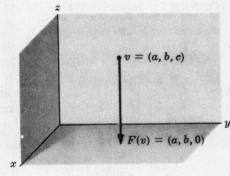

Fig. 9-2

(b)　Let V be the vector space of polynomials over $\mathbf{R}$ and let $\mathbf{T} : V \to V$ be the third derivative operator, that is,

$$\mathbf{T}[f(t)] = d^3f/dt^3$$

[Sometimes the notation $\mathbf{T} = D^3$, where D is the derivative mapping in Example 9.3(f), is used.] Then

$$\text{Ker } \mathbf{T} = \{\text{polynomials of degree} \leq 2\}$$

[since $\mathbf{T}(at^2 + bt + c) = 0$ but $\mathbf{T}(t^n) \neq 0$ for $n > 3$]. On the other hand,

$$\text{Im } \mathbf{T} = V$$

since every polynomial $f(t)$ in V is the third derivative of some polynomial.

(c)   Consider an arbitrary 4×3 matrix A over a field K:

$$A = \begin{pmatrix} a_1 & a_2 & a_3 \\ b_1 & b_2 & b_3 \\ c_1 & c_2 & c_3 \\ d_1 & d_2 & d_3 \end{pmatrix}$$

which we view as a linear mapping $A : K^3 \to K^4$. Now the usual basis $\{e_1, e_2, e_3\}$ of K^3 span K^3 and so their values Ae_1, Ae_2, Ae_3 under A span the image of A. But the vectors Ae_1, Ae_2, and Ae_3 are the columns of A:

$$Ae_1 = \begin{pmatrix} a_1 & a_2 & a_3 \\ b_1 & b_2 & b_3 \\ c_1 & c_2 & c_3 \\ d_1 & d_2 & d_3 \end{pmatrix} \begin{pmatrix} 1 \\ 0 \\ 0 \end{pmatrix} = \begin{pmatrix} a_1 \\ b_1 \\ c_1 \\ d_1 \end{pmatrix}, \qquad Ae_2 = \begin{pmatrix} a_1 & a_2 & a_3 \\ b_1 & b_2 & b_3 \\ c_1 & c_2 & c_3 \\ d_1 & d_2 & d_3 \end{pmatrix} \begin{pmatrix} 0 \\ 1 \\ 0 \end{pmatrix} = \begin{pmatrix} a_2 \\ b_2 \\ c_2 \\ d_2 \end{pmatrix}$$

$$Ae_3 = \begin{pmatrix} a_1 & a_2 & a_3 \\ b_1 & b_2 & b_3 \\ c_1 & c_2 & c_3 \\ d_1 & d_2 & d_3 \end{pmatrix} \begin{pmatrix} 0 \\ 0 \\ 1 \end{pmatrix} = \begin{pmatrix} a_3 \\ b_3 \\ c_3 \\ d_3 \end{pmatrix}$$

thus the image of A is precisely the column space of A.

On the other hand, the kernel of A consists of all vectors v for which $Av = 0$. This means that the kernel of A is the solution space of the homogeneous system $AX = 0$.

> **Remark:**   The above result is true in general. That is, if A is any $m \times n$ matrix viewed as a linear mapping $A : K^n \to K^m$ and $E = \{e_i\}$ is the usual basis of K^n, then $Ae_1, \ldots, Ae_n$ are the columns of A and
>
> $$\text{Ker } A = \text{nullsp } A \qquad \text{and} \qquad \text{Im } A = \text{colsp } A$$
>
> Here colsp A means the column space of A and nullsp A means the *null space of A*, i.e., the solution space of the homogeneous system $AX = 0$.

Rank and Nullity of a Linear Mapping

So far we have not related the notion of dimension to that of a linear mapping $F : V \to U$. In the case that V is of finite dimension, we have the following fundamental relationship.

Theorem 9.5:   Let V be of finite dimension and let $F : V \to U$ be a linear mapping. Then

$$\dim V = \dim (\text{Ker } F) + \dim (\text{Im } F) \tag{9.1}$$

That is, the sum of the dimensions of the image and kernel of a linear mapping is equal to the dimension of its domain.

Equation (9.1) is easily seen to hold for the projection mapping F in Example 9.5(a). There the image (xy plane) and the kernel (z axis) of F have dimensions 2 and 1, respectively, whereas the domain $\mathbf{R}^3$ of F has dimension 3.

> **Remark:**   Let $F : V \to U$ be a linear mapping. Then the rank of F is defined to be the dimension of its image, and the nullity of F is defined to be the dimension of its kernel: that is,
>
> $$\text{rank } F = \dim (\text{Im } F) \qquad \text{and} \qquad \text{nullity } F = \dim (\text{Ker } F)$$

Thus Theorem 9.5 yields the following formula for F when V has finite dimension:

$$\text{rank } F + \text{nullity } F = \dim V$$

Recall that the rank of a matrix A was originally defined to be the dimension of its column space and of its row space. Observe that if we now view A as a linear mapping, then both definitions correspond since the image of A is precisely its column space.

Example 9.6. Let $F : \mathbf{R}^4 \to \mathbf{R}^3$ be the linear mapping defined by

$$F(x, y, s, t) = (x - y + s + t, \; x + 2s - t, \; x + y + 3s - 3t)$$

(a) Find a basis and the dimension of the image of F.

Find the image of the usual basis vectors of $\mathbf{R}^4$:

$$F(1, 0, 0, 0) = (1, 1, 1) \qquad F(0, 0, 1, 0) = (1, 2, 3)$$
$$F(0, 1, 0, 0) = (-1, 0, 1) \qquad F(0, 0, 0, 1) = (1, -1, -3)$$

By Proposition 9.4, the image vectors span Im F; hence form the matrix whose rows are these image vectors and row reduce to echelon form:

$$\begin{pmatrix} 1 & 1 & 1 \\ -1 & 0 & 1 \\ 1 & 2 & 3 \\ 1 & -1 & -3 \end{pmatrix} \sim \begin{pmatrix} 1 & 1 & 1 \\ 0 & 1 & 2 \\ 0 & 1 & 2 \\ 0 & -2 & -4 \end{pmatrix} \sim \begin{pmatrix} 1 & 1 & 1 \\ 0 & 1 & 2 \\ 0 & 0 & 0 \\ 0 & 0 & 0 \end{pmatrix}$$

Thus $(1, 1, 1)$ and $(0, 1, 2)$ form a basis of Im F; hence dim (Im F) = 2 or, in other words, rank $F = 2$.

(b) Find a basis and the dimension of the kernel of the map F.

Set $F(v) = 0$ where $v = (x, y, z, t)$:

$$F(x, y, s, t) = (x - y + s + t, \; x + 2s - t, \; x + y + 3s - 3t) = (0, 0, 0)$$

Set corresponding components equal to each other to form the following homogeneous system whose solution space is Ker F:

$$\begin{array}{ll}
x - y + s + t = 0 & \\
x \quad\;\; + 2s - t = 0 & \text{or} \\
x + y + 3s - 3t = 0 &
\end{array}
\qquad
\begin{array}{l}
x - y + s + t = 0 \\
y + s - 2t = 0 \qquad \text{or} \\
2y + 2s - 4t = 0
\end{array}
\qquad
\begin{array}{l}
x - y + s + t = 0 \\
y + s - 2t = 0
\end{array}$$

The free variables are s and t; hence dim (Ker F) = 2 or nullity $F = 2$. Set:

  (i) $s = -1, t = 0$, to obtain the solution $(2, 1, -1, 0)$,

  (ii) $s = 0, t = 1$, to obtain the solution $(1, 2, 0, 1)$.

Thus $(2, 1, -1, 0)$ and $(1, 2, 0, 1)$ form a basis for Ker F. (Observe that rank F + nullity $F = 2 + 2 = 4$, which is the dimension of the domain $\mathbf{R}^4$ of F.)

Application to Systems of Linear Equations

Consider a system of m linear equations in n unknowns over a field K:

$$a_{11}x_1 + a_{12}x_2 + \cdots + a_{1n}x_n = b_1$$
$$a_{21}x_1 + a_{22}x_2 + \cdots + a_{2n}x_n = b_2$$
$$\dots\dots\dots\dots\dots\dots\dots\dots\dots\dots\dots\dots\dots\dots\dots$$
$$a_{m1}x_1 + a_{m2}x_2 + \cdots + a_{mn}x_n = b_m$$

which is equivalent to the matrix equation

$$Ax = b$$

where $A = (a_{ij})$ is the coefficient matrix, and $x = (x_i)$ and $b = (b_i)$ are the column vectors of the unknowns and of the constants, respectively. Now the matrix A may also be viewed as the linear mapping

$$A : K^n \to K^m$$

Thus the solution of the equation $Ax = b$ may be viewed as the preimage of $b \in K^m$ under the linear mapping $A : K^n \to K^m$. Furthermore, the solution of the associated homogeneous equation $Ax = 0$ may be viewed as the kernel of the linear mapping $A : K^n \to K^m$.

Theorem 9.5 on linear mappings gives us the following relation:

$$\dim (\text{Ker } A) = \dim K^n - \dim (\text{Im } A) = n - \text{rank } A$$

But n is exactly the number of unknowns in the homogeneous system $Ax = 0$. Thus we have the following theorem on linear equations appearing in Chapter 5.

Theorem 9.6: The dimension of the solution space W of the homogeneous system of linear equations $AX = 0$ is $n - r$ where n is the number of unknowns and r is the rank of the coefficient matrix A.

9.5 SINGULAR AND NONSINGULAR LINEAR MAPPINGS, ISOMORPHISMS

A linear mapping $F : V \to U$ is said to be singular if the image of some nonzero vector under F is 0, i.e., if there exists $v \in V$ for which $v \neq 0$ but $F(v) = 0$. Thus $F : V \to U$ is nonsingular if only $0 \in V$ maps into $0 \in U$ or, equivalently, if its kernel consists only of the zero vector: $\text{Ker } F = \{0\}$.

One fundamental property of nonsingular mappings follows (see proof in Problem 9.29).

Theorem 9.7: Suppose a linear mapping $F : V \to U$ is nonsingular. Then the image of any linearly independent set is linearly independent.

Isomorphisms

Suppose a linear mapping $F : V \to U$ is one-to-one. Then only $0 \in V$ can map into $0 \in U$ and so F is nonsingular. The converse is also true. For suppose F is nonsingular and $F(v) = F(w)$; then $F(v - w) = F(w) = 0$ and hence $v - w = 0$ or $v = w$. Thus $F(v) = F(w)$ implies $v = w$, that is, F is one-to-one. Thus we have proven

Proposition 9.8: A linear mapping $F : V \to U$ is one-to-one if and only if it is nonsingular.

Recall that a mapping $F : V \to U$ is called an isomorphism if F is linear and if F is bijective, i.e., if F is one-to-one and onto. Also, recall that a vector space V is said to be isomorphic to a vector space U, written $V \simeq U$, if there is an isomorphism $F : V \to U$.

The following theorem, proved in Problem 9.30, applies.

Theorem 9.9: Suppose V has finite dimension and $\dim V = \dim U$. Suppose $F : V \to U$ is linear. Then F is an isomorphism if and only if F is nonsingular.

9.6 OPERATIONS WITH LINEAR MAPPINGS

We are able to combine linear mappings in various ways to obtain new linear mappings. These operations are very important and shall be used throughout the text.

Suppose $F : V \to U$ and $G : V \to U$ are linear mappings of vector spaces over a field K. We define the sum $F + G$ to be the mapping from V into U which assigns $F(v) + G(v)$ to $v \in V$:

$$(F + G)(v) = F(v) + G(v)$$

Furthermore, for any scalar $k \in K$, we define the product kF to be the mapping from V into U which assigns $kF(v)$ to $v \in V$:

$$(kF)(v) = kF(v)$$

We show that if F and G are linear, then $F + G$ and kF are also linear. We have, for any vectors v, $w \in V$ and any scalars $a, b \in K$,

$$\begin{aligned}(F + G)(av + bw) &= F(av + bw) + G(av + bw) \\ &= aF(v) + bF(w) + aG(v) + bG(w) \\ &= a(F(v) + G(v)) + b(F(w) + G(w)) \\ &= a(F + G)(v) + b(F + G)(w)\end{aligned}$$

and

$$\begin{aligned}(kF)(av + bw) &= kF(av + bw) = k(aF(v) + bF(w)) \\ &= akF(v) + bkF(w) = a(kF)(v) + b(kF)(w)\end{aligned}$$

Thus $F + G$ and kF are linear.

The following theorem applies.

Theorem 9.10: Let V and U be vector spaces over a field K. Then the collection of all linear mappings from V into U with the above operations of addition and scalar multiplication form a vector space over K.

The space in Theorem 9.10 is usually denoted by

$$\text{Hom } (V, U)$$

Here Hom comes from the word homomorphism. In the case that V and U are of finite dimension, we have the following theorem, proved in Problem 9.36.

Theorem 9.11: Suppose dim $V = m$ and dim $U = n$. Then dim Hom $(V, U) = mn$.

Composition of Linear Mappings

Now suppose that V, U, and W are vector spaces over the same field K, and that $F : V \to U$ and $G : U \to W$ are linear mappings:

$$V \xrightarrow{\;F\;} U \xrightarrow{\;G\;} W$$

Recall that the composition function $G \circ F$ is the mapping from V into W defined by $(G \circ F)(v) = G(F(v))$. We show that $G \circ F$ is linear whenever F and G are linear. We have, for any vectors $v, w \in V$ and any scalars $a, b \in K$,

$$\begin{aligned}(G \circ F)(av + bw) &= G(F(av + bw)) = G(aF(v) + bF(w)) \\ &= aG(F(v)) + bG(F(w)) = a(G \circ F)(v) + b(G \circ F)(w)\end{aligned}$$

That is, $G \circ F$ is linear.

The composition of linear mappings and that of addition and scalar multiplication are related as follows (see proof in Problem 9.37):

Theorem 9.12: Let V, U, and W be vector spaces over K. Let F, F' be linear mappings from V into U and G, G' linear mappings from U into W, and let $k \in K$. Then

(i) $G \circ (F + F') = G \circ F + G \circ F'$

(ii) $(G + G') \circ F = G \circ F + G' \circ F$

(iii) $k(G \circ F) = (kG) \circ F = G \circ (kF)$

9.7   ALGEBRA $A(V)$ OF LINEAR OPERATORS

Let V be a vector space over a field K. We now consider the special case of linear mappings $T : V \to V$, i.e., from V into itself. They are also called linear operators or linear transformations on V. We will write $A(V)$, instead of Hom (V, V), for the space of all such mappings.

By Theorem 9.10, $A(V)$ is a vector space over K; it is of dimension n^2 if V is of dimension n. Now if $T, S \in A(V)$, then the composition $S \circ T$ exists and is also a linear mapping from V into itself, i.e., $S \circ T \in A(V)$. Thus we have a "multiplication" defined in $A(V)$. [We shall write ST for $S \circ T$ in the space $A(V)$.]

> **Remark:** An algebra A over a field K is a vector space over K in which an operation of multiplication is defined satisfying, for every $F, G, H \in A$ and every $k \in K$,
>
> (i) $F(G + H) = FG + FH$
>
> (ii) $(G + H)F = GF + HF$
>
> (iii) $k(GF) = (kG)F = G(kF)$
>
> If the associative law also holds for the multiplication, i.e., if for every $F, G, H \in A$,
>
> (iv) $(FG)H = F(GH)$
>
> then the algebra A is said to be associative.

The above definition of an algebra and Theorems 9.10, 9.11, and 9.12 give us the following basic result.

Theorem 9.13:   Let V be a vector space over K. Then $A(V)$ is an associative algebra over K with respect to composition of mappings. If dim $V = n$, then dim $A(V) = n^2$.

In view of the above theorem, $A(V)$ is frequently called the algebra of linear operators on V.

Polynomials and Linear Operators

Observe that the identity mapping $I : V \to V$ belongs to $A(V)$. Also, for any $T \in A(V)$, we have $TI = IT = T$. We note that we can also form "powers" of T; we use the notation $T^2 = T \circ T$, $T^3 = T \circ T \circ T, \ldots$. Furthermore, for any polynomial

$$p(x) = a_0 + a_1 x + a_2 x^2 + \cdots + a_n x^n \qquad a_i \in K$$

we can form the operator $p(T)$ defined by

$$p(T) = a_0 I + a_1 T + a_2 T^2 + \cdots + a_n T^n$$

(For a scalar $k \in K$, the operator kI is frequently denoted by simply k.) In particular, if $p(T) = 0$, the zero mapping, then T is said to be a zero of the polynomial $p(x)$.

Example 9.7.   Let $T : \mathbf{R}^3 \to \mathbf{R}^3$ be defined by $T(x, y, z) = (0, x, y)$. Now if (a, b, c) is any element of $\mathbf{R}^3$, then

$$(T + I)(a, b, c) = (0, a, b) + (a, b, c) = (a, a + b, b + c)$$

and $T^3(a, b, c) = T^2(0, a, b) = T(0, 0, a) = (0, 0, 0)$

Thus we see that $T^3 = 0$, the zero mapping from V into itself. In other words, T is a zero of the polynomial $p(x) = x^3$.

9.8 INVERTIBLE OPERATORS

A linear operator $T : V \to V$ is said to be invertible if it has an inverse, i.e., if there exists $T^{-1} \in A(V)$ such that $TT^{-1} = T^{-1}T = I$.

Now T is invertible if and only if it is one-to-one and onto. Thus in particular, if T is invertible, then only $0 \in V$ can map into itself, i.e., T is nonsingular. The converse is not true in general as seen by the following example.

Example 9.8. Let V be the vector space of polynomials over K, and let T be the operator on V defined by

$$T(a_0 + a_1 t + \cdots + a_n t^n) = a_0 t + a_1 t^2 + \cdots + a_n t^{n+1}$$

i.e., T increases the exponent of t in each term by 1. Now T is a linear mapping and is nonsingular. However, T is not onto and so is not invertible.

The above situation changes significantly when V has finite dimension. Specifically, the following theorem applies.

Theorem 9.14: Suppose T is a linear operator on a finite-dimensional vector space V. Then the following four conditions are equivalent:

  (i) T is nonsingular, i.e., Ker $T = \{0\}$.

  (ii) T is injective, i.e., one-to-one.

  (iii) T is surjective, i.e., onto.

  (iv) T is invertible, i.e., one-to-one and onto.

Proposition 9.8 tells us that (i) and (ii) are equivalent. Thus, to prove the theorem, we need only show that (i) and (iii) are equivalent. [It will then follow that (iv) is equivalent to the others.] By Theorem 9.6,

$$\dim V = \dim (\mathrm{Im}\ T) + \dim (\mathrm{Ker}\ T)$$

If T is nonsingular, then dim (Ker T) = 0 and so dim V = dim (Im T). This means that $V = \mathrm{Im}\ T$ or T is surjective. Thus (i) implies (iii). Conversely, suppose T is surjective. Then $V = \mathrm{Im}\ T$ and so dim V = dim (Im T). This means that dim (Ker T) = 0 and hence T is nonsingular. Thus (iii) implies (i). Accordingly, the theorem is proved. (The proof of Theorem 9.9 is identical to this proof.)

Example 9.9. Let T be the operator on $\mathbf{R}^2$ defined by $T(x, y) = (y, 2x - y)$. The kernel of T is $\{(0, 0)\}$; hence T is nonsingular and, by Theorem 9.14, invertible. We now find a formula for T^{-1}. Suppose (s, t) is the image of (x, y) under T; hence (x, y) is the image of (s, t) under T^{-1}: $T(x, y) = (s, t)$ and $T^{-1}(s, t) = (x, y)$. We have

$$T(x, y) = (y, 2x - y) = (s, t) \qquad \text{and so} \qquad y = s, \ 2x - y = t$$

Solving for x and y in terms of s and t, we obtain $x = \frac{1}{2}s + \frac{1}{2}t$, $y = s$. Thus T^{-1} is given by the formula $T^{-1}(s, t) = (\frac{1}{2}s + \frac{1}{2}t, s)$.

Applications to Systems of Linear Equations

Consider a system of linear equations over K and suppose the system has the same number of equations as unknowns, say n. We can represent this system by the matrix equation

$$Ax = b \tag{*}$$

where A is an n-square matrix over K which we view as a linear operator on K^n. Suppose the matrix A is nonsingular, i.e., the matrix equation $Ax = 0$ has only the zero solution. Then the linear mapping A is one-to-one and onto. This means that the system (*) has a unique solution for any $b \in K^n$. On the other hand, suppose the matrix A is singular, i.e., the matrix equation $Ax = 0$ has a nonzero solution. Then the linear mapping A is not onto. This means that there exist $b \in K^n$ for which (*) does not have a solution. Furthermore, if a solution exists it is not unique. Thus we have proven the following fundamental result:

Theorem 9.15: Consider the following system of linear equations with the same number of equations as unknowns:

$$a_{11}x_1 + a_{12}x_2 + \cdots + a_{1n}x_n = b_1$$
$$a_{21}x_1 + a_{22}x_2 + \cdots + a_{2n}x_n = b_2$$
$$\dotfill$$
$$a_{n1}x_1 + a_{n2}x_2 + \cdots + a_{nn}x_n = b_b$$

(a) If the corresponding homogeneous system has only the zero solution, then the above system has a unique solution for any values of the b_i.

(b) If the corresponding homogeneous system has a nonzero solution, then: (i) there are values for the b_i for which the above system does not have a solution; (ii) whenever a solution of the above system exists, it is not unique.

Solved Problems

MAPPINGS

9.1. State whether or not each diagram in Fig. 9-3 defines a mapping from $A = \{a, b, c\}$ into $B = \{x, y, z\}$.

(a) No. There is nothing assigned to the element $b \in A$.

(b) No. Two elements, x and z, are assigned to $c \in A$.

(c) Yes.

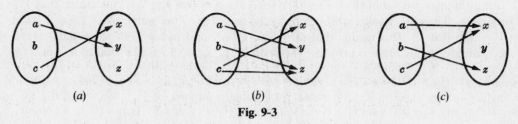

$\qquad$ (a) $\qquad\qquad\qquad\qquad$ (b) $\qquad\qquad\qquad\qquad$ (c)

Fig. 9-3

9.2. Let the mappings $f: A \to B$ and $g: B \to C$ be defined by the diagram in Fig. 9-4. Find: (a) the composition mapping $(g \circ f): A \to C$, and (b) the image of the mappings f, g, and $g \circ f$.

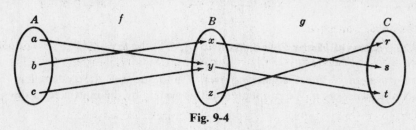

Fig. 9-4

(a) We use the definition of the composition mapping to compute:

$$(g \circ f)(a) = g(f(a)) = g(y) = t$$
$$(g \circ f)(b) = g(f(b)) = g(x) = s$$
$$(g \circ f)(c) = g(f(c)) = g(y) = t$$

Observe that we arrive at the same answer if we "follow the arrows" in Fig. 9-4:

$$a \to y \to t \qquad b \to x \to s \qquad c \to y \to t$$

(b) By Fig. 9-4, the image values under the mapping f are x and y, and the image values under g are r, s, and t; hence

$$\text{Im } f = \{x, y\} \qquad \text{and} \qquad \text{Im } g = \{r, s, t\}$$

Also, by (a), the image values under the composition mapping $g \circ f$ are t and s; accordingly, Im $(g \circ f) = \{s, t\}$. Note that the images of g and $g \circ f$ are different.

9.3. Consider the mapping $F : \mathbf{R}^3 \to \mathbf{R}^2$ defined by $F(x, y, z) = (yz, x^2)$. Find:

(a) $F(2, 3, 4)$,   (b) $F(5, -2, 7)$,   (c) $F^{-1}(0, 0)$, i.e., all vectors $v \in \mathbf{R}^3$ such that $F(v) = 0$.

(a) Substitute in the formula for F to get $F(2, 3, 4) = (3 \cdot 4, 2^2) = (12, 4)$.

(b) $F(5, -2, 7) = (-2 \cdot 7, 5^2) = (-14, 25)$.

(c) Set $F(v) = 0$ where $v = (x, y, z)$, and then solve for x, y, z:

$$F(x, y, z) = (yz, x^2) = (0, 0) \qquad \text{or} \qquad yz = 0 \text{ and } x^2 = 0$$

Thus $x = 0$ and either $y = 0$ or $z = 0$. In other words, $x = 0$, $y = 0$ or $x = 0$, $z = 0$. Accordingly, v lies on the z axis or the y axis.

9.4. Consider the mapping $G : \mathbf{R}^3 \to \mathbf{R}^2$ defined by $G(x, y, z) = (x + 2y - 4z,\ 2x + 3y + z)$. Find $G^{-1}(3, 4)$.

Set $G(x, y, z) = (3, 4)$ to get the system

$$\begin{array}{ccc} x + 2y - 4z = 3 & & x + 2y - 4z = 3 & & x + 2y - 4z = 3 \\ 2x + 3y + z = 4 & \text{or} & -y + 9z = -2 & \text{or} & y - 9z = 2 \end{array}$$

Here z is a free variable. Set $z = a$ to obtain the general solution

$$x = -14a - 1 \qquad y = 9a + 2 \qquad z = a$$

In other words, $G^{-1}(3, 4) = \{(-14a - 1, 9a + 2, a)\}$.

9.5. Consider the mapping $F : \mathbf{R}^2 \to \mathbf{R}^2$ defined by $F(x, y) = (3y, 2x)$. Let S be the unit circle in $\mathbf{R}^2$, that is, the solution set of $x^2 + y^2 = 1$. (a) Describe $F(S)$. (b) Find $F^{-1}(S)$.

(a) Let (a, b) be an element of $F(S)$. Then there exists $(x, y) \in S$ such that $F(x, y) = (a, b)$. Hence:

$$(3y, 2x) = (a, b) \qquad \text{or} \qquad 3y = a, 2x = b \qquad \text{or} \qquad y = a/3, x = b/2$$

Since $(x, y) \in S$, that is, $x^2 + y^2 = 1$, we have

$$(b/2)^2 + (a/3)^2 = 1 \qquad \text{or} \qquad a^2/9 + b^2/4 = 1$$

Thus $F(S)$ is an elipse.

(b) Let $F(x, y) = (a, b)$ where $(a, b) \in S$. Then $(3y, 2x) = (a, b)$ or $3y = a, 2x = b$. Since $(a, b) \in S$, we have $a^2 + b^2 = 1$. Thus $(3y)^2 + (2x)^2 = 1$. Accordingly, $F^{-1}(S)$ is the ellipse $4x^2 + 9y^2 = 1$.

9.6. Let the mappings f and g be defined by $f(x) = 2x + 1$ and $g(x) = x^2 - 2$. Compute formulas for the mappings (a) $g \circ f$, (b) $f \circ g$, and (c) $g \circ g$ (sometimes denoted by g^2).

(a) Compute the formula for $g \circ f$ as follows:

$$(g \circ f)(x) = g(f(x)) = g(2x + 1) = (2x + 1)^2 - 2 = 4x^2 + 4x - 1$$

Observe that the same answer can be found by writing

$$y = f(x) = 2x + 1 \qquad \text{and} \qquad z = g(y) = y^2 - 2$$

and then eliminating y as follows: $z = y^2 - 2 = (2x + 1)^2 - 2 = 4x^2 + 4x - 1$.

(b) $(f \circ g)(x) = f(g(x)) = f(x^2 - 2) = 2(x^2 - 2) + 1 = 2x^2 - 3$

(c) $(g \circ g)(x) = g(g(x)) = g(x^2 - 2) = (x^2 - 2)^2 - 2 = x^4 - 4x^2 + 2$

9.7. Suppose $f : A \rightarrow B$ and $g : B \rightarrow C$; hence the composition $(g \circ f) : A \rightarrow C$ exists. Prove:

(a) If f and g are one-to-one, then $g \circ f$ is one-to-one.

(b) If f and g are onto mappings, then $g \circ f$ is an onto mapping.

(c) If $g \circ f$ is one-to-one, then f is one-to-one.

(d) If $g \circ f$ is an onto mapping, then g is an onto mapping.

(a) Suppose $(g \circ f)(x) = (g \circ f)(y)$. Then $g(f(x)) = g(f(y))$. Since g is one-to-one, $f(x) = f(y)$. Since f is one-to-one, $x = y$. We have proven that $(g \circ f)(x) = (g \circ f)(y)$ implies $x = y$; hence $g \circ f$ is one-to-one.

(b) Suppose $c \in C$. Since g is onto, there exists $b \in B$ for which $g(b) = c$. Since f is onto, there exists $a \in A$ for which $f(a) = b$. Thus $(g \circ f)(a) = g(f(a)) = g(b) = c$; hence $g \circ f$ is onto.

(c) Suppose f is not one-to-one. Then there exists distinct elements $x, y \in A$ for which $f(x) = f(y)$. Thus $(g \circ f)(x) = g(f(x)) = g(f(y)) = (g \circ f)(y)$; hence $g \circ f$ is not one-to-one. Therefore, if $g \circ f$ is one-to-one, then f must be one-to-one.

(d) If $a \in A$, then $(g \circ f)(a) = g(f(a)) \in g(B)$; hence $(g \circ f)(A) \subseteq g(B)$. Suppose g is not onto. Then $g(B)$ is properly contained in C and so $(g \circ f)(A)$ is properly contained in C; thus $g \circ f$ is not onto. Accordingly if $g \circ f$ is onto, then g must be onto.

9.8. Prove that a mapping $f : A \rightarrow B$ has an inverse if and only if it is one-to-one and onto.

Suppose f has an inverse function $f^{-1} : B \rightarrow A$ and hence $f^{-1} \circ f = 1_A$ and $f \circ f^{-1} = 1_B$. Since 1_A is one-to-one, f is one-to-one by Problem 9.7(c), and since 1_B is onto, f is onto by Problem 9.7(d). That is, f is both one-to-one and onto.

Now suppose f is both one-to-one and onto. Then each $b \in B$ is the image of a unique element in A, say $\hat{b}$. Thus if $f(a) = b$, then $a = \hat{b}$; hence $f(\hat{b}) = b$. Now let g denote the mapping from B to A defined by $b \mapsto \hat{b}$. We have:

(i) $(g \circ f)(a) = g(f(a)) = g(b) = \hat{b} = a$, for every $a \in A$; hence $g \circ f = 1_A$.

(ii) $(f \circ g)(b) = f(g(b)) = f(\hat{b}) = b$, for every $b \in B$; hence $f \circ g = 1_B$.

Accordingly, f has an inverse. Its inverse is the mapping g.

LINEAR MAPPINGS

9.9. Show that the following mapping F is linear: $F : \mathbf{R}^3 \rightarrow \mathbf{R}$ defined by $F(x, y, z) = 2x - 3y + 4z$.

Let $v = (a, b, c)$ and $w = (a', b', c')$; hence

$$v + w = (a + a', b + b', c + c') \qquad \text{and} \qquad kv = (ka, kb, kc) \qquad k \in \mathbf{R}$$

We have $F(v) = 2a - 3b + 4c$ and $F(w) = 2a' - 3b' + 4c'$. Thus

$$F(v + w) = F(a + a', b + b', c + c') = 2(a + a') - 3(b + b') + 4(c + c')$$
$$= (2a - 3b + 4c) + (2a' - 3b' + 4c') = F(v) + F(w)$$

and
$$F(kv) = F(ka, kb, kc) = 2ka - 3kb + 4kc = k(2a - 3b + 4c) = kF(v)$$

Accordingly, F is linear.

9.10. Show that the following mapping F is not linear: $F : \mathbf{R}^2 \to \mathbf{R}^3$ defined by $F(x, y) = (x + 1, 2y, x + y)$.

Since $F(0, 0) = (1, 0, 0) \neq (0, 0, 0)$, F cannot be linear.

9.11. Let V be the vector space of n-square matrices over K. Let M be an arbitrary matrix in V. Let $T : V \to V$ be defined by $T(A) = AM + MA$, where $A \in V$. Show that T is linear.

For any $A, B \in V$ and any $k \in K$, we have

$$T(A + B) = (A + B)M + M(A + B) = AM + BM + MA + MB$$
$$= (AM + MA) + (BM + MB) = T(A) + T(B)$$

and

$$T(kA) = (kA)M + M(kA) = k(AM) + k(MA) = k(AM + MA) = kT(A)$$

Accordingly, T is linear.

9.12. Prove Theorem 9.2.

There are three steps to the proof of the theorem: (1) Define the mapping $F : V \to U$ such that $F(v_i) = u_i$, $i = 1, \ldots, n$. (2) Show that F is linear. (3) Show that F is unique.

Step 1. Let $v \in V$. Since $\{v_1, \ldots, v_n\}$ is a basis of V, there exist unique scalars $a_1, \ldots, a_n \in K$ for which $v = a_1 v_1 + a_2 v_2 + \cdots + a_n v_n$. We define $F : V \to U$ by

$$F(v) = a_1 u_1 + a_2 u_2 + \cdots + a_n u_n$$

(Since the a_i are unique, the mapping F is well-defined.) Now, for $i = 1, \ldots, n$,

$$v_i = 0v_1 + \cdots + 1v_i + \cdots + 0v_n$$

Hence

$$F(v_i) = 0u_1 + \cdots + 1u_i + \cdots + 0u_n = u_i$$

Thus the first step of the proof is complete.

Step 2. Suppose $v = a_1 v_1 + a_2 v_2 + \cdots + a_n v_n$ and $w = b_1 v_1 + b_2 v_2 + \cdots + b_n v_n$. Then

$$v + w = (a_1 + b_1)v_1 + (a_2 + b_2)v_2 + \cdots + (a_n + b_n)v_n$$

and, for any $k \in K$, $kv = ka_1 v_1 + ka_2 v_2 + \cdots + ka_n v_n$. By definition of the mapping F,

$$F(v) = a_1 u_1 + a_2 u_2 + \cdots + a_n u_n \quad \text{and} \quad F(w) = b_1 u_1 + b_2 v_2 + \cdots + b_n v_n$$

Hence

$$F(v + w) = (a_1 + b_1)u_1 + (a_2 + b_2)u_2 + \cdots + (a_n + b_n)u_n$$
$$= (a_1 u_1 + a_2 u_2 + \cdots + a_n u_n) + (b_1 u_1 + b_2 u_2 + \cdots + b_n u_n)$$
$$= F(v) + F(w)$$

and

$$F(kv) = k(a_1 u_1 + a_2 u_2 + \cdots + a_n u_n) = kF(v)$$

Thus F is linear.

Step 3. Suppose $G : V \to U$ is linear and $G(v_i) = u_i$, $i = 1, \ldots, n$. If

$$v = a_1 v_1 + a_2 v_2 + \cdots + a_n v_n$$

then

$$G(v) = G(a_1 v_1 + a_2 v_2 + \cdots + a_n v_n) = a_1 G(v_1) + a_2 G(v_2) + \cdots + a_n G(v_n)$$
$$= a_1 u_1 + a_2 u_2 + \cdots + a_n u_n = F(v)$$

Since $G(v) = F(v)$ for every $v \in V$, $G = F$. Thus F is unique and the theorem is proved.

9.13. Let $F : \mathbf{R}^2 \to \mathbf{R}^2$ be the linear mapping for which

$$F(1, 2) = (2, 3) \quad \text{and} \quad F(0, 1) = (1, 4)$$

[Since $(1, 2)$ and $(0, 1)$ form a basis of $\mathbf{R}^2$, such a linear map F exists and is unique by Theorem 9.2.] Find a formula for F, that is, find $F(a, b)$.

Write (a, b) as a linear combination of $(1, 2)$ and $(0, 1)$ using unknowns x and y:

$$(a, b) = x(1, 2) + y(0, 1) = (x, 2x + y) \qquad \text{so} \qquad a = x, \; b = 2x + y$$

Solve for x and y in terms of a and b to get $x = a$, $y = -2a + b$. Then

$$F(a, b) = xF(1, 2) + yF(0, 1) = a(2, 3) + (-2a + b)(1, 4) = (b, -5a + 4b)$$

9.14. Let $T : V \to U$ be linear, and suppose $v_1, \ldots, v_n \in V$ have the property that their images $T(v_1), \ldots, T(v_n)$ are linearly independent. Show that the vectors $v_1, \ldots, v_n$ are also linearly independent.

Suppose that, for scalars $a_1, \ldots, a_n$, $a_1 v_1 + a_2 v_2 + \cdots + a_n v_n = 0$. Then

$$0 = T(0) = T(a_1 v_1 + a_2 v_2 + \cdots + a_n v_n) = a_1 T(v_1) + a_2 T(v_2) + \cdots + a_n T(v_n)$$

Since the $T(v_i)$ are linearly independent, all the $a_i = 0$. Thus the vectors $v_1, \ldots, v_n$ are linearly independent.

9.15. Suppose the linear mapping $F : V \to U$ is one-to-one and onto. Show that the inverse mapping $F^{-1} : U \to V$ is also linear.

Suppose $u, u' \in U$. Since F is one-to-one and onto, there exist unique vectors $v, v' \in V$ for which $F(v) = u$ and $F(v') = u'$. Since F is linear, we also have

$$F(v + v') = F(v) + F(v') = u + u' \qquad \text{and} \qquad F(kv) = kF(v) = ku$$

By definition of the inverse mapping, $F^{-1}(u) = v$, $F^{-1}(u') = v'$, $F^{-1}(u + u') = v + v'$, and $F^{-1}(ku) = kv$. Then

$$F^{-1}(u + u') = v + v' = F^{-1}(u) + F^{-1}(u') \qquad \text{and} \qquad F^{-1}(ku) = kv = kF^{-1}(u)$$

and thus F^{-1} is linear.

IMAGE AND KERNEL OF LINEAR MAPPINGS

9.16. Let $F : \mathbf{R}^5 \to \mathbf{R}^3$ be the linear mapping defined by

$$F(x, y, z, s, t) = (x + 2y + z - 3s + 4t, \; 2x + 5y + 4z - 5s + 5t, \; x + 4y + 5z - s - 2t)$$

Find a basis and the dimension of the image of F.

Find the image of the usual basis vectors of $\mathbf{R}^5$:

$$F(1, 0, 0, 0, 0) = (1, 2, 1) \qquad F(0, 1, 0, 0, 0) = (2, 5, 4) \qquad F(0, 0, 1, 0, 0) = (1, 4, 5)$$

$$F(0, 0, 0, 1, 0) = (-3, -5, -1) \qquad F(0, 0, 0, 0, 1) = (4, 5, -2)$$

By Proposition 9.4, the image vectors span Im F; hence form the matrix whose rows are these image vectors, and row reduce to echelon form:

$$\begin{pmatrix} 1 & 2 & 1 \\ 2 & 5 & 4 \\ 1 & 4 & 5 \\ -3 & -5 & -1 \\ 4 & 5 & -2 \end{pmatrix} \sim \begin{pmatrix} 1 & 2 & 1 \\ 0 & 1 & 2 \\ 0 & 2 & 4 \\ 0 & 1 & 2 \\ 0 & -3 & -6 \end{pmatrix} \sim \begin{pmatrix} 1 & 2 & 1 \\ 0 & 1 & 2 \\ 0 & 0 & 0 \\ 0 & 0 & 0 \\ 0 & 0 & 0 \end{pmatrix}$$

9.17. Let $G : \mathbf{R}^3 \to \mathbf{R}^3$ be the linear mapping defined by

$$G(x, y, z) = (x + 2y - z, \; y + z, \; x + y - 2z)$$

Find a basis and the dimension of the kernel of T.

Set $G(v) = 0$ where $v = (x, y, z)$:

$$G(x, y, z) = (x + 2y - z, \ y + z, \ x + y - 2z) = (0, 0, 0)$$

Set corresponding components equal to each other to form the homogeneous system whose solution space is the kernel W of G:

$$\begin{matrix} x + 2y - \ z = 0 \\ y + \ z = 0 \\ x + \ y - 2z = 0 \end{matrix} \quad \text{or} \quad \begin{matrix} x + 2y - z = 0 \\ y + z = 0 \\ -y - z = 0 \end{matrix} \quad \text{or} \quad \begin{matrix} x + 2y - z = 0 \\ y + z = 0 \end{matrix}$$

The only free variable is z; hence dim $W = 1$. Let $z = 1$; then $y = -1$ and $x = 3$. Thus $(3, -1, 1)$ forms a basis for Ker G.

9.18. Consider the matrix mapping $A: \mathbf{R}^4 \to \mathbf{R}^3$ where $A = \begin{pmatrix} 1 & 2 & 3 & 1 \\ 1 & 3 & 5 & -2 \\ 3 & 8 & 13 & -3 \end{pmatrix}$. Find a basis and the dimension of (a) the image of A, and (b) the kernel of A.

(a) The column space of A is equal to Im A. Thus reduce A^T to echelon form:

$$A^T = \begin{pmatrix} 1 & 1 & 3 \\ 2 & 3 & 8 \\ 3 & 5 & 13 \\ 1 & -2 & -3 \end{pmatrix} \sim \begin{pmatrix} 1 & 1 & 3 \\ 0 & 1 & 2 \\ 0 & 2 & 4 \\ 0 & -3 & -6 \end{pmatrix} \sim \begin{pmatrix} 1 & 1 & 3 \\ 0 & 1 & 2 \\ 0 & 0 & 0 \\ 0 & 0 & 0 \end{pmatrix}$$

Thus $\{(1, 1, 3), (0, 1, 2)\}$ is a basis of Im A and dim (Im A) = 2.

(b) Here Ker A is the solution space of the homogeneous system $AX = 0$ where $X = (x, y, z, t)^T$. Thus reduce the matrix A of coefficients to echelon form:

$$\begin{pmatrix} 1 & 2 & 3 & 1 \\ 0 & 1 & 2 & -3 \\ 0 & 2 & 4 & -6 \end{pmatrix} \sim \begin{pmatrix} 1 & 2 & 3 & 1 \\ 0 & 1 & 2 & -3 \\ 0 & 0 & 0 & 0 \end{pmatrix} \sim \begin{cases} x + 2y + 3z + \ t = 0 \\ y + 2z - 3t = 0 \end{cases}$$

The free variables are z and t. Thus dim (Ker A) = 2. Set:

  (i) $z = 1, t = 0$, to get the solution $(1, -2, 1, 0)$,

  (ii) $z = 0, t = 1$, to get the solution $(-7, 3, 0, 1)$.

Thus $(1, -2, 1, 0)$ and $(-7, 3, 0, 1)$ form a basis for Ker A.

9.19. Consider the matrix map $B: \mathbf{R}^3 \to \mathbf{R}^3$ where $B = \begin{pmatrix} 1 & 2 & 5 \\ 3 & 5 & 13 \\ -2 & -1 & -4 \end{pmatrix}$. Find the dimension and a basis for (a) the kernel of B, and (b) the image of B.

(a) Reduce B to echelon form to get the homogeneous system corresponding to Ker B:

$$B = \begin{pmatrix} 1 & 2 & 5 \\ 3 & 5 & 13 \\ -2 & -1 & -4 \end{pmatrix} \sim \begin{pmatrix} 1 & 2 & 5 \\ 0 & -1 & -2 \\ 0 & 3 & 6 \end{pmatrix} \sim \begin{pmatrix} 1 & 2 & 5 \\ 0 & 1 & 2 \\ 0 & 0 & 0 \end{pmatrix} \sim \begin{cases} x + 2y + 5z = 0 \\ y + 2z = 0 \end{cases}$$

There is one free variable z so dim (Ker B) = 1. Set $z = 1$ to get the solution $(-1, -2, 1)$ which forms a basis of Ker B.

(b)   Reduce B^T to echelon form:

$$B^T = \begin{pmatrix} 1 & 3 & -2 \\ 2 & 5 & -1 \\ 5 & 13 & -4 \end{pmatrix} \sim \begin{pmatrix} 1 & 3 & -2 \\ 0 & -1 & 3 \\ 0 & -2 & 6 \end{pmatrix} \sim \begin{pmatrix} 1 & 3 & -2 \\ 0 & 1 & -3 \\ 0 & 0 & 0 \end{pmatrix}$$

Thus $(1, 3, -2)$ and $(0, 1, -3)$ form a basis of Im B.

9.20.   Find a linear map $F : \mathbf{R}^3 \to \mathbf{R}^4$ whose image is spanned by $(1, 2, 0, -4)$ and $(2, 0, -1, -3)$.

Method 1.   Consider the usual basis of $\mathbf{R}^3$: $e_1 = (1, 0, 0)$, $e_2 = (0, 1, 0)$, $e_3 = (0, 0, 1)$. Set

$$F(e_1) = (1, 2, 0, -4) \qquad F(e_2) = (2, 0, -1, -3) \qquad \text{and} \qquad F(e_3) = (0, 0, 0, 0)$$

By Theorem 9.2, such a linear map F exists and is unique. Furthermore, the image of F is spanned by the $F(e_i)$; hence F has the required property. We find a general formula for $F(x, y, z)$:

$$\begin{aligned} F(x, y, z) &= F(xe_1 + ye_2 + ze_3) = xF(e_1) + yF(e_1) + zF(e_3) \\ &= x(1, 2, 0, -4) + y(2, 0, -1, -3) + z(0, 0, 0, 0) \\ &= (x + 2y, 2x, -y, -4x - 3y) \end{aligned}$$

Method 2.   Form a 4×3 matrix A whose columns consist only of the given vectors; say,

$$A = \begin{pmatrix} 1 & 2 & 2 \\ 2 & 0 & 0 \\ 0 & -1 & -1 \\ -4 & -3 & -3 \end{pmatrix}$$

Recall that A determines a linear map $A : \mathbf{R}^3 \to \mathbf{R}^4$ whose image is spanned by the columns of A. Thus A satisfies the required condition.

9.21.   Let V be the vector space of 2 by 2 matrices over $\mathbf{R}$ and let $M = \begin{pmatrix} 1 & 2 \\ 0 & 3 \end{pmatrix}$. Let $F : V \to V$ be the linear map defined by $F(A) = AM - MA$. Find a basis and the dimension of the kernel W of F.

We seek the set of $\begin{pmatrix} x & y \\ s & t \end{pmatrix}$ such that $F\begin{pmatrix} x & y \\ s & t \end{pmatrix} = \begin{pmatrix} 0 & 0 \\ 0 & 0 \end{pmatrix}$.

$$\begin{aligned} F\begin{pmatrix} x & y \\ s & t \end{pmatrix} &= \begin{pmatrix} x & y \\ s & t \end{pmatrix}\begin{pmatrix} 1 & 2 \\ 0 & 3 \end{pmatrix} - \begin{pmatrix} 1 & 2 \\ 0 & 3 \end{pmatrix}\begin{pmatrix} x & y \\ s & t \end{pmatrix} \\ &= \begin{pmatrix} x & 2x + 3y \\ s & 2s + 3t \end{pmatrix} - \begin{pmatrix} x + 2s & y + 2t \\ 3s & 3t \end{pmatrix} \\ &= \begin{pmatrix} -2s & 2x + 2y - 2t \\ -2s & 2s \end{pmatrix} = \begin{pmatrix} 0 & 0 \\ 0 & 0 \end{pmatrix} \end{aligned}$$

Thus $\qquad \begin{cases} 2x + 2y - 2t = 0 \\ \qquad\qquad 2s = 0 \end{cases}$ or $\begin{cases} x + y - t = 0 \\ \qquad\quad s = 0 \end{cases}$

The free variables are y and t; hence dim $W = 2$. To obtain a basis of W set

(a) $y = -1, t = 0$ to obtain the solution $x = 1, y = -1, s = 0, t = 0$;

(b) $y = 0, t = 1$ to obtain the solution $x = 1, y = 0, s = 0, t = 1$.

Thus $\left\{ \begin{pmatrix} 1 & -1 \\ 0 & 0 \end{pmatrix}, \begin{pmatrix} 1 & 0 \\ 0 & 1 \end{pmatrix} \right\}$ is a basis of W.

9.22. Prove Theorem 9.3.

(a) Since $F(0) = 0$, we have $0 \in \text{Im } F$. Now suppose $u, u' \in \text{Im } F$ and $a, b \in K$. Since u and u' belong to the image of F, there exist vectors $v, v' \in V$ such that $F(v) = u$ and $F(v') = u'$. Then

$$F(av + bv') = aF(v) + bF(v') = au + bu' \in \text{Im } F$$

Thus the image of F is a subspace of U.

(b) Since $F(0) = 0$, we have $0 \in \text{Ker } F$. Now suppose $v, w \in \text{Ker } F$ and $a, b \in K$. Since v and w belong to the kernel of F, $F(v) = 0$ and $F(w) = 0$. Thus

$$F(av + bw) = aF(v) + bF(w) = a0 + b0 = 0 \qquad \text{and so} \qquad av + bw \in \text{Ker } F$$

Thus the kernel of F is a subspace of V.

9.23. Prove Theorem 9.5.

Suppose $\dim (\text{Ker } F) = r$ and $\{w_1, \ldots, w_r\}$ is a basis of $\text{Ker } F$, and suppose $\dim (\text{Im } F) = s$ and $\{u_1, \ldots, u_2\}$ is a basis of $\text{Im } F$. (By Problem 9.30, $\text{Im } F$ has finite dimension.) Since $u_j \in \text{Im } F$, there exist vectors $v_1, \ldots, v_s$ in V such that $F(v_1) = u_1, \ldots, F(V_s) = u_s$. We claim that the set

$$B = \{w_1, \ldots, w_r, v_1, \ldots, v_s\}$$

is a basis of V, that is, (i) B spans V, and (ii) B is linearly independent. Once we prove (i) and (ii), then $\dim V = r + s = \dim (\text{Ker } F) + \dim (\text{Im } F)$.

(i) B spans V.

Let $v \in V$. Then $F(v) \in \text{Im } F$. Since the u_j span $\text{Im } F$, there exist scalars $a_1, \ldots, a_s$ such that $F(v) = a_1 u_1 + \cdots + a_s u_s$. Set $\hat{v} = a_1 v_1 + \cdots + a_s v_s - v$. Then

$$F(\hat{v}) = F(a_1 v_1 + \cdots + a_s v_s - v) = a_1 F(v_1) + \cdots + a_s F(v_s) - F(v)$$
$$= a_1 u_1 + \cdots + a_s u_s - F(v) = 0$$

Thus $\hat{v} \in \text{Ker } F$. Since the w_i span $\text{Ker } F$, there exists scalars $b_1, \ldots, b_r$ such that

$$\hat{v} = b_1 w_1 + \cdots + b_r w_r = a_1 v_1 + \cdots + a_s v_s - v$$

Accordingly,

$$v = a_1 v_1 + \cdots + a_s v_s - b_1 w_1 - \cdots - b_r w_r$$

Thus B spans V.

(ii) B is linearly independent.

Suppose

$$x_1 w_1 + \cdots + x_r w_r + y_1 v_1 + \cdots + y_s v_s = 0 \tag{1}$$

where $x_i, y_j \in K$. Then

$$0 = F(0) = F(x_1 w_1 + \cdots + x_r w_r + y_1 v_1 + \cdots + y_s v_s)$$
$$= x_1 F(w_1) + \cdots + x_r F(w_r) + y_1 F(v_1) + \cdots + y_s F(v_s) \tag{2}$$

But $F(w_i) = 0$ since $w_i \in \text{Ker } F$, and $F(v_j) = u_j$. Substitution in (2) gives $y_1 u_1 + \cdots + y_s u_s = 0$. Since the u_j are linearly independent, each $y_j = 0$. Substitution in (1) gives $x_1 w_1 + \cdots + x_r w_r = 0$. Since the w_i are linearly independent, each $x_i = 0$. Thus B is linearly independent.

9.24. Suppose $F : V \to U$ and $G : U \to W$ are linear. Prove:

$$\text{(a) rank } (G \circ F) \le \text{rank } G. \text{ (b) rank } (G \circ F) \le \text{rank } F.$$

(a) Since $F(V) \subseteq U$, we also have $G(F(V)) \subseteq G(U)$ and so $\dim G(F(V)) \le \dim G(U)$. Then

$$\text{rank } (G \circ F) = \dim ((G \circ F)(V)) = \dim (G(F(V))) \le \dim G(U) = \text{rank } G$$

(b) We have $\dim (G(F(V))) \le \dim F(V)$. Hence

$$\text{rank } (G \circ F) = \dim ((G \circ F)(V)) = \dim (G(F(V))) \le \dim F(V) = \text{rank } F$$

9.25. Suppose $f: V \rightarrow U$ is linear with kernel W, and that $f(v) = u$. Show that the "coset" $v + W = \{v + w: w \in W\}$ is the preimage of u, that is, $f^{-1}(u) = v + W$.

We must prove that (i) $f^{-1}(u) \subseteq v + W$ and (ii) $v + W \subseteq f^{-1}(u)$. We first prove (i). Suppose $v' \in f^{-1}(u)$. Then $f(v') = u$ and so

$$f(v' - v) = f(v') - f(v) = u - u = 0$$

that is, $v' - v \in W$. Thus $v' = v + (v' - v) \in v + W$ and hence $f^{-1}(u) \subseteq v + W$.

Now we prove (ii). Suppose $v' \in v + W$. Then $v' = v + w$ where $w \in W$. Since W is the kernel of f, $f(w) = 0$. Accordingly,

$$f(v') = f(v + w) + f(v) + f(w) = f(v) + 0 = f(v) = u$$

Thus $v' \in f^{-1}(u)$ and so $v + W \subseteq f^{-1}(u)$.

SINGULAR AND NONSINGULAR LINEAR MAPPINGS, ISOMORPHISMS

9.26. Determine whether or not each linear map is nonsingular. If not, find a nonzero vector v whose image is 0.

(a) $F: \mathbf{R}^2 \rightarrow \mathbf{R}^2$ defined by $F(x, y) = (x - y, x - 2y)$.

(b) $G: \mathbf{R}^2 \rightarrow \mathbf{R}^2$ defined by $G(x, y) = (2x - 4y, 3x - 6y)$.

(a) Find Ker F by setting $F(v) = 0$ where $v = (x, y)$:

$$(x - y, x - 2y) = (0, 0) \quad \text{or} \quad \begin{cases} x - y = 0 \\ x - 2y = 0 \end{cases} \quad \text{or} \quad \begin{cases} x - y = 0 \\ -y = 0 \end{cases}$$

The only solution is $x = 0$, $y = 0$; hence F is nonsingular.

(b) Set $G(x, y) = (0, 0)$ to find Ker G:

$$(2x - 4y, 3x - 6y) = (0, 0) \quad \text{or} \quad \begin{cases} 2x - 4y = 0 \\ 3x - 6y = 0 \end{cases} \quad \text{or} \quad x - 2y = 0$$

The system has nonzero solutions since y is a free variable; hence G is singular. Let $y = 1$ to obtain the solution $v = (2, 1)$ which is a nonzero vector such that $G(v) = 0$.

9.27. Let $H: \mathbf{R}^3 \rightarrow \mathbf{R}^3$ be defined by $H(x, y, z) = (x + y - 2z, x + 2y + z, 2x + 2y - 3z)$. (a) Show that H is nonsingular. (b) Find a formula for H^{-1}.

(a) Set $H(x, y, z) = (0, 0, 0)$; that is, set

$$(x + y - 2z, x + 2y + z, 2x + 2y - 3z) = (0, 0, 0)$$

This yields the homogeneous system

$$\begin{cases} x + y - 2z = 0 \\ x + 2y + z = 0 \\ 2x + 2y - 3z = 0 \end{cases} \quad \text{or} \quad \begin{cases} x + y - 2z = 0 \\ y + 3z = 0 \\ z = 0 \end{cases}$$

The echelon system is in triangular form so the only solution is $x = 0$, $y = 0$, $z = 0$. Thus H is nonsingular.

(b) Set $H(x, y, z) = (a, b, c)$ and then solve for x, y, z in terms of a, b, c:

$$\begin{cases} x + y - 2z = a \\ x + 2y + z = b \\ 2x + 2y - 3z = c \end{cases} \quad \text{or} \quad \begin{cases} x + y - 2z = a \\ y + 3z = b - a \\ z = c - 2a \end{cases}$$

Solving for x, y, z yields $x = -8a - b + 5c$, $y = 5a + b - 3c$, $z = -2a + c$. Thus

$$H^{-1}(a, b, c) = (-8a - b + 5c, 5a + b - 3c, -2a + c)$$

or, replacing a, b, c by x, y, z, respectively,

$$H^{-1}(x, y, z) = (-8x - y + 5z, 5x + y - 3z, -2x + z)$$

9.28. Suppose $F : V \to U$ is linear and that V is of finite dimension. Show that V and the image of F have the same dimension if and only if F is nonsingular. Determine all nonsingular linear mappings $T : \mathbf{R}^4 \to \mathbf{R}^3$.

By Theorem 9.5, dim V = dim (Im F) + dim (Ker F). Hence V and Im F have the same dimension if and only if dim (Ker F) = 0 or Ker $F = \{0\}$, i.e., if and only if F is nonsingular.

Since dim $\mathbf{R}^3$ is less than dim $\mathbf{R}^4$, we have dim (Im T) is less than the dimension of the domain $\mathbf{R}^4$ of T. Accordingly, no linear mapping $T : \mathbf{R}^4 \to \mathbf{R}^3$ can be nonsingular.

9.29. Prove Theorem 9.7.

Suppose $v_1, v_2, \dots, v_n$ are linearly independent vectors in V. We claim that $F(v_1), F(v_2), \dots, F(v_n)$ are also linearly independent. Suppose $a_1 F(v_1) + a_2 F(v_2) + \cdots + a_n F(v_n) = 0$, where $a_i \in K$. Since F is linear, $F(a_1 v_1 + a_2 v_2 + \cdots + a_n v_n) = 0$; hence

$$a_1 v_1 + a_2 v_2 + \cdots + a_n v_n \in \text{Ker } F$$

But F is nonsingular, i.e., Ker $F = \{0\}$; hence $a_1 v_1 + a_2 v_2 + \cdots + a_n v_n = 0$. Since the v_i are linearly independent, all the a_i are 0. Accordingly, the $F(v_i)$ are linearly independent. Thus the theorem is proved.

9.30. Prove Theorem 9.9.

If F is an isomorphism then only 0 maps to 0 so F is nonsingular. Suppose F is nonsingular. Then dim (Ker F) = 0. By Theorem 9.5, dim V = dim (Ker F) + dim (Im F). Thus dim U = dim V = dim (Im F). Since U has finite dimension, Im $F = U$ and so F is surjective. Thus F is both one-to-one and onto, i.e., F is an isomorphism.

OPERATIONS WITH LINEAR MAPPINGS

9.31. Let $F : \mathbf{R}^3 \to \mathbf{R}^2$ and $G : \mathbf{R}^3 \to \mathbf{R}^2$ be defined by $F(x, y, z) = (2x, y + z)$ and $G(x, y, z) = (x - z, y)$, respectively. Find formulas defining the maps (a) $F + G$, (b) $3F$, and (c) $2F - 5G$.

(a) $(F + G)(x, y, z) = F(x, y, z) + (G(x, y, z)$
$= (2x, y + z) + (x - z, y) = (3x - z, 2y + z)$

(b) $(3F)(x, y, z) = 3F(x, y, z) = 3(2x, y + z) = (6x, 3y + 3z)$

(c) $(2F - 5G)(x, y, z) = 2F(x, y, z) - 5G(x, y, z) = 2(2x, y + z) - 5(x - z, y)$
$= (4x, 2y + 2z) + (-5x + 5z, -5y) = (-x + 5z, -3y + 2z)$

9.32. Let $F : \mathbf{R}^3 \to \mathbf{R}^2$ and $G : \mathbf{R}^2 \to \mathbf{R}^2$ be defined by $F(x, y, z) = (2x, y + z)$ and $G(x, y) = (y, x)$, respectively. Derive formulas defining the mappings (a) $G \circ F$, (b) $F \circ G$.

(a) $(G \circ F)(x, y, z) = G(F(x, y, z)) = G(2x, y + z) = (y + z, 2x)$

(b) The mapping $F \circ G$ is not defined since the image of G is not contained in the domain of F.

9.33. Prove: (a) The zero mapping $\mathbf{0}$, defined by $\mathbf{0}(v) = \mathbf{0}$ for every $v \in V$, is the zero element of Hom (V, U). (b) The negative of $F \in$ Hom (V, U) is the mapping $(-1)F$, i.e., $-F = (-1)F$.

(a) Let $F \in$ Hom (V, U). Then, for every $v \in V$,

$$(F + \mathbf{0})(v) = F(v) + \mathbf{0}(v) = F(v) + \mathbf{0} = F(v)$$

Since $(F + \mathbf{0})(v) = F(v)$ for every $v \in V$, $F + \mathbf{0} = F$.

(b)   For every $v \in V$,

$$(F + (-1)F)(v) = F(v) + (-1)F(v) = F(v) - F(v) = \mathbf{0} = \mathbf{0}(v)$$

Since $(F + (-1)F)(v) = \mathbf{0}(v)$ for every $v \in V$, $F + (-1)F = \mathbf{0}$. Thus $(-1)F$ is the negative of F.

9.34.   Suppose $F_1, F_2, \ldots, F_n$ are linear maps from V into U. Show that, for any scalars $a_1, a_2, \ldots, a_n$, and for any $v \in V$,

$$(a_1 F_1 + a_2 F_2 + \cdots + a_n F_n)(v) = a_1 F_1(v) + a_2 F_2(v) + \cdots + a_n F_n(v)$$

By definition of the mapping $a_1 F_1$, $(a_1 F_1)(v) = a_1 F_1(v)$; hence the theorem holds for $n = 1$. Thus by induction,

$$(a_1 F_1 + a_2 F_2 + \cdots + a_n F_n)(v) = (a_1 F_1)(v) + (a_2 F_2 + \cdots + a_n F_n)(v)$$
$$= a_1 F_1(v) + a_2 F_2(v) + \cdots + a_n F_n(v)$$

9.35.   Consider linear mappings $F : \mathbf{R}^3 \to \mathbf{R}^2$, $G : \mathbf{R}^3 \to \mathbf{R}^2$, $H : \mathbf{R}^3 \to \mathbf{R}^2$ defined by

$$F(x, y, z) = (x + y + z, x + y) \qquad G(x, y, z) = (2x + z, x + t) \qquad H(x, y, z) = (2y, x)$$

Show that F, G, H are linearly independent [as elements of Hom $(\mathbf{R}^3, \mathbf{R}^2)$].

Suppose, for scalars $a, b, c \in K$,

$$aF + bG + cH = \mathbf{0} \tag{1}$$

(Here $\mathbf{0}$ is the zero mapping.) For $e_1 = (1, 0, 0) \in \mathbf{R}^3$, we have

$$(aF + bG + cH)(e_2) = aF(0, 1, 0) + bG(0, 1, 0) + cH(0, 1, 0)$$
$$= a(1, 1) + b(0, 1) + c(2, 0) = (a + 2c, a + b) = \mathbf{0}(e_2) = (0, 0)$$

and $\mathbf{0}(e_1) = (0, 0)$. Thus by (1), $(a + 2b, a + b + c) = (0, 0)$ and so

$$a + 2b = 0 \qquad \text{and} \qquad a + b + c = 0 \tag{2}$$

Similarly for $e_2 = (0, 1, 0) \in \mathbf{R}^3$, we have

$$(aF + bG + cH)(e_2) = aF(0, 1, 0) + bG(0, 1, 0) + cH(0, 1, 0)$$
$$= a(1, 1) + b(0, 1) + c(2, 0) = (a + 2c, a + b) = \mathbf{0}(e_2) = (0, 0)$$

Thus $a + 2c = 0 \qquad \text{and} \qquad a + b = 0$ $\tag{3}$

Using (2) and (3) we obtain $a = 0 \qquad b = 0 \qquad c = 0$ $\tag{4}$

Since (1) implies (4), the mappings $F, G,$ and H are linearly independent.

9.36.   Prove Theorem 9.11.

Suppose $\{v_1, \ldots, v_m\}$ is a basis of V and $\{u_1, \ldots, u_n\}$ is a basis of U. By Theorem 9.2, a linear mapping in Hom (V, U) is uniquely determined by arbitrarily assigning elements of U to the basis elements v_i of V. We define

$$F_{ij} \in \text{Hom } (V, U) \qquad i = 1, \ldots, m, j = 1, \ldots, n$$

to be the linear mapping for which $F_{ij}(v_i) = u_j$, and $F_{ij}(v_k) = 0$ for $k \neq i$. That is, F_{ij} maps v_i into u_j and the other v's into 0. Observe that $\{F_{ij}\}$ contains exactly mn elements; hence the theorem is proved if we show that it is a basis of Hom (V, U).

Proof that $\{F_{ij}\}$ generates Hom (V, U). Consider an arbitrary function $F \in$ Hom (V, U). Suppose $F(v_1) = w_1, F(v_2) = w_2, \ldots, F(v_m) = w_m$. Since $w_k \in U$, it is a linear combination of the u's; say,

$$w_k = a_{k1} u_1 + a_{k2} u_2 + \cdots + a_{kn} u_n \qquad k = 1, \ldots, m, \quad a_{ij} \in K \tag{1}$$

Consider the linear mapping $G = \sum_{i=1}^{m} \sum_{j=1}^{n} a_{ij} F_{ij}$. Since G is a linear combination of the F_{ij}, the proof that $\{F_{ij}\}$ generates Hom (V, U) is complete if we show that $F = G$.

We now compute $G(v_k)$, $k = 1, \ldots, m$. Since $F_{ij}(v_k) = 0$ for $k \neq i$ and $F_{ki}(v_k) = u_i$,

$$G(v_k) = \sum_{i=1}^{m} \sum_{j=1}^{n} a_{ij} F_{ij}(v_k) = \sum_{j=1}^{n} a_{kj} F_{kj}(v_k) = \sum_{j=1}^{n} a_{kj} u_j$$
$$= a_{k1} u_1 + a_{k2} u_2 + \cdots + a_{kn} u_n$$

Thus by (1), $G(v_k) = w_k$ for each k. But $F(v_k) = w_k$ for each k. Accordingly, by Theorem 9.2, $F = G$; hence $\{F_{ij}\}$ generates Hom (V, U).

Proof that $\{F_{ij}\}$ is linearly independent. Suppose, for scalars $a_{ij} \in K$,

$$\sum_{i=1}^{m} \sum_{j=1}^{n} a_{ij} F_{ij} = 0$$

For v_k, $k = 1, \ldots, m$,

$$0 = 0(v_k) = \sum_{i=1}^{m} \sum_{j=1}^{n} a_{ij} F_{ij}(v_k) = \sum_{j=1}^{n} a_{kj} F_{kj}(v_k) = \sum_{j=1}^{n} a_{kj} u_j$$
$$= a_{k1} u_1 + a_{k2} u_2 + \cdots + a_{kn} u_n$$

But the u_i are linearly independent; hence for $k = 1, \ldots, m$, we have $a_{k1} = 0, a_{k2} = 0, \ldots, a_{kn} = 0$. In other words, all the $a_{ij} = 0$ and so $\{F_{ij}\}$ is linearly independent.

Thus $\{F_{ij}\}$ is a basis of Hom (V, U); hence dim Hom $(V, U) = mn$.

9.37. Prove Theorem 9.12.

(i) For every $v \in V$,

$$(G \circ (F + F'))(v) = G((F + F')(v)) = G(F(v) + F'(v))$$
$$= G(F(v)) + G(F'(v)) = (G \circ F)(v) + (G \circ F')(v) = (G \circ F + G \circ F')(v)$$

Thus $G \circ (F + F') = G \circ F + G \circ F'$.

(ii) For every $v \in V$,

$$((G + G') \circ F)(v) = (G + G')(F(v)) = G(F(v)) + G'(F(v))$$
$$= (G \circ F)(v) + (G' \circ F)(v) = (G \circ F + G' \circ F)(v)$$

Thus $(G + G') \circ F = G \circ F + G' \circ F$.

(iii) For every $v \in V$,

$$(k(G \circ F))(v) = k(G \circ F)(v) = k(G(F(v))) = (kG)(F(v)) = (kG \circ F)(v)$$

and

$$(k(G \circ F))(v) = k(G \circ F)(v) = k(G(F(v))) = G(kF(v)) = G((kF)(v)) = (G \circ kF)(v)$$

Accordingly, $k(G \circ F) = (kG) \circ F = G \circ (kF)$. (We emphasize that two mappings are shown to be equal by showing that they assign the same image to each point in the domain.)

ALGEBRA OF LINEAR OPERATORS

9.38. Let S and T be the linear operators on $\mathbf{R}^2$ defined by $S(x, y) = (y, x)$ and $T(x, y) = (0, x)$. Find formulas defining the operators (a) $S + T$, (b) $2S - 3T$, (c) ST, (d) TS, (e) S^2, (f) T^2.

(a) $(S + T)(x, y) = S(x, y) + T(x, y) = (y, x) + (0, x) = (y, 2x)$.

(b) $(2S - 3T)(x, y) = 2S(x, y) - 3T(x, y) = 2(y, x) - 3(0, x) = (2y, -x)$.

(c) $(ST)(x, y) = S(T(x, y)) = S(0, x) = (x, 0)$.

(d) $(TS)(x, y) = T(S(x, y)) = T(y, x) = (0, y)$.

(e) $S^2(x, y) = S(S(x, y)) = S(y, x) = (x, y)$. Note $S^2 = I$, the identity mapping.

(f) $T^2(x, y) = T(T(x, y)) = T(0, x) = (0, 0)$. Note $T^2 = 0$, the zero mapping.

9.39. Consider the linear operator T on $\mathbf{R}^3$ defined by $T(x, y, z) = (2x, 4x - y, 2x + 3y - z)$. (a) Show that T is invertible. Find formulas for: (b) T^{-1}, (c) T^2, and (d) T^{-2}.

(a) Let $W = \text{Ker } T$. We need only show that T is nonsingular, i.e., that $W = \{0\}$. Set $T(x, y, z) = (0, 0, 0)$ which yields

$$T(x, y, z) = (2x, 4x - y, 2x + 3y - z) = (0, 0, 0)$$

Thus W is the solution space of the homogeneous system

$$2x = 0 \qquad 4x - y = 0 \qquad 2x + 3y - z = 0$$

which has only the trivial solution $(0, 0, 0)$. Thus $W = \{0\}$; hence T is nonsingular and so T is invertible.

(b) Set $T(x, y, z) = (r, s, t)$ [and so $T^{-1}(r, s, t) = (x, y, z)$]. We have

$$(2x, 4x - y, 2x + 3y - z) = (r, s, t) \qquad \text{or} \qquad 2x = r, \ 4x - y = s, \ 2x + 3y - z = t$$

Solve for x, y, z in terms of r, s, t to get $x = \frac{1}{2}r$, $y = 2r - s$, $z = 7r - 3s - t$. Thus

$$T^{-1}(r, s, t) = (\tfrac{1}{2}r, 2r - s, 7r - 3s - t) \qquad \text{or} \qquad T^{-1}(x, y, z) = (\tfrac{1}{2}x, 2x - y, 7x - 3y - z)$$

(c) Apply T twice to get

$$\begin{aligned} T^2(x, y, z) &= T(2x, 4x - y, 2x + 3y - z) \\ &= [4x, 4(2x) - (4x - y), 2(2x) + 3(4x - y) - (2x + 3y - z)] \\ &= (4x, 4x + y, 14x - 6y + z) \end{aligned}$$

(d) Apply T^{-1} twice to get

$$\begin{aligned} T^{-2}(x, y, z) &= T^{-2}(\tfrac{1}{2}x, 2x - y, 7x - 3y - z) \\ &= [\tfrac{1}{4}x, 2(\tfrac{1}{2}x) - (2x - y), 7(\tfrac{1}{2}x) - 3(2x - y) - (7x - 3y - z)] \\ &= (\tfrac{1}{4}x, -x + y, -\tfrac{19}{2}x + 6y + z) \end{aligned}$$

9.40. Let V be of finite dimension and let T be a linear operator on V for which $TS = I$, for some operator S on V. (We call S a right inverse of T.) (a) Show that T is invertible. (b) Show that $S = T^{-1}$. (c) Give an example showing that the above need not hold if V is of infinite dimension.

(a) Let $\dim V = n$. By Theorem 9.14, T is invertible if and only if T is onto; hence T is invertible if and only if $\text{rank } T = n$. We have $n = \text{rank } I = \text{rank } TS \leq \text{rank } T \leq n$. Hence $\text{rank } T = n$ and T is invertible.

(b) $TT^{-1} = T^{-1}T = I$. Then $S = IS = (T^{-1}T)S = T^{-1}(TS) = T^{-1}I = T^{-1}$.

(c) Let V be the space of polynomials in t over K; say, $p(t) = a_0 + a_1 t + a_2 t^2 + \cdots + a_n t^n$. Let T and S be the operators on V defined by

$$T(p(t)) = 0 + a_1 + a_2 t + \cdots + a_n t^{n-1} \qquad \text{and} \qquad S(p(t)) = a_0 t + a_1 t^2 + \cdots + a_n t^{n+1}$$

We have

$$(TS)(p(t)) = T(S(p(t))) = T(a_0 t + a_1 t^2 + \cdots + a_n t^{n+1}) = a_0 + a_1 t + \cdots + a_n t^n = p(t)$$

and so $TS = I$, the identity mapping. On the other hand, if $k \in K$ and $k \neq 0$, then

$$(ST)(k) = S(T(k)) = S(0) = 0 \neq k$$

Accordingly, $ST \neq I$.

9.41. Let S and T be the linear operators on $\mathbf{R}^2$ defined by $S(x, y) = (0, x)$ and $T(x, y) = (x, 0)$. Show that $TS = 0$ but $ST \neq 0$. Also show that $T^2 = T$.

$(TS)(x, y) = T(S(x, y)) = T(0, x) = (0, 0)$. Since TS assigns $0 = (0, 0)$ to every $(x, y) \in \mathbf{R}^2$, it is the zero mapping: $TS = 0$.

$(ST)(x, y) = S(T(x, y)) = S(x, 0) = (0, x)$. For example, $(ST)(4, 2) = (0, 4)$. Thus $ST \neq 0$, since it does not assign $0 = (0, 0)$ to every element of $\mathbf{R}^2$.

For any $(x, y) \in \mathbf{R}^2$, $T^2(x, y) = T(T(x, y)) = T(x, 0) = (x, 0) = T(x, y)$. Hence $T^2 = T$.

9.42. Consider the linear operator T on $\mathbf{R}^2$ defined by $T(x, y) = (2x + 4y, \ 3x + 6y)$. Find: (a) a formula for T^{-1}, (b) $T^{-1}(8, 12)$, and (c) $T^{-1}(1, 2)$. (d) Is T an onto mapping?

(a) T is singular, e.g., $T(2, 1) = (0, 0)$; hence the linear operator $T^{-1} : \mathbf{R}^2 \to \mathbf{R}^2$ does not exist.

(b) $T^{-1}(8, 12)$ means the preimage of $(8, 12)$ under T. Set $T(x, y) = (8, 12)$ to get the system

$$\begin{cases} 2x + 4y = 8 \\ 3x + 6y = 12 \end{cases} \quad \text{or} \quad x + 2y = 4$$

Here y is a free variable. Set $y = a$, where a is a parameter, to get the solution $x = -2a + 4$, $y = a$. Thus $T^{-1}(8, 12) = \{(-2a + 4, a) : a \in \mathbf{R}\}$.

(c) Set $T(x, y) = (1, 2)$ to get the system

$$2x + 4y = 1 \qquad 3x + 6y = 2$$

The system has no solution. Thus $T^{-1}(1, 2) = \varnothing$, the empty set.

(d) No, since, e.g., $(1, 2)$ has no preimage.

9.43. Let $B = \{v_1, v_2, v_3\}$ be a basis of V and let $B' = \{u_1, u_2\}$ be a basis of U. Let $T : V \to U$ be linear. Also, suppose

$$\begin{array}{ll} T(v_1) = a_1 u_1 + a_2 u_2 \\ T(v_2) = b_1 u_1 + b_2 u_2 & \text{and} \qquad A = \begin{pmatrix} a_1 & b_1 & c_1 \\ a_2 & b_2 & c_2 \end{pmatrix} \\ T(v_3) = c_1 u_1 + c_2 u_2 \end{array}$$

Show that, for any $v \in V$, $A[v]_B = [T(v)]_{B'}$ (where the vectors in K^2 and K^3 are column vectors).

Suppose $v = k_1 v_1 + k_2 v_2 + k_3 v_3$; then $[v]_B = [k_1, k_2, k_3]^T$. Also

$$\begin{aligned} T(v) &= k_1 T(v_1) + k_2 T(v_2) + k_3 T(v_3) \\ &= k_1(a_1 u_1 + a_2 u_2) + k_2(b_1 u_1 + b_2 u_2) + k_3(c_1 u_1 + c_2 u_2) \\ &= (a_1 k_1 + b_1 k_2 + c_1 k_3)u_1 + (a_2 k_1 + b_2 k_2 + c_2 k_3)u_2 \end{aligned}$$

Accordingly,

$$[T(v)]_{B'} = \begin{pmatrix} a_1 k_1 + b_1 k_2 + c_1 k_3 \\ a_2 k_1 + b_2 k_2 + c_2 k_3 \end{pmatrix}$$

Computing, we obtain

$$A[v]_B = \begin{pmatrix} a_1 & b_1 & c_1 \\ a_2 & b_2 & c_2 \end{pmatrix} \begin{pmatrix} k_1 \\ k_2 \\ k_3 \end{pmatrix} = \begin{pmatrix} a_1 k_1 + b_1 k_2 + c_1 k_3 \\ a_2 k_1 + b_2 k_2 + c_2 k_3 \end{pmatrix} = [T(v)]_{B'}$$

9.44. Let k be a nonzero scalar. Show that a linear map T is singular if and only if kT is singular. Hence T is singular if and only if $-T$ is singular.

Suppose T is singular. Then $T(v) = 0$ for some vector $v \neq 0$. Hence

$$(kT)(v) = kT(v) = k0 = 0$$

and so kT is singular.

Now suppose kT is singular. Then $(kT)(w) = 0$ for some vector $w \neq 0$; hence

$$T(kw) = kT(w) = (kT)(w) = 0$$

But $k \neq 0$ and $w \neq 0$ implies $kw \neq 0$; thus T is also singular.

9.45. Let E be a linear operator on V for which $E^2 = E$. (Such an operator is termed a *projection*.) Let U be the image of E and W the kernel. Show that: (a) if $u \in U$, then $E(u) = u$, i.e., E is the identity map on U; (b) if $E \neq I$, then E is singular, i.e., $E(v) = 0$ for some $v \neq 0$; (c) $V = U \oplus W$.

(a) If $u \in U$, the image of E, then $E(v) = u$ for some $v \in V$. Hence using $E^2 = E$, we have

$$u = E(v) = E^2(v) = E(E(v)) = E(u)$$

(b) If $E \neq I$ then, for some $v \in V$, $E(v) = u$ where $v \neq u$. By (i), $E(u) = u$. Thus

$$E(v - u) = E(v) - E(u) = u - u = 0 \qquad \text{where} \quad v - u \neq 0$$

(c) We first show that $V = U + W$. Let $v \in V$. Set $u = E(v)$ and $w = v - E(v)$. Then

$$v = E(v) + v - E(v) = u + w$$

By definition, $u = E(v) \in U$, the image of E. We now show that $w \in W$, the kernel of E:

$$E(w) = E(v - E(v)) = E(v) - E^2(v) = E(v) - E(v) = 0$$

and thus $w \in W$. Hence $V = U + W$.

We next show that $U \cap W = \{0\}$. Let $v \in U \cap W$. Since $v \in U$, $E(v) = v$ by (a). Since $v \in W$, $E(v) = 0$. Thus $v = E(v) = 0$ and so $U \cap W = \{0\}$.

The above two properties imply that $V = U \oplus W$.

9.46. Find the dimension d of (a) Hom $(\mathbf{R}^3, \mathbf{R}^2)$, (b) Hom $(\mathbf{C}^3, \mathbf{R}^2)$, (c) Hom $(V, \mathbf{R}^2)$ where $V = \mathbf{C}^3$ viewed as a vector space over $\mathbf{R}$, (d) $A(\mathbf{R}^3)$, (e) $A(\mathbf{C}^3)$, (f) $A(V)$ where $V = \mathbf{C}^3$ viewed as a vector space over $\mathbf{R}$.

(a) Since dim $\mathbf{R}^3 = 3$ and dim $\mathbf{R}^2 = 2$, we have (Theorem 9.11) $d = 3 \cdot 2 = 6$.

(b) $\mathbf{C}^3$ is a vector space over $\mathbf{C}$ and $\mathbf{R}^2$ is a vector space over $\mathbf{R}$; hence Hom $(\mathbf{C}^3, \mathbf{R}^2)$ does not exist.

(c) As a vector space over $\mathbf{R}$, $V = \mathbf{C}^3$ has dimension 6. Hence (Theorem 9.11) $d = 6 \cdot 2 = 12$.

(d) $A(\mathbf{R}^3) = $ Hom $(\mathbf{R}^3, \mathbf{R}^3)$ and dim $\mathbf{R}^3 = 3$; hence $d = 3^2 = 9$.

(e) $A(\mathbf{C}^3) = $ Hom $(\mathbf{C}^3, \mathbf{C}^3)$ and dim $\mathbf{C}^3 = 3$; hence $d = 3^2 = 9$.

(f) Since dim $V = 6$, $d = $ dim $A(V) = 6^2 = 36$.

Supplementary Problems

MAPPINGS

9.47. Determine the number of different mappings from $\{a, b\}$ into $\{1, 2, 3\}$.

9.48. Let the mapping g assign to each name in the set {Betty, Martin, David, Alan, Rebecca} the number of different letters needed to spell the name. Find (a) the graph of g, and (b) the image of g.

9.49. Figure 9-5 is a diagram of mappings $f : A \to B$, $g : B \to A$, $h : C \to B$, $F : B \to C$ and $G : A \to C$. Determine whether each of the following defines a composition mapping and, if it does, find its domain and codomain: (a) $g \circ f$, (b) $h \circ f$, (c) $F \circ f$, (d) $G \circ f$, (e) $g \circ h$, (f) $h \circ G \circ g$.

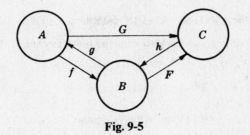

Fig. 9-5

9.50. Let $f : \mathbf{R} \to \mathbf{R}$ and $g : \mathbf{R} \to \mathbf{R}$ be defined by $f(x) = x^2 + 3x + 1$ and $g(x) = 2x - 3$. Find formulas defining the composition mappings $(a)\, f \circ g$, $(b)\, g \circ f$, $(c)\, g \circ g$, $(d)\, f \circ f$.

9.51. For each of the following mappings $f : \mathbf{R} \to \mathbf{R}$ find a formula for the inverse mapping: $(a)\, f(x) = 3x - 7$, and $(b)\, f(x) = x^3 + 2$.

9.52. For any mapping $f : A \to B$, show that $1_B \circ f = f = f \circ 1_A$.

LINEAR MAPPINGS

9.53. Let V be the vector space of polynomials in t over K. Show that the mappings $T : V \to V$ and $S : V \to V$ defined below are linear:

(a) $T(a_0 + a_1 t + \cdots + a_n t^n) = a_0 t + a_1 t^2 + \cdots + a_n t^{n+1}$

(b) $S(a_0 + a_1 t + \cdots + a_n t^n) = 0 + a_1 + a_2 t + \cdots + a_n t^{n-1}$

9.54. Let V be the vector space of $n \times n$ matrices over K; and let M be an arbitrary matrix in V. Show that the first two mappings $T : V \to V$ are linear, but the third is not linear (unless $M = 0$): $(a)\, T(A) = MA$, $(b)\, T(A) = MA - AM$, $(c)\, T(A) = M + A$.

9.55. Find $T(a, b)$ where $T : \mathbf{R}^2 \to \mathbf{R}^3$ is defined by $T(1, 2) = (3, -1, 5)$ and $T(0, 1) = (2, 1, -1)$.

9.56. Give an example of a nonlinear map $F : V \to U$ such that $F^{-1}(0) = \{0\}$ but F is not one-to-one.

9.57. Show that if $F : V \to U$ is linear and maps independent sets into independent sets then F is nonsingular.

9.58. Find a 2×2 matrix A which maps u_1 and u_2 into v_1 and v_2, respectively, where: $u_1 = (2, -4)^T$, $u_2 = (-1, 2)^T$ and $v_1 = (1, 1)^T$, $v_2 = (1, 3)^T$.

9.59. Find a 2×2 singular matrix B which maps $(1, 1)^T$ into $(1, 3)^T$.

9.60. Find a 2×2 matrix C with an eigenvalue $\lambda = 3$ and maps $(1, 1)^T$ into $(1, 3)^T$.

9.61. Let $T : \mathbf{C} \to \mathbf{C}$ be the conjugate mapping on the complex field $\mathbf{C}$. That is, $T(z) = \bar{z}$ where $z \in \mathbf{C}$, or $T(a + bi) = a - bi$ where $a, b \in \mathbf{R}$. (a) Show that T is not linear if $\mathbf{C}$ is viewed as a vector space over itself. (b) Show that T is linear if $\mathbf{C}$ is viewed as a vector space over the real field $\mathbf{R}$.

9.62. Let $F : \mathbf{R}^2 \to \mathbf{R}^2$ be defined by $F(x, y) = (3x + 5y, 2x + 3y)$, and let S be the unit circle in $\mathbf{R}^2$. (S consists of all points satisfying $x^2 + y^2 = 1$.) Find: (a) the image $F(S)$, and (b) the preimage $F^{-1}(S)$.

9.63. Consider the linear map $G : \mathbf{R}^3 \to \mathbf{R}^3$ defined by $G(x, y, z) = (x + y + z, y - 2z, y - 3z)$ and the unit sphere S_2 in $\mathbf{R}^3$ which consists of the points satisfying $x^2 + y^2 + z^2 = 1$. Find: (a) $G(S_2)$, and (b) $G^{-1}(S_2)$.

9.64. Let H be the plane $x + 2y - 3z = 4$ in $\mathbf{R}^3$ and let G be the linear map in Problem 9.63. Find: (a) $G(H)$, and (b) $G^{-1}(H)$.

KERNEL AND IMAGE OF LINEAR MAPPINGS

9.65. For the following linear map G, find a basis and the dimension of (i) the image of G, (ii) the kernel of G:
$G : \mathbf{R}^3 \to \mathbf{R}^2$ defined by $G(x, y, z) = (x + y, y + z)$.

9.66. Find a linear mapping $F : \mathbf{R}^3 \to \mathbf{R}^3$ whose image is spanned by $(1, 2, 3)$ and $(4, 5, 6)$.

9.67. Find a linear mapping $F : \mathbf{R}^4 \to \mathbf{R}^3$ whose kernel is spanned by $(1, 2, 3, 4)$ and $(0, 1, 1, 1)$.

9.68. Let $F : V \to U$ be linear. Show that (a) the image of any subspace of V is a subspace of U and (b) the preimage of any subspace of U is a subspace of V.

9.69. Each of the following matrices determines a linear map from $\mathbf{R}^4$ into $\mathbf{R}^3$:

$$(a) \quad A = \begin{pmatrix} 1 & 2 & 0 & 1 \\ 2 & -1 & 2 & -1 \\ 1 & -3 & 2 & -2 \end{pmatrix} \qquad (b) \quad B = \begin{pmatrix} 1 & 0 & 2 & -1 \\ 2 & 3 & -1 & 1 \\ -2 & 0 & -5 & 3 \end{pmatrix}$$

Find a basis and the dimension of the image U and the kernel W of each map.

9.70. Consider the vector space V of real polynomials $f(t)$ of degree 10 or less and the linear map $\mathbf{D}^4 : V \to V$ defined by $d^4 f/dt^4$, i.e., the fourth derivative. Find a basis and the dimension of (a) the image of $\mathbf{D}^4$, and (b) the kernel of $\mathbf{D}^4$.

OPERATIONS WITH LINEAR MAPPINGS

9.71. Let $F : \mathbf{R}^3 \to \mathbf{R}^2$ and $G : \mathbf{R}^3 \to \mathbf{R}^2$ be defined by $F(x, y, z) = (y, x + z)$ and $G(x, y, z) = (2z, x - y)$. Find formulas defining the mappings $F + G$ and $3F - 2G$.

9.72. Let $H : \mathbf{R}^2 \to \mathbf{R}^2$ be defined by $H(x, y) = (y, 2x)$. Using the mappings F and G in Problem 9.71, find formulas defining the mappings: (a) $H \circ F$ and $H \circ G$, (b) $F \circ H$ and $G \circ H$, (c) $H \circ (F + G)$ and $H \circ F + H \circ G$.

9.73. Show that the following mappings F, G, and H are linearly independent:

(a) $F, G, H \in \text{Hom } (\mathbf{R}^2, \mathbf{R}^2)$ defined by $F(x, y) = (x, 2y)$, $G(x, y) = (y, x + y)$, $H(x, y) = (0, x)$.

(b) $F, G, H \in \text{Hom } (\mathbf{R}^3, \mathbf{R})$ defined by $F(x, y, z) = x + y + z$, $G(x, y, z) = y + z$, $H(x, y, z) = x - z$.

9.74. For $F, G \in \text{Hom } (V, U)$, show that $\text{rank } (F + G) \leq \text{rank } F + \text{rank } G$. (Here V has finite dimension.)

9.75. Let $F : V \to U$ and $G : U \to V$ be linear. Show that if F and G are nonsingular then $G \circ F$ is nonsingular. Give an example where $G \circ F$ is nonsingular but G is not.

9.76. Prove that $\text{Hom } (V, U)$ does satisfy all the required axioms of a vector space. That is, prove Theorem 9.10.

ALGEBRA OF LINEAR OPERATORS

9.77. Suppose S and T are linear operators on V and that S is nonsingular. Assume V has finite dimension. Show that $\text{rank } (ST) = \text{rank } (TS) = \text{rank } T$.

9.78. Suppose $V = U \oplus W$. Let E_1 and E_2 be the linear operators on V defined by $E_1(v) = u$, $E_2(v) = w$, where $v = u + w$, $u \in U$, $w \in W$. Show that: (a) $E_1^2 = E_1$ and $E_2^2 = E_2$, i.e., that E_1 and E_2 are "projections"; (b) $E_1 + E_2 = I$, the identity mapping; (c) $E_1 E_2 = 0$ and $E_2 E_1 = 0$.

9.79. Let E_1 and E_2 be linear operators on V satisfying (a), (b), (c) of Problem 9.78. Show that V is the direct sum of the image of $E_2 : V = \text{Im } E_1 \oplus \text{Im } E_2$.

9.80. Show that if the linear operators S and T are invertible, then ST is invertible and $(ST)^{-1} = T^{-1}S^{-1}$.

9.81. Let V have finite dimension, and let T be a linear operator on V such that rank $T^2 = $ rank T. Show that Ker $T \cap$ Im $T = \{0\}$.

9.82. Which of the following integers can be the dimension of an algebra $A(V)$ of linear maps: 5, 9, 18, 25, 31, 36, 44, 64, 88, 100?

9.83. An algebra A is said to have an identity element 1 if $1 \cdot a = a \cdot 1 = a$ for every $a \in A$. Show that $A(V)$ has an identity element.

9.84. Find the dimension of $A(V)$ where: (a) $V = \mathbf{R}^4$, (b) $V = \mathbf{C}^4$, (c) $V = \mathbf{C}^4$ viewed as a vector space over $\mathbf{R}$, (d) $V = $ polynomials of degree ≤ 10.

MISCELLANEOUS PROBLEMS

9.85. Suppose $T : K^n \to K^m$ is a linear mapping. Let $\{e_1, \ldots, e_n\}$ be the usual basis of K^n and let A be the $m \times n$ matrix whose columns are the vectors $T(e_1), \ldots, T(e_n)$, respectively. Show that, for every vector $v \in K^n$, $T(v) = Av$, where v is written as a column vector.

9.86. Suppose $F : V \to U$ is linear and k is a nonzero scalar. Show that the maps F and kF have the same kernel and the same image.

9.87. Show that if $F : V \to U$ is onto, then dim $U \leq$ dim V. Determine all linear maps $T : \mathbf{R}^3 \to \mathbf{R}^4$ which are onto.

9.88. Let $T : V \to U$ be linear and let W be a subspace of V. The restriction of T to W is the map $T_W : W \to U$ defined by $T_W(w) = T(w)$, for every $w \in W$. Prove the following: (a) T_W is linear. (b) Ker $T_W = $ Ker $T \cap W$. (c) Im $T_W = T(w)$.

9.89. Two operators $S, T \in A(V)$ are said to be similar if there exists an invertible operator $P \in A(V)$ for which $S = P^{-1}TP$. Prove the following: (a) Similarity of operators is an equivalence relation. (b) Similar operators have the same rank (when V has finite dimension).

9.90. Let v and w be elements of a real vector space V. The line segment L from v to $v + w$ is defined to be the set of vectors $v + tw$ for $0 \leq t \leq 1$. (See Fig. 9-6.)

(a) Show that the line segment L between vectors v and u consists of the points:

    (i) $(1 - t)v + tu$ for $0 \leq t \leq 1$,   (ii) $t_1 v + t_2 u$ for $t_1 + t_2 = 1$, $t_1 \geq 0$, $t_2 \geq 0$.

(b) Let $F : V \to U$ be linear. Show that the image $F(L)$ of a line segment L in V is a line segment in U.

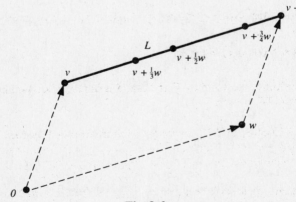

Fig. 9-6

9.91. A subset X of a vector space V is said to be convex if the line segment L between any two points (vectors) $P, Q \in X$ is contained in X.

    (a) Show that the intersection of convex sets is convex.

    (b) Suppose $F : V \to U$ is linear and X is convex. Show that $F(X)$ is convex.

Answers to Supplementary Problems

9.47. Nine.

9.48. (a) {(Betty, 4), (Martin, 6), (David, 4), (Alan, 3), (Rebecca, 5)}.   (b) Image of $g = \{3, 4, 5, 6\}$.

9.49. (a) $(g \circ f) : A \to A$,   (b) No,   (c) $(F \circ f) : A \to C$,   (d) No,   (e) $(g \circ h) : C \to A$,

    (f) $(h \circ G \circ g) : B \to B$.

9.50. (a) $(f \circ g)(x) = 4x^2 - 6x + 1$,   (c) $(g \circ g)(x) = 4x - 9$,

    (b) $(g \circ f)(x) = 2x^2 + 6x - 1$,   (d) $(f \circ f)(x) = x^4 + 6x^3 + 14x^2 + 15x + 5$.

9.51. (a) $f^{-1}(x) = (x + 7)/3$,   (b) $f^{-1}(x) = \sqrt[3]{x - 2}$.

9.55. $T(a, b) = (-a + 2b, -3a + b, 7a - b)$.

9.56. $f(x) = x^2$.

9.58. (a) $\begin{pmatrix} -12 & 7 \\ -27 & 17 \end{pmatrix}$,   (b) $\begin{pmatrix} -17 & 5 \\ 23 & -6 \end{pmatrix}$,   (c) Does not exist.

9.59. $\begin{pmatrix} 1 & 0 \\ 3 & 0 \end{pmatrix}$. [*Hint:* Send $(0, 1)^T$ into $(0, 0)^T$.]

9.60. $\begin{pmatrix} -2 & 3 \\ 0 & 3 \end{pmatrix}$. [*Hint:* Send $(0, 1)^T$ into $(0, 3)^T$.]

9.62. (a) $13x^2 - 42xy + 34y^2 = 1$,   (b) $13x^2 + 42xy + 24y^2 = 1$.

9.63. (a) $x^2 - 8xy + 26y^2 + 6xz - 38yz + 14z^2 = 1$,   (b) $x^2 + 2xy + 3y^2 + 2xz - 8yz + 14z^2 = 1$.

9.64. (a) $x - y + 2z = 4$,   (b) $x - 12z = 4$.

9.65. (i) $(1, 0), (0, 1)$, rank $G = 2$;   (ii) $(1, -1, 1)$, nullity $G = 1$.

9.66. $F(x, y, z) = (x + 4y, 2x + 5y, 3x + 6y)$.

9.67. $F(x, y, z, t) = (x + y - z, 3x + y - t, 0)$.

9.69. (a) $\{(1, 2, 1), (0, 1, 1)\}$ basis of Im A; dim (Im A) = 2.
             $\{(4, -2, -5, 0), (1, -3, 0, 5)\}$ basis of Ker A; dim (Ker A) = 2.

    (b) Im $B = \mathbf{R}^3$; $\{(-1, \frac{2}{3}, 1, 1)\}$ basis of Ker B; dim (Ker B) = 1.

9.70. (a) $1, t, \ldots, t^6$; rank $\mathbf{D}^4 = 7$;   (b) $1, t, t^2, t^3$; nullity $\mathbf{D}^4 = 4$.

9.71. $(F + G)(x, y, z) = (y + 2z, 2x - y + z), (3F - 2G)(x, y, z) = (3y - 4z, x + 2y + 3z)$.

9.72. (a) $(H \circ F)(x, y, z) = (x + y, 2y), (H \circ G)(x, y, z) = (x - y, 4z)$.   (b) Not defined.

  (c) $(H \circ (F + G))(x, y, z) = (H \circ F + H \circ G)(x, y, z) = (2x - y + z, 2y + 4z)$.

9.82. Squares: 9, 25, 36, 64, 100.

9.84. (a) 16,   (b) 16,   (c) 64,   (d) 121.

Chapter 10

Matrices and Linear Mappings

10.1 INTRODUCTION

Suppose $S = \{u_1, u_2, \ldots, u_n\}$ is a basis of a vector space V over a field K and, for $v \in V$, suppose

$$v = a_1 u_1 + a_2 u_2 + \cdots + a_n u_n$$

Then the coordinate vector of v relative to the basis S, which we write as a column vector unless otherwise specified or implied, is denoted and defined by

$$[v]_S = \begin{pmatrix} a_1 \\ a_2 \\ \ldots \\ a_n \end{pmatrix} = [a_1, a_2, \ldots, a_n]^T$$

Recall that the mapping $v \to [v]_S$, determined by the basis S, is an isomorphism between V and the space K^n.

This chapter shows that there is also an isomorphism, determined by the basis S, between the algebra $A(V)$ of linear operators on V and the algebra $\mathbf{M}$ of n-square matrices over K. Thus every linear operator $T : V \to V$ will correspond to an n-square matrix $[T]_S$ determined by the basis S.

The question of whether or not a linear operator T can be represented by a diagonal matrix will also be addressed in this chapter.

10.2 MATRIX REPRESENTATION OF A LINEAR OPERATOR

Let T be a linear operator on a vector space V over a field K and suppose $S = \{u_1, u_2, \ldots, u_n\}$ is a basis of V. Now $T(u_1), \ldots, T(u_n)$ are vectors in V and so each is a linear combination of the vectors in the basis S; say,

$$\begin{aligned} T(u_1) &= a_{11} u_1 + a_{12} u_2 + \cdots + a_{1n} u_n \\ T(u_2) &= a_{21} u_1 + a_{22} u_2 + \cdots + a_{2n} u_n \\ &\cdots\cdots\cdots\cdots\cdots\cdots\cdots\cdots\cdots\cdots \\ T(u_n) &= a_{n1} u_1 + a_{n2} u_2 + \cdots + a_{nn} u_n \end{aligned}$$

The following definition applies.

Definition: The transpose of the above matrix of coefficients, denoted by $m_S(T)$ or $[T]_S$, is called the matrix representation of T relative to the basis S or simply the matrix of T in the basis S; that is,

$$[T]_S = \begin{pmatrix} a_{11} & a_{21} & \cdots & a_{n1} \\ a_{12} & a_{22} & \cdots & a_{n2} \\ \cdots\cdots\cdots\cdots\cdots\cdots\cdots \\ a_{1n} & a_{2n} & \cdots & a_{nn} \end{pmatrix}$$

(The subscript S may be omitted if the basis S is understood.)

Remark: Using the coordinate (column) vector notation, the matrix representation of T may also be written in the form

$$m(T) = [T] = ([T(u_1)], [T(u)2)], \ldots, [T(u_n)])$$

that is, the columns of $m(T)$ are the coordinate vectors $[T(u_1)], \ldots, [T(u_n)]$.

Example 10.1

(a) Let V be the vector space of polynomials in t over $\mathbf{R}$ of degree ≤ 3, and let $\mathbf{D} : V \to V$ be the differential operator defined by $\mathbf{D}(p(t)) = d(p(t))/dt$. We compute the matrix of $\mathbf{D}$ in the basis $\{1, t, t^2, t^3\}$. We have:

$$\begin{aligned}
\mathbf{D}(1) &= 0 &&= 0 - 0t + 0t^2 + 0t^3 \\
\mathbf{D}(t) &= 1 &&= 1 + 0t + 0t^2 + 0t^3 \\
\mathbf{D}(t^2) &= 2t &&= 0 + 2t + 0t^2 + 0t^3 \\
\mathbf{D}(t^3) &= 3t^2 &&= 0 + 0t + 3t^2 + 0t^3
\end{aligned}
\qquad \text{and} \qquad
[\mathbf{D}] = \begin{pmatrix} 0 & 1 & 0 & 0 \\ 0 & 0 & 2 & 0 \\ 0 & 0 & 0 & 3 \\ 0 & 0 & 0 & 0 \end{pmatrix}$$

[Note the coordinate vectors of $\mathbf{D}(1)$, $\mathbf{D}(t)$, $\mathbf{D}(t^2)$, and $\mathbf{D}(t^3)$ are the columns, not rows, in $[\mathbf{D}]$.]

(b) Consider the linear operator $F : \mathbf{R}^2 \to \mathbf{R}^2$ defined by $F(x, y) = (4x - 2y, 2x + y)$ and the following bases of $\mathbf{R}^2$:

$$S = \{u_1 = (1, 1), u_2 = (-1, 0)\} \qquad \text{and} \qquad E = \{e_1 = (1, 0), e_2 = (0, 1)\}$$

We have

$$\begin{aligned}
F(u_1) &= F(1, 1) = (2, 3) = 3(1, 1) + (-1, 0) = 3u_1 + u_2 \\
F(u_2) &= F(-1, 0) = (-4, -2) = -2(1, 1) + 2(-1, 0) = -2u_1 + 2u_2
\end{aligned}$$

Therefore, $[F]_S = \begin{pmatrix} 3 & -2 \\ 1 & 2 \end{pmatrix}$ is the matrix representation of F in the basis S. We also have

$$\begin{aligned}
F(e_1) &= F(1, 0) = (4, 2) = 4e_1 + 2e_2 \\
F(e_2) &= F(0, 1) = (-2, 1) = -2e_1 + e_2
\end{aligned}$$

Accordingly, $[F]_E = \begin{pmatrix} 4 & -2 \\ 2 & 1 \end{pmatrix}$ is the matrix representation of F relative to the usual basis E.

(c) Consider any n-square matrix A over K (which defines a matrix map $A : K^n \to K^n$) and the usual basis $E = \{e_i\}$ of K^n. Then $Ae_1, Ae_2, \ldots, Ae_n$ are precisely the columns of A (Example 9.5) and their coordinates relative to the usual basis E are the vectors themselves. Accordingly,

$$[A]_E = A$$

that is, relative to the usual basis E, the matrix representation of a matrix map A is the matrix A itself.

The following algorithm will be used to compute matrix representations.

Algorithm 10.2

Given a linear operator T on V and a basis $S = \{u_1, \ldots, u_n\}$ of V, this algorithm finds the matrix $[T]_S$ which represents T relative to basis S.

Step 1. Repeat for each basis vector u_k in S:

    (a) Find $T(u_k)$.

    (b) Write $T(u_k)$ as a linear combination of the basis vectors $u_1, \ldots, u_n$ to obtain the coordinates of $T(u_k)$ in the basis S.

Step 2. Form the matrix $[T]_S$ whose columns are the coordinate vectors $[T(u_k)]_S$ obtained in Step 1(*b*).

Step 3. Exit.

> **Remark:** Observe that Step 2(*b*) is repeated for each basis vector u_k. Accordingly, it may be useful to first apply:
>
> *Step 0.* Find a formula for the coordinates of an arbitrary vector v relative to the basis S.

Our first theorem, proved in Problem 10.10, tells us that the "action" of an operator T on a vector v is preserved by its matrix representation:

Theorem 10.1: Let $S = \{u_1, u_2, \ldots, u_n\}$ be a basis for V and let T be any linear operator on V. Then, for any vector $v \in V$, $[T]_S[v]_S = [T(v)]_S$.

That is, if we multiply the coordinate vector of v by the matrix representation of T, then we obtain the coordinate vector of $T(v)$.

Example 10.2. Consider the differential operator $D : V \to V$ in Example 10.1(*a*). Let

$$p(t) = a + bt + ct^2 + dt^3 \qquad \text{and so} \qquad D(p(t)) = b + 2ct + 3dt^2$$

Hence, relative to the basis $\{1, t, t^2, t^3\}$,

$$[p(t)] = [a, b, c, d]^T \qquad \text{and} \qquad [D(p(t))] = [b, 2c, 3d, 0]^T$$

We show that Theorem 10.1 does hold here:

$$[D][p(t)] = \begin{pmatrix} 0 & 1 & 0 & 0 \\ 0 & 0 & 2 & 0 \\ 0 & 0 & 0 & 3 \\ 0 & 0 & 0 & 0 \end{pmatrix} \begin{pmatrix} a \\ b \\ c \\ d \end{pmatrix} = \begin{pmatrix} b \\ 2c \\ 3d \\ 0 \end{pmatrix} = [D(p(t))]$$

Now we have associated a matrix $[T]$ to each T in $A(V)$, the algebra of linear operators on V. By our first theorem, the action of an individual operator T is preserved by this representation. The next two theorems, (proved in Problems 10.11 and 10.12, respectively), tell us that the three basic operations with these operators,

  (i) Addition,

 (ii) Scalar multiplication,

(iii) Composition

are also preserved.

Theorem 10.2: Let $S = \{u_1, u_2, \ldots, u_n\}$ be a basis for a vector space V over K, and let $\mathbf{M}$ be the algebra of n-square matrices over K. Then the mapping $m : A(V) \to \mathbf{M}$ defined by $m(T) = [T]_S$ is a vector space isomorphism. That is, for any $F, G \in A(V)$ and any $k \in K$, we have

  (i) $m(F + G) = m(F) + m(G)$, i.e., $[F + G] = [F] + [G]$,

 (ii) $m(kF) = km(F)$, i.e., $[kF] = k[F]$,

(iii) m is one-to-one and onto.

Theorem 10.3:   For any linear operators $G, F \in A(v)$,

$$m(G \circ F) = m(G)m(F) \qquad \text{i.e.,} \qquad [G \circ F] = [G][F]$$

(Here $G \circ F$ denotes the composition of the maps G and F.)

We illustrate the above theorems in the case dim $V = 2$. Suppose $\{u_1, u_2\}$ is a basis for V, and F and G are linear operators on V for which

$$F(u_1) = a_1 u_1 + a_2 u_2 \qquad\qquad G(u_1) = c_1 u_1 + c_2 u_2$$
$$F(u_2) = b_1 u_1 + b_2 u_2 \qquad\qquad G(u_2) = d_1 u_1 + d_2 u_2$$

Then
$$[F] = \begin{pmatrix} a_1 & b_1 \\ a_2 & b_2 \end{pmatrix} \qquad \text{and} \qquad [G] = \begin{pmatrix} c_1 & d_1 \\ c_2 & d_2 \end{pmatrix}$$

We have

$$(F + G)(u_1) = F(u_1) + G(u_1) = a_1 u_1 + a_2 u_2 + c_1 u_1 + c_2 u_2$$
$$= (a_1 + c_1)u_1 + (a_2 + c_2)u_2$$
$$(F + G)(u_2) = F(u_2) + G(u_2) = b_1 u_1 + b_2 u_2 + d_1 u_1 + d_2 u_2$$
$$= (b_1 + d_1)u_1 + (b_2 + d_2)u_2$$

Thus
$$[F + G] = \begin{pmatrix} a_1 + c_1 & b_1 + d_1 \\ a_2 + c_2 & b_2 + d_2 \end{pmatrix} = \begin{pmatrix} a_1 & b_1 \\ a_2 & b_2 \end{pmatrix} + \begin{pmatrix} c_1 & d_1 \\ c_2 & d_2 \end{pmatrix} = [F] + [G]$$

Also, for $k \in K$, we have

$$(kF)(u_1) = kF(u_1) = k(a_1 u_1 + a_2 u_2) = ka_1 u_1 + ka_2 u_2$$
$$(kF)(u_2) = kF(u_2) = k(b_1 u_1 + b_2 u_2) = kb_1 u_1 + kb_2 u_2$$

Thus

$$[kF] = \begin{pmatrix} ka_1 & kb_1 \\ ka_2 & kb_2 \end{pmatrix} = k\begin{pmatrix} a_1 & b_1 \\ a_2 & b_2 \end{pmatrix} = k[F]$$

Finally, we have

$$(G \circ F)(u_1) = G(F(u_1)) = G(a_1 u_1 + a_2 u_2) = a_1 G(u_1) + a_2 G(u_2)$$
$$= a_1(c_1 u_1 + c_2 u_2) + a_2(d_1 u_1 + d_2 u_2)$$
$$= (a_1 c_1 + a_2 d_1)u_1 + (a_1 c_2 + a_2 d_2)u_2$$

$$(G \circ F)(u_2)) = G(F(u_2) = G(b_1 u_1 + b_2 u_2) = b_1 G(u_1) + b_2 G(u_2)$$
$$= b_1(c_1 u_1 + c_2 u_2) + b_2(d_1 u_1 + d_2 u_2)$$
$$= (b_1 c_1 + b_2 d_1)u_1 + (b_1 c_2 + b_2 d_2)u_2$$

Accordingly,

$$[G \circ F] = \begin{pmatrix} a_1 c_1 + a_2 d_1 & b_1 c_1 + b_2 d_1 \\ a_1 c_2 + a_2 d_2 & b_1 c_2 + b_2 d_2 \end{pmatrix} = \begin{pmatrix} c_1 & d_1 \\ c_2 & d_2 \end{pmatrix}\begin{pmatrix} a_1 & b_1 \\ a_2 & b_2 \end{pmatrix} = [G][F]$$

10.3   CHANGE OF BASIS AND LINEAR OPERATORS

The above discussion shows that we can represent a linear operator by a matrix once we have chosen a basis. We ask the following natural question: How does our representation change if we select another basis? In order to answer this question, we first recall a definition and some facts.

Definition: Let $S = \{u_1, u_2, \ldots, u_n\}$ be a basis of V and let $S' = \{v_1, v_2, \ldots, v_n\}$ be another basis. Suppose, for $i = 1, 2, \ldots, n$,

$$v_i = a_{i1}u_1 + a_{i2}u_2 + \cdots + a_{in}u_n$$

The transpose P of the above matrix of coefficients is termed the change-of-basis (or transition) matrix from the "old" basis S to the "new" basis S'.

Fact 1. The above change-of-basis matrix P is invertible and its inverse P^{-1} is the change-of-basis matrix from S' back to S.

Fact 2. Let P be the change-of-basis matrix from the usual basis E of K^n to another basis S. Then P is the matrix whose columns are precisely the elements of S.

Fact 3. Let P be the change-of-basis matrix from a basis S to a basis S' in V. Then (Theorem 5.27), for any vector $v \in V$,

$$P[v]_{S'} = [v]_S \quad \text{and} \quad P^{-1}[v]_S = [v]_{S'}$$

(Thus P^{-1} transforms the coordinates of v in the "old" basis S to the "new" basis S'.)

The following theorem, proved in Problem 10.19, answers the above question, that is, shows how the matrix representation of a linear operator is affected by a change of basis.

Theorem 10.4: Let P be the change-of-basis matrix from a basis S to a basis S' in a vector space V. Then, for any linear operator T on V,

$$[T]_{S'} = P^{-1}[T]_S P$$

In other words, if A is the matrix representing T in a basis S, then $B = P^{-1}AP$ is the matrix which represents T in a new basis S' where P is the change-of-basis matrix from S to S'.

Example 10.3. Consider the following bases for $\mathbf{R}^2$:

$$E = \{e_1 = (1, 0), e_2 = (0, 1)\} \quad \text{and} \quad S = \{u_1 = (1, -2), u_2 = (2, -5)\}$$

Since E is the usual basis of $\mathbf{R}^2$, we write the basis vector in S as columns to obtain the change-of-basis matrix P from E to S:

$$P = \begin{pmatrix} 1 & 2 \\ -2 & -5 \end{pmatrix}$$

Consider the linear operator F on $\mathbf{R}^2$ defined by $F(x, y) = (2x - 3y, 4x + y)$. We have

$$f(e_1) = F(1, 0) = (2, 4) = 2e_1 + 4e_2$$
$$f(e_2) = F(0, 1) = (-3, 1) = -3e_1 + e_2$$

and hence $\quad A = \begin{pmatrix} 2 & -3 \\ 4 & 1 \end{pmatrix}$

is the matrix representation of F relative to the usual basis E. By Theorem 10.4,

$$B = P^{-1}AP = \begin{pmatrix} 5 & 2 \\ -2 & -1 \end{pmatrix}\begin{pmatrix} 2 & -3 \\ 4 & 1 \end{pmatrix}\begin{pmatrix} 1 & 2 \\ -2 & -5 \end{pmatrix} = \begin{pmatrix} 44 & 101 \\ -18 & -41 \end{pmatrix}$$

is the matrix representation of F relative to the basis S.

Remark: Suppose $P = (a_{ij})$ is any n-square invertible matrix over a field K, and suppose $S = \{u_1, u_2, \ldots, u_n\}$ is a basis for a vector space V over K. Then the n vectors,

$$v_i = a_{1i}u_1 + a_{2i}u_2 + \cdots + a_{ni}u_n \qquad i = 1, 2, \ldots, n$$

are linearly independent and hence they form another basis S' for V. Furthermore, P is the change-of-basis matrix from the basis S to the basis S'. Accordingly, if A is any matrix representation of a linear operator T on V, then the matrix $B = P^{-1}AP$ is also a matrix representation of T.

Similarity and Linear Operators

Suppose A and B are square matrices for which there exists an invertible matrix P such that $B = P^{-1}AP$. Recall that B is said to be similar to A or is said to be obtained from A by a similarity transformation. By Theorem 10.4 and the above remark, we have the following basic result.

Theorem 10.5:   Two matrices A and B represent the same linear operator T if and only if they are similar to each other.

That is, all the matrix representations of the linear operator T form an equivalence class of similar matrices.

Now suppose f is a function on square matrices which assigns the same value to similar matrices; that is, $f(A) = f(B)$ whenever A is similar to B. Then f induces a function, also denoted by f, on linear operators T in the following natural way: $f(T) = f([T]_S)$ where S is any basis. The function is well-defined by Theorem 10.5. Three important examples of such functions are:

(1)   determinant,   (2)   trace,   and   (3)   characteristic polynomial

Thus the determinant, trace, and characteristic polynomial of a linear operator T are well-defined.

Example 10.4.   Let F be the linear operator on $\mathbf{R}^2$ defined by $F(x, y) = (2x - 3y, 4x + y)$. By Example 10.34, the matrix representation of T relative to the usual basis for $\mathbf{R}^2$ is

$$A = \begin{pmatrix} 2 & -3 \\ 4 & 1 \end{pmatrix}$$

Accordingly:

(i) $\det(T) = \det(A) = 2 + 12 = 14$ is the determinant of T.

(ii) $\operatorname{tr} T = \operatorname{tr} A = 2 + 1 = 3$ is the trace of T.

(iii) $\Delta_T(t) = \Delta_A(t) = t^2 - 3t + 14$ is the characteristic polynomial of T.

By Example 10.3, another matrix representation of T is the matrix

$$B = \begin{pmatrix} 44 & 101 \\ -18 & -41 \end{pmatrix}$$

Using this matrix, we obtain:

(i) $\det(T) = \det(A) = -1804 + 1818 = 14$ is the determinant of T.

(ii) $\operatorname{tr} T = \operatorname{tr} A = 44 - 41 = 3$ is the trace of T.

(iii) $\Delta_T(t) = \Delta_B(t) = t^2 - 3t + 14$ is the characteristic polynomial of T.

As expected, both matrices yield the same results.

10.4 DIAGONALIZATION OF LINEAR OPERATORS

A linear operator T on a vector space V is said to be *diagonalizable* if T can be represented by a diagonal matrix D. Thus T is diagonalizable if and only if there exists a basis $S = \{u_1, u_2, \ldots, u_n\}$ of V for which

$$
\begin{aligned}
T(u_1) &= k_1 u_1 \\
T(u_2) &= k_2 u_2 \\
&\cdots\cdots\cdots\cdots\cdots\cdots\cdots\cdots\cdots\cdots \\
T(u_n) &= k_n u_n
\end{aligned}
$$

In such a case, T is represented by the diagonal matrix

$$D = \operatorname{diag}(k_1, k_2, \ldots, k_n)$$

relative to the basis S.

The above observation leads us to the following definitions and theorems which are analogous to the definitions and theorems for matrices discussed in Chapter 8.

A scalar $\lambda \in K$ is called an *eigenvalue* of T if there exists a nonzero vector $v \in V$ for which

$$T(v) = \lambda v$$

Every vector satisfying this relation is called an *eigenvector* of T *belonging* to the eigenvalue λ. The set E_λ of all such vectors is a subspace of V called the *eigenspace* of λ. (Alternatively, λ is an eigenvalue of T if $\lambda I - T$ is singular and, in this case, E_λ is the kernel of $\lambda I - T$.)

The following theorems apply.

Theorem 10.6: T can be represented by a diagonal matrix D (or T is diagonalizable) if and only if there exists a basis S of V consisting of eigenvectors of T. In this case, the diagonal elements of D are the corresponding eigenvalues.

Theorem 10.7: Nonzero eigenvectors $u_1, u_2, \ldots, u_r$ of T, belonging, respectively, to distinct eigenvalues $\lambda_1, \lambda_2, \ldots, \lambda_r$, are linearly independent. (See Problem 10.26 for the proof.)

Theorem 10.8: T is a root of its characteristic polynomial $\Delta(t)$.

Theorem 10.9: The scalar λ is an eigenvalue of T if and only if λ is a root of the characteristic polynomial $\Delta(t)$ of T.

Theorem 10.10: The geometric multiplicity of an eigenvalue λ of T does not exceed its algebraic multiplicity. (See Problem 10.27 for the proof.)

Theorem 10.11: Suppose A is a matrix representation of T. Then T is diagonalizable if and only if A is diagonalizable.

> **Remark:** Theorem 10.11 reduces the investigation of the diagonalization of a linear operator T to the diagonalization of a matrix A which was discussed in detail in Chapter 8.

Example 10.5

(a) Let V be the vector space of real functions for which $S = \{\sin \theta, \cos \theta\}$ is a basis, and let D be the differential operator on V. Then

$$D(\sin \theta) = \quad \cos \theta = \quad 0(\sin \theta) + 1(\cos \theta)$$
$$D(\cos \theta) = -\sin \theta = -1(\sin \theta) + 0(\cos \theta)$$

Hence $A = \begin{pmatrix} 0 & 1 \\ -1 & 0 \end{pmatrix}$ is the matrix representation of D in the basis S. Therefore,

$$\Delta(t) = t^2 - (\text{tr } A)t + |A| = t^2 + 1$$

is the characteristic polynomial of both A and D. Thus A and D have no (real) eigenvalues and, in particular, D is not diagonalizable.

(b) Consider the functions $e^{a_1 t}, e^{a_2 t}, \ldots, e^{a_r t}$ where $a_1, a_2, \ldots, a_r$ are distinct real numbers. Let D be the differential operator; hence $D(e^{a_k t}) = a_k e^{a_k t}$. Accordingly, the functions $e^{a_k t}$ are eigenvectors of D belonging to distinct eigenvalues. Thus, by Theorem 10.7, the functions are linearly independent.

(c) Let $T : \mathbf{R}^2 \to \mathbf{R}^2$ be the linear operator which rotates each vector $v \in \mathbf{R}^2$ by an angle $\theta = 90°$ (as shown in Fig. 10-1). Note that no nonzero vector is a multiple of itself. Hence T has no eigenvalues and so no eigenvectors.

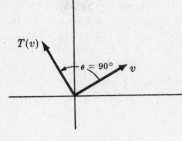

Fig. 10-1

Now the minimum polynomial $m(t)$ of a linear operator T is defined independently of the theory of matrices, as the monic polynomial of lowest degree which has T as a zero. However, for any polynomial $f(t)$,

$$f(T) = 0 \quad \text{if and only if} \quad f(A) = 0$$

where A is any matrix representation of T. Accordingly, T and A have the same minimum polynomial. Thus all theorems in Chapter 8 on the minimum polynomial of a matrix also hold for the minimum polynomial of a linear operator T.

10.5 MATRICES AND GENERAL LINEAR MAPPINGS

Lastly, we consider the general case of linear mappings from one space into another. Let V and U be vector spaces over the same field K and, say, dim $V = m$ and dim $U = n$. Furthermore, let $\mathbf{e} = \{v_1, v_2, \ldots, v_m\}$ and $\mathbf{f} = \{u_1, u_2, \ldots, u_n\}$ be arbitrary but fixed bases of V and U, respectively.

Suppose $F : V \to U$ is a linear mapping. Then the vectors $F(v_1), \ldots, F(v_m)$ belong to U and so each is a linear combination of the u_k; say

$$F(v_1) = a_{11}u_1 + a_{12}u_2 + \cdots + a_{1n}u_n$$
$$F(v_2) = a_{21}u_1 + a_{22}u_2 + \cdots + a_{2n}u_n$$
$$\cdots\cdots\cdots\cdots\cdots\cdots\cdots\cdots\cdots\cdots\cdots\cdots$$
$$F(v_m) = a_{m1}u_n + a_{m2}u_2 + \cdots + a_{mn}u_n$$

The transpose of the above matrix of coefficients, denoted by $[F]_e^f$, is called the *matrix representation* of F relative to the bases $\mathbf{e}$ and $\mathbf{f}$:

$$[F]_e^f = \begin{pmatrix} a_{11} & a_{21} & \cdots & a_{m1} \\ a_{12} & a_{22} & \cdots & a_{m2} \\ \cdots\cdots\cdots\cdots\cdots\cdots\cdots \\ a_{1n} & a_{2n} & \cdots & a_{mn} \end{pmatrix}$$

(We will use the simple notation $[F]$ when the bases are understood.)

The following theorems apply.

Theorem 10.12:   For any vector $v \in V$, $[F]_e^f[v]_e = [F(v)]_f$.

That is, multiplying the coordinate vector of v in the basis $\mathbf{e}$ by the matrix $[F]_e^f$, we obtain the coordinate vector of $F(v)$ in the basis $\mathbf{f}$.

Theorem 10.13:   The mapping $F \mapsto [F]$ is an isomorphism from Hom (V, U) onto the vector space of $n \times m$ matrices over K. That is, the mapping is one-to-one and onto and, for any $F, G \in$ Hom (V, U) and any $k \in K$,

$$[F + G] = [F] + [G] \quad \text{and} \quad [kF] = k[F]$$

Remark: Recall that any $n \times m$ matrix A over K has been identified with the linear mapping from K^m into K^n given by $v \mapsto Av$. Now suppose V and U are vector spaces over K of dimensions m and n, respectively, and suppose $\mathbf{e}$ is a basis of V and $\mathbf{f}$ is a basis of U. Then in view of the preceding theorem, we shall also identify A with the linear mapping $F : V \to U$ given by $[F(v)]_f = A[v]_e$. We comment that if other bases of V and U are given, then A is identified with another linear mapping from V into U.

Theorem 10.14: Let $\mathbf{e}$, $\mathbf{f}$, and $\mathbf{g}$ be bases of V, U, and W, respectively. Let $F : V \to U$ and $G : U \to W$ be linear mappings. Then

$$[G \circ F]_e^g = [G]_f^g [F]_e^f$$

That is, relative to the appropriate bases, the matrix representation of the composition of two linear mappings is equal to the product of the matrix representations of the individual mappings.

Our next theorem shows how the matrix representation of a linear mapping $F : V \to U$ is affected when new bases are selected.

Theorem 10.15: Let P be the change-of-basis matrix from a basis $\mathbf{e}$ to a basis $\mathbf{e}'$ in V, and let Q be the change-of-basis matrix from a basis $\mathbf{f}$ to a basis $\mathbf{f}'$ in U. Then, for any linear mapping $F : V \to U$,

$$[F]_{e'}^{f'} = Q^{-1} [F]_e^f P$$

In other words, if A represents the linear mapping F relative to the bases $\mathbf{e}$ and $\mathbf{f}$, then

$$B = Q^{-1}AP$$

represents F relative to the new bases $\mathbf{e}'$ and $\mathbf{f}'$.

Our last theorem, proved in Problem 10.34, shows that every linear mapping from one space into another can be represented by a very simple matrix.

Theorem 10.16: Let $F : V \to U$ be linear and, say, rank $F = r$. Then there exist bases of V and of U such that the matrix representation of F has the form

$$A = \begin{pmatrix} I & 0 \\ 0 & 0 \end{pmatrix}$$

where I is the r-square identity matrix.

The above matrix A is called the *normal* or *canonical* form of the linear map F.

Solved Problems

MATRIX REPRESENTATIONS OF LINEAR OPERATORS

10.1. Suppose $F : R^2 \to R^2$ is defined by $F(x, y) = (2y, 3x - y)$. Find the matrix representation of F relative to the usual basis $E = \{e_1 = (1, 0), e_2 = (0, 1)\}$:

Note first that if $(a, b) \in \mathbf{R}^2$, then $(a, b) = ae_1 + be_2$.

$$\begin{aligned} F(e_1) &= F(1, 0) = (0, 3) &= 0e_1 + 3e_2 \\ F(e_2) &= F(0, 1) = (2, -1) = 2e_1 - e_2 \end{aligned} \quad \text{and} \quad [F]_E = \begin{pmatrix} 0 & 2 \\ 3 & -1 \end{pmatrix}$$

It is seen that the rows of $[F]_E$ are given directly by the coefficients in the components of $F(x, y)$. This generalizes to any space K^n.

It is seen that the rows of $[F]_E$ are given directly by the coefficients in the components of $F(x, y)$. This generalizes to any space K^n.

10.2. Find the matrix representation of the linear operator F in Problem 10.1 relative to the basis $S = \{u_1 = (1, 3), u_2 = (2, 5)\}$.

First find the coordinates of an arbitrary vector $(a, b) \in \mathbf{R}^2$ with respect to the basis S. We have

$$\binom{a}{b} = x\binom{1}{3} + y\binom{2}{5} \qquad \text{or} \qquad \begin{array}{l} x + 2y = a \\ 3x + 5y = b \end{array}$$

Solve for x and y in terms of a and b to get $x = 2b - 5a$ and $y = 3a - b$. Thus

$$(a, b) = (-5a + 2b)u_1 + (3a - b)u_2$$

We have $F(x, y) = (2y, 3x - y)$. Hence

$$\begin{array}{ll} F(u_1) = F(1, 3) = (6, 0) = -30u_1 + 18u_2 \\ F(u_2) = F(2, 5) = (10, 1) = -48u_1 + 29u_2 \end{array} \quad \text{and} \quad [F]_S = \begin{pmatrix} -30 & -48 \\ 18 & 29 \end{pmatrix}$$

(**Remark:** We emphasize that the coefficients of u_1 and u_2 are written as columns, not rows, in each matrix representation.)

10.3. Let G be the linear operator on $\mathbf{R}^3$ defined by $G(x, y, z) = (2y + z, x - 4y, 3x)$.

(a) Find the matrix representation of G relative to the basis

$$S = \{w_1 = (1, 1, 1), w_2 = (1, 1, 0), w_3 = (1, 0, 0)\}$$

(b) Verify that $[G][v] = [G(v)]$ for any vector $v \in \mathbf{R}^3$.

First find the coordinates of an arbitrary vector $(a, b, c) \in \mathbf{R}^3$ with respect to the basis S. Write (a, b, c) as a linear combination of w_1, w_2, w_3 using unknown scalars $x, y,$ and z:

$$(a, b, c) = x(1, 1, 1) + y(1, 1, 0) + z(1, 0, 0) = (x + y + z, x + y, x)$$

Set corresponding components equal to each other to obtain the system of equations

$$x + y + z = a \qquad x + y = b \qquad x = c$$

Solve the system for $x, y,$ and z in terms of $a, b,$ and c to find $x = c, y = b - c, z = a - b$. Thus

$$(a, b, c) = cw_1 + (b - c)w_2 + (a - b)w_3 \qquad \text{or equivalently} \qquad [(a, b, c)] = [c, b - c, a - b]^T$$

(a) Since $G(x, y, z) = (2y + z, x - 4y, 3x)$,

$$\begin{array}{l} G(w_1) = G(1, 1, 1) = (3, -3, 3) = 3w_1 - 6w_2 + 6w_3 \\ G(w_2) = G(1, 1, 0) = (2, -3, 3) = 3w_1 - 6w_2 + 5w_3 \\ G(w_3) = G(1, 0, 0) = (0, 1, 3) = 3w_1 - 2w_2 - w_3 \end{array}$$

Write the coordinates of $G(w_1), G(w_2), G(w_3)$ as columns to get

$$[G] = \begin{pmatrix} 3 & 3 & 3 \\ -6 & -6 & -2 \\ 6 & 5 & -1 \end{pmatrix}$$

(b) Write $G(v)$ as a linear combination of w_1, w_2, w_3 where $v = (a, b, c)$ is an arbitrary vector in $\mathbf{R}^3$:

$$G(v) = G(a, b, c) = (2b + c, a - 4b, 3a) = 3aw_1 + (-2a - 4b)w_2 + (-a + 6b + c)w_3$$

or, equivalently,

$$[G(v)] = [3a, -2a - 4b, -a + 6b + c]^T$$

Accordingly,

$$[G][v] = \begin{pmatrix} 3 & 3 & 3 \\ -6 & -6 & -2 \\ 6 & 5 & -1 \end{pmatrix} \begin{pmatrix} c \\ b-c \\ a-b \end{pmatrix} = \begin{pmatrix} 3a \\ -2a-4b \\ -a+6b+c \end{pmatrix} = [G(v)]$$

10.4. Let $A = \begin{pmatrix} 1 & 2 \\ 3 & 4 \end{pmatrix}$ and let T be the linear operator on $\mathbf{R}^2$ defined by $T(v) = Av$ (where v is written as a column vector). Find the matrix of T in each of the following bases:

(a) $E = \{e_1 = (1, 0), e_2 = (0, 1)\}$, i.e., the usual basis;   and   (b) $S = \{u_1 = (1, 3), u_2 = (2, 5)\}$.

(a) $$T(e_1) = \begin{pmatrix} 1 & 2 \\ 3 & 4 \end{pmatrix}\begin{pmatrix} 1 \\ 0 \end{pmatrix} = \begin{pmatrix} 1 \\ 3 \end{pmatrix} = 1e_1 + 3e_2 \quad \text{and thus} \quad [T]_E = \begin{pmatrix} 1 & 2 \\ 3 & 4 \end{pmatrix}$$

$$T(e_2) = \begin{pmatrix} 1 & 2 \\ 3 & 4 \end{pmatrix}\begin{pmatrix} 0 \\ 1 \end{pmatrix} = \begin{pmatrix} 2 \\ 4 \end{pmatrix} = 2e_1 + 4e_2$$

Observe that the matrix of T in the usual basis is precisely the original matrix A which defined T. This is not unusual. In fact, this is true for any matrix A when using the usual basis.

(b) From Problem 10.2, $(a, b) = (-5a + 2b)u_1 + (3a - b)u_2$. Hence

$$T(u_1) = \begin{pmatrix} 1 & 2 \\ 3 & 4 \end{pmatrix}\begin{pmatrix} 1 \\ 3 \end{pmatrix} = \begin{pmatrix} 7 \\ 15 \end{pmatrix} = -5u_1 + 6u_2 \quad \text{and thus} \quad [T]_S = \begin{pmatrix} -5 & -8 \\ 6 & 10 \end{pmatrix}$$

$$T(u_2) = \begin{pmatrix} 1 & 2 \\ 3 & 4 \end{pmatrix}\begin{pmatrix} 2 \\ 5 \end{pmatrix} = \begin{pmatrix} 12 \\ 26 \end{pmatrix} = -8u_1 + 10u_2$$

10.5. Each of the sets (a) $\{1, t, e^t, te^t\}$ and (b) $\{e^{3t}, te^{3t}, t^2e^{3t}\}$ is a basis of a vector space V of functions $f: \mathbf{R} \to \mathbf{R}$. Let D be the differential operator on V, that is, $D(f) = df/dt$. Find the matrix of D in each given basis.

(a)
$$\begin{aligned} D(1) &= 0 &&= 0(1) + 0(t) + 0(e^t) + 0(te^t) \\ D(t) &= 1 &&= 1(1) + 0(t) + 0(e^t) + 0(te^t) \\ D(e^t) &= e^t &&= 0(1) + 0(t) + 1(e^t) + 0(te^t) \\ D(te^t) &= e^t + te^t &&= 0(1) + 0(t) + 1(e^t) + 1(te^t) \end{aligned}$$
and thus $\quad [D] = \begin{pmatrix} 0 & 1 & 0 & 0 \\ 0 & 0 & 0 & 0 \\ 0 & 0 & 1 & 1 \\ 0 & 0 & 0 & 1 \end{pmatrix}$

(b)
$$\begin{aligned} D(e^{3t}) &= 3e^{3t} &&= 3(e^{3t}) + 0(te^{3t}) + 0(t^2e^{3t}) \\ D(te^{3t}) &= e^{3t} + 3te^{3t} &&= 1(e^{3t}) + 3(te^{3t}) + 0(t^2e^{3t}) \\ D(t^2e^{3t}) &= 2te^{3t} + 3t^2e^{3t} &&= 0(e^{3t}) + 2(te^{3t}) + 3(t^2e^{3t}) \end{aligned}$$
and thus $\quad [D] = \begin{pmatrix} 3 & 1 & 0 \\ 0 & 3 & 2 \\ 0 & 0 & 3 \end{pmatrix}$

10.6. Let V be the vector space of 2×2 matrices with the usual basis

$$\left\{ E_1 = \begin{pmatrix} 1 & 0 \\ 0 & 0 \end{pmatrix}, \; E_2 = \begin{pmatrix} 0 & 1 \\ 0 & 0 \end{pmatrix}, \; E_3 = \begin{pmatrix} 0 & 0 \\ 1 & 0 \end{pmatrix}, \; E_4 = \begin{pmatrix} 0 & 0 \\ 0 & 1 \end{pmatrix} \right\}$$

Let $M = \begin{pmatrix} 1 & 2 \\ 3 & 4 \end{pmatrix}$ and T be the linear operator on V defined by $T(A) = MA$. Find the matrix representation of T relative to the above usual basis of V.

We have

$$T(E_1) = ME_1 = \begin{pmatrix} 1 & 2 \\ 3 & 4 \end{pmatrix}\begin{pmatrix} 1 & 0 \\ 0 & 0 \end{pmatrix} = \begin{pmatrix} 1 & 0 \\ 3 & 0 \end{pmatrix} = 1E_1 + 0E_2 + 3E_3 + 0E_4$$

$$T(E_2) = ME_2 = \begin{pmatrix} 1 & 2 \\ 3 & 4 \end{pmatrix}\begin{pmatrix} 0 & 1 \\ 0 & 0 \end{pmatrix} = \begin{pmatrix} 0 & 1 \\ 0 & 3 \end{pmatrix} = 0E_1 + 1E_2 + 0E_3 + 3E_4$$

$$T(E_3) = ME_3 = \begin{pmatrix} 1 & 2 \\ 3 & 4 \end{pmatrix}\begin{pmatrix} 0 & 0 \\ 1 & 0 \end{pmatrix} = \begin{pmatrix} 2 & 0 \\ 4 & 0 \end{pmatrix} = 2E_1 + 0E_2 + 4E_3 + 0E_4$$

$$T(E_4) = ME_4 = \begin{pmatrix} 1 & 2 \\ 3 & 4 \end{pmatrix}\begin{pmatrix} 0 & 0 \\ 0 & 1 \end{pmatrix} = \begin{pmatrix} 0 & 2 \\ 0 & 4 \end{pmatrix} = 0E_1 + 2E_2 + 0E_3 + 4E_4$$

Hence

$$[T] = \begin{pmatrix} 1 & 0 & 2 & 0 \\ 0 & 1 & 0 & 2 \\ 3 & 0 & 4 & 0 \\ 0 & 3 & 0 & 4 \end{pmatrix}$$

(Since dim $V = 4$, any matrix representation of a linear operator on V must be a 4-square matrix.)

10.7. Consider the basis $S = \{(1, 0), (1, 1)\}$ of $\mathbf{R}^2$. Let $L : \mathbf{R}^2 \to \mathbf{R}^2$ be defined by $L(1, 0) = (6, 4)$ and $L(1, 1) = (1, 5)$. (Recall that a linear map is completely defined by its action on a basis.) Find the matrix representation of L with respect to the basis S.

Write $(6, 4)$ and then $(1, 5)$ each as a linear combination of the basis vectors to get

$$\begin{aligned} L(1, 0) &= (6, 4) = 2(1, 0) + 4(1, 1) \\ L(1, 1) &= (1, 5) = -4(1, 0) + 5(1, 1) \end{aligned} \quad \text{and thus} \quad [L] = \begin{pmatrix} 2 & -4 \\ 4 & 5 \end{pmatrix}$$

10.8. Consider the usual basis $E = \{e_1, e_2, \ldots, e_n\}$ of K^n. Let $L : K^n \to K^n$ be defined by $L(e_i) = v_i$. Show that the matrix A representing L relative to the usual basis E is obtained by writing the image vectors $v_1, v_2, \ldots, v_n$ as columns.

Suppose $v_i = (a_{i1}, a_{i2}, \ldots, a_{in})$. Then $L(e_i) = v_i = a_{i1}e_1 + a_{i2}e_2 + \cdots + a_{in}e_n$. Thus

$$[L] = \begin{pmatrix} a_{11} & a_{21} & \cdots & a_{n1} \\ a_{12} & a_{22} & \cdots & a_{n2} \\ \cdots\cdots\cdots\cdots\cdots\cdots \\ a_{1n} & a_{2n} & \cdots & a_{nn} \end{pmatrix}$$

as claimed.

10.9. For each of the following linear operators L on $\mathbf{R}^2$, find the matrix A which represents L (relative to the usual basis of $\mathbf{R}^2$):

(a) L is defined by $L(1, 0) = (2, 4)$ and $L(0, 1) = (5, 8)$.

(b) L is the rotation in $\mathbf{R}^2$ counterclockwise by $90°$.

(c) L is the reflection in $\mathbf{R}^2$ about the line $y = -x$.

(a) Since $(1, 0)$ and $(0, 1)$ do form the usual basis of $\mathbf{R}^2$, write their images under L as columns to get

$$A = \begin{pmatrix} 2 & 5 \\ 4 & 8 \end{pmatrix}$$

(b) Under the rotation L, we have $L(1, 0) = (0, 1)$ and $L(0, 1) = (-1, 0)$. Thus $A = \begin{pmatrix} 0 & -1 \\ 1 & 0 \end{pmatrix}$.

(c) Under the reflection L, we have $L(1, 0) = (0, -1)$ and $L(0, 1) = (-1, 0)$. Thus $A = \begin{pmatrix} 0 & -1 \\ -1 & 0 \end{pmatrix}$.

10.10. Prove Theorem 10.1.

Suppose, for $i = 1, \ldots, n$,

$$T(u_i) = a_{i1}u_1 + a_{i2}u_2 + \cdots + a_{in}u_n = \sum_{j=1}^{n} a_{ij}u_j$$

Then $[T]_S$ is the n-square matrix whose jth row is

$$(a_{1j}, a_{2j}, \ldots, a_{nj}) \qquad\qquad (1)$$

Now suppose

$$v = k_1u_1 + k_2u_2 + \cdots + k_nu_n = \sum_{i=1}^{n} k_iu_i$$

Writing a column vector as the transpose of a row vector,

$$[v]_S = (k_1, k_2, \ldots, k_n)^T \qquad\qquad (2)$$

Furthermore, using the linearity of T,

$$T(v) = T\left(\sum_{i=1}^{n} k_iu_i\right) = \sum_{i=1}^{n} k_i T(u_i) = \sum_{i=1}^{n} k_i\left(\sum_{j=1}^{n} a_{ij}u_j\right)$$

$$= \sum_{j=1}^{n} \left(\sum_{i=1}^{n} a_{ij}k_i\right)u_j = \sum_{j=1}^{n} (a_{1j}k_1 + a_{2j}k_2 + \cdots + a_{nj}k_n)u_j$$

Thus $[T(v)]_S$ is the column vector whose jth entry is

$$a_{1j}k_1 + a_{2j}k_2 + \cdots + a_{nj}k_n \qquad\qquad (3)$$

On the other hand, the jth entry of $[T]_S[v]_S$ is obtained by multiplying the jth row of $[T]_S$ by $[v]_S$, i.e., (1) by (2). But the product of (1) and (2) is (3); hence $[T]_S[v]_S$ and $[T(v)]_S$ have the same entries. Thus $[T]_S[v]_S = [T(v)]_S$.

10.11. Prove Theorem 10.2.

Suppose, for $i = 1, \ldots, n$,

$$F(u_i) \sum_{j=1}^{n} a_{ij}u_j \qquad \text{and} \qquad G(u_i) = \sum_{j=1}^{n} b_{ij}u_j$$

Consider the matrices $A = (a_{ij})$ and $B = (b_{ij})$. Then $[F] = A^T$ and $[G] = B^T$. We have, for $i = 1, \ldots, n$,

$$(F + G)(u_i) = F(u_i) + G(u_i) = \sum_{j=1}^{n} (a_{ij} + b_{ij})u_j$$

Since $A + B$ is the matrix $(a_{ij} + b_{ij})$, we have

$$[A + B] = (A + B)^T = A^T + B^T = [A] + [B]$$

Also, for $i = 1, \ldots, n$,

$$(kF)(u_i) = kF(u_i) = k\sum_{j=1}^{n} a_{ij}u_j = \sum_{j=1}^{n} (ka_{ij})u_j$$

Since kA is the matrix (ka_{ij}), we have

$$[kF] = (kA)^T = kA^T = k[F]$$

Lastly, m is one-to-one since a linear mapping is completely determined by its values on a basis and map m is onto since each matrix $A = (a_{ij})$ in M is the image of the linear operator

$$F(u_i) = \sum_{j=1}^{n} a_{ij}u_j \qquad i = 1, \ldots, n$$

Thus the theorem is proved.

10.12. Prove Theorem 10.3.

Using the notation in Problem 10.11, we have

$$(G \circ F)(u_i) = G(F(u_i)) = G\left(\sum_{j=1}^{n} a_{ij} u_j\right) = \sum_{j=1}^{n} a_{ij} G(u_j)$$

$$= \sum_{j=1}^{n} a_{ij}\left(\sum_{k=1}^{n} b_{jk} u_k\right) = \sum_{k=1}^{n}\left(\sum_{j=1}^{n} a_{ij} b_{jk}\right) u_k$$

Recall that AB is the matrix $AB = (c_{ik})$ where $c_{ik} = \sum_{j=1}^{n} a_{ij} b_{jk}$. Accordingly,

$$[G \circ F] = (AB)^T = B^T A^T = [G][F]$$

Thus the theorem is proved.

10.13. Let A be a matrix representation of an operator T. Show that $f(A)$ is the matrix representation of $f(T)$, for any polynomial $f(t)$. [Thus $f(T) = 0$ if and only if $f(A) = 0$.]

Let ϕ be the mapping $T \mapsto A$, i.e., which sends the operator T into its matrix representation A. We need to prove that $\phi(f(T)) = f(A)$. Suppose $f(t) = a_n t^n + \cdots + a_1 t + a_0$. The proof is by induction on n, the degree of $f(t)$.

Suppose $n = 0$. Recall that $\phi(I') = I$ where I' is the identity mapping and I is the identity matrix. Thus

$$\phi(f(T)) = \phi(a_0 I') = a_0 \phi(I') = a_0 I = f(A)$$

and so the theorem holds for $n = 0$.

Now assume the theorem holds for polynomials of degree less than n. Then, since ϕ is an algebra isomorphism,

$$\begin{aligned}
\phi(f(T)) &= \phi(a_n T^n + a_{n-1} T^{n-1} + \cdots + a_1 T + a_0 I') \\
&= a_n \phi(T)\phi(T^{n-1}) + \phi(a_{n-1} T^{n-1} + \cdots + a_1 T + a_0 I') \\
&= a_n A A^{n-1} + (a_{n-1} A^{n-1} + \cdots + a_1 A + a_0 I) = f(A)
\end{aligned}$$

and the theorem is proved.

10.14. Let V be the vector space of functions which has $\{\sin \theta, \cos \theta\}$ as a basis, and let D be the differential operator on V. Show that D is a zero of $f(t) = t^2 + 1$.

Apply $f(D)$ to each basis vector:

$$f(D)(\sin \theta) = (D^2 + I)(\sin \theta) = D^2(\sin \theta) + I(\sin \theta) = -\sin \theta + \sin \theta = 0$$
$$f(D)(\cos \theta) = (D^2 + I)(\cos \theta) = D^2(\cos \theta) + I(\cos \theta) = -\cos \theta + \cos \theta = 0$$

Since each basis vector is mapped into 0, every vector $v \in V$ is also mapped into 0 by $f(D)$. Thus $f(D) = 0$.
[This result is expected since, by Example 10.5(a), $f(t)$ is the characteristic polynomial of D.]

CHANGE OF BASIS, SIMILAR MATRICES

10.15. Let $F : \mathbf{R}^2 \to \mathbf{R}^2$ be defined by $F(x, y) = (4x - y, 2x + y)$ and consider the following bases for $\mathbf{R}^2$:

$$E = \{e_1 = (1, 0), e_2 = (0, 1)\} \qquad \text{and} \qquad S = \{u_1 = (1, 3), u_2 = (2, 5)\}$$

(a) Find the change-of-basis matrix P from E to S, the change-of-basis matrix Q from S back to E, and verify that $Q = P^{-1}$.

(b)   Find the matrix A that represents F in the basis E, the matrix B that represents F in the basis S, and verify that $B = P^{-1}AP$.

(c)   Find the trace $\text{tr } F$, the determinant $\det (F)$, and the characteristic polynomial $\Delta(t)$ of F.

(a)   Since E is the usual basis, write the elements of S as columns to obtain the change-of-basis matrix

$$P = \begin{pmatrix} 1 & 2 \\ 3 & 5 \end{pmatrix}$$

Solving the system $u_1 = e_1 + 3e_2$, $u_2 = 2e_1 + 5e_2$, for e, and e_2 yields

$$\begin{aligned} e_1 &= -5u_1 + 3u_2 \\ e_2 &= 2u_1 - u_2 \end{aligned} \qquad \text{and thus} \qquad Q = \begin{pmatrix} -5 & 2 \\ 3 & -1 \end{pmatrix}$$

We have

$$PQ = \begin{pmatrix} 1 & 2 \\ 3 & 5 \end{pmatrix} \begin{pmatrix} -5 & 2 \\ 3 & -1 \end{pmatrix} = \begin{pmatrix} 1 & 0 \\ 0 & 1 \end{pmatrix} = I$$

(b)   Write the coefficients of x and y as rows (Problem 10.3) to obtain

$$A = \begin{pmatrix} 4 & -1 \\ 2 & 1 \end{pmatrix}$$

Since $F(x, y) = (4x - y, 2x + y)$ and $(a, b) = (-5a + 2b)u_1 + (3a - b)u_2$,

$$\begin{aligned} F(u_1) &= F(1, 3) = (1, 5) = 5u_1 - 2u_2 \\ F(u_2) &= F(2, 5) = (3, 9) = 3u_1 \end{aligned} \qquad \text{and thus} \qquad B = \begin{pmatrix} 5 & 3 \\ -2 & 0 \end{pmatrix}$$

We have

$$P^{-1}AP = \begin{pmatrix} -5 & 2 \\ 3 & -1 \end{pmatrix} \begin{pmatrix} 4 & -1 \\ 2 & 1 \end{pmatrix} \begin{pmatrix} 1 & 2 \\ 3 & 5 \end{pmatrix} = \begin{pmatrix} 5 & 3 \\ -2 & 0 \end{pmatrix} = B$$

(c)   Use A (or B) to obtain

$$\text{tr } F = \text{tr } A = 4 + 1 = 5 \qquad \det (F) = \det (A) = 4 + 2 = 6$$

$$\Delta(t) = t^2 - (\text{tr } A)t + |A| = t^2 - 5t + 6$$

10.16. Let G be the linear operator on $\mathbf{R}^3$ defined by $G(x, y, z) = (2y + z, x - 4y, 3x)$ and consider the usual basis E of $\mathbf{R}^3$ and the following basis S of $\mathbf{R}^3$:

$$S = \{w_1 = (1, 1, 1), w_2 = (1, 1, 0), w_3 = (1, 0, 0)\}$$

(a)   Find the change-of-basis matrix P from E to S, the change-of-basis matrix Q from S back to E, and verify that $Q = P^{-1}$.

(b)   Verify that $[G]_S = P^{-1}[G]_E P$.

(c)   Find the trace, determinant, and characteristic polynomial of G.

(a)   Since E is the usual basis, write the elements of S as columns to obtain the change-of-basis matrix

$$P = \begin{pmatrix} 1 & 1 & 1 \\ 1 & 1 & 0 \\ 1 & 0 & 0 \end{pmatrix}$$

By the familiar inversion process (see Problem 10.3) we obtain

$$\begin{aligned} e_1 &= 0w_1 + 0w_2 + 1w_3 \\ e_2 &= 0w_1 + 1w_2 - 1w_3 \\ e_3 &= 1w_1 - 1w_2 + 0w_3 \end{aligned} \qquad \text{and thus} \qquad Q = \begin{pmatrix} 0 & 0 & 1 \\ 0 & 1 & -1 \\ 1 & -1 & 0 \end{pmatrix}$$

We have

$$PQ = \begin{pmatrix} 1 & 1 & 1 \\ 1 & 1 & 0 \\ 1 & 0 & 0 \end{pmatrix}\begin{pmatrix} 0 & 0 & 1 \\ 0 & 1 & -1 \\ 1 & -1 & 0 \end{pmatrix} = \begin{pmatrix} 1 & 0 & 0 \\ 0 & 1 & 0 \\ 0 & 0 & 1 \end{pmatrix} = I$$

(b) From Problems 10.1 and 10.3, $[G]_E = \begin{pmatrix} 0 & 2 & 1 \\ 1 & -4 & 0 \\ 3 & 0 & 0 \end{pmatrix}$ and $[G]_S = \begin{pmatrix} 3 & 3 & 3 \\ -6 & -6 & -2 \\ 6 & 5 & -1 \end{pmatrix}$. Thus

$$P^{-1}[G]_E P = \begin{pmatrix} 0 & 0 & 1 \\ 0 & 1 & -1 \\ 1 & -1 & 0 \end{pmatrix}\begin{pmatrix} 0 & 2 & 1 \\ 1 & -4 & 0 \\ 3 & 0 & 0 \end{pmatrix}\begin{pmatrix} 1 & 1 & 1 \\ 1 & 1 & 0 \\ 1 & 0 & 0 \end{pmatrix} = \begin{pmatrix} 3 & 3 & 3 \\ -6 & -6 & -2 \\ 6 & 5 & -1 \end{pmatrix} = [G]_S$$

(c) Use $[G]_E$ (the simpler matrix) to obtain

$$\text{tr } G = 0 - 4 + 0 = -4 \qquad \det (G) = 12 \qquad \text{and} \qquad \Delta(t) = t^3 + 4t^2 - 5t - 12$$

10.17. Find the trace and determinant of the following operator on $\mathbf{R}^3$:

$$T(x, y, z) = (a_1 x + a_2 y + a_3 z, b_1 x + b_2 y + b_3 z, c_1 x + c_2 y + c_3 z)$$

First find a matrix representation A of T. Choosing the usual basis E,

$$A = \begin{pmatrix} a_1 & a_2 & a_3 \\ b_1 & b_2 & b_3 \\ c_1 & c_2 & c_3 \end{pmatrix}$$

Then

$$\text{tr } T = \text{tr } A = a_1 + b_2 + c_3$$

and

$$\det (T) = \det (A) = a_1 b_2 c_3 + a_2 b_3 c_1 + a_3 b_1 c_2 - a_3 b_b c_3 - a_2 b_1 c_3 - a_1 b_3 c_2$$

10.18. Let V be the space of 2×2 matrices over $\mathbf{R}$, and let $M = \begin{pmatrix} 1 & 2 \\ 3 & 4 \end{pmatrix}$. Let T be the linear operator on V defined by $T(A) = MA$. Find the trace and determinant of T.

We must first find a matrix representation of T. Choose the usual basis of V to obtain (Problem 10.6) the following matrix representation:

$$[T] = \begin{pmatrix} 1 & 0 & 2 & 0 \\ 0 & 1 & 0 & 2 \\ 3 & 0 & 4 & 0 \\ 0 & 3 & 0 & 4 \end{pmatrix}$$

Then $\text{tr } T = 1 + 1 + 4 + 4 = 10$ and $\det (T) = 4$.

10.19. Prove Theorem 10.4.

Let v be any vector in V. Then, by Theorem 5.27, $P[v]_{S'} = [v]_S$. Therefore,

$$P^{-1}[T]_S P[v]_{S'} = P^{-1}[T]_S[v]_S = P^{-1}[T(v)]_S = [T(v)]_{S'}$$

But $[T]_{S'}[v]_{S'} = [T(v)]_{S'}$; hence

$$P^{-1}[T]_S P[v]_{S'} = [T]_{S'}[v]_{S'}$$

Since the mapping $v \mapsto [v]_{S'}$ is onto K^n, we have $P^{-1}[T]_S PX = [T]_{S'} X$ for every $X \in K^n$. Thus $P^{-1}[T]_S P = [T]_{S'}$, as claimed.

DIAGONALIZATION OF LINEAR OPERATORS, EIGENVALUES AND EIGENVECTORS

10.20. Find the eigenvalues and linearly independent eigenvectors of the following linear operator on $\mathbf{R}^2$, and, if it is diagonalizable, find a diagonal representation D: $F(x, y) = (6x - y, 3x + 2y)$.

First find the matrix A which represents F in the usual basis of $\mathbf{R}^2$ by writing down the coefficients of x and y as rows:

$$A = \begin{pmatrix} 6 & -1 \\ 3 & 2 \end{pmatrix}$$

The characteristic polynomial $\Delta(t)$ of F is then

$$\Delta(t) = t^2 - (\operatorname{tr} A)t + |A| = t^2 - 8t + 15 = (t - 3)(t - 5)$$

Thus $\lambda_1 = 3$ and $\lambda_2 = 5$ are eigenvalues of F. We find the corresponding eigenvectors as follows:

(i)   Subtract $\lambda_1 = 3$ down the diagonal of A to obtain the matrix $M = \begin{pmatrix} 3 & -1 \\ 3 & -1 \end{pmatrix}$ which corresponds to the homogeneous system $3x - y = 0$. Here $v_1 = (1, 3)$ is a nonzero solution and hence an eigenvector of F belonging to $\lambda_1 = 3$.

(ii) Subtract $\lambda_2 = 5$ down the diagonal of A to obtain $M = \begin{pmatrix} 1 & -1 \\ 3 & -3 \end{pmatrix}$ which corresponds to the system $x - y = 0$. Here $v_2 = (1, 1)$ is a nonzero solution and hence an eigenvector of F belonging to $\lambda_2 = 5$.

Then $S = \{v_1, v_2\}$ is a basis of $\mathbf{R}^2$ consisting of eigenvectors of F. Thus F is diagonalizable with the matrix representation $D = \begin{pmatrix} 3 & 0 \\ 0 & 5 \end{pmatrix}$.

10.21. Let L be the linear operator on $\mathbf{R}^2$ which reflects points across the line $y = kx$ (where $k \neq 0$). See Fig. 10-2.

(a)   Show that $v_1 = (k, 1)$ and $v_2 = (1, -k)$ are eigenvectors of L.

(b)   Show that L is diagonalizable, and find such a diagonal representation D.

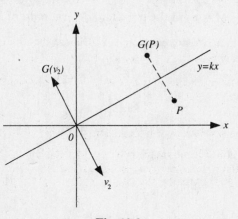

Fig. 10-2

(a)   The vector $v_1 = (k, 1)$ lies on the line $y = kx$ and hence is left fixed by L, that is, $L(v_1) = v_1$. Thus v_1 is an eigenvector of L belonging to the eigenvalue $\lambda_1 = 1$. The vector $v_2 = (1, -k)$ is perpendicular to the line $y = kx$ and hence L reflects v_2 into its negative, that is, $L(v_2) = -v_2$. Thus v_2 is an eigenvector of L belonging to the eigenvalue $\lambda_2 = -1$.

(b) Here $S = \{v_1, v_2\}$ is a basis of $\mathbf{R}^2$ consisting of eigenvectors of L. Thus L is diagonalizable with the diagonal representation (relative to S) $D = \begin{pmatrix} 1 & 0 \\ 0 & -1 \end{pmatrix}$.

10.22. Find all eigenvalues and a basis of each eigenspace of the operator $T : \mathbf{R}^3 \to \mathbf{R}^3$ defined by $T(x, y, z) = (2x + y, y - z, 2y + 4z)$. Is T diagonalizable? If so, find such a representation D.

First find the matrix A which represents T in the usual basis of $\mathbf{R}^3$ by writing down the coefficients of x, y, z as rows:

$$A = [T] = \begin{pmatrix} 2 & 1 & 0 \\ 0 & 1 & -1 \\ 0 & 2 & 4 \end{pmatrix}$$

The characteristic polynomial $\Delta(t)$ of T is then

$$\Delta(t) = |tI - A| = \begin{vmatrix} t - 2 & -1 & 0 \\ 0 & t - 1 & 1 \\ 0 & -2 & t - 4 \end{vmatrix} = (t - 2)^2(t - 3)$$

Thus 2 and 3 are the eigenvalues of T.

We find a basis of the eigenspace E_2 of the eigenvalue 2. Substitute $t = 2$ into $tI - A$ to obtain the homogeneous system

$$\begin{pmatrix} 0 & -1 & 0 \\ 0 & 1 & 1 \\ 0 & -2 & -2 \end{pmatrix}\begin{pmatrix} x \\ y \\ z \end{pmatrix} = \begin{pmatrix} 0 \\ 0 \\ 0 \end{pmatrix} \quad \text{or} \quad \begin{cases} -y & = 0 \\ y + z = 0 \\ -2y - 2z = 0 \end{cases} \quad \text{or} \quad \begin{cases} y & = 0 \\ y + z = 0 \end{cases}$$

The system has only one independent solution, e.g., $x = 1$, $y = 0$, $z = 0$. Thus $u = (1, 0, 0)$ forms a basis of the eigenspace of E_2.

We find a basis of the eigenspace E_3 of the eigenvalue 3. Substitute $t = 3$ into $tI - A$ to obtain the homogeneous system

$$\begin{pmatrix} 1 & -1 & 0 \\ 0 & 2 & 1 \\ 0 & -2 & -1 \end{pmatrix}\begin{pmatrix} x \\ y \\ z \end{pmatrix} = \begin{pmatrix} 0 \\ 0 \\ 0 \end{pmatrix} \quad \text{or} \quad \begin{cases} x - y = 0 \\ 2y + z = 0 \\ -2y - z = 0 \end{cases} \quad \text{or} \quad \begin{cases} x - y = 0 \\ 2y + z = 0 \end{cases}$$

The system has only one independent solution, e.g., $x = 1$, $y = 1$, $z = -2$. Thus $v = (1, 1, -2)$ forms a basis of the eigenspace E_3.

Observe that T is not diagonalizable, since T has only two linearly independent eigenvectors.

10.23. Show that 0 is an eigenvalue of T if and only if T is singular.

We have that 0 is an eigenvalue of T if and only if there exists a nonzero vector v such that $T(v) = 0v = 0$, i.e., that T is singular.

10.24. Suppose λ is an eigenvalue of an invertible operator T. Show that λ^{-1} is an eigenvalue of T^{-1}.

Since T is invertible, it is also nonsingular; hence, by Problem 10.23 $\lambda \neq 0$.

By definition of an eigenvalue, there exists a nonzero vector v for which $T(v) = \lambda v$. Applying T^{-1} to both sides, we obtain $v = T^{-1}(\lambda v) = \lambda T^{-1}(v)$. Hence $T^{-1}(v) = \lambda^{-1}v$; that is, λ^{-1} is an eigenvalue of T^{-1}.

10.25. Suppose dim $V = n$. Let $T : V \to V$ be an invertible operator. Show that T^{-1} is equal to a polynomial in T of degree not exceeding n.

Let $m(t)$ be the minimum polynomial of T. Then $m(t) = t^r + a_{r-1}t^{r-1} + \cdots + a_1t + a_0$, where $r \leq n$. Since T is invertible, $a_0 \neq 0$. We have

$$m(T) = T^r + a_{r-1}T^{r-1} + \cdots + a_1T + a_0I = 0$$

Hence

$$-\frac{1}{a_0}(T^{r-1} + a_{r-1}T^{r-2} + \cdots + a_1 I)T = I \qquad \text{and} \qquad T^{-1} = -\frac{1}{a_0}(T^{r-1} + a_{r-1}T^{r-2} + \cdots + a_1 I)$$

10.26. Prove Theorem 10.7.

The proof is by induction on n. If $n = 1$, then u_1 is linearly independent since $u_1 \neq 0$. Assume $n > 1$. Suppose

$$a_1 u_1 + a_2 u_2 + \cdots + a_n u_n = 0 \tag{1}$$

where the a_i are scalars. Applying T to the above relation, we obtain by linearity

$$a_1 T(u_1) + a_2 T(u_2) + \cdots + a_n T(u_n) = T(0) = 0$$

But by hypothesis $T(u_i) = \lambda_i u_i$; hence

$$a_1 \lambda_1 u_1 + a_2 \lambda_2 u_2 + \cdots + a_n \lambda_n u_n = 0 \tag{2}$$

On the other hand, multiplying (1) by λ_n.

$$a_1 \lambda_n u_1 + a_2 \lambda_n u_2 + \cdots + a_n \lambda_n u_n = 0 \tag{3}$$

Now subtracting (3) from (2),

$$a_1(\lambda_1 - \lambda_n)u_1 + a_2(\lambda_2 - \lambda_n)u_2 + \cdots + a_{n-1}(\lambda_{n-1} - \lambda_n)u_{n-1} = 0$$

By induction, $u_1, u_2, \ldots, u_{n-1}$ are linearly independent; hence each of the above coefficients is 0. Since the λ_i are distinct, $\lambda_i - \lambda_n \neq 0$ for $i \neq n$. Hence $a_1 = \cdots = a_{n-1} = 0$. Substituting this into (1) we get $a_n u_n = 0$, and hence $a_n = 0$. Thus the u_i are linearly independent.

10.27. Prove Theorem 10.10.

Suppose the geometric multiplicity of λ is r. Then λ contains r linearly independent eigenvectors $v_1, \ldots, v_r$. Extend the set $\{v_i\}$ to a basis of V: $\{v_1, \ldots, v_r, w_1, \ldots, w_s\}$. We have

$$
\begin{aligned}
T(v_1) &= \lambda v_1 \\
T(v_2) &= \qquad \lambda v_2 \\
&\cdots\cdots\cdots\cdots\cdots\cdots\cdots\cdots\cdots\cdots\cdots\cdots \\
T(v_r) &= \qquad\qquad\quad \lambda v_r \\
T(w_1) &= a_{11}v_1 + \cdots + a_{1r}v_r + b_{11}w_1 + \cdots + b_{1s}w_s \\
T(w_2) &= a_{21}v_1 + \cdots + a_{2r}v_r + b_{21}w_1 + \cdots + b_{2s}w_s \\
&\cdots\cdots\cdots\cdots\cdots\cdots\cdots\cdots\cdots\cdots\cdots\cdots \\
T(w_s) &= a_{s1}v_1 + \cdots + a_{sr}v_r + b_{s1}w_1 + \cdots + b_{ss}w_s
\end{aligned}
$$

The matrix of T in the above basis is

$$
M = \left(
\begin{array}{cccc:cccc}
\lambda & 0 & \cdots & 0 & a_{11} & a_{21} & \cdots & a_{s1} \\
0 & \lambda & \cdots & 0 & a_{12} & a_{22} & \cdots & a_{s2} \\
\multicolumn{8}{c}{\cdots\cdots\cdots\cdots\cdots\cdots\cdots\cdots} \\
0 & 0 & \cdots & \lambda & a_{1r} & a_{2r} & \cdots & a_{sr} \\
\hdashline
0 & 0 & \cdots & 0 & b_{11} & b_{21} & \cdots & b_{r1} \\
0 & 0 & \cdots & 0 & b_{12} & b_{22} & \cdots & b_{r2} \\
\multicolumn{8}{c}{\cdots\cdots\cdots\cdots\cdots\cdots\cdots\cdots} \\
0 & 0 & \cdots & 0 & b_{1s} & b_{2s} & \cdots & b_{ss}
\end{array}
\right)
= \left(
\begin{array}{c:c}
\lambda I_r & A \\
\hdashline
0 & B
\end{array}
\right)
$$

where $A = (a_{ij})^T$ and $B = (b_{ij})^T$.

Since M is a block triangular matrix, the characteristic polynomial of λI_r, which is $(t - \lambda)^r$, must divide the characteristic polynomial of M and hence T. Thus the algebraic multiplicity of λ for the operator T is at least r, as required.

10.28. Let $\{v_1, \ldots, v_n\}$ be a basis of V. Let $T : V \to V$ be an operator for which $T(v_1) = 0$, $T(v_2) = a_{21}v_1$, $T(v_3) = a_{31}v_1 + a_{32}v_2, \ldots, T(v_n) = a_{n1}v_1 + \cdots + a_{n, n-1}v_{n-1}$. Show that $T^n = 0$.

It suffices to show that

$$T^j(v_j) = 0 \qquad\qquad (*)$$

for $j = 1, \ldots, n$. For then it follows that

$$T^n(v_j) = T^{n-j}(T^j(v_j)) = T^{n-j}(0) = 0, \qquad \text{for } j = 1, \ldots, n$$

and, since $\{v_1, \ldots, v_n\}$ is a basis, $T^n = 0$.

We prove $(*)$ by induction on j. The case $j = 1$ is true by hypothesis. The inductive step follows (for $j = 2, \ldots, n$) from

$$
\begin{aligned}
T^j(v_j) &= T^{j-1}(T(v_j)) = T^{j-1}(a_{j1}v_1 + \cdots + a_{j, j-1}v_{j-1}) \\
&= a_{j1}T^{j-1}(v_1) + \cdots + a_{j, j-1}T^{j-1}(v_{j-1}) \\
&= a_{j1}0 + \cdots + a_{j, j-1}0 = 0
\end{aligned}
$$

Remark: Observe that the matrix representation of T in the above basis is triangular with diagonal elements 0:

$$
\begin{pmatrix}
0 & a_{21} & a_{31} & \cdots & a_{n1} \\
0 & 0 & a_{32} & \cdots & a_{n2} \\
\multicolumn{5}{c}{\dotfill} \\
0 & 0 & 0 & \cdots & a_{n, n-1} \\
0 & 0 & 0 & \cdots & 0
\end{pmatrix}
$$

MATRIX REPRESENTATIONS OF LINEAR MAPPINGS

10.29. Let $F : \mathbf{R}^3 \to \mathbf{R}^2$ be the linear mapping defined by $F(x, y, z) = (3x + 2y - 4z, x - 5y + 3z)$.

(a) Find the matrix of F in the following bases of $\mathbf{R}^3$ and $\mathbf{R}^2$:

$$S = \{w_1 = (1, 1, 1),\ w_2 = (1, 1, 0),\ w_3 = (1, 0, 0)\} \qquad S' = \{u_1 = (1, 3),\ u_2 = (2, 5)\}$$

(b) Verify that the action of F is preserved by its matrix representation; that is, for any $v \in \mathbf{R}^3$, $[F]_S^{S'}[v]_S = [F(v)]_{S'}$

(a) From Problem 10.2, $(a, b) = (-5a + 2b)u_1 + (3a - b)u_2$. Thus

$$
\begin{aligned}
F(w_1) &= F(1, 1, 1) = (1, -1) = -7u_1 + 4u_2 \\
F(w_2) &= F(1, 1, 0) = (5, -4) = -33u_1 + 19u_2 \\
F(w_3) &= F(1, 0, 0) = (3, 1) = -13u_1 + 8u_2
\end{aligned}
$$

Write the coordinates of $F(w_1)$, $F(w_2)$, $F(w_3)$ as columns to get

$$[F]_S^{S'} = \begin{pmatrix} -7 & -33 & -13 \\ 4 & 19 & 8 \end{pmatrix}$$

(b) If $v = (x, y, z)$ then, by Problem 10.3, $v = zw_1 + (y - z)w_2 + (x - y)w_3$. Also,

$$F(v) = (3x + 2y - 4z,\ x - 5y + 3z) = (-13x - 20y + 26z)u_1 + (8x + 11y - 15z)u_2$$

Hence $\qquad [v]_S = (z,\ y - z,\ x - y)^T \quad$ and $\quad [F(v)]_{S'} = \begin{pmatrix} -13x - 20y + 26z \\ 8x + 11y - 15z \end{pmatrix}$

Thus $\qquad [F]_S^{S'}[v]_S = \begin{pmatrix} -7 & -33 & -13 \\ 4 & 19 & 8 \end{pmatrix} \begin{pmatrix} z \\ y - z \\ x - y \end{pmatrix} = \begin{pmatrix} -13x - 20y + 26z \\ 8x + 11y - 15z \end{pmatrix} = [F(v)]_{S'}$

10.30. Let $F : K^n \to K^m$ be the linear mapping defined by

$$F(x_1, x_2, \ldots, x_n) = (a_{11}x_1 + \cdots + a_{1n}x_n, a_{21}x_1 + \cdots + a_{2n}x_n, \ldots, a_{m1}x_1 + \cdots + a_{mn}x_n)$$

Show that the matrix representation of F relative to the usual bases of K^n and of K^m is given by

$$[F] = \begin{pmatrix} a_{11} & a_{12} & \cdots & a_{1n} \\ a_{21} & a_{22} & \cdots & a_{2n} \\ \hdotsfor{4} \\ a_{m1} & a_{m2} & \cdots & a_{mn} \end{pmatrix}$$

That is, the rows of $[F]$ are obtained from the coefficients of the x_i in the components of $F(x_1, \ldots, x_n)$, respectively.

We have

$$\begin{aligned} F(1, 0, \ldots, 0) &= (a_{11}, a_{21}, \ldots, a_{m1}) \\ F(0, 1, \ldots, 0) &= (a_{12}, a_{22}, \ldots, a_{m2}) \\ &\cdots\cdots\cdots\cdots\cdots \\ F(0, 0, \ldots, 1) &= (a_{1n}, a_{2n}, \ldots, a_{mn}) \end{aligned} \quad \text{and thus} \quad [F] = \begin{pmatrix} a_{11} & a_{12} & \cdots & a_{1n} \\ a_{21} & a_{22} & \cdots & a_{2n} \\ \hdotsfor{4} \\ a_{m1} & a_{m2} & \cdots & a_{mn} \end{pmatrix}$$

10.31. Find the matrix representation of each of the following linear mappings relative to the usual bases of $\mathbf{R}^n$:

$$F : \mathbf{R}^2 \to \mathbf{R}^3 \text{ defined by } F(x, y) = (3x - y, 2x + 4y, 5x - 6y)$$
$$F : \mathbf{R}^4 \to \mathbf{R}^2 \text{ defined by } F(x, y, s, t) = (3x - 4y + 2s - 5t, 5x + 7y - s - 2t)$$
$$F : \mathbf{R}^3 \to \mathbf{R}^4 \text{ defined by } F(x, y, z) = (2x + 3y - 8z, x + y + z, 4x - 5z, 6y)$$

By Problem 10.30, we need only look at the coefficients of the unknowns in $F(x, y, \ldots)$. Thus

$$[F] = \begin{pmatrix} 3 & -1 \\ 2 & 4 \\ 5 & -6 \end{pmatrix} \qquad [F] = \begin{pmatrix} 3 & -4 & 2 & -5 \\ 5 & 7 & -1 & -2 \end{pmatrix} \qquad [F] = \begin{pmatrix} 2 & 3 & -8 \\ 1 & 1 & 1 \\ 4 & 0 & -5 \\ 0 & 6 & 0 \end{pmatrix}$$

10.32. Let $T : \mathbf{R}^2 \leftarrow \mathbf{R}^2$ be defined by $T(x, y) = (2x - 3y, x + 4y)$. Find the matrix of T relative, respectively, to the following bases of $\mathbf{R}^2$:

$$E = \{e_1 = (1, 0), e_2 = (0, 1)\} \qquad \text{and} \qquad S = \{u_1 = (1, 3), u_2 = (2, 5)\}$$

(We can view T as a linear mapping from one space into another, each having its own basis.)

From Problem 10.2, $(a, b) = (-5a + 2b)u_1 + (3a - b)u_2$. Hence

$$\begin{aligned} T(e_1) &= T(1, 0) = (2, 1) &= -8u_1 + 5u_2 \\ T(e_2) &= T(0, 1) = (-3, 4) = 23u_1 - 13u_2 \end{aligned} \quad \text{and thus} \quad [T]_E^S = \begin{pmatrix} -8 & 23 \\ 5 & -13 \end{pmatrix}$$

10.33. Let $A = \begin{pmatrix} 2 & 5 & -3 \\ 1 & -4 & 7 \end{pmatrix}$. Recall that A determines a linear mapping $F : \mathbf{R}^3 \to \mathbf{R}^2$ defined by $F(v) = Av$ where v is written as a column vector. Find the matrix representation of F relative to the following bases of $\mathbf{R}^3$ and $\mathbf{R}^2$.

$$S = \{w_1 = (1, 1, 1), w_2 = (1, 1, 0), w_3 = (1, 0, 0)\} \qquad S' = \{u_1 = (1, 3), u_2 = (2, 5)\}$$

From Problem 10.2, $(a, b) = (-5a + 2b)u_1 + (3a - b)u_2$. Thus

$$F(w_1) = \begin{pmatrix} 2 & 5 & -3 \\ 1 & -4 & 7 \end{pmatrix} \begin{pmatrix} 1 \\ 1 \\ 1 \end{pmatrix} = \begin{pmatrix} 4 \\ 4 \end{pmatrix} = -12u_1 + 8u_2$$

$$F(w_2) = \begin{pmatrix} 2 & 5 & -3 \\ 1 & -4 & 7 \end{pmatrix} \begin{pmatrix} 1 \\ 1 \\ 0 \end{pmatrix} = \begin{pmatrix} 7 \\ -3 \end{pmatrix} = -41u_1 + 24u_2$$

$$F(w_3) = \begin{pmatrix} 2 & 5 & -3 \\ 1 & -4 & 7 \end{pmatrix} \begin{pmatrix} 1 \\ 0 \\ 0 \end{pmatrix} = \begin{pmatrix} 2 \\ 1 \end{pmatrix} = -8u_1 + 5u_2$$

Writing the coefficients of $F(w_1)$, $F(w_2)$, $F(w_3)$ as columns yields

$$[F]_S^{S'} = \begin{pmatrix} -12 & -41 & -8 \\ 8 & 24 & 5 \end{pmatrix}$$

10.34. Prove Theorem 10.16.

Suppose dim $V = m$ and dim $U = n$. Let W be the kernel of F and U' the image of F. We are given that rank $F = r$; hence the dimension of the kernel of F is $m - r$. Let $\{w_1, \ldots, w_{m-r}\}$ be a basis of the kernel of F and extend this to a basis of V:

$$\{v_1, \ldots, v_r, w_1, \ldots, w_{m-r}\}$$

Set $\qquad\qquad u_1 = F(v_1),\ u_2 = F(v_2),\ \ldots,\ u_r = F(v_r)$

We note that $\{u_1, \ldots, u_r\}$ is a basis of U', the image of F. Extend this to a basis

$$\{u_1, \ldots, u_r, u_{r+1}, \ldots, u_n\}$$

of U. Observe that

$$
\begin{aligned}
F(v_1) \quad &= u_1 = 1u_1 + 0u_2 + \cdots + 0u_r + 0u_{r+1} + \cdots + 0u_n \\
F(v_2) \quad &= u_2 = 0u_1 + 1u_2 + \cdots + 0u_r + 0u_{r+1} + \cdots + 0u_n \\
&\cdots\cdots\cdots\cdots\cdots\cdots\cdots\cdots\cdots\cdots\cdots\cdots\cdots\cdots \\
F(v_r) \quad &= u_r = 0u_1 + 0u_2 + \cdots + 1u_r + 0u_{r+1} + \cdots + 0u_n \\
F(w_1) \quad &= 0 = 0u_1 + 0u_2 + \cdots + 0u_r + 0u_{r+1} + \cdots + 0u_n \\
&\cdots\cdots\cdots\cdots\cdots\cdots\cdots\cdots\cdots\cdots\cdots\cdots\cdots\cdots \\
F(w_{m-r}) &= 0 = 0u_1 + 0u_2 + \cdots + 0u_r + 0u_{r+1} + \cdots + 0u_n
\end{aligned}
$$

Thus the matrix of F in the above bases has the required form.

Supplementary Problems

MATRIX REPRESENTATIONS OF LINEAR OPERATORS

10.35. Find the matrix representation of each of the following linear operators T on $\mathbf{R}^3$ relative to the usual basis:

(a) $T(x, y, z) = (x, y, 0)$

(b) $T(x, y, z) = (2x - 7y - 4z, 3x + y + 4z, 6x - 8y + z)$

(c) $T(x, y, z) = (z, y + z, x + y + z)$

10.36. Find the matrix of each operator T in Problem 10.35 with respect to the basis

$$S = \{u_1 = (1, 1, 0),\ u_2 = (1, 2, 3),\ u_3 = (1, 3, 5)\}$$

10.37. Let D be the differential operator, i.e., $D(f) = df/dt$. Each of the following sets is a basis of a vector space V of functions $f : \mathbf{R} \to \mathbf{R}$. Find the matrix of D in each basis:

(a) $\{e^t, e^{2t}, te^{2t}\}$,   (b) $\{\sin t, \cos t\}$,   (c) $\{e^{5t}, te^{5t}, t^2 e^{5t}\}$,   (d) $\{1, t, \sin 3t, \cos 3t\}$

10.38. Consider the complex field $\mathbf{C}$ as a vector space over the real field $\mathbf{R}$. Let T be the conjugation operator on $\mathbf{C}$, that is, $T(z) = \bar{z}$. Find the matrix of T in each basis: (a) $\{1, i\}$, and (b) $\{1 + i, 1 + 2i\}$.

10.39. Let V be the vector space of 2×2 matrices over $\mathbf{R}$ and let $M = \begin{pmatrix} a & b \\ c & d \end{pmatrix}$. Find the matrix of each of the following linear operators T on V in the usual basis (see Problem 10.18):

(a) $T(A) = MA$,   (b) $T(A) = AM$,   (c) $T(A) = MA - AM$

10.40. Let 1_V and 0_V denote the identity and zero operators, respectively, on a vector space V. Show that, for any basis S of V, (a) $[1_V]_S = I$, the identity matrix, (b) $[0_V]_S = 0$, the zero matrix.

CHANGE OF BASIS, SIMILAR MATRICES

10.41. Consider the following bases of $\mathbf{R}^2$: $E = \{e_1 = (1, 0),\ e_2 = (0, 1)\}$ and $S = \{u_1 = (1, 2),\ u_2 = (2, 3)\}$.

(a) Find the change-of-basis matrices P and Q from E to S and from S to E, respectively. Verify $Q = P^{-1}$.

(b) Show that $[v]_E = P[v]_S$ for any vector $v \in \mathbf{R}^2$.

(c) Verify that $[T]_S = P^{-1}[T]_E P$ for the linear operator $T(x, y) = (2x - 3y, x + y)$.

10.42. Find the trace and determinant of each linear map on $\mathbf{R}^3$:

(a) $F(x, y, z) = (x + 3y, 3x - 2z, x - 4y - 3z)$,   (b) $G(x, y, z) = (x + y - z, x + 3y, 4y + 3z)$

10.43. Suppose $S = \{u_1, u_2\}$ is a basis of V and $T : V \to V$ is a linear operator for which $T(u_1) = 3u_1 - 2u_2$ and $T(u_2) = u_1 + 4u_2$. Suppose $S' = \{w_1, w_2\}$ is a basis of V for which $w_1 = u_1 + u_2$ and $w_2 = 2u_1 + 3u_2$. Find the matrix of T in the basis S'.

10.44. Consider the bases $\{1, i\}$ and $\{1 + i, 1 + 2i\}$ of the complex field $\mathbf{C}$ over the real field $\mathbf{R}$. (a) Find the change-of-basis matrices P and Q from S to S' and from S' to S, respectively. Verify that $Q = P^{-1}$. (b) Show that $[T]_{S'} = P^{-1}[T]_S P$ for the conjugate operator T in Problem 10.38.

DIAGONALIZATION OF LINEAR OPERATORS, EIGENVALUES AND EIGENVECTORS

10.45. Suppose v is an eigenvector of operators S and T. Show that v is also an eigenvector of the operator $aS + bT$ where a and b are any scalars.

10.46. Suppose v is an eigenvector of an operator T belonging to the eigenvalue λ. Show that for $n > 0$, v is also an eigenvector of T^n belonging to λ^n.

10.47. Suppose λ is an eigenvalue of an operator T. Show that $f(\lambda)$ is an eigenvalue of $f(T)$.

10.48. Show that a linear operator T is diagonalizable if and only if its minimum polynomial is a product of distinct linear factors.

10.49. Let S and T be linear operators such that $ST = TS$. Let λ be an eigenvalue of T and let W be its eigenspace. Show that W is invariant under S, i.e., $S(W) \subseteq W$.

MATRIX REPRESENTATIONS OF LINEAR MAPPINGS

10.50. Find the matrix representation relative to the usual bases for $\mathbf{R}^n$ of the linear mapping $F: \mathbf{R}^3 \to \mathbf{R}^2$ defined by $F(x, y, z) = (2x - 4y + 9z, 5x + 3y - 2z)$.

10.51. Let $F: \mathbf{R}^3 \to \mathbf{R}^2$ be the linear mapping defined by $F(x, y, z) = (2x + y - z, 3x - 2y + 4z)$. Find the matrix of F in the following bases of $\mathbf{R}^3$ and $\mathbf{R}^2$:

$$S = \{w_1 = (1, 1, 1), w_2 = (1, 1, 0), w_3 = (1, 0, 0)\} \qquad S' = \{v_1 = (1, 3), v_2 = (1, 4)\}$$

Verify that, for any vector $v \in \mathbf{R}^3$, $[F]_S^{S'}[v]_S = [F(v)]_{S'}$.

10.52. Let S and S' be bases of V, and let 1_V be the identity mapping on V. Show that the matrix of 1_V relative to the bases S and S' is the inverse of the change-of-basis matrix P from S to S'; that is, $[1_V]_S^{S'} = P^{-1}$.

10.53. Prove Theorem 10.12.

10.54. Prove Theorem 10.13.

10.55. Prove Theorem 10.15.

Answers to Supplementary Problems

10.35. (a) $\begin{pmatrix} 1 & 0 & 0 \\ 0 & 1 & 0 \\ 0 & 0 & 0 \end{pmatrix}$, (b) $\begin{pmatrix} 2 & -7 & -4 \\ 3 & 1 & 4 \\ 6 & -8 & 1 \end{pmatrix}$, (c) $\begin{pmatrix} 0 & 0 & 1 \\ 0 & 1 & 1 \\ 1 & 1 & 1 \end{pmatrix}$

10.36. (a) $\begin{pmatrix} 1 & 3 & 5 \\ 0 & -5 & -10 \\ 0 & 3 & 6 \end{pmatrix}$, (b) $\begin{pmatrix} 15 & 51 & 104 \\ -49 & -191 & -351 \\ 29 & 116 & 208 \end{pmatrix}$, (c) $\begin{pmatrix} 0 & 1 & 2 \\ -1 & 2 & 3 \\ 1 & 0 & 0 \end{pmatrix}$

10.37. (a) $\begin{pmatrix} 1 & 0 & 0 \\ 0 & 2 & 1 \\ 0 & 0 & 2 \end{pmatrix}$, (b) $\begin{pmatrix} 0 & -1 \\ 1 & 0 \end{pmatrix}$, (c) $\begin{pmatrix} 5 & 1 & 0 \\ 0 & 5 & 2 \\ 0 & 0 & 5 \end{pmatrix}$, (d) $\begin{pmatrix} 0 & 1 & 0 & 0 \\ 0 & 0 & 0 & 0 \\ 0 & 0 & 0 & -3 \\ 0 & 0 & 3 & 0 \end{pmatrix}$

10.38. (a) $\begin{pmatrix} 1 & 0 \\ 0 & -1 \end{pmatrix}$, (b) $\begin{pmatrix} -3 & 4 \\ -2 & -3 \end{pmatrix}$

10.39. (a) $\begin{pmatrix} a & 0 & b & 0 \\ 0 & a & 0 & b \\ c & 0 & d & 0 \\ 0 & c & 0 & d \end{pmatrix}$, (b) $\begin{pmatrix} a & c & 0 & 0 \\ b & d & 0 & 0 \\ 0 & 0 & a & c \\ 0 & 0 & b & d \end{pmatrix}$, (c) $\begin{pmatrix} 0 & -c & b & 0 \\ -b & a-d & 0 & b \\ c & 0 & d-a & -c \\ 0 & c & -b & 0 \end{pmatrix}$

10.41. $P = \begin{pmatrix} 1 & 2 \\ 2 & 3 \end{pmatrix}$, $Q = \begin{pmatrix} -3 & 2 \\ 2 & -1 \end{pmatrix}$

10.42. (*a*) $-2, 13, t^3 + 2t^2 - 20t - 13$;   (*b*) $7, 2, t^3 - 7t^2 + 14t - 2$

10.43. $\begin{pmatrix} 8 & 11 \\ -2 & -1 \end{pmatrix}$

10.44. $P = \begin{pmatrix} 1 & 1 \\ 1 & 2 \end{pmatrix}, \qquad Q = \begin{pmatrix} 2 & -1 \\ -1 & 1 \end{pmatrix}$

10.50. $\begin{pmatrix} 2 & -4 & 9 \\ 5 & 3 & -2 \end{pmatrix}$

10.51. $\begin{pmatrix} 3 & 11 & 5 \\ -1 & -8 & -3 \end{pmatrix}$

Chapter 11

Canonical Forms

11.1 INTRODUCTION

Let T be a linear operator on a vector space of finite dimension. As seen in Chapter 10, T may not have a diagonal matrix representation. However, it is still possible to "simplify" the matrix representation of T in a number of ways. This is the main topic of this chapter. In particular, we obtain the primary decomposition theorem, and the triangular, Jordan and rational canonical forms.

We comment that the triangular and Jordan canonical forms exist for T if and only if the characteristic polynomial $\Delta(t)$ of T has all its roots in the base field K. This is always true if K is the complex field $\mathbf{C}$ but may not be true if K is the real field $\mathbf{R}$.

We also introduce the idea of a quotient space. This is a very powerful tool and will be used in the proof of the existence of the triangular and rational canonical forms.

11.2 TRIANGULAR FORM

Let T be a linear operator on an n-dimensional vector space V. Suppose T can be represented by the triangular matrix

$$A = \begin{pmatrix} a_{11} & a_{12} & \cdots & a_{1n} \\ & a_{22} & \cdots & a_{2n} \\ & & \cdots\cdots\cdots \\ & & & a_{nn} \end{pmatrix}$$

Then the characteristic polynomial of T,

$$\Delta(t) = |tI - A| = (t - a_{11})(t - a_{22}) \cdots (t - a_{nn})$$

is a product of linear factors. The converse is also true and is an important theorem, namely (see Problem 11.28 for the proof),

Theorem 11.1: Let $T : V \to V$ be a linear operator whose characteristic polynomial factors into linear polynomials. Then there exists a basis of V in which T is represented by a triangular matrix.

Theorem 11.1 (Alternate Form): Let A be a square matrix whose characteristic polynomial factors into linear polynomials. Then A is similar to a triangular matrix, i.e., there exists an invertible matrix P such that $P^{-1}AP$ is triangular.

We say that an operator T can be brought into triangular form if it can be represented by a triangular matrix. Note that in this case the eigenvalues of T are precisely those entries appearing on the main diagonal. We give an application of this remark.

Example 11.1. Let A be a square matrix over the complex field $\mathbf{C}$. Suppose λ is an eigenvalue of A^2. Show that $\sqrt{\lambda}$ or $-\sqrt{\lambda}$ is an eigenvalue of A. We know by Theorem 11.1 that A is similar to a triangular matrix

$$B = \begin{pmatrix} \mu_1 & * & \cdots & * \\ & \mu_2 & \cdots & * \\ & & \cdots\cdots \\ & & & \mu_n \end{pmatrix}$$

Hence A^2 is similar to the matrix

$$B^2 = \begin{pmatrix} \mu_1^2 & * & \cdots & * \\ & \mu_2^2 & \cdots & * \\ & & \cdots\cdots \\ & & & \mu_n^2 \end{pmatrix}$$

Since similar matrices have the same eigenvalues, $\lambda = \mu_i^2$ for some i. Hence $\mu_i = \sqrt{\lambda}$ or $\mu_i = -\sqrt{\lambda}$; that is, $\sqrt{\lambda}$ or $-\sqrt{\lambda}$ is an eigenvalue of A.

11.3 INVARIANCE

Let $T : V \to V$ be linear. A subspace W of V is said to be invariant under T or T-invariant if T maps W into itself, i.e., if $v \in W$ implies $T(v) \in W$. In this case T restricted to W defines a linear operator on W; that is, T induces a linear operator $\hat{T} : W \to W$ defined by $\hat{T}(w) = T(w)$ for every $w \in W$.

Example 11.2

(a) Let $T : \mathbf{R}^3 \to \mathbf{R}^3$ be the linear operator which rotates each vector about the z axis by an angle θ (Fig. 11-1); that is, defined by

$$T(x, y, z) = (x \cos \theta - y \sin \theta,\ x \sin \theta + y \cos \theta,\ z)$$

Observe that each vector $w = (a, b, 0)$ in the xy plane W remains in W under the mapping T, hence, W is T-invariant. Observe also that the z axis U is invariant under T. Furthermore, the restriction of T to W rotates each vector about the origin O, and the restriction of T to U is the identity mapping of U.

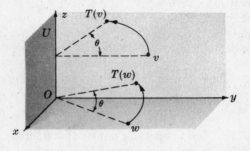

Fig. 11-1

(b) Nonzero eigenvectors of a linear operator $T : V \to V$ may be characterized as generators of T-invariant 1-dimensional subspaces. For suppose $T(v) = \lambda v$, $v \neq 0$. Then $W = \{kv,\ k \in K\}$, the 1-dimensional subspace generated by v, is invariant under T because

$$T(kv) = kT(v) = k(\lambda v) = k\lambda v \in W$$

Conversely, suppose $\dim U = 1$ and $u \neq 0$ spans U, and U is invariant under T. Then $T(u) \in U$ and so $T(u)$ is a multiple of u, i.e., $T(u) = \mu u$. Hence u is an eigenvector of T.

The next theorem, proved in Problem 11.3, gives us an important class of invariant subspaces.

Theorem 11.2: Let $T : V \to V$ be any linear operator, and let $f(t)$ be any polynomial. Then the kernel of $f(T)$ is invariant under T.

The notion of invariance is related to matrix representations (Problem 11.5) as follows.

Theorem 11.3: Suppose W is an invariant subspace of $T : V \to V$. Then T has a block matrix representation $\begin{pmatrix} A & B \\ 0 & C \end{pmatrix}$ where A is a matrix representation of the restriction $\hat{T}$ of T to W.

11.4 INVARIANT DIRECT-SUM DECOMPOSITIONS

A vector space V is termed the direct sum of its subspaces $W_1, \ldots, W_r$, written

$$V = W_1 \oplus W_2 \oplus \cdots \oplus W_r$$

if every vector $v \in V$ can be written uniquely in the form

$$v = w_1 + w_2 + \cdots + w_r \qquad \text{with} \qquad w_i \in W_i$$

The following theorem, proved in Problem 11.7, applies.

Theorem 11.4: Suppose $W_1, \ldots, W_r$ are subspaces of V, and suppose

$$\{w_{11}, \ldots, w_{1n_1}\}, \ldots, \{w_{r1}, \ldots, w_{rn_r}\}$$

are bases of $W_1, \ldots, W_r$, respectively. Then V is the direct sum of the W_i if and only if the union $B = \{w_{11}, \ldots, w_{1n_1}, \ldots, w_{r1}, \ldots, w_{rn_r}\}$ is a basis of V.

Now suppose $T : V \to V$ is linear and V is the direct sum of (nonzero) T-invariant subspaces $W_1, \ldots, W_r$:

$$V = W_1 \oplus \cdots \oplus W_r \qquad \text{and} \qquad T(W_i) \subseteq W_i \qquad i = 1, \ldots, r$$

Let T_i denote the restriction of T to W_i. Then T is said to be decomposable into the operators T_i or T is said to be the direct sum of the T_i, written $T = T_1 \oplus \cdots \oplus T_r$. Also, the subspaces $W_1, \ldots, W_r$ are said to reduce T or to form a T-invariant direct-sum decomposition of V.

Consider the special case where two subspaces U and W reduce an operator $T : V \to V$; say, $\dim U = 2$ and $\dim W = 3$ and suppose $\{u_1, u_2\}$ and $\{w_1, w_2, w_3\}$ are bases of U and W, respectively. If T_1 and T_2 denote the restrictions of T to U and W, respectively, then

$$
\begin{aligned}
T_1(u_1) &= a_{11}u_1 + a_{12}u_2 & T_2(w_1) &= b_{11}w_1 + b_{12}w_2 + b_{13}w_3 \\
T_1(u_2) &= a_{21}u_1 + a_{22}u_2 & T_2(w_2) &= b_{21}w_1 + b_{22}w_2 + b_{23}w_3 \\
& & T_2(w_3) &= b_{31}w_1 + b_{32}w_2 + b_{33}w_3
\end{aligned}
$$

Hence

$$A = \begin{pmatrix} a_{11} & a_{21} \\ a_{12} & a_{22} \end{pmatrix} \qquad \text{and} \qquad B = \begin{pmatrix} b_{11} & b_{21} & b_{31} \\ b_{12} & b_{22} & b_{32} \\ b_{13} & b_{23} & b_{33} \end{pmatrix}$$

are matrix representations of T_1 and T_2, respectively. By Theorem 11.4, $\{u_1, u_2, w_1, w_2, w_3\}$ is a basis of V. Since $T(u_i) = T_1(u_i)$ and $T(w_j) = T_2(w_j)$, the matrix of T in this basis is the block diagonal matrix $\begin{pmatrix} A & 0 \\ 0 & B \end{pmatrix}$.

A generalization of the above argument gives us the following theorem.

Theorem 11.5: Suppose $T : V \to V$ is linear and V is the direct sum of T-invariant subspaces, say $W_1, \ldots, W_r$. If A_i is a matrix representation of the restriction of T to W_i, then T can be represented by the block diagonal matrix

$$M = \begin{pmatrix} A_1 & 0 & \dots & 0 \\ 0 & A_2 & \dots & 0 \\ \dots\dots\dots\dots\dots\dots \\ 0 & 0 & \dots & A_r \end{pmatrix}$$

The block diagonal matrix M with diagonal entries $A_1, \dots, A_r$ is sometimes called the direct sum of the matrices $A_1, \dots, A_r$ and denoted by $M = A_1 \oplus \cdots \oplus A_r$.

11.5 PRIMARY DECOMPOSITION

The following theorem shows that any operator $T : V \to V$ is decomposable into operators whose minimum polynomials are powers of irreducible polynomials. This is the first step in obtaining a canonical form for T,

Primary Decomposition Theorem 11.6: Let $T : V \to V$ be a linear operator with minimal polynomial

$$m(t) = f_1(t)^{n_1} f_2(t)^{n_2} \cdots f_r(t)^{n_r}$$

where the $f_i(t)$ are distinct monic irreducible polynomials. Then V is the direct sum of T-invariant subspaces $W_1, \dots, W_r$ where W_i is the kernel of $f_i(T)^{n_i}$. Moreover, $f_i(t)^{n_i}$ is the minimum polynomial of the restriction of T to W_i.

Since the polynomials $f_1(t)^{n_i}$ are relatively prime, the above fundamental result follows (Problem 11.11) from the next two theorems.

Theorem 11.7: Suppose $T : V \to V$ is linear, and suppose $f(t) = g(t)h(t)$ are polynomials such that $f(T) = 0$ and $g(t)$ and $h(t)$ are relatively prime. Then V is the direct sum of the T-invariant subspaces U and W, where $U = \text{Ker } g(T)$ and $W = \text{Ker } h(T)$.

Theorem 11.8: In Theorem 11.7, if $f(t)$ is the minimum polynomial of T [and $g(t)$ and $h(t)$ are monic], then $g(t)$ and $h(t)$ are the minimum polynomials of the restrictions of T to U and W, respectively.

We will also use the primary decomposition theorem to prove the following useful characterization of diagonalizable operators (see Problem 11.12 for the proof).

Theorem 11.9 (Alternate Form): A matrix A is similar to a diagonal matrix if and only if its minimum polynomial is a product of distinct linear polynomials.

Theorem 11.9 (Alternate Form): A matrix A is similar to a diagonal matrix if and only if its minimum polynomial is a product of distinct linear polynomials.

Example 11.3. Suppose $A \neq I$ is a square matrix for which $A^3 = I$. Determine whether or not A is similar to a diagonal matrix if A is a matrix over (i) the real field $\mathbf{R}$, (ii) the complex field $\mathbf{C}$.
Since $A^3 = I$, A is a zero of the polynomial $f(t) = t^3 - 1 = (t - 1)(t^2 + t + 1)$. The minimum polynomial $m(t)$ of A cannot be $t - 1$, since $A \neq I$. Hence

$$m(t) = t^2 + t + 1 \qquad \text{or} \qquad m(t) = t^3 - 1$$

Since neither polynomial is a product of linear polynomials over $\mathbf{R}$, A is not diagonalizable over $\mathbf{R}$. On the other hand, each of the polynomials is a product of distinct linear polynomials over $\mathbf{C}$. Hence A is diagonalizable over $\mathbf{C}$.

11.6 NILPOTENT OPERATORS

A linear operator $T : V \to V$ is termed nilpotent if $T^n = 0$ for some positive integer n; we call k the index of nilpotency of T if $T^k = 0$ but $T^{k-1} \neq 0$. Analogously, a square matrix A is termed nilpotent if $A^n = 0$ for some positive integer n, and of index k if $A^k = 0$ but $A^{k-1} \neq 0$. Clearly the minimum polynomial of a nilpotent operator (matrix) of index k is $m(t) = t^k$; hence 0 is its only eigenvalue.

The fundamental result on nilpotent operators follows.

Theorem 11.10:   Let $T : V \to V$ be a nilpotent operator of index k. Then T has a block diagonal matrix representation whose diagonal entries are of the form

$$N = \begin{pmatrix} 0 & 1 & 0 & \dots & 0 & 0 \\ 0 & 0 & 1 & \dots & 0 & 0 \\ \multicolumn{6}{c}{\dotfill} \\ 0 & 0 & 0 & \dots & 0 & 1 \\ 0 & 0 & 0 & \dots & 0 & 0 \end{pmatrix}$$

(i.e., all entries of N are 0 except those just above the main diagonal where they are 1). There is at least one N of order k and all other N are of orders $\leq k$. The number of N of each possible order is uniquely determined by T. Moreover, the total number of N of all orders is equal to the nullity of T.

In the proof of the above theorem (Problem 11.16), we shall show that the number of N of order i is equal to $2m_i - m_{i+1} - m_{i-1}$, where m_i is the nullity of T^i.

We remark that the above matrix N is itself nilpotent and that its index of nilpotency is equal to its order (Problem 11.13). Note that the matrix N of order 1 is just the 1×1 zero matrix (0).

11.7 JORDAN CANONICAL FORM

An operator T can be put into Jordan canonical form if its characteristic and minimum polynomials factor into linear polynomials. This is always true if K is the complex field $\mathbf{C}$. In any case, we can always extend the base field K to a field in which the characteristic and minimum polynomials do factor into linear factors; thus in a broad sense every operator has a Jordan canonical form. Analogously, every matrix is similar to a matrix in Jordan canonical form.

Theorem 11.11:   Let $T : V \to V$ be a linear operator whose characteristic and minimum polynomials are, respectively,

$$\Delta(t) = (t - \lambda_1)^{n_1} \cdots (t - \lambda_r)^{n_r} \qquad \text{and} \qquad m(t) = (t - \lambda_1)^{m_1} \cdots (t - \lambda_r)^{m_r}$$

where the λ_i are distinct scalars. Then T has a block diagonal matrix representation J whose diagonal entries are of the form

$$J_{ij} = \begin{pmatrix} \lambda_i & 1 & 0 & \dots & 0 & 0 \\ 0 & \lambda_i & 1 & \dots & 0 & 0 \\ \multicolumn{6}{c}{\dotfill} \\ 0 & 0 & 0 & \dots & \lambda_i & 1 \\ 0 & 0 & 0 & \dots & 0 & \lambda_i \end{pmatrix}$$

For each λ_i the corresponding blocks J_{ij} have the following properties:

(i)   There is at least one J_{ij} of order m_i; all other J_{ij} are of order $\leq m_i$.

(ii)   The sum of the orders of the J_{ij} is n_i.

(iii)   The number of J_{ij} equals the geometric multiplicity of λ_i.

(iv)   The number of J_{ij} of each possible order is uniquely determined by T.

The matrix J appearing in the above theorem is called the Jordan canonical form of the operator T. A diagonal block J_{ij} is called a Jordan block belonging to the eigenvalue λ_i. Observe that

$$
\begin{pmatrix}
\lambda_i & 1 & 0 & \dots & 0 & 0 \\
0 & \lambda_i & 1 & \dots & 0 & 0 \\
\multicolumn{6}{c}{\dotfill} \\
0 & 0 & 0 & \dots & \lambda_i & 1 \\
0 & 0 & 0 & \dots & 0 & \lambda_i
\end{pmatrix}
=
\begin{pmatrix}
\lambda_i & 0 & \dots & 0 & 0 \\
0 & \lambda_i & \dots & 0 & 0 \\
\multicolumn{5}{c}{\dotfill} \\
0 & 0 & \dots & 0 & 0 \\
0 & 0 & \dots & 0 & \lambda_i
\end{pmatrix}
+
\begin{pmatrix}
0 & 1 & 0 & \dots & 0 & 0 \\
0 & 0 & 1 & \dots & 0 & 0 \\
\multicolumn{6}{c}{\dotfill} \\
0 & 0 & 0 & \dots & 0 & 1 \\
0 & 0 & 0 & \dots & 0 & 0
\end{pmatrix}
$$

That is,
$$J_{ij} = \lambda_i I + N$$

where N is the nilpotent block appearing in Theorem 11.10. In fact, we will prove Theorem 11.11 (Problem 11.18) by showing that T can be decomposed into operators, each the sum of a scalar and a nilpotent operator.

Example 11.4. Suppose the characteristic and minimum polynomials of an operator T are, respectively,

$$\Delta(t) = (t - 2)^4(t - 3)^3 \quad \text{and} \quad m(t) = (t - 2)^2(t - 3)^2$$

Then the Jordan canonical form of T is one of the following matrices:

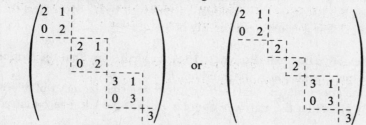

The first matrix occurs if T has two independent eigenvectors belonging to its eigenvalue 2; and the second matrix occurs if T has three independent eigenvectors belonging to 2.

11.8 CYCLIC SUBSPACES

Let T be a linear operator on a vector space V of finite dimension over K. Suppose $v \in V$ and $v \neq 0$. The set of all vectors of the form $f(T)(v)$, where $f(t)$ ranges over all polynomials over K, is a T-invariant subspace of V called the T-cyclic subspace of V generated by v; we denote it by $Z(v, T)$ and denote the restriction of T to $Z(v, T)$ by T_v. We could equivalently define $Z(v, T)$ as the intersection of all T-invariant subspaces of V containing v.

Now consider the sequence

$$v, \ T(v), \ T^2(v), \ T^3(v), \ \dots$$

of powers of T acting on v. Let k be the lowest integer such that $T^k(v)$ is a linear combination of these vectors which precede it in the sequence; say,

$$T^k(v) = -a_{k-1}T^{k-1}(v) - \cdots - a_1 T(v) - a_0 v$$

Then
$$m_v(t) = t^k + a_{k-1}t^{k-1} + \cdots + a_1 t + a_0$$

is the unique monic polynomial of lowest degree for which $m_v(T)(v) = 0$. We call $m_v(t)$ the T-annihilator of v and $Z(v, T)$.

The following theorem (proved in Problem 11.29) applies.

Theorem 11.12: Let $Z(v, T)$, T_v, and $m_v(t)$ be defined as above. Then:

    (i)   The set $\{v, T(v), \dots, T^{k-1}(v)\}$ is a basis of $Z(v, T)$; hence dim $Z(v, T) = k$.

   (ii)   The minimum polynomial of T_v is $m_v(t)$.

   (iii) The matrix representation of T_v in the above basis is

$$
C = \begin{pmatrix}
0 & 0 & 0 & \ldots & 0 & -a_0 \\
1 & 0 & 0 & \ldots & 0 & -a_1 \\
0 & 1 & 0 & \ldots & 0 & -a_2 \\
\ldots\ldots\ldots\ldots\ldots\ldots\ldots\ldots \\
0 & 0 & 0 & \ldots & 0 & +a_{k-2} \\
0 & 0 & 0 & \ldots & 1 & -a_{k-1}
\end{pmatrix}
$$

The above matrix C is called the companion matrix of the polynomial $m_v(t)$.

11.9 RATIONAL CANONICAL FORM

In this section we present the rational canonical form for a linear operator $T : V \to V$. We emphasize that this form exists even when the minimum polynomial cannot be factored into linear polynomials. (Recall that this is not the case for the Jordan canonical form.)

Lemma 11.13: Let $T : V \to V$ be a linear operator whose minimum polynomial is $f(t)^n$ where $f(t)$ is a monic irreducible polynomial. Then V is the direct sum

$$V = Z(v_1, T) \oplus \cdots \oplus Z(v_r, T)$$

of T-cyclic subspaces $Z(v_i, T)$ with corresponding T-annihilators

$$f(t)^{n_1}, f(t)^{n_2}, \ldots, f(t)^{n_r} \qquad n = n_1 \geq n_2 \geq \cdots \geq n_r$$

Any other decomposition of V into T-cyclic subspaces has the same number of components and the same set of T-annihilators.

We emphasize that the above lemma, proved in Problem 11.31, does not say that the vectors v_i or the T-cyclic subspaces $Z(v_i, T)$ are uniquely determined by T; but it does say that the set of T-annihilators are uniquely determined by T. Thus T has a unique matrix representation

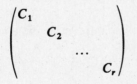

where the C_i are companion matrices. In fact, the C_i are the companion matrices to the polynomials $f(t)^{n_i}$.

Using the primary decomposition theorem and Lemma 11.13, we obtain the following fundamental result.

Theorem 11.14: Let $T : V \to V$ be a linear operator with minimum polynomial

$$m(t) = f_1(t)^{m_1} f_2(t)^{m_2} \cdots f_s(t)^{m_s}$$

where the $f_i(t)$ are distinct monic irreducible polynomials. Then T has a unique block diagonal matrix representation

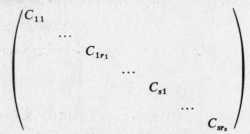

where the C_{ij} are companion matrices. In particular, the C_{ij} are the companion matrices of the polynomials $f_1(t)^{n_{ij}}$ where

$$m_1 = n_{11} \geq n_{12} \geq \cdots \geq n_{1r_1}, \ldots, m_s = n_{s1} \geq n_{s2} \geq \cdots \geq n_{sr_s}$$

The above matrix representation of T is called its rational canonical form. The polynomials $f_i(t)^{n_{ij}}$ are called the elementary divisors of T.

Example 11.5. Let V be a vector space of dimension 6 over $\mathbf{R}$, and let T be a linear operator whose minimum polynomial is $m(t) = (t^2 - t + 3)(t - 2)^2$. Then the rational canonical form of T is one of the following direct sums of companion matrices:

(i) $C(t^2 - t + 3) \oplus C(t^2 - t + 3) \oplus C((t - 2)^2)$

(ii) $C(t^2 - t + 3) \oplus C((t - 2)^2) \oplus C((t - 2)^2)$

(iii) $C(t^2 - t + 3) \oplus C((t - 2)^2) \oplus C(t - 2) \oplus C(t - 2)$

where $C(f(t))$ is the companion matrix of $f(t)$; that is,

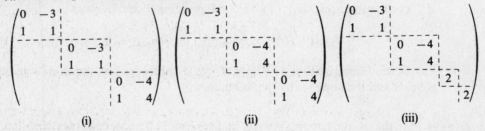

$$\text{(i)} \qquad\qquad \text{(ii)} \qquad\qquad \text{(iii)}$$

11.10 QUOTIENT SPACES

Let V be a vector space over a field K and let W be a subspace of V. If v is any vector in V, we write $v + W$ for the set of sums $v + w$ with $w \in W$:

$$v + W = \{v + w : w \in W\}$$

These sets are called the cosets of W in V. We show (Problem 11.22) that these cosets partition V into mutually disjoint subsets.

Example 11.6. Let W be the subspace of $\mathbf{R}^2$ defined by

$$W = \{(a, b): a = b\}$$

That is, W is the line given by the equation $x - y = 0$. We can view $v + W$ as a translation of the line, obtained by adding the vector v to each point in W. As shown in Fig. 11-2, $v + W$ is also a line and is parallel to W. Thus the cosets of W in $\mathbf{R}^2$ are precisely all the lines parallel to W.

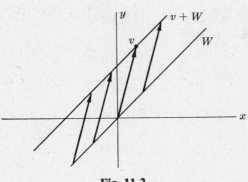

Fig. 11-2

In the following theorem we use the cosets of a subspace W of a vector space V to define a new vector space; it is called the quotient space of V by W and is denoted by V/W.

Theorem 11.15: Let W be a subspace of a vector space over a field K. Then the cosets of W in V form a vector space over K with the following operations of addition and scalar multiplication:

      (i) $(u + W) + (v + W) = (u + v) + W$

      (ii) $k(u + W) = ku + W$, where $k \in K$.

We note that, in the proof of Theorem 11.15 (Problem 11.24), it is first necessary to show that the operations are well defined; that is, whenever $u + W = u' + W$ and $v + W = v' + W$, then

      (i) $(u + v) + W = (u' + v') + W$ and   (ii) $ku + W = ku' + W$, for any $k \in K$

In the case of an invariant subspace, we have the following useful result, proved in Problem 11.27.

Theorem 11.16: Suppose W is a subspace invariant under a linear operator $T : V \to V$. Then T induces a linear operator $\bar{T}$ on V/W defined by $\bar{T}(v + W) = T(v) + W$. Moreover, if T is a zero of any polynomial, then so is $\bar{T}$. Thus the minimum polynomial of $\bar{T}$ divides the minimum polynomial of T.

Solved Problems

INVARIANT SUBSPACES

11.1. Suppose $T : V \to V$ is linear. Show that each of the following is invariant under T: (a) $\{0\}$, (b) V, (c) kernel of T, and (d) image of T.

    (a)   We have $T(0) = 0 \in \{0\}$; hence $\{0\}$ is invariant under T.

    (b)   For every $v \in V$, $T(v) \in V$; hence V is invariant under T.

    (c)   Let $u \in \operatorname{Ker} T$. Then $T(u) = 0 \in \operatorname{Ker} T$ since the kernel of T is a subspace of V. Thus $\operatorname{Ker} T$ is invariant under T.

    (d)   Since $T(v) \in \operatorname{Im} T$ for every $v \in V$, it is certainly true when $v \in \operatorname{Im} T$. Hence the image of T is invariant under T.

11.2. Suppose $\{W_i\}$ is a collection of T-invariant subspaces of a vector space V. Show that the intersection $W = \cap_i W_i$ is also T-invariant.

Suppose $v \in W$; then $v \in W_i$ for every i. Since W_i is T-invariant, $T(v) \in W_i$ for every i. Thus $T(v) \in W = \cap_i W_i$ and so W is T-invariant.

11.3. Prove Theorem 11.2.

Suppose $v \in \mathrm{Ker}\, f(T)$, i.e., $f(T)(v) = 0$. We need to show that $T(v)$ also belongs to the kernel of $f(T)$, i.e., $f(T)(T(v)) = 0$. Since $f(t)t = tf(t)$, we have $f(T)T = Tf(T)$. Thus

$$f(T)T(v) = Tf(T)(v) = T(0) = 0$$

as required.

11.4. Find all invariant subspaces of $A = \begin{pmatrix} 2 & -5 \\ 1 & -2 \end{pmatrix}$ viewed as an operator on $\mathbf{R}^2$.

First of all, we have that $\mathbf{R}^2$ and $\{0\}$ are invariant under A. Now if A has any other invariant subspaces, then it must be 1-dimensional. However, the characteristic polynomial of A is

$$\Delta(t) = |tI - A| = \begin{vmatrix} t - 2 & 5 \\ -1 & t + 2 \end{vmatrix} = t^2 + 1$$

Hence A has no eigenvalues (in $\mathbf{R}$) and so A has no eigenvectors. But the 1-dimensional invariant subspaces correspond to the eigenvectors; thus $\mathbf{R}^2$ and $\{0\}$ are the only subspaces invariant under A.

11.5. Prove Theorem 11.3.

We choose a basis $\{w_1, \ldots, w_r\}$ of W and extend it to a basis $\{w_1, \ldots, w_r, v_1, \ldots, v_s\}$ of V. We have

$$\begin{aligned}
\hat{T}(w_1) &= T(w_1) = a_{11}w_1 + \cdots + a_{1r}w_r, \\
\hat{T}(w_2) &= T(w_2) = a_{21}w_1 + \cdots + a_{2r}w_r, \\
&\cdots\cdots\cdots\cdots\cdots\cdots\cdots\cdots\cdots\cdots\cdots\cdots \\
\hat{T}(w_r) &= T(w_r) = a_{r1}w_1 + \cdots + a_{rr}w_r, \\
T(v_1) &= b_{11}w_1 + \cdots + b_{1r}w_r + c_{11}v_1 + \cdots + c_{1s}v_s \\
T(v_2) &= b_{21}w_1 + \cdots + b_{2r}w_r + c_{21}v_1 + \cdots + c_{2s}v_s \\
&\cdots\cdots\cdots\cdots\cdots\cdots\cdots\cdots\cdots\cdots\cdots\cdots \\
T(v_s) &= b_{s1}w_1 + \cdots + b_{sr}w_r + c_{s1}v_1 + \cdots + c_{ss}v_s
\end{aligned}$$

But the matrix of T in this basis is the transpose of the matrix of coefficients in the above system of equations. (See page 344.) Therefore it has the form $\begin{pmatrix} A & B \\ 0 & C \end{pmatrix}$ where A is the transpose of the matrix of coefficients for the obvious subsystem. By the same argument, A is the matrix of $\hat{T}$ relative to the basis $\{w_i\}$ of W.

11.6. Let $\hat{T}$ denote the restriction of an operator T to an invariant subspace W, i.e., $\hat{T}(w) = T(w)$ for every $w \in W$. Prove:

(i)   For any polynomial $f(t), f(\hat{T})(w) = f(T)(w)$.

(ii) The minimum polynomial of $\hat{T}$ divides the minimum polynomial of T.

(i)   If $f(t) = 0$ or if $f(t)$ is a constant, i.e., of degree 1, then the result clearly holds.

Assume $\deg f = n > 1$ and that the result holds for polynomials of degree less than n. Suppose that

$$f(t) = a_n t^n + a_{n-1}t^{n-1} + \cdots + a_1 t + a_0$$

Then
$$f(\hat{T})(w) = (a_n \hat{T}^n + a_{n-1}\hat{T}^{n-1} + \cdots + a_0 I)(w)$$
$$= (a_n \hat{T}^{n-1})(\hat{T}(w)) + (a_{n-1}\hat{T}^{n-1} + \cdots + a_0 I)(w)$$
$$= (a_n T^{n-1})(T(w)) + (a_{n-1}T^{n-1} + \cdots + a_0 I)(w)$$
$$= f(T)(w)$$

(ii)   Let $m(t)$ denote the minimum polynomial of T. Then by (i), $m(\hat{T})(w) = m(T)(w) = 0(w) = 0$ for every $w \in W$; that is, $\hat{T}$ is a zero of the polynomial $m(t)$. Hence the minimum polynomial of $\hat{T}$ divides $m(t)$.

INVARIANT DIRECT-SUM DECOMPOSITIONS

11.7.   Prove Theorem 11.4.

Suppose B is a basis of V. Then, for any $v \in V$,

$$v = a_{11}w_{11} + \cdots + a_{1n_1}w_{1n_1} + \cdots + a_{r1}w_{r1} + \cdots + a_{rn_r}w_{rn_r} = w_1 + w_2 + \cdots + w_r$$

where $w_i = a_{i1}w_{i1} + \cdots + a_{in_i}w_{in_i} \in W_i$. We next show that such a sum is unique. Suppose

$$v = w_1' + w_2' + \cdots + w_r' \qquad \text{where } w_i' \in W_i$$

Since $\{w_{i1}, \ldots, w_{in_i}\}$ is a basis of W_i, $w_i' = b_{i1}w_{i1} + \cdots + b_{in_i}w_{in_i}$ and so

$$v = b_{11}w_{11} + \cdots + b_{1n_1}w_{1n_1} + \cdots + b_{r1}w_{r1} + \cdots + b_{rn_r}w_{rn_r}$$

Since B is a basis of V, $a_{ij} = b_{ij}$, for each i and each j. Hence $w_i = w_i'$ and so the sum for v is unique. Accordingly, V is the direct sum of the W_i.

Conversely, suppose V is the direct sum of the W_i. Then for any $v \in V$, $v = w_1 + \cdots + w_r$ where $w_i \in W_i$. Since $\{w_{ij_i}\}$ is a basis of W_i, each w_i is a linear combination of the w_{ij_i} and so v is a linear combination of the elements of B. Thus B spans V. We now show that B is linearly independent. Suppose

$$a_{11}w_{11} + \cdots + a_{1n_1}w_{1n_1} + \cdots + a_{r1}w_{r1} + \cdots + a_{rn_r}w_{rn_r} = 0$$

Note that $a_{i1}w_{i1} + \cdots + a_{in_i}w_{in_i} \in W_i$. We also have that $0 = 0 + 0 + \cdots + 0$ where $0 \in W_i$. Since such a sum for 0 is unique,

$$a_{i1}w_{i1} + \cdots + a_{in_i}w_{in_i} = 0 \qquad \text{for } i = 1, \ldots, r$$

The independence of the bases $\{w_{ij_i}\}$ imply that all the a's are 0. Thus B is linearly independent and hence is a basis of V.

11.8.   Suppose $T : V \to V$ is linear and suppose $T = T_1 \oplus T_2$ with respect to a T-invariant direct-sum decomposition $V = U \oplus W$. Show that:

(a) $m(t)$ is the least common multiple of $m_1(t)$ and $m_2(t)$ where $m(t)$, $m_1(t)$, and $m_2(t)$ are the minimum polynomials of T, T_1, and T_2, respectively.

(b) $\Delta(t) = \Delta_1(t)\, \Delta_2(t)$, where $\Delta(t)$, $\Delta_1(t)$ and $\Delta_2(t)$ are the characteristic polynomials of T, T_1 and T_2, respectively.

(a)   By Problem 11.6, each of $m_1(t)$ and $m_2(t)$ divides $m(t)$. Now suppose $f(t)$ is a multiple of both $m_1(t)$ and $m_2(t)$; then $f(T_1)(U) = 0$ and $f(T_2)(W) = 0$. Let $v \in V$; then $v = u + w$ with $u \in U$ and $w \in W$. Now

$$f(T)v = f(T)u + f(T)w = f(T_1)u + f(T_2)w = 0 + 0 = 0$$

That is, T is a zero of $f(t)$. Hence $m(t)$ divides $f(t)$, and so $m(t)$ is the least common multiple of $m_1(t)$ and $m_2(t)$.

(b)   By Theorem 11.5, T has a matrix representation $M = \begin{pmatrix} A & 0 \\ 0 & B \end{pmatrix}$ where A and B are matrix representations of T_1 and T_2, respectively. Then,

$$\Delta(t) = |tI - M| = \begin{vmatrix} tI - A & 0 \\ 0 & tI - B \end{vmatrix} = |tI - A||tI - B| = \Delta_1(t)\, \Delta_2(t)$$

as required.

11.9. Prove Theorem 11.7.

Note first that U and W are T-invariant by Theorem 11.2. Now since $g(t)$ and $h(t)$ are relatively prime, there exist polynomials $r(t)$ and $s(t)$ such that

$$r(t)g(t) + s(t)h(t) = 1$$

Hence for the operator T

$$r(T)g(T) + s(T)h(T) = I \tag{$*$}$$

Let $v \in V$; then by $(*)$,

$$v = r(T)g(T)v + s(T)h(T)v$$

But the first term in this sum belongs to $W = \operatorname{Ker} h(T)$ since

$$h(T)r(T)g(T) = r(T)g(T)h(T)v = r(T)f(T)v = r(T)0v = 0$$

Similarly, the second term belongs to U. Hence V is the sum of U and W.

To prove that $V = U \oplus W$, we must show that a sum $v = u + w$ with $u \in U$, $w \in W$, is uniquely determined by v. Applying the operator $r(T)g(T)$ to $v = u + w$ and using $g(T)u = 0$, we obtain

$$r(T)g(T)v = r(T)g(T)u + r(T)g(T)w = r(T)g(T)w$$

Also, applying $(*)$ to w alone and using $h(T)w = 0$, we obtain

$$w = r(T)g(T)w + s(T)h(T)w = r(T)g(T)w$$

Both of the above formulas give us $w = r(T)g(T)v$ and so w is uniquely determined by v. Similarly u is uniquely determined by v. Hence $V = U \oplus W$, as required.

11.10. Prove Theorem 11.8: In Theorem 11.7 (Problem 11.9), if $f(t)$ is the minimum polynomial of T [and $g(t)$ and $h(t)$ are monic], then $g(t)$ is the minimum polynomial of the restriction $\hat{T}_1$ of T to U and $h(t)$ is the minimum polynomial of the restriction $\hat{T}_2$ of T to W.

Let $m_1(t)$ and $m_2(t)$ be the minimum polynomials of $\hat{T}_1$ and $\hat{T}_2$, respectively. Note that $g(\hat{T}_1) = 0$ and $h(\hat{T}_2) = 0$ because $U = \operatorname{Ker} g(\hat{T})$ and $W = \operatorname{Ker} h(T)$. Thus

$$m_1(t) \text{ divides } g(t) \qquad \text{and} \qquad m_2(t) \text{ divides } h(t) \tag{1}$$

By Problem 11.9, $f(t)$ is the least common multiple of $m_1(t)$ and $m_2(t)$. But $m_1(t)$ and $m_2(t)$ are relatively prime since $g(t)$ and $h(t)$ are relatively prime. Accordingly, $f(t) = m_1(t)m_2(t)$. We also have that $f(t) = g(t)h(t)$. These two equations together with (1) and the fact that all the polynomials are monic, imply that $g(t) = m_1(t)$ and $h(t) = m_2(t)$, as required.

11.11. Prove the Primary Decomposition Theorem 11.6.

The proof is by induction on r. The case $r = 1$ is trivial. Suppose that the theorem has been proved for $r - 1$. By Theorem 11.7 we can write V as the direct sum of T-invariant subspaces W_1 and V_1 where W_1 is the kernel of $f_1(T)^{n_1}$ and where V_1 is the kernel of $f_2(T)^{n_2} \cdots f_r(T)^{n_r}$. By Theorem 11.8, the minimum polynomial of the restrictions of T to W_1 and V_1 are, respectively, $f_1(t)^{n_1}$ and $f_2(t)^{n_2} \cdots f_r(t)^{n_r}$.

Denote the restriction of T to V_1 by $\hat{T}_1$. By the inductive hypothesis, V_1 is the direct sum of subspaces $W_2, \ldots, W_r$ such that W_i is the kernel of $f_i(T_1)^{n_i}$ and such that $f_i(t)^{n_i}$ is the minimum polynomial for the restriction of $\hat{T}_1$ to W_i. But the kernel of $f_i(T)^{n_i}$, for $i = 2, \ldots, r$ is necessarily contained in V_1 since $f_i(t)^{n_i}$ divides $f_2(t)^{n_2} \cdots f_r(t)^{n_r}$. Thus the kernel of $f_i(T)^{n_i}$ is the same as the kernel of $f_i(T_1)^{n_i}$ which is W_i. Also, the restriction of T to W_i is the same as the restriction of $\hat{T}_1$ to W_i (for $i = 2, \ldots, r$); hence $f_i(t)^{n_i}$ is also the minimum polynomial for the restriction of T to W_i. Thus $V = W_1 \oplus W_2 \oplus \cdots \oplus W_r$ is the desired decomposition of T.

11.12. Prove Theorem 11.9.

Suppose $m(t)$ is a product of distinct linear polynomials; say,

$$m(t) = (t - \lambda_1)(t - \lambda_2) \cdots (t - \lambda_r)$$

where the λ_i are distinct scalars. By Theorem 11.6, V is the direct sum of subspaces $W_1, \ldots, W_r$ where $W_i = \text{Ker }(T - \lambda_i I)$. Thus if $v \in W_i$, then $(T - \lambda_i I)(v) = 0$ or $T(v) = \lambda_i v$. In other words, every vector in W_i is an eigenvector belonging to the eigenvalue λ_i. By Theorem 11.4, the union of bases for $W_1, \ldots, W_r$ is a basis of V. This basis consists of eigenvectors and so T is diagonalizable.

Conversely, suppose T is diagonalizable, i.e., V has a basis consisting of eigenvectors of T. Let $\lambda_1, \ldots, \lambda_s$ be the distinct eigenvalues of T. Then the operator

$$f(T) = (T - \lambda_1 I)(T - \lambda_2 I) \cdots (T - \lambda_3 I)$$

maps each basis vector into 0. Thus $f(T) = 0$ and hence the minimum polynomial $m(t)$ of T divides the polynomial

$$f(t) = (t - \lambda_1)(t - \lambda_2) \cdots (t - \lambda_3 I)$$

Accordingly, $m(t)$ is a product of distinct linear polynomials.

NILPOTENT OPERATORS, JORDAN CANONICAL FORM

11.13. Let $T : V \to V$ be linear. Suppose, for $v \in V$, $T^k(v) = 0$ but $T^{k-1}(v) \neq 0$. Prove:

(a) The set $S = \{v, T(v), \ldots, T^{k-1}(v)\}$ is linearly independent.

(b) The subspace W generated by S is T-invariant.

(c) The restriction $\hat{T}$ of T to W is nilpotent of index k.

(d) Relative to the basis $\{T^{k-1}(v), \ldots, T(v), v\}$ of W, the matrix of T is of the form

$$\begin{pmatrix} 0 & 1 & 0 & \ldots & 0 & 0 \\ 0 & 0 & 1 & \ldots & 0 & 0 \\ \multicolumn{6}{c}{\dotfill} \\ 0 & 0 & 0 & \ldots & 0 & 1 \\ 0 & 0 & 0 & \ldots & 0 & 0 \end{pmatrix}$$

Hence the above k-square matrix is nilpotent of index k.

(a) Suppose

$$av + a_1 T(v) + a_2 T^2(v) + \cdots + a_{k-1} T^{k-1}(v) = 0 \qquad (*)$$

Applying T^{k-1} to $(*)$ and using $T^k(v) = 0$, we obtain $aT^{k-1}(v) = 0$; since $T^{k-1}(v) \neq 0$, $a = 0$. Now applying T^{k-2} to $(*)$ and using $T^k(v) = 0$ and $a = 0$, we find $a_1 T^{k-1}(v) = 0$; hence $a_1 = 0$. Next applying T^{k-3} to $(*)$ and using $T^k(v) = 0$ and $a = a_1 = 0$, we obtain $a_2 T^{k-1}(v) = 0$; hence $a_2 = 0$. Continuing this process, we find that all the a's are 0; hence S is independent.

(b) Let $v \in W$. Then

$$v = bv + b_1 T(v) + b_2 T^2(v) + \cdots + b_{k-1} T^{k-1}(v)$$

Using $T^k(v) = 0$, we have that

$$T(v) = bT(v) + b_1 T^2(v) + \cdots + b_{k-2} T^{k-1}(v) \in W$$

Thus W is T-invariant.

(c) By hypothesis, $T^k(v) = 0$. Hence, for $i = 0, \ldots, k-1$,

$$\hat{T}^k(T^i(v)) = T^{k+i}(v) = 0$$

That is, applying $\hat{T}^k$ to each generator of W, we obtain 0; hence $\hat{T}^k = 0$ and so $\hat{T}$ is nilpotent of index at most k. On the other hand, $\hat{T}^{k-1}(v) = T^{k-1}(v) \neq 0$; hence T is nilpotent of index exactly k.

(d) For the basis $\{T^{k-1}(v), T^{k-2}(v), \ldots, T(v), v\}$ of W,

$$\hat{T}(T^{k-1}(v)) = T^k(v) = 0$$
$$\hat{T}(T^{k-2}(v)) \quad = \quad\quad\quad T^{k-1}(v)$$
$$\hat{T}(T^{k-3}(v)) \quad = \quad\quad\quad\quad\quad\quad T^{k-2}(v)$$
$$\cdots\cdots\cdots\cdots\cdots\cdots\cdots\cdots\cdots\cdots\cdots\cdots\cdots\cdots\cdots\cdots\cdots\cdots$$
$$\hat{T}(T(v)) \quad\quad = \quad\quad\quad\quad\quad\quad\quad\quad T^2(v)$$
$$\hat{T}(v) \quad\quad\quad = \quad\quad\quad\quad\quad\quad\quad\quad\quad\quad T(v)$$

Hence the matrix of T in this basis is

$$\begin{pmatrix} 0 & 1 & 0 & \ldots & 0 & 0 \\ 0 & 0 & 1 & \ldots & 0 & 0 \\ \cdots & \cdots & \cdots & \cdots & \cdots & \cdots \\ 0 & 0 & 0 & \ldots & 0 & 1 \\ 0 & 0 & 0 & \ldots & 0 & 0 \end{pmatrix}$$

11.14. Let $T : V \to V$ be linear. Let $U = \operatorname{Ker} T^i$ and $W = \operatorname{Ker} T^{i+1}$. Show that (a) $U \subseteq W$, and (b) $T(W) \subseteq U$.

(a) Suppose $u \in U = \operatorname{Ker} T^i$. Then $T^i(u) = 0$ and so $T^{i+1}(u) = T(T^i(u)) = T(0) = 0$. Thus $u \in \operatorname{Ker} T^{i+1} = W$. But this is true for every $u \in U$; hence $U \subseteq W$.

(b) Similarly, if $w \in W = \operatorname{Ker} T^{i+1}$, then $T^{i+1}(w) = 0$. Thus $T^{i+1}(w) = T^i(T(w)) = T^i(0) = 0$ and so $T(W) \subseteq U$.

11.15. Let $T : V \to V$ be linear. Let $X = \operatorname{Ker} T^{i-1}$, $Y = \operatorname{Ker} T^{i-1}$, and $Z = \operatorname{Ker} T^i$. By Problem 11.14, $X \subseteq Y \subseteq Z$. Suppose

$$\{u_1, \ldots, u_r\} \quad\quad \{u_1, \ldots, u_r, v_1, \ldots, v_s\} \quad\quad \{u_1, \ldots, u_r, v_1, \ldots, v_s\, w_1, \ldots, w_t\}$$

are bases of X, Y, and Z, respectively. Show that

$$S = \{u_1, \ldots, u_r, T(w_1), \ldots, T(w_t)\}$$

is contained in Y and is linearly independent.

By Problem 11.14, $T(Z) \subseteq Y$ and hence $S \subseteq Y$. Now suppose S is linearly dependent. Then there exists a relation

$$a_1 u_1 + \cdots + a_r u_r + b_1 T(w_1) + \cdots + b_t T(w_t) = 0$$

where at least one coefficient is not zero. Furthermore, since $\{u_i\}$ is independent, at least one of the b_k must be nonzero. Transposing, we find

$$b_1 T(w_1) + \cdots + b_t T(w_t) = -a_1 u_1 - \cdots - a_r u_r \in X = \operatorname{Ker} T^{i-2}$$

Hence $$T^{i-2}(b_1 T(w_1) + \cdots + b_t T(w_t)) = 0$$

Thus $$T^{i-1}(b_1 w_1 + \cdots + b_t w_t) = 0 \quad\quad \text{and so} \quad\quad b_1 w_1 + \cdots + b_t w_t \in Y = \operatorname{Ker} T^{i-1}$$

Since $\{u_i, v_j\}$ generates Y, we obtain a relation among the u_i, v_j and w_k where one of the coefficients, i.e., one of the b_k, is not zero. This contradicts the fact that $\{u_i, v_j, w_k\}$ is independent. Hence S must also be independent.

11.16. Prove Theorem 11.10.

Suppose $\dim V = n$. Let $W_1 = \operatorname{Ker} T$, $W_2 = \operatorname{Ker} T^2, \ldots, W_k = \operatorname{Ker} T^k$. Let us set $m_i = \dim W_i$, for $i = 1, \ldots, k$. Since T is of index k, $W_k = V$ and $W_{k-1} \neq V$ and so $m_{k-1} < m_k = n$. By Problem 11.17,

$$W_1 \subseteq W_2 \subseteq \cdots \subseteq W_k = V$$

Thus, by induction, we can choose a basis $\{u_1, \ldots, u_n\}$ of V such that $\{u_1, \ldots, u_{m_i}\}$ is a basis of W_i.

We now choose a new basis for V with respect to which T has the desired form. It will be convenient to label the members of this new basis by pairs of indices. We begin by setting

$$v(1, k) = u_{m_{k-1}+1}, \, v(2, k) = u_{m_{k-1}+2}, \ldots, v(m_k - m_{k-1}, k) = u_{m_k}$$

and setting

$$v(1, k - 1) = Tv(1, k), \, v(2, k - 1) = Tv(2, k), \ldots, v(m_k - m_{k-1}, k - 1) = Tv(m_k - m_{k-1}, k)$$

By Problem 11.15, $\quad S_1 = \{u_1, \ldots, u_{m_{k-2}}, v(1, k - 1), \ldots, v(m_k - m_{k-1}, k - 1)\}$

is a linearly independent subset of W_{k-1}. We extend S_1 to a basis of W_{k-1} by adjoining new elements (if necessary) which we denote by

$$v(m_k - m_{k-1} + 1, k - 1), \, v(m_k - m_{k-1} + 2, k - 1), \ldots, v(m_{k-1} - m_{k-2}, k - 1)$$

Next we set

$$v(1, k - 2) = Tv(1, k - 1), \, v(2, k - 2) = Tv(2, k - 1), \ldots, v(m_{k-1} - m_{k-2}, k - 2) = Tv(m_{k-1} - m_{k-2}, k - 1)$$

Again by Problem 11.15,

$$S_2 = \{u_1, \ldots, u_{m_{k-3}}, v(1, k - 2), \ldots, v(m_{k-1} - m_{k-2}, k - 2)\}$$

is a linearly independent subset of W_{k-2} which we can extend to a basis of W_{k-2} by adjoining elements

$$v(m_{k-1} - m_{k-2} + 1, k - 2), \, v(m_{k-1} - m_{k-2} + 2, k - 2), \ldots, v(m_{k-2} - m_{k-3}, k - 2)$$

Continuing in this manner we get a new basis for V which for convenient reference we arrange as follows:

$$v(1, k), \quad \ldots, v(m_k - m_{k-1}, k)$$
$$v(1, k - 1), \ldots, v(m_k - m_{k-1}, k - 1), \ldots, v(m_{k-1} - m_{k-2}, k - 1)$$

. . . .

$$v(1, 2), \quad \ldots, v(m_k - m_{k-1}, 2), \quad \ldots, v(m_{k-1} - m_{k-2}, 2), \ldots, v(m_2 - m_1, 2)$$
$$v(1, 1), \quad \ldots, v(m_k - m_{k-1}, 1), \quad \ldots, v(m_{k-1} - m_{k-2}, 1), \ldots, v(m_2 - m_1, 1), \ldots, v(m_1, 1)$$

The bottom row forms a basis of W_1, the bottom two rows form a basis of W_2, etc. But what is important for us is that T maps each vector into the vector immediately below it in the table or into 0 if the vector is in the bottom row. That is,

$$Tv(i, j) = \begin{cases} v(i, j - 1) & \text{for } j > 1 \\ 0 & \text{for } j = 1 \end{cases}$$

Now it is clear [see Problem 11.13(d)] that T will have the desired form if the $v(i, j)$ are ordered lexicographically: beginning with $v(1, 1)$ and moving up the first column to $v(1, k)$, then jumping to $v(2, 1)$ and moving up the second column as far as possible, etc.

Moreover, there will be exactly

$$m_k - m_{k-1} \qquad \text{diagonal entries of order } k$$
$$(m_{k-1} - m_{k-2}) - (m_k - m_{k-1}) = 2m_{k-1} - m_k - m_{k-2} \quad \text{diagonal entries of order } k - 1$$

. . . .

$$2m_2 - m_1 - m_3 \qquad \text{diagonal entries of order } 2$$
$$2m_1 - m_2 \qquad \text{diagonal entries of order } 1$$

as can be read off directly from the table. In particular, since the numbers $m_1, \ldots, m_k$ are uniquely determined by T, the number of diagonal entries of each order is uniquely determined by T. Finally, the identity

$$m_1 = (m_k - m_{k-1}) + (2m_{k-1} - m_k - m_{k-2}) + \cdots + (2m_2 - m_1 - m_3) + (2m_1 - m_2)$$

shows that the nullity m_1 of the total number of diagonal entries of T.

11.17. Let $A = \begin{pmatrix} 0 & 1 & 1 & 0 & 1 \\ 0 & 0 & 1 & 1 & 1 \\ 0 & 0 & 0 & 0 & 0 \\ 0 & 0 & 0 & 0 & 0 \\ 0 & 0 & 0 & 0 & 0 \end{pmatrix}$. Then $A^2 = \begin{pmatrix} 0 & 0 & 1 & 1 & 1 \\ 0 & 0 & 0 & 0 & 0 \\ 0 & 0 & 0 & 0 & 0 \\ 0 & 0 & 0 & 0 & 0 \\ 0 & 0 & 0 & 0 & 0 \end{pmatrix}$ and $A^3 = 0$; hence A is nilpotent of

index 3. Find the nilpotent matrix M in canonical form which is similar to A.

Since A is nilpotent of index 3, M contains a diagonal block of order 3 and none greater than 3. Note that rank $A = 2$; hence nullity $A = 5 - 2 = 3$. Thus M contains 3 diagonal blocks. Accordingly M must contain one diagonal block of order 3 and two of order 1; that is,

$$M = \begin{pmatrix} 0 & 1 & 0 & 0 & 0 \\ 0 & 0 & 1 & 0 & 0 \\ 0 & 0 & 0 & 0 & 0 \\ 0 & 0 & 0 & 0 & 0 \\ 0 & 0 & 0 & 0 & 0 \end{pmatrix}$$

11.18. Prove Theorem 11.11.

By the primary decomposition theorem, T is decomposable into operators $T_1, \ldots, T_r$, that is, $T = T_1 \oplus \cdots \oplus T_r$, where $(t - \lambda_i)^{m_i}$ is the minimum polynomial of T_i. Thus in particular,

$$(T_1 - \lambda_1 I)^{m_1} = 0, \ldots, (T_r - \lambda_r I)^{m_r} = 0$$

Set $N_i = T_i - \lambda_i I$. Then, for $i = 1, \ldots, r$,

$$T_i = N_i + \lambda_i I \qquad \text{where } N_i^{m_i} = 0$$

That is, T_i is the sum of the scalar operator $\lambda_i I$ and a nilpotent operator N_i, which is of index m_i since $(t - \lambda_i)^{m_i}$ is the minimum polynomial of T_i.

Now by Theorem 11.10 on nilpotent operators, we can choose a basis so that N_i is in canonical form. In this basis, $T_i = N_i + \lambda_i I$ is represented by a block diagonal matrix M_i whose diagonal entries are the matrices J_{ij}. The direct sum J of the matrices M_i is in Jordan canonical form and, by Theorem 11.5, is a matrix representation of T.

Lastly we must show that the blocks J_{ij} satisfy the required properties. Property (i) follows from the fact that N_i is of index m_i. Property (ii) is true since T and J have the same characteristic polynomial. Property (iii) is true since the nullity of $N_i = T_i - \lambda_i I$ is equal to the geometric multiplicity of the eigenvalue λ_i. Property (iv) follows from the fact that the T_i and hence the N_i are uniquely determined by T.

11.19. Determine all possible Jordan canonical forms for a linear operator $T : V \to V$ whose characteristic polynomial is $\Delta(t) = (t - 2)^3 (t - 5)^2$.

Since $t - 2$ has exponent 3 in $\Delta(t)$, 2 must appear three times on the main diagonal. Similarly 5 must appear twice. Thus the possible Jordan canonical forms are

$$\begin{pmatrix} 2 & 1 & & & \\ & 2 & 1 & & \\ & & 2 & & \\ \hline & & & 5 & 1 \\ & & & & 5 \end{pmatrix} \qquad \begin{pmatrix} 2 & 1 & & & \\ & 2 & & & \\ & & 2 & & \\ \hline & & & 5 & 1 \\ & & & & 5 \end{pmatrix} \qquad \begin{pmatrix} 2 & & & & \\ & 2 & & & \\ & & 2 & & \\ \hline & & & 5 & 1 \\ & & & & 5 \end{pmatrix}$$

$$\text{(i)} \qquad\qquad\qquad \text{(ii)} \qquad\qquad\qquad \text{(iii)}$$

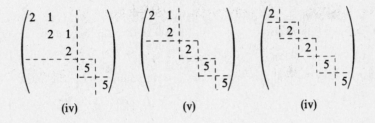

(iv)   (v)   (iv)

11.20. Determine all possible Jordan canonical forms J for a matrix of order 5 whose minimal polynomial is $m(t) = (t - 2)^2$.

J must have one Jordan block of order 2 and the others must be of order 2 or 1. Thus there are only two possibilities:

$$J = \begin{pmatrix} 2 & 1 & & & \\ & 2 & & & \\ & & 2 & 1 & \\ & & & 2 & \\ & & & & 2 \end{pmatrix} \quad \text{or} \quad J = \begin{pmatrix} 2 & 1 & & & \\ & 2 & & & \\ & & 2 & & \\ & & & 2 & \\ & & & & 2 \end{pmatrix}$$

Note that all the diagonal entries must be 2 since 2 is the only eigenvalue.

QUOTIENT SPACE AND TRIANGULAR FORM

11.21. Let W be a subspace of a vector space V. Show that the following are equivalent: (i) $u \in v + W$, (ii) $u - v \in W$, (iii) $v \in u + W$.

Suppose $u \in v + W$. Then there exists $w_0 \in W$ such that $u = v + w_0$. Hence $u - v = w_0 \in W$. Conversely, suppose $u - v \in W$. Then $u - v = w_0$ where $w_0 \in W$. Hence $u = v + w_0 \in W$. Thus (i) and (ii) are equivalent.

We also have: $u - v \in W$ iff $-(u - v) = v - u \in W$ iff $v \in u + W$. Thus (ii) and (iii) are also equivalent.

11.22. Prove: The cosets of W in V partition V into mutually disjoint sets. That is:

(i)   any two cosets $u + W$ and $v + W$ are either identical or disjoint; and

(ii)   each $v \in V$ belongs to a coset; in fact, $v \in v + W$.

Furthermore, $u + W = v + W$ if and only if $u - v \in W$, and so $(v + w) + W = v + W$ for any $w \in W$.

Let $v \in V$. Since $0 \in W$, we have $v = v + 0 \in v + W$ which proves (ii).

Now suppose the cosets $u + W$ and $v + W$ are not disjoint; say, the vector x belongs to both $u + W$ and $v + W$. Then $u - x \in W$ and $x - v \in W$. The proof of (i) is complete if we show that $u + W = v + W$. Let $u + w_0$ be any element in the coset $u + W$. Since $u - x$, $x - v$, and w_0 belong to W,

$$(u + w_0) - v = (u - x) + (x - v) + w_0 \in W$$

Thus $u + w_0 \in v + W$ and hence the coset $u + W$ is contained in the coset $v + W$. Similarly $v + W$ is contained in $u + W$ and so $u = W = v + W$.

The last statement follows from the fact that $u + W = v + W$ if and only if $u \in v + W$, and, by Problem 11.21, this is equivalent to $u - v \in W$.

11.23. Let W be the solution space of the homogeneous equation $2x + 3y + 4z = 0$. Describe the cosets of W in $\mathbf{R}^3$.

W is a plane through the origin $O = (0, 0, 0)$, and the cosets of W are the planes parallel to W as shown in Fig. 11-3. Equivalently, the cosets of W are the solution sets of the family of equations

$$2x + 3y + 4z = k \qquad k \in \mathbf{R}$$

In particular the coset $v + W$, where $v = (a, b, c)$, is the solution set of the linear equation

$$2x + 3y + 4z = 2a + 3b + 4c \qquad \text{or} \qquad 2(x - a) + 3(y - b) + 4(z - c) = 0$$

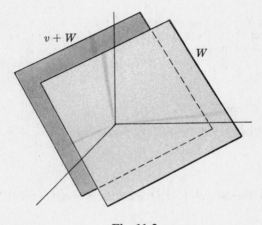

Fig. 11-3

11.24. Suppose W is a subspace of a vector space V. Show that the operations in Theorem 11.15 are well defined; namely, show that if $u + W = u' + W$ and $v + W = v' + W$, then:

  (a) $(u + v) + W = (u' + v') + W$ and   (b) $ku + W = ku' + W$,   for any $k \in K$.

  (a) Since $u + W = u' + W$ and $v + W = v' + W$, both $u - u'$ and $v - v'$ belong to W. But then $(u + v) - (u' + v') = (u - u') + (v - v') \in W$. Hence $(u + v) + W = (u' + v') + W$.

  (b) Also, since $u - u' \in W$ implies $k(u - u') \in W$, then $ku - ku' = k(u - u') \in W$; accordingly $ku + W = ku' + W$.

11.25. Let V be a vector space and W a subspace of V. Show that the natural map $\eta : V \to V/W$, defined by $\eta(v) = v + W$, is linear.

  For any $u, v \in V$ and any $k \in K$, we have

$$\eta(u + v) = u + v + W = u - W + v + W = \eta(u) + \eta(v)$$

  and

$$\eta(kv) = kv + W = k(v + W) = k\eta(v)$$

  Accordingly, η is linear.

11.26. Let W be a subspace of a vector space V. Suppose $\{w_1, \ldots, w_r\}$ is a basis of W and the set of cosets $\{\bar{v}_1, \ldots, \bar{v}_s\}$, where $\bar{v}_j = v_j + W$, is a basis of the quotient space. Show that B is a basis of V where $B = \{v_1, \ldots, v_s, w_1, \ldots, w_r\}$. Thus dim V = dim W + dim (V/W).

  Suppose $u \in V$. Since $\{\bar{v}_j\}$ is a basis of V/W,

$$\bar{u} = u + W = a_1\bar{v}_1 + a_2\bar{v}_2 + \cdots + a_s\bar{v}_s$$

  Hence $u = a_1v_1 + \cdots + a_sv_s + w$ where $w \in W$. Since $\{w_i\}$ is a basis of W,

$$u = a_1v_1 + \cdots + a_sv_s + b_1w_1 + \cdots + b_rw_r$$

  Accordingly, B spans V.

We now show that B is linearly independent. Suppose

$$c_1 v_1 + \cdots + c_s v_s + d_1 w_1 + \cdots + d_r w_r = 0 \qquad (1)$$

Then

$$c_1 \bar{v}_1 + \cdots + c_s \bar{v}_s = \bar{0} = W$$

Since $\{\bar{v}_j\}$ is independent, the c's are all 0. Substituting into (1), we find $d_1 w_1 + \cdots + d_r w_r = 0$. Since $\{w_i\}$ is independent, the d's are all 0. Thus B is linearly independent and therefore a basis of V.

11.27. Prove Theorem 11.16.

We first show that $\bar{T}$ is well defined, i.e., if $u + W = v + W$ then $\bar{T}(u + W) = \bar{T}(v + W)$. If $u + W = v + W$ then $u - v \in W$ and, since W is T-invariant, $T(u - v) = T(u) - T(v) \in W$. Accordingly,

$$\bar{T}(u + W) = T(u) + W = T(v) + W = \bar{T}(v + W)$$

as required.

We next show that $\bar{T}$ is linear. We have

$$\bar{T}((u + W) + (v + W)) = \bar{T}(u + v + W) = T(u + v) + W = T(u) + T(v) + W$$
$$= T(u) + W + T(v) + W = \bar{T}(u + W) + \bar{T}(v + W)$$

and

$$\bar{T}(k(u + W)) = \bar{T}(ku + W) = T(ku) + W = kT(u) + W = k(T(u) + W) = k\bar{T}(u + W)$$

Thus $\bar{T}$ is linear.

Now, for any coset $u + W$ in V/W,

$$\overline{T^2}(u + W) = T^2(u) + W = T(T(u)) + W = \bar{T}(T(u) + W) = \bar{T}(\bar{T}(u + W)) = \bar{T}^2(u + W)$$

Hence $\overline{T^2} = \bar{T}^2$. Similarly $\overline{T^n} = \bar{T}^n$ for any n. Thus for any polynomial

$$f(t) = a_n t^n + \cdots + a_0 = \sum a_t t^i$$

$$\overline{f(T)}(u + W) = f(T)(u) + W = \sum a_i T^i(u) + W = \sum a_i(T^i(u) + W)$$
$$= \sum a_i \overline{T^i}(u + W) = \sum a_i \bar{T}^i(u + W) = (\sum a_i \bar{T}^i)(u + W) = f(\bar{T})(u + W)$$

and so $\overline{f(T)} = f(\bar{T})$. Accordingly, if T is a root of $f(t)$ then $\overline{f(T)} = \bar{0} = W = f(\bar{T})$, i.e. $\bar{T}$ is also a root of $f(t)$. Thus the theorem is proved.

11.28. Prove Theorem 11.1.

The proof is by induction on the dimension of V. If $\dim V = 1$, then every matrix representation of T is a 1×1 matrix, which is triangular.

Now suppose $\dim V = n > 1$ and that the theorem holds for spaces of dimension less than n. Since the characteristic polynomial of T factors into linear polynomials, T has at least one eigenvalue and so at least one nonzero eigenvector v, say $T(v) = a_{11} v$. Let W be the 1-dimensional subspace spanned by v. Set $\bar{V} = V/W$. Then (Problem 11.26) $\dim \bar{V} = \dim V - \dim W = n - 1$. Note also that W is invariant under T. By Theorem 11.16, T induces a linear operator $\bar{T}$ on $\bar{V}$ whose minimum polynomial divides the minimum polynomial of T. Since the characteristic polynomial of T is a product of linear polynomials, so is its minimum polynomial; hence so are the minimum and characteristic polynomials of $\bar{T}$. Thus $\bar{V}$ and $\bar{T}$ satisfy the hypothesis of the theorem. Hence, by induction, there exists a basis $\{\bar{v}_2, \ldots, \bar{v}_n\}$ of $\bar{V}$ such that

$$\bar{T}(\bar{v}_2) = a_{22} \bar{v}_2$$
$$\bar{T}(\bar{v}_3) = a_{32} \bar{v}_2 + a_{33} \bar{v}_3$$
$$\dots\dots\dots\dots\dots\dots\dots\dots\dots\dots\dots\dots\dots\dots\dots$$
$$\bar{T}(\bar{v}_n) = a_{n2} \bar{v}_2 + a_{n3} \bar{v}_3 + \cdots + a_{nn} \bar{v}_n$$

Now let $v_2, \ldots, v_n$ be elements of V which belong to the cosets $\bar{v}_2, \ldots, \bar{v}_n$, respectively. Then $\{v, v_2, \ldots, v_n\}$ is a basis of V (Problem 11.26). Since $\bar{T}(\bar{v}_2) = a_{22} \bar{v}_2$, we have

$$\bar{T}(\bar{v}_2) - a_{22} \bar{v}_2 = 0 \qquad \text{and so} \qquad T(v_2) - a_{22} v_2 \in W$$

But W is spanned by v; hence $T(v_2) - a_{22} v_2$ is a multiple of v, say

$$T(v_2) - a_{22} v_2 = a_{21} v \quad \text{and so} \quad T(v_2) = a_{21} v + a_{22} v_2$$

Similarly, for $i = 3, \ldots, n$,

$$T(v_i) - a_{i2} v_2 - a_{i3} v_3 - \cdots - a_{ii} v_i \in W \quad \text{and so} \quad T(v_i) = a_{i1} v + a_{i2} v_2 + \cdots + a_{ii} v_i$$

Thus

$$T(v) = a_{11} v$$
$$T(v_2) = a_{21} v + a_{22} v_2$$
$$\cdots\cdots\cdots\cdots\cdots\cdots\cdots\cdots\cdots$$
$$T(v_n) = a_{n1} v + a_{n2} v_2 + \cdots + a_{nn} v_n$$

and hence the matrix of T in this basis is triangular.

CYCLIC SUBSPACES, RATIONAL CANONICAL FORM

11.29. Prove Theorem 11.12.

(i) By definition of $m_v(t)$, $T^k(v)$ is the first vector in the sequence v, $T(v)$, $T^2(v)$, ... which is a linear combination of those vectors which precede it in the sequence; hence the set $B = \{v, T(v), \ldots, T^{k-1}(v)\}$ is linearly independent. We now only have to show that $Z(v, T) = L(B)$, the linear span of B. By the above, $T^k(v) \in L(B)$. We prove by induction that $T^n(v) \in L(B)$ for every n. Suppose $n > k$ and $T^{n-1}(v) \in L(B)$, i.e., $T^{n-1}(v)$ is a linear combination of $v, \ldots, T^{k-1}(v)$. Then $T^n(v) = T(T^{n-1}(v))$ is a linear combination of $T(v), \ldots, T^k(v)$. But $T^k(v) \in L(B)$; hence $T^n(v) \in L(B)$ for every n. Consequently $f(T)(v) \in L(B)$ for any polynomial $f(t)$. Thus $Z(v, T) = L(B)$ and so B is a basis as claimed.

(ii) Suppose $m(t) = t^s + b_{s-1} t^{s-1} + \cdots + b_0$ is the minimal polynomial of T_v. Then, since $v \in Z(v, T)$,

$$0 = m(T_v)(v) = m(T)(v) = T^s(v) + b_{s-1} T^{s-1}(v) + \cdots + b_0 v$$

Thus $T^s(v)$ is a linear combination of $v, T(v), \ldots, T^{s-1}(v)$, and therefore $k \le s$. However, $m_v(T) = 0$ and so $m_v(T_v) = 0$. Then $m(t)$ divides $m_v(t)$ and so $s \le k$. Accordingly $k = s$ and hence $m_v(t) = m(t)$.

(iii)
$$T_v(v) = T(v)$$
$$T_v(T(v)) = T^2(v)$$
$$\cdots\cdots\cdots\cdots\cdots\cdots\cdots\cdots\cdots\cdots\cdots\cdots\cdots\cdots\cdots\cdots\cdots\cdots$$
$$T_v(T^{k-2}(v)) = T^{k-1}(v)$$
$$T_v(T^{k-1}(v)) = T^k(v) = -a_0 v - a_1 T(v) - a_2 T^2(v) - \cdots - a_{k-1} T^{k-1}(v)$$

By definition, the matrix of T_v in this basis is the transpose of the matrix of coefficients of the above system of equations; hence it is C, as required.

11.30. Let $T : V \to V$ be linear. Let W be a T-invariant subspace of V and $\bar{T}$ the induced operator on V/W. Prove: (a) The T-annihilator of $v \in V$ divides the minimum polynomial of T. (b) The $\bar{T}$-annihilator of $\bar{v} \in V/W$ divides the minimum polynomial of T.

(a) The T-annihilator of $v \in V$ is the minimum polynomial of the restriction of T to $Z(v, T)$ and therefore, by Problem 11.6, it divides the minimum polynomial of T.

(b) The $\bar{T}$-annihilator of $\bar{v} \in V/W$ divides the minimum polynomial of $\bar{T}$, which divides the minimum polynomial of T by Theorem 11.16.

Remark. In case the minimum polynomial of T is $f(t)^n$ where $f(t)$ is a monic irreducible polynomial, then the T-annihilator of $v \in V$ and the $\bar{T}$-annihilator of $\bar{v} \in V/W$ are of the form $f(t)^m$ where $m \le n$.

11.31. Prove Lemma 11.13.

The proof is by induction on the dimension of V. If $\dim V = 1$, then V is itself T-cyclic and the lemma holds. Now suppose $\dim V > 1$ and that the lemma holds for those vector spaces of dimension less than that of V.

Since the minimum polynomial of T is $f(t)^n$, there exists $v_1 \in V$ such that $f(T)^{n-1}(v_1) \neq 0$; hence the T-annihilator of v_1 is $f(t)^n$. Let $Z_1 = Z(v_1, T)$ and recall that Z_1 is T-invariant. Let $\bar{V} = V/Z_1$ and let $\bar{T}$ be the linear operator on $\bar{V}$ induced by T. By Theorem 11.16, the minimum polynomial of $\bar{T}$ divides $f(t)^n$; hence the hypothesis holds for $\bar{V}$ and $\bar{T}$. Consequently, by induction, $\bar{V}$ is the direct sum of $\bar{T}$-cyclic subspaces; say,

$$\bar{V} = Z(\bar{v}_2, \bar{T}) \oplus \cdots \oplus Z(\bar{v}_r, \bar{T})$$

where the corresponding $\bar{T}$-annihilators are $f(t)^{n_2}, \ldots, f(t)^{n_r}, n \geq n_2 \geq \cdots \geq n_r$.

We claim that there is a vector v_2 in the coset $\bar{v}_2$ whose T-annihilator is $f(t)^{n_2}$, the $\bar{T}$-annihilator of $\bar{v}_2$. Let w be any vector in $\bar{v}_2$. Then $f(T)^{n_2}(w) \in Z_1$. Hence there exists a polynomial $g(t)$ for which

$$f(T)^{n_2}(w) = g(T)(v_1) \tag{1}$$

Since $f(t)^n$ is the minimum polynomial of T, we have by (1),

$$0 = f(T)^n(w) = f(T)^{n-n_2}g(T)(v_1)$$

But $f(T)^n$ is the T-annihilator of v_1; hence $f(t)^n$ divides $f(t)^{n-n_2}g(t)$ and so $g(t) = f(t)^{n_2}h(t)$ for some polynomial $h(t)$. We set

$$v_2 = w - h(T)(v_1)$$

Since $w - v_2 = h(T)(v_1) \in Z_1$, v_2 also belongs to the coset $\bar{v}_2$. Thus the T-annihilator of v_2 is a multiple of the $\bar{T}$-annihilator of $\bar{v}_2$. On the other hand, by (1)

$$f(T)^{n_2}(v_2) = f(T)^{n_2}(w - h(T)(v_1)) = f(T)^{n_2}(w) - g(T)(v_1) = 0$$

Consequently the T-annihilator of v_2 is $f(t)^{n_2}$ as claimed.

Similarly, there exist vectors $v_3, \ldots, v_r \in V$ such that $v_i \in \bar{v}_i$ and that the T-annihilator of v_i is $f(t)^{n_i}$, the $\bar{T}$-annihilator of $\bar{v}_i$. We set

$$Z_2 = Z(v_2, T), \ldots, Z_r = Z(v_r, T)$$

Let d denote the degree of $f(t)$ so that $f(t)^{n_i}$ has degree dn_i. Then since $f(t)^{n_i}$ is both the T-annihilator of v_i and the $\bar{T}$-annihilator of $\bar{v}_i$, we know that

$$\{v_i, T(v_i), \ldots, T^{dn_i - 1}(v_i)\} \qquad \text{and} \qquad \{\bar{v}_i, \bar{T}(\bar{v}_i), \ldots, \bar{T}^{dn_i - 1}(\bar{v}_i)\}$$

are bases for $Z(v_i, T)$ and $Z(\bar{v}_i, \bar{T})$, respectively, for $i = 2, \ldots, r$. But $\bar{V} = Z(\bar{v}_2, \bar{T}) \oplus \cdots \oplus Z(\bar{v}_r, \bar{T})$; hence

$$\{\bar{v}_2, \ldots, \bar{T}^{dn_2 - 1}(\bar{v}_2), \ldots, \bar{v}_r, \ldots, \bar{T}^{dn_r - 1}(\bar{v}_r)\}$$

is a basis for $\bar{V}$. Therefore by Problem 11.26 and the relation $\bar{T}^i(\bar{v}) = \overline{T^i(v)}$ (see Problem 11.27),

$$\{v_1, \ldots, T^{dn_1 - 1}(v_1), v_2, \ldots, T^{dn_2 - 1}(v_2), \ldots, v_r, \ldots, T^{dn_r - 1}(v_r)\}$$

is a basis for V. Thus by Theorem 11.4, $V = Z(v_1, T) \oplus \cdots \oplus Z(v_r, T)$, as required.

It remains to show that the exponents $n_1, \ldots, n_r$ are uniquely determined by T. Since d denotes the degree of $f(t)$,

$$\dim V = d(n_1 + \cdots + n_r) \qquad \text{and} \qquad \dim Z_i = dn_i \qquad i = 1, \ldots, r$$

Also, if s is any positive integer then (Problem 11.59) $f(T)^s(Z_i)$ is a cyclic subspace generated by $f(T)^s(v_i)$ and it has dimension $d(n_i - s)$ if $n_i > s$ and dimension 0 if $n_i \leq s$.

Now any vector $v \in V$ can be written uniquely in the form $v = w_1 + \cdots + w_r$ where $w_i \in Z_i$. Hence any vector in $f(T)^s(V)$ can be written uniquely in the form

$$f(T)^s(v) = f(T)^s(w_1) + \cdots + f(T)^s(w_r)$$

where $f(T)^s(w_i) \in f(T)^s(Z_i)$. Let t be the integer, dependent on s, for which

$$n_1 > s, \ldots, n_t > s, n_{t+1} \geq s$$

Then

$$f(T)^s(V) = f(T)^s(Z_1) \oplus \cdots \oplus f(T)^s(Z_t)$$

and so

$$\dim (f(T)^s(V)) = d[(n_1 - s) + \cdots + (n_t - s)] \tag{*}$$

The numbers on the left of (∗) are uniquely determined by T. Set $s = n - 1$ and (∗) determines the number of n_i equal to n. Next set $s = n - 2$ and (∗) determines the number of n_i (if any) equal to $n - 1$. We repeat the process until we set $s = 0$ and determine the number of n_i equal to 1. Thus the n_i are uniquely determined by T and V, and the lemma is proved.

11.32. Let V be a vector space of dimension 7 over $\mathbf{R}$, and let $T : V \to V$ be a linear operator with minimum polynomial $m(t) = (t^2 + 2)(t + 3)^3$. Find all the possible rational canonical forms for T.

The sum of the degrees of the companion matrices must add up to 7. Also, one companion matrix must be $t^2 + 2$ and one must be $(t + 3)^3$. Thus the rational canonical form of T is exactly one of the following direct sums of companion matrices:

(i) $C(t^2 + 2) \oplus C(t^2 + 2) \oplus C((t + 3)^3)$

(ii) $C(t^2 + 2) \oplus C((t + 3)^3) \oplus C((t + 3)^2)$

(iii) $C(t^2 + 2) \oplus C((t + 3)^3) \oplus C(t + 3) \oplus C(t + 3)$

That is,

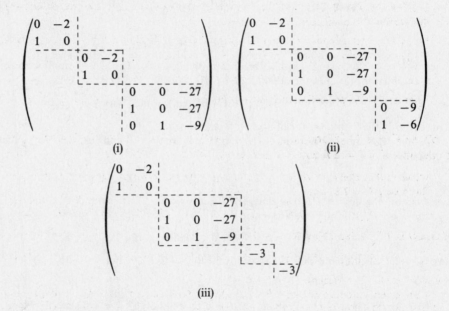

(i)   (ii)

(iii)

PROJECTIONS

11.33. Suppose $V = W_1 \oplus \cdots \oplus W_r$. The projection of V into its subspace W_k is the mapping $E : V \to V$ defined by $E(v) = w_k$ where $v = w_1 + \cdots + w_r$, $w_i \in W_i$. Show that (a) E is linear, and (b) $E^2 = E$.

(a)   Since the sum $v = w_1 + \cdots + w_r$, $w_i \in W$ is uniquely determined by v, the mapping E is well defined. Suppose, for $u \in V$, $u = w'_1 + \cdots + w'_r$, $w'_i \in W_i$. Then

$$v + u = (w_1 + w'_1) + \cdots + (w_r + w'_r) \quad \text{and} \quad kv = kw_1 + \cdots + kw_r \qquad kw_i, w_i + w'_i \in W_i$$

are the unique sums corresponding to $v + u$ and kv. Hence

$$E(v + u) = w_k + w'_k = E(v) + E(u) \quad \text{and} \quad E(kv) = kw_k + kE(v)$$

and therefore E is linear.

(b)   We have that $$w_k = 0 + \cdots + 0 + w_k + 0 + \cdots + 0$$

is the unique sum corresponding to $w_k \in W_k$; hence $E(w_k) = w_k$. Then for any $v \in V$,

$$E^2(v) = E(E(v)) = E(w_k) = w_k = E(v)$$

Thus $E^2 = E$, as required.

11.34. Suppose $E : V \to V$ is linear and $E^2 = E$. Show that: (a) $E(u) = u$ for any $u \in \text{Im } E$, i.e., the restriction of E to its image is the identity mapping; (b) V is the direct sum of the image and kernel of E: that is, $V = \text{Im } E \oplus \text{Ker } E$; (c) E is the projection of V into $\text{Im } E$, its image. Thus, by Problem 11.33, a linear mapping $T : V \to V$ is a projection if and only if $T^2 = T$; this characterization of a projection is frequently used as its definition.

(a) If $u \in \text{Im } E$, then there exists $v \in V$ for which $E(v) = u$; hence

$$E(u) = E(E(v)) = E^2(v) = E(v) = u$$

as required.

(b) Let $v \in V$. We can write v in the form $v = E(v) + v - E(v)$. Now $E(v) \in \text{Im } E$ and, since

$$E(v - E(v)) = E(v) - E^2(v) = E(v) - E(v) = 0$$

$v - E(v) \in \text{Ker } E$. Accordingly, $V = \text{Im } E + \text{Ker } E$.

Now suppose $w \in \text{Im } E \cap \text{Ker } E$. By (i), $E(w) = w$ because $w \in \text{Im } E$. On the other hand, $E(w) = 0$ because $w \in \text{Ker } E$. Thus $w = 0$ and so $\text{Im } E \cap \text{Ker } E = \{0\}$. These two conditions imply that V is the direct sum of the image and kernel of E.

(c) Let $v \in V$ and suppose $v = u + w$ where $u \in \text{Im } E$ and $w \in \text{Ker } E$. Note that $E(u) = u$ by (i), and $E(w) = 0$ because $w \in \text{Ker } E$. Hence

$$E(v) = E(u + w) = E(u) + E(w) = u + 0 = u$$

That is, E is the projection of V into its image.

11.35. Suppose $V = U \oplus W$ and suppose $T : V \to V$ is linear. Show that U and W are both T-invariant if and only if $TE = ET$ where E is the projection of V into U.

Observe that $E(v) \in U$ for every $v \in V$, and that (i) $E(v) = v$ iff $v \in U$, (ii) $E(v) = 0$ iff $v \in W$.

Suppose $ET = TE$. Let $u \in U$. Since $E(u) = u$,

$$T(u) = T(E(u)) = (TE)(u) = (ET)(u) = E(T(u)) \in U$$

Hence U is T-invariant. Now let $w \in W$. Since $E(w) = 0$,

$$E(T(w)) = (ET)(w) = (TE)(w) = T(E(w)) = T(0) = 0 \qquad \text{and so} \qquad T(w) \in W$$

Hence W is also T-invariant.

Conversely, suppose U and W are both T-invariant. Let $v \in V$ and suppose $v = u + w$ where $u \in T$ and $w \in W$. Then $T(u) \in U$ and $T(w) \in W$; hence $E(T(u)) = T(u)$ and $E(T(w)) = 0$. Thus

$$(ET)(v) = (ET)(u + w) = (ET)(u) + (ET)(w) = E(T(u)) + E(T(w)) = T(u)$$

and

$$(TE)(v) = (TE)(u + w) = T(E(u + w)) = T(u)$$

That is, $(ET)(v) = (TE)(v)$ for every $v \in V$; therefore $ET = TE$ as required.

Supplementary Problems

INVARIANT SUBSPACES

11.36. Suppose W is invariant under $T : V \to V$. Show that W is invariant under $f(T)$ for any polynomial $f(t)$.

11.37. Show that every subspace of V is invariant under I and 0, the identity and zero operators.

11.38. Suppose W is invariant under $T_1 : V \to V$ and $T_2 : V \to V$. Show that W is also invariant under $T_1 + T_2$ and ST.

11.39. Let $T : V \to V$ be linear and let W be the eigenspace belonging to an eigenvalue λ of T. Show that W is T-invariant.

11.40. Let V be a vector space of odd dimension (greater than 1) over the real field $\mathbf{R}$. Show that any linear operator on V has an invariant subspace other than V or $\{0\}$.

11.41. Determine the invariant subspaces of $A = \begin{pmatrix} 2 & -4 \\ 5 & -2 \end{pmatrix}$ viewed as a linear operator on (i) $\mathbf{R}^2$, (ii) $\mathbf{C}^2$.

11.42. Suppose $\dim V = n$. Show that $T : V \to V$ has a triangular matrix representation if and only if there exist T-invariant subspaces $W_1 \subset W_2 \subset \cdots \subset W_n = V$ for which $\dim W_k = k$, $k = 1, \ldots, n$.

INVARIANT DIRECT SUMS

11.43. The subspaces $W_1, \ldots, W_r$ are said to be independent if $w_1 + \cdots + w_r = 0$, $w_i \in W_i$, implies that each $w_i = 0$. Show that span $(W_i) = W_1 \oplus \cdots \oplus W_r$ if and only if the W_i are independent. [Here span (W_i) denotes the linear span of the W_i.]

11.44. Show that $V = W_1 \oplus \cdots \oplus W_r$ if and only if (i) $V = $ span (W_i) and (ii) $W_k \cap$ span $(W_1, \ldots, W_{k-1}, W_{k+1}, \ldots, W_r) = \{0\}$, $k = 1, \ldots, r$.

11.45. Show that span $(W_i) = W_1 \oplus \cdots \oplus W_r$ if and only if $\dim$ span $(W_i) = \dim W_1 + \cdots + \dim W_r$.

11.46. Suppose the characteristic polynomial of $T : V \to V$ is $\Delta(t) = f_1(t)^{n_1} f_2(t)^{n_2} \cdots f_r(t)^{n_r}$ where the $f_i(t)$ are distinct monic irreducible polynomials. Let $V = W_1 \oplus \cdots \oplus W_r$ be the primary decomposition of V into T-invariant subspaces. Show that $f_i(t)^{n_i}$ is the characteristic polynomial of the restriction of T to W_i.

NILPOTENT OPERATORS

11.47. Suppose T_1 and T_2 are nilpotent operators which commute, i.e., $T_1 T_2 = T_2 T_1$. Show that $T_1 + T_2$ and $T_1 T_2$ are also nilpotent.

11.48. Suppose A is a supertriangular matrix, i.e., all entries on and below the main diagonal are 0. Show that A is nilpotent.

11.49. Let V be the vector space of polynomials of degree $\leq n$. Show that the differential operator on V is nilpotent of index $n + 1$.

11.50. Show that the following nilpotent matrices of order n are similar:

$$\begin{pmatrix} 0 & 1 & 0 & \ldots & 0 \\ 0 & 0 & 1 & \ldots & 0 \\ \cdots\cdots\cdots\cdots\cdots \\ 0 & 0 & 0 & \ldots & 1 \\ 0 & 0 & 0 & \ldots & 0 \end{pmatrix} \quad \text{and} \quad \begin{pmatrix} 0 & 0 & \ldots & 0 & 0 \\ 1 & 0 & \ldots & 0 & 0 \\ 0 & 1 & \ldots & 0 & 0 \\ \cdots\cdots\cdots\cdots\cdots \\ 0 & 0 & \ldots & 1 & 0 \end{pmatrix}$$

11.51. Show that two nilpotent matrices of order 3 are similar if and only if they have the same index of nilpotency. Show by example that the statement is not true for nilpotent matrices of order 4.

JORDAN CANONICAL FORM

11.52. Find all possible Jordan canonical forms for those matrices whose characteristic polynomial $\Delta(t)$ and minimum polynomial $m(t)$ are as follows:

  (a) $\Delta(t) = (t-2)^4(t-3)^2$, $m(t) = (t-2)^2(t-3)^2$

  (b) $\Delta(t) = (t-7)^5$, $m(t) = (t-7)^2$

  (c) $\Delta(t) = (t-2)^7$, $m(t) = (t-2)^3$

  (d) $\Delta(t) = (t-3)^4(t-5)^4$, $m(t) = (t-3)^2(t-5)^2$

11.53. Show that every complex matrix is similar to its transpose. (*Hint:* Use its Jordan canonical form and Problem 11.50.)

11.54. Show that all $n \times n$ complex matrices A for which $A^n = I$ but $A^k \neq I$ for $k < n$ are similar.

11.55. Suppose A is a complex matrix with only real eigenvalues. Show that A is similar to a matrix with only real entries.

CYCLIC SUBSPACES

11.56. Suppose $T : V \to V$ is linear. Prove that $Z(v, T)$ is the intersection of all T-invariant subspaces containing v.

11.57. Let $f(t)$ and $g(t)$ be the T-annihilators of u and v, respectively. Show that if $f(t)$ and $g(t)$ are relatively prime, then $f(t)g(t)$ is the T-annihilator of $u + v$.

11.58. Prove that $Z(u, T) = Z(v, T)$ if and only if $g(T)(u) = v$ where $g(t)$ is relatively prime to the T-annihilator of u.

11.59. Let $W = Z(v, T)$, and suppose the T-annihilator of v is $f(t)^n$ where $f(t)$ is a monic irreducible polynomial of degree d. Show that $f(T)^s(W)$ is a cyclic subspace generated by $f(T)^s(v)$ and it has dimension $d(n-s)$ if $n > s$ and dimension 0 if $n \leq s$.

RATIONAL CANONICAL FORM

11.60. Find all possible rational canonical forms for:

  (a) 6×6 matrices with minimum polynomial $m(t) = (t^2 + 3)(t + 1)^2$

  (b) 6×6 matrices with minimum polynomial $m(t) = (t + 1)^3$

  (c) 8×8 matrices with minimum polynomial $m(t) = (t^2 + 2)^2(t + 3)^2$

11.61. Let A be a 4×4 matrix with minimum polynomial $m(t) = (t^2 + 1)(t^2 - 3)$. Find the rational canonical form for A if A is a matrix over (a) the rational field $\mathbf{Q}$, (b) the real field $\mathbf{R}$, (c) the complex field C.

11.62. Find the rational canonical form for the Jordan block $\begin{pmatrix} \lambda & 1 & 0 & 0 \\ 0 & \lambda & 1 & 0 \\ 0 & 0 & \lambda & 1 \\ 0 & 0 & 0 & \lambda \end{pmatrix}$.

11.63. Prove that the characteristic polynomial of an operator $T : V \to V$ is a product of its elementary divisors.

11.64. Prove that two 3×3 matrices with the same minimum and characteristic polynomials are similar.

11.65. Let $C(f(t))$ denote the companion matrix to an arbitrary polynomial $f(t)$. Show that $f(t)$ is the characteristic polynomial of $C(f(t))$.

PROJECTIONS

11.66. Suppose $V = W_1 \oplus \cdots \oplus W_r$. Let E_i denote the projection of V into W_i. Prove: (i) $E_i E_j = 0$, $i \neq j$; and (ii) $I = E_1 + \cdots + E_r$.

11.67. Let $E_1, \ldots, E_r$ be linear operators on V such that: (i) $E_i^2 = E_i$, i.e., the E_i are projections; (ii) $E_i E_j = 0$, $i \neq j$; (iii) $I = E_1 + \cdots + E_r$. Prove that $V = \text{Im } E_1 \oplus \cdots \oplus \text{Im } E_r$.

11.68. Suppose $E : V \to V$ is a projection, i.e., $E^2 = E$. Prove that E has a matrix representation of the form $\begin{pmatrix} I_r & 0 \\ 0 & 0 \end{pmatrix}$ where r is the rank of E and I_r is the r-square identity matrix.

11.69. Prove that any two projections of the same rank are similar. (*Hint:* Use the result of Problem 11.68.)

11.70. Suppose $E : V \to V$ is a projection. Prove: (i) $I - E$ is a projection and $V = \text{Im } E \oplus \text{Im } (I - E)$; and (ii) $I + E$ is invertible (if $1 + 1 \neq 0$).

QUOTIENT SPACES

11.71. Let W be a subspace of V. Suppose the set of cosets $\{v_1 + W, v_2 + W, \ldots, v_n + W\}$ in V/W is linearly independent. Show that the set of vectors $\{v_1, v_2, \ldots, v_n\}$ in V is also linearly independent.

11.72. Let W be a subspace of V. Suppose the set of vectors $\{u_1, u_2, \ldots, u_n\}$ in V is linearly independent, and that span $(u_i) \cap W = \{0\}$. Show that the set of cosets $\{u_1 + W, \ldots, u_n + W\}$ in V/W is also linearly independent.

11.73. Suppose $V = U \oplus W$ and that $\{u_1, \ldots, u_n\}$ is a basis of U. Show that $\{u_1 + W, \ldots, u_n + W\}$ is a basis of the quotient space V/W. (Observe that no condition is placed on the dimensionality of V or W.)

11.74. Let W be the solution space of the linear equation

$$a_1 x_1 + a_2 x_2 + \cdots + a_n x_n = 0 \qquad a_i \in K$$

and let $v = (b_1, b_2, \ldots, b_n) \in K^n$. Prove that the coset $v + W$ of W in K^n is the solution set of the linear equation

$$a_1 x_1 + a_2 x_2 + \cdots + a_n x_n = b \qquad \text{where} \qquad b = a_1 b_1 + \cdots + a_n b_n$$

11.75. Let V be the vector space of polynomials over $\mathbf{R}$ and let W be the subspace of polynomials divisible by t^4, that is, of the form $a_0 t^4 + a_1 t_5 + \cdots + a_{n-4} t^n$. Show that the quotient space V/W is of dimension 4.

11.76. Let U and W be subspaces of V such that $W \subseteq U \subseteq V$. Note that any coset $u + W$ of W in U may also be viewed as a coset of W in V since $u \in U$ implies $u \in V$; hence U/W is a subset of V/W. Prove that (i) U/W is a subspace of V/W, and (ii) dim $(V/W) - $ dim $(U/W) = $ dim V/U.

11.77. Let U and W be subspaces of V. Show that the cosets of $U \cap W$ in V can be obtained by intersecting each of the cosets of U in V by each of the cosets of W in V:

$$V/(U \cap W) = \{(u + U) \cap (v' + W) : v, v' \in V\}$$

11.78. Let $T : V \to V'$ be linear with kernel W and image U. Show that the quotient space V/W is isomorphic to U under the mapping $\theta : V/W \to U$ defined by $\theta(v + W) = T(v)$. Furthermore, show that $T = i \circ \theta \circ \eta$

where $\eta : V \to V/W$ is the natural mapping of V into V/W, i.e. $\eta(v) = v + W$, and $i : U \to V'$ is the inclusion mapping, i.e. $i(u) = u$. (See Fig. 11-4.)

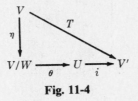

Fig. 11-4

Answers to Supplementary Problems

11.41. (a) $\mathbf{R}^2$ and $\{0\}$,   (b) $\mathbf{C}^2$, $\{0\}$, $W_1 = L((2, 1 - 2i))$, $W_2 = L((2, 1 + 2i))$.

11.52. (a)

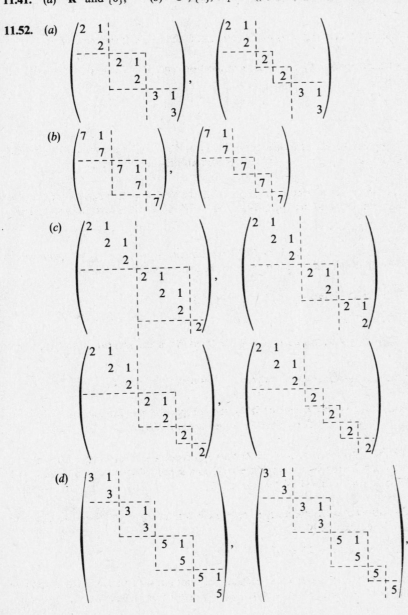

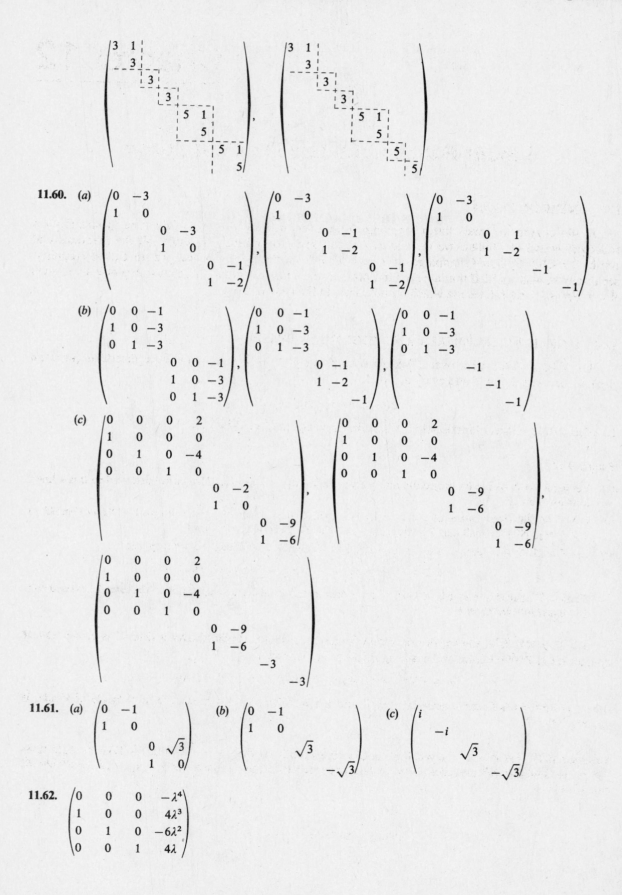

$$\begin{pmatrix} 3 & 1 & & & & & & \\ & 3 & & & & & & \\ & & 3 & & & & & \\ & & & 3 & & & & \\ & & & & 5 & 1 & & \\ & & & & & 5 & & \\ & & & & & & 5 & 1 \\ & & & & & & & 5 \end{pmatrix}, \quad \begin{pmatrix} 3 & 1 & & & & & & \\ & 3 & & & & & & \\ & & 3 & & & & & \\ & & & 3 & & & & \\ & & & & 5 & 1 & & \\ & & & & & 5 & & \\ & & & & & & 5 & \\ & & & & & & & 5 \end{pmatrix}$$

11.60. (a)
$$\begin{pmatrix} 0 & -3 & & & & \\ 1 & 0 & & & & \\ & & 0 & -3 & & \\ & & 1 & 0 & & \\ & & & & 0 & -1 \\ & & & & 1 & -2 \end{pmatrix}, \quad \begin{pmatrix} 0 & -3 & & & & \\ 1 & 0 & & & & \\ & & 0 & -1 & & \\ & & 1 & -2 & & \\ & & & & 0 & -1 \\ & & & & 1 & -2 \end{pmatrix}, \quad \begin{pmatrix} 0 & -3 & & & \\ 1 & 0 & & & \\ & & 0 & 1 & \\ & & 1 & -2 & \\ & & & & -1 \\ & & & & & -1 \end{pmatrix}$$

(b)
$$\begin{pmatrix} 0 & 0 & -1 & & & \\ 1 & 0 & -3 & & & \\ 0 & 1 & -3 & & & \\ & & & 0 & 0 & -1 \\ & & & 1 & 0 & -3 \\ & & & 0 & 1 & -3 \end{pmatrix}, \quad \begin{pmatrix} 0 & 0 & -1 & & & \\ 1 & 0 & -3 & & & \\ 0 & 1 & -3 & & & \\ & & & 0 & -1 & \\ & & & 1 & -2 & \\ & & & & & -1 \end{pmatrix}, \quad \begin{pmatrix} 0 & 0 & -1 & & & \\ 1 & 0 & -3 & & & \\ 0 & 1 & -3 & & & \\ & & & -1 & & \\ & & & & -1 & \\ & & & & & -1 \end{pmatrix}$$

(c)
$$\begin{pmatrix} 0 & 0 & 0 & 2 & & & & \\ 1 & 0 & 0 & 0 & & & & \\ 0 & 1 & 0 & -4 & & & & \\ 0 & 0 & 1 & 0 & & & & \\ & & & & 0 & -2 & & \\ & & & & 1 & 0 & & \\ & & & & & & 0 & -9 \\ & & & & & & 1 & -6 \end{pmatrix}, \quad \begin{pmatrix} 0 & 0 & 0 & 2 & & & & \\ 1 & 0 & 0 & 0 & & & & \\ 0 & 1 & 0 & -4 & & & & \\ 0 & 0 & 1 & 0 & & & & \\ & & & & 0 & -9 & & \\ & & & & 1 & -6 & & \\ & & & & & & 0 & -9 \\ & & & & & & 1 & -6 \end{pmatrix},$$

$$\begin{pmatrix} 0 & 0 & 0 & 2 & & & & \\ 1 & 0 & 0 & 0 & & & & \\ 0 & 1 & 0 & -4 & & & & \\ 0 & 0 & 1 & 0 & & & & \\ & & & & 0 & -9 & & \\ & & & & 1 & -6 & & \\ & & & & & & -3 & \\ & & & & & & & -3 \end{pmatrix}$$

11.61. (a)
$$\begin{pmatrix} 0 & -1 & & \\ 1 & 0 & & \\ & & 0 & \sqrt{3} \\ & & 1 & 0 \end{pmatrix}$$
(b)
$$\begin{pmatrix} 0 & -1 & & \\ 1 & 0 & & \\ & & \sqrt{3} & \\ & & & -\sqrt{3} \end{pmatrix}$$
(c)
$$\begin{pmatrix} i & & & \\ & -i & & \\ & & \sqrt{3} & \\ & & & -\sqrt{3} \end{pmatrix}$$

11.62.
$$\begin{pmatrix} 0 & 0 & 0 & -\lambda^4 \\ 1 & 0 & 0 & 4\lambda^3 \\ 0 & 1 & 0 & -6\lambda^2 \\ 0 & 0 & 1 & 4\lambda \end{pmatrix}$$

Chapter 12

Linear Functionals and the Dual Space

12.1 INTRODUCTION

In this chapter we study linear mappings from a vector space V into its field K of scalars. (Unless otherwise stated or implied, we view K as a vector space over itself.) Naturally all the theorems and results for arbitrary linear mappings on V hold for this special case. However, we treat these mappings separately because of their fundamental importance and because the special relationship of V to K gives rise to new notions and results which do not apply in the general case.

12.2 LINEAR FUNCTIONALS AND THE DUAL SPACE

Let V be a vector space over a field K. A mapping $\phi : V \to K$ is termed a *linear functional* (or *linear form*) if, for every $u, v \in V$ and every $a, b \in K$,

$$\phi(au + bv) = a\phi(u) + b\phi(v)$$

In other words, a linear functional on V is a linear mapping from V into K.

Example 12.1

(a) Let $\pi_i : K^n \to K$ be the ith projection mapping, i.e., $\pi_i(a_1, a_2, \ldots, a_n) = a_i$. Then π_i is linear and so it is a linear functional on K^n.

(b) Let V be the vector space of polynomials in t over **R**. Let $\mathbf{J} : V \to \mathbf{R}$ be the integral operator defined by $\mathbf{J}(p(t)) = \int_0^1 p(t) \, dt$. Recall that $\mathbf{J}$ is linear; and hence it is a linear functional on V.

(c) Let V be the vector space of n-square matrices over K. Let $T : V \to K$ be the trace mapping

$$T(A) = a_{11} + a_{22} + \cdots + a_{nn} \qquad \text{where } A = (a_{ij})$$

That is, T assigns to a matrix A the sum of its diagonal elements. This map is linear (Problem 12.24) and so it is a linear functional on V.

By Theorem 9.10, the set of linear functionals on a vector space V over a field K is also a vector space over K with addition and scalar multiplication defined by

$$(\phi + \sigma)(v) = \phi(v) + \sigma(v) \qquad \text{and} \qquad (k\phi)(v) = k\phi(v)$$

where ϕ and σ are linear functionals on V and $k \in K$. This space is called the dual space of V and is denoted by V^*.

Example 12.2. Let $V = K^n$, the vector space of n-tuples which we write as column vectors. Then the dual space V^* can be identified with the space of row vectors. In particular, any linear functional $\phi = (a_1, \ldots, a_n)$ in V^* has the representation

$$\phi(x_1, \ldots, x_n) = (a_1, a_2, \ldots, a_n) \begin{pmatrix} x_1 \\ x_2 \\ \cdots \\ x_n \end{pmatrix}$$

or simply

$$\phi(x_1, \ldots, x_n) = a_1 x_1 + a_2 x_2 + \cdots + a_n x_n$$

Historically, the above formal expression was termed a linear form.

12.3 DUAL BASIS

Suppose V is a vector space of dimension n over K. By Theorem 9.11 the dimension of the dual space V^* is also n (since K is of dimension 1 over itself). In fact, each basis of V determines a basis of V^* as follows (see Problem 12.3 for the proof):

Theorem 12.1:   Suppose $\{v_1, \ldots, v_n\}$ is a basis of V over K. Let $\phi_1, \ldots, \phi_n \in V^*$ be the linear functionals defined by

$$\phi_i(v_j) = \delta_{ij} = \begin{cases} 1 & \text{if } i = j \\ 0 & \text{if } i \neq j \end{cases}$$

Then $\{\phi_1, \ldots, \phi_n\}$ is a basis of V^*.

The above basis $\{\phi_i\}$ is termed the basis *dual* to $\{v_i\}$ or the *dual basis*. The above formula, which uses the Kronecker delta δ_{ij}, is a short way of writing

$$\phi_1(v_1) = 1, \ \phi_1(v_2) = 0, \ \phi_1(v_3) = 0, \ \ldots, \ \phi_1(v_n) = 0$$
$$\phi_2(v_1) = 0, \ \phi_2(v_2) = 1, \ \phi_2(v_3) = 0, \ \ldots, \ \phi_2(v_n) = 0$$
$$\cdots\cdots\cdots\cdots\cdots\cdots\cdots\cdots\cdots\cdots\cdots\cdots\cdots\cdots\cdots$$
$$\phi_n(v_1) = 0, \ \phi_n(v_2) = 0, \ \ldots, \ \phi_n(v_{n-1}) = 0, \ \phi_n(v_n) = 1$$

By Theorem 9.2 these linear mappings ϕ_i are unique and well-defined.

Example 12.3.   Consider the following basis of $\mathbf{R}^2$: $\{v_1 = (2, 1), v_2 = (3, 1)\}$. Find the dual basis $\{\phi_1, \phi_2\}$.

We seek linear functionals $\phi_1(x, y) = ax + by$ and $\phi_2(x, y) = cx + dy$ such that

$$\phi_1(v_1) = 1 \qquad \phi_1(v_2) = 0 \qquad \phi_2(v_1) = 0 \qquad \phi_2(v_2) = 1$$

Thus

$$\left. \begin{array}{l} \phi_1(v_1) = \phi_1(2, 1) = 2a + b = 1 \\ \phi_1(v_2) = \phi_1(3, 1) = 3a + b = 0 \end{array} \right\} \quad \text{or} \quad a = -1, b = 3$$

$$\left. \begin{array}{l} \phi_2(v_1) = \phi_2(2, 1) = 2c + d = 0 \\ \phi_2(v_2) = \phi_2(3, 1) = 3c + d = 1 \end{array} \right\} \quad \text{or} \quad c = 1, d = -2$$

Hence the dual basis is $\{\phi_1(x, y) = -x + 3y, \ \phi_2(x, y) = x - 2y\}$.

The next theorems give relationships between bases and their duals.

Theorem 12.2:   Let $\{v_1, \ldots, v_n\}$ be a basis of V and let $\{\phi_1, \ldots, \phi_n\}$ be the dual basis of V^*. Then for any vector $u \in V$,

$$u = \phi_1(u)v_1 + \phi_2(u)v_2 + \cdots + \phi_n(u)v_n \tag{12.1}$$

and, for any linear functional $\sigma \in V^*$,

$$\sigma = \sigma(v_1)\phi_1 + \sigma(v_2)\phi_2 + \cdots + \sigma(v_n)\phi_n \tag{12.2}$$

Theorem 12.3:   Let $\{v_1, \ldots, v_n\}$ and $\{w_1, \ldots, w_n\}$ be bases of V and let $\{\phi_1, \ldots, \phi_n\}$ and $\{\sigma_1, \ldots, \sigma_n\}$ be the bases of V^* dual to $\{v_i\}$ and $\{w_i\}$, respectively. Suppose P is the change-of-basis matrix from $\{v_i\}$ to $\{w_i\}$. Then $(P^{-1})^T$ is the change-of-basis matrix from $\{\theta_i\}$ to $\{\sigma_i\}$.

12.4 SECOND DUAL SPACE

We repeat: Every vector space V has a dual space V^* which consists of all the linear functionals on V. Thus V^* itself has a dual space V^{**}, called the second dual of V, which consists of all the linear functionals on V^*.

We now show that each $v \in V$ determines a specific element $\hat{v} \in V^{**}$. First of all, for any $\phi \in V^*$ we define

$$\hat{v}(\phi) = \phi(v)$$

It remains to be shown that this map $\hat{v} : V^* \to K$ is linear. For any scalars $a, b \in K$ and any linear functionals $\phi, \sigma \in V^*$, we have

$$\hat{v}(a\phi + b\sigma) = (a\phi + b\sigma)(v) = a\phi(v) + b\sigma(v) = a\hat{v}(\phi) + b\hat{v}(\sigma)$$

That is, $\hat{v}$ is linear and so $\hat{v} \in V^{**}$. The following theorem, proved in Problem 12.7, applies.

Theorem 12.4:   If V has finite dimension, then the mapping $v \mapsto \hat{v}$ is an isomorphism of V onto V^{**}.

The above mapping $v \mapsto \hat{v}$ is called the natural mapping of V into V^{**}. We emphasize that this mapping is never onto V^{**} if V is not finite-dimensional. However, it is always linear and, moreover, it is always one-to-one.

Now suppose V does have finite dimension. By Theorem 12.4, the natural mapping determines an isomorphism between V and V^{**}. Unless otherwise stated we shall identify V with V^{**} by this mapping. Accordingly, we shall view V as the space of linear functionals on V^* and shall write $V = V^{**}$. We remark that if $\{\phi_i\}$ is the basis of V^* dual to a basis $\{v_i\}$ of V, then $\{v_i\}$ is the basis of $V = V^{**}$ which is dual to $\{\phi_i\}$.

12.5 ANNIHILATORS

Let W be a subset (not necessarily a subspace) of a vector space V. A linear functional $\phi \in V^*$ is called an annihilator of W if $\phi(w) = 0$ for every $w \in W$, i.e., if $\phi(W) = \{0\}$. We show that the set of all such mappings, denoted by W^0 and called the annihilator of W, is a subspace of V^*. Clearly $0 \in W^0$. Now suppose $\phi, \sigma \in W^0$. Then, for any scalars $a, b \in K$ and for any $w \in W$,

$$(a\phi + b\sigma)(w) = a\phi(w) + b\sigma(w) = a0 + b0 = 0$$

Thus $a\phi + b\sigma \in W^0$ and so W^0 is a subspace of V^*.

In the case that W is a subspace of V, we have the following relationship between W and its annihilator W^0. (See proof in Problem 12.11.)

Theorem 12.5:   Suppose V has finite dimension and W is a subspace of V. Then:

$$\text{(i) dim } W + \text{dim } W^0 = \text{dim } V \quad \text{and (ii) } W^{00} = W.$$

Here $W^{00} = \{v \in V : \phi(v) = 0 \text{ for every } \phi \in W^0\}$ or, equivalently, $W^{00} = (W^0)^0$ where W^{00} is viewed as a subspace of V under the identification of V and V^{**}.

The concept of an annihilator enables us to give another interpretation of a homogeneous system of linear equations

$$\begin{aligned}
a_{11}x_1 + a_{12}x_2 + \cdots + a_{1n}x_n &= 0 \\
a_{21}x_1 + a_{22}x_2 + \cdots + a_{2n}x_n &= 0 \\
&\cdots\cdots\cdots\cdots\cdots\cdots\cdots\cdots\cdots \\
a_{m1}x_1 + a_{m2}x_2 + \cdots + a_{mn}x_n &= 0
\end{aligned} \qquad (*)$$

Here each row $(a_{i1}, a_{i2}, \ldots, a_{in})$ of the coefficient matrix $A = (a_{ij})$ is viewed as an element of K^n and each solution vector $\phi = (x_1, x_2, \ldots, x_n)$ is viewed as an element of the dual space. In this context, the solution space S of $(*)$ is the annihilator of the rows of A and hence of the row space of A. Consequently, using Theorem 12.5, we again obtain the following fundamental result on the dimension of the solution space of a homogeneous system of linear equations:

$$\dim S = \dim K^n - \dim (\text{row space of } A) = n - \text{rank } A$$

12.6 TRANSPOSE OF A LINEAR MAPPING

Let $T : V \to U$ be an arbitrary linear mapping from a vector space V into a vector space U. Now for any linear functional $\phi \in U^*$, the composition $\phi \circ T$ is a linear mapping from V into K:

$$V \xrightarrow{\ T\ } U \xrightarrow{\ \phi\ } K$$
$$\underbrace{\qquad\qquad\qquad}_{\phi \, \circ \, T}$$

That is, $\phi \circ T \in V^*$. Thus the correspondence

$$\phi \mapsto \phi \circ T$$

is a mapping from U^* into V^*; we denote it by T^t and call it the transpose of T. In other words, $T^t : U^* \to V^*$ is defined by

$$T^t(\phi) = \phi \circ T$$

Thus $(T^t(\phi))(v) = \phi(T(v))$ for every $v \in V$.

Theorem 12.6: The transpose mapping T^t defined above is linear.

Proof. For any scalars $a, b \in K$ and any linear functionals $\phi, \sigma \in U^*$,

$$T^t(a\phi + b\sigma) = (a\phi + b\sigma) \circ T = a(\phi \circ T) + b(\sigma \circ T) = aT^t(\phi) + bT^t(\sigma)$$

That is, T^t is linear as claimed.

We emphasize that if T is a linear mapping from V into U, then T^t is a linear mapping from U^* into V^*:

$$V \xrightarrow{\ T\ } U \qquad\qquad V^* \xleftarrow{\ T^t\ } U^*$$

The name "transpose" for the mapping T^t no doubt derives from the following theorem, proved in Problem 12.16.

Theorem 12.7: Let $T : V \to U$ be linear, and let A be the matrix representation of T relative to bases $\{v_i\}$ of V and $\{u_i\}$ of U. Then the transpose matrix A^T is the matrix representation of $T^t : U^* \to V^*$ relative to the bases dual to $\{u_i\}$ and $\{v_i\}$.

Solved Problems

DUAL SPACES AND BASES

12.1. Consider the following basis of $\mathbf{R}^3$: $\{v_1 = (1, -1, 3), v_2 = (0, 1, -1), v_3 = (0, 3, -2)\}$. Find the dual basis $\{\phi_1, \phi_2, \phi_3\}$.

We seek linear functionals

$$\phi_1(x, y, z) = a_1 x + a_2 y + a_3 z \qquad \phi_2(x, y, z) = b_1 x + b_2 y + b_3 z \qquad \phi_3(x, y, z) = c_1 x + c_2 y + c_3 z$$

such that
$$
\begin{array}{lll}
\phi_1(v_1) = 1 & \phi_1(v_2) = 0 & \phi_1(v_3) = 0 \\
\phi_2(v_1) = 0 & \phi_2(v_2) = 1 & \phi_2(v_3) = 0 \\
\phi_3(v_1) = 0 & \phi_3(v_2) = 0 & \phi_3(v_3) = 1
\end{array}
$$

We find ϕ_1 as follows:

$$
\begin{aligned}
\phi_1(v_1) &= \phi_1(1, -1, 3) = a_1 - a_2 + 3a_3 = 1 \\
\phi_1(v_2) &= \phi_1(0, 1, -1) = \phantom{a_1 - {}} a_2 - a_3 = 0 \\
\phi_1(v_3) &= \phi_1(0, 3, -2) = \phantom{a_1 - {}} 3a_2 - 2a_3 = 0
\end{aligned}
$$

Solving the system of equations, we obtain $a_1 = 1, a_2 = 0, a_3 = 0$. Thus $\phi_1(x, y, z) = x$.
   We next find ϕ_2 :

$$
\begin{aligned}
\phi_2(v_1) &= \phi_2(1, -1, 3) = b_1 - b_2 + 3b_3 = 0 \\
\phi_2(v_2) &= \phi_2(0, 1, -1) = \phantom{b_1 - {}} b_2 - b_3 = 1 \\
\phi_2(v_3) &= \phi_2(0, 3, -2) = \phantom{b_1 - {}} 3b_2 - 2b_3 = 0
\end{aligned}
$$

Solving the system, we obtain $b_1 = 7, b_2 = -2, b_3 = -3$. Hence $\phi_2(x, y, z) = 7x - 2y - 3z$.
   Finally, we find ϕ_3 :

$$
\begin{aligned}
\phi_3(v_1) &= \phi_3(1, -1, 3) = c_1 - c_2 + 3c_3 = 0 \\
\phi_3(v_2) &= \phi_3(0, 1, -1) = \phantom{c_1 - {}} c_2 - c_3 = 0 \\
\phi_3(v_3) &= \phi_3(0, 3, -2) = \phantom{c_1 - {}} 3c_2 - 2c_3 = 1
\end{aligned}
$$

Solving the system, we obtain $c_1 = -2, c_2 = 1, c_3 = 1$. Thus $\phi_3(x, y, z) = -2x + y + z$.

12.2. Let V be the vector space of polynomials over $\mathbf{R}$ of degree ≤ 1, i.e., $V = \{a + bt : a, b \in \mathbf{R}\}$. Let $\phi_1 : V \to \mathbf{R}$ and $\phi_2 : V \to \mathbf{R}$ be defined by

$$\phi_1(f(t)) = \int_0^1 f(t)\, dt \qquad \text{and} \qquad \phi_2(f(t)) = \int_0^2 f(t)\, dt$$

(We remark that ϕ_1 and ϕ_2 are linear and so belong to the dual space V^*.) Find the basis $\{v_1, v_2\}$ of V which is dual to $\{\phi_1, \phi_2\}$.

   Let $v_1 = a + bt$ and $v_2 = c + dt$. By definition of the dual basis,

$$\phi_1(v_1) = 1, \; \phi_2(v_1) = 0 \qquad \text{and} \qquad \phi_1(v_2) = 0, \; \phi_2(v_2) = 1$$

Thus

$$
\left.
\begin{aligned}
\phi_1(v_1) &= \int_0^1 (a + bt)\, dt = a + \tfrac{1}{2}b = 1 \\
\phi_2(v_1) &= \int_0^2 (a + bt)\, dt = 2a + 2b = 0
\end{aligned}
\right\} \qquad \text{or} \qquad a = 2, \, b = -2
$$

$$
\left.
\begin{aligned}
\phi_1(v_2) &= \int_0^1 (c + dt)\, dt = c + \tfrac{1}{2}d = 0 \\
\phi_2(v_2) &= \int_0^2 (c + dt)\, dt = 2c + 2d = 1
\end{aligned}
\right\} \qquad \text{or} \qquad c = -1/2, \, d = 1
$$

In other words, $\{2 - 2t, \, -\tfrac{1}{2} + t\}$ is the basis of V which is dual to $\{\phi_1, \phi_2\}$.

12.3. Prove Theorem 12.1.

We first show that $\{\phi_1, \ldots, \phi_n\}$ spans V^*. Let ϕ be an arbitrary element of V^*, and suppose

$$\phi(v_1) = k_1, \; \phi(v_2) = k_2, \ldots, \phi(v_n) = k_n$$

Set $\sigma = k_1\phi_1 + \cdots + k_n\phi_n$. Then

$$\sigma(v_1) = (k_1\phi_1 + \cdots + k_n\phi_n)(v_1) = k_1\phi_1(v_1) + k_2\phi_2(v_1) + \cdots + k_n\phi_n(v_1)$$
$$= k_1 \cdot 1 + k_2 \cdot 0 + \cdots + k_n \cdot 0 = k_1$$

Similarly, for $i = 2, \ldots, n$,

$$\sigma(v_i) = (k_1\phi_1 + \cdots + k_n\phi_n)(v_i) = k_1\phi_1(v_i) + \cdots + k_i\phi_i(v_i) + \cdots + k_n\phi_n(v_i) = k_i$$

Thus $\phi(v_i) = \sigma(v_i)$ for $i = 1, \ldots, n$. Since ϕ and σ agree on the basis vectors, $\phi = \sigma = k_1\phi_1 + \cdots + k_n\phi_n$. Accordingly, $\{\phi_1, \ldots, \phi_n\}$ spans V^*.

It remains to be shown that $\{\phi_1, \ldots, \phi_n\}$ is linearly independent. Suppose

$$a_1\phi_1 + a_2\phi_2 + \cdots + a_n\phi_n = 0$$

Applying both sides to v_1, we obtain

$$0 = 0(v_1) = (a_1\phi_1 + \cdots + a_n\phi_n)(v_1) = a_1\phi_1(v_1) + a_2\phi_2(v_1) + \cdots + a_n\phi_n(v_1)$$
$$= a_1 \cdot 1 + a_2 \cdot 0 + \cdots + a_n \cdot 0 = a_1$$

Similarly, for $i = 2, \ldots, n$,

$$0 = 0(v_i) = (a_1\phi_1 + \cdots + a_n\phi_n)(v_i) = a_1\phi_1(v_i) + \cdots + a_i\phi_i(v_i) + \cdots + a_n\phi_n(v_i) = a_i$$

That is, $a_1 = 0, \ldots, a_n = 0$. Hence $\{\phi_1, \ldots, \phi_n\}$ is linearly independent and so it is a basis of V^*.

12.4. Prove Theorem 12.2.

Suppose

$$u = a_1v_1 + a_2v_2 + \cdots + a_nv_n \qquad\qquad (1)$$

Then

$$\phi_1(u) = a_1\phi_1(v_1) + a_2\phi_1(v_2) + \cdots + a_n\phi_1(v_n) = a_1 \cdot 1 + a_2 \cdot 0 + \cdots + a_n \cdot 0 = a_1$$

Similarly, for $i = 2, \ldots, n$,

$$\phi_i(u) = a_1\phi_i(v_1) + \cdots + a_i\phi_i(v_i) + \cdots + a_n\phi_i(v_n) = a_i$$

That is, $\phi_1(u) = a_1$, $\phi_2(u) = a_2, \ldots, \phi_n(u) = a_n$. Substituting these results into (1), we obtain (12.1).

Next we prove (12.2). Applying the linear functional σ to both sides of (12.1),

$$\sigma(u) = \phi_1(u)\sigma(v_1) + \phi_2(u)\sigma(v_2) + \cdots + \phi_n(u)\sigma(v_n)$$
$$= \sigma(v_1)\phi_1(u) + \sigma(v_2)\phi_2(u) + \cdots + \sigma(v_n)\phi_n(u)$$
$$= (\sigma(v_1)\phi_1 + \sigma(v_2)\phi_2 + \cdots + \sigma(v_n)\phi_n)(u)$$

Since the above holds for every $u \in V$, $\sigma = \sigma(v_1)\phi_1 + \sigma(v_2)\phi_2 + \cdots + \sigma(v_n)\phi_n$ as claimed.

12.5. Prove Theorem 12.3.

Suppose

$$w_1 = a_{11}v_1 + a_{12}v_2 + \cdots + a_{1n}v_n \qquad\qquad \sigma_1 = b_{11}\phi_1 + b_{12}\phi_2 + \cdots + b_{1n}\phi_n$$
$$w_2 = a_{21}v_1 + a_{22}v_2 + \cdots + a_{2n}v_n \qquad\qquad \sigma_2 = b_{21}\phi_1 + b_{22}\phi_2 + \cdots + b_{2n}\phi_n$$
$$\cdots\cdots\cdots\cdots\cdots\cdots\cdots\cdots\cdots\cdots\cdots\cdots\cdots \qquad\qquad \cdots\cdots\cdots\cdots\cdots\cdots\cdots\cdots\cdots\cdots\cdots\cdots\cdots$$
$$w_n = a_{n1}v_1 + a_{n2}v_2 + \cdots + a_{nn}v_n \qquad\qquad \sigma_n = b_{n1}\phi_1 + b_{n2}\phi_2 + \cdots + b_{nn}\phi_n$$

where $P = (a_{ij})$ and $Q = (b_{ij})$. We seek to prove that $Q = (P^{-1})^T$.

Let R_i denote the ith row of Q and let C_j denote the jth column of P^T. Then

$$R_i = (b_{i1}, b_{i2}, \ldots, b_{in}) \quad \text{and} \quad C_j = (a_{j1}, a_{j2}, \ldots, a_{jn})^T$$

By definition of the dual basis,

$$\sigma_i(w_j) = (b_{i1}\phi_1 + b_{i2}\phi_2 + \cdots + b_{in}\phi_n)(a_{j1}v_1 + a_{j2}v_2 + \cdots + a_{jn}v_n)$$
$$= b_{i1}a_{j1} + b_{i2}a_{j2} + \cdots + b_{in}a_{jn} = R_i C_j = \delta_{ij}$$

where δ_{ij} is the Kronecker delta. Thus

$$QP^T = \begin{pmatrix} R_1 C_1 & R_1 C_2 & \ldots & R_n C_n \\ R_2 C_1 & R_2 C_2 & \ldots & R_2 C_n \\ \hdotsfor{4} \\ R_n C_1 & R_n C_2 & \ldots & R_n C_n \end{pmatrix} = \begin{pmatrix} 1 & 0 & \ldots & 0 \\ 0 & 1 & \ldots & 0 \\ \hdotsfor{4} \\ 0 & 0 & \ldots & 1 \end{pmatrix} = I$$

and hence $Q = (P^T)^{-1} = (P^{-1})^T$ as claimed.

12.6. Suppose V has finite dimension. Show that if $v \in V$, $v \neq 0$, then there exists $\phi \in V^*$ such that $\phi(v) \neq 0$.

We extend $\{v\}$ to a basis $\{v, v_2, \ldots, v_n\}$ of V. By Theorem 9.2, there exists a unique linear mapping $\phi : V \to K$ such that $\phi(v) = 1$ and $\phi(v_i) = 0$, $i = 2, \ldots, n$. Hence ϕ has the desired property.

12.7. Prove Theorem 12.4.

We first prove that the map $v \mapsto \hat{v}$ is linear, i.e., for any vectors v, $w \in V$ and any scalars a, $b \in K$, $\widehat{av + bw} = a\hat{v} + b\hat{w}$. For any linear functional $\phi \in V^*$,

$$\widehat{av + bw}(\phi) = \phi(av + bw) = a\phi(v) + b\phi(w) = a\hat{v}(\phi) + b\hat{w}(\phi) = (a\hat{v} + b\hat{w})(\phi)$$

Since $\widehat{av + bw}(\phi) = (a\hat{v} + b\hat{w})(\phi)$ for every $\phi \in V^*$, we have $\widehat{av + bw} = a\hat{v} + b\hat{w}$. Thus the map $v \mapsto \hat{v}$ is linear.

Now suppose $v \in V$, $v \neq 0$. Then, by Problem 12.6, there exists $\phi \in V^*$ for which $\phi(v) \neq 0$. Hence $\hat{v}(\phi) = \phi(v) \neq 0$ and thus $\hat{v} \neq 0$. Since $v \neq 0$ implies $\hat{v} \neq 0$, the map $v \mapsto \hat{v}$ is nonsingular and hence an isomorphism (Theorem).

Now dim $V = $ dim $V^* = $ dim V^{**} because V has finite dimension. Accordingly, mapping $v \mapsto \hat{v}$ is an isomorphism of V onto V^{**}.

ANNIHILATORS

12.8. Show that if $\phi \in V^*$ annihilates a subset S of V, then ϕ annihilates the linear span $L(S)$ of S. Hence $S^0 = (\text{span }(S))^0$.

Suppose $v \in \text{span }(S)$. Then there exist $w_1, \ldots, w_r \in S$ for which $v = a_1 w_1 + a_2 w_2 + \cdots + a_r w_r$.

$$\phi(v) = a_1 \phi(w_1) + a_2 \phi(w_2) + \cdots + a_r \phi(w_r) = a_1 0 + a_2 0 + \cdots + a_r 0 = 0$$

Since v was an arbitrary element of span (S), ϕ annihilates span (S) as claimed.

12.9. Let W be the subspace of $\mathbf{R}^4$ spanned by $v_1 = (1, 2, -3, 4)$ and $v_2 = (0, 1, 4, -1)$. Find a basis of the annihilator of W.

By Problem 12.8, it suffices to find a basis of the set of linear functionals of the form $\phi(x, y, z, w) = ax + by + cz + dw$ for which $\phi(v_1) = 0$ and $\phi(v_2) = 0$:

$$\phi(1, 2, -3, 4) = a + 2b - 3c + 4d = 0$$
$$\phi(0, 1, 4, -1) = \qquad b + 4c - d = 0$$

The system of equations in unknowns a, b, c, d is in echelon form with free variables c and d.

Set $c = 1$, $d = 0$ to obtain the solution $a = 11$, $b = -4$, $c = 1$, $d = 0$ and hence the linear functional $\phi_1(x, y, z, w) = 11x - 4y + z$.

Set $c = 0$, $d = -1$ to obtain the solution $a = 6$, $b = -1$, $c = 0$, $d = -1$ and hence the linear functional $\phi_2(x, y, z, w) = 6x - y - w$.

The set of linear functionals $\{\phi_1, \phi_2\}$ is a basis of W^0, the annihilator of W.

12.10. Show that: (a) for any subset S of V, $S \subseteq S^{00}$; and (b) if $S_1 \subseteq S_2$, then $S_2^0 \subseteq S_1^0$.

(a) Let $v \in S$. Then for every linear functional $\phi \in S^0$, $\hat{v}(\phi) = \phi(v) = 0$. Hence $\hat{v} \in (S^0)^0$. Therefore, under the identification of V and V^{**}, $v \in S^{00}$. Accordingly, $S \subseteq S^{00}$.

(b) Let $\phi \in S_2^0$. Then $\phi(v) = 0$ for every $v \in S_2$. But $S_1 \subseteq S_2$; hence ϕ annihilates every element of S_1, i.e., $\phi \in S_1$. Therefore $S_2^0 \subseteq S_1^0$.

12.11. Prove Theorem 12.5.

(i) Suppose dim $V = n$ and dim $W = r \leq n$. We want to show that dim $W^0 = n - r$. We choose a basis $\{w_1, \ldots, w_r\}$ of W and extend it to the following basis of V, say $\{w_1, \ldots, w_r, v_1, \ldots, v_{n-r}\}$. Consider the dual basis

$$\{\phi_1, \ldots, \phi_r, \sigma_1, \ldots, \sigma_{n-r}\}$$

By definition of the dual basis, each of the above σ's annihilates each w_i; hence $\sigma_1, \ldots, \sigma_{n-r} \in W^0$. We claim that $\{\sigma_j\}$ is a basis of W^0. Now $\{\sigma_j\}$ is part of a basis of V^* and so it is linearly independent.

We next show that $\{\phi_j\}$ spans W^0. Let $\sigma \in W^0$. By Theorem 12.2,

$$\sigma = \sigma(w_1)\phi_1 + \cdots + \sigma(w_r)\phi_r + \sigma(v_1)\sigma_1 + \cdots + \sigma(v_{n-r})\sigma_{n-r}$$
$$= 0\phi_1 + \cdots + 0\phi_r + \sigma(v_1)\sigma_1 + \cdots + \sigma(v_{n-r})\sigma_{n-r}$$
$$= \sigma(v_1)\sigma_1 + \cdots + \sigma(v_{n-r})\sigma_{n-r}$$

Consequently $\{\sigma_1, \ldots, \sigma_{n-r}\}$ spans W^0 and so it is a basis of W^0. Accordingly, as required

$$\dim W^0 = n - r = \dim V - \dim W.$$

(ii) Suppose dim $V = n$ and dim $W = r$. Then dim $V^* = n$ and, by (i), dim $W^0 = n - r$. Thus by (i), dim $W^{00} = n - (n - r) = r$; therefore dim $W = \dim W^{00}$. By Problem 12.10, $W \subseteq W^{00}$. Accordingly, $W = W^{00}$.

12.12. Let U and W be subspaces of V. Prove: $(U + W)^0 = U^0 \cap W^0$.

Let $\phi \in (U + W)^0$. Then ϕ annihilates $U + W$ and so, in particular, ϕ annihilates U and V. That is, $\phi \in U^0$ and $\phi \in W^0$; hence $\phi \in U^0 \cap W^0$. Thus $(U + W)^0 \subseteq U^0 \cap W^0$.

On the other hand, suppose $\sigma \in U^0 \cap W^0$. Then σ annihilates U and also W. If $v \in U + W$, then $v = u + w$ where $u \in U$ and $w \in W$. Hence $\sigma(v) = \sigma(u) + \sigma(w) = 0 + 0 = 0$. Thus σ annihilates $U + W$, i.e., $\sigma \in (U + W)^0$. Accordingly, $U^0 + W^0 \subseteq (U + W)^0$.

Both inclusion relations give us the desired equality.

Remark: Observe that no dimension argument is employed in the proof; hence the result holds for spaces of finite or infinite dimension.

TRANSPOSE OF A LINEAR MAPPING

12.13. Let ϕ be the linear functional on $\mathbf{R}^2$ defined by $\phi(x, y) = x - 2y$. For each of the following linear operators T on $\mathbf{R}^2$, find $(T^t(\phi))(x, y)$:

(a) $T(x, y) = (x, 0)$,   (b) $T(x, y) = (y, x + y)$,   (c) $T(x, y) = (2x - 3y, 5x + 2y)$

By definition of the transpose mapping, $T^t(\phi) = \phi \circ T$, that is, $(T^t(\phi))(v) = \phi(T(v))$ for every vector v. Hence:

(a) $(T^t(\phi))(x, y) = \phi(T(x, y)) = \phi(x, 0) = x$

(b) $(T^t(\phi))(x, y) = \phi(T(x, y)) = \phi(y, x + y) = y - 2(x + y) = -2x - y$

(c) $(T^t(\phi))(x, y) = \phi(T(x, y)) = \phi(2x - 3y, 5x + 2y) = (2x - 3y) - 2(5x + 2y) = -8x - 7y$

12.14. Let $T : V \to U$ be linear and let $T^t : U^* \to V^*$ be its transpose. Show that the kernel of T^t is the annihilator of the image of T, i.e., Ker $T^t = (\text{Im } T)^0$.

Suppose $\phi \in$ Ker T^t; that is, $T^t(\phi) = \phi \circ T = 0$. If $u \in$ Im T, then $u = T(v)$ for some $v \in V$; hence

$$\phi(u) = \phi(T(v)) = (\phi \circ T)(v) = 0(v) = 0$$

We have that $\phi(u) = 0$ for every $u \in$ Im T; hence $\phi \in (\text{Im } T)^0$. Thus Ker $T^t \subset (\text{Im } T)^0$.

On the other hand, suppose $\sigma \in (\text{Im } T)^0$; that is, $\sigma(\text{Im } T) = \{0\}$. Then, for every $v \in V$,

$$(T^t(\sigma))(v) = (\sigma \circ T)(v) = \sigma(T(v)) = 0 = 0(v)$$

We have $(T^t(\sigma))(v) = 0(v)$ for every $v \in V$; hence $T^t(\sigma) = 0$. Thus $\sigma \in$ Ker T^t and so $(\text{Im } T)^0 \subseteq$ Ker T^t. Both inclusion relations give us the required equality.

12.15. Suppose V and U have finite dimension and suppose $T : V \to U$ is linear. Prove: rank $T =$ rank T^t.

Suppose dim $V = n$ and dim $U = m$. Also suppose rank $T = r$. Then, by Theorem 12.5,

$$\dim (\text{Im } T)^0 = \dim U - \dim (\text{Im } T) = m - \text{rank } T = m - r$$

By Problem 12.14, Ker $T^t = (\text{Im } T)^0$. Hence nullity $T^t = m - r$. It then follows that, as claimed,

$$\text{rank } T^t = \dim U^* - \text{nullity } T^t = m - (m - r) = r = \text{rank } T$$

12.16. Prove Theorem 12.7.

Suppose

$$\begin{aligned}
T(v_1) &= a_{11} u_1 + a_{12} u_2 + \cdots + a_{1n} u_n \\
T(v_2) &= a_{21} u_1 + a_{22} u_2 + \cdots + a_{2n} u_n \\
&\cdots\cdots\cdots\cdots\cdots\cdots\cdots\cdots\cdots\cdots \\
T(v_m) &= a_{m1} u_1 + a_{m2} u_2 + \cdots + a_{mn} u_n
\end{aligned} \tag{1}$$

We want to prove that

$$\begin{aligned}
T^t(\sigma_1) &= a_{11} \phi_1 + a_{21} \phi_2 + \cdots + a_{m1} \phi_m \\
T^t(\sigma_2) &= a_{12} \phi_1 + a_{22} \phi_2 + \cdots + a_{m2} \phi_m \\
&\cdots\cdots\cdots\cdots\cdots\cdots\cdots\cdots\cdots\cdots \\
T^t(\sigma_n) &= a_{1n} \phi_1 + a_{2n} \phi_2 + \cdots + a_{mn} \phi_m
\end{aligned} \tag{2}$$

where $\{\sigma_i\}$ and $\{\phi_j\}$ are the bases dual to $\{u_i\}$ and $\{v_j\}$, respectively.

Let $v \in V$ and suppose $v = k_1 v_1 + k_2 v_2 + \cdots + k_m v_m$. Then, by (1),

$$\begin{aligned}
T(v) &= k_1 T(v_1) + k_2 T(v_2) + \cdots + k_m T(v_m) \\
&= k_1(a_{11} u_1 + \cdots + a_{1n} u_n) + k_2(a_{21} u_1 + \cdots + a_{2n} u_n) + \cdots + k_m(a_{m1} u_1 + \cdots + a_{mn} u_n) \\
&= (k_1 a_{11} + k_2 a_{21} + \cdots + k_m a_{m1})u_1 + \cdots + (k_1 a_{1n} + k_2 a_{2n} + \cdots + k_m a_{mn})u_n \\
&= \sum_{i=1}^{n} (k_1 a_{1i} + k_2 a_{2i} + \cdots + k_m a_{mi})u_i
\end{aligned}$$

Hence for $j = 1, \ldots, n$,

$$\begin{aligned}
(T^t(\sigma_j)(v)) &= \sigma_j(T(v)) = \sigma_j\left(\sum_{i=1}^{n} (k_1 a_{1i} + k_2 a_{2i} + \cdots + k_m a_{mi})u_i \right) \\
&= k_1 a_{1j} + k_2 a_{2j} + \cdots + k_m a_{mj}
\end{aligned} \tag{3}$$

On the other hand, for $j = 1, \ldots, n$,

$$(a_{1j}\phi_1 + a_{2j}\phi_2 + \cdots + a_{mj}\phi_m)(v) = (a_{1j}\phi_1 + a_{2j}\phi_2 + \cdots + a_{mj}\phi_m)(k_1 v_1 + k_2 v_2 + \cdots + k_m v_m)$$
$$= k_1 a_{1j} + k_2 a_{2j} + \cdots + k_m a_{mj} \tag{4}$$

Since $v \in V$ was arbitrary, (3) and (4) imply that

$$T^t(\sigma_j) = a_{1j}\phi_1 + a_{2j}\phi_2 + \cdots + a_{mj}\phi_m \qquad j = 1, \ldots, n$$

which is (2). Thus the theorem is proved.

12.17. Let A be an arbitrary $m \times n$ matrix over a field K. Prove that the row rank and the column rank of A are equal.

Let $T : K^n \to K^m$ be the linear map defined by $T(v) = Av$, where the elements of K^n and K^m are written as column vectors. Then A is the matrix representation of T relative to the usual bases of K^n and K^m, and the image of T is the column space of A. Hence

$$\text{rank } T = \text{column rank of } A$$

By Theorem 12.7, A^T is the matrix representation of T^t relative to the dual bases. Hence

$$\text{rank } T^t = \text{column rank of } A^T = \text{row rank of } A$$

But by Problem 12.16, rank $T =$ rank T^t; hence the row rank and the column rank of A are equal. (This result was stated earlier as Theorem 5.18, and was proved in a direct way in Problem 5.53.)

Supplementary Problems

DUAL SPACES AND DUAL BASES

12.18. Let $\phi : \mathbf{R}^3 \to \mathbf{R}$ and $\sigma : \mathbf{R}^3 \to \mathbf{R}$ be the linear functionals defined by $\phi(x, y, z) = 2x - 3y + z$ and $\sigma(x, y, z) = 4x - 2y + 3z$. Find: (a) $\phi + \sigma$, (b) 3ϕ, (c) $2\phi - 5\sigma$.

12.19. Let V be the vector space of polynomials over $\mathbf{R}$ of degree ≤ 2. Let ϕ_1, ϕ_2, and ϕ_3 be the linear functionals on V defined by

$$\phi_1(f(t)) = \int_0^1 f(t)\, dt \qquad \phi_2(f(t)) = f'(1) \qquad \phi_3(f(t)) = f(0)$$

Here $f(t) = a + bt + ct^2 \in V$ and $f'(t)$ denotes the derivative of $f(t)$. Find the basis $\{f_1(t), f_2(t), f_3(t)\}$ of V which is dual to $\{\phi_1, \phi_2, \phi_3\}$.

12.20. Suppose $u, v \in V$ and that $\phi(u) = 0$ implies $\phi(v) = 0$ for all $\phi \in V^*$. Show that $v = ku$ for some scalar k.

12.21. Suppose $\phi, \sigma \in V^*$ and that $\phi(v) = 0$ implies $\sigma(v) = 0$ for all $v \in V$. Show that $\sigma = k\phi$ for some scalar k.

12.22. Let V be the vector space of polynomials over K. For $a \in K$, define $\phi_a : V \to K$ by $\phi_a(f(t)) = f(a)$. Show that: (a) ϕ_a is linear; (b) if $a \neq b$, then $\phi_a \neq \phi_b$.

12.23. Let V be the vector space of polynomials of degree ≤ 2. Let $a, b, c \in K$ be distinct scalars. Let ϕ_a, ϕ_b, and ϕ_c be the linear functionals defined by $\phi_a(f(t)) = f(a)$, $\phi_b(f(t)) = f(b)$, $\phi_c(f(t)) = f(c)$. Show that $\{\phi_a, \phi_b, \phi_c\}$ is linearly independent, and find the basis $\{f_1(t), f_2(t), f_3(t)\}$ of V which is its dual.

12.24. Let V be the vector space of square matrices of order n. Let $T : V \to K$ be the trace mapping: that is $T(A) = a_{11} + a_{22} + \cdots + a_{nn}$, where $A = (a_{ij})$. Show that T is linear.

12.25. Let W be a subspace of V. For any linear functional ϕ on W, show that there is a linear functional σ on V such that $\sigma(w) = \phi(w)$ for any $w \in W$, i.e., ϕ is the restriction of σ to W.

12.26. Let $\{e_1, \ldots, e_n\}$ be the usual basis of K^n. Show that the dual basis is $\{\pi_1, \ldots, \pi_n\}$ where π_i is the ith projection mapping: that is, $\pi_i(a_1, \ldots, a_n) = a_i$.

12.27. Let V be a vector space over $\mathbf{R}$. Let $\phi_1, \phi_2 \in V^*$ and suppose $\sigma : V \to R$ defined by $\sigma(v) = \phi_1(v)\phi_2(v)$ also belongs to V^*. Show that either $\phi_1 = 0$ or $\phi_2 = 0$.

ANNIHILATORS

12.28. Let W be the subspace of $\mathbf{R}^4$ spanned by $(1, 2, -3, 4)$, $(1, 3, -2, 6)$, and $(1, 4, -1, 8)$. Find a basis of the annihilator of W.

12.29. Let W be the subspace of $\mathbf{R}^3$ spanned by $(1, 1, 0)$ and $(0, 1, 1)$. Find a basis of the annihilator of W.

12.30. Show that, for any subset S of V, span $(S) = S^{00}$ where span (S) is the linear span of S.

12.31. Let U and W be subspaces of a vector space V of finite dimension. Prove: $(U \cap W)^0 = U^0 + W^0$.

12.32. Suppose $V = U \oplus W$. Prove that $V^* = U^0 \oplus W^0$.

TRANSPOSE OF A LINEAR MAPPING

12.33. Let ϕ be the linear functional on $\mathbf{R}^2$ defined by $\phi(x, y) = 3x - 2y$. For each of the following linear mapping $T : \mathbf{R}^3 \to \mathbf{R}^2$, find $(T^t(\phi))(x, y, z)$.

    (a) $T(x, y, z) = (x + y, y + z)$;   (b) $T(x, y, z) = (x + y + z, 2x - y)$

12.34. Suppose $T_1 : U \to V$ and $T_2 : V \to W$ are linear. Prove that $(T_2 \circ T_1)^t = T_1^t \circ T_2^t$.

12.35. Suppose $T : V \to U$ is linear and V has finite dimension. Prove that Im $T^t = (\text{Ker } T)^0$.

12.36. Suppose $T : V \to U$ is linear and $u \in U$. Prove that $u \in$ Im T or there exists $\phi \in V^*$ such that $T^t(\phi) = 0$ and $\phi(u) = 1$.

12.37. Let V be of finite dimension. Show that the mapping $T \mapsto T^t$ is an isomorphism from Hom (V, V) onto Hom (V^*, V^*). (Here T is any linear operator on V.)

MISCELLANEOUS PROBLEMS

12.38. Let V be a vector space over $\mathbf{R}$. The line segment $\overline{uv}$ joining points $u, v \in V$ is defined by

$$\overline{uv} = \{tu + (1 - t)v : 0 \le t \le 1\}.$$

A subset S of V is termed convex if $u, v \in S$ implies $\overline{uv} \subseteq S$. Let $\phi \in V^*$ and let

$$W^+ = \{v \in V : \phi(v) > 0\} \qquad W = \{v \in V : \phi(v) = 0\} \qquad W^- = \{v \in V : \phi(v) < 0\}$$

12.39. Let V be a vector space of finite dimension. A hyperplane H of V is defined to be the kernel of a nonzero linear functional ϕ on V. Show that every subspace of V is the intersection of a finite number of hyperplanes.

Answers to Supplementary Problems

12.18. (a) $6x - 5y + 4z$, (b) $6x - 9y + 3z$, (c) $-16x + 4y - 13z$

12.22. (b) Let $f(t) = t$. Then $\phi_a(f(t)) = a \neq b = \phi_b(f(t))$, and therefore $\phi_a \neq \phi_b$.

12.23. $\left\{ f_1(t) = \dfrac{t^2 - (b+c)t + bc}{(a-b)(a-c)}, \quad f_2(t) = \dfrac{t^2 - (a+c)t + ac}{(b-a)(b-c)}, \quad f_3(t) = \dfrac{t^2 - (a+b)t + ab}{(c-a)(c-b)} \right\}$

12.28. $\{\phi_1(x, y, z, t) = 5x - y + z, \ \phi_2(x, y, z, t) = 2y - t\}$

12.29. $\{\phi(x, y, z) = x - y + z\}$

12.33. (a) $(T^t(\phi))(x, y, z) = 3x + y - 2z$, (b) $(T^t(\phi))(x, y, z) = -x + 5y + 3z$

Chapter 13

Bilinear, Quadratic, and Hermitian Forms

13.1 INTRODUCTION

This chapter generalizes the notions of linear mappings and linear functionals. Specifically, we introduce the notion of a bilinear form. (Actually, the notion of a general multilinear mappings did appear in Chapter 7.) These bilinear maps also give rise to quadratic and Hermitian forms. Although quadratic forms did appear previously in the context of matrices, this chapter is treated independently of the previous results. (Thus there may be some overlap in the discussion and some examples and problems.)

13.2 BILINEAR FORMS

Let V be a vector space of finite dimension over a field K. A bilinear form on V is a mapping $f: V \times V \to K$ which satisfies:

   (i) $f(au_1 + bu_2, v) = af(u_1, v) + bf(u_2, v)$

  (ii) $f(u, av_1 + bv_2) = af(u, v_1) + bf(u, v_2)$

for all $a, b \in K$ and all $u_i, v_i \in V$. We express condition (i) by saying f is linear in the first variable, and condition (ii) by saying f is linear in the second variable.

Example 13.1

(a) Let ϕ and σ be arbitrary linear functionals on V. Let $f: V \times V \to K$ be defined by $f(u, v) = \phi(u)\sigma(v)$. Then f is bilinear because ϕ and σ are each linear. (Such a bilinear form f turns out to be the "tensor product" of ϕ and σ and so is sometimes written $f = \phi \otimes \sigma$.)

(b) Let f be the dot product on $\mathbf{R}^n$; that is,

$$f(u, v) = u \cdot v = a_1 b_1 + a_2 b_2 + \cdots + a_n b_n$$

where $u = (a_i)$ and $v = (b_i)$. Then f is a bilinear form on $\mathbf{R}^n$.

(c) Let $A = (a_{ij})$ be any $n \times n$ matrix over K. Then A may be viewed as a bilinear form f on K^n by defining

$$f(X, Y) = X^T A Y = (x_1, x_2, \ldots, x_n) \begin{pmatrix} a_{11} & a_{12} & \cdots & a_{1n} \\ a_{21} & a_{22} & \cdots & a_{2n} \\ \cdots\cdots\cdots\cdots\cdots\cdots \\ a_{n1} & a_{n2} & \cdots & a_{nn} \end{pmatrix} \begin{pmatrix} y_1 \\ y_2 \\ \cdots \\ y_n \end{pmatrix}$$

$$= \sum_{i,\,j=1}^{n} a_{ij} x_i y_i = a_{11} x_1 y_1 + a_{12} x_1 y_2 + \cdots + a_{nn} x_n y_n$$

The above formal expression in variables x_i, y_i is termed the bilinear polynomial corresponding to the matrix A. Equation (13.1) below shows that, in a certain sense, every bilinear form is of this type.

We will let $B(V)$ denote the set of bilinear forms on V. A vector space structure is placed on $B(V)$ by defining $f + g$ and kf as follows:

$$(f + g)(u, v) = f(u, v) + g(u, v)$$

$$(kf)(u, v) = kf(u, v)$$

for any $f, g \in B(V)$ and any $k \in K$. In fact,

Theorem 13.1: Let V be a vector space of dimension n over K. Let $\{\phi_1, \ldots, \phi_n\}$ be any basis of the dual space V^*. Then $\{f_{ij} : i, j = 1, \ldots, n\}$ is a basis of $B(V)$ where f_{ij} is defined by $f_{ij}(u, v) = \phi_i(u)\phi_j(v)$. Thus, in particular, dim $B(V) = n^2$.

(See Problem 13.4 for the proof.)

13.3 BILINEAR FORMS AND MATRICES

Let f be a bilinear form on V, and let $S = \{u_1, u_2, \ldots, u_n\}$ be a basis of V. Suppose $u, v \in V$ and suppose

$$u = a_1 u_1 + \cdots + a_n u_n \quad \text{and} \quad v = b_1 u_1 + \cdots + b_n u_n$$

Then

$$f(u, v) = f(a_1 u_1 + \cdots + a_n u_n, \, b_1 u_1 + \cdots + b_n u_n)$$

$$= a_1 b_1 f(u_1, u_1) + a_1 b_2 f(u_1, u_2) + \cdots + a_n b_n f(u_n, u_n)$$

$$= \sum_{i, j = 1}^{n} a_i b_j f(u_i, u_j)$$

Thus f is completely determined by the n^2 values $f(u_i, u_j)$.

The matrix $A = (a_{ij})$ where $a_{ij} = f(u_i, u_j)$ is called the matrix representation of f relative to the basis S or, simply, the matrix of f in S. It "represents" f in the sense that

$$f(u, v) = \sum a_i b_j f(u_i, u_j) = (a_1, \ldots, a_n) A \begin{pmatrix} b_1 \\ b_2 \\ \cdots \\ b_n \end{pmatrix} = [u]_S^T A [v]_S \qquad (13.1)$$

for all $u, v \in V$. [As usual, $[u]_S$ denotes the coordinate (column) vector of $u \in V$ in the basis S.]

We next ask, how does a matrix representing a bilinear form transform when a new basis is selected? The answer is given in the following theorem, proved in Problem 13.6. (Recall Theorem 10.4 that the change-of-basis matrix P from one basis S to another basis S' has the property that $[u]_S = P[u]_{S'}$, for every $u \in V$.)

Theorem 13.2: Let P be the change-of-basis matrix from one basis S to another basis S'. If A is the matrix of f in the original basis S, then

$$B = P^T A P$$

is the matrix of f in the new basis S'.

The above theorem motivates the following definition.

Definition: A matrix B is said to be congruent to a matrix A if there exists an invertible (or: nonsingular) matrix P such that $B = P^T A P$.

Thus by Theorem 13.2 matrices representing the same bilinear form are congruent. We remark that congruent matrices have the same rank because P and P^t are nonsingular; hence the following definition is well defined.

Definition: The rank of a bilinear form f on V, written rank f, is defined to be the rank of any matrix representation. We say that f is degenerate or nondegenerate according as to whether rank $f < \dim V$ or rank $f = \dim V$.

13.4   ALTERNATING BILINEAR FORMS

A bilinear form f on V is said to be alternating if

$$\text{(i)} \quad f(v, v) = 0$$

for every $v \in V$. If f is alternating, then

$$0 = f(u + v, u + v) = f(u, u) + f(u, v) + f(v, u) + f(v, v)$$

and so

$$\text{(ii)} \quad f(u, v) = -f(v, u)$$

for every $u, v \in V$. A bilinear form which satisfies condition (ii) is said to be skew symmetric (or anti-symmetric). If $1 + 1 \neq 0$ in K, then condition (ii) implies $f(v, v) = -f(v, v)$ which implies condition (i). In other words, alternating and skew symmetric are equivalent when $1 + 1 \neq 0$.

The main structure theorem of alternating bilinear forms, proved in Problem 13.19, follows.

Theorem 13.3:   Let f be an alternating bilinear form on V. Then there exists a basis of V in which f is represented by a matrix of the form

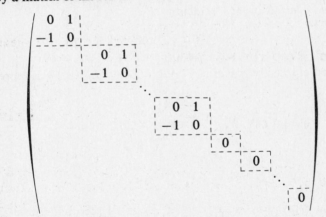

Moreover, the number of $\begin{pmatrix} 0 & 1 \\ -1 & 0 \end{pmatrix}$ is uniquely determined by f (because it is equal to $\frac{1}{2}$ rank f).

In particular, the above theorem shows that an alternating bilinear form must have even rank.

13.5   SYMMETRIC BILINEAR FORMS, QUADRATIC FORMS

A bilinear form f on V is said to be symmetric if

$$f(u, v) = f(v, u)$$

for every $u, v \in V$. If A is a matrix representation of f, we can write

$$f(X, Y) = X^T A Y = (X^T A Y)^T = Y^T A^T X$$

(We use the fact that $X^T A Y$ is a scalar and therefore equals its transpose.) Thus if f is symmetric,

$$Y^T A^T X = f(X, Y) = f(Y, X) = Y^T A X$$

and since this is true for all vectors X, Y it follows that $A = A^T$ or A is symmetric. Conversely if A is symmetric, then f is symmetric.

The main result for symmetric bilinear forms, proved in Problem 13.11, is given in

Theorem 13.4:   Let f be a symmetric bilinear form on V over K (in which $1 + 1 \neq 0$). Then V has a basis $\{v_1, \ldots, v_n\}$ in which f is represented by a diagonal matrix, i.e., $f(v_i, v_j) = 0$ for $i \neq j$.

Theorem 13.4 (Alternate Form):   Let A be a symmetric matrix over K (in which $1 + 1 \neq 0$). Then there exists an invertible (or nonsingular) matrix P such that $P^T A P$ is diagonal. That is, A is congruent to a diagonal matrix.

Since an invertible matrix P is a product of elementary matrices (Theorem 4.10), one way of obtaining the diagonal form $P^t A P$ is by a sequence of elementary row operations and the same sequence of elementary column operations. These same elementary row operations on I will yield P^T. This method is illustrated in the next example.

Definition: A mapping $q : V \to K$ is called a quadratic form if $q(v) = f(v, v)$ for some symmetric bilinear form f on V.

We call q the quadratic form associated with the symmetric bilinear form f. If $1 + 1 \neq 0$ in K, then f is obtainable from q according to the identity

$$f(u, v) = \tfrac{1}{2}[q(u + v) - q(u) - q(v)]$$

The above formula is called the polar form of f.

Now if f is represented by a symmetric matrix $A = (a_{ij})$, then q is represented in the form

$$q(X) = f(X, X) = X^T A X = (x_1, \ldots, x_n) \begin{pmatrix} a_{11} & a_{12} & \cdots & a_{1n} \\ a_{21} & a_{22} & \cdots & a_{2n} \\ \cdots\cdots\cdots\cdots\cdots\cdots \\ a_{n1} & a_{n2} & \cdots & a_{nn} \end{pmatrix} \begin{pmatrix} x_1 \\ x_2 \\ \cdots \\ x_n \end{pmatrix}$$

$$= \sum_{i,j} a_{ij} x_i x_j = a_{11} x_1^2 + a_{22} x_2^2 + \cdots + a_{nn} x_n^2 + 2 \sum_{i<j} a_{ij} x_i x_j$$

The above formal expression in variables x_i is termed the quadratic polynomial corresponding to the symmetric matrix A. Observe that if the matrix A is diagonal, then q has the diagonal representation

$$q(X) = X^t A X = a_{11} x_1^2 + a_{22} x_2^2 + \cdots + a_{nn} x_n^2$$

that is, the quadratic polynomial representing q will contain no "cross product" terms. Moreover, by Theorem 13.4, every quadratic form has such a representation (when $1 + 1 \neq 0$).

13.6   REAL SYMMETRIC BILINEAR FORMS, LAW OF INERTIA

In this section we treat symmetric bilinear forms and quadratic forms on vector spaces over the real field **R**. These forms appear in many branches of mathematics and physics. The special nature of **R** permits an independent theory. The main result, proved in Problem 13.13, follows.

Theorem 13.5:   Let f be a symmetric bilinear form on V over **R**. Then there is a basis of V in which f is represented by a diagonal matrix; every other diagonal representation has the same number P of positive entries and the same number N of negative entries. The difference $S = P - N$ is called the *signature* of f.

A real symmetric bilinear form f is said to be nonnegative semidefinite if

$$q(v) = f(v, v) \geq 0$$

for every vector v; and is said to be positive definite if

$$q(v) = f(v, v) > 0$$

for every vector $v \neq 0$. By Theorem 13.5,

(i) f is nonnegative semidefinite if and only if $S = \text{rank}\,(f)$,

(ii) f is positive definite if and only if $S = \dim V$,

where S is the signature of f.

Example 13.2. Let f be the dot product on $\mathbf{R}^n$; that is,

$$f(u, v) = u \cdot v = a_1 b_1 + a_2 b_2 + \cdots + a_n b_n$$

where $u = (a_i)$ and $v = (b_i)$. Note that f is symmetric since

$$f(u, v) = u \cdot v = v \cdot u = f(v, u)$$

Furthermore, f is positive definite because

$$f(u, u) = a_1^2 + a_2^2 + \cdots + a_n^2 > 0$$

when $u \neq 0$.

In Chapter 14 we will see how a real quadratic form q transforms when the transition matrix P is "orthogonal." If no condition is placed on P, then q can be represented in diagonal form with only 1s and -1s as nonzero coefficients. Specifically,

Corollary 13.6: Any real quadratic from q has a unique representation in the form

$$q(x_1, \ldots, x_n) = x_1^2 + \cdots + x_s^2 - x_{s+1}^2 - \cdots - x_r^2$$

The above result for real quadratic forms is sometimes referred to as the Law of Inertia or Sylvester's Theorem.

13.7 HERMITIAN FORMS

Let V be a vector space of finite dimension over the complex field $\mathbf{C}$. Let $f : V \times V \to \mathbf{C}$ be such that

(i) $f(au_1 + bu_2, v) = af(u_1, v) + bf(u_2, v)$

(ii) $f(u, v) = \overline{f(v, u)}$

where $a, b \in \mathbf{C}$ and $u_i, v \in V$. Then f is called a Hermitian form on V. (As usual, $\bar{k}$ denotes the complex conjugate of $k \in \mathbf{C}$.) By (i) and (ii),

$$\begin{aligned} f(u, av_1 + bv_2) &= \overline{f(av_1 + bv_2, u)} = \overline{af(v_1, u) + bf(v_2, u)} \\ &= \bar{a}\overline{f(v_1, u)} + \bar{b}\overline{f(v_2, u)} = \bar{a}f(u, v_1) + \bar{b}f(u, v_2) \end{aligned}$$

That is,

(iii) $f(u, av_1 + bv_2) = \bar{a}f(u, v_1) + \bar{b}f(u, v_2)$

As before, we express condition (i) by saying f is linear in the first variable. On the other hand, we express condition (iii) by saying f is conjugate linear in the second variable. Note that, by (ii), we have $f(v, v) = \overline{f(v, v)}$ and so $f(v, v)$ is real for every $v \in V$.

The mapping $q : V \to \mathbf{R}$, defined by $q(v) = f(v, v)$, is called the Hermitian quadratic form or complex quadratic form associated with the Hermitian form f. We can obtain f from q according to the following identity called the polar form of f:

$$f(u, v) = \tfrac{1}{4}[q(u + v) - q(u - v)] + \tfrac{1}{4}[q(u + iv) - q(u - iv)]$$

Now suppose $S = \{u_1, \ldots, u_n\}$ is a basis of V. The matrix $H = (h_{ij})$ where $h_{ij} = f(u_i, u_j)$ is called the *matrix representation* of f in the basis S. By (ii), $f(u_i, u_j) = \overline{f(u_j, u_i)}$; hence H is Hermitian and, in particular, the diagonal entries of H are real. Thus any diagonal representation of f contains only real entries. The next theorem, to be proved in Problem 13.33, is the complex analog of Theorem 13.5 on real symmetric bilinear forms.

Theorem 13.7: Let f be a Hermitian form on V. Then there exists a basis $S = \{u_1, \ldots, u_n\}$ of V in which f is represented by a diagonal matrix, i.e., $f(u_i, u_j) = 0$ for $i \neq j$. Moreover, every diagonal representation of f has the same number P of positive entries, and the same number N of negative entries. The difference $S = P - N$ is called the signature of f.

Analogously, a Hermitian form f is said to be nonnegative semidefinite if

$$q(v) = f(v, v) \geq 0$$

for every $v \in V$, and is said to be positive definite if

$$q(v) = f(v, v) > 0$$

for every $v \neq 0$.

Example 13.3.   Let f be the dot product on $\mathbf{C}^n$; that is,

$$f(u, v) = u \cdot v = z_1 \bar{w}_1 + z_2 \bar{w}_2 + \cdots + z_n \bar{w}_n$$

where $u = (z_i)$ and $v = (w_i)$. Then f is a Hermitian form on $\mathbf{C}^n$. Moreover, f is positive definite since, for any $v \neq 0$,

$$f(u, u) = z_1 \bar{z}_1 + z_2 \bar{z}_2 + \cdots + z_n \bar{z}_n = |z_1|^2 + |z_2|^2 + \cdots + |z_n|^2 > 0$$

Solved Problems

BILINEAR FORMS

13.1.   Let $u = (x_1, x_2, x_3)$ and $v = (y_1, y_2, y_3)$, and let

$$f(u, v) = 3x_1 y_1 - 2x_1 y_2 + 5x_2 y_1 + 7x_2 y_2 - 8x_2 y_3 + 4x_3 y_2 - x_3 y_3$$

Express f in matrix notation.

Let A be the 3×3 matrix whose ij-entry is the coefficient of $x_i y_j$. Then

$$f(u, v) = X^T A Y = (x_1, x_2, x_3) \begin{pmatrix} 3 & -2 & 0 \\ 5 & 7 & -8 \\ 0 & 4 & -1 \end{pmatrix} \begin{pmatrix} y_1 \\ y_2 \\ y_3 \end{pmatrix}$$

13.2.   Let A be an $n \times n$ matrix over K. Show that the mapping f defined by

$$f(X, Y) = X^T A Y$$

is a bilinear form on K^n.

For any $a, b \in K$ and any $X_i, Y_i \in K^n$,

$$f(aX_1 + bX_2, Y) = (aX_1 + bX_2)^T AY = (aX_1^T + bX_2^T)AY$$
$$= aX_1^T AY + bX_2^T AY = af(X_1, Y) + bf(X_2, Y)$$

Hence f is linear in the first variable. Also,

$$f(X, aY_1 + bY_2) = X^T A(aY_1 + bY_2) = aX^T AY_1 + bX^T AY_2 = af(X, Y_1) + bf(X, Y_2)$$

Hence f is linear in the second variable, and so f is a bilinear form on K^n.

13.3. Let f be the bilinear form on $\mathbf{R}^2$ defined by

$$f((x_1, x_2), (y_1, y_2)) = 2x_1 y_1 - 3x_1 y_2 + x_2 y_2$$

(a) Find the matrix A of f in the basis $\{u_1 = (1, 0), u_2 = (1, 1)\}$.

(b) Find the matrix B of f in the basis $\{v_1 = (2, 1), v_2 = (1, -1)\}$.

(c) Find the change-of-basis matrix P from the basis $\{u_i\}$ to the basis $\{v_i\}$, and clarify that $B = P^T AP$.

(a) Set $A = (a_{ij})$ where $a_{ij} = f(u_i, u_j)$:

$$\begin{aligned}
a_{11} &= f(u_1, u_1) = f((1, 0), (1, 0)) = 2 - 0 + 0 = 2 \\
a_{12} &= f(u_1, u_2) = f((1, 0), (1, 1)) = 2 - 3 + 0 = -1 \\
a_{21} &= f(u_2, u_1) = f((1, 1), (1, 0)) = 2 - 0 + 0 = 2 \\
a_{22} &= f(u_2, u_2) = f((1, 1), (1, 1)) = 2 - 3 + 1 = 0
\end{aligned}$$

Thus $A = \begin{pmatrix} 2 & -1 \\ 2 & 0 \end{pmatrix}$ is the matrix of f in the basis $\{u_1, u_2\}$.

(b) Set $B = (b_{ij})$ where $b_{ij} = f(v_i, v_j)$:

$$\begin{aligned}
b_{11} &= f(v_1, v_1) = f((2, 1), (2, 1)) & = 8 - 6 + 1 = 3 \\
b_{12} &= f(v_1, v_2) = f((2, 1), (1, -1)) & = 4 + 6 - 1 = 9 \\
b_{21} &= f(v_2, v_1) = f((1, -1), (2, 1)) & = 4 - 3 - 1 = 0 \\
b_{22} &= f(v_2, v_2) = f((1, -1), (1, -1)) & = 2 + 3 + 1 = 6
\end{aligned}$$

Thus $B = \begin{pmatrix} 3 & 9 \\ 0 & 6 \end{pmatrix}$ is the matrix of f in the basis $\{v_1, v_2\}$.

(c) We must write v_1 and v_2 in terms of the u_i:

$$\begin{aligned}
v_1 &= (2, 1) = (1, 0) + (1, 1) = u_1 + u_2 \\
v_2 &= (1, -1) = 2(1, 0) - (1, 1) = 2u_1 - u_2
\end{aligned}$$

Then $P = \begin{pmatrix} 1 & 2 \\ 1 & -1 \end{pmatrix}$ and so $P^T = \begin{pmatrix} 1 & 1 \\ 2 & -1 \end{pmatrix}$. Thus

$$P^T AP = \begin{pmatrix} 1 & 1 \\ 2 & -1 \end{pmatrix}\begin{pmatrix} 2 & -1 \\ 2 & 0 \end{pmatrix}\begin{pmatrix} 1 & 2 \\ 1 & -1 \end{pmatrix} = \begin{pmatrix} 3 & 9 \\ 0 & 6 \end{pmatrix} = B$$

13.4. Prove Theorem 13.1.

Let $\{u_1, \dots, u_n\}$ be the basis of V dual to $\{\phi_i\}$. We first show that $\{f_{ij}\}$ spans $B(V)$. Let $f \in B(V)$ and suppose $f(u_i, u_j) = a_{ij}$. We claim that $f = \sum a_{ij} f_{ij}$. It suffices to show that

$$f(u_s, u_t) = (\sum a_{ij} f_{ij})(u_s, u_t) \qquad \text{for} \quad s, t = 1, \dots, n$$

We have

$$\left(\sum a_{ij} f_{ij}\right)(u_s, u_t) = \sum a_{ij} f_{ij}(u_s, u_t) = \sum a_{ij} \phi_i(u_s)\phi_j(u_t) = \sum a_{ij} \delta_{is} \delta_{jt} = a_{st} = f(e_s, e_t)$$

as required. Hence $\{f_{ij}\}$ spans $B(V)$. Next, suppose $\sum a_{ij} f_{ij} = 0$. Then for $s, t = 1, \ldots, n$,

$$0 = 0(u_s, u_t) = \left(\sum a_{ij} f_{ij}\right)(u_s, u_t) = a_{rs}$$

The last step follows as above. Thus $\{f_{ij}\}$ is independent and hence is a basis of $B(V)$.

13.5. Let $[f]$ denote the matrix representation of a bilinear form f on V relative to a basis $\{u_1, \ldots, u_n\}$ of V. Show that the mapping $f \mapsto [f]$ is an isomorphism of $B(V)$ onto the vector space of n-square matrices.

Since f is completely determined by the scalars $f(u_i, u_j)$, the mapping $f \mapsto [f]$ is one-to-one and onto. It suffices to show that the mapping $f \mapsto [f]$ is a homomorphism; that is, that

$$[af + bg] = a[f] + b[g] \qquad \qquad (*)$$

However, for $i, j = 1, \ldots, n$,

$$(af + bg)(u_i, u_j) = af(u_i, u_j) + bg(u_i, u_j)$$

which is a restatement of $(*)$. Thus the result is proved.

13.6. Prove Theorem 13.2.

Let $u, v \in V$. Since P is the change-of-basis matrix from S to S', we have $P[u]_{S'} = [u]_S$ and also $P[v]_{S'} = [v]_S$; hence $[u]_S^T = [u]_{S'}^T P^T$. Thus

$$f(u, v) = [u]_S^T A[v]_S = [u]_{S'}^T P^T A P[v]_{S'}$$

Since u and v are arbitrary elements of V, $P^T A P$ is the matrix of f in the basis S'.

SYMMETRIC BILINEAR FORMS, QUADRATIC FORMS

13.7. Find the symmetric matrix which corresponds to each of the following quadratic polynomials:

(a) $q(x, y, z) = 3x^2 + 4xy - y^2 + 8xz - 6yz + z^2$ (b) $q(x, y, z) = x^2 - 2yz + xz$

The symmetric matrix $A = (a_{ij})$ representing $q(x_1, \ldots, x_n)$ has the diagonal entry a_{ii} equal to the coefficient of x_i^2 and has the entries a_{ij} and a_{ji} each equal to half the coefficient of $x_i x_j$. Thus

$$\begin{pmatrix} 3 & 2 & 4 \\ 2 & -1 & -3 \\ 4 & -3 & 1 \end{pmatrix} \qquad \begin{pmatrix} 2 & 0 & \frac{1}{2} \\ 0 & 0 & -1 \\ \frac{1}{2} & -1 & 0 \end{pmatrix}$$

$$\qquad (a) \qquad\qquad\qquad (b)$$

13.8. For the following real symmetric matrix A, find a nonsingular matrix P such that $P^T A P$ is diagonal and also find its signature:

$$A = \begin{pmatrix} 1 & -3 & 2 \\ -3 & 7 & -5 \\ 2 & -5 & 8 \end{pmatrix}$$

First form the block matrix (A, I):

$$(A, I) = \begin{pmatrix} 1 & -3 & 2 & \vdots & 1 & 0 & 0 \\ -3 & 7 & -5 & \vdots & 0 & 1 & 0 \\ 2 & -5 & 8 & \vdots & 0 & 0 & 1 \end{pmatrix}$$

Apply the row operations $3R_1 + R_2 \to R_2$ and $-2R_1 + R_3 \to R_3$ to (A, I) and then the corresponding column operations $3C_1 + C_2 \to C_2$ and $-2C_1 + C_3 \to C_3$ to A to obtain

$$
\begin{pmatrix}
1 & -3 & 2 & \vdots & 1 & 0 & 0 \\
0 & -2 & 1 & \vdots & 3 & 1 & 0 \\
0 & 1 & 4 & \vdots & -2 & 0 & 1
\end{pmatrix}
\quad \text{and then} \quad
\begin{pmatrix}
1 & 0 & 0 & \vdots & 1 & 0 & 0 \\
0 & -2 & 1 & \vdots & 3 & 1 & 0 \\
0 & 1 & 4 & \vdots & -2 & 0 & 1
\end{pmatrix}
$$

Next apply row operation $R_2 + 2R_3 \to R_3$ and then the corresponding column operation $C_2 + 2C_3 \to C_3$ to obtain

$$
\begin{pmatrix}
1 & 0 & 0 & \vdots & 1 & 0 & 0 \\
0 & -2 & 1 & \vdots & 3 & 1 & 0 \\
0 & 0 & 9 & \vdots & -1 & 1 & 2
\end{pmatrix}
\quad \text{and then} \quad
\begin{pmatrix}
1 & 0 & 0 & \vdots & 1 & 0 & 0 \\
0 & -2 & 0 & \vdots & 3 & 1 & 0 \\
0 & 0 & 18 & \vdots & -1 & 1 & 2
\end{pmatrix}
$$

Now A has been diagonalized. Set $P = \begin{pmatrix} 1 & 3 & -1 \\ 0 & 1 & 1 \\ 0 & 0 & 2 \end{pmatrix}$; then $P^T A P = \begin{pmatrix} 1 & 0 & 0 \\ 0 & -2 & 0 \\ 0 & 0 & 18 \end{pmatrix}$.

The signature S of A is $S = 2 - 1 = 1$.

13.9. Suppose $1 + 1 \neq 0$ in K. Give a formal algorithm to diagonalize (under congruence) a symmetric matrix $A = (a_{ij})$ over K.

Case (i): $a_{11} \neq 0$. Apply the row operations $-a_{i1}R_1 + a_{11}R_i \to R_i$, $i = 2, \ldots, n$, and then the corresponding column operations $-a_{i1}C_1 + a_{11}C_i \to C_i$ to reduce A to the form $\begin{pmatrix} a_{11} & 0 \\ 0 & B \end{pmatrix}$.

Case (ii): $a_{11} = 0$ but $a_{ii} \neq 0$, for some $i > 1$. Apply the row operation $R_1 \leftrightarrow R_i$ and then the corresponding column operation $C_1 \leftrightarrow C_i$ to bring a_{ii} into the first diagonal position. This reduces the matrix to (i).

Case (iii): All diagonal entries $a_{ii} = 0$. Choose i, j such that $a_{ij} \neq 0$, and apply the row operation $R_j + R_i \to R_i$ and the corresponding column operation $C_j + C_i \to C_i$ to bring $2a_{ij} \neq 0$ into the ith diagonal position. This reduces the matrix to (ii).

In each of the cases, we can finally reduce A to the form $\begin{pmatrix} a_{11} & 0 \\ 0 & B \end{pmatrix}$ where B is a symmetric matrix of order less than A. By induction we can finally bring A into diagonal form.

Remark: The hypothesis that $1 + 1 \neq 0$ in K, is used in (iii) where we state that $2a_{ij} \neq 0$.

13.10. Let q be the quadratic form associated with the symmetric bilinear form f. Verify the polar form of f: that is, $f(u, v) = \frac{1}{2}[q(u + v) - q(u) - q(v)]$. (Assume that $1 + 1 \neq 0$.)

We have

$$
\begin{aligned}
q(u + v) - q(u) - q(v) &= f(u + v, u + v) - f(u, u) - f(v, v) \\
&= f(u, u) + f(u, v) + f(v, u) + f(v, v) - f(u, u) - f(v, v) \\
&= 2f(u, v)
\end{aligned}
$$

If $1 + 1 \neq 0$, we can divide by 2 to obtain the required identity.

13.11. Prove Theorem 13.4.

Method 1.   If $f = 0$ or if $\dim V = 1$, then the theorem clearly holds. Hence we can suppose $f \neq 0$ and $\dim V = n > 1$. If $q(v) = f(v, v) = 0$ for every $v \in V$, then the polar form of f (see Problem 13.10) implies that $f = 0$. Hence we can assume there is a vector $v_1 \in V$ such that $f(v_1, v_1) \neq 0$. Let U be the subspace spanned by v_1 and let W consist of those vectors $v \in V$ for which $f(v_1, v) = 0$. We claim that $V = U \oplus W$.

(i) Proof that $U \cap W = \{0\}$: Suppose $u \in U \cap W$. Since $u \in U$, $u = kv_1$ for some scalar $k \in K$. Since $u \in W$, $0 = f(u, u) = f(kv_1, kv_1) = k^2 f(v_1, v_1)$. But $f(v_1, v_1) \neq 0$; hence $k = 0$ and therefore $u = kv_1 = 0$. Thus $U \cap W = \{0\}$.

(ii) Proof that $V = U + W$: let $v \in V$. Set

$$w = v - \frac{f(v_1, v)}{f(v_1, v_1)} \, v_1 \qquad (1)$$

Then

$$f(v_1, w) = f(v_1, v) - \frac{f(v_1, v)}{f(v_1, v_1)} \, f(v_1, v_1) = 0$$

Thus $w \in W$. By (1), v is the sum of an element of U and an element of W. Thus $V = U + W$. By (i) and (ii), $V = U \oplus W$.

Now f restricted to W is a symmetric bilinear form on W. But $\dim W = n - 1$; hence by induction there is a basis $\{v_2, \ldots, v_n\}$ of W such that $f(v_i, v_j) = 0$ for $i \neq j$ and $2 \leq i, j \leq n$. But by the very definition of W, $f(v_1, v_j) = 0$ for $j = 2, \ldots, n$. Therefore the basis $\{v_1, \ldots, v_n\}$ of V has the required property that $f(v_i, v_j) = 0$ for $i \neq j$.

Method 2. The algorithm in Problem 13.9 shows that every symmetric matrix over K is congruent to a diagonal matrix. This is equivalent to the statement that f has a diagonal matrix representation.

13.12. Let $A = \begin{pmatrix} a_1 & & & \\ & a_2 & & \\ & & \ddots & \\ & & & a_n \end{pmatrix}$, a diagonal matrix over K. Show that:

(a) For any nonzero scalars $k_1, \ldots, k_n \in K$, A is congruent to a diagonal matrix with diagonal entries $a_i k_i^2$;

(b) If K is the complex field $\mathbf{C}$, then A is congruent to a diagonal matrix with only 1s and 0s as diagonal entries;

(c) If K is the real field $\mathbf{R}$, then A is congruent to a diagonal matrix with only 1s, -1s, and 0s as diagonal entries.

(a) Let P be the diagonal matrix with diagonal entries k_i. Then

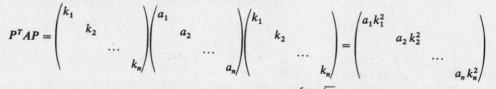

(b) Let P be the diagonal matrix with diagonal entries $b_i = \begin{cases} 1/\sqrt{a_i} & \text{if } a_i \neq 0 \\ 1 & \text{if } a_i = 0 \end{cases}$. Then $P^T A P$ has the required form.

(c) Let P be the diagonal matrix with diagonal entries $b_i = \begin{cases} 1/\sqrt{|a_i|} & \text{if } a_i \neq 0 \\ 1 & \text{if } a_i = 0 \end{cases}$. Then $P^T A P$ has the required form.

Remark: We emphasize that (b) is no longer true if congruence is replaced by Hermitian congruence (see Problems 13.32 and 13.33).

13.13. Prove Theorem 13.5.

By Theorem 13.4, there is a basis $\{u_1, \ldots, u_n\}$ of V in which f is represented by a diagonal matrix, say, with P positive and N negative entries. Now suppose $\{w_1, \ldots, w_n\}$ is another basis of V in which f is represented by a diagonal matrix, say, with P' positive and N' negative entries. We can assume without loss in generality that the positive entries in each matrix appear first. Since rank $f = P + N = P' + N'$, it suffices to prove that $P = P'$.

Let U be the linear span of $u_1, \ldots, u_P$ and let W be the linear span of $w_{P'+1}, \ldots, w_n$. Then $f(v, v) > 0$ for every nonzero $v \in U$, and $f(v, v) \le 0$ for every nonzero $v \in W$. Hence $U \cap W = \{0\}$. Note that $\dim U = P$ and $\dim W = n - P'$. Thus

$$\dim (U + W) = \dim U + \dim W - \dim (U \cap W) = P + (n - P') - 0 = P - P' + n$$

But $\dim (U + W) \le \dim V = n$; hence $P - P' + n \le n$ or $P \le P'$. Similarly, $P' \le P$ and therefore $P = P'$, as required.

Remark: The above theorem and proof depend only on the concept of positivity. Thus the theorem is true for any subfield K of the real field $\mathbf{R}$ such as the rational field $\mathbf{Q}$.

13.14. An $n \times n$ real symmetric matrix A is said to be positive definite if $X^T A X > 0$ for every nonzero (column) vector $X \in \mathbf{R}^n$, i.e., if A is positive definite viewed as a bilinear form. Let B be any real nonsingular matrix. Show that (a) $B^T B$ is symmetric and (b) $B^T B$ is positive definite.

(a) $(B^T B)^T = B^T B^{TT} = B^T B$; hence $B^t B$ is symmetric.

(b)   Since B is nonsingular, $BX \ne 0$ for any nonzero $X \in \mathbf{R}^n$. Hence the dot product of BX with itself, $BX \cdot BX = (BX)^T (BX)$, is positive. Thus $X^T(B^T B)X = (X^T B^T) = (BX)^T(BX) > 0$ as required.

HERMITIAN FORMS

13.15. Determine which of the following matrices are Hermitian:

$$\begin{pmatrix} 2 & 2+3i & 4-5i \\ 2-3i & 5 & 6+2i \\ 4+5i & 6-2i & -7 \end{pmatrix} \qquad \begin{pmatrix} 3 & 2-i & 4+i \\ 2-i & 6 & i \\ 4+i & i & 3 \end{pmatrix} \qquad \begin{pmatrix} 4 & -3 & 5 \\ -3 & 2 & 1 \\ 5 & 1 & -6 \end{pmatrix}$$

$$\qquad\qquad (a) \qquad\qquad\qquad\qquad\qquad\qquad (b) \qquad\qquad\qquad\qquad\qquad (c)$$

A matrix $A = (a_{ij})$ is Hermitian iff $A = A^*$, i.e., iff $a_{ij} = \overline{a_{ij}}$.

(a)   The matrix is Hermitian, since it is equal to its conjugate transpose.

(b)   The matrix is not Hermitian, even though it is symmetric.

(c)   The matrix is Hermitian. In fact, a real matrix is Hermitian if and only if it is symmetric.

13.16. Let A be a Hermitian matrix. Show that f is a Hermitian form on $\mathbf{C}^n$ where f is defined by $f(X, Y) = X^T A \bar{Y}$.

For all $a, b \in \mathbf{C}$ and all $X_1, X_2, Y \in \mathbf{C}^n$,

$$f(aX_1 + bX_2, Y) = (aX_1 + bX_2)^T A \bar{Y} = (aX_1^T + bX_2^T)A\bar{Y}$$
$$= aX_1^T A \bar{Y} + bX_2^T A \bar{Y} = af(X_1, Y) + bf(X_2, Y)$$

Hence f is linear in the first variable. Also,

$$\overline{f(X, Y)} = \overline{X^T A \bar{Y}} = \overline{(X^T A \bar{Y})^T} = \overline{\bar{Y}^T A^T X} = Y^T A^* \bar{X} = Y^T A \bar{X} = f(Y, X)$$

Hence f is a Hermitian form on $\mathbf{C}^n$. (*Remark:* We use the fact that $X^t A \bar{Y}$ is a scalar and so it is equal to its transpose.)

13.17. Let f be a Hermitian form on V. Let H be the matrix of f in a basis $S = \{u_1, \ldots, u_n\}$ of V. Show that:

(a) $f(u, v) = [u]_S^T H \overline{[v]_S}$ for all $u, v \in V$;

(b) If P is the change-of-basis matrix from S to a new basis S' of V, then $B = P^t H \bar{P}$ (or $B = Q^*HQ$ where $Q = \bar{P}$) is the matrix of f in the new basis S'.

Note that (b) is the complex analog of Theorem 13.2.

(a) Let $u, v \in V$ and suppose $u = a_1 u_1 + a_2 u_2 + \cdots + a_n u_n$ and $v = b_1 u_1 + b_2 u_2 + \cdots + b_n u_n$. Then

$$f(u, v) = f(a_1 u_1 + \cdots + a_n u_n, b_1 u_1 + \cdots + b_n u_n)$$

$$= \sum_{i,j} a_i \bar{b}_j f(u_i u_j) = (a_1, \ldots, a_n) H \begin{pmatrix} \bar{b}_1 \\ \bar{b}_2 \\ \cdots \\ \bar{b}_n \end{pmatrix} = [u]_S^T H \overline{[v]_S}$$

as required.

(b) Since P is the change-of-basis matrix from S to S', we have $P[u]_{S'} = [u]_S$ and $P[v]_{S'} = [v]_S$; hence $[u]_S^T = [u]_{S'}^T P^T$ and $\overline{[v]_S} = \bar{P}\overline{[v]_{S'}}$. Thus by (a),

$$f(u, v) = [u]_S^T H \overline{[v]_S} = [u]_{S'}^T P^T H \bar{P}\overline{[v]_{S'}}$$

But u and v are arbitrary elements of V; hence $P^T H \bar{P}$ is the matrix of f in the basis S'.

13.18. Let $H = \begin{pmatrix} 1 & 1+i & 2i \\ 1-i & 4 & 2-3i \\ -2i & 2+3i & 7 \end{pmatrix}$, a Hermitian matrix. Find a nonsingular matrix P such that $P^T H \bar{P}$ is diagonal.

First form the block matrix (H, I):

$$\begin{pmatrix} 1 & 1+i & 2i & \vdots & 1 & 0 & 0 \\ 1-i & 4 & 2-3i & \vdots & 0 & 1 & 0 \\ -2i & 2+3i & 7 & \vdots & 0 & 0 & 1 \end{pmatrix}$$

Apply the row operations $(-1 + i)R_1 + R_2 \to R_2$ and $2iR_1 + R_3 \to R_3$ to (A, I) and then the corresponding "Hermitian column operations" (see Problem 13.42) $(-1 - i)C_1 + C_2 \to C_2$ and $-2iC_1 + C_3 \to C_3$ to A to obtain

$$\begin{pmatrix} 1 & 1+i & 2i & \vdots & 1 & 0 & 0 \\ 0 & 2 & -5i & \vdots & -1+i & 1 & 0 \\ 0 & 5i & 3 & \vdots & 2i & 0 & 1 \end{pmatrix} \quad \text{and then} \quad \begin{pmatrix} 1 & 0 & 0 & \vdots & 1 & 0 & 0 \\ 0 & 2 & -5i & \vdots & -1+i & 1 & 0 \\ 0 & 5i & 3 & \vdots & 2i & 0 & 1 \end{pmatrix}$$

Next apply the row operation $R_3 \to -5iR_2 + 2R_3$ and the corresponding Hermitian column operation $C_3 \to 5iC_2 + 2C_3$ to obtain

$$\begin{pmatrix} 1 & 0 & 0 & \vdots & 1 & 0 & 0 \\ 0 & 2 & -5i & \vdots & -1+i & 1 & 0 \\ 0 & 0 & -19 & \vdots & 5+9i & -5i & 2 \end{pmatrix} \quad \text{and then} \quad \begin{pmatrix} 1 & 0 & 0 & \vdots & 1 & 0 & 0 \\ 0 & 2 & 0 & \vdots & -1+i & 1 & 0 \\ 0 & 0 & -38 & \vdots & 5+9i & -5i & 2 \end{pmatrix}$$

Now H has been diagonalized. Set

$$P = \begin{pmatrix} 1 & -1+i & 5+9i \\ 0 & 1 & -5i \\ 0 & 0 & 2 \end{pmatrix} \quad \text{and then} \quad P^T H \bar{P} = \begin{pmatrix} 1 & 0 & 0 \\ 0 & 2 & 0 \\ 0 & 0 & -38 \end{pmatrix}$$

Note that the signature S of H is $S = 2 - 1 = 1$.

MISCELLANEOUS PROBLEMS

13.19. Prove Theorem 13.3.

If $f = 0$, then the theorem is obviously true. Also, if dim $V = 1$, then $f(k_1 u, k_2 u) = k_1 k_2 f(u, u) = 0$ and so $f = 0$. Accordingly we can assume that dim $V > 1$ and $f \neq 0$.

Since $f \neq 0$, there exist (nonzero) $u_1, u_2 \in V$ such that $f(u_1, u_2) \neq 0$. In fact, multiplying u_1 by an appropriate factor, we can assume that $f(u_1, u_2) = 1$ and so $f(u_2, u_1) = -1$. Now u_1 and u_2 are linearly independent; because if, say, $u_2 = ku_1$, then $f(u_1, u_2) = f(u_1, ku_1) = kf(u_1, u_1) = 0$. Let U be the subspace spanned by u_1 and u_2, i.e., $U = L(u_1, u_2)$. Note:

(i) The matrix representation of the restriction of f to U in the basis $\{u_1, u_2\}$ is $\begin{pmatrix} 0 & 1 \\ -1 & 0 \end{pmatrix}$;

(ii) If $u \in U$, say $u = au_1 + bu_2$, then

$$f(u, u_1) = f(au_1 + bu_2, u_1) = -b$$
$$f(u, u_2) = f(au_1 + bu_2, u_2) = \quad a$$

Let W consist of those vectors $w \in V$ such that $f(w, u_1) = 0$ and $f(w, u_2) = 0$. Equivalently,

$$W = \{w \in V : f(w, u) = 0 \text{ for every } u \in U\}$$

We claim that $V = U \oplus W$. It is clear that $U \cap W = \{0\}$, and so it remains to show that $V = U + W$. Let $v \in V$. Set

$$u = f(v, u_2)u_1 - f(v, u_1)u_2 \quad \text{and} \quad w = v - u \tag{1}$$

Since u is a linear combination of u_1 and u_2, $u \in U$. We show that $w \in W$. By (1) and (ii), $f(u, u_1) = f(v, u_1)$; hence

$$f(w, u_1) = f(v - u, u_1) = f(v, u_1) - f(u, u_1) = 0$$

Similarly, $f(u, u_2) = f(v, u_2)$ and so

$$f(w, u_2) + f(v - u, u_2) = f(v, u_2) - f(u, u_2) = 0$$

Then $w \in W$ and so, by (1), $v = u + w$ where $u \in W$. This shows that $V = U + W$; and therefore $V = U \oplus W$.

Now the restriction of f to W is an alternating bilinear form on W. By induction, there exists a basis $u_3, \ldots, u_n$ of W in which the matrix representing f restricted to W has the desired form. Accordingly, $u_1, u_2, u_3, \ldots, u_n$ is a basis of V in which the matrix representing f has the desired form.

Supplementary Problems

BILINEAR FORMS

13.20. Let V be the vector space of 2×2 matrices over $\mathbf{R}$. Let $M = \begin{pmatrix} 1 & 2 \\ 3 & 5 \end{pmatrix}$, and let $f(A, B) = \operatorname{tr} A^t M B$, where $A, B \in V$ and "tr" denotes trace. (a) Show that f is a bilinear form on V. (b) Find the matrix of f in the basis $\left\{ \begin{pmatrix} 1 & 0 \\ 0 & 0 \end{pmatrix}, \begin{pmatrix} 0 & 1 \\ 0 & 0 \end{pmatrix}, \begin{pmatrix} 0 & 0 \\ 1 & 0 \end{pmatrix}, \begin{pmatrix} 0 & 0 \\ 0 & 1 \end{pmatrix} \right\}$.

13.21. Let $B(V)$ be the set of bilinear forms on V over K. Prove:

(a) If $f, g \in B(V)$, then $f + g$ and kf, for $k \in K$, also belong to $B(V)$, and so $B(V)$ is a subspace of the vector space of functions from $V \times V$ into K;

(b) If ϕ and σ are linear functionals on V, then $f(u, v) = \phi(u)\sigma(v)$ belongs to $B(V)$.

13.22. Let f be a bilinear form on V. For any subset S of V, we write

$$S^\perp = \{v \in V : f(u, v) = 0 \text{ for every } u \in S\} \qquad S^\perp = \{v \in V : f(v, u) = 0 \text{ for every } u \in S\}$$

Show that: (a) $S^\perp$ and $S^\perp$ are subspaces of V; (b) $S_1 \subseteq S_2$ implies $S_2^\perp \subseteq S_1^\perp$ and $S_2^\perp \subseteq S_1^\perp$; and (c) $\{0\}^\perp = \{0\}^\perp = V$.

13.23. Prove: if f is a bilinear form on V, then rank $f = \dim V - \dim V^\perp = \dim V - \dim V^\top$ and hence $\dim V^\perp = \dim V^\top$.

13.24. Let f be a bilinear form on V. For each $u \in V$, let $\hat{u} : V \to K$ and $\tilde{u} : V \to K$ be defined by $\hat{u}(x) = f(x, u)$ and $\tilde{u}(x) = f(u, x)$. Prove:

(a) $\hat{u}$ and $\tilde{u}$ are each linear, i.e. $\hat{u}, \tilde{u} \in V^*$;

(b) $u \mapsto \hat{u}$ and $u \mapsto \tilde{u}$ are each linear mappings from V into V^*;

(c) rank $f =$ rank $(u \mapsto \hat{u}) =$ rank $(u \mapsto \tilde{u})$.

13.25. Show that congruence of matrices is an equivalence relation, i.e., (i) A is congruent to A; (ii) if A is congruent to B, then B is congruent to A; (iii) if A is congruent to B and B is congruent to C, then A is congruent to C.

SYMMETRIC BILINEAR FORMS, QUADRATIC FORMS

13.26. Find the symmetric matrix belonging to each of the following quadratic polynomials:

(a) $q(x, y, z) = 2x^2 - 8xy + y^2 - 16xz + 14yz + 5z^2$ (c) $q(x, y, z) = xy + y^2 + 4xz + z^2$

(b) $q(x, y, z) = x^2 - xz + y^2$ (d) $q(x, y, z) = xy + yz$

13.27. For each of the following matrices A, find a nonsingular matrix P such that $P^T A P$ is diagonal:

(a) $A = \begin{pmatrix} 2 & 3 \\ 3 & 4 \end{pmatrix}$, (b) $A = \begin{pmatrix} 1 & -2 & 3 \\ -2 & 6 & -9 \\ 3 & -9 & 4 \end{pmatrix}$, (c) $A = \begin{pmatrix} 1 & 1 & -2 & -3 \\ 1 & 2 & -5 & -1 \\ -2 & -5 & 6 & 9 \\ -3 & -1 & 9 & 11 \end{pmatrix}$

In each case find the rank and signature.

13.28. Let $S(V)$ be the set of symmetric bilinear forms on V. Show that:

(i) $S(V)$ is a subspace of $B(V)$;   (ii) if $\dim V = n$, then $\dim S(V) = \frac{1}{2}n(n + 1)$.

13.29. Suppose A is a real symmetric positive definite matrix. Show that there exists a nonsingular matrix P such that $A = P^T P$.

13.30. Consider a real quadratic polynomial $q(x_1, \ldots, x_n) = \sum_{i, j=1}^{n} a_{ij} x_i x_j$, where $a_{ij} = a_{ji}$.

(i) If $a_{11} \neq 0$, show that the substitution

$$x_1 = y_1 - \frac{1}{a_{11}} (a_{12} y_2 + \cdots + a_{1n} y_n), \quad x_2 = y_2, \ldots, x_n = y_n$$

yields the equation $q(x_1, \ldots, x_n) = a_{11} y_1^2 + q'(y_2, \ldots, y_n)$, where q' is also a quadratic polynomial.

(ii) If $a_{11} = 0$ but, say, $a_{12} \neq 0$, show that the substitution

$$x_1 = y_1 + y_2, \quad x_2 = y_1 - y_2, \quad x_3 = y_3, \ldots, x_n = y_n$$

yields the equation $q(x_1, \ldots, x_n) = \sum b_{ij} y_i y_j$, where $b_{11} \neq 0$, i.e., reduces this case to case (i). This method of diagonalizing q is known as "completing the square."

HERMITIAN FORMS

13.31. Let A be any complex nonsingular matrix. Show that $H = A^*A$ is Hermitian and positive definite.

13.32. We say that B is Hermitian congruent to A if there exists a nonsingular matrix Q such that $B = Q^*AQ$. Show that Hermitian congruence is an equivalence relation.

13.33. Prove Theorem 13.7. [Note that the second part of the theorem does not hold for complex symmetric bilinear forms, as is shown by Problem 13.12(ii). However, the proof of Theorem 13.5 in Problem 13.13 does carry over to the Hermitian case.]

MISCELLANEOUS PROBLEMS

13.34. Let V and W be vector spaces over K. A mapping $f: V \times W \to K$ is called a bilinear form on V and W if:

(i) $f(av_1 + bv_2, w) = af(v_1, w) + bf(v_2, w)$

(ii) $f(v, aw_1 + bw_2) = af(v, w_1) + bf(v, w_2)$

for every $a, b \in K$, $v_i \in V$, $w_j \in W$. Prove the following:

(a) The set $B(V, W)$ of bilinear forms on V and W is a subspace of the vector space of functions from $V \times W$ into K.

(b) If $\{\phi_1, \ldots, \phi_m\}$ is a basis of V^* and $\{\sigma_1, \ldots, \sigma_n\}$ is a basis of W^*, then $\{f_{ij}: i = 1, \ldots, m, j = 1, \ldots, n\}$ is a basis of $B(V, W)$ where f_{ij} is defined by $f_{ij}(v, w) = \phi_i(v)\sigma_j(w)$. Thus dim $B(V, W) = $ dim $V \cdot$ dim W.

> **Remark:** Observe that if $V = W$, then we obtain the space $B(V)$ investigated in this chapter.

13.35. Let V be a vector space over K. A mapping $f: \overbrace{V \times V \times \cdots \times V}^{m \text{ times}} \to K$ is called a *multilinear* (or *m-linear*) *form* on V if f is linear in each variable, i.e., for $i = 1, \ldots, m$,

$$f(\ldots, \widehat{au + bv}, \ldots) = af(\ldots, \hat{u}, \ldots) + bf(\ldots, \hat{v}, \ldots)$$

where $\hat{}$ denotes the ith component, and other components are held fixed. An m-linear form f is said to be alternating if

$$f(v_1, \ldots, v_m) = 0 \qquad \text{whenever} \qquad v_i = v_k, \, i \neq k$$

Prove:

(a) The set $B_m(V)$ of m-linear forms on V is a subspace of the vector space of functions from $V \times V \times \cdots \times V$ into K.

(b) The set $A_m(V)$ of alternating m-linear forms on V is a subspace of $B_m(V)$.

> **Remark 1:** If $m = 2$, then we obtain the space $B(V)$ investigated in this chapter.

> **Remark 2:** If $V = K^m$, then the determinant function is a particular alternating m-linear form on V.

Answers to Supplementary Problems

13.20. (b) $\begin{pmatrix} 1 & 0 & 2 & 0 \\ 0 & 1 & 0 & 2 \\ 3 & 0 & 4 & 0 \\ 0 & 3 & 0 & 4 \end{pmatrix}$

13.26. (a) $\begin{pmatrix} 2 & -4 & -8 \\ -4 & 1 & 7 \\ -8 & 7 & 5 \end{pmatrix}$,   (b) $\begin{pmatrix} 1 & 0 & -\frac{1}{2} \\ 0 & 1 & 0 \\ -\frac{1}{2} & 0 & 0 \end{pmatrix}$,   (c) $\begin{pmatrix} 0 & \frac{1}{2} & 2 \\ \frac{1}{2} & 1 & 0 \\ 2 & 0 & 1 \end{pmatrix}$,   (d) $\begin{pmatrix} 0 & \frac{1}{2} & 0 \\ \frac{1}{2} & 0 & \frac{1}{2} \\ 0 & \frac{1}{2} & 0 \end{pmatrix}$

13.27. (a) $P = \begin{pmatrix} 1 & -3 \\ 0 & 2 \end{pmatrix}$, $P^T A P \begin{pmatrix} 2 & 0 \\ 0 & -2 \end{pmatrix}$, $S = 0$

(b) $P = \begin{pmatrix} 1 & 2 & 0 \\ 0 & 1 & 3 \\ 0 & 0 & 2 \end{pmatrix}$, $P^T A P = \begin{pmatrix} 1 & 0 & 0 \\ 0 & 2 & 0 \\ 0 & 0 & -38 \end{pmatrix}$, $S = 1$

(c) $P = \begin{pmatrix} 1 & -1 & -1 & 26 \\ 0 & 1 & 3 & 13 \\ 0 & 0 & 1 & 9 \\ 0 & 0 & 0 & 7 \end{pmatrix}$, $P^T A P = \begin{pmatrix} 1 & 0 & 0 & 0 \\ 0 & 1 & 0 & 1 \\ 0 & 0 & -7 & 0 \\ 0 & 0 & 0 & 469 \end{pmatrix}$, $S = 2$

Chapter 14

Linear Operators on Inner Product Spaces

14.1 INTRODUCTION

This chapter investigates the space $A(V)$ of linear operators T on an inner product space V. (See Chapter 6.) Thus the base field K is either the real field $\mathbf{R}$ or the complex field $\mathbf{C}$. In fact, different terminology will be used for the real case and for the complex case. We also use the fact that the inner product on Euclidean space $\mathbf{R}^n$ may be defined by

$$\langle u, v \rangle = u^T v$$

and that the inner product on complex Euclidean space $\mathbf{C}^n$ may be defined by

$$\langle u, v \rangle = u^T \bar{v}$$

where u and v are column vectors.

The reader should review the material in Chapter 6. In particular, the reader should be very familiar with the notions of norm (length), orthogonality, and orthonormal bases.

Lastly, we mention that Chapter 6 mainly dealt with real inner product spaces. On the other hand, here we assume that V is a complex inner product space unless otherwise stated or implied.

14.2 ADJOINT OPERATORS

We begin with the following basic definition.

Definition: A linear operator T on an inner product space V is said to have an adjoint operator T^* on V if $\langle T(u), v \rangle = \langle u, T^*(v) \rangle$ for every $u, v \in V$.

The following example shows that the adjoint operator has a simple description within the context of matrix mappings.

Example 14.1

(a) Let A be a real n-square matrix viewed as a linear operator on $\mathbf{R}^n$. Then, for every $u, v \in \mathbf{R}^n$,

$$\langle Au, v \rangle = (Au)^T v = u^T A^T v = \langle u, A^T v \rangle$$

Thus the transposed matrix A^T is the adjoint of A.

(b) Let B be a complex n-square matrix viewed as a linear on $\mathbf{C}^n$. Then, for every $u, v \in \mathbf{C}^n$,

$$\langle Bu, v \rangle = (Bu)^T v = u^T B^T v = u^T \overline{\bar{B}^T} \bar{v} = u^T B^* v = \langle u, B^* v \rangle$$

Thus the conjugate transposed matrix $B^* = \bar{B}^T$ is the adjoint of B.

> **Remark:** The notation B^* is used to denote the adjoint of B and, previously, to denote the conjugate transpose of B. The above example shows that they both give the same result.

The following theorem, proved in Problem 14.4, is the main result in this section.

Theorem 14.1:   Let T be a linear operator on a finite-dimensional inner product space V over K. Then:

(i)   There exists a unique linear operator T^* on V such that $\langle T(u), v \rangle = \langle u, T^*(v) \rangle$ for every $u, v \in V$. (That is, T has an adjoint T^*.)

(ii)   If A is the matrix representation of T with respect to any orthonormal basis $S = \{u_i\}$ of V, then the matrix representation of T^* in the basis S is the conjugate transpose A^* of A (or the transpose A^T of A when K is real).

We emphasize that no such simple relationship exists between the matrices representing T and T^* if the basis is not orthonormal. Thus we see one useful property of orthonormal bases. We also emphasize that this theorem is not valid if V has infinite dimension (Problem 14.31).

Example 14.2.   Let T be the linear operator on $\mathbf{C}^3$ defined by

$$T(x, y, z) = (2x + iy, \; y - 5iz, \; x + (1 - i)y + 3z)$$

We find a similar formula for the adjoint T^* of T. Note (Problem 10.1) that the matrix of T in the usual basis of $\mathbf{C}^3$ is

$$[T] = \begin{pmatrix} 2 & i & 0 \\ 0 & 1 & -5i \\ 1 & 1-i & 3 \end{pmatrix}$$

Recall that the usual basis is orthonormal. Thus by Theorem 14.1, the matrix of T^* in this basis is the conjugate transpose of $[T]$:

$$[T^*] = \begin{pmatrix} 2 & 0 & 1 \\ -i & 1 & 1+i \\ 0 & 5i & 3 \end{pmatrix}$$

Accordingly,

$$T^*(x, y, z) = (2x + z, \; -ix + y + (1 + i)z, \; 5iy + 3z)$$

The following theorem, proved in Problem 14.5, summarizes some of the properties of the adjoint.

Theorem 14.2:   Let T, T_1, T_2 be linear operators on V and let $k \in K$. Then:

(i) $(T_1 + T_2)^* = T_1^* + T_2^*$ (iii) $(T_1 T_2)^* = T_2^* T_1^*$

(ii) $(kT)^* = \bar{k}T^*$ (iv) $(T^*)^* = T$

Observe the similarity between the above theorem and Theorem 3.3 on properties of the transpose operation on matrices.

Linear Functionals and Inner Product Spaces

Recall (Chapter 12) that a linear functional ϕ on a vector space V is a linear mapping from V into the base field K. This subsection contains an important result (Theorem 14.3) which is used in the proof of the above basic Theorem 14.1.

Let V be an inner product space. Each $u \in V$ determines a mapping $\hat{u} : V \to K$ defined by

$$\hat{u}(v) = \langle v, u \rangle$$

Now for any $a, b \in K$ and any $v_1, v_2 \in V$,

$$\hat{u}(av_1 + bv_2) = \langle av_1 + bv_2, u \rangle = a\langle v_1, u \rangle + b\langle v_2, u \rangle = a\hat{u}(v_1) + b\hat{u}(v_2)$$

That is, $\hat{u}$ is a linear functional on V. The converse is also true for spaces of finite dimension and is an important theorem (proved in Problem 14.3). Namely,

Theorem 14.3: Let ϕ be a linear functional on a finite-dimensional inner product space V. Then there exists a unique vector $u \in V$ such that $\phi(v) = \langle v, u \rangle$ for every $v \in V$.

We remark that the above theorem is not valid for spaces of infinite dimension (Problem 14.24), although some general results in this direction are known. (One such famous result is the Riesz representation theorem.)

14.3 ANALOGY BETWEEN $A(V)$ AND C, SPECIAL OPERATORS

Let $A(V)$ denote the algebra of all linear operators on a finite-dimensional inner product space V. The adjoint mapping $T \mapsto T^*$ on $A(V)$ is quite analogous to the conjugation mapping $z \mapsto \bar{z}$ on the complex field C. To illustrate this analogy we identify in Table 14-1 certain classes of operators $T \in A(V)$ whose behavior under the adjoint map imitates the behavior under conjugation of familiar classes of complex numbers.

Table 14-1

Class of complex numbers	Behavior under conjugation	Class of operators in $A(V)$	Behavior under the adjoint map
Unit circle ($|z| = 1$)	$\bar{z} = 1/z$	Orthogonal operators (real case) Unitary operators (complex case)	$T^* = T^{-1}$

Real axis	$\bar{z} = z$	Self-adjoint operators
		 Also called:
		     symmetric (real case)
		     Hermitian (complex case)
			$T^* = T$

Imaginary axis	$\bar{z} = -z$	Skew-adjoint operators
		 Also called:
		     skew-symmetric (real case)
		     skew-Hermitian (complex case)
			$T^* = -T$

Positive half axis $(0, \infty)$	$z = \bar{w}w$, $w \neq 0$	Positive definite operators	$T = S^*S$ with S nonsingular

The analogy between these classes of operators T and complex numbers z is reflected in the following theorem.

Theorem 14.4: Let λ be an eigenvalue of a linear operator T on V.

    (i)   If $T^* = T^{-1}$ (i.e., T is orthogonal or unitary), then $|\lambda| = 1$.

    (ii) If $T^* = T$ (i.e., T is self-adjoint), then λ is real.

    (iii) If $T^* = -T$ (i.e., T is skew-adjoint), then λ is pure imaginary.

    (iv) If $T = S^*S$ with S nonsingular (i.e., T is positive definite), then λ is real and positive.

Proof. In each case let v be a nonzero eigenvector of T belonging to λ, that is, $T(v) = \lambda v$ with $v \neq 0$; hence $\langle v, v \rangle$ is positive.

Proof of (i):   We show that $\lambda\bar{\lambda}\langle v, v\rangle = \langle v, v\rangle$:

$$\lambda\bar{\lambda}\langle v, v\rangle = \langle \lambda v, \lambda v\rangle = \langle T(v), T(v)\rangle = \langle v, T^*T(v)\rangle = \langle v, I(v)\rangle = \langle v, v\rangle$$

But $\langle v, v\rangle \neq 0$; hence $\lambda\bar{\lambda} = 1$ and so $|\lambda| = 1$.

Proof of (ii):   We show that $\lambda\langle v, v\rangle = \bar{\lambda}\langle v, v\rangle$:

$$\lambda\langle v, v\rangle = \langle \lambda v, v\rangle = \langle T(v), v\rangle = \langle v, T^*(v)\rangle = \langle v, T(v)\rangle = \langle v, \lambda v\rangle = \bar{\lambda}\langle v, v\rangle$$

But $\langle v, v\rangle \neq 0$; hence $\lambda = \bar{\lambda}$ and so λ is real.

Proof of (iii):   We show that $\lambda\langle v, v\rangle = -\bar{\lambda}\langle v, v\rangle$:

$$\lambda\langle v, v\rangle = \langle \lambda v, v\rangle = \langle T(v), v\rangle = \langle v, T^*(v)\rangle = \langle v, -T(v)\rangle = \langle v, -\lambda v\rangle = -\bar{\lambda}\langle v, v\rangle$$

But $\langle v, v\rangle \neq 0$; hence $\lambda = -\bar{\lambda}$ or $\bar{\lambda} = -\lambda$, and so λ is pure imaginary.

Proof of (iv):   Note first that $S(v) \neq 0$ because S is nonsingular; hence $\langle S(v), S(v)\rangle$ is positive. We show that $\lambda\langle v, v\rangle = \langle S(v), S(v)\rangle$:

$$\lambda\langle v, v\rangle = \langle \lambda v, v\rangle = \langle T(v), v\rangle = \langle S^*S(v), v\rangle = \langle S(v), S(v)\rangle$$

But $\langle v, v\rangle$ and $\langle S(v), S(v)\rangle$ are positive; hence λ is positive.

Remark: Each of the above operators T commute with their adjoint, that is, $TT^* = T^*T$. Such operators are called normal operators.

14.4 SELF-ADJOINT OPERATORS

Let T be a self-adjoint operator on an inner product space V, that is, suppose

$$T^* = T$$

(In case, T is defined by a matrix A, then A is symmetric or Hermitian according as A is real or complex.) By Theorem 14.4, the eigenvalues of T are real. Another important property of T follows.

Theorem 14.5:   Let T be a self-adjoint operator on V. Suppose u and v are eigenvectors of T belonging to distinct eigenvalues. Then u and v are orthogonal, i.e., $\langle u, v\rangle = 0$.

Proof.   Suppose $T(u) = \lambda_1 u$ and $T(v) = \lambda_2 v$ where $\lambda_1 \neq \lambda_2$. We show that $\lambda_1\langle u, v\rangle = \lambda_2\langle u, v\rangle$:

$$\lambda_1\langle u, v\rangle = \langle \lambda_1 u, v\rangle = \langle T(u), v\rangle = \langle u, T^*(v)\rangle = \langle u, T(v)\rangle$$
$$= \langle u, \lambda_2 v\rangle = \bar{\lambda_2}\langle u, v\rangle = \lambda_2\langle u, v\rangle$$

(The fourth equality uses the fact that $T^* = T$ and the last equality uses the fact that the eigenvalue λ_2 is real.) Since $\lambda_1 \neq \lambda_2$, we get $\langle u, v\rangle = 0$. Thus the theorem is proved.

14.5 ORTHOGONAL AND UNITARY OPERATORS

Let U be a linear operator on a finite-dimensional inner product space V. Recall that if

$$U^* = U^{-1} \qquad \text{or equivalently} \qquad UU^* = U^*U = I$$

then U is said to be orthogonal or unitary according as the underlying field is real or complex. The next theorem, proved in Problem 14.10, gives alternative characterizations of these operators.

Theorem 14.6:   The following conditions on an operator U are equivalent:

  (i) $U^* = U^{-1}$, that is, $UU^* = U^*U = I$.

  (ii) U preserves inner products, i.e., for every $v, w \in V$,

$$\langle U(v), U(w) \rangle = \langle v, w \rangle$$

  (iii) U preserves lengths, i.e., for every $v \in V$, $\|U(v)\| = \|v\|$.

Example 14.3

(a) Let $T : \mathbf{R}^3 \to \mathbf{R}^3$ be the linear operator which rotates each vector about the z axis by a fixed angle θ as shown in Fig. 14-1, that is, defined by

$$T(x, y, z) = (x \cos \theta - y \sin \theta, \, x \sin \theta + y \cos \theta, \, z)$$

Note that lengths (distances from the origin) are preserved under T. Thus T is an orthogonal operator.

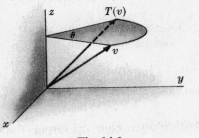

Fig. 14-1

(b) Let V be the l_2-space of Example 6.3. Let $T : V \to V$ be the linear operator defined by

$$T(a_1, a_2, \ldots) = (0, a_1, a_2, \ldots).$$

Clearly, T preserves inner products and lengths. However, T is not surjective since, for example, $(1, 0, 0, \ldots)$ does not belong to the image of T; hence T is not invertible. Thus we see that Theorem 14.6 is not valid for spaces of infinite dimension.

An isomorphism from one inner product space into another is a bijective mapping which preserves the three basic operations of an inner product space: vector addition, scalar multiplication, and inner products. Thus the above mappings (orthogonal and unitary) may also be characterized as the isomorphisms of V into itself. Note that such a mapping U also preserves distances, since

$$\|U(v) - U(w)\| = \|U(v - w)\| = \|v - w\|$$

and so U is also called an isometry.

14.6 ORTHOGONAL AND UNITARY MATRICES

Let U be a linear operator on an inner product space V. By Theorem 14.1, we obtain the following result when the base field K is complex.

Theorem 14.7A: A matrix A with complex entries represents a unitary operator U (relative to an orthonormal basis) if and only if $A^* = A^{-1}$.

On the other hand, if the base field K is real then $A^* = A^T$; hence we have the following corresponding theorem for real inner product spaces.

Theorem 14.7B: A matrix A with real entries represents an orthogonal operator U (relative to an orthonormal basis) if and only if $A^T = A^{-1}$.

The above theorems motivate the following definitions.

Definition: A complex matrix A for which $A^* = A^{-1}$, or equivalently $AA^* = A^*A = I$, is called a unitary matrix.

Definition: A real matrix A for which $A^T = A^{-1}$, or equivalently $AA^T = A^TA = I$, is called an orthogonal matrix.

Observe that a unitary matrix with real entries is orthogonal.

Example 14.4. Suppose $A = \begin{pmatrix} a_1 & a_2 \\ b_1 & b_2 \end{pmatrix}$ is a unitary matrix. Then $AA^* = I$ and hence

$$AA^* = \begin{pmatrix} a_1 & a_2 \\ b_1 & b_2 \end{pmatrix}\begin{pmatrix} \bar{a}_1 & \bar{b}_1 \\ \bar{a}_2 & \bar{b}_2 \end{pmatrix} = \begin{pmatrix} |a_1|^2 + |a_2|^2 & a_1\bar{b}_1 + a_2\bar{b}_2 \\ \bar{a}_1 b_1 + \bar{a}_2 b_2 & |b_1|^2 + |b_2|^2 \end{pmatrix} = \begin{pmatrix} 1 & 0 \\ 0 & 1 \end{pmatrix} = I$$

Thus

$$|a_1|^2 + |a_2|^2 = 1 \qquad |b_1|^2 + |b_2|^2 = 1 \qquad \text{and} \qquad a_1\bar{b}_1 + a_2\bar{b}_2 = 0$$

Accordingly, the rows of A form an orthonormal set. Similarly, $A^*A = I$ forces the columns of A to form an orthonormal set.

The result in the above example holds true in general; namely,

Theorem 14.8: The following conditions for a matrix A are equivalent:

   (i) A is unitary (orthogonal).

   (ii) The rows of A form an orthonormal set.

   (iii) The columns of A form an orthonormal set.

14.7 CHANGE OF ORTHONORMAL BASIS

In view of the special role of orthonormal bases in the theory of inner product spaces, we are naturally interested in the properties of the change-of-basis matrix from one such basis into another. The following theorem, proved in Problem 14.12, applies.

Theorem 14.9: Let $\{u_1, \ldots, u_n\}$ be an orthonormal basis of an inner product space V. Then the change-of-basis matrix from $\{u_i\}$ into another orthonormal basis is unitary (orthogonal). Conversely, if $P = (a_{ij})$ is a unitary (orthogonal) matrix, then the following is an orthonormal basis:

$$\{u_i' = a_{1i}u_1 + a_{2i}u_2 + \cdots + a_{ni}u_n : i = 1, \ldots, n\}$$

Recall that matrices A and B representing the same linear operator T are similar, i.e., $B = P^{-1}AP$ where P is the (nonsingular) change-of-basis matrix. On the other hand, if V is an inner product space, we are usually interested in the case when P is unitary (or orthogonal) as suggested by Theorem 14.9. (Recall that P is unitary if $P^* = P^{-1}$, and P is orthogonal if $P^T = P^{-1}$.) This leads to the following definition.

Definition: Complex matrices A and B are unitarily equivalent if there is a unitary matrix P for which $B = P^*AP$. Analogously, real matrices A and B are orthogonally equivalent if there is an orthogonal matrix P for which $B = P^TAP$.

Note that orthogonally equivalent matrices are necessarily congruent.

14.8 POSITIVE OPERATORS

Let P be a linear operator on an inner product space V. P is said to be *positive* (or *semidefinite*) if

$$P = S^*S \qquad \text{for some operator } S$$

and is said to be *positive definite* if S is also nonsingular. The next theorems give alternative characterizations of these operators. (Theorem 14.10A is proved in Problem 14.21.)

Theorem 14.10A: The following conditions on an operator P are equivalent:

    (i) $P = T^2$ for some self-adjoint operator T.

    (ii) $P = S^*S$ for some operator S.

    (iii) P is self-adjoint and $\langle P(u), u \rangle \geq 0$ for every $u \in V$.

The corresponding theorem for positive definite operators is

Theorem 14.10B: The following conditions on an operator P are equivalent:

    (i) $P = T^2$ for some nonsingular self-adjoint operator T.

    (ii) $P = S^*S$ for some nonsingular operator S.

    (iii) P is self-adjoint and $\langle P(u), u \rangle > 0$ for every $u \neq 0$ in V.

14.9 DIAGONALIZATION AND CANONICAL FORMS IN EUCLIDEAN SPACES

Let T be a linear operator on a finite-dimensional inner product space V over K. Representing T by a diagonal matrix depends upon the eigenvectors and eigenvalues of T, and hence upon the roots of the characteristic polynomial $\Delta(t)$ of T. Now $\Delta(t)$ always factors into linear polynomials over the complex field $\mathbf{C}$, but may not have any linear polynomials over the real field $\mathbf{R}$. Thus the situation for Euclidean spaces (where $K = \mathbf{R}$) is inherently different than that for unitary spaces (where $K = \mathbf{C}$); hence we treat them separately. We investigate Euclidean spaces below (see Problem 14.14 for proof of Theorem 14.11), and unitary spaces in the next section.

Theorem 14.11: Let T be a symmetric (self-adjoint) operator on a real finite-dimensional inner product space V. Then there exists an orthonormal basis of V consisting of eigenvectors of T; that is, T can be represented by a diagonal matrix relative to an orthonormal basis.

We give the corresponding statement for matrices.

Theorem 14.11 (Alternate Form): Let A be a real symmetric matrix. Then there exists an orthogonal matrix P such that $B = P^{-1}AP = P^TAP$ is diagonal.

We can choose the columns of the above matrix P to be normalized orthogonal eigenvectors of A; then the diagonal entries of B are the corresponding eigenvalues.

Canonical Form for Orthogonal Operators

An orthogonal operator T need not be symmetric, and so it may not be represented by a diagonal matrix relative to an orthonormal basis. However, such an operator T does have a simple canonical representation, as described in the following theorem, proved in Problem 14.16.

Theorem 14.12: Let T be an orthogonal operator on a real inner product space V. Then there is an orthonormal basis with respect to which T has the following form:

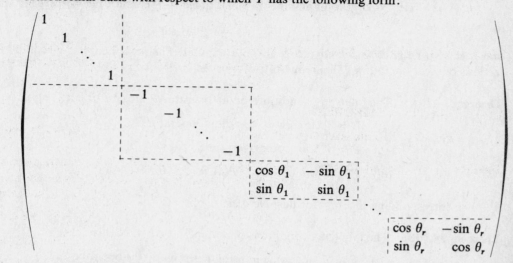

The reader may recognize that each of the above 2×2 diagonal blocks represents a rotation in the corresponding two-dimensional subspace and that each diagonal entry -1 represents a reflection in the corresponding one-dimensional subspace.

14.10 DIAGONALIZATION AND CANONICAL FORMS IN UNITARY SPACES

We now present the fundamental diagonalization theorem for complex inner product spaces, i.e., for unitary spaces. Recall that an operator T is said to be normal if it commutes with its adjoint, i.e., if $TT^* = T^*T$. Analogously, a complex matrix A is said to be normal if it commutes with its conjugate transpose, i.e., if $AA^* = A^*A$.

Example 14.5. Let $A = \begin{pmatrix} 1 & 1 \\ i & 3+2i \end{pmatrix}$. Then

$$AA^* = \begin{pmatrix} 1 & 1 \\ i & 3+2i \end{pmatrix}\begin{pmatrix} 1 & -i \\ 1 & 3-2i \end{pmatrix} = \begin{pmatrix} 2 & 3-3i \\ 3+3i & 14 \end{pmatrix}$$

$$A^*A = \begin{pmatrix} 1 & -i \\ 1 & 3-2i \end{pmatrix}\begin{pmatrix} 1 & 1 \\ i & 3+2i \end{pmatrix} = \begin{pmatrix} 2 & 3-3i \\ 3+3i & 14 \end{pmatrix}$$

Thus A is a normal matrix.

The following theorem applies.

Theorem 14.13: Let T be a normal operator on a complex finite-dimensional inner product space V. Then there exists an orthonormal basis of V consisting of eigenvectors of T; that is, T can be represented by a diagonal matrix relative to an orthonormal basis.

We give the corresponding statement for matrices.

Theorem 14.13 (Alternate Form): Let A be a normal matrix. Then there exists a unitary matrix P such that $B = P^{-1}AP = P^*AP$ is diagonal.

The following theorem shows that even nonnormal operators on unitary spaces have a relatively simple form.

Theorem 14.14: Let T be an arbitrary operator on a complex finite-dimensional inner product space V. Then T can be represented by a triangular matrix relative to an orthonormal basis of V.

Theorem 14.14 (Alternate Form): Let A be an arbitrary complex matrix. Then there exists a unitary matrix P such that $B = P^{-1}AP = P^*AP$ is triangular.

14.11 SPECTRAL THEOREM

The Spectral Theorem is a reformulation of the diagonalization Theorems 14.11 and 14.13.

Theorem 14.15 (Spectral Theorem): Let T be a normal (symmetric) operator on a complex (real) finite-dimensional inner product space V. Then there exist linear operators $E_1, \ldots, E_r$ on V and scalars $\lambda_1, \ldots, \lambda_r$ such that

(i) $T = \lambda_1 E_1 + \lambda_2 E_2 + \cdots + \lambda_r E_r$ (iii) $E_1^2 = E_1, \ldots, E_r^2 = E_r$

(ii) $E_1 + E_2 + \cdots + E_r = I$ (iv) $E_i E_j = 0$ for $i \neq j$.

The above linear operators $E_1, \ldots, E_r$ are *projections* in the sense that $E_i^2 = E_i$. Moreover, they are said to be *orthogonal projections* since they have the additional property that $E_i E_j = 0$ for $i \neq j$.

The following example shows the relationship between a diagonal matrix representation and the corresponding orthogonal projections.

Example 14.6. Consider a diagonal matrix, say $A = \begin{pmatrix} 2 & & & \\ & 3 & & \\ & & 3 & \\ & & & 5 \end{pmatrix}$. Let

$$E_1 = \begin{pmatrix} 1 & & & \\ & 0 & & \\ & & 0 & \\ & & & 0 \end{pmatrix} \quad E_2 = \begin{pmatrix} 0 & & & \\ & 1 & & \\ & & 1 & \\ & & & 0 \end{pmatrix} \quad E_3 = \begin{pmatrix} 0 & & & \\ & 0 & & \\ & & 0 & \\ & & & 1 \end{pmatrix}$$

The reader can verify that:

(i) $A = 2E_1 + 3E_2 + 5E_3$,   (ii) $E_1 + E_2 + E_3 = I$,   (iii) $E_i^2 = E_i$,   and   (iv) $E_i E_j = 0$ for $i \neq j$

Solved Problems

ADJOINTS

14.1. Find the adjoint of the matrix $B = \begin{pmatrix} 3 - 7i & 18 & 4 + i \\ -7i & 6 - i & 2 - 3i \\ 8 + i & 7 + 9i & 6 + 3i \end{pmatrix}$.

The conjugate transpose gives us $B^* = \begin{pmatrix} 3 + 7i & 7i & 8 - i \\ 18 & 6 + i & 7 - 9i \\ 4 - i & 2 + 3i & 6 - 3i \end{pmatrix}$.

14.2. Find the adjoint of each linear operator:

(a) $F : \mathbf{R}^3 \to \mathbf{R}^3$ defined by $F(x, y, z) = (3x + 4y - 5z, 2x - 6y + 7z, 5x - 9y + z)$.

(b) $G : \mathbf{C}^3 \to \mathbf{C}^3$ defined by

$$G(x, y, z) = (2x + (1 - i)y, (3 + 2i)x - 4iz, 2ix + (4 - 3i)y - 3z)$$

(a) First find the matrix A representing T in the usual basis of $\mathbf{R}^3$. (Recall the rows of A are the coefficients of x, y, z.) Thus

$$A = \begin{pmatrix} 3 & 4 & -5 \\ 2 & -6 & 7 \\ 5 & -9 & 1 \end{pmatrix}$$

Since the base field is $\mathbf{R}$, the adjoint F^* is represented by the transpose A^T of A. Thus form

$$A^T = \begin{pmatrix} 3 & 2 & 5 \\ 4 & -6 & -9 \\ -5 & 7 & 1 \end{pmatrix}$$

Then $F^*(x, y, z) = (3x + 2y + 5z, 4x - 6y - 9z, -5x + 7y + z)$.

(b) First find the matrix B representing T in the usual basis of $\mathbf{C}^3$:

$$B = \begin{pmatrix} 2 & 1 - i & 0 \\ 3 + 2i & 0 & -4i \\ 2i & 4 - 3i & -3 \end{pmatrix}$$

Form the conjugate transpose B^* of B:

$$B^* = \begin{pmatrix} 2 & 3 - 2i & -2i \\ 1 + i & 0 & 4 + 3i \\ 0 & 4i & -3 \end{pmatrix}$$

Thus $G^*(x, y, z) = (2x + (3 - 2i)y - 2iz, (1 + i)x + (4 + 3i)z, 4iy - 3z)$.

14.3. Prove Theorem 14.3.

Let $\{w_1, \ldots, w_n\}$ be an orthonormal basis of V. Set

$$u = \overline{\phi(w_1)}w_1 + \overline{\phi(w_2)}w_2 + \cdots + \overline{\phi(w_n)}w_n$$

Let $\hat{u}$ be the linear functional on V defined by $\hat{u}(v) = \langle v, u \rangle$, for every $v \in V$. Then for $i = 1, \ldots, n$,

$$\hat{u}(w_i) = \langle w_i, u \rangle = \langle w_i, \overline{\phi(w_1)}w_1 + \cdots + \overline{\phi(w_n)}w_n \rangle = \theta(w_i)$$

Since $\hat{u}$ and ϕ agree on each basis vector, $\hat{u} = \phi$.

Now suppose u' is another vector in V for which $\phi(v) = \langle v, u' \rangle$ for every $v \in V$. Then $\langle v, u \rangle = \langle v, u' \rangle$ or $\langle v, u - u' \rangle = 0$. In particular this is true for $v = u - u'$ and so $\langle u - u', u - u' \rangle = 0$. This yields $u - u' = 0$ and $u = u'$. Thus such a vector u is unique as claimed.

14.4. Prove Theorem 14.1.

Proof of (i):

We first define the mapping T^*. Let v be an arbitrary but fixed element of V. The map $u \mapsto \langle T(u), v \rangle$ is a linear functional on V. Hence by Theorem 14.3, there exists a unique element $v' \in V$ such that $\langle T(u), v \rangle = \langle u, v' \rangle$ for every $u \in V$. We define $T^* V \to V$ by $T^*(v) = v'$. Then $\langle T(u), v \rangle = \langle u, T^*(v) \rangle$ for every $u, v \in V$.

We next show that T^* is linear. For any $u, v_i \in V$, and any $a, b \in K$,

$$\langle u, T^*(av_1 + bv_2) \rangle = \langle T(u), av_1 + bv_2 \rangle = \bar{a}\langle T(u), v_1 \rangle + \bar{b}\langle T(u), v_2 \rangle$$
$$= \bar{a}\langle u, T^*(v_1) \rangle + \bar{b}\langle u, T^*(v_2) \rangle = \langle u, aT^*(v_1) + bT^*(v_2) \rangle$$

But this is true for every $u \in V$; hence $T^*(av_1 + bv_2) = aT^*(v_1) + bT^*(v_2)$. Thus T^* is linear.

Proof of (ii):

The matrices $A = (a_{ij})$ and $B = (b_{ij})$ representing T and T^*, respectively, in the basis $S = \{u_i\}$ are given by $a_{ij} = \langle T(u_j), u_i \rangle$ and $b_{ij} = \langle T^*(u_j), u_i \rangle$ (see Problem 6.23). Hence

$$b_{ij} = \langle T^*(u_j), u_i \rangle = \overline{\langle u_i, T^*(u_j) \rangle} = \overline{\langle T(u_i), u_j \rangle} = \bar{a}_{ji}$$

Thus $B = A^*$, as claimed.

14.5. Prove Theorem 14.2.

(i)   For any $u, v \in V$,

$$\langle (T_1 + T_2)(u), v \rangle = \langle T_1(u) + T_2(u), v \rangle = \langle T_1(u), v \rangle + \langle T_2(u), v \rangle$$
$$= \langle u, T_1^*(v) \rangle + \langle u, T_2^*(v) \rangle = \langle u, T_1^*(v) + T_2^*(v) \rangle$$
$$= \langle u, (T_1^* + T_2^*)(v) \rangle$$

The uniqueness of the adjoint implies $(T_1 + T_2)^* = T_1^* + T_2^*$.

(ii) For any $u, v \in V$,

$$\langle (kT)(u), v \rangle = \langle kT(u), v \rangle = k\langle T(u), v \rangle = k\langle u, T^*(v) \rangle = \langle u, \bar{k}T^*(v) \rangle = \langle u, (\bar{k}T^*)(v) \rangle$$

The uniqueness of the adjoint implies $(kT)^* = \bar{k}T^*$.

(iii) For any $u, v \in V$,

$$\langle (T_1 T_2)(u), v \rangle = \langle T_1(T_2)(u), v \rangle = \langle T_2(u), T_1^*(v) \rangle$$
$$= \langle u, T_2^*(T_1^*(v)) \rangle = \langle u, (T_2^* T_1^*)(v) \rangle$$

The uniqueness of the adjoint implies $(T_1 T_2)^* = T_2^* T_1^*$.

(iv) For any $u, v \in V$,

$$\langle T^*(u), v \rangle = \overline{\langle v, T^*(u) \rangle} = \overline{\langle T(v), u \rangle} = \langle u, T(v) \rangle$$

The uniqueness of the adjoint implies $(T^*)^* = T$.

14.6. Show that: (*a*) $I^* = I$, and (*b*) $0^* = 0$.

(*a*) For every $u, v \in V$, $\langle I(u), v \rangle = \langle u, v \rangle = \langle u, I(v) \rangle$; hence $I^* = I$.

(*b*) For every $u, v \in V$, $\langle 0(u), v \rangle = \langle 0, v \rangle = 0 = \langle u, 0 \rangle = \langle u, 0(v) \rangle$; hence $0^* = 0$.

14.7. Suppose T is invertible. Show that $(T^{-1})^* = (T^*)^{-1}$.

$I = I^* = (TT^{-1})^* = (T^{-1})^* T^*$; hence $(T^{-1})^* = T^{*-1}$.

14.8. Let T be a linear operator on V, and let W be a T-invariant subspace of V. Show that $W^\perp$ is invariant under T^*.

Let $u \in W^\perp$. If $w \in W$, then $T(w) \in W$ and so $\langle w, T^*(u) \rangle = \langle T(w), u \rangle = 0$. Thus $T^*(u) \in W^\perp$ since it is orthogonal to every $w \in W$. Hence $W^\perp$ is invariant under T^*.

14.9. Let T be a linear operator on V. Show that each of the following conditions implies $T = 0$:

(i) $\langle T(u), v \rangle = 0$ for every $u, v \in V$;

(ii) V is a complex space, and $\langle T(u), u \rangle = 0$ for every $u \in V$;

(iii) T is self-adjoint and $\langle T(u), u \rangle = 0$ for every $u \in V$.

Give an example of an operator T on a real space V for which $\langle T(u), u \rangle = 0$ for every $u \in V$ but $T \neq 0$. [Thus (ii) need not hold for a real space V.]

(i) Set $v = T(u)$. Then $\langle T(u), T(u) \rangle = 0$ and hence $T(u) = 0$, for every $u \in V$. Accordingly, $T = 0$.

(ii) By hypothesis, $\langle T(v + w), v + w \rangle = 0$ for any $v, w \in V$. Expanding and setting $\langle T(v), v \rangle = 0$ and $\langle T(w), w \rangle = 0$,

$$\langle T(v), w \rangle + \langle T(w), v \rangle = 0 \tag{1}$$

Note w is arbitrary in (1). Substituting iw for w, and using $\langle T(v), iw \rangle = \bar{i}\langle T(v), w \rangle = -i\langle T(v), w \rangle$ and $\langle T(iw), v \rangle = \langle iT(w), v \rangle = i\langle T(w), v \rangle$,

$$-i\langle T(v), w \rangle + i\langle T(w), v \rangle = 0$$

Dividing through by i and adding to (1), we obtain $\langle T(w), v \rangle = 0$ for any $v, w \in V$. By (i), $T = 0$.

(iii) By (ii), the result holds for the complex case; hence we need only consider the real case. Expanding $\langle T(v + w), v + w \rangle = 0$, we again obtain (1). Since T is self-adjoint and since it is a real space, we have $\langle T(w), v \rangle = \langle w, T(v) \rangle = \langle T(v), w \rangle$. Substituting this into (1), we obtain $\langle T(v), w \rangle = 0$ for any $v, w \in V$. By (i), $T = 0$.

For our example, consider the linear operator T on $\mathbf{R}^2$ defined by $T(x, y) = (y, -x)$. Then $\langle T(u), u \rangle = 0$ for every $u \in V$, but $T \neq 0$.

ORTHOGONAL AND UNITARY OPERATORS AND MATRICES

14.10. Prove Theorem 14.6.

Suppose (i) holds. Then, for every $v, w \in V$,

$$\langle U(v), U(w) \rangle = \langle v, U^*U(w) \rangle = \langle v, I(w) \rangle = \langle v, w \rangle$$

Thus (i) implies (ii). Now if (ii) holds, then

$$\| U(v) \| = \sqrt{\langle U(v), U(v) \rangle} = \sqrt{\langle v, v \rangle} = \| v \|$$

Hence (ii) implies (iii). It remains to show that (iii) implies (i).

Suppose (iii) holds. Then for every $v \in V$,

$$\langle U^*U(v), v \rangle = \langle U(v), U(v) \rangle = \langle v, v \rangle = \langle I(v), v \rangle$$

Hence $\langle (U^*U - I)(v), v \rangle = 0$ for every $v \in V$. But $U^*U - I$ is self-adjoint (Prove!); then by Problem 14.9 we have $U^*U - I = 0$ and so $U^*U = I$. Thus $U^* = U^{-1}$ as claimed.

14.11. Let U be a unitary (orthogonal) operator on V, and let W be a subspace invariant under U. Show that $W^\perp$ is also invariant under U.

Since U is nonsingular, $U(W) = W$; that is, for any $w \in W$ there exists $w' \in W$ such that $U(w') = w$. Now let $v \in W^\perp$. Then for any $w \in W$,

$$\langle U(v), w \rangle = \langle U(v), U(w') \rangle = \langle v, w' \rangle = 0$$

Thus $U(v)$ belongs to $W^\perp$. Therefore $W^\perp$ is invariant under U.

14.12. Prove Theorem 14.9.

Suppose $\{v_i\}$ is another orthonormal basis and suppose

$$v_i = b_{i1}u_1 + b_{i2}u_2 + \cdots + b_{in}u_n \qquad i = 1, \dots, n \tag{1}$$

Since $\{v_i\}$ is orthonormal,

$$\delta_{ij} = \langle v_i, v_j \rangle = b_{i1}\overline{b_{j1}} + b_{i2}\overline{b_{j2}} + \cdots + b_{in}\overline{b_{jn}} \tag{2}$$

Let $B = (b_{ij})$ be the matrix of coefficients in (1). (Then B^T is the change-of-basis matrix from $\{u_i\}$ to $\{v_i\}$.) Then $BB^* = (c_{ij})$ where $c_{ij} = b_{i1}\overline{b_{j1}} + b_{i2}\overline{b_{j2}} + \cdots + b_{in}\overline{b_{jn}}$. By (2), $c_{ij} = \delta_{ij}$ and therefore $BB^* = I$. Accordingly, B, and hence B^T, are unitary.

It remains to prove that $\{u_i'\}$ is orthonormal. By Problem 6.23,

$$\langle u_i', u_j' \rangle = a_{1i}\overline{a_{1j}} + a_{2i}\overline{a_{2j}} + \cdots + a_{ni}\overline{a_{nj}} = \langle C_i, C_j \rangle$$

where C_i denotes the ith column of the unitary (orthogonal) matrix $P = (a_{ij})$. Since P is unitary (orthogonal), its columns are orthonormal; hence $\langle u_i', u_j' \rangle = \langle C_i, C_j \rangle = \delta_{ij}$. Thus $\{u_i'\}$ is an orthonormal basis.

SYMMETRIC OPERATORS AND CANONICAL FORMS IN EUCLIDEAN SPACES

14.13. Let T be a symmetric operator. Show that: (a) the characteristic polynomial $\Delta(t)$ of T is a product of linear polynomials (over **R**); (b) T has a nonzero eigenvector.

(a) Let A be a matrix representing T relative to an orthonormal basis of V; then $A = A^T$. Let $\Delta(t)$ be the characteristic polynomial of A. Viewing A as a complex self-adjoint operator, A has only real eigenvalues by Theorem 14.4. Thus

$$\Delta(t) = (t - \lambda_1)(t - \lambda_2) \cdots (t - \lambda_n)$$

where the λ_i are all real. In other words, $\Delta(t)$ is a product of linear polynomials over **R**.

(b) By (a), T has at least one (real) eigenvalue. Hence T has a nonzero eigenvector.

14.14. Prove Theorem 14.11.

The proof is by induction on the dimension of V. If $\dim V = 1$, the theorem trivially holds. Now suppose $\dim V = n > 1$. By Problem 14.13, there exists a nonzero eigenvector v_1 of T. Let W be the space spanned by v_1, and let u_1 be a unit vector in W, e.g., let $u_1 = v_1/\|v_1\|$.

Since v_1 is an eigenvector of T, the subspace W of V is invariant under T. By Problem 14.8, $W^\perp$ is invariant under $T^* = T$. Thus the restriction $\hat{T}$ of T to $W^\perp$ is a symmetric operator. By Theorem 6.4, $V = W \oplus W^\perp$. Hence $\dim W^\perp = n - 1$ since $\dim W = 1$. By induction, there exists an orthonormal basis $\{u_2, \ldots, u_n\}$ of $W^\perp$ consisting of eigenvectors of $\hat{T}$ and hence of T. But $\langle u_1, u_i \rangle = 0$ for $i = 2, \ldots, n$ because $u_i \in W^\perp$. Accordingly $\{u_1, u_2, \ldots, u_n\}$ is an orthonormal set and consists of eigenvectors of T. Thus the theorem is proved.

14.15. Find an orthogonal change of coordinates which diagonalizes the real quadratic form defined by $q(x, y) = 2x^2 + 2xy + 2y^2$.

First find the symmetric matrix A representing q and then its characteristic polynomial $\Delta(t)$:

$$A = \begin{pmatrix} 2 & 1 \\ 1 & 2 \end{pmatrix} \quad \text{and} \quad \Delta(t) = |tI - A| = \begin{vmatrix} t - 2 & -1 \\ -1 & t - 2 \end{vmatrix} = (t - 1)(t - 3)$$

The eigenvalues of A are 1 and 3; hence the diagonal form of q is

$$q(x', y') = x'^2 + 3y'^2$$

We find the corresponding transformation of coordinates by obtaining a corresponding orthonormal set of eigenvectors of A.

Set $t = 1$ into the matrix $tI - A$ to obtain the corresponding homogeneous system

$$-x - y = 0 \qquad -x - y = 0$$

A nonzero solution is $v_1 = (1, -1)$. Now set $t = 3$ into the matrix $tI - A$ to find the corresponding homogeneous system

$$x - y = 0 \qquad -x + y = 0$$

A nonzero solution is $v_2 = (1, 1)$. As expected from Theorem 14.5, v_1 and v_2 are orthogonal. Normalize v_1 and v_2 to obtain the orthonormal basis

$$\{u_1 = (1/\sqrt{2}, -1/\sqrt{2}), u_2 = (1/\sqrt{2}, 1/\sqrt{2})\}$$

The transition matrix P and the required transformation of coordinates follow:

$$P = \begin{pmatrix} 1/\sqrt{2} & 1/\sqrt{2} \\ -1/\sqrt{2} & 1/\sqrt{2} \end{pmatrix} \quad \text{and} \quad \begin{pmatrix} x \\ y \end{pmatrix} = P\begin{pmatrix} x' \\ y' \end{pmatrix} \quad \text{or} \quad \begin{aligned} x &= (x' + y')/\sqrt{2} \\ y &= (-x' + y')/\sqrt{2} \end{aligned}$$

Note that the columns of P are u_1 and u_2. We can also express x' and y' in terms of x and y by using $P^{-1} = P^T$; that is,

$$x' = (x - y)/\sqrt{2} \qquad y' = (x + y)/\sqrt{2}$$

14.16. Prove Theorem 14.12.

Let $S = T + T^{-1} = T + T^*$. Then $S^* = (T + T^*)^* = T^* + T = S$. Thus S is a symmetric operator on V. By Theorem 14.11, there exists an orthonormal basis of V consisting of eigenvectors of S. If $\lambda_1, \ldots, \lambda_m$ denote the distinct eigenvalues of S, then V can be decomposed into the direct sum $V = V_1 \oplus V_2 \oplus \cdots \oplus V_m$ where the V_i consists of the eigenvectors of S belonging to λ_i. We claim that each V_i is invariant under T. For suppose $v \in V_i$; then $S(v) = \lambda_i v$ and

$$S(T(v)) = (T + T^{-1})T(v) = T(T + T^{-1})(v) = TS(v) = T(\lambda_i v) = \lambda_i T(v)$$

That is, $T(v) \in V_i$. Hence V_i is invariant under T. Since the V_i are orthogonal to each other, we can restrict our investigation to the way that T acts on each individual V_i.

On a given V_i, $(T + T^{-1})v = S(v) = \lambda_i v$. Multiplying by T,

$$(T^2 - \lambda_i T + I)(v) = 0$$

We consider the cases $\lambda_i = \pm 2$ and $\lambda_i \neq \pm 2$ separately. If $\lambda_i = \pm 2$, then $(T \pm I)^2(v) = 0$ which leads to $(T \pm I)(v) = 0$ or $T(v) = \pm v$. Thus T restricted to this V_i is either I or $-I$.

If $\lambda_i \neq \pm 2$, then T has no eigenvectors in V_i since by Theorem 14.4 the only eigenvalues of T are 1 or -1. Accordingly, for $v \neq 0$ the vectors v and $T(v)$ are linearly independent. Let W be the subspace spanned by v and $T(v)$. Then W is invariant under T, since

$$T(T(v)) = T^2(v) = \lambda_i T(v) - v$$

By Theorem 6.4, $V_i = W \oplus W^\perp$. Furthermore, by Problem 14.8, $W^\perp$ is also invariant under T. Thus we can decompose V_i into the direct sum of two-dimensional subspaces W_j where the W_j are orthogonal to each other and each W_j is invariant under T. Thus we can now restrict our investigation to the way T acts on each individual W_j.

Since $T^2 - \lambda_i T + I = 0$, the characteristic polynomial $\Delta(t)$ of T acting on W_j is $\Delta(t) = t^2 - \lambda_i t + 1$. Thus the determinant of T is 1, the constant term in $\Delta(t)$. By Theorem 4.6, the matrix A representing T acting on W_j relative to any orthonormal basis of W_j must be of the form

$$\begin{pmatrix} \cos\theta & -\sin\theta \\ \sin\theta & \cos\theta \end{pmatrix}$$

The union of the basis of the W_j gives an orthonormal basis of V_i, and the union of the basis of the V_i gives an orthonormal basis of V in which the matrix representing T is of the desired form.

NORMAL OPERATORS AND CANONICAL FORMS IN UNITARY SPACES

14.17. Determine which matrix is normal: (a) $A = \begin{pmatrix} 1 & i \\ 0 & 1 \end{pmatrix}$, and (b) $B = \begin{pmatrix} 1 & i \\ 1 & 2+i \end{pmatrix}$.

(a) $AA^* = \begin{pmatrix} 1 & i \\ 0 & 1 \end{pmatrix}\begin{pmatrix} 1 & 0 \\ -i & 1 \end{pmatrix} = \begin{pmatrix} 2 & i \\ -i & 1 \end{pmatrix}$, $\quad A^*A = \begin{pmatrix} 1 & 0 \\ -i & 1 \end{pmatrix}\begin{pmatrix} 1 & i \\ 0 & 1 \end{pmatrix} = \begin{pmatrix} 1 & i \\ -i & 2 \end{pmatrix}$

Since $AA^* \neq A^*A$, the matrix A is not normal.

(b) $BB^* = \begin{pmatrix} 1 & i \\ 1 & 2+i \end{pmatrix}\begin{pmatrix} 1 & 1 \\ -i & 2-i \end{pmatrix} = \begin{pmatrix} 2 & 2+2i \\ 2-2i & 6 \end{pmatrix}$,

$B^*B = \begin{pmatrix} 1 & 1 \\ -i & 2-i \end{pmatrix}\begin{pmatrix} 1 & i \\ 1 & 2+i \end{pmatrix} = \begin{pmatrix} 2 & 2+2i \\ 2-2i & 6 \end{pmatrix}$

Since $BB^* = B^*B$, the matrix B is normal.

14.18. Let T be a normal operator. Prove:

   (a) $T(v) = 0$ if and only if $T^*(v) = 0$.

   (b) $T - \lambda I$ is normal.

   (c) If $T(v) = \lambda v$, then $T^*(v) = \bar{\lambda} v$; hence any eigenvector of T is also an eigenvector of T^*.

   (d) If $T(v) = \lambda_1 v$ and $T(w) = \lambda_2 w$ where $\lambda_1 \neq \lambda_2$, then $\langle v, w \rangle = 0$; that is, eigenvectors of T belonging to distinct eigenvalues are orthonormal.

   (a) We show that $\langle T(v), T(v) \rangle = \langle T^*(v), T^*(v) \rangle$:

$$\langle T(v), T(v) \rangle = \langle v, T^*T(v) \rangle = \langle v, TT^*(v) \rangle = \langle T^*(v), T^*(v) \rangle$$

Hence by $[I_3]$, $T(v) = 0$ if and only if $T^*(v) = 0$.

   (b) We show that $T - \lambda I$ commutes with its adjoint:

$$(T - \lambda I)(T - \lambda I)^* = (T - \lambda I)(T^* - \bar{\lambda} I) = TT^* - \lambda T^* - \bar{\lambda} T + \lambda \bar{\lambda} I$$
$$= T^*T - \bar{\lambda} T - \lambda T^* + \bar{\lambda} \lambda I = (T^* - \bar{\lambda} I)(T - \lambda I)$$
$$= (T - \lambda I)^*(T - \lambda I)$$

Thus $T - \lambda I$ is normal.

   (c) If $T(v) = \lambda v$, then $(T - \lambda I)(v) = 0$. Now $T - \lambda I$ is normal by (b); therefore, by (a), $(T - \lambda I)^*(v) = 0$. That is, $(T^* - \bar{\lambda} I)(v) = 0$; hence $T^*(v) = \bar{\lambda} v$.

   (d) We show that $\lambda_1 \langle v, w \rangle = \lambda_2 \langle v, w \rangle$:

$$\lambda_1 \langle v, w \rangle = \langle \lambda_1 v, w \rangle = \langle T(v), w \rangle = \langle v, T^*(w) \rangle = \langle v, \bar{\lambda}_2 w \rangle = \lambda_2 \langle v, w \rangle$$

But $\lambda_1 \neq \lambda_2$; hence $\langle v, w \rangle = 0$.

14.19. Prove Theorem 14.13.

The proof is by induction on the dimension of V. If dim $V = 1$, then the theorem trivially holds. Now suppose dim $V = n > 1$. Since V is a complex vector space, T has at least one eigenvalue and hence a nonzero eigenvector v. Let W be the subspace of V spanned by v and let u_1 be a unit vector in W.

Since v is an eigenvector of T, the subspace W is invariant under T. However, v is also an eigenvector of T^* by Problem 14.18; hence W is also invariant under T^*. By Problem 14.8, $W^\perp$ is invariant under $T^{**} = T$. The remainder of the proof is identical with the latter part of the proof of Theorem 14.11.

14.20. Prove Theorem 14.14.

The proof is by induction on the dimension of V. If dim $V = 1$, then the theorem trivially holds. Now suppose dim $V = n > 1$. Since V is a complex vector space, T has at least one eigenvalue and hence at least one nonzero eigenvector v. Let W be the subspace of V spanned by v and let u_1 be a unit vector in W. Then u_1 is an eigenvector of T and, say, $T(u_1) = a_{11} u_1$.

By Theorem 6.4, $V = W \oplus W^\perp$. Let E denote the orthogonal projection of V into $W^\perp$. Clearly $W^\perp$ is invariant under the operator ET. By induction, there exists an orthonormal basis $\{u_2, \ldots, u_n\}$ of $W^\perp$ such that, for $i = 2, \ldots, n$,

$$ET(u_i) = a_{i2} u_2 + a_{i3} u_3 + \cdots + a_{ii} u_i$$

(Note that $\{u_1, u_2, \ldots, u_n\}$ is an orthonormal basis of V.) But E is the orthogonal projection of V onto $W^\perp$; hence we must have

$$T(u_i) = a_{i1} u_1 + a_{i2} u_2 + \cdots + a_{ii} u_i$$

for $i = 2, \ldots, n$. This with $T(u_1) = a_{11} u_1$ gives us the desired result.

MISCELLANEOUS PROBLEMS

14.21. Prove Theorem 14.10A.

Suppose (i) holds, that is, $P = T^2$ where $T = T^*$. Then $P = TT = T^*T$ and so (i) implies (ii). Now suppose (ii) holds. Then $P^* = (S^*S)^* = S^*S^{**} = S^*S = P$ and so P is self-adjoint. Furthermore,

$$\langle P(u), u \rangle = \langle S^*S(u), u \rangle = \langle S(u), S(u) \rangle \geq 0$$

Thus (ii) implies (iii), and so it remains to prove that (iii) implies (i).

Now suppose (iii) holds. Since P is self-adjoint, there exists an orthonormal basis $\{u_1, \ldots, u_n\}$ of V consisting of eigenvectors of P; say, $P(u_i) = \lambda_i u_i$. By Theorem 14.4, the λ_i are real. Using (iii), we show that the λ_i are nonnegative. We have, for each i,

$$0 \leq \langle P(u_i), u_i \rangle = \langle \lambda_i u_i, u_i \rangle = \lambda_i \langle u_i, u_i \rangle$$

Thus $\langle u_i, u_i \rangle \geq 0$ forces $\lambda_i \geq 0$, as claimed. Accordingly, $\sqrt{\lambda_i}$ is a real number. Let T be the linear operator defined by

$$T(u_i) = \sqrt{\lambda_i}\, u_i \qquad \text{for } i = 1, \ldots, n$$

Since T is represented by a real diagonal matrix relative to the orthonormal basis $\{u_i\}$, T is self-adjoint. Moreover, for each i,

$$T^2(u_i) = T(\sqrt{\lambda_i}\, u_i) = \sqrt{\lambda_i}\, T(u_i) = \sqrt{\lambda_i}\sqrt{\lambda_i}\, u_i = \lambda_i u_i = P(u_i)$$

Since T^2 and P agree on a basis of V, $P = T^2$. Thus the theorem is proved.

Remark: The above operator T is the unique positive operator such that $P = T^2$ it is called the positive square root of P.

14.22. Show that any operator T is the sum of a self-adjoint operator and skew-adjoint operator.

Set $S = \frac{1}{2}(T + T^*)$ and $U = \frac{1}{2}(T - T^*)$. Then $T = S + U$ where

$$S^* = (\tfrac{1}{2}(T + T^*))^* = \tfrac{1}{2}(T^* + T^{**}) = \tfrac{1}{2}(T^* + T) = S$$

and

$$U^* = (\tfrac{1}{2}(T - T^*))^* = \tfrac{1}{2}(T^* - T) = -\tfrac{1}{2}(T - T^*) = -U$$

i.e., S is self-adjoint and U is skew adjoint.

14.23. Prove: Let T be an arbitrary linear operator on a finite-dimensional inner product space V. Then T is a product of a unitary (orthogonal) operator U and a unique positive operator P, that is, $T = UP$. Furthermore, if T is invertible, then U is also uniquely determined.

By Theorem 14.10, T^*T is a positive operator and hence there exists a (unique) positive operator P such that $P^2 = T^*T$ (Problem 14.43). Observe that

$$\|P(v)\|^2 = \langle P(v), P(v) \rangle = \langle P^2(v), v \rangle = \langle T^*T(v), v \rangle = \langle T(v), T(v) \rangle = \|T(v)\|^2 \qquad (I)$$

We now consider separately the cases when T is invertible and non-invertible.

If T is invertible, then we set $\hat{U} = PT^{-1}$. We show that $\hat{U}$ is unitary:

$$\hat{U}^* = (PT^{-1})^* = T^{-1*}P^* = (T^*)^{-1}P \qquad \text{and} \qquad \hat{U}^*\hat{U} = (T^*)^{-1}PPT^{-1} = (T^*)^{-1}T^*TT^{-1} = I$$

Thus $\hat{U}$ is unitary. We next set $U = \hat{U}^{-1}$. Then U is also unitary and $T = UP$ as required.

To prove uniqueness, we assume $T = U_0 P_0$ where U_0 is unitary and P_0 is positive. Then

$$T^*T = P_0^* U_0^* U_0 P_0 = P_0 I P_0 = P_0^2$$

But the positive square root of T^*T is unique (Problem 14.43) hence $P_0 = P$. (Note that the invertibility of T is not used to prove the uniqueness of P.) Now if T is invertible, then P is also by (I). Multiplying $U_0 P = UP$ on the right by P^{-1} yields $U_0 = U$. Thus U is also unique when T is invertible.

Now suppose T is not invertible. Let W be the image of P, i.e., $W = \text{Im } P$. We define $U_1 : W \to V$ by

$$U_1(w) = T(v) \qquad \text{where} \qquad P(v) = w \tag{2}$$

We must show that U_1 is well defined, that is, that $P(v) = P(v')$ implies $T(v) = T(v')$. This follows from the fact that $P(v - v') = 0$ is equivalent to $\|P(v - v')\| = 0$ which forces $\|T(v - v')\| = 0$ by (1). Thus U_1 is well defined. We next define $U_2 : W \to V$. Note by (1) that P and T have the same kernels. Hence the images of P and T have the same dimension, i.e., dim (Im P) = dim W = dim (Im T). Consequently, $W^\perp$ and (Im T)$^\perp$ also have the same dimension. We let U_2 be any isomorphism between $W^\perp$ and (Im T)$^\perp$.

We next set $U = U_1 \oplus U_2$. [Here U is defined as follows: if $v \in V$ and $v = w + w'$ where $w \in W$, $w' \in W^\perp$, then $U(v) = U_1(w) + U_2(w')$.] Now U is linear (Problem 14.69) and, if $v \in V$ and $P(v) = w$, then by (2)

$$T(v) = U_1(w) = U(w) = UP(v)$$

Thus $T = UP$ as required.

It remains to show that U is unitary. Now every vector $x \in V$ can be written in the form $x = P(v) + w'$ where $w' \in W^\perp$. Then $U(x) = UP(v) + U_2(w') = T(v) + U_2(w')$ where $\langle T(v), U_2(w') \rangle = 0$ by definition of U_2. Also, $\langle T(v), T(v) \rangle = \langle P(v), P(v) \rangle$ by (1). Thus

$$\begin{aligned}
\langle U(x), U(x) \rangle &= \langle T(v) + U_2(w'), T(v) + U_2(w') \rangle \\
&= \langle T(v), T(v) \rangle + \langle U_2(w'), U_2(w') \rangle \\
&= \langle P(v), P(v) \rangle + \langle w', w' \rangle = \langle P(v) + w', P(v) + w' \rangle \\
&= \langle x, x \rangle
\end{aligned}$$

[We also used the fact that $\langle P(v), w' \rangle = 0$.] Thus U is unitary and the theorem is proved.

14.24. Let V be the vector space of polynomials over **R** with inner product defined by

$$\langle f, g \rangle = \int_0^1 f(t)g(t)\, dt.$$

Give an example of a linear functional ϕ on V for which Theorem 14.3 does not hold, i.e., there does not exist a polynomial $h(t)$ for which $\phi(f) = \langle f, h \rangle$ for every $f \in V$.

Let $\phi : V \to \mathbf{R}$ be defined by $\phi(f) = f(0)$, that is, ϕ evaluates $f(t)$ at 0 and hence maps $f(t)$ into its constant term. Suppose a polynomial $h(t)$ exists for which

$$\phi(f) = f(0) = \int_0^1 f(t)h(t)\, dt \tag{1}$$

for every polynomial $f(t)$. Observe that ϕ maps the polynomial $tf(t)$ into 0; hence by (1),

$$\int_0^1 tf(t)h(t)\, dt = 0 \tag{2}$$

for every polynomial $f(t)$. In particular, (2) must hold for $f(t) = th(t)$, that is,

$$\int_0^1 t^2 h^2(t)\, dt = 0$$

This integral forces $h(t)$ to be the zero polynomial; hence $\phi(f) = \langle f, h \rangle = \langle f, 0 \rangle = 0$ for every polynomial $f(t)$. This contradicts the fact that ϕ is not the zero functional; hence the polynomial $h(t)$ does not exist.

Supplementary Problems

ADJOINT OPERATORS

14.25. Find the adjoint of each of the following matrices:

(a) $\quad A = \begin{pmatrix} 5 - 2i & 3 + 7i \\ 4 - 6i & 8 + 3i \end{pmatrix}$, $\qquad$ (b) $\quad B = \begin{pmatrix} 3 & 5i \\ i & -2i \end{pmatrix}$ $\qquad$ (c) $\quad C = \begin{pmatrix} 1 & 1 \\ 2 & 3 \end{pmatrix}$

14.26. Let $T : \mathbf{R}^3 \to \mathbf{R}^3$ be defined by $T(x, y, z) = (x + 2y, 3x - 4z, y)$. Find $T^*(x, y, z)$.

14.27. Let $T : \mathbf{C}^3 \to \mathbf{C}^3$ be defined by

$$T(x, y, z) = (ix + (2 + 3i)y, 3x + (3 - i)z, (2 - 5i)y + iz)$$

Find $T^*(x, y, z)$.

14.28. For each of the following linear functions ϕ on V find a vector $u \in V$ such that $\phi(v) = \langle v, u \rangle$ for every $v \in V$:

   (a) $\phi : \mathbf{R}^3 \to \mathbf{R}$ defined by $\phi(x, y, z) = x + 2y - 3z$.

   (b) $\phi : \mathbf{C}^3 \to \mathbf{C}$ defined by $\phi(x, y, z) = ix + (2 + 3i)y + (1 - 2i)z$.

   (c) $\phi : V \to \mathbf{R}$ defined by $\phi(f) = f(1)$ where V is the vector space of Problem 14.24.

14.29. Suppose V has finite dimension. Prove that the image of T^* is the orthogonal complement of the kernel of T, i.e., Im $T^* = (\text{Ker } T)^\perp$. Hence rank $T =$ rank T^*.

14.30. Show that $T^*T = 0$ implies $T = 0$.

14.31. Let V be the vector space of polynomials over $\mathbf{R}$ with inner product defined by $\langle f, g \rangle = \int_0^1 f(t)g(t)\, dt$. Let D be the derivative operator on V, i.e., $D(f) = df/dt$. Show that there is no operator D^* on V such that $\langle D(f), g \rangle = \langle f, D^*(g) \rangle$ for every $f, g \in V$. That is, D has no adjoint.

UNITARY AND ORTHOGONAL OPERATORS AND MATRICES

14.32. Find a unitary (orthogonal) matrix whose first row is:

   (a) $(2/\sqrt{13}, 3/\sqrt{13})$,   (b) a multiple of $(1, 1 - i)$,   (c) $(\frac{1}{2}, \frac{1}{2}i, \frac{1}{2} - \frac{1}{2}i)$

14.33. Prove: The product and inverses of orthogonal matrices are orthogonal. (Thus the orthogonal matrices form a group under multiplication called the orthogonal group.)

14.34. Prove: The product and inverses of unitary matrices are unitary. (Thus the unitary matrices form a group under multiplication called the unitary group.)

14.35. Show that if an orthogonal (unitary) matrix is triangular, then it is diagonal.

14.36. Recall that the complex matrices A and B are unitarily equivalent if there exists a unitary matrix P such that $B = P^*AP$. Show that this relation is an equivalence relation.

14.37. Recall that the real matrices A and B are orthogonally equivalent if there exists an orthogonal matrix P such that $B = P^T A P$. Show that this relation is an equivalence relation.

14.38. Let W be a subspace of V. For any $v \in V$ let $v = w + w'$ where $w \in W$, $w' \in W^\perp$. (Such a sum is unique because $V = W \oplus W^\perp$.) Let $T : V \to V$ be defined by $T(v) = w - w'$. Show that T is a self-adjoint unitary operator on V.

14.39. Let V be an inner product space, and suppose $U : V \to V$ (not necessarily linear) is surjective (onto) and preserves inner products, i.e., $\langle U(v), U(w) \rangle = \langle u, w \rangle$ for every $v, w \in V$. Prove that U is linear and hence unitary.

POSITIVE AND POSITIVE DEFINITE OPERATORS

14.40. Show that the sum of two positive (positive definite) operators is positive (positive definite).

14.41. Let T be a linear operator on V and let $f: V \times V \to K$ be defined by $f(u, v) = \langle T(u), v \rangle$. Show that f is itself an inner product on V if and only if T is positive definite.

14.42. Suppose E is an orthogonal projection onto some subspace W of V. Prove that $kI + E$ is positive (positive definite) if $k \geq 0$ ($k > 0$).

14.43. Consider the operator T defined by $T(u_i) = \sqrt{\lambda_i} u_i$, $i = 1, \ldots, n$, in the proof of Theorem 14.10A. Show that T is positive and that it is the only positive operator for which $T^2 = P$.

14.44. Suppose P is both positive and unitary. Prove that $P = I$.

14.45. Determine which of the following matrices are positive (positive definite):

$$\begin{pmatrix} 1 & 1 \\ 1 & 1 \end{pmatrix} \qquad \begin{pmatrix} 0 & i \\ -i & 0 \end{pmatrix} \qquad \begin{pmatrix} 0 & 1 \\ -1 & 0 \end{pmatrix} \qquad \begin{pmatrix} 1 & 1 \\ 0 & 1 \end{pmatrix} \qquad \begin{pmatrix} 2 & 1 \\ 1 & 2 \end{pmatrix} \qquad \begin{pmatrix} 1 & 2 \\ 2 & 1 \end{pmatrix}$$

$$\text{(i)} \qquad\qquad \text{(ii)} \qquad\qquad \text{(iii)} \qquad\qquad \text{(iv)} \qquad\qquad \text{(v)} \qquad\qquad \text{(vi)}$$

14.46. Prove that a 2×2 complex matrix $A = \begin{pmatrix} a & b \\ c & d \end{pmatrix}$ is positive if and only if (i) $A = A^*$, and (ii) a, d, and $ad - bc$ are nonnegative real numbers.

14.47. Prove that a diagonal matrix A is positive (positive definite) if and only if every diagonal entry is a nonnegative (positive) real number.

SELF-ADJOINT AND SYMMETRIC OPERATORS

14.48. For any operator T, show that $T + T^*$ is self-adjoint and $T - T^*$ is skew-adjoint.

14.49. Suppose T is self-adjoint. Show that $T^2(v) = 0$ implies $T(v) = 0$. Use this to prove that $T^n(v) = 0$ also implies $T(v) = 0$ for $n > 0$.

14.50. Let V be a complex inner product space. Suppose $\langle T(v), v \rangle$ is real for every $v \in V$. Show that T is self-adjoint.

14.51. Suppose S and T are self-adjoint. Show that ST is self-adjoint if and only if S and T commute, i.e., $ST = TS$.

14.52. For each of the following symmetric matrices A, find an orthogonal matrix P for which $P^T A P$ is diagonal:

$$\text{(a)} \quad A = \begin{pmatrix} 1 & 2 \\ 2 & -2 \end{pmatrix}, \qquad \text{(b)} \quad A = \begin{pmatrix} 5 & 4 \\ 4 & -1 \end{pmatrix}, \qquad \text{(c)} \quad A = \begin{pmatrix} 7 & 3 \\ 3 & -1 \end{pmatrix}$$

14.53. Find an orthogonal transformation of coordinates which diagonalizes each of the following quadratic forms: (a) $q(x, y) = 2x^2 - 6xy + 10y^2$, (b) $q(x, y) = x^2 + 8xy - 5y^2$

14.54. Find an orthogonal transformation of coordinates which diagonalizes the quadratic form

$$q(x, y, z) = 2xy + 2xz + 2yz.$$

NORMAL OPERATORS AND MATRICES

14.55. Verify that $A = \begin{pmatrix} 2 & i \\ i & 2 \end{pmatrix}$ is normal. Find a unitary matrix P such that P^*AP is diagonal, and find P^*AP.

14.56. Show that a triangular matrix is normal if and only if it is diagonal.

14.57. Prove that if T is normal on V, then $\|T(v)\| = \|T^*(v)\|$ for every $v \in V$. Prove that the converse holds in complex inner product spaces.

14.58. Show that self-adjoint, skew-adjoint and unitary (orthogonal) operators are normal.

14.59. Suppose T is normal. Prove that:

    (a) T is self-adjoint if and only if its eigenvalues are real.

    (b) T is unitary if and only if its eigenvalues have absolute value 1.

    (c) T is positive if and only if its eigenvalues are nonnegative real numbers.

14.60. Show that if T is normal, then T and T^* have the same kernel and the same image.

14.61. Suppose S and T are normal and commute. Show that $S + T$ and ST are also normal.

14.62. Suppose T is normal and commutes with S. Show that T also commutes with S^*.

14.63. Prove: Let S and T be normal operators on a complex finite-dimensional vector space V. Then there exists an orthonormal basis of V consisting of eigenvectors of both S and T. (That is, S and T can be simultaneously diagonalized.)

INNER PRODUCT SPACE ISOMORPHISM PROBLEMS

14.64. Let $S = \{u_1, \ldots, u_n\}$ be an orthonormal basis of an inner product space V over K. Show that the map $v \mapsto [v]_S$ is an (inner product space) isomorphism between V and K^n. (Here $[v]_S$ denotes the coordinate vector of v in the basis S.)

14.65. Show that inner product spaces V and W over K are isomorphic if and only if V and W have the same dimension.

14.66. Suppose $\{u_1, \ldots, u_n\}$ and $\{u'_1, \ldots, u'_n\}$ are orthonormal basis of V and W, respectively. Let $T : V \to W$ be the linear map defined by $T(u_i) = u'_i$, for each i. Show that T is an isomorphism.

14.67. Let V be an inner product space. Recall (page 426) that each $u \in V$ determines a linear functional $\hat{u}$ in the dual space V^* by the definition $\hat{u}(v) = \langle v, u \rangle$ for every $v \in V$. Show that the map $u \mapsto \hat{u}$ is linear and nonsingular, and hence an isomorphism from V onto V^*.

MISCELLANEOUS PROBLEMS

14.68. Show that there exists an orthonormal basis $\{u_1, \ldots, u_n\}$ of V consisting of eigenvectors of T if and only if there exist orthogonal projections $E_1, \ldots, E_r$ and scalars $\lambda_1, \ldots, \lambda_r$ such that:

    (i) $T = \lambda_1 E_1 + \cdots + \lambda_r E_r$; (ii) $E_1 + \cdots + E_r = I$; (iii) $E_i E_j = 0$ for $i \neq j$.

14.69. Suppose $V = U \oplus W$ and suppose $T_1 : U \to V$ and $T_2 : W \to V$ are linear. Show that $T = T_1 \oplus T_2$ is also linear. [Here T is defined as follows: If $v \in V$ and $v = u + w$ where $u \in U$, $w \in W$, then

$$T(v) = T_1(u) + T_2(w).]$$

14.70. Suppose U is an orthogonal operator on $\mathbf{R}^3$ with positive determinant. Show that U is either a rotation or a reflection through a plane.

Answers to Supplementary Problems

14.25. (a) $\begin{pmatrix} 5+2i & 4+6i \\ 3-7i & 8-3i \end{pmatrix}$,   (b) $\begin{pmatrix} 3 & -i \\ -5i & 2i \end{pmatrix}$,   (c) $\begin{pmatrix} 1 & 2 \\ 1 & 3 \end{pmatrix}$

14.26. $T^*(x, y, z) = (x + 3y, 2x + z, -4y)$

14.27. $T^*(x, y, z) = (-ix + 3y, (2 - 3i)x + (2 + 5i)z, (3 + i)y - iz)$

14.28. (a) $u = (1, 2, -3)$,   (b) $u = (-i, 2 - 3i, 1 + 2i)$,   (c) $u = (18t^2 - 8t + 13)/15$

14.32. (a) $\begin{pmatrix} 2/\sqrt{13} & 3/\sqrt{13} \\ 3/\sqrt{13} & -2/\sqrt{13} \end{pmatrix}$,   (b) $\begin{pmatrix} 1/\sqrt{3} & (1-i)/\sqrt{3} \\ (1+i)/\sqrt{3} & -1/\sqrt{3} \end{pmatrix}$,   (c) $\begin{pmatrix} \frac{1}{2} & \frac{1}{2}i & \frac{1}{2} - \frac{1}{2}i \\ i/\sqrt{2} & -1/\sqrt{2} & 0 \\ \frac{1}{2} & -\frac{1}{2}i & -\frac{1}{2} + \frac{1}{2}i \end{pmatrix}$

14.45. Only (i) and (v) are positive. Moreover, (v) is positive definite.

14.52. (a) $P = \begin{pmatrix} 2/\sqrt{5} & -1/\sqrt{5} \\ -1/\sqrt{5} & 2/\sqrt{5} \end{pmatrix}$,   (b) $P = \begin{pmatrix} 2/\sqrt{5} & -1/\sqrt{5} \\ -1/\sqrt{5} & 2/\sqrt{5} \end{pmatrix}$,   (c) $P = \begin{pmatrix} 3/\sqrt{10} & -1/\sqrt{10} \\ -1/\sqrt{10} & 3/\sqrt{10} \end{pmatrix}$

14.53. (a) $x = (3x' - y')/\sqrt{10}$, $y = (x' + 3y')/\sqrt{10}$,   (b) $x = (2x' - y')/\sqrt{5}$, $y = (x' + 2y')/\sqrt{5}$

14.54. $x = x'/\sqrt{3} + y'/\sqrt{2} + z'/\sqrt{6}$, $y = x'/\sqrt{3} - y'/\sqrt{2} + z'/\sqrt{6}$, $z = x'/\sqrt{3} - 2z'/\sqrt{6}$

14.55. $P = \begin{pmatrix} 1/\sqrt{2} & -1/\sqrt{2} \\ 1/\sqrt{2} & 1/\sqrt{2} \end{pmatrix}$, $P^*AP = \begin{pmatrix} 2+i & 0 \\ 0 & 2-i \end{pmatrix}$

Appendix

Polynomials Over a Field

A.1 INTRODUCTION

The ring $K[t]$ of polynomials over a field K has many properties analogous to properties of the integers. These play an important role in obtaining canonical forms for a linear operator T on a vector space V over K.

Any polynomial in $K[t]$ may be written in the form

$$f(t) = a_n t^n + \cdots + a_1 t + a_0$$

The entry a_k is called the kth *coefficient* of f. If n is the largest integer for which $a_n \neq 0$, then we say that the degree of f is n, written

$$\deg f = n$$

We also call a_n the *leading coefficient* of f, and if $a_n = 1$ we call f a *monic polynomial*. On the other hand, if every coefficient of f is 0 then f is called the *zero polynomial*, written $f = 0$. The degree of the zero polynomial is not defined.

A.2 DIVISIBILITY; GREATEST COMMON DIVISOR

The following theorem formalizes the process known as "long division."

Theorem A.1 (Euclidean Algorithm): Let f and g be polynomials over a field K with $g \neq 0$. Then there exist polynomials q and r such that

$$f = qg + r$$

where either $r = 0$ or $\deg r < \deg g$.

Proof: If $f = 0$ or if $\deg f < \deg g$, then we have the required representation

$$f = 0g + f$$

Now suppose $\deg f \geq \deg g$, say

$$f = a_n t^n + \cdots + a_1 t + a_0 \qquad \text{and} \qquad g = b_m t^m + \cdots + b_1 t + b_0$$

where $a_n, b_m \neq 0$ and $n \geq m$. We form the polynomial

$$f_1 = f - \frac{a_n}{b_m} t^{n-m} g \tag{1}$$

Then $\deg f_1 < \deg f$. By induction, there exist polynomials q_1 and r such that

$$f_1 = q_1 g + r$$

where either $r = 0$ or $\deg r < \deg g$. Substituting this into (1) and solving for f,

$$f = \left(q_1 + \frac{a_n}{b_m} t^{n-m} \right) g + r$$

which is the desired representation.

Theorem A.2: The ring $K[t]$ of polynomials over a field K is a principal ideal ring. If I is an ideal in $K[t]$, then there exists a unique monic polynomial d which generates I, that is, such that d divides every polynomial $f \in I$.

Proof: Let d be a polynomial of lowest degree in I. Since we can multiply d by a nonzero scalar and still remain in I, we can assume without loss in generality that d is a monic polynomial. Now suppose $f \in I$. By Theorem A.1 there exist polynomials q and r such that

$$f = qd + r \qquad \text{where either } r = 0 \text{ or deg } r < \text{deg } d$$

Now f, $d \in I$ implies $qd \in I$ and hence $r = f - qd \in I$. But d is a polynomial of lowest degree in I. Accordingly, $r = 0$ and $f = qd$, that is, d divides f. It remains to show that d is unique. If d' is another monic polynomial which generates I, then d divides d' and d' divides d. This implies that $d = d'$, because d and d' are monic. Thus the theorem is proved.

Theorem A.3: Let f and g be nonzero polynomials in $K[t]$. Then there exists a unique monic polynomial d such that: (i) d divides f and g; and (ii) if d' divides f and g, then d' divides d.

Definition: The above polynomial d is called the greatest common divisor of f and g. If $d = 1$, then f and g are said to be relatively prime.

Proof of Theorem A.3: The set $I = \{mf + ng : m, n \in K[t]\}$ is an ideal. Let d be the monic polynomial which generates I. Note f, $g \in I$; hence d divides f and g. Now suppose d' divides f and g. Let J be the ideal generated by d'. Then f, $g \in J$ and hence $I \subset J$. Accordingly, $d \in J$ and so d' divides d as claimed. It remains to show that d is unique. If d_1 is another (monic) greatest common divisor of f and g, then d divides d_1 and d_1 divides d. This implies that $d = d_1$ because d and d_1 are monic. Thus the theorem is proved.

Corollary A.4: Let d be the greatest common divisor of the polynomials f and g. Then there exist polynomials m and n such that $d = mf + ng$. In particular, if f and g are relatively prime then there exist polynomials m and n such that $mf + ng = 1$.

The corollary follows directly from the fact that d generates the ideal

$$I = \{mf + ng : m, n \in K[t]\}$$

A.3 FACTORIZATION

A polynomial $p \in K[t]$ of positive degree is said to be irreducible if $p = fg$ implies f or g is a scalar.

Lemma A.5: Suppose $p \in K[t]$ is irreducible. If p divides the product fg of polynomials f, $g \in K[t]$, then p divides f or p divides g. More generally, if p divides the product of n polynomials $f_1 f_2 \cdots f_n$, then p divides one of them.

Proof: Suppose p divides fg but not f. Since p is irreducible, the polynomials f and p must then be relatively prime. Thus there exist polynomials m, $n \in K[t]$ such that $mf + np = 1$. Multiplying this equation by g, we obtain $mfg + npg = g$. But p divides fg and so mfg, and p divides npg; hence p divides the sum $g = mfg + npg$.

Now suppose p divides $f_1 f_2 \cdots f_n$. If p divides f_1, then we are through. If not, then by the above result p divides the product $f_2 \cdots f_n$. By induction on n, p divides one of the polynomials $f_2, \ldots, f_n$. Thus the lemma is proved.

Theorem A.6 (Unique Factorization Theorem): Let f be a nonzero polynomial in $K[t]$. Then f can be written uniquely (except for order) as a product

$$f = kp_1p_2 \cdots p_n$$

where $k \in K$ and the p_i are monic irreducible polynomials in $K[t]$.

Proof. We prove the existence of such a product first. If f is irreducible or if $f \in K$, then such a product clearly exists. On the other hand, suppose $f = gh$ where f and g are nonscalars. Then g and h have degrees less than that of f. By induction, we can assume

$$g = k_1g_1g_2 \cdots g_r \qquad \text{and} \qquad h = k_2h_1h_2 \cdots h_s$$

where $k_1, k_2 \in K$ and the g_i and h_j are monic irreducible polynomials. Accordingly,

$$f = (k_1k_2)g_1g_2 \cdots g_rh_1h_2 \cdots h_s$$

is our desired representation.

We next prove uniqueness (except for order) of such a product for f. Suppose

$$f = kp_1p_2 \cdots p_n = k'q_1q_2 \cdots q_m$$

where $k, k' \in K$ and the $p_1, \ldots, p_n, q_1, \ldots, q_m$ are monic irreducible polynomials. Now p_1 divides $k'q_1 \cdots q_m$. Since p_1 is irreducible it must divide one of the q_i, by Lemma A.5. Say p_1 divides q_1. Since p_1 and q_1 are both irreducible and monic, $p_1 = q_1$. Accordingly,

$$kp_2 \cdots p_n = k'q_2 \cdots q_m$$

By induction, we have that $n = m$ and $p_2 = q_2, \ldots, p_n = q_n$ for some rearrangement of the q_i. We also have that $k = k'$. Thus the theorem is proved.

If the field K is the complex field $\mathbf{C}$, then we have the following result which is known as the fundamental theorem of algebra; its proof lies beyond the scope of this text.

Theorem A.7 (Fundamental Theorem of Algebra): Let $f(t)$ be a nonzero polynomial over the complex field $\mathbf{C}$. Then $f(t)$ can be written uniquely (except for order) as a product

$$f(t) = k(t - r_1)(t - r_2) \cdots (t - r_n)$$

where $k, r_i \in \mathbf{C}$, i.e. as a product of linear polynomials.

In the case of the real field $\mathbf{R}$ we have the following result.

Theorem A.8: Let $f(t)$ be a nonzero polynomial over the real field $\mathbf{R}$. Then $f(t)$ can be written uniquely (except for order) as a product

$$f(t) = kp_1(t)p_2(t) \cdots p_m(t)$$

where $k \in \mathbf{R}$ and the $p_i(t)$ are monic irreducible polynomials of degree one or two.

Index

A(V), 322
Absolute value, 52
Adjoint, classical, 253
Adjoint operator, 425, 433
  matrix representation, 426
Algebra of linear maps, 322
Algebraic multiplicity, 285
Algorithm:
  basis-finding, 154
  determinant, 253
  diagonalization under:
    congruence, 102
    similarity, 102, 286
  Elimination (Gaussian), 11
  Euclidean, 446
  Gauss-Jordan,
  inverse matrix, 100
  orthogonal diagonalization, 289
Alternating bilinear form, 411
Angle, 45, 206,
Annihilators, 399, 403
Augmented matrix, 78

$B(V)$, 409
Back-substitution, 9
Basis:
  dual, 398, 400
  general solution, 19
  orthogonal, 208
  orthonormal, 209
  second dual, 399
  usual, 150
  vector space, 150, 169
Basis-finding algorithm, 154
Bessel inequality, 212
Bijective mapping, 314
Bilinear form, 409, 414
  alternating, 411
  matrix representation of, 410
  polar form of, 412
  rank of, 411
  real symmetric, 412
  symmetric, 411
Block matrix, 79, 86
  determinant of, 256
  diagonal, 98, 291
  Jordan, 374
  square, 98, 123
  triangular, 98

C^n, 36, 52
Caley-Hamilton theorem, 282

Cancellation law, 142
Canonical form, 14, 352
  Jordan, 373, 381
  rational, 375, 388
  row, 14, 16, 28
Cauchy-Schwarz inequality, 45, 58, 205, 218
Cells, 79
Change of basis, 160, 187
Change-of-basis matrix, 160, 348, 357
Change-of-variable matrix, 105
Characteristic polynomial, 281, 349
Classical adjoint, 253
Codomain, 312
Coefficient, 1
  Fourier, 211
Coefficient matrix, 17, 78
Cofactor, 252
Column, 13, 74
  operations, 100
  rank, 152
  space, 146
  vector, 40, 75
Commuting matrices, 90
Companion matrix, 375
Complement, orthogonal, 207, 226
Complex:
  conjugate, 51
  inner product, 216
  matrix, 96, 121
  n-space, 52
  numbers, 51
Component, 40, 211
Composition of mappings, 313
Congruent:
  diagonalization, 102
  matrices, 102, 410
  symmetric matrices, 102, 124
Conjugate, complex, 51
Conjugate matrix, 96
Consistent systems, 13
Coordinate vector, 157, 186
Coordinates, 40, 157
Cosets, 376
Cramer's rule, 255, 263
Cross product, 49, 65
Curves, 48
Cyclic subspaces, 374, 388

Degenerate linear equations, 2
Degree of polynomial, 446
Dependence, linear, 147, 166

Determinant, 246
  block matrix, 256
  computation of, 252, 260
  linear equations, 254
  linear operators, 349
  multiplinearity, 257
  order n, 250
  order 3, 248
  properties, 251
  volume, 257
Diagonal (of a matrix), 90
Diagonal matrix, 93
  block, 98
Diagonalization, 349, 360
  algorithm, 102, 286
Diagonalizable matrices, 280, 298
Dimension of solution spaces, 19
Dimension of vector spaces, 150
  subspaces, 151, 171
Direct sum, 156, 181
  decomposition, 371, 379
Directed line segment, 46
Distance, 45, 206, 216
Domain, 312
Dot product, 43, 52, 56
Dual:
  basis, 398
  space, 397

Echelon:
  form, 10, 23, 14
  matrices, 14
Eigenspace, 284, 350
Eigenvalue, 284, 292, 350
  computing, 286
Eigenvector, 284, 292, 350
  computing, 286
Elementary operations, 7
  column, 100
  row, 14, 98
Elementary divisors, 376
Elementary matrix, 99, 101, 117
Elimination algorithm, 6
Elimination, Gaussian, 7, 100
Equal:
  matrices, 74
  vectors, 40
Equations (see Linear equations),
Equivalence:
  matrix, 100, 102
  row, 14
Equivalent systems, 7
Euclidean:
  algorithm, 446
  space, 202
Evaluation map, 92

$F(X)$, 143
Factorization:
  polynomial, 448
  LU, 109
Field, 74
Finite dimension, 150
Fourier, coefficient, 211
Free variable, 3, 10
Function,
  square matrix, 89
  spaces, 143
Functional, linear, 397

Gauss–Jordan algorithm, 29
Gaussian elimination, 8, 24, 100, 261
General solution, 1, 7
Geometric multiplicity, 285
Gram–Schmidt orthogonalization, 213
Graph, 4
Greatest common divisor, 447

Hermitian:
  form, 413
  matrix, 96
  quadratic form, 414
Homogeneous systems, 18, 32, 399
Hilbert space, 205
Hyperplane, 46

ijk notation, 49
Im F, 316
Im(z), 51
Identity:
  mapping, 314
  matrix, 91
Image of linear mapping, 316, 328
Imaginary part, 51
Inconsistent systems, 13
Independence, linear, 147
Infinite dimension, 150
Infinity-norm, 219
Injective mapping, 314
Inner product, 202, 221
  complex, 216
  usual, 202, 218
Inner product spaces, 202
  linear operators on, 425
Invariance, 370
Invariant subspaces, 370, 377,
Inverse matrix, 92
  computing algorithm, 100
Invertible matrices, 92, 115
  linear substitutions, 105
  operators, 323

Isomorphic vector spaces, 316
Isomorphism, 158, 316, 320

Jordan canonical form, 373
Jordan block, 374

K^n, 142
Ker F, 316
Kernel of a linear map, 316, 328
Kronecker delta, 91

Laplace expansion, 252
Law of Inertia, 102, 413
Leading nonzero entry, 13
Leading unknown, 3
Length, 44, 203
Line, 47,
Linear:
    combination, 19, 41, 145, 149, 165
    dependence, 42, 147, 166
    functional, 397, 426
    independence, 42, 147
    span, 145
Linear equation, 1
    degenerate, 2
    one unknown, 2
Linear equations (system), 21, 152, 319, 323, 399
    consistent, 13
    echelon form, 23
    two unknowns, 4
    triangular form, 9, 23
Linear mapping, 314, 326, 319
    image, 316, 328
    kernel, 316, 328
    matrix representation, 351
    nullity, 318,
    rank, 318
    transpose, 400
Linear operator, 322
    adjoint, 425
    characteristic polynomial, 349
    determinant, 349,
    inner product spaces, 425
    invertible, 323
    matrix representation, 344
    nilpotent, 373
Located vectors, 46
LU factorization, 109, 130

Mappings (maps), 312, 324
    composition of, 313
    linear, 314, 326
    matrix, 315

Matrices, 13, 74
    augmented, 78
    Block, 79, 86
    change-of-basis, 160, 348, 430
    change-of-variable, 105
    coefficient, 17, 78
    companion, 374
    complex, 96
    diagonal, 93
    diagonizable, 280, 298
    echelon, 14
    equivalent, 100
    invertible, 92
    nonsingular, 92
    normal, 97
    similar, 357
    square, 89
    triangular, 94
Matrix mappings, 315
Matrix representation:
    adjoint operator, 426
    bilinear form, 410
    linear maps, 344, 351, 363
    quadratic form, 102
Matrix space, 142
Minkowski's inequality, 45
Minimal polynomial, 289, 301, 350
Minor, 252, 255
    principle, 256
Monic polynomial, 446
Multiplication of matrices, 76
Multilinearity, 257
Multiplicity, 285

n-space, 40
Nilpotent, 373
Nondegenerate linear equations, 3
Nonnegative semidefinite, 413
Nonsingular:
    linear maps, 320
    matrices, 92
Norm, 44, 53, 203, 219
Normal:
    matrix, 97
    operator, 428, 432
Normal vector, 46
Normalizing, 44
Normed vector spaces, 219
Nullity, 318

One-norm, 219
One-to-one:
    correspondence, 314
    mapping, 314
Onto mapping, 314

Operations on linear maps, 320
Operators (*see* Linear operators)
Orthogonal:
   basis, 208
   complement, 207, 226
   matrix, 95, 216, 429
   operator, 427, 431
   projections, 433
   sets, 208
Orthogonality, 43, 280
Orthogonalization (Gram–Schmidt), 213
Orthonormal basis, 209
Outer product, 49

$P(t)$, 143
Parallelogram law, 39
Parameters, 4, 10, 47
Parity of permutations, 249
Perpendicular, 43
Permutations, 249, 269
Pivot entries, 16
Pivoting (row reduction), 27
Planes, 61
Point, 40, 52
Polar form, 412
Polynomial, 446
   characteristic, 281
   degree, 446
   Legendre, 214
   minimal, 350
   monic, 446
Positive definite:
   matrices, 108, 127, 214
   operators, 413, 427
Positive operators, 427
Powers of matrices, 91
Primary decomposition, 372
Principal minor, 256
Product:
   cross, 49
   dot, 43, 52
   inner, 202, 221
Projections, 45, 211, 315, 338, 397, 433
   orthogonal, 433
Pythagorean theorem, 208

Quadratic forms, 102, 124, 289, 295, 412
   diagonalizing, 102
   Hermitian, 414
   matrix representation, 102
Quotient spaces, 376, 385

R^n, 40, 53
Re z, 51

Rank, 102, 107
Rational canonical form, 375
Real part, 51
Real symmetric bilinear form, 412
Reduction algorithm:
   linear equations, 11
   determinants, 252
Ring of polynomials, 446
Row, 13, 74
   canonical form, 14, 16, 28
   equivalence, 14, 98
   operations, 14, 98
   rank, 152
   reducing, 14, 27
   vector, 75

Scalar, 40, 141
   matrix, 91
   product, 43
Schwarz (Cauchy–Schwarz) inequality, 45, 58
Second dual space, 399
Self-adjoint operator, 427
Signature, 102, 107
Similarity:
   matrices, 109, 128, 357
   operators, 349
Singular, 320, 332
Size of a matrix, 13, 74
Skew-adjoint operator, 427
Skew-Hermitian, 96
Skew-symmetric, 94
Solutions, 1, 21
   general, 1, 7
   simultaneous, 5
Spacial vectors, 39, 49
Span, 145
Spanning sets, 145
Spectral theorem, 433
Square matrix, 89, 111
   block, 98, 123
Standard form, 7
Sums of vector spaces, 155
Surjective map, 314
Sylvester's theorem, 413
Symmetric bilinear form, 411, 416
Symmetric matrices, 94
   congruent, 102, 124
Systems of linear equations,
   (*see* Linear equations)

Tangent vector, 48
Trace, 90, 349
Transpose:
   matrix, 77, 85
   linear mapping, 400

Transition matrix, 160, 348
Triangle inequality, 205
Triangular form:
   linear equations, 9, 23
   linear operators, 369, 385
Triangular matrix, 94
   block, 98
Two-norm, 219

Unknown, leading, 3
Unit vector, 44, 206
Unitary:
   matrix, 97, 430
   operator, 427
Usual:
   basis, 150
   inner product, 202, 218

V^*, 397
V^{**}, 399
Variable, free, 3

Vector, 39, 40
   addition, 41, 52
   column, 75
   coordinate, 157, 186
   located, 46
   normal, 46
   row, 75
   scalar multiplication, 41, 52
   spacial, 39, 40
   tangent, 48
   unit, 44, 206
   zero, 40
Vector space, 141, 162
   basis, 150, 169
   dimension, 150
   normed, 219
   sums, 155

Zero:
   mapping, 315
   matrix, 76
   polynomial, 447
   vector, 40